The Most Challenging Problems Book
by an Indian Author Perfectly Suited for JEE

500 SELECTED Problems in Physics for JEE MAIN & ADVANCED

The Most Challenging Problems Book
by an Indian Author Perfectly Suited for JEE

500

SELECTED Problems in Physics

for JEE MAIN & ADVANCED

DC Pandey

arihant
ARIHANT PRAKASHAN (Series), MEERUT

arihant

Arihant Prakashan (Series), Meerut

Administrative & Production Offices

Regd. Office

'Ramchhaya' 4577/15, Agarwal Road, Darya Ganj, New Delhi -110002
Tele: 011- 47630600, 43518550; Fax: 011- 23280316

Head Office

Kalindi, TP Nagar, Meerut (UP) - 250002
Tele: 0121-2401479, 2512970, 4004199; Fax: 0121-2401648

Sales & Support Offices

Agra, Ahmedabad, Bengaluru, Bhubaneswar, Bareilly, Chennai, Delhi, Guwahati, Hyderabad, Jaipur, Jhansi, Kolkata, Lucknow, Meerut, Nagpur & Pune

ISBN : 978-93-12147-37-5

Price : ₹ 290.00

PRINTED & BOUND BY
ARIHANT PUBLICATIONS (I) LTD. (PRESS UNIT)

For further information about the books published by Arihant log on to www.arihantbooks.com or email to info@arihantbooks.com

PREFACE

The primary objective of preparing this book, Problems in Physics, is to help students develop a command over the basic concepts and principles of Physics through problem solving. With more than a decade of distinguished career in teaching Physics to JEE aspirants, I can firmly say that the best way to grapple with the concepts of the Physics is to learn with the problems themselves.

The problems included in the book with their solutions, aim to give you the mastery required for solving the intricate problems asked in the exams such as JEE. Critical analysis of the situation is required whatever be format of the questions. This book just gives emphasis on this, and it is hoped that it will give a boost to all the meritorious and hard-working students.

I am thankful to my student, Vineet Jain for the help in editing of this book. I would also like to thank Ankit Kapoor and Bhaskar Sharma for their valuable suggestions.

Above all, I shall like to thank Mr Yogesh Chand Jain, Chairman, Arihant Group, for bringing the book in this nice form.

Suggestions for further improvement of the book are welcome.

DC Pandey

This book is dedicated to my honourable grandfather

(LATE) SH. PITAMBER PANDEY

a Kumaoni poet, and a resident of village

Dhaura (Almora), Uttarakhand

CONTENTS

PROBLEMS

CHAPTER 1

KINEMATICS

Problem 1. The coordinates of a particle moving in a plane are given by

$$x = 4\cos 6t \quad \text{and} \quad y = 6\sin 6t$$

(a) find the equation of the path of the particle

(b) find the angle between position vector $\vec{r}$ and velocity vector $\vec{v}$ at time $t = \pi/12$

(c) prove that the acceleration of the particle is always directed toward a fixed point.

Problem 2. A particle is projected vertically upwards with speed u in a medium in which the resistance to motion is proportional to the square of the speed. Initially net force on the particle is two times its weight downwards. Find the time of ascent and distance ascended by the particle.

Problem 3. A car starts from rest with an acceleration of 6 m/s^2 which decreases to zero linearly with time, in 10 second, after which the car continues at a constant speed. Find the time required for the car to travel 400 m from the start.

Problem 4. A charged particle of mass m and charge q is moved from rest at $x = 0$ by an electric field $\vec{E} = (E_0 - \alpha x)\,\hat{i}$ where α is a positive constant and x is the displacement of the particle in time t along x-axis. Find the distance moved by the particle when it again is brought to rest and the acceleration of the particle at that instant. Describe the motion of the particle.

Hint : Force on a charge 'q' placed in an electric field $\vec{E}$ is given by

$$\vec{F_e} = q\vec{E}$$

Problem 5. A particle of mass m moving in a straight line is acted upon by a force F which varies with its velocity as $F = -kv^n$. Here k is a constant. For what values of n the average value of velocity of the particle averaged over the time, till it stops, is one third the initial velocity. ($n \neq 1$ or 2)

Problem 6. A lift of total mass M kg is raised by cables from rest to rest through a height h. The greatest tension which the cables can safely bear is nMg newtons. Find the shortest interval of time in which the ascent can be made ($n > 1$).

Problem 7. A particle moves in a straight line with constant acceleration 'a'. The displacements of particle from origin in times t_1, t_2 and t_3 are s_1, s_2 and s_3 respectively. If times are in A.P. with common difference d and displacements are in G.P. Then prove that $a = \dfrac{(\sqrt{s_1} - \sqrt{s_3})^2}{d^2}$.

Problem 8. A ball of mass 2 kg is dropped from a height of 80 m on a floor. At each collision with the floor the ball loses half of its speed. Plot the velocity-time, speed-time, and kinetic energy-time graphs of its motion till first two collisions with the floor (Take $g = 10$ m/s^2).

Problem 9. The acceleration-displacement graph of a particle moving in a straight line is as shown alongside. Initial velocity of particle is zero. Find the velocity of the particle when displacement of the particle is, $s = 12$ m.

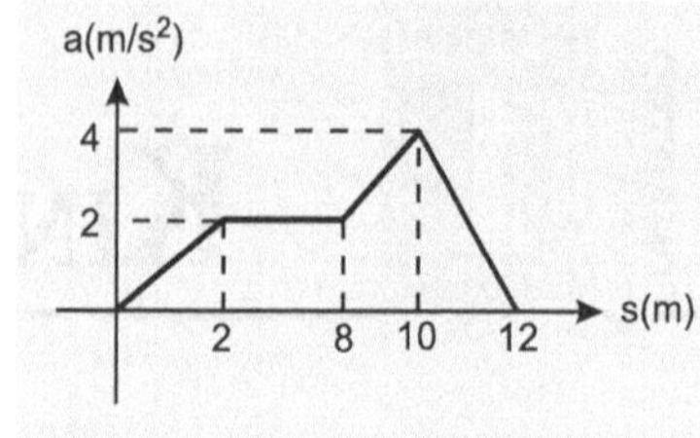

Problem 10. An aircraft flies with constant airspeed (speed of aircraft in still air) 200 km/hr from position A to position B, which is 100 km north-east of A and then flies back to A. Throughout the whole flight the wind velocity is 60 km/hr from the west. Find total time of flight from A to B and back.

Problem 11. A motor boat going downstream overcame a raft at point A. After one hour it turned back and after some time it met the raft again at a distance 6 km from point A. Find the river velocity.

Problem 12. A particle A is projected with an initial velocity of 60 m/s at an angle 30° to the horizontal. At the same time a second particle B is projected in opposite direction with initial speed of 50 m/s from a point at a distance of 100 m from A. If the particles collide in air, find the angle of projection α of B, time when collision occurs and the distance of point P from A, where collision takes place. ($g = 10$ m/s^2)

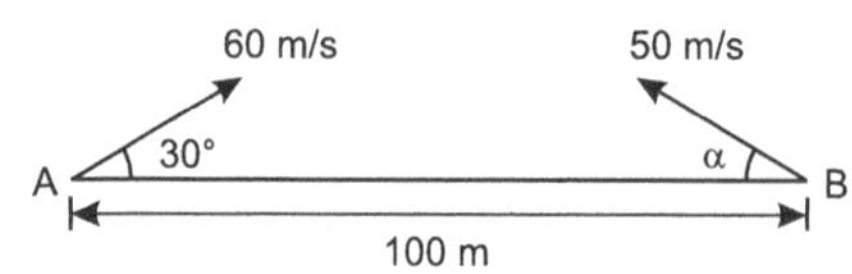

Problem 13. A particle is projected with speed v at an angle θ to the horizontal from the point $x = 0, y = 0$. If x and y-axes are horizontal and vertically upwards respectively and point of projection is the lowest point on the inner surface of a bowl formed by rotating the curve $x^2 = 4ay$, where a is a positive constant. Show that the particle strikes the bowl at a horizontal distance

$$x = \frac{4av^2 \tan\theta}{v^2 + 2ag + 2ag\tan^2\theta}$$

Problem 14. A particle is released from a certain height $H = 400$ m. Due to the wind the particle gathers the horizontal velocity component $v_x = ay$ where $a = \sqrt{5}$ s^{-1} and y is the vertical displacement of the particle from point of release, then find :

(a) the horizontal drift of the particle when it strikes the ground

(b) the speed with which the particle strikes the ground (Take $g = 10$ m/s^2)

Problem 15. A particle of mass m is attached by a light inextensible string of length 2R to a fixed point O. When vertically below O at point A, the particle is given a horizontal velocity u. When the string becomes horizontal, it hits a small smooth nail C, at a distance R from O and the particle continues to rotate about C. Find the minimum value of u so that the particle just describes complete circle abou

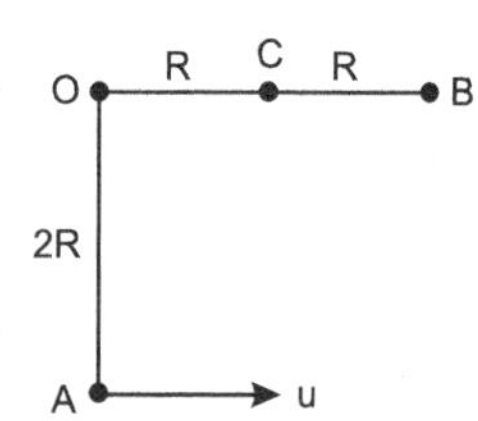

Problem 16. A particle moves in a vertical circle. Its velocity at topmost point is half of its velocity at bottommost point. Find the magnitude of acceleration of the particle at the moment when its velocity is directed vertically upwards. ($g = 10$ m/s^2)

Problem 17. A uniform electric field of strength 10^6 N/C is directed vertically downwards. A particle of mass 0.01 kg and charge 10^{-6} C is suspended by an inextensible thread of length 1 m. The particle is displaced slightly from its mean position and released. Calculate the time period of its oscillation. What minimum velocity should be given to the particle at bottom so that it completes a full circle. Calculate the maximum and minimum tensions in the thread. ($g = 9.8 \text{ m/s}^2$)

Hint : Electrostatic force $\vec{F_e}$ on a charge q in electric field $\vec{E}$ is $\vec{F_e} = q\vec{E}$

Problem 18. A particle moves along a straight line and its velocity depends on time as $v = 3t - t^2$. Here v is in m/s and t in second. Find:

(a) average velocity and

(b) average speed for first five seconds.

Problem 19. At the initial moment three points A, B and C are on a horizontal straight line at equal distances from one another. Point A begins to move vertically upward with a constant velocity v and point c vertically downward without any initial velocity but with a constant acceleration a. How should point B move vertically for all the three points to be constantly on one straight line. The points begin to move simultaneously.

Problem 20. Two particles A and B start from positions shown in figure and move with constant speeds v and u ($> v$). A moves along x-axis and B moves such that its velocity is always aimed at A. Let r be the distance between them and θ be the angle made by the trajectory of B with x-axis at sometime t. Prove that,

$$\frac{r}{d} = \frac{(\sin\theta)^{\frac{u}{v}+1}}{(1-\cos\theta)^{u/v}}$$

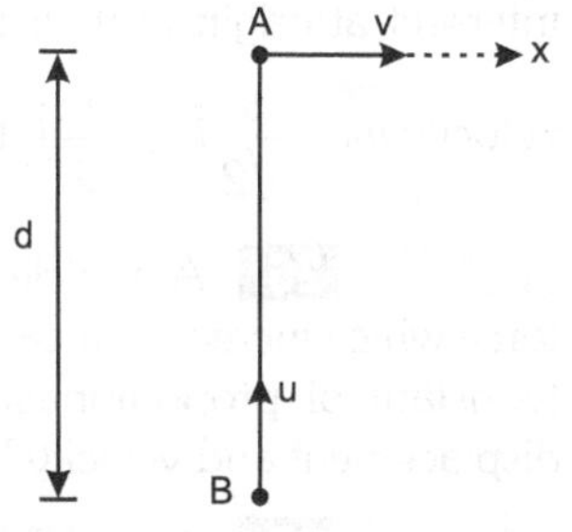

Problem 21. A boat is moving in a river with a speed v w.r.t. water. The water is flowing with a speed u. At time $t = 0$ the boat is at the origin of a co-ordinate system with x-y axes in the horizontal plane and positive x-axis in the opposite direction of the flow of water. The boat has to reach the point $P(x, y)$ as shown in the figure.

Show that the boat has to start in a direction inclined at an angle $\sin^{-1}\left\{\frac{u}{v}\frac{y}{\sqrt{x^2+y^2}}\right\}$ to the line joining O to P.

Also find the time taken by the boat to reach the point P.

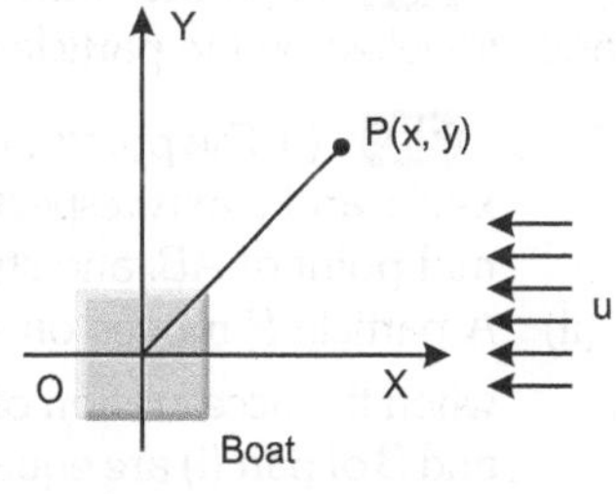

Problem 22. A fighter plane enters inside the enemy territory, at time $t = 0$, with velocity $v_o = 250$ m/s and moves horizontally with constant acceleration $a = 20 \text{ m/s}^2$ (see figure). An enemy tank at the border, spot the plane and fire shots at an angle $\theta = 60°$ with the horizontal and with velocity $u = 600$ m/s. At what altitude H of the plane it can be hit by the shot ?

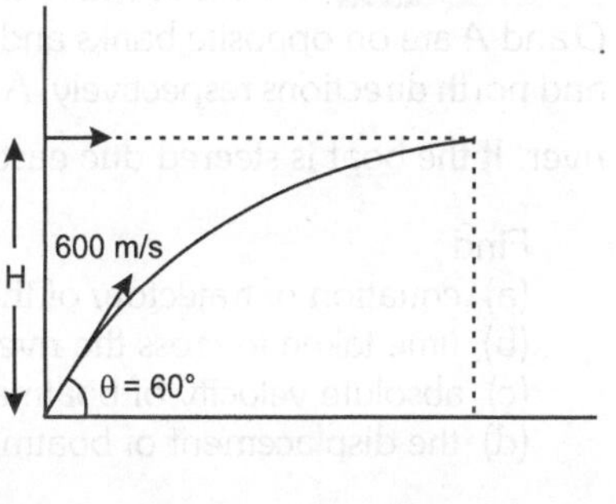

Problem 23. A 3 m long arm OA rotates in a plane such that $\theta = 0.5\, t^2$ where θ is the angle with x-axis in radian and t is in second. A slider collar B slides along the arm in such a way that its distance from the hinge O is given by $r = 3 - 0.4\, t^2$ where r is in meters. Find the velocity of the collar at an instant the arm has turned to $\theta = 30°$.

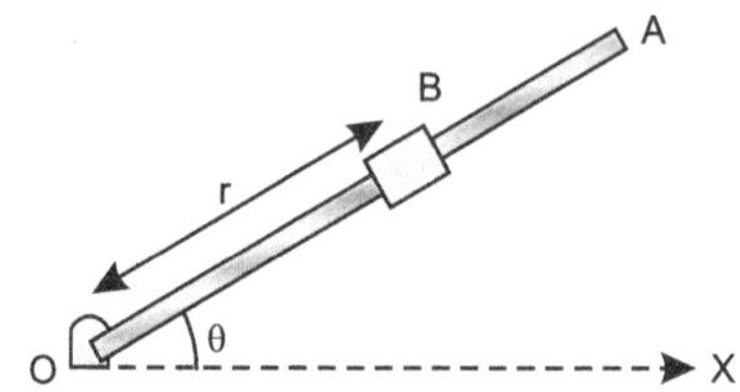

Problem 24. A rod follower AB is subjected to a vertical up and down movement while resting on the circular contour of radius 30 cm of a cam. The cam moves to the right with a velocity of 5 cm/s and an acceleration of $10\,\text{cm/s}^2$. Find the velocity and acceleration of point B on the rod at the instant of interest as shown in figure.

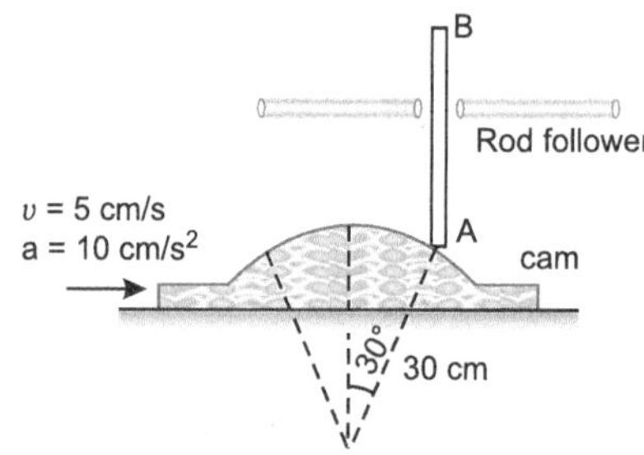

Problem 25. Two rods of equal length are lying one along x-axis and the other along line $x = y$. They intersect at origin at their mid point. The first rod moves with velocity $\vec{v}_1 = v\,\hat{j}$ and the second with velocity $\vec{v}_2 = \frac{v}{\sqrt{2}}\hat{i} - \frac{v}{\sqrt{2}}\hat{j}$. Find the velocity of point of intersection of two rods.

Problem 26. A particle of mass 1 kg which moves along the x-axis is subjected to an accelerating force which increases linearly with time and a retarding force which increases directly with displacement (constant of proportionality being one with proper dimensions in both the cases). At time $t = 0$, displacement and velocity both are zero. Find the displacement as a function of time t.

Problem 27. A particle moves along the x-axis according to the equation $x = A \cos \omega t$. Find the distance travelled by the particle during the time interval $t = 0$ to $t = t$.

Problem 28. (i) The points A and B are moving with the same speed u in the positive direction of the x-axis and y-axis respectively. Find the magnitude of velocity relative to A of a point C, which is mid point of AB, and show that it is reverse of the velocity of C relative to B.

(ii) A particle P moves on the circle $x^2 + y^2 = 1$ with constant speed v. Show that each instant when the acceleration of P is parallel to the line $x + y = 0$, the velocities of P relative to points A and B of part (i) are equal in magnitude. Find v in terms of u if the maximum value of the velocity of P relative to C is u.

Problem 29. A river of width 'a' with straight parallel banks flows due north with speed u. The points O and A are on opposite banks and A is due east of O. Coordinate axes Ox and Oy are taken in the east and north directions respectively. A boat, whose speed is v relative to water, starts from O and crosses the river. If the boat is steered due east and u varies with x as : $u = x\,(a - x)\,\frac{v}{a^2}$.

Find :

(a) equation of trajectory of the boat

(b) time taken to cross the river

(c) absolute velocity of boatman when he reaches the opposite bank

(d) the displacement of boatman when he reaches the opposite bank from the initial position.

Problem 30. A river of width w is flowing such that the stream velocity varies with y as

$$v_R = v_o \left[1 + \frac{\sqrt{3}-1}{w} y\right]$$

where y is the perpendicular distance from one bank. A boat starts rowing from the bank with constant velocity $v = 2\,v_o$ in such a way that it always moves along a straight line perpendicular to the banks.

(a) at what time will he reach the other bank?

(b) what will be the velocity of the boat along the straight line when he reaches the other bank ?

Problem 31. Two points A and B move with speeds v and $2v$ in two concentric circles with centre O and radii $2r$ and r respectively. If the points move in the same sense and if $\angle OAB = \alpha$, when the relative motion is along AB, find the value of α.

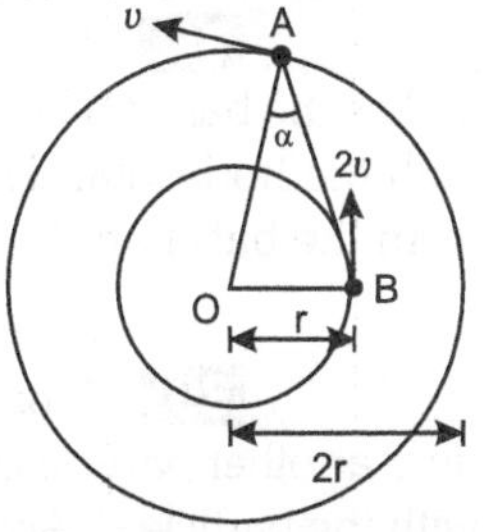

Problem 32. Two parallel straight lines are inclined to the horizon at an angle α. A particle is projected from a point midway between them so as to graze one of the lines and strikes the other at right angles. Show that if θ is the angle between the direction of projection and either of the lines, then

$$\tan\theta = (\sqrt{2}-1)\cot\alpha$$

Problem 33. A regular hexagon stands with one side on the ground and a particle projected so as to graze its four upper vertices. Show that the ratio of its maximum velocity to that of its minimum velocity is $\sqrt{\frac{31}{3}}$.

Problem 34. Two stones are projected simultaneously with equal speeds from a point on an inclined plane along the line of its greatest slope upwards and downwards respectively. The maximum distance between their points of striking the plane is double that of when they are projected on a horizontal ground with same speed. If one strikes the plane after two second of the other, find :

(a) the angle of inclination of plane.

(b) the speeds of their projection (Take $g = 9.8 \text{ m/s}^2$).

Problem 35. A river of width ω is flowing with a uniform velocity v. A boat starts moving from point P also with velocity v relative to the river. The direction of resultant velocity is always perpendicular to the line joining boat and the fixed point R. Point Q is on the opposite side of the river and P, Q and R are in a straight line. If $PQ = QR = \omega$, find :

(a) the trajectory of the boat,

(b) the drifting of the boat and

(c) the time taken by the boat to cross the river.

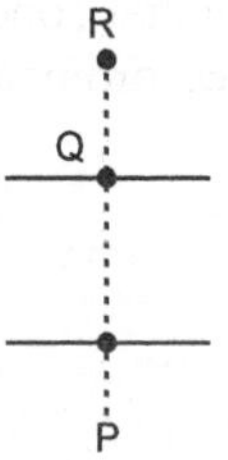

Problem 36. AB is an inclined roof and a body is projected from origin towards the roof as shown in figure. Find 'h' for which body will just touch the roof.

Given : $\theta = \alpha = 45°$ and $u = 10 \text{ m/s}$, $g = 10 \text{ m/s}^2$.

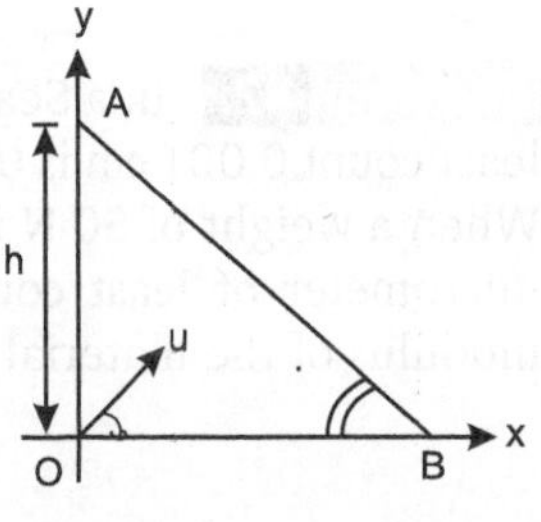

Problem 37. A particle is projected from point G, such that it touches the points B, C, D and F of a regular hexagon of side 'a'. Find its horizontal range GH.

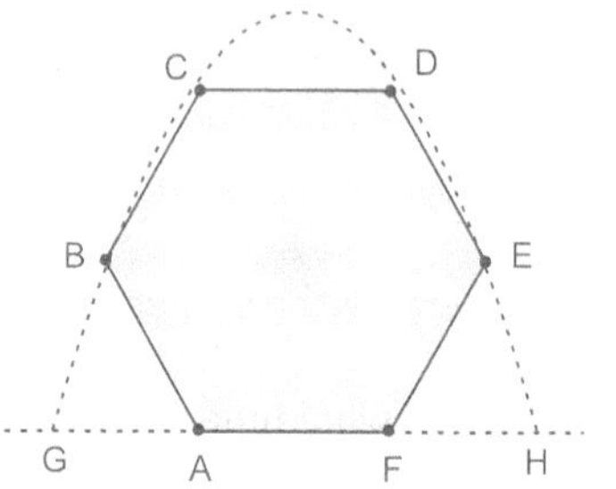

Problem 38. The benches of a gallery in a cricket stadium are 1 m high and 1 m wide. A batsman strikes the ball at a level 1 m about the ground and hits a ball. The ball starts at 35 m/s at an angle of 53° with the horizontal. The benches are perpendicular to the plane of motion and the first bench is 110 m from the batsman. On which bench will the ball hit.

Problem 39. A particle is dropped from point P at time $t = 0$. At the same time another particle is thrown from point O as shown in the figure and it collides with the particle P. Acceleration due to gravity is along the negative y-axes. If the two particles collide 2 sec after they start, find the initial velocity of particle which was projected from O.

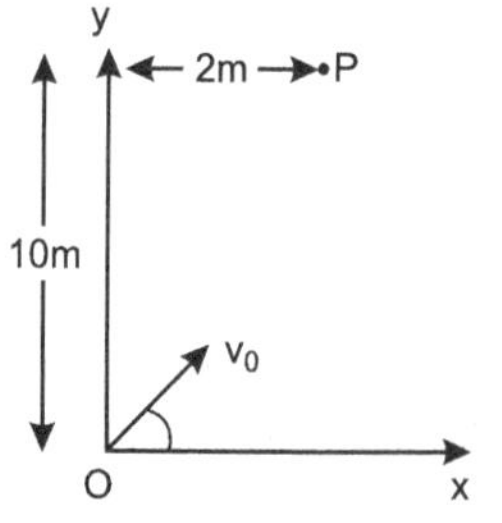

Problem 40. In the vertical plane shown two particles 'P' and 'Q' are located at points 'A' and 'B'. At $t = 0$, the particle 'P' is projected perpendicular to the inclined plane 'OA' with velocity $2v_o$ and simultaneously the particle Q is projected horizontally in 'BO' direction. What is the necessary value of v_o (in terms of 'd' and θ) so that both the particles meet each other between the points 'O' and 'B'.

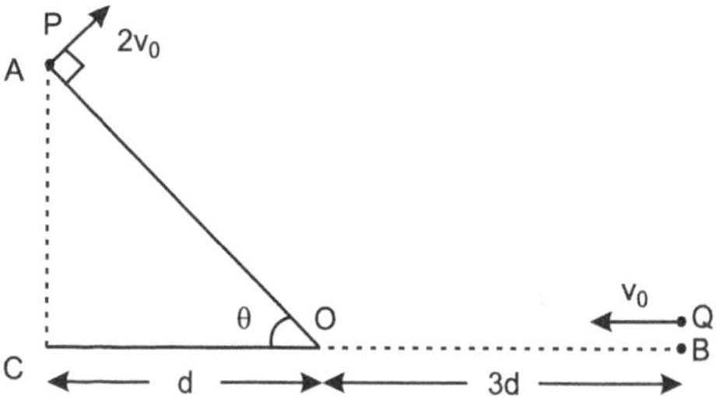

Problem 41. A projectile is launched from point 'A' with the initial conditions shown in the figure. Determine the 'x' and 'y' co-ordinates of the point of impact.

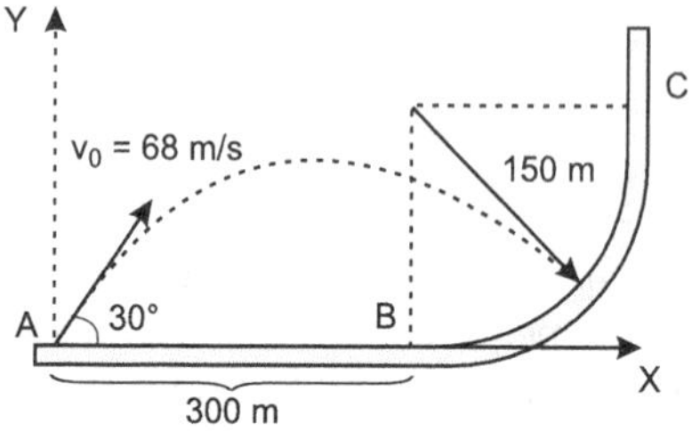

IIT JEE PROBLEMS

Problem 42. In a Searle's experiment, the diameter of the wire as measured by a screw gauge of least count 0.001 cm is 0.050 cm. The length, measured by a scale of least count 0.1 cm, is 110.0 cm. When a weight of 50 N is suspended from the wire, the extension is measured to be 0.125 cm by a micrometer of least count 0.001 cm. Find the maximum error in the measurement of Young's modulus of the material of the wire from these data. **(JEE 2004)**

Problem 43. The pitch of a screw gauge is 1 mm and there are 100 divisions on the circular scale. While measuring the diameter of a wire, the linear scale reads 1 mm and 47th division on the circular scale coincides with the reference line. The length of the wire is 5.6 cm. Find the curved surface area (in cm^2) of the wire in appropriate number of significant figures. **(JEE 2004)**

Problem 44. N-divisions on the main scale of a vernier callipers coincide with $N + 1$ divisions on the vernier scale. If each division on the main scale is of a units, determine the least count of the instrument. **(JEE 2003)**

Problem 45. On a frictionless horizontal surface, assumed to be the x-y plane, a small trolley A is moving along a straight line parallel to the y-axis (see figure) with a constant velocity of $(\sqrt{3} - 1)$m/s. At a particular instant, when the line OA makes an angle of 45° with the x-axis, a ball is thrown along the surface from the origin O. Its velocity makes an angle ϕ with the x-axis and it hits the trolley.

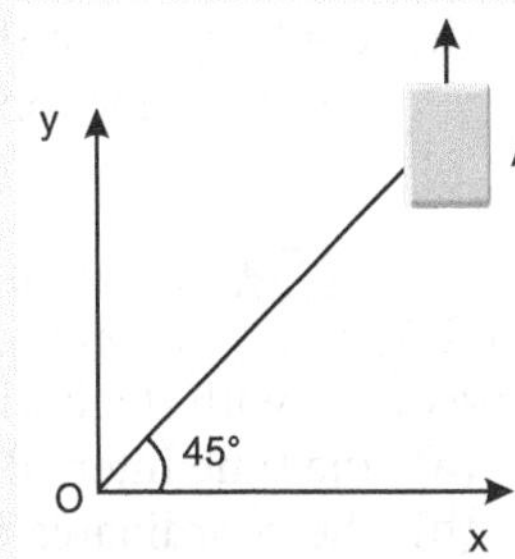

(a) The motion of the ball is observed from the frame of the trolley. Calculate the angle θ made by the velocity vector of the ball with the x-axis in this frame.

(b) Find the speed of the ball with respect to the surface, if $\phi = \frac{4\theta}{3}$. **(JEE 2002)**

Problem 46. An object A is kept fixed at the point $x = 3$ m and $y = 1.25$ m on a plank P raised above the ground. At time $t = 0$ the plank starts moving along the $+X$ direction with an acceleration 1.5 m/s^2. At the same instant a stone is projected from the origin with a velocity $\vec{u}$ as shown. A stationary person on the ground observes the stone hitting the object during its downward motion at an angle of 45° to the horizontal. All the motions are in X-Y plane. Find $\vec{u}$ and the time after which the stone hits the object. Take $g = 10$ m/s^2. **(JEE 2000)**

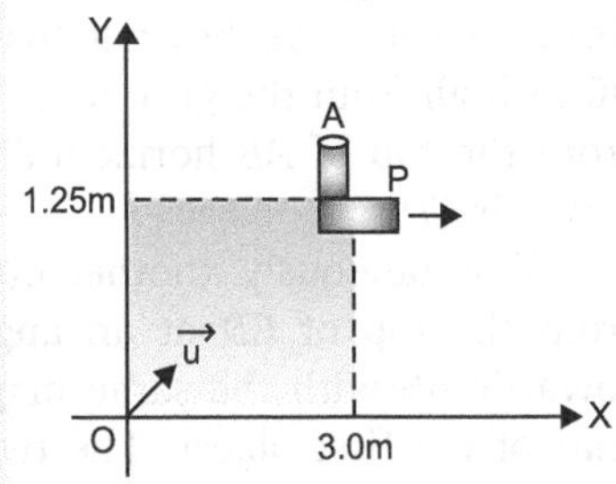

Problem 47. A particle of mass 10^{-2} kg is moving along the positive X-axis under the influence of a force $F(x) = -\frac{k}{2x^2}$ where $k = 10^{-2}$ Nm^2. At time $t = 0$, it is at $x = 1.0$ m and its velocity is $v = 0$.

(a) find its velocity when it reaches $x = 0.5$ m

(b) find the time at which it reaches $x = 0.25$ m **(JEE 1998)**

Problem 48. A large heavy box is sliding without friction down a smooth plane of inclindation θ. From a point P on the bottom of the box, a particle is projected inside the box. The inital speed of the particle with respect to the box is u and the direction of projection makes an angle α with the bottom as shown in the figure.

(a) find the distance along the bottom of the box between the point of projection P and the point Q where the particle lands (assume that the particle does not hit any other surface of the box. Neglect air resistance.)

(b) if the horizontal displacement of the partcle as seen by an observeer on the ground is zero, find the speed of the box with respect to the ground at the instant when the particle was projected. **(JEE 1998)**

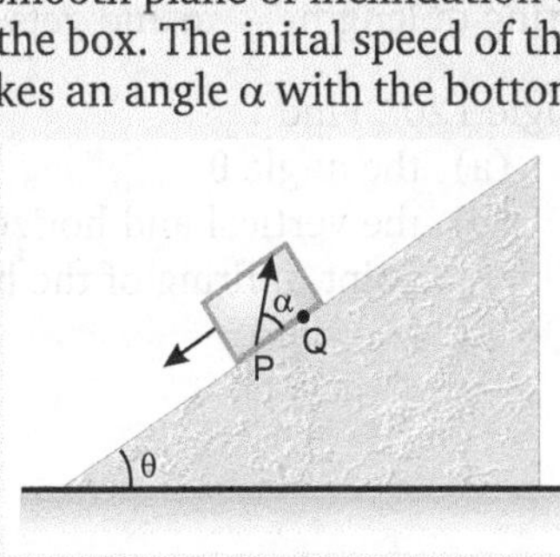

Problem 49. A cart is moving along X-direction with a velocity of 4 m/s. A person on the cart throws a stone with a velocity of 6 m/s relative to himself. In the frame of reference of the cart the stone is thrown in Y-Z plane making an angle of 30° with vertical z-axis. At the highest point of its trajectory, the stone hits an object of equal mass hung vertically from branch of a tree by means of a string of length L. A completely inelastic collision occurs in which the stone gets embedded in the object. Determine :

(i) the speed of the combined mass immediately after the collision with respect to an observer on the ground.

(ii) the length L of the string such that the tension in the string becomes zero when the string becomes horizontal during the subsequent motion of the combined mass. ($g = 9.8\ m/s^2$)

(JEE 1997)

Problem 50. Two guns situated on the top of a hill of height 10 m fire one shot each with the same speed $5\sqrt{3}$ m/s at some interval of time. One gun fires horizontally and other fires upwards at an angle of 60° with the horizontal. The shots collide in air at point P. Find

(a) the time interval between the firings and

(b) the coordinates of the point P. Take origin of the coordinate system at the foot of the hill right below the muzzle and trajectories in x-y plane. ($g = 10\ m/s^2$) **(JEE 1996)**

Problem 51. Two towers AB and CD are situated a distance d apart as shown in figure. AB is 20 m high and CD is 30 m high from the ground. An object of mass m is thrown from the top of AB horizontally with a velocity of 10 m/s towards CD.

Simultaneously another object of mass 2 m is thrown from the top of CD at an angle of 60° to the horizontal towards AB with the same magnitude of initial velocity as that of the first object. The two objects move in the same vertical plane, collide in mid-air and stick to each other.

(i) calculate the distance d between the towers,

(ii) find the position where the objects hit the ground. ($g = 9.8\ m/s^2$) **(JEE 1994)**

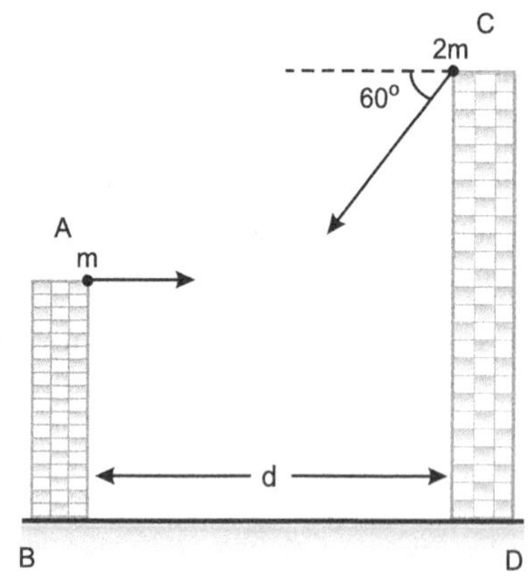

Problem 52. A bullet of mass M is fired with a velocity 50 m/s at an angle θ with the horizontal. At the highest point of its trajectory, it collides head on with a bob of mass 3 M suspended by a massless string of length $\frac{10}{3}$ m and gets embedded in the bob. After the collision the string moves through an angle 120°. Find :

(a) the angle θ

(b) the vertical and horizontal coordinates of the initial position of the bob with respect to the point of firing of the bullet ($g = 10\ m/s^2$). **(JEE 1988)**

CHAPTER 2

Laws Of Motion

Problem 53. Two men of masses M and $M + m$ start simultaneously from the ground and climb with uniform accelerations up from the free ends of a massless inextensible rope which passes over a smooth pulley at a height h from the ground.

(a) Which man reaches the pulley first.

(b) If the man who reaches first takes time t to reach the pulley. Find the distance of the second man from the pulley at that instant.

Problem 54. In the arrangement shown in figure, all pulleys are smooth and massless. When the system is released from rest, accelerations of blocks 2 and 3 relative to 1 are 1 m/s^2 downwards and 5 m/s^2 downwards. Acceleration of block 3 relative to 4 is zero.

Find the absolute accelerations of blocks 1, 2, 3 and 4.

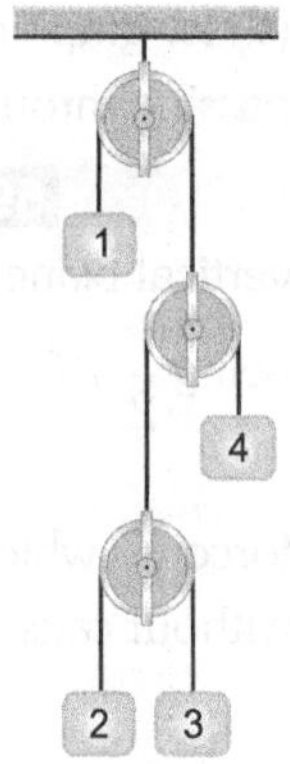

Problem 55. The system shown in figure is in equilibrium. Find the force F and magnitude of total force on the bearing of pulley C. Each pulley is free to rotate about its bearing, and the weights of all parts are small compared with the load. ($g = 10$ m/s^2)

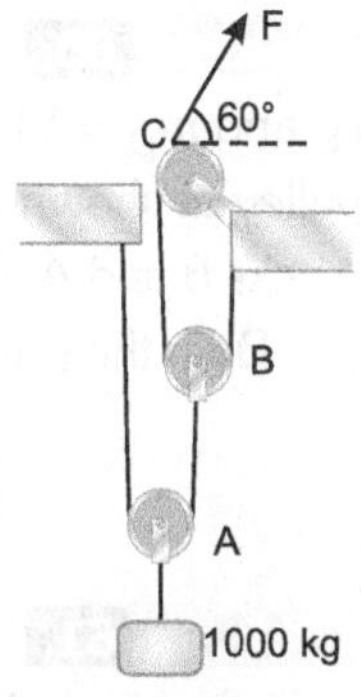

Problem 56. Determine the constraint equation which relates the accelerations of bodies A, B and C.

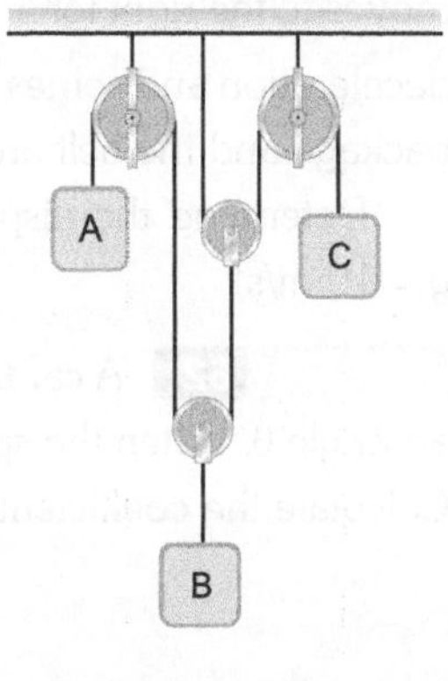

Problem 57. Under the action of a force F the constant acceleration of block B is 3 m/s^2 to the right. At the instant when the velocity of B is 2 m/s to the right, determine the velocity of B relative to A; the acceleration of B relative to A and the absolute velocity of point C of the cable.

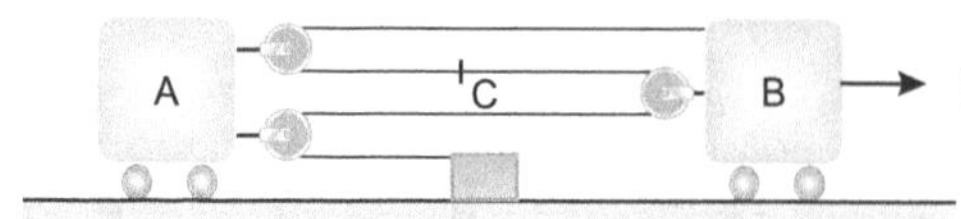

Problem 58. A small bead of mass m is free to move inside a smooth vertical semicircular ring of radius R. The bead is attached to one end of a massless spring of force constant $k = \dfrac{2mg}{R}$ and natural lenght $l_o = \dfrac{3R}{4}$. The other end of the spring is fixed at the centre of the ring. How does the normal reaction on the bead and tangential acceleration of it varies with angle θ (show graphically), where θ is the angle which the spring makes with a vertical line passing through the centre of the ring. The bead is released from the position shown in figure.

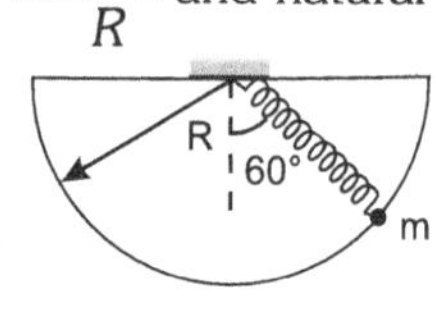

Problem 59. O is the centre of a circular disc of mass 50 kg which rests in a vertical plane on two rough pegs A and B, the coefficient of friction with each being $\frac{1}{2}$. AO makes 60° and BO makes 30° with the vertical. Find the maximum force F, which can be applied tangentially at the highest point of the disc without causing rotation of the disc. ($g = 10$ m/s^2)

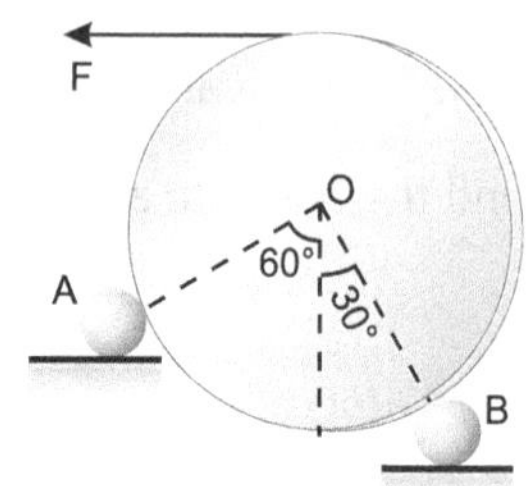

Problem 60. Three blocks shown in figure have the masses $m_A = 1$ kg, $m_B = 2$ kg and $m_C = 1$ kg. A time varying force (in newtons) $F = 20\,t$ is applied on the pulley as shown in figure (here t is in seconds). Find the relative velocity between blocks B and A, when block C has acquired a velocity of 2.5 m/s. ($g = 10$ m/s^2)

Both the pulleys are massless and friction is absent everywhere.

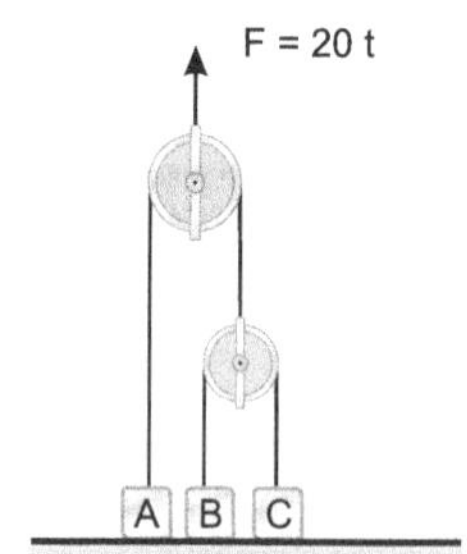

Problem 61. A package is at rest on a conveyor belt which is initially at rest. The belt is started and moves to the right for 1.3 s with a constant acceleration of 2 m/s^2. The belt then moves with a constant deceleration and comes to stop after a total displacement of 2.2 m. The coefficient of friction between the package and the belt are $\mu_s = 0.35$ and $\mu_k = 0.25$.

Determine the displacement of the package relative to the belt as the belt comes to stop. Take $g = 10$ m/s^2

Problem 62. A car travelling at 28 m/s has no tendency to slip on a track of radius 200 m banked at an angle θ. When the speed is increased to 35 m/s, the car is just on the point of slipping up the track. Calculate the coefficient of friction between the car and the track. (g = 9.8 m/s^2)

Problem 63. Two blocks of mass 10 kg and 2 kg respectively are connected by an ideal string passing over a fixed smooth pulley as shown in figure. A monkey of 8 kg started climbing the string with a constant acceleration $2\ m/s^2$ with respect to the string at $t = 0$. Initially the system is in equilibrium and the monkey is at a distance of 2 m from the pulley. Find the time taken by the monkey to reach the pulley.

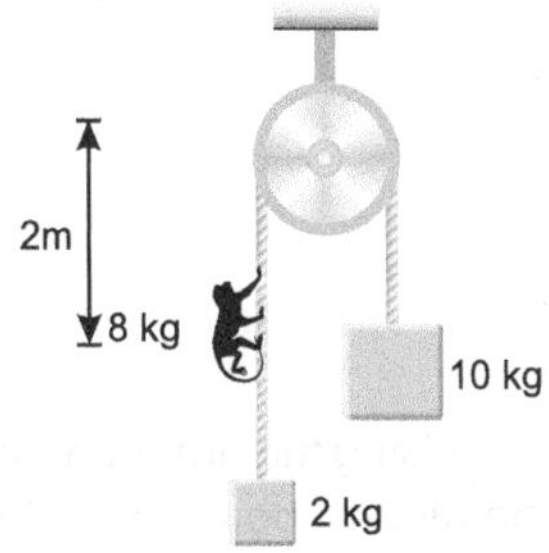

Problem 64. If masses of the blocks A and B shown in figure (a) and (b) are 10 kg and 5 kg respectively, find the acceleration of the two masses. Assume all pulleys and strings are ideal.

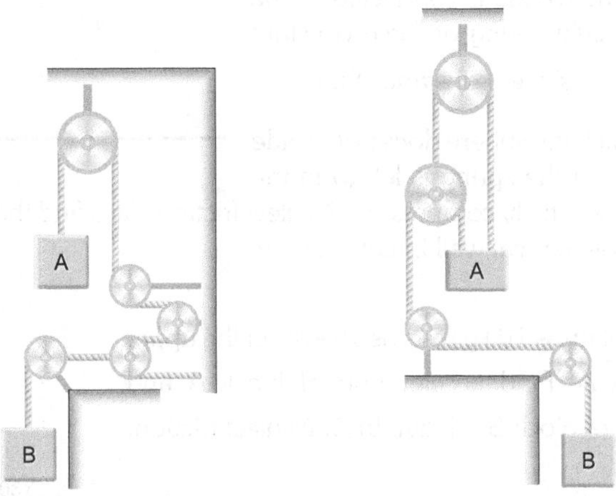

(a) (b)

Problem 65. Two identical blocks each having a mass of 20 kg are connected to each other by a light inextensible string as shown and are placed over a rough surface. Pulleys are connected to the blocks.

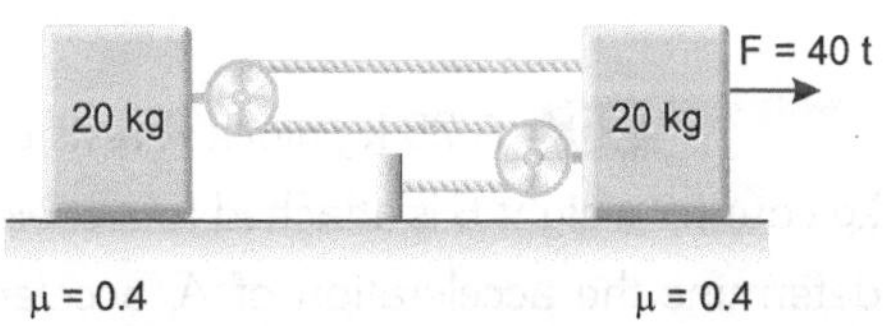

Find acceleration of the blocks after one second after the application of the time varying force of $40t$ N, where t is in second.

Problem 66. An insect lying on the bottom of the hemi-spherical bowl tries to come out from it. The coefficient of static friction between insect and bowl is 0.5. How high up does the insect go without slipping? Now if the bowl starts rotating about axis as shown in figure. At what angular speed ω will the insect just be able to come out of the bowl? (Radius of the bowl 5 cm)

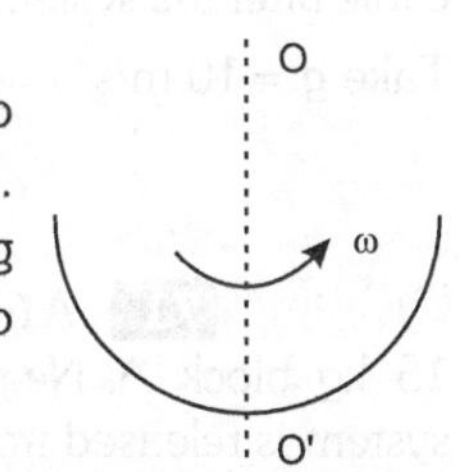

Problem 67. There is a parabolic-shaped bridge across a river of width 100 m. The highest point of the bridge is 5 m above the level of the banks. A car of mass 1000 kg is crossing the bridge at a constant speed of $20\ ms^{-1}$.

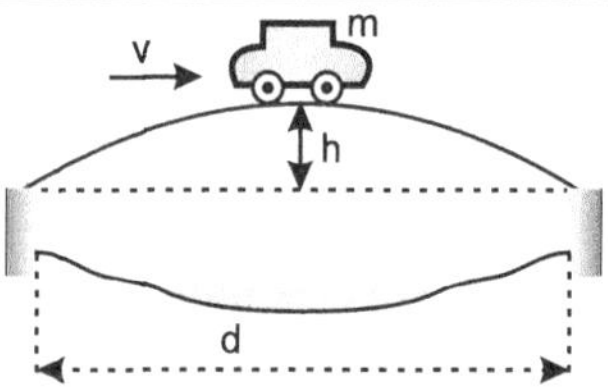

Using the notation indicated in the figure, find the force exerted on the bridge by the car when it is at the highest point of the bridge. (Ignore air resistance and take g as 10 ms^{-2})

Problem 68. A sphere rests between a smooth wall and a smooth wedge of mass M whose inclination to the horizontal is $\alpha = 60°$. Mass of sphere is m and its radius is R. The wedge initially touches the right wall. The vertical side of the wedge is connected to the side wall with the help of light spring of force constant $k = \eta \frac{mg}{R}$, where η is a positive constant. Find the minimum value of η for which the sphere does not collide with the horizontal surface; if the spring is let go in the position shown and spring is initially compressed. Neglect friction. Also find the normal reaction between the sphere and the right side vertical wall in critical case.

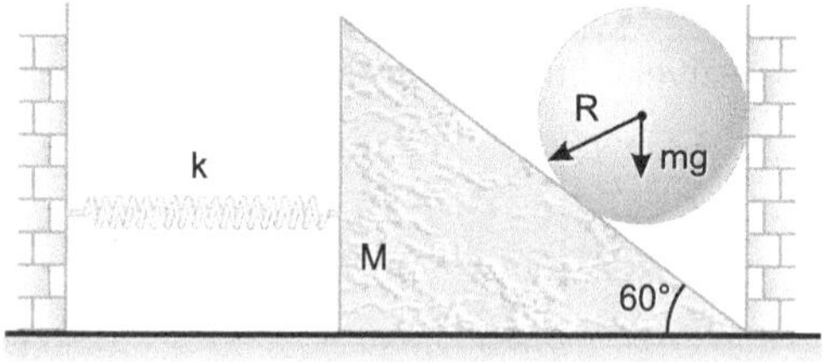

Problem 69. Block B of mass 10 kg rests as shown on the upper surface of a 22 kg block A. Find acceleration of block A and magnitude of acceleration of block B relative to A. Neglect friction. Wedge is fixed (g = 10 m/s^2).

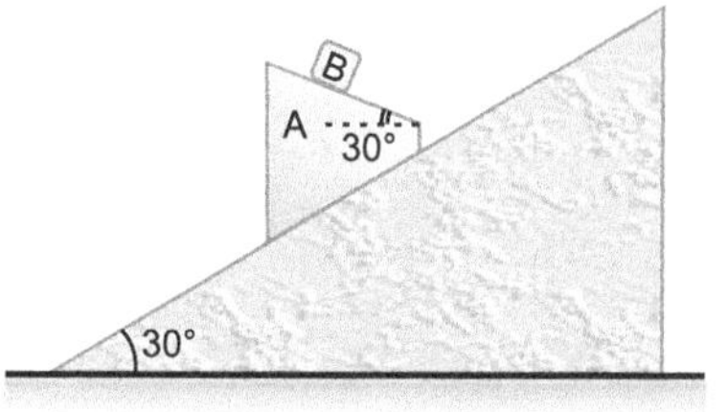

Problem 70. A 25 kg block A rests on an inclined surface and a 15 kg counter weight B is attached to a cable as shown. Neglecting friction, determine the acceleration of A, acceleration of B and tension in the cable after the system is released from rest. Cable is parallel to the plane. Take g = 10 m/s^2.

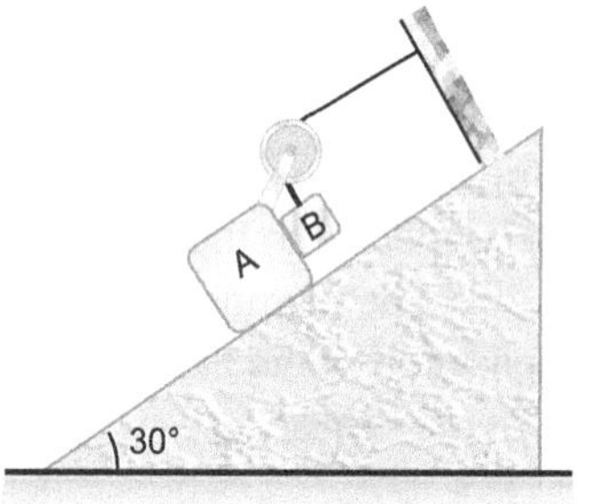

Problem 71. A 6 kg block B rests as shown on the upper surface of a 15 kg block A. Neglecting friction, determine immediately after the system is released from rest

(a) the acceleration of A

(b) the acceleration of B relative to A.

Take g = 10 m/s^2. (wedge is fixed)

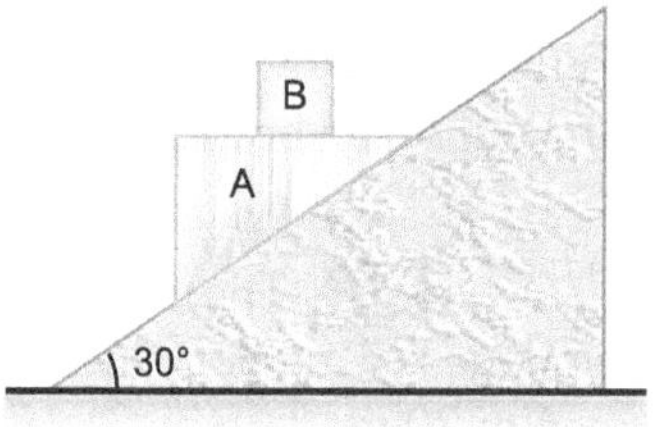

Problem 72. The 10 kg solid cylinder is resting in the inclined V-block. If the coefficient of static friction between the cylinder and the block is 0.5, determine

(a) the frictional force F acting on the cylinder at each side before force P is applied

(b) the value of P required to start sliding the cylinder up the incline ($g = 9.8 \text{ m/s}^2$)

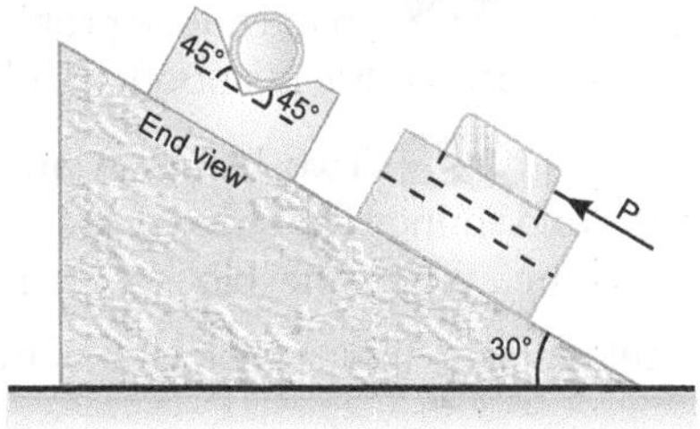

Problem 73. In the given figure, assume that there is no friction between block B and the surface on which it moves and the coefficient of friction between blocks A and B is μ.

(a) find the minimum value of M so that block A starts sliding over B.

(b) if M is two times that obtained in part (a), find the time when the block A topples from B.

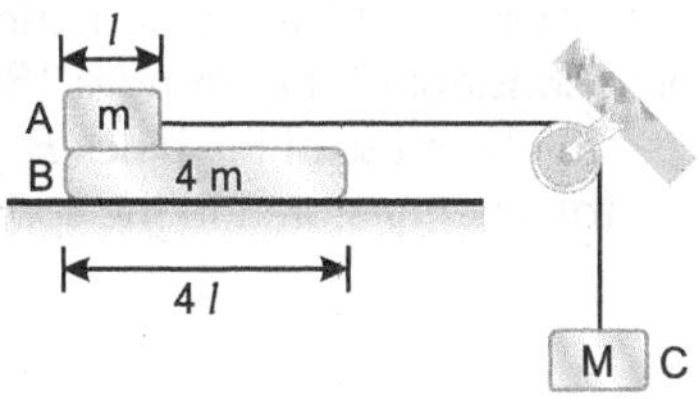

Problem 74. A wedge of mass $M = 4$ kg with a smooth quarter circular plane is kept on a rough horizontal plane. A particle of mass $m = 2$ kg is released from rest from the top of the wedge as shown in figure. Find the minimum value of coefficient of friction between the wedge and the horizontal plane so that the wedge does not move during complete journey of the particle.

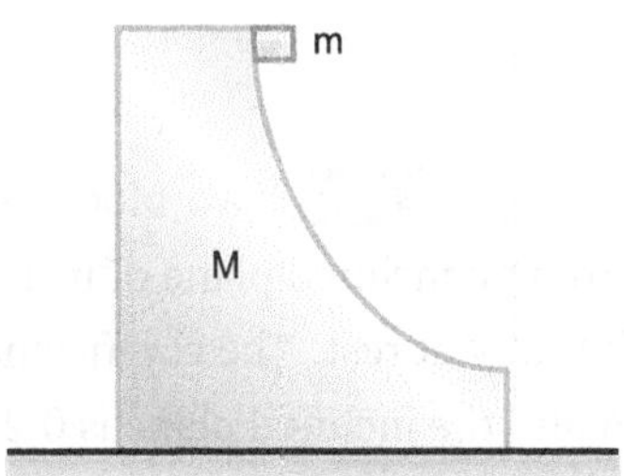

Problem 75. A block A of mass m and length l is placed on a horizontal floor. A rectangular box B is used to cover A. The distance between interior of the walls of B is $L (> l)$ and the mass of B is also m. The coefficient of friction between A and floor is μ_1 and that between B and floor is μ_2 ($\mu_2 > \mu_1$). Initially the left end of A touches the left wall of B as shown in figure and both A and B move with velocity v_0 towards the right. All collisions between A ad B are elastic and contact time during each collision is very short. Find an expression for the period between two consecutive collisions.

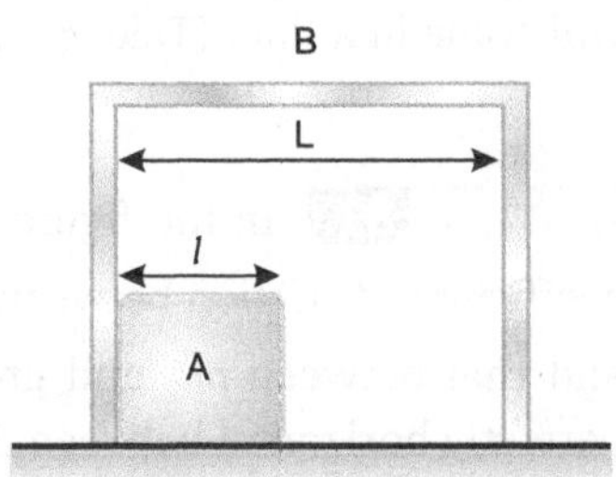

Problem 76. A point mass of 0.5 kg moving with a constant speed of 5 m/s on an elliptical track experiences an inward force of 10 N when at either end point of the major axis and a similar force of 1.25 N at each end of the minor axis. How long are the axes of the ellipse.

Problem 77. A disc of mass m and radius 1 m is hinged at its centre on a frictionless horizontal surface. It has a massless wall of short height around the circumference. A small particle of mass $\frac{m}{2}$ is projected with velocity 10 m/s keeping it in contact with the wall and base of the disc. If coefficient of friction between the small particle and the base of the disc in 0.5 and the wall is smooth. Find the angular displacement of the mass after 2 sec.

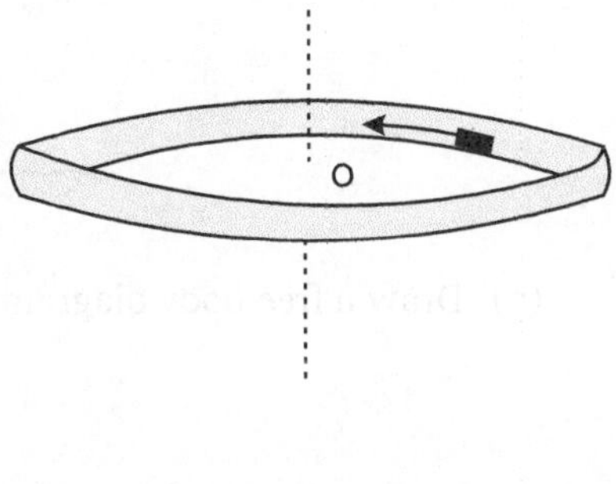

Problem 78. In the above problem, the coefficient of friction between the particle and the wall is 0.5 and the base of disc is smooth. Find the time after which relative motion between the two is stopped.

Problem 79. Two blocks A and B connected to each other by a string and spring. The string passes over a frictionless pulley as shown in the figure. When the block C is moving on an inclined plane with acceleration $\frac{g}{2}$ upward, block B of mass 2 kg slides over the top surface of block C and block A slides over the side wall of C, both with the same uniform speed. Coefficient of friction between all the blocks is 0.2. The force constant of the spring is 1800 N/m. Find:

(a) the mass of the block B.

(b) energy stored in the spring.

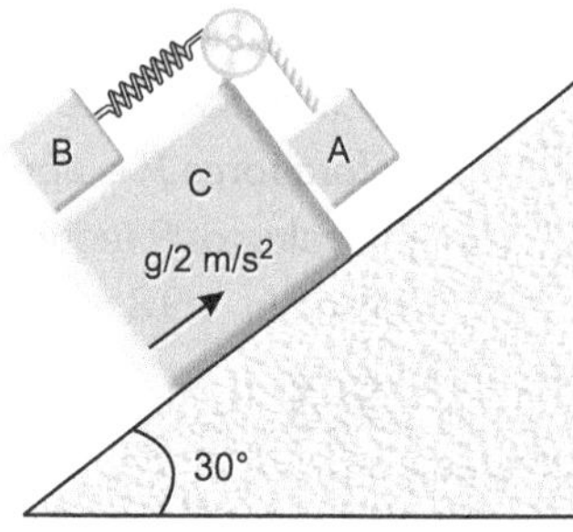

IIT JEE PROBLEMS

Problem 80. Two blocks A and B of equal masses are released from an inclined plane of inclination 45° at $t = 0$. Both the blocks are initially at rest. The coefficient of kinetic friction between the block A and the inclined plane is 0.2 while it is 0.3 for block B. Initially the block A is $\sqrt{2}$ m behind the block B. When and where their front faces will come in a line. (Take $g = 10 \text{ m/s}^2$) **(JEE 2004)**

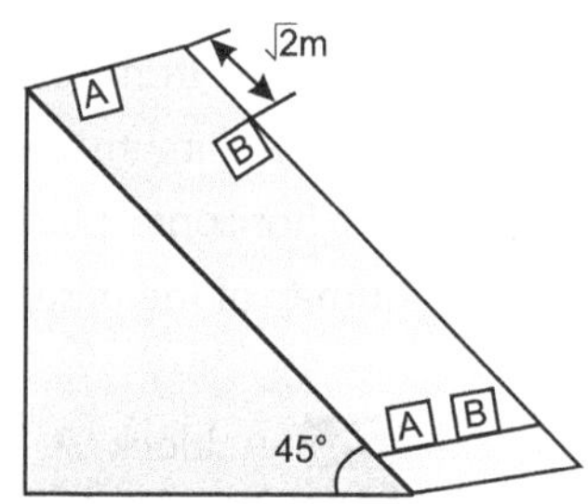

Problem 81. In the figure, masses m_1, m_2 and M are 20 kg, 5 kg and 50 kg respectively. The coefficient of friction between M and ground is zero. The coefficient of friction between m_1 and M and that between m_2 and ground is 0.3. The pulleys and the strings are massless. The string is perfectly horizontal between P_1 and m_1 and also between P_2 and m_2. The string is perfectly vertical between P_1 and P_2. An external horizontal force F is applied to the mass M. Take $g = 10 \text{ m/s}^2$.

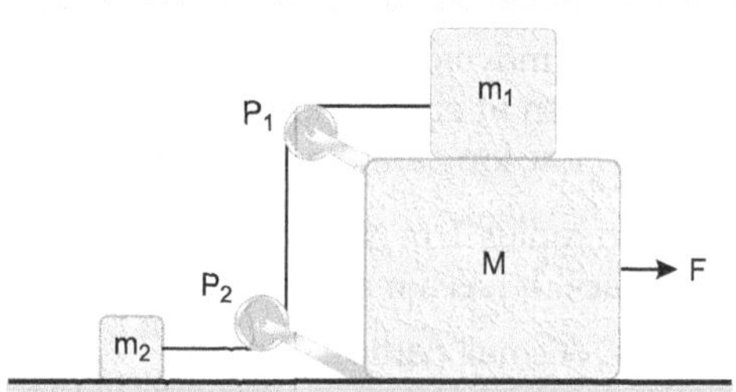

(a) Draw a free body diagram of mass M, clearly showing all the forces.

(b) Let the magnitude of the force of friction between m_1 and M be f_1 and that between m_2 and ground be f_2. For a particular F it is found that $f_1 = 2f_2$. Find f_1 and f_2. Write equations of motion of all the masses. Find F, tension in the string and accelerations of the masses.

(JEE 2000)

Problem 82. Two blocks of mass $m_1 = 10\,\text{kg}$ and $m_2 = 5\,\text{kg}$ connected to each other by a massless inextensible string of length 0.3 m are placed along a diameter of turn table. The coefficient of friction between the table and m_1 is 0.5 while there is no friction between m_2 and the table. The table is rotating with an angular velocity of 10 rad/s about a vertical axis passing through its centre O. The masses are placed along the diameter of the table on either side of the centre O such that the mass m_1 is at a distance of 0.124 m from O. The masses are observed to be at rest with respect to an observer on the turn table. ($g = 9.8\,\text{m/s}^2$)

(i) Calculate the frictional force on m_1.

(ii) What should be the minimum angular speed of the turn table so that the masses will slip from this position.

(iii) How should the masses be placed with the string remaining taut so that there is no frictional force acting on the mass m_1.

(JEE 1997)

Problem 83. A smooth semicircular wire track of radius R is fixed in a vertical plane. One end of a massless spring of natural length $3R/4$ is attached to the lowest point O of the wire track. A small ring of mass m which can slide on the track is attached to the other end of the spring. The ring is held stationary at point P such that the spring makes an angle 60° with the vertical. The spring constant $k = mg/R$. Consider the instant when the ring is released.

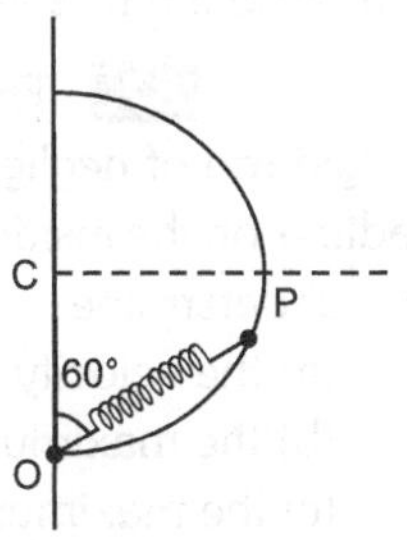

(i) Draw the free body diagram of the ring.

(ii) Determine the tangential acceleration of the ring and the normal reaction.

(JEE 1996)

Problem 84. A hemispherical bowl of radius $R = 0.1$ m is rotating about its own axis (which is vertical) with an angular velocity ω. A particle of mass 10^{-2} kg on the frictionless inner surface of the bowl is also rotating with the same ω.

The particle is at a height h from the bottom of the bowl.

(a) Obtain the relation between h and ω. What is the minimum value of ω needed, in order to have a non-zero value of h?

(b) It is desired to measure g (acceleration due to gravity) using this set-up, by measuring h accurately. Assuming that R and ω are known precisely, and that the least count in the measurement of h is 10^{-4} m, what is the minimum possible error Δg in the measured value of g?

(JEE 1993)

CHAPTER 3 WORK, POWER AND ENERGY

Problem 85. A thin circular rod is supported in a vertical plane by a bracket at A. Attached to the bracket and loosely wound around the rod is a spring of constant $k = 40$ N/m and undeformed length equal to arc of the circle AB. A 0.2 kg collar C, not attached to the spring, can slide without friction along the rod. The collar is released from rest at an angle θ with the vertical.

(a) Make the equation for minimum value of θ for which the collar will pass through D and reach point A.

(b) Determine the velocity of collar as it reaches point A for minimum value of θ. (Take $g = 10 \text{ m/s}^2$)

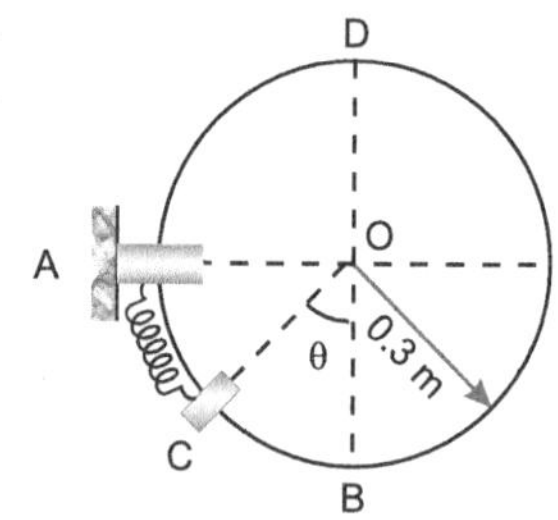

Problem 86. A body of mass m slides down a plane inclined at an angle α. The coefficient of friction is μ. Find the rate at which kinetic plus gravitational potential energy is dissipated at any time t.

Problem 87. The two particles of mass m and $2m$ respectively are connected by a rigid rod of negligible mass and slide with negligible friction in a circular path of radius r on the inside of the vertical circular ring. If the system is released from rest at $\theta = 0°$ determine

(a) the velocity v of the particles when the rod passes the horizontal position,

(b) the maximum velocity v_{max} of the particles and

(c) the maximum value of θ.

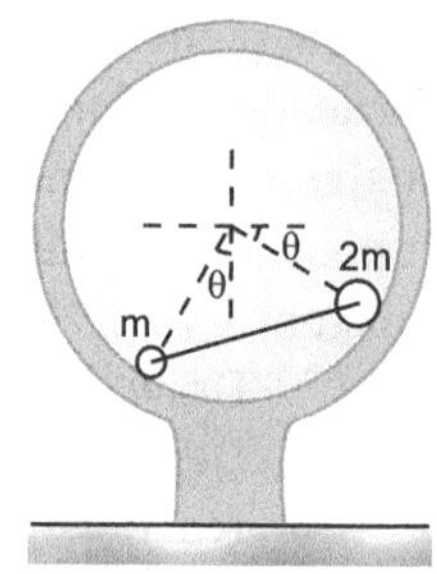

Problem 88. A person rolls a small ball with speed u along the floor from point A. If $x = 3R$, determine the required speed u so that the ball returns to A after rolling on the circular surface in the vertical plane from B to C and becoming a projectile at C. What is the minimum value of x for which the game could be played if contact must be maintained at point C. Neglect friction.

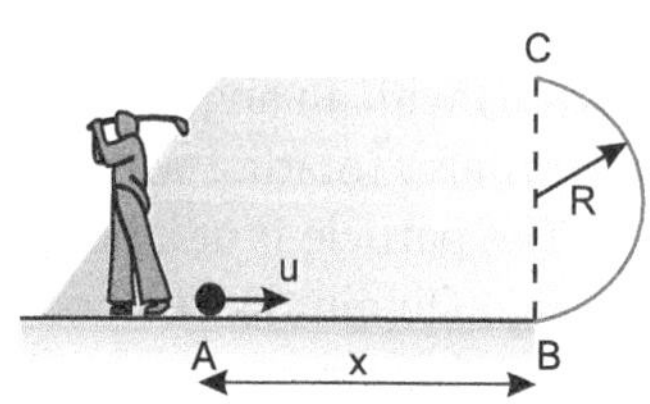

Problem 89. If the system is released from rest, determine the speeds of both masses after B has moved 1 m. Neglect friction and masses of the pulleys. ($g = 10 \text{ m/s}^2$)

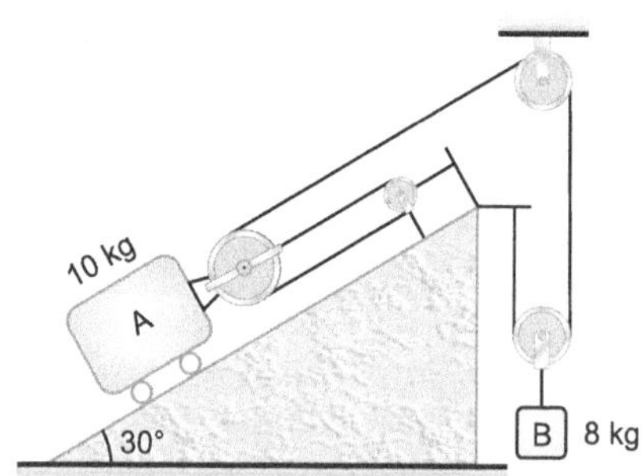

Problem 90. The free end of a flexible rope of length L and mass λ per unit length is released from rest in the position shown in the (a)-part of the figure. Determine the velocity v of the moving portion of the rope in terms of y.

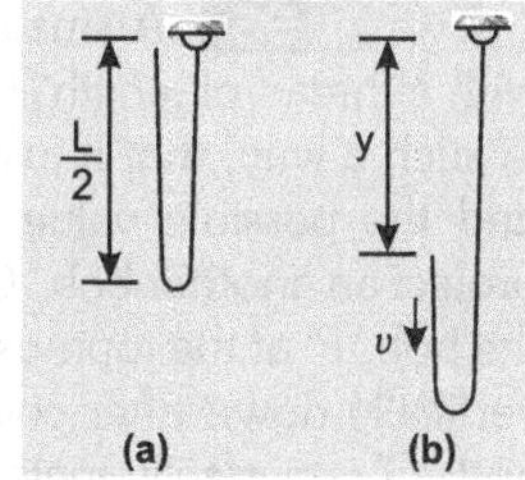

Problem 91. A disc slides up without friction a hill of height h whose profile depends only on the x-coordinate (figure *a*). At the bottom the disc has the velocity v_1 whose direction forms the angle α_1 with the X-axis (see figure *b*, top view). Find the direction of motion of the disc after it reaches the top, *i.e.*, find the angle α_2. Also describe the condition that the disc cannot overcome the hill.

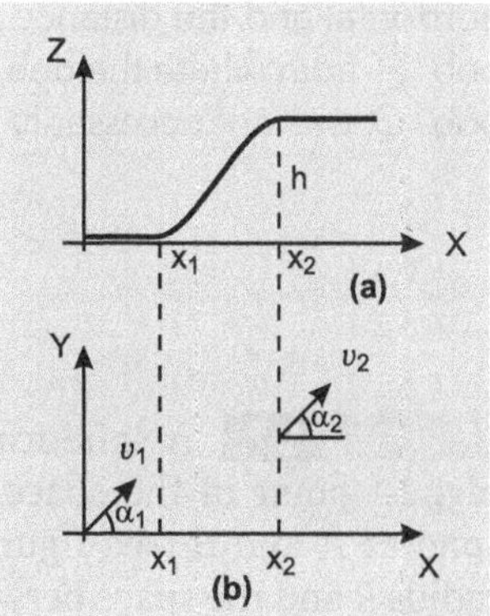

Problem 92. A small object loops a vertical loop of radius R in which a symmetrical section of 2α has been removed. Find the maximum and minimum heights from which the object after loosing contact with the loop at point A and flying through the air, will reach point B. Find the corresponding angles of the section removed for which this is possible.

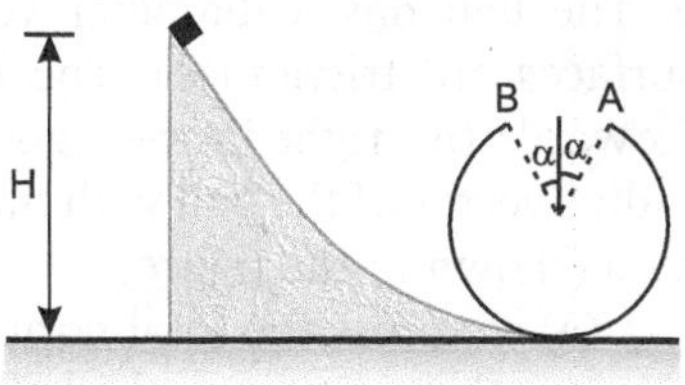

Problem 93. A car of mass $4m$ holds a block of mass m which is attached to the former by means of a spring of spring constant k, as shown in the diagram. All surfaces are frictionless and the wheels are massless. The system is at rest. A bullet of mass m is fired at the first block with a horizontal velocity v_0 and sticks to it. Find:

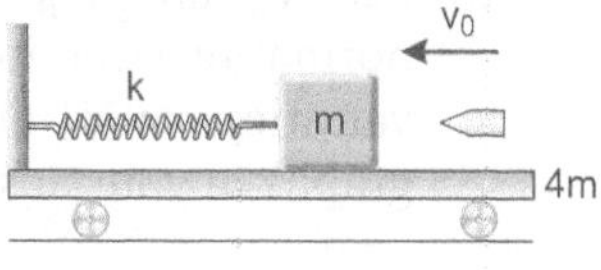

(a) the velocity of the car at the moment when the spring undergoes maximum compression.

(b) the maximum compression of the spring.

Problem 94. A block of mass 2 kg approaches a spring of stiffness $k = 80$ N/m on a smooth horizontal plane with a speed of 10 m/s. Find the time in which kinetic energy of the block reduces to 50% of its initial kinetic energy. Assume $t = 0$ when the block just touches the spring, for the first time.

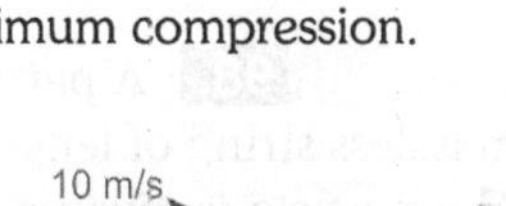

Problem 95. A pendulum bob of mass m is suspended at rest. A constant horizontal force $F = mg$ starts acting on it. Find:

(a) the maximum angular deflection of the string.

(b) the maximum possible tension in the string.

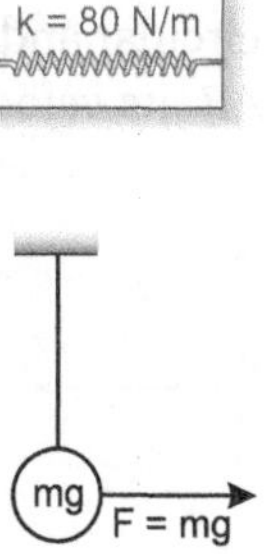

Problem 96. A small bob '*P*' of mass '*m*' is attached with a thread of length '*l*'. If the bob '*P*' is given a velocity in such a way, that it just completes a vertical circle, then find the possible value of velocity, with which we can project an another bob '*Q*' of same mass '*m*', which can hit the bob '*P*' at the topmost point of their trajectory and falls vertically down after collision (The collision between '*P*' and '*Q*' is perfectly inelastic). Also find the angle with the horizontal and the distance from the bottommost point of bob '*P*' from where the bob '*Q*' should be projected. (The bob '*Q*' was not necessarily projected, when '*P*' was at the bottom of its trajectory).

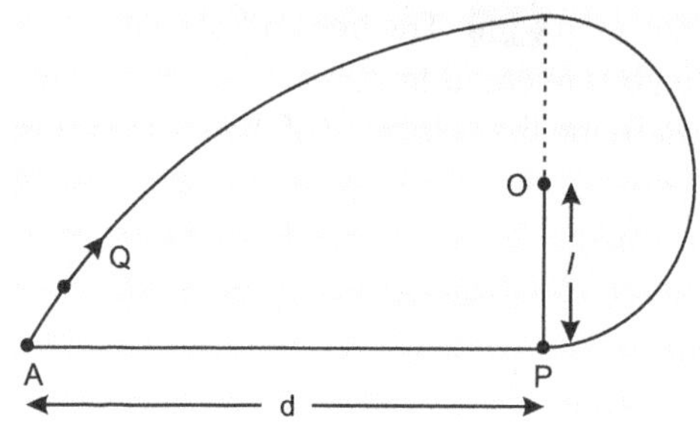

IIT JEE PROBLEMS

Problem 97. A spherical ball of mass m is kept at the highest point in the space between two fixed, concentric spheres A and B (see figure). The smaller sphere A has a radius R and the space between the two spheres has a width d. The ball has a diameter very slightly less then d. All surfaces are frictionless. The ball is given a gentle push (towards the right in the figure). The angle made by the radius vector of the ball with the upward vertical is denoted by θ (shown in the figure).

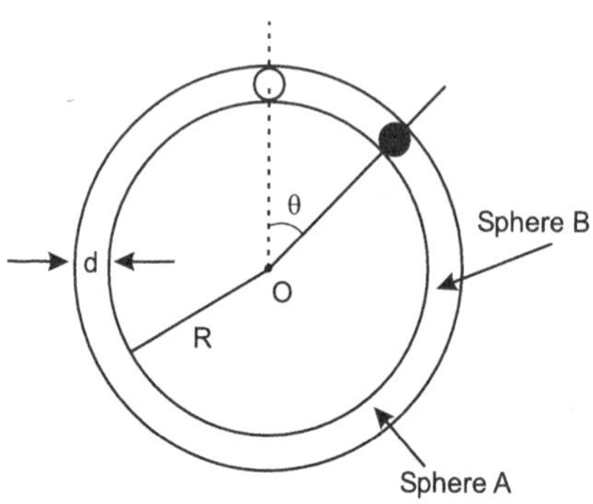

(a) Express the total normal reaction force exerted by the spheres on the ball as a function of angle θ.

(b) Let N_A and N_B denote the magnitudes of the normal reaction forces on the ball exerted by the spheres A and B, respectively. Sketch the variations of N_A and N_B as function of $\cos\theta$ in the range $0 \le \theta \le \pi$ by drawing two separate graphs in your answer book, taking $\cos\theta$ on the horizontal axes. **(JEE 2002)**

Problem 98. A particle is suspended vertically from a point O by an inextensible massless string of length L. A vertical line AB is at a distance of $L/8$ from O as shown. The particle is given a horizontal velocity u. At some point, its motion ceases to be circular and eventually the object passes through the line AB. At the instant of crossing AB, its velocity is horizontal. Find u. **(JEE 1999)**

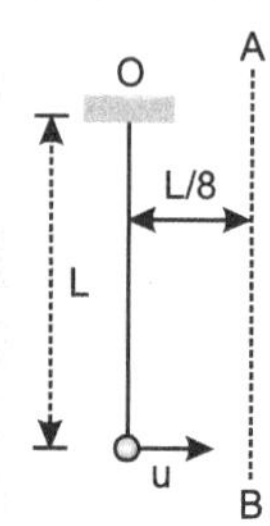

CHAPTER 4

Centre Of Mass, Conservation Of Momentum, Collision, Impulse

Problem 99. A 60 kg man and a 50 kg woman are standing on opposite ends of a platform of mass 20 kg. The platform is placed on a smooth horizontal ground. The man and woman begin to approach each other. Find the displacement of the platform when the two meet in terms of the displacement x_0 of the man relative to the platform. The length of the platform is 6 m.

Problem 100. A drinking straw of length $3a/2$ and mass $2m$ is placed on a square table of side 'a' parallel to one of its sides such that one third of its length extends beyond the table. An insect of mass $m/2$ lands on the inner end of the straw (*i.e.,* the end which lies on the table) and walks along the straw until it reaches the outer end. It does not topple even when another insect lands on top of the first one. Find the largest mass of the second insect that can have without toppling the straw. Neglect friction.

Problem 101. A cannon of mass m slides down a smooth inclined plane forming the angle α with the horizontal. At the moment when the velocity of cannon reaches $\vec{v}$ it fires a shell in a horizontal direction with the result that the cannon stops and the shell carries away the momentum $\vec{p}$. Suppose that the firing duration is Δt. What is the reaction force of the inclined plane averaged over the time Δt.

Problem 102. A car of mass 1000 kg and running at 25 m/s holds three men each of mass 75 kg. Each man runs with a speed of 5 m/s relative to the car and jumps off from the back end. Find the speed of the car if the three men jump off

(i) in succession

(ii) all together. Neglect friction between the car and the ground.

Problem 103. A ring of mass m can slide on a smooth horizontal wire. The ring is attached to a particle of mass $3m$ by a string of length l. A horizontal velocity v_o is given to the ring. Find the maximum angle the string will make with the vertical in subsequent motion.

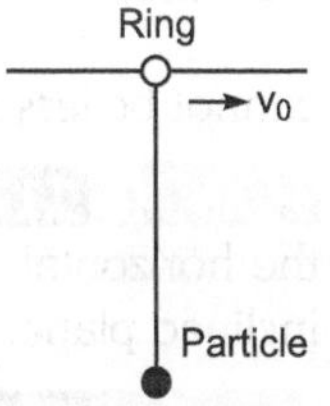

Problem 104. Two bodies of mass $m_1 = 1$ kg and $m_2 = 2$ kg move towards each other in mutually perpendicular directions at velocities $v_1 = 3$ m/s and $v_2 = 2$ m/s. As a result of collision, the bodies stick together. Determine the amount of heat liberated as a result of collision.

Problem 105. A particle of mass m moving with a velocity $(3\hat{i} + 2\hat{j})$ m/s collides with a stationary body of mass M and finally moves with a velocity $(-2\hat{i} + \hat{j})$ m/s. If $\frac{m}{M} = \frac{1}{13}$, find:

(a) the impulse received by each.
(b) the velocity of the M.
(c) the coefficient of restitution.

Problem 106. A hemisphere of mass M and radius R is at rest. One solid sphere of mass $2M$ and radius R, moving with a velocity v_o, collides with the hemisphere. If e is the coefficient of restitution, find the speed of the hemisphere and the solid sphere after collision.

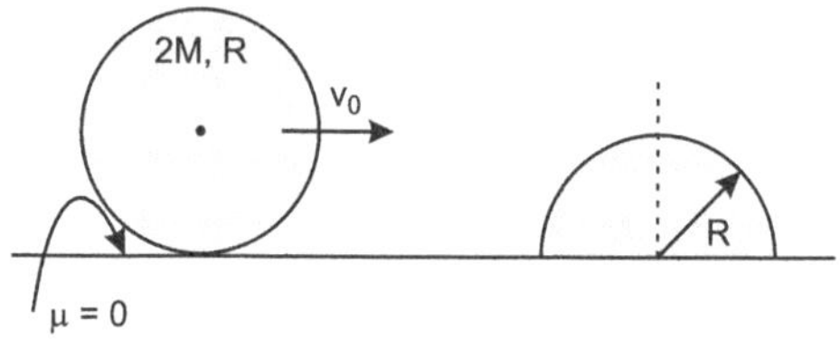

Problem 107. You are given three billiards-tables of different lengths and the same width. Balls are struck simultaneously from the edge of one of the long sides of each table (fig.) with velocities which are equal in direction and magnitude. Is it possible that these balls should not return to the side from which they started at exactly the same moment?

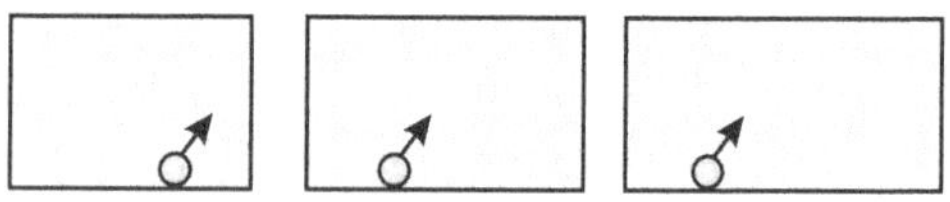

Problem 108. A rope thrown over a pulley has a ladder with a man A on one of its ends and a counterbalancing mass M on its other end. The man whose mass is m, climbs upwards by $\Delta\vec{r}$ relative to the ladder and then stops. Ignoring the masses of the pulley and the rope, as well as the friction in the pulley axis, find the displacement of the centre of mass of this system.

Problem 109. Two identical particles are projected at the same instant from points A and B at the same level, the first from A towards B with velocity u at $45°$ above AB and the second from B towards A with velocity v at $60°$ above BA. If the particles collide directly when each reaches its greatest height, find the ratio $\frac{v^2}{u^2}$ and prove that $u^2 = ag\,(3-\sqrt{3})$ where a is the distance AB. After the collision the first particle falls vertically. Show that the coefficient of restitution between the particles is $(\sqrt{3}-1)/(\sqrt{3}+1)$.

Problem 110. A boy throws a ball with initial speed $2\sqrt{ag}$ at an angle θ to the horizontal. It strikes a smooth vertical wall and returns to his hand. Show that if the boy is standing at a distance 'a' from the wall, the coefficient of restitution between the ball and the wall equals $\frac{1}{(4\sin 2\theta - 1)}$. Also show that θ cannot be less than 15°.

Problem 111. A ball falls freely from a height onto an smooth inclined plane forming an angle α with the horizontal. Find the ratio of the distance between the points at which the jumping ball strikes the inclined plane. Assume the impacts to be elastic.

Problem 112. A ball is projected from a point A on a smooth inclined plane which makes an angle α to the horizontal. The velocity of projection makes an angle θ with the plane upwards. If on the second bounce the ball is moving perpendicular to the plane, find e in terms of α and θ. Here e is the coefficient of restitution between the ball and the plane.

Problem 113. Two identical smooth balls are projected towards each other from points A and B on the horizontal ground with same speed of projection. The angle of projection in each case is 30°. The distance between A and B is 100 m. The balls collide in air and return to their respective points of projection. If coefficient of restitution is $e = 0.7$, find

(a) the speed of projection of either ball.
(b) coordinates of point with respect to A where the balls collide. (Take $g = 10$ m/s^2)

Problem 114. A ball strikes the wall at point A ($e_1 = 0.5$) and then hits the ground at B ($e_2 = 0.3$). Find the distance x from the point O, where should a fielder be to catch the ball at a height of 1.2 m from the ground. Here e is coefficient of restitution. Take $g = 10$ m/s^2.

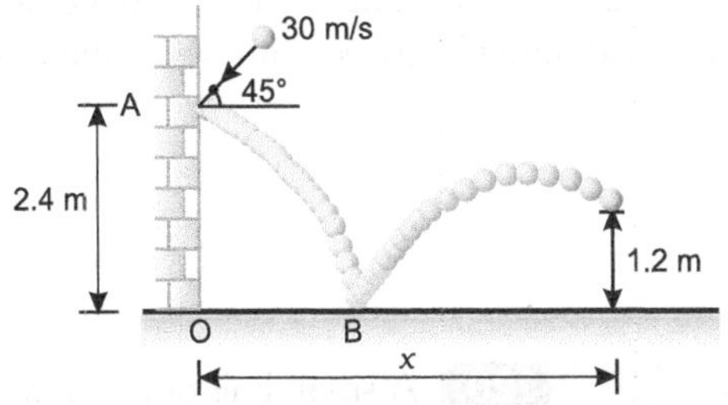

Problem 115. Two smooth spheres A and B of equal radius but of masses m and M respectively are free to move on a horizontal table. A is projected with speed u towards B which is at rest. On impact, the line joining their centres is inclined at an angle θ to the velocity of A before impact. If e is the coefficient of restitution between the spheres, find the speed with which B begins to move. If A deviates by 90° from its initial path, find the angle θ.

Problem 116. A shell of mass $m = 1$ kg is fired from a gun of mass $M = 10$ kg from $x = 0$ with an initial speed u relative to the gun as shown in figure. The gun is situated on a horizontal floor. The gun strikes a wall at $x = L$ moving towards negative x-direction with velocity v. Find the coefficient of restitution between the shell and the wall so that the shell returns to the point of the gun from where it was started. Take $u = 10$ m/s, $L = 4$ m and $v = 2$ m/s. Neglect friction. ($g = 10$ m/s^2)

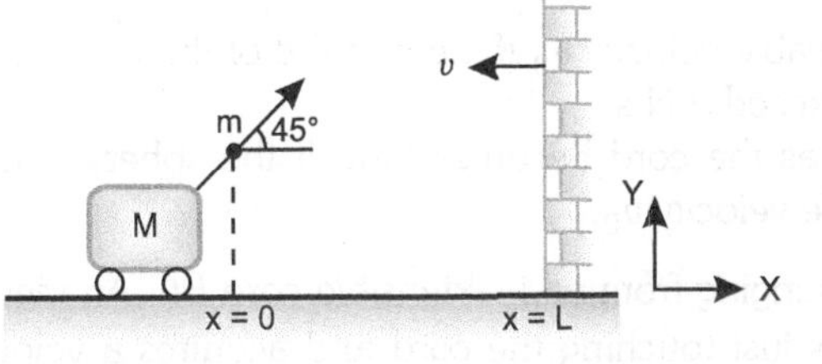

Problem 117. Two identical cubes A and B of same mass $2m$ and side $2d$ are placed one over the other as shown in figure. B is attached to one end of a spring of force constant k. The other end of the spring is attached to a wall. The system is resting on a smooth horizontal surface with the spring in the relaxed state. A small object of mass m moving horizontally with speed v at a height d above the horizontal surface hits elastically the block B along the line of their centre of mass. There is no friction between A and B :

(a) find the minimum value of v (say v_0) such that block A will topple over block B
(b) if $v = v_0/2$, find the amplitude of oscillation of block spring system.

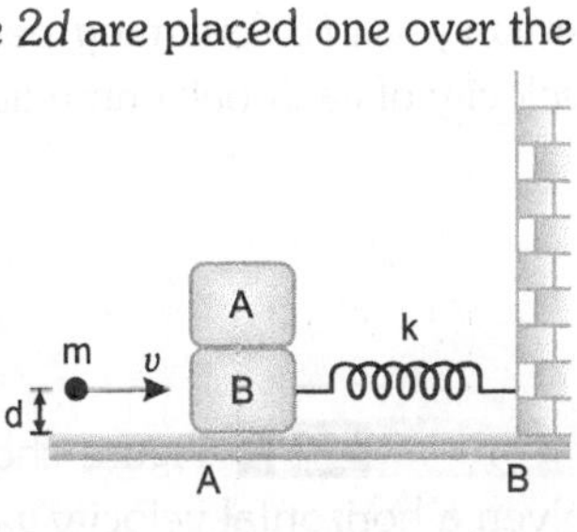

Problem 118. Three identical particles A, B and C lie on a smooth horizontal table. Light inextensible strings which are just taut connect AB and BC and $\angle ABC$ is 135°. An impulse J is applied to the particle C in the direction BC. Find the initial speed of each particle. The mass of each particle is m.

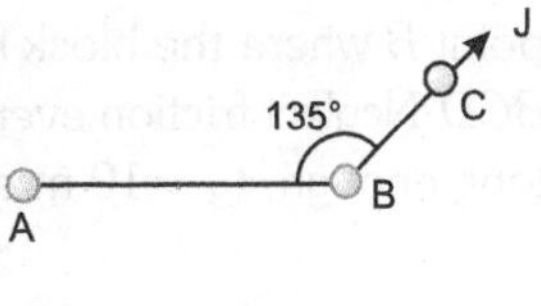

Problem 119. A mass $2m$ rests on a horizontal table. It is attached to a light inextensible string which passes over a smooth pulley and carries a mass m at the other end. If the mass m is raised vertically through a distance h and is then dropped, find the speed with which the mass $2m$ begins to rise.

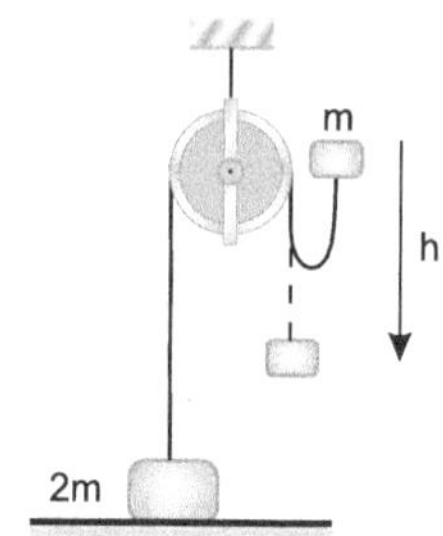

Problem 120. A small ball of mass m is connected by an inextensible massless string of length l with an another ball of mass $M = 4m$. They are released with zero tension in the string from a height h as shown in figure. Find the time when the string becomes taut for the first time after the mass M collides with the ground. Take all collisions to be elastic.

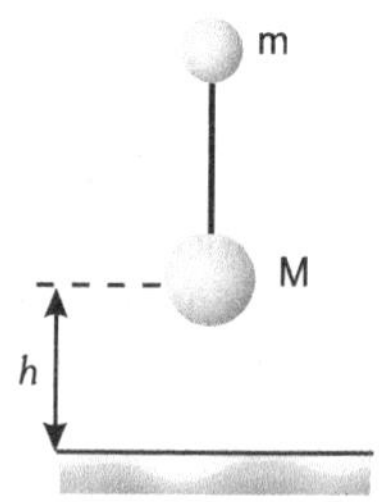

Problem 121. A 2 kg sphere A is connected to a fixed point O by an inextensible cord of length 1.2 m. The sphere is resting on a frictionless horizontal surface at a distance of 0.5 m from O when it is given a velocity v_0 in a direction perpendicular to the line OA. It moves freely until it reaches position A' when the cord becomes taut. Determine

(a) the maximum allowable velocity v_0 if the impulse of the force exerted on the cord is not to exceed 3 N-s

(b) the loss of energy as the cord becomes taut, if the sphere is given the maximum allowable velocity v_0.

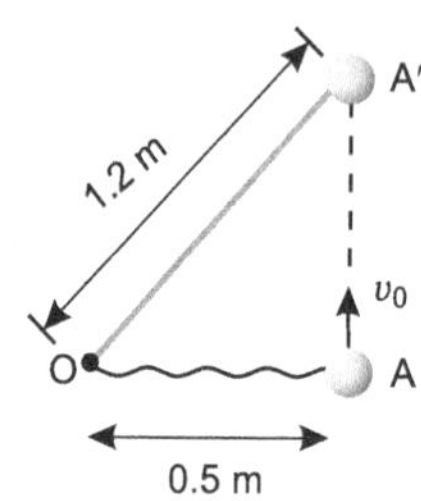

Problem 122. Ball B is hanging from an inextensible cord BC. An identical ball A is released from rest when it is just touching the cord and acquires a velocity v_0 before striking ball B. Assuming perfectly elastic impact ($e = 1$) and no friction, determine the velocity of each ball immediately after impact.

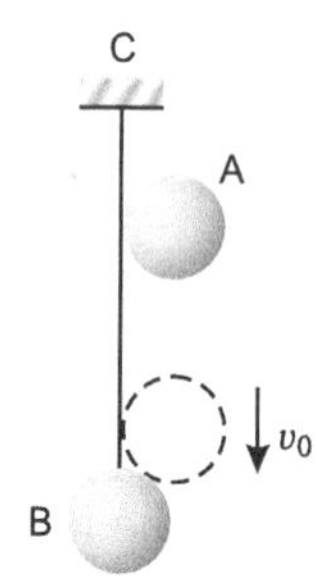

Problem 123. Figure shows a small block of mass $m = 1$ kg which is given a horizontal velocity $v_0 = 10$ m/s on the horizontal part of the bigger block of mass $M = 9$ kg placed on a horizontal floor. The curved part of the surface shown is semicircular of radius $R = 1$ m. Find the distance from point B where the block m lands finally after looping the semicircular part BCD. Neglect friction everywhere. Assume that the horizontal portion AB is long enough. ($g = 10$ m/s^2)

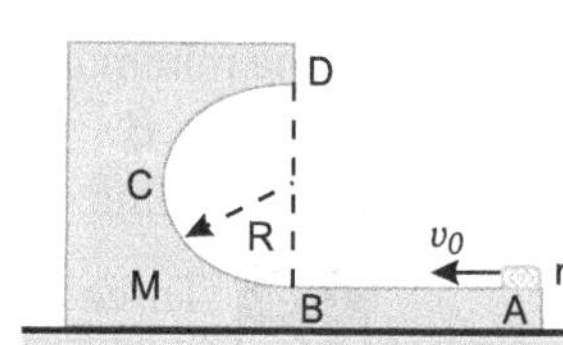

Problem 124. A railroad car of length L and mass m_0 when empty is moving freely on a smooth horizontal track while being loaded with sand from a stationary chute at a rate $\frac{dm}{dt} = q$. Knowing that the car was approaching the chute at a speed v_0. Determine

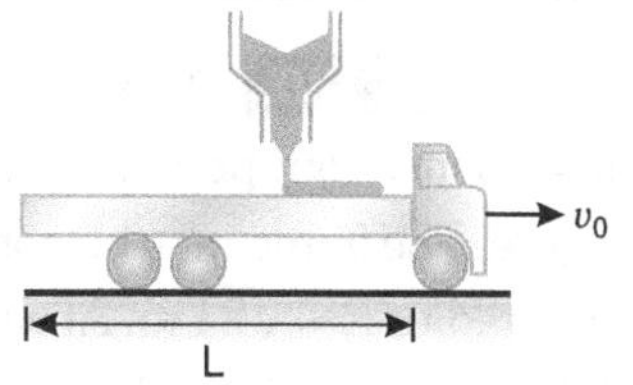

(a) the speed of the car v_f at the instant when the car has cleared the chute.
(b) the mass of the car and its load at that instant.

Problem 125. A particle whose initial mass is m_0 is projected vertically upwards at time $t = 0$ with speed gT, where T is a constant. At time t the mass of the particle has increased to $m_0 e^{t/T}$. If the added mass is at rest relative to particle when it is acquired, find the time when it is at highest point and mass at that instant.

Problem 126. A rain drop falls from rest in an atmosphere saturated with water vapour. As it falls, water vapour condenses on the drop at the rate of mass μ per second. If initial mass of drop is m_0, how much distance the drop falls in time t.

Problem 127. A cannon with bullets of total mass M_0 is kept on a rough horizontal surface. The coefficient of friction between the cannon and the surface is μ. If the cannon fires bullets with constant frequency with a relative velocity u, find the velocity of the cannon when its mass with remaining bullets become M after time t. Initially cannon was at rest. Assume that the thrust force is greater than the limiting friction right from the beginning.

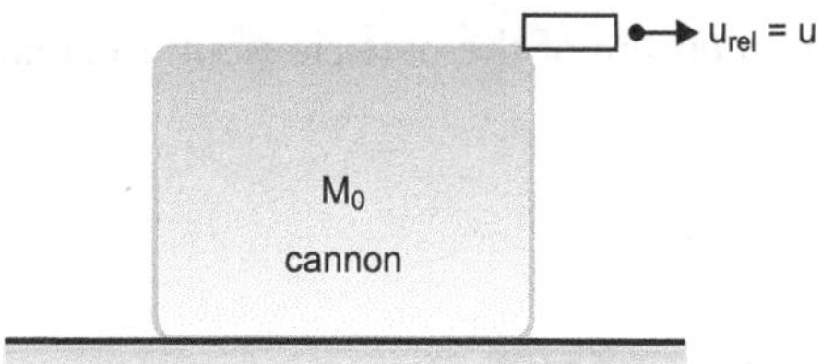

Problem 128. A long thin carpet is laid on the floor. One end of the carpet is bent back and then pulled backwards with constant velocity $v = 1$ m/s, just above the part of the carpet which is still at rest on the floor.

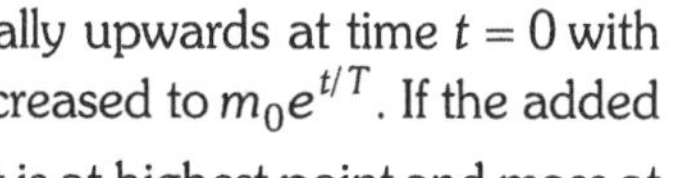

(a) Find the speed of centre of mass of the moving part.
(b) What is the minimum force needed to pull the moving part, if the carpet has unit length and unit mass.

Problem 129. A light inextensible thread passes over a small frictionless pulley. Two blocks of masses $m = 1$ kg and $M = 3$ kg respectively are attached with the thread as shown in the figure.

The heavier block rests on a horizontal surface. A shell of mass 1 kg moving upward with a velocity 10 ms^{-1} collides and sticks with the block of mass m as shown in the figure at $t = 0$. If the inclined plane is smooth. Calculate:

(a) maximum height ascended by M.
(b) time 't' at that instant.

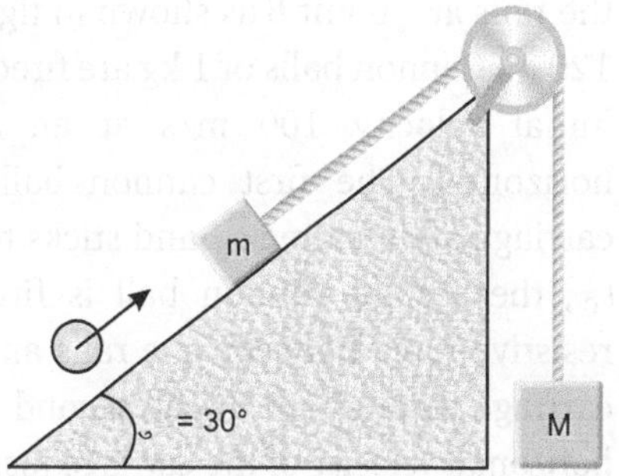

Problem 130. A particle is projected horizontally with a speed of $u = 10$ m/s from point A. A steel frame is rigidly fixed as shown in the figure. The frame may be considered as an arc of circle with centre at A and radius $R = \sqrt{21}$ m. At which point will the particle strike the frame? If the particle rebounds elastically from the frame will it again strike it? Take $g = 10$ m/s^2

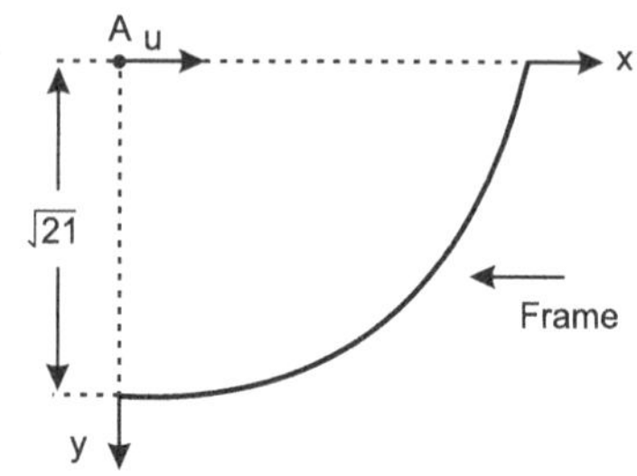

IIT JEE PROBLEMS

Problem 131. A particle of mass m, moving in a circular path of radius R with a constant speed v_2 is located a point $(2R, 0)$ at time $t = 0$ and a man starts moving with a velocity v_1 along the positive y-axis from origin at time $t = 0$. Calculate the linear momentum of the particle w.r.t. the man as a function of time.
(JEE 2003)

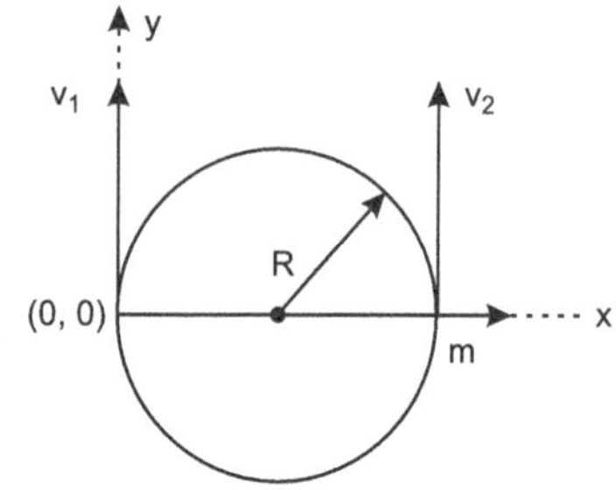

Problem 132. Two point masses m_1 and m_2 are connected by a spring of natural length l_0. The spring is compressed such that the two point masses touch each other and then they are fastened by a string. Then the system is moved with a velocity v_0 along positive x-axis. When the system reaches the origin the string breaks $(t = 0)$. The position of the point mass m_1 is given by

$$x_1 = v_0 t - A(1 - \cos \omega t) \qquad \text{where } A \text{ and } \omega \text{ are constants.}$$

Find the position of the second block as a function of time. Also find the relation between A and l_0.
(JEE 2003)

Problem 133. A car P is moving with a uniform speed of $5\sqrt{3}$ m/s towards a carriage of mass 9 kg at rest kept on the rails at a point B as shown in figure. The height AC is 120 m. Cannon balls of 1 kg are fired from the car with an initial velocity 100 m/s at an angle 30° with the horizontal. The first cannon ball hits the stationary carriage after a time t_0 and sticks to it. Determine t_0. At t_0, the second cannon ball is fired. Assume that the resistive force between the rails and the carriage is constant and ignore the vertical motion of the carriage throughout. If the second cannon ball also hits and sticks to the carriage. What will be the horizontal velocity of the carriage just after the second impact ? Take $g = 10$ m/s^2 **(JEE 2001)**

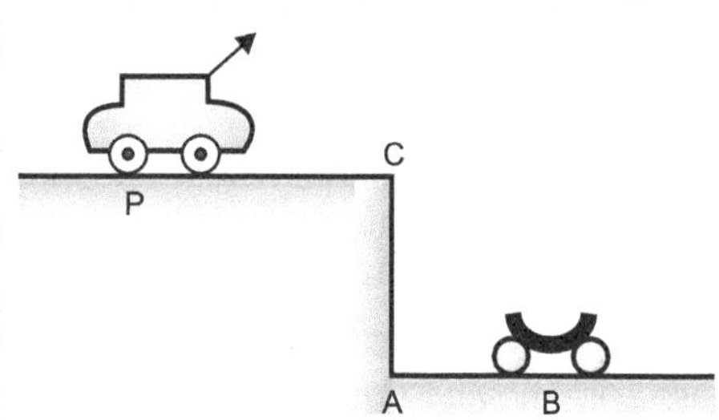

Problem 134. Two blocks of mass 2 kg and M are at rest on an inclined plane and are separated by a distance of 6.0 m as shown. The coefficient of friction between each block and the inclined plane is 0.25. The 2 kg block is given a velocity of 10.0 m/s up the inclined plane. It collides with M, comes back and has a velocity of 1.0 m/s when it reaches its initial position. The other block M after the collison moves 0.5 m up and comes to rest. Calculate the coefficient of restitution between the blocks and the mass of the block M.

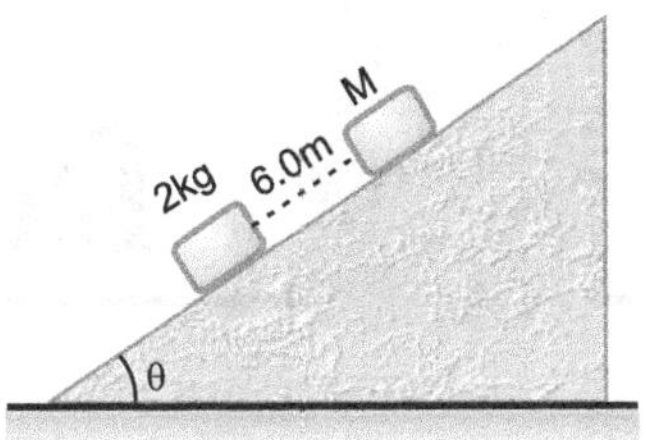

[take $\sin\theta \approx \tan\theta = 0.05$ and $g = 10\ \text{m/s}^2$] **(JEE 1999)**

Problem 135. A large open top container of negligible mass and uniform cross-sectional area A has a small hole of cross-sectional area $A/100$ in its side wall near the bottom. The container is kept on a smooth horizontal floor and contains a liquid of density ρ and mass m_0. Assuming that the liquid starts flowing out horizontally through the hole at $t = 0$, calculate

(i) the acceleration of the container, and

(ii) its velocity when 75% of the liquid has drained out. **(JEE 1997)**

Problem 136. A small sphere of radius R is held against the inner surface of larger sphere of radius $6R$ (as shown in figure). The masses of large and small spheres are $4M$ and M respectively. This arrangement is placed on a horizontal table. There is no friction between any surfaces of contact. The small sphere is now released. Find the coordinates of the centre of the large sphere, when the smaller sphere reaches the other extreme position. **(JEE 1996)**

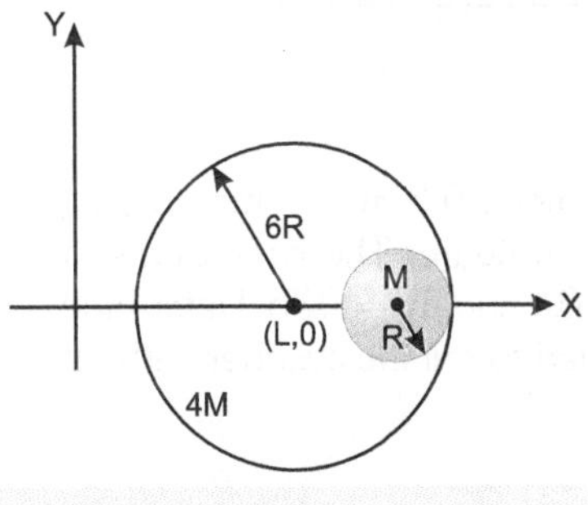

Problem 137. A cylindrical solid of mass 10^{-2} kg and cross-secional area $10^{-4}\ \text{m}^2$ is moving parallel to its axis (the X-axis) with a uniform speed of 10^3 m/s in the positive direction. At $t = 0$, its front face passes the plane $x = 0$. The region to the right of this plane is filled with stationary dust particles of uniform density $10^{-3}\ \text{kg/m}^3$. When a dust particle collides with the face of the cylinder, it sticks to its surface. Assuming that the dimensions of the cylinder remain practically unchanged, and that the dust sticks only to the front face of the cylinder, find the x-coordinate of the front of the cylinder at $t = 150$ s. **(JEE 1993)**

CHAPTER 5

ROTATION

Problem 138. A ring of radius R rolls without slipping on a horizontal plane with constant velocity υ. Find the position, velocity and acceleration of any point P on the circumference of the ring at any time t. Assuming that point P was at origin at time $t = 0$.

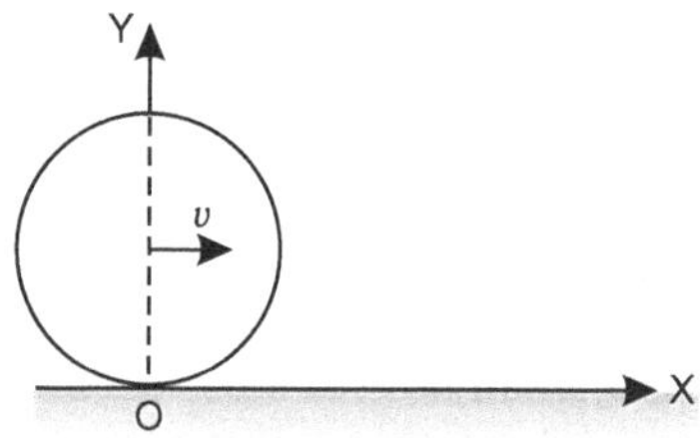

Problem 139. A 4m long rod AB slides down the plane with $\upsilon_A = 4$ m/s to the left and $a_A = 5$ m/s^2 to the right. Angle θ at this moment is 30°. Determine the angular velocity and angular acceleration of the rod at this instant of time.

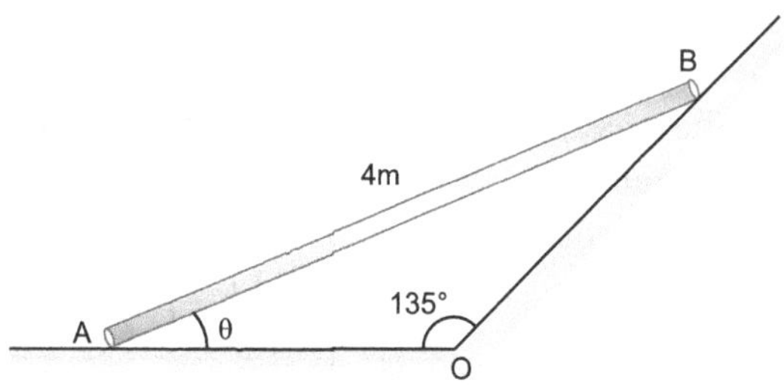

Problem 140. A cylinder of diameter 0.3 m and mass 25 kg rests on a rough surface as shown in figure. The coefficients of static and kinetic frictions are $\mu_s = 0.4$ and $\mu_k = 0.35$. Determine the minimum value of P to be applied to roll the cylinder without slip over the step ($g = 10$ m/s^2).

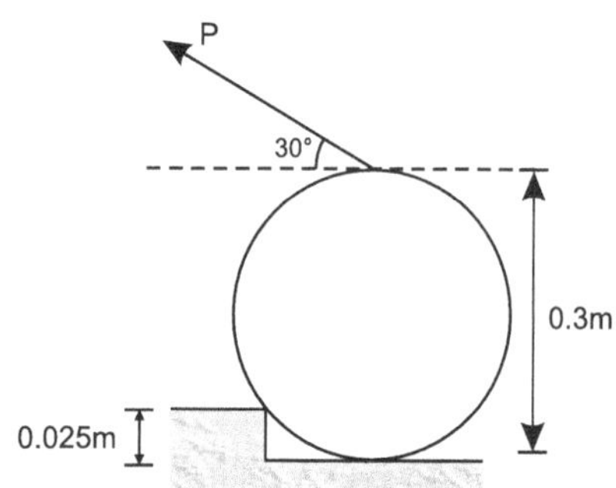

Problem 141. Two heavy and light cylindrical rollers of diameters D and d respectively rest on a horizontal plane as shown in figure. The larger roller has a string wound round it to which a horizontal force P can be applied as shown. Assuming that the coefficient of friction μ has the same value for all surfaces of contact, determine the limits of μ so that the larger roller can be pulled over the smaller one.

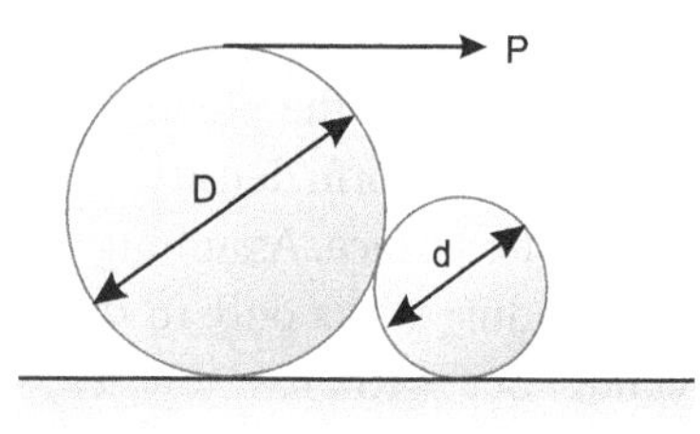

Problem 142. A wheel of radius r rolls to the left without slipping and at the instant considered the centre O has a velocity v_0 and acceleration a_0 to the left. Determine the magnitude of acceleration of points A and C on the wheel for the instant considered.

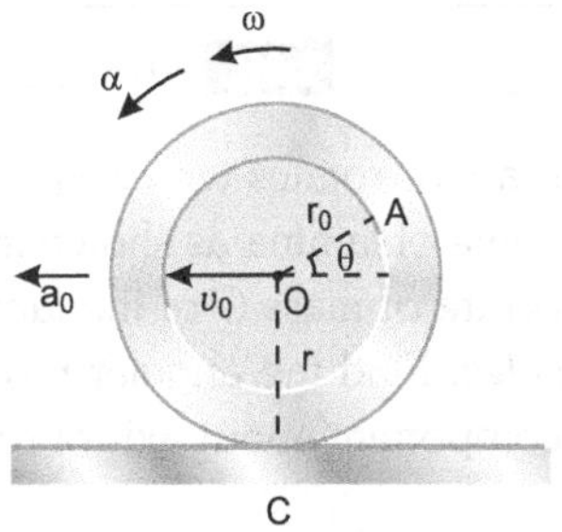

Problem 143. A sphere of radius r and mass m has a linear velocity v_0 directed to the left and no angular velocity as it is placed on a belt moving to the right with a constant velocity v_1. If after sliding on the belt the sphere is to have no linear velocity relative to the ground as it starts rolling on the belt without sliding. In terms of v_1 and the coefficient of kinetic friction μ_k between the sphere and the belt, determine

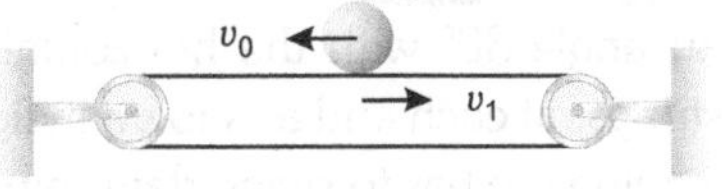

(a) the required value of v_0
(b) the time t_1 at which the sphere will start rolling on the belt
(c) the distance the sphere will have moved relative to the ground at time t_1.

Problem 144. Two identical rods of same mass and length 'l' are hinged at P as shown. Initially rod 2 was perfectly horizontal. Find the angular acceleration of rod 1 at the moment the string connecting the rod 2 in cut.

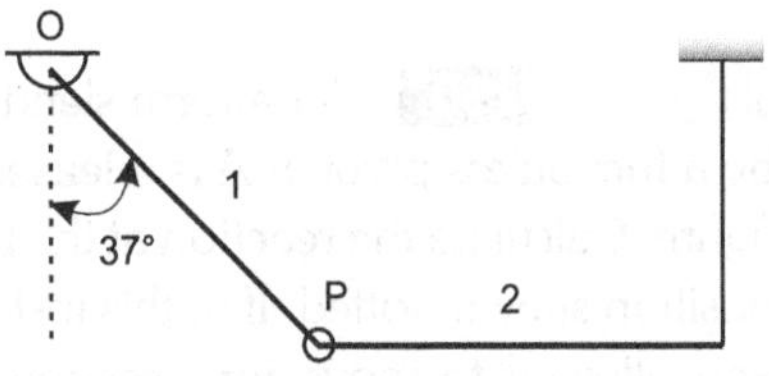

Problem 145. A ring of radius R is rolling (without slipping) over a rough horizontal surface with a velocity v_0. Two points are located at A and B on the rim of the ring. Find the velocity of A w.r.t. B.

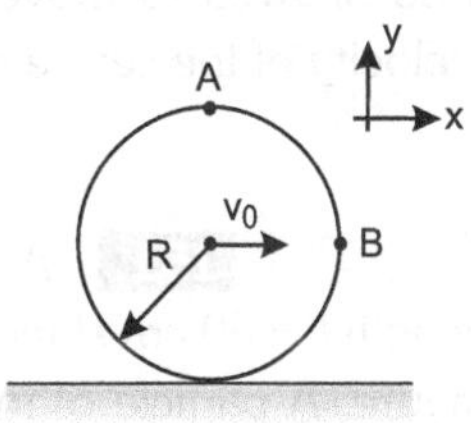

Problem 146. A disc of radius R is given a linear velocity v_o and an angular velocity $\frac{v_o}{R}$ and placed on a rough surface as shown in the figure. Assume coefficient of kinetic friction = μ.

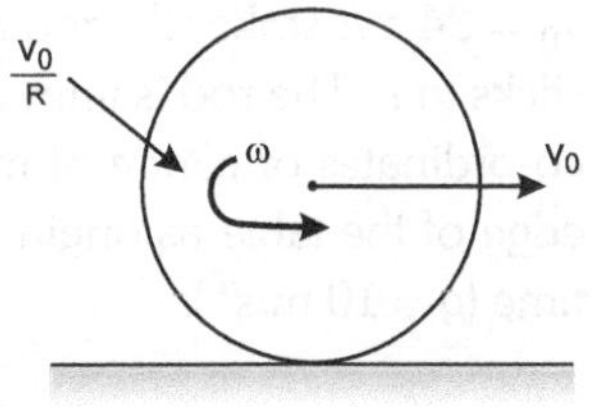

(a) Will the disc return?
(b) Plot a curve for linear velocity of disc with respect to time, indicating appropriate values of velocities and time.

Problem 147. A uniform bar of length L and weight W stands vertically touching a vertical wall (y-axis). When slightly displaced, its lower end begins to slide along the floor (x-axis). Obtain an expression for the angular velocity (ω) of the bar as a function of θ. Determine the distance moved by the lower end at which the bar no longer touches the vertical wall. Neglect friction everywhere.

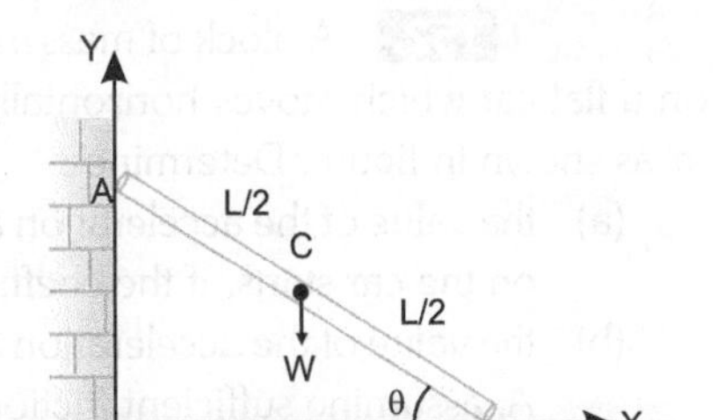

Problem 148. A rectangular lamina of length l breath b, and mass M rotates with angular velocity $\vec{\omega}$ about an axis ACB in the plane of lamina as shown in figure. Axis of rotation passes through centre of mass C of the lamina making an angle θ with one of the sides. Find the angular momentum of the lamina about C and its component about axis of rotation.

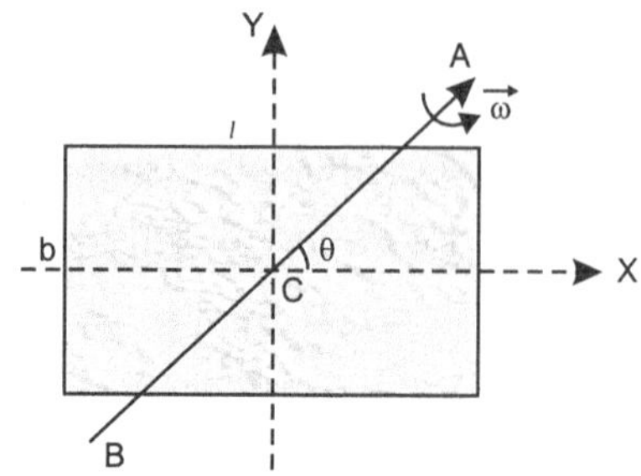

Problem 149. A thin uniform bar of mass m and length $2L$ is held at an angle 30° with the horizontal by means of two vertical inextensible strings, at each end as shown in figure. If the string at the right end breaks, leaving the bar to swing determine the tension in the string at the left end and the angular acceleration of the bar immediately after string breaks.

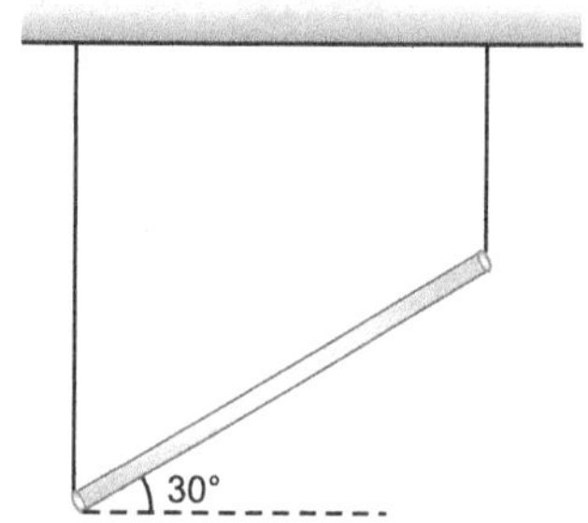

Problem 150. A uniform slender bar AB of mass m and length L supported by a frictionless pivot at A is released from rest at its vertical position as shown in figure. Calculate the reaction at the pivot when the bar just acquires the horizontal position shown dotted. If at this instant, the bar is released from its support gently and allowed to move for t second further, estimate its angular speed and the velocity of the centre of mass at that instant.

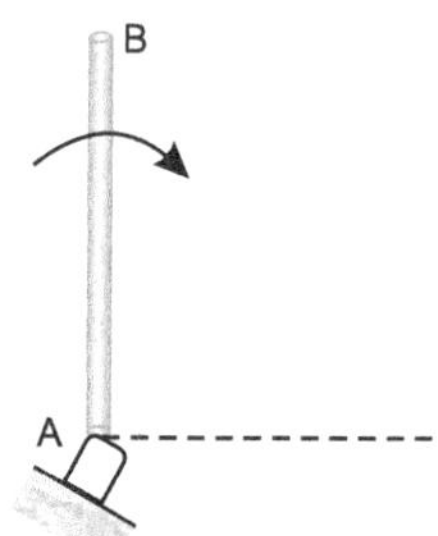

Problem 151. A uniform thin rod AB of mass $M = 0.6$ kg and length $l = 60$ cm stands at the edge of a frictionless table as shown in figure. A particle of mass $m = 0.3$ kg flying horizontally with velocity $v_0 = 24$ m/s strikes the rod at point P at a height 45 cm from base and sticks to it. The rod is immediately driven off the table. Determine the co-ordinates of centre of mass (COM) of the combined system with edge of the table as origin when rod becomes horizontal for the first time ($g = 10$ m/s^2).

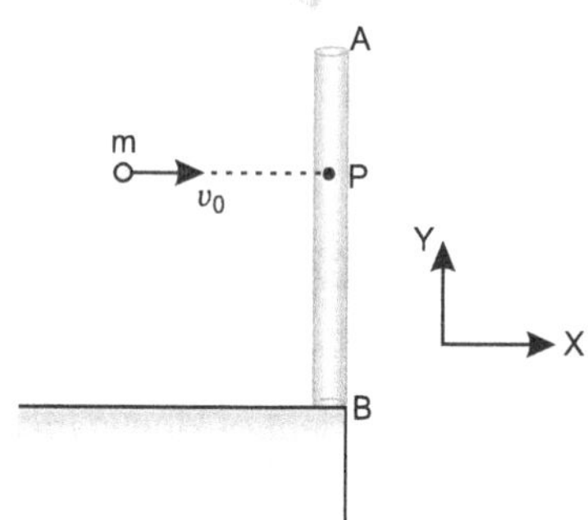

Problem 152. A block of mass m height $2h$ and width $2b$ rests on a flat car which moves horizontally with constant acceleration 'a' as shown in figure. Determine :

(a) the value of the acceleration at which slipping of the block on the car starts, if the coefficient of friction is μ

(b) the value of the acceleration at which block topples about A, assuming sufficient friction to prevent slipping and

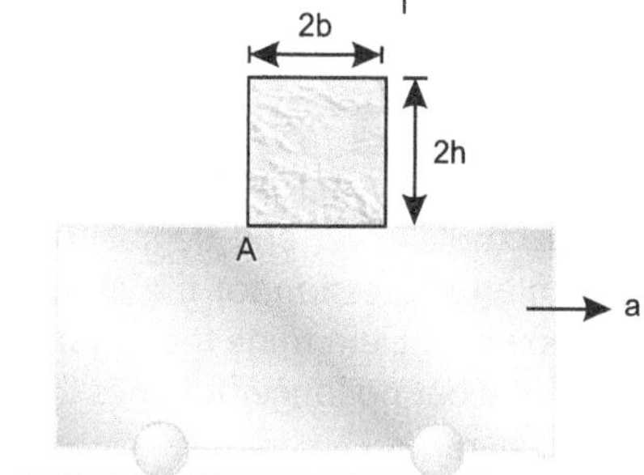

(c) the shortest distance in which it can be stopped from a speed of 72 km/hr with constant deceleration so that the block is not disturbed.
The following data are given

$$b = 0.6 \text{ m}, \quad h = 0.9 \text{ m}$$
$$\mu = 0.5 \quad \text{and} \quad g = 9.8 \text{ m/s}^2$$

Problem 153. A block of mass $M = 4$ kg of height 'h' and breadth 'b' is placed on a rough plank of same mass M. A light inextensible string is connected to the upper end of the block and passed through a light smooth pulley as shown in figure. A mass $m = 1$ kg is hung to the other end of the string.

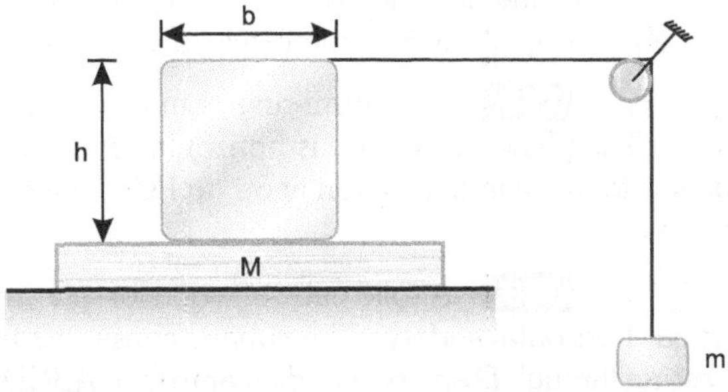

(a) What should be the minimum value of coefficient of friction between the block and the plank so that, there is no slipping between the block and the wedge ?
(b) Find the minimum value of b/h so that the block does not topple over the plank, friction is absent between the plank and the ground.

Problem 154. A beam of mass $m_1 = 100$ kg supported on two solid cylindrical rollers each of mass $m_2 = m_3 = 20$ kg and radius $R = 0.1$ m are moved up the inclined plane by a force F applied at an angle 30° with the incline as shown in figure. The inclined plane makes an angle of 30° with horizontal. Find the magnitude of F if the beam is moving up with an acceleration of $a = 1 \text{ m/s}^2$. There is no slipping at points of contact. ($g = 10 \text{ m/s}^2$).

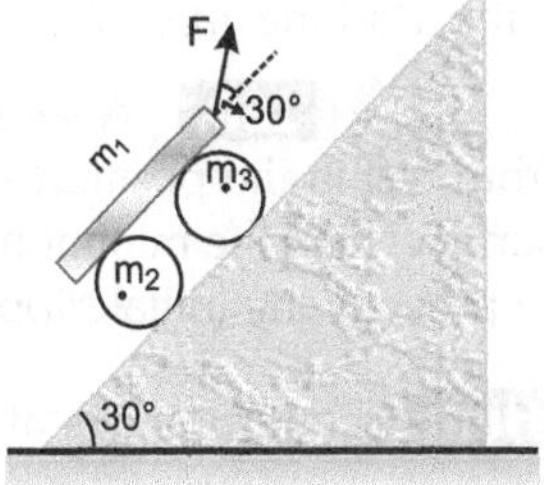

Problem 155. A homogeneous cylinder and a homogeneous sphere of equal mass $m = 20$ kg and equal radii R are connected together by a light frame and are free to roll without slipping down the plane inclined at 30° with the horizontal. Determine the force in the frame. Assume that the bearings are frictionless. Take $g = 10 \text{ m/s}^2$.

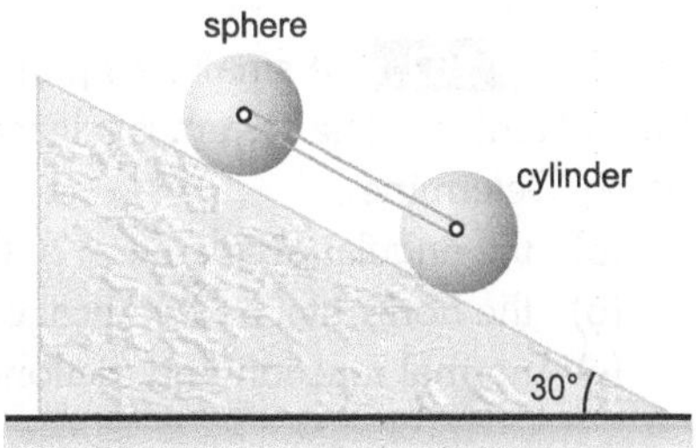

Problem 156. A cylindrical pipe of diameter 1 m is kept on the truck as shown in figure. If the truck now starts moving with constant acceleration of 1 m/s^2 the pipe rolls backward without slipping on the floor of the truck and finally falls on the road. If the pipe moves a total length of 4m on the floor of the truck. Find how much distance the pipe moves on the road before it finally stops. The coefficient of friction between the pipe and the road is 0.4. ($g = 10 \text{ m/s}^2$)

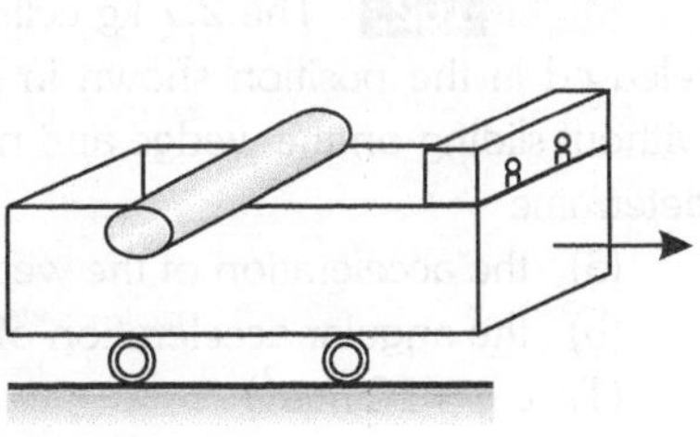

Problem 157. A rod AB of mass 3 m and length 4 a is falling freely in a horizontal position and C is a point distant a from A. When the speed of the rod is u, the point C collides with a particle of mass m which is moving vertically upwards with speed u. If the impact between the particle and the rod is perfectly elastic find

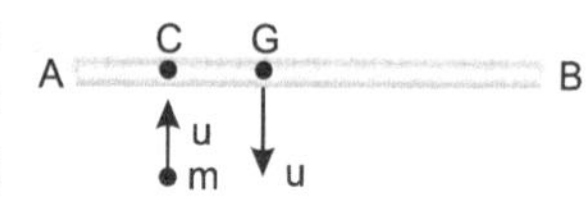

(a) the velocity of particle immediately after impact
(b) the angular velocity of the rod immediately after impact.
(c) the speed of B immediately after impact.

Problem 158. A uniform ring of mass m, radius a and centre C lies at rest on a smooth horizontal table. The plane of the ring is horizontal. A point P on the circumference is struck horizontally and it begins to move in a direction at 60° to PC. If the magnitude of impulse is $mv\sqrt{7}$, find the initial speed of point P.

Problem 159. A solid cube of wood of side $2a$ and mass M is resting on a horizontal surface. The cube is constrained to rotate about an axis passing through D and perpendicular to face $ABCD$. A bullet of mass m and speed v is shot at a height of $\frac{4a}{3}$ as shown in figure. The bullet becomes embedded in the cube. Find the minimum value of v required to topple the cube. Assume $m << M$.

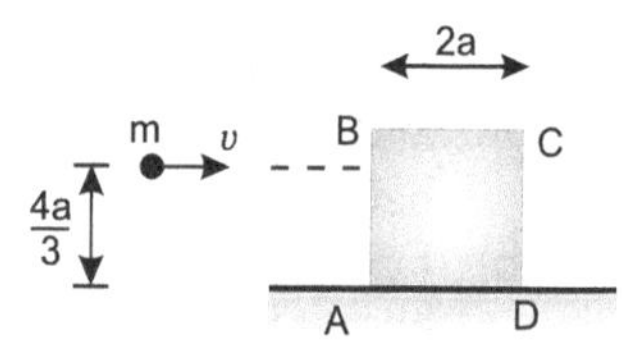

Problem 160. A semicircular ring of mass m and radius r is released from rest in the position shown with its lower edge resting on a horizontal surface. Find the minimum coefficient of static friction μ_s which is necessary to prevent any initial slipping of the ring.

Hint: centre of mass of a semicircular ring lies at a distance of $\frac{2r}{\pi}$ from centre

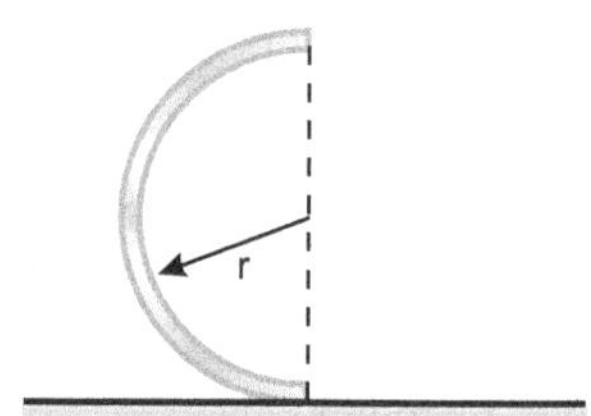

Problem 161. A small clamp of mass m is attached at B to a hoop of mass $3m$ and radius r. The system is released from rest with $\theta = 90°$ and rolls without sliding. Determine

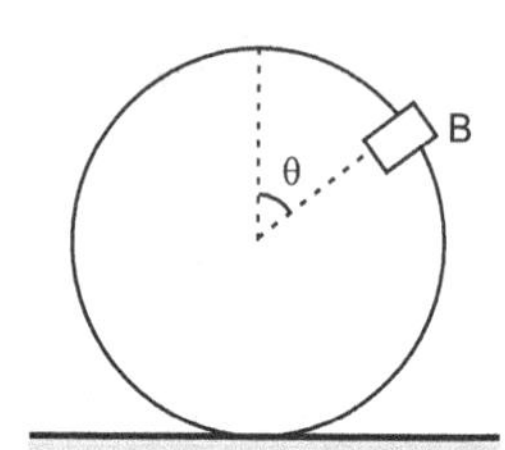

(a) the angular acceleration of the hoop
(b) the horizontal and vertical components of the acceleration of B.
(c) normal reaction and frictional force just after the release.

Problem 162. The 2.7 kg cylinder B and the 1.8 kg wedge A are released in the position shown in figure. Assuming that cylinder rolls without sliding on the wedge and neglecting friction between A and C, determine

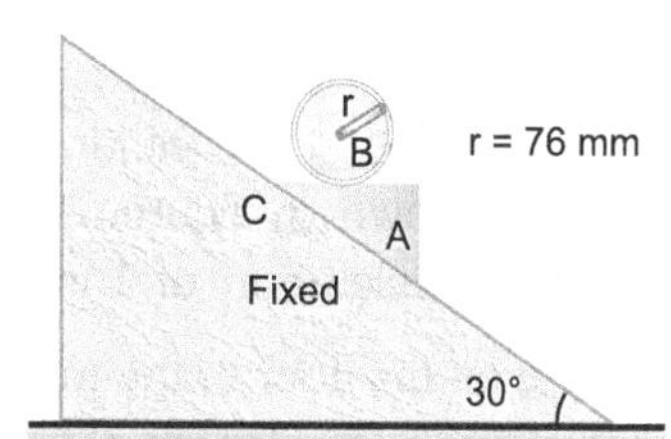

(a) the acceleration of the wedge
(b) the angular acceleration of the cylinder.
(Take $g = 10$ m/s^2)

Problem 163. A spherical ball of radius r and mass m collides with a plank of mass M kept on a smooth horizontal surface. Before impact, the centre of the ball has a velocity v_0 and angular velocity ω_0 as shown. The normal velocity is reversed with same magnitude and the ball stops rotating after the impact. Find the distance on the plank between first two impacts of the ball. The coefficient of friction between the ball and the plank is μ. Assume that plank is large enough.

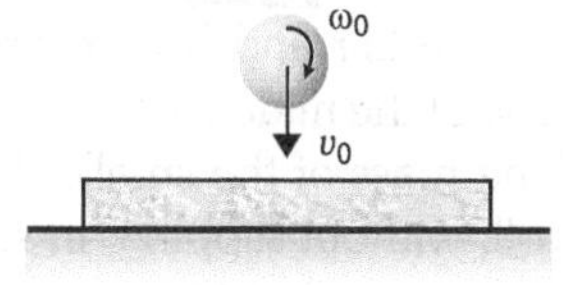

Problem 164. A ring shown in figure is made up of two semicircular rings A and B of masses 2 kg and 4 kg respectively. The ring has the diameter of 1 m. The ring rolls without slipping. Find :

(a) angular acceleration
(b) normal reaction and
(c) frictional force

[COM of a semicircular ring lies at a distance of $\frac{2r}{\pi}$ from centre]

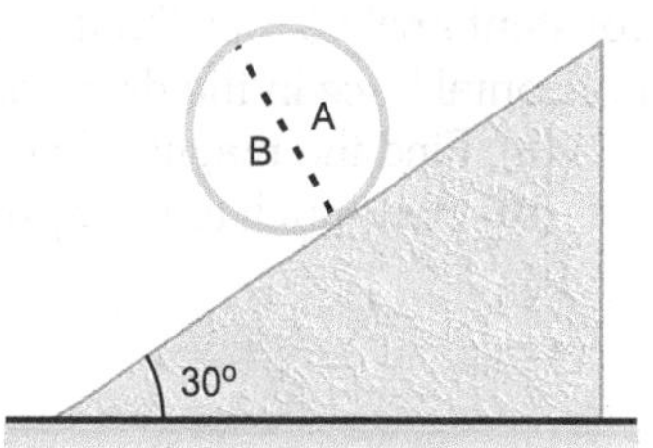

Problem 165. Two spools of masses m_1 and m_2 rest on rough inclined planes. Their internal and external radii are R and $2R$ respectively. Spool of mass m_1 on left hand side is connected to a block A of mass m_A and spool of mass m_2 on right hand side is connected to block B of mass m_B through a spring of force constant k. The coefficient of friction between blocks A and B is μ, whereas the horizontal part of the wedge is smooth. The whole system is in equilibrium. Find

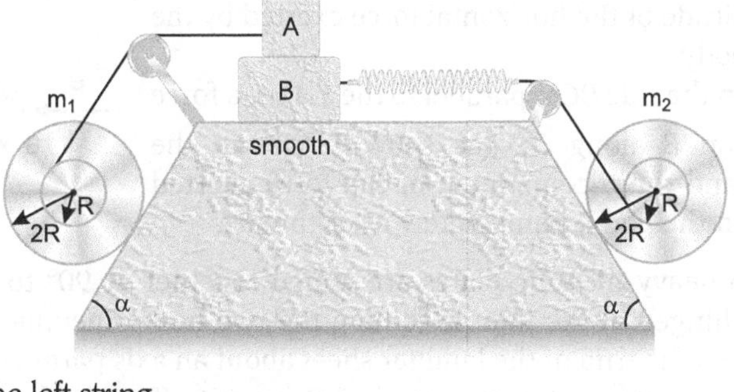

(a) the tension in the left string
(b) the minimum value of μ for equilibrium
(c) the extension in the spring
(d) the ratio of m_1 and m_2.

Problem 166. A uniform rod of length L and mass m is placed on a smooth horizontal surface. The rod is hinged at the end A and is free to rotate in horizontal plane about a vertical axis passing through A. As shown in the figure, there is a nail N at a perpendicular distance $\frac{L}{4}$ from the end A of the rod. At $t = 0$, an impulse P_0 is applied at a distance $\frac{3L}{4}$ from the end A. The impulse is in the horizontal plane and is perpendicular to the rod. At $t = \frac{8\pi mL}{27P_0}$, the rod returns to its initial position. Find the impulse due to the force exerted by the nail on the rod.

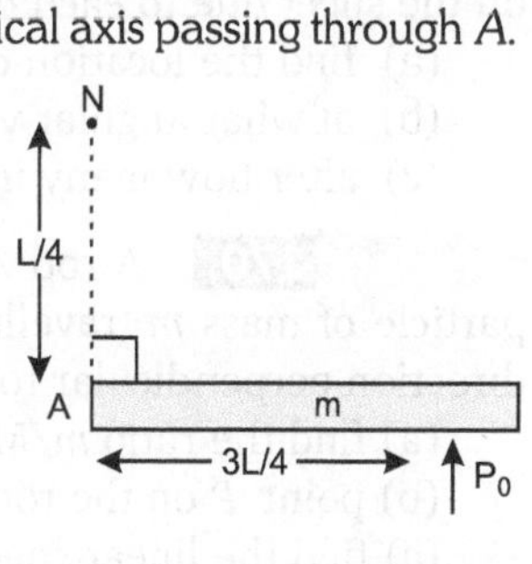

Problem 167. In a physics stunt, two balls of equal density and radii r and $R = 2r$, are placed with the centre of the larger one at the middle of a cart of mass $M = 6$ kg and length $L = 2$ m. The mass of the smaller ball is $m = 1$ kg. The balls are made to roll, without slipping, in such a way that the larger ball rests on the cart and a straight line connecting their centres remains at a constant angle $\phi = 60°$ to the horizontal. The cart is pulled by a horizontal force in the direction shown in the figure.

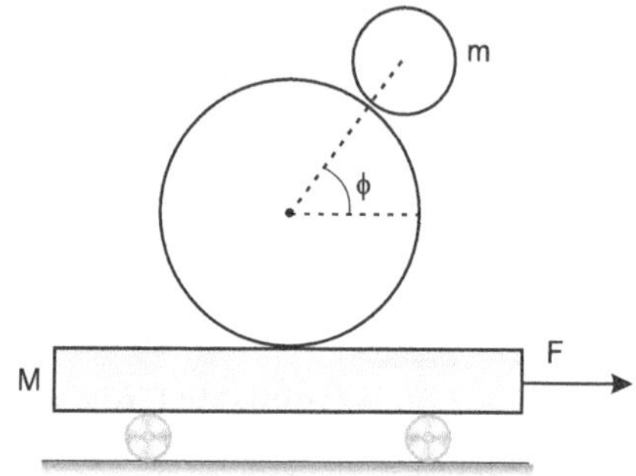

(a) Find the magnitude of the force F.

(b) How much time elapses before the balls fall off the cart?

IIT JEE PROBLEMS

Problem 168. Three particles A, B and C, each of mass m, are connected to each other by three massless rigid rods to form a rigid, equilateral triangular body of side l. This body is placed on a horizontal frictionless table (x-y plane) and is hinged to it at the point A so that it can move without friction about the vertical axis through A (see figure). The body is set into rotational motion on the table about A with a constant angular velocity ω

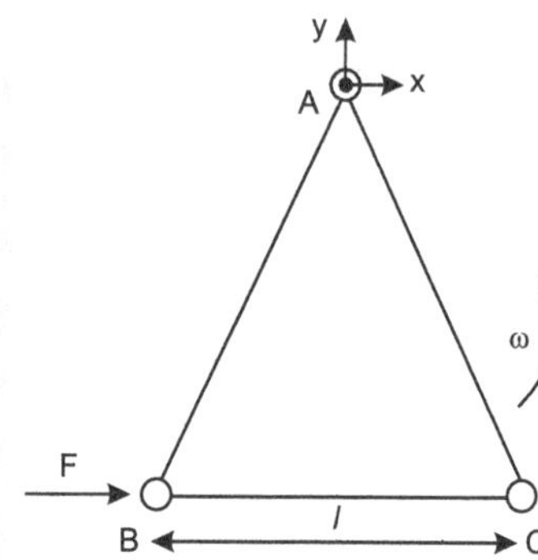

(a) Find the magnitude of the horizontal force exerted by the hinge on the body.

(b) At time T, when the side BC is parallel to the x-axis, a force F is applied on B along BC (as shown). Obtain the x-component and the y-component of the force exerted by the hinge on the body, immediately after time T. **(JEE 2002)**

Problem 169. Two heavy metallic plates are joined together at 90° to each other. A laminar sheet of mass 30 kg is hinged at the line AB joining the two heavy metallic plates. The hinges are frictionless. The moment of inertia of the laminar sheet about an axis parallel to AB and passing through its centre of mass is 1.2 kg-m^{-2}. Two rubber obstacles P and Q are fixed, one on each metallic plate at a distance 0.5 m from the line AB. This distance is chosen so that the reaction due to the hinges on the laminar sheet is zero during the impact. Initially the laminar sheet hits one of the obstacles with an angular velocity 1 rad/s and turns back. If the impulse on the sheet due to each obstacle is 6 N-s.

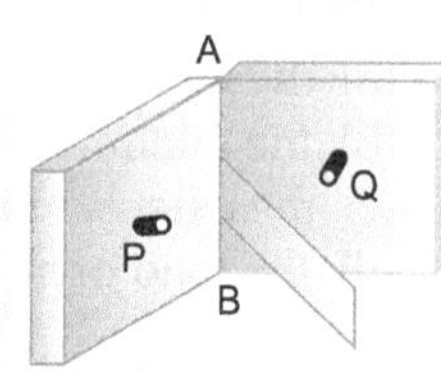

(a) find the location of the centre of mass of the laminar sheet from AB.

(b) at what angular velocity does the laminar sheet come back after the first impact ?

(c) after how many impacts, does the laminar sheet come to rest ? **(JEE 2001)**

Problem 170. A rod AB of mass M and length L is lying on a horizontal frictionless surface. A particle of mass m travelling along the surface hits the end `A' of the rod with a velocity v_0 in a direction perpendicular to AB. The collision is elastic. After the collision the particle comes to rest.

(a) find the ratio m/M.

(b) point P on the rod is at rest immediately after collision. Find the distance AP.

(c) find the linear speed of the point P after a time $\pi L/3v_0$ after the collision. **(JEE 2000)**

Problem 171. A man pushes a cylinder of mass m_1 with the help of a plank of mass m_2 as shown. There is no slipping at any contact. The horizontal component of the force applied by the man is F. Find :

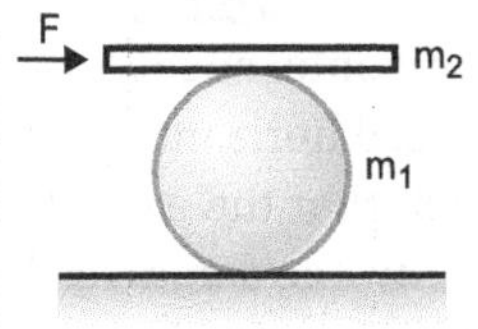

(a) the accelerations of the plank and the centre of mass of the cylinder, and

(b) the magnitudes and directions of frictional forces at contact points.

(JEE 1999)

Problem 172. A uniform circular disc has radius R and mass m. A particle, also of mass m, is fixed at a point A on the edge of the disc as shown in the figure. The disc can rotate freely about a fixed horizontal chord PQ that is at a distance $R/4$ from the centre C of the disc. The line AC is perpendicular to PQ.

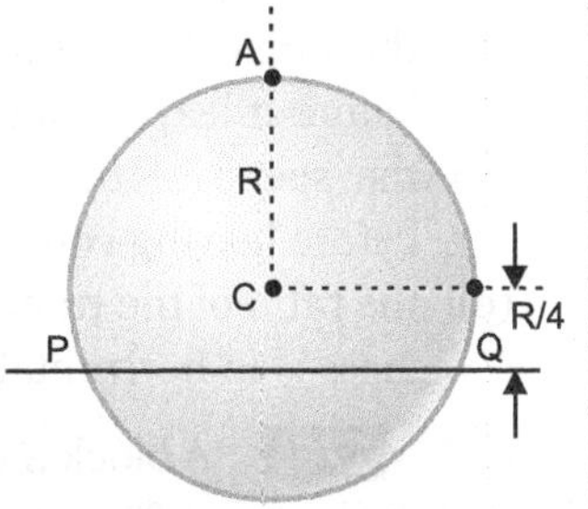

Initially the disc is held vertical with the point A at its highest position. It is then allowed to fall so that it starts rotating about PQ. Find the linear speed of the particle as it reaches its lowest position.

(JEE 1998)

Problem 173. Two thin circular discs of mass 2 kg and radius 10 cm each are joined by a rigid massless rod of length 20 cm. The axis of the rod is along the perpendicular to the planes of the disc through their centres. This object is kept on a truck in such a way that the axis of the object is horizontal and perpendicular to the direction of motion of the truck. Its friction with the floor of the truck is large enough so that the object can roll on the truck without slipping. Take X-axis as the direction of motion of the truck and Z-axis as the vertically upwards direction. If the truck has an acceleration $9 \text{ m}/\text{s}^2$, Calculate :

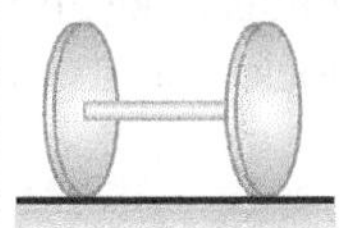

(i) the force of friction on each disc.

(ii) the magnitude and direction of the frictional torque acting on each disc about the centre of mass O of the object. Express the torque in the vector form in terms of unit vectors $\hat{i}$, $\hat{j}$ and $\hat{k}$ in X, Y and Z directions.

(JEE 1997)

Problem 174. A uniform disc of mass m and radius R is projected horizontally with velocity v_0 on a rough horizontal floor so that it starts off with a purely sliding motion at $t = 0$. After t_0 second, it acquires a purely rolling motion as shown in figure.

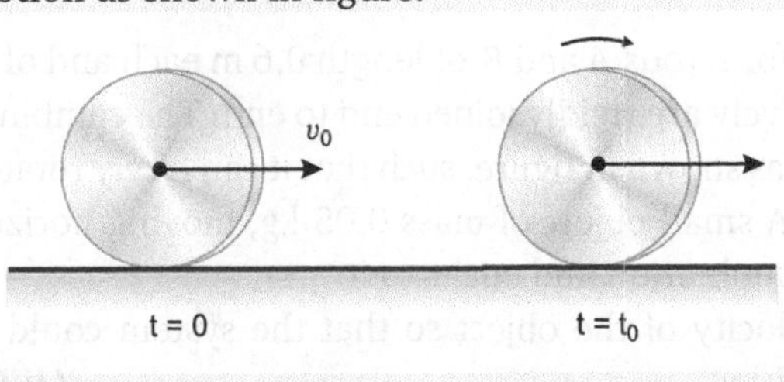

(i) calculate the velocity of the centre of mass of the disc at t_0.

(ii) assuming the coefficient of friction to be μ, calculate t_0. Also calculate the work done by the frictional force as a function of time and the total work done by it over a time t much longer than t_0.

(JEE 1997, Cancelled)

Problem 175. A rectangular rigid fixed block has a long horizontal edge. A solid homogeneous cylinder of radius R is placed horizontally at rest with its length parallel to the edge such that the axis of the cylinder and the edge of the block are in the same vertical plane as shown in figure. There is sufficient friction present at the edge so that a very small displacement causes the cylinder to roll of the edge without slipping. Determine :

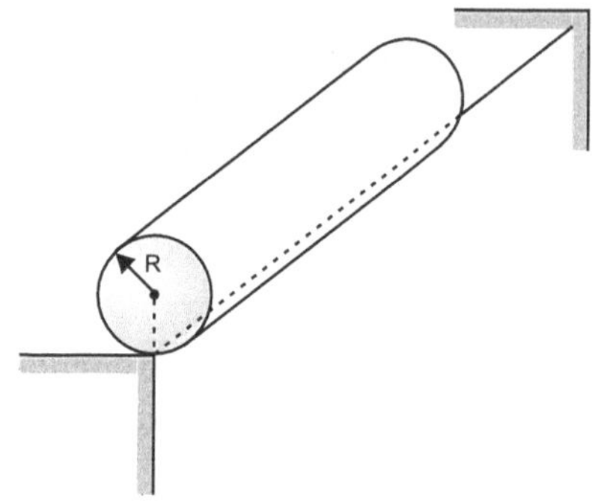

(a) the angle θ_c through which the cylinder rotates before it leaves contact with the edge.

(b) the speed of the centre of mass of the cylinder before leaving contact with the edge, and

(c) the ratio of the translational to rotational kinetic energies of the cylinder when its centre of mass is in horizontal line with the edge. **(JEE 1995)**

Problem 176. A block X of mass 0.5 kg is held by a long massless string on a frictionless inclined plane of inclination 30° to the horizontal. The string is wound on a uniform soilid cylindrical drum Y of mass 2 kg and of radius 0.2 m as shown in Figure. The drum is given an initial velocity such that the block X starts moving up the plane.

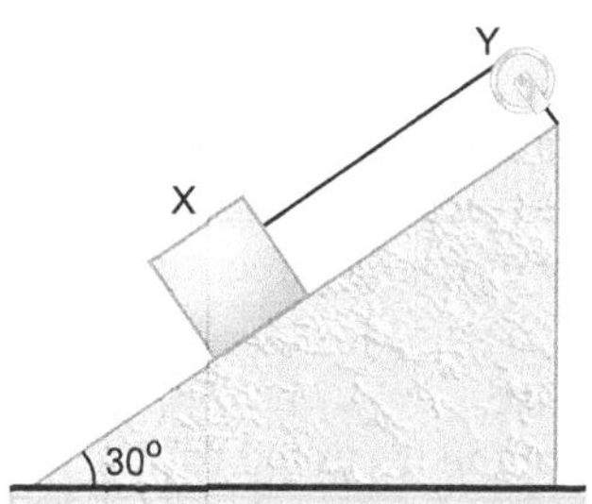

(i) Find the tension in the string during the motion. Take $g = 9.8 \text{ m/s}^2$

(ii) At a certain instant of time the magnitude of the angular velocity of Y is 10 rad/s^{-1}. Calculate the distance travelled by X from the instant of time until it comes to rest. **(JEE 1994)**

Problem 177. Two uniform rods A and B of length 0.6 m each and of masses 0.01 kg and 0.02 kg respectively are rigidly joined end to end. The combination is pivoted at the lighter end, P as shown in figure. such that it can freely rotate about point P in a vertical plane. A small object of mass 0.05 kg, moving horizontally, hits the lower end of the combination and sticks to it.

What should be the velocity of the object so that the system could just be raised to the horizontal position ? **(JEE 1994)**

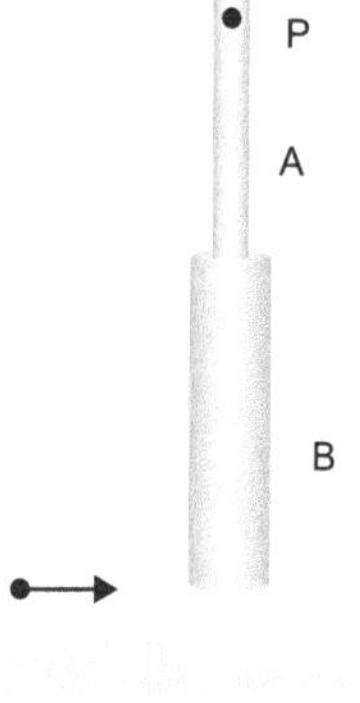

CHAPTER 6

GRAVITATION

Problem 178. Two planets A and B describe circles of radii r_1 and r_2 round the sun as centre with speed varying inversely as the square root of their radii. Find the angle between the radii of these two planets when their relative angular velocity is zero.

Problem 179. If a planet was suddenly stopped in its orbit supposed to be circular, show that it would fall onto the sun in a time $\frac{\sqrt{2}}{8}$ times the period of the planet's revolution.

Problem 180. A thin spherical shell of total mass M and radius R is held fixed. There is a small hole in the shell. A mass m is released from rest a distance R from the hole along a line that passes through the hole and also through the centre of the shell. This mass subsequently moves under gravitational force of the shell. How long does the mass take to travel from the hole to the point diametrically opposite ?

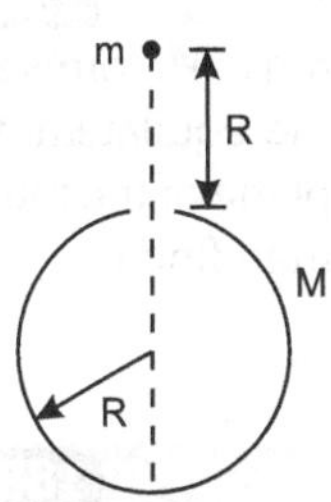

Problem 181. A satellite is revolving round the earth in a circular orbit of radius r and velocity v_o. A particle is projected from the satellite in forward direction with relative velocity $v = (\sqrt{5/4} - 1)v_o$. Calculate its minimum and maximum distances from earth's centre during subsequent motion of the particle.

Problem 182. Two earth's satellites move in a common plane along circular orbits. The orbital radius of one satellite is $r = 7000$ km while that of the other is 70 km less. What time interval separates the periodic approaches of the satellites to each other over the minimum distance ?

$$(G = 6.67 \times 10^{-11} \frac{m^3}{kg\text{-}s^2} \text{ and Mass of earth, } M = 6 \times 10^{24} \text{ kg}).$$

Problem 183. A body is launched from the earth's surface at an angle $\alpha = 30°$ to the horizontal at a speed $v_0 = \sqrt{\frac{1.5\, GM}{R}}$. Neglecting air resistance and earth's rotation. Find the height to which the body will rise. Here M is mass of earth and R the radius of earth.

Problem 184. An earth satellite is revolving in a circular orbit of radius `a' with velocity v_o. A gun is in the satellite and is aimed directly towards the earth. A bullet is fired from the gun with muzzle velocity $\frac{v_o}{2}$. Neglecting resistance offered by cosmic dust and recoil of gun, calculate maximum and minimum distance of bullet from the centre of earth during its subsequent motion.

Problem 185. One of the star of a binary (double) stars system is rotating in a circular orbit of radius r_1 with time period T. If the mass of this star is m_1. Find the mass m_2 of the other star. Also find the distance between the two stars. Take $m_1 = 10^{30}$ kg, $r_1 = 2.455 \times 10^6$ km and $T = 20$ hrs, $G = 6.67 \times 10^{-11}\ \dfrac{\text{N-m}^2}{\text{kg}^2}$.

Hint : For the equation $(1 + x)^2 = 0.59\, x^3$, $x \approx 3$

Problem 186. Binary stars of comparable masses m_1 and m_2 rotate under the influence of each other's gravity with a time period T. If they are stopped suddenly in their motions, find their relative velocity when they collide with each other. The radii of the stars are R_1 and R_2 respectively. G is the universal constant of gravitation.

Problem 187. The line joining the positions of three identical stars, each of mass M, forms an equilateral triangle of side a. A particle, located at the centroid of the equilateral triangle, is given a velocity v_o in a direction perpendicular to the plane of the triangle. If the particle stops momentarily after travelling a distance $3a$, then find v_o.

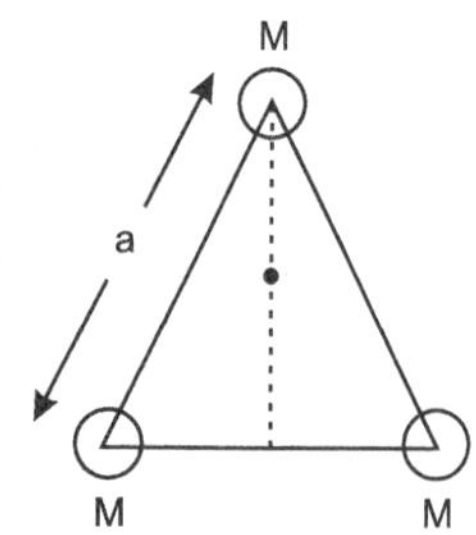

Problem 188. If the earth (supposed spherical) was covered by an ocean of uniform depth h, prove that the value of the gravity at the bottom of the ocean would exceed that at the top by $4\pi G h\,(2\rho - 3\sigma)/3$ approximately, where σ is density of ocean, ρ is mean density of the earth. ($h << R$, radius of the earth)

Problem 189. A rocket is launched from and returns to a spherical planet of radius R in such a way that its velocity vector on return is parallel to its launch vector. The angular separation at the centre of the planet between the launch and arrival points is θ. How long does the flight of the rocket take, if the period of a satellite flying around the planet just above its surface is T_0? What is the maximum distance of the rocket above the surface of the planet? Consider whether your analysis also applies to the limiting case of $\theta \to 0$.

IIT JEE PROBLEMS

Problem 190. There is a crater of depth $R/100$ on the surface of the moon (radius R). A projectile is fired vertically upward from the crater with a velocity, which is equal to the escape velocity v from the surface of the moon. Find the maximum height attained by the projectile.

(JEE 2003)

Problem 191. Distance between the centres of two stars is $10a$. The masses of these stars are M and $16M$ and their radii a and $2a$ respectively. A body of mass m is fired straight from the surface of the larger star towards the smaller star. What should be its minimum initial speed to reach the surface fo the smaller star? Obtain the expression in terms of G, M and a. **(JEE 1996)**

CHAPTER 7 SIMPLE HARMONIC MOTION

Problem 192. A simple pendulum, which is meant to beat seconds (*i.e.*, each half oscillation takes 1 second), gains 1 minute per day. By what percentage of its length should be increased to make it accurate ?

Problem 193. Consider a spring that exerts the following restoring force :

$$F = \begin{cases} -kx & for\ x > 0 \\ -2kx & for\ x < 0 \end{cases}$$

A mass m on a frictionless surface is attached to this spring, displaced to $x = A$ by stretching the spring and released.

(a) find the period of motion.

(b) what is the most negative value of x that the mass m reaches ?

Problem 194. A load of mass M is on horizontal rails. A pendulum made of ball of mass m tied to a weightless inextensible thread is suspended to the load. The load can move only along the rails. Determine the ratio of the periods T_2/T_1 of small oscillations of the pendulum in vertical planes parallel and perpendicular to rails. Neglect friction everywhere.

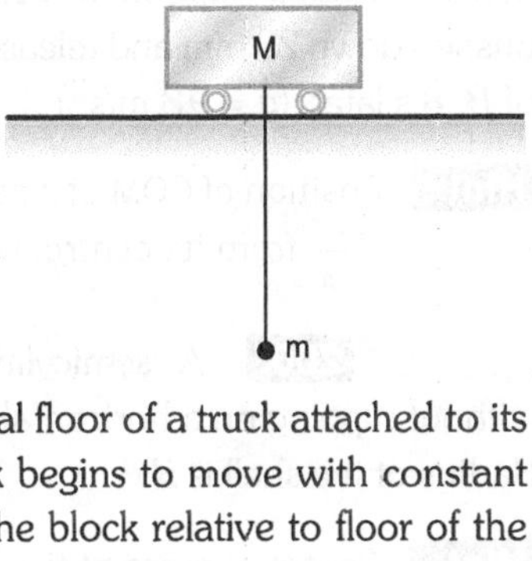

Problem 195. A block of mass 2 kg is resting on a smooth horizontal floor of a truck attached to its front by a spring of force constant $k = 40$ N/m. At time $t = 0$, the truck begins to move with constant acceleration 4 m/s^2. Find the amplitude and period of oscillations of the block relative to floor of the truck.

Problem 196. As a submerged body moves through a fluid, the particles of the fluid flow around the body and thus acquire kinetic energy. In the case of a sphere moving in an ideal fluid, the total kinetic energy acquired by the fluid is $\frac{1}{4}\rho V v^2$ where ρ is the mass density of fluid, V the volume of sphere and v is the velocity of the sphere. Consider a 0.5 kg hollow spherical shell of radius 8 cm which is held submerged in a tank of water by a spring of force constant 500 N/m.

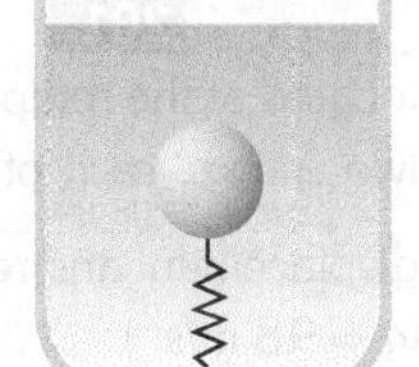

(a) Neglecting fluid friction, determine the period of vibration of the shell when it is displaced vertically and then released.

(b) Solve part (a) assuming that the tank is accelerated upward at the constant rate of 8 m/s^2. Density of water is 10^3 kg/m^3.

Problem 197. Find the natural angular frequency of the system shown in figure. Pulleys are massless and friction is absent everywhere.

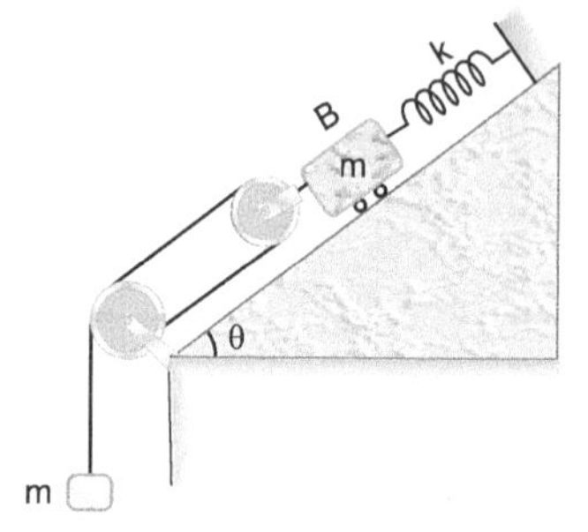

Problem 198. A 7 kg uniform cylinder can roll without sliding on an incline and is attached to a spring AB as shown. If the centre of the cylinder is moved 10 mm down the incline and released. Determine

(a) the period of oscillation.

(b) the maximum velocity of the centre of the cylinder. $(g = 10 \text{ m/s}^2)$

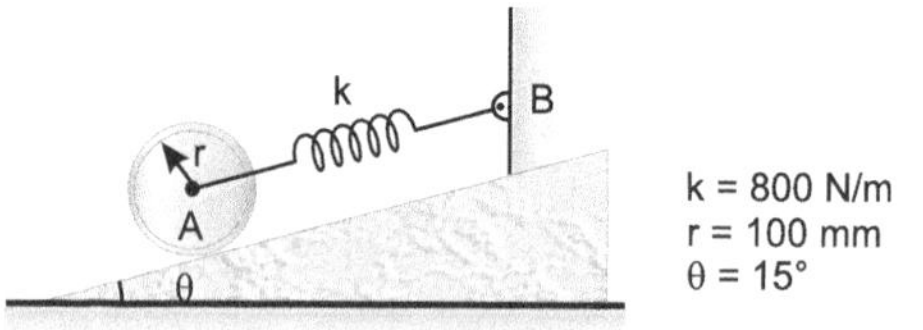

Problem 199. A homogeneous wire bent to form the figure shown is attached to a pin support at A. Knowing that $r = 220$ mm and that point B is pushed down 20 mm and released, determine the magnitude of the velocity of B, 8 s later. $(g = 9.8 \text{ m/s}^2)$

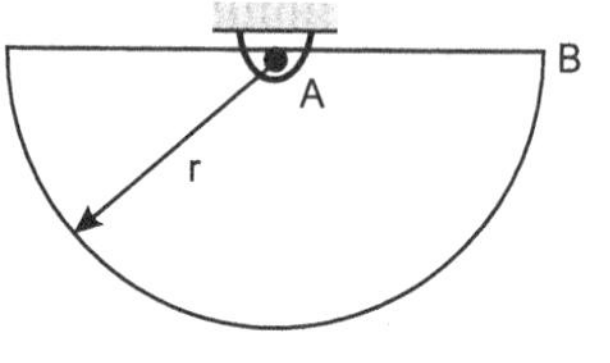

Hint : Position of COM of a semicircular wire of shape ∪ is at a distance of $\frac{2r}{\pi}$ from its centre, where r is the radius of semicircle

Problem 200. A semicylindrical shell with negligible thickness oscillates without slipping on a horizontal surface. Find the time period of small oscillations. Radius of the shell is R.

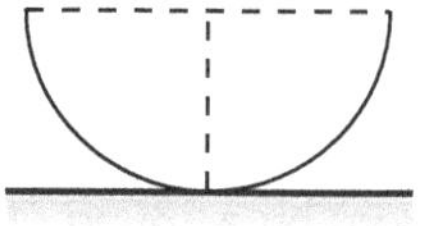

Hint : Centre of mass of the shell lies at $\frac{2R}{\pi}$ from centre

Problem 201. A 20 kg uniform square plate is suspended from a pin located at the midpoint A of one of its edges of length 0.4 m and is attached to two springs, each of constant $k = 1.4 \times 10^3$ N/m. If corner B is given a small displacement and released. Determine the frequency of the resulting vibration. $(g = 9.8 \text{ m/s}^2)$

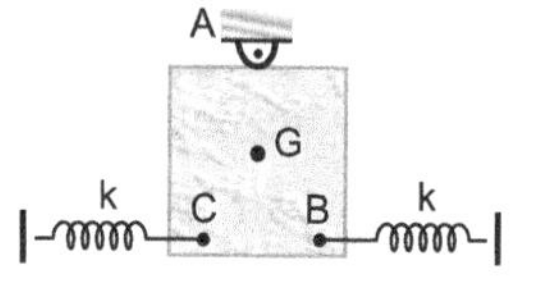

Problem 202. A 3 kg block is dropped from a height of 2 m onto the initially stationary 28 kg block which is supported by four springs, each of which has a constant $k = 800$ N/m. The two blocks stick after collision. Determine the displacement x as a function of time during the resulting vibration, where x is measured from the initial position of the block as shown. Two springs which are not shown in the figure are behind the two springs shown in figure. $(g = 9.8 \text{ m/s}^2)$

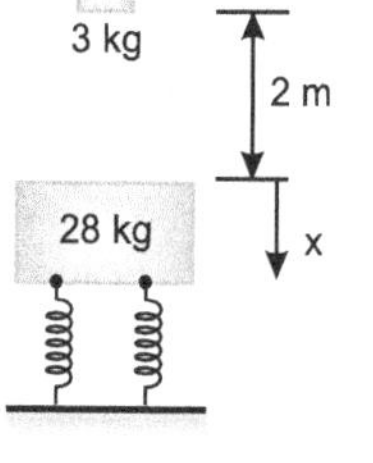

Problem 203. Two identical circular discs each of radius R and their connecting links form a single rigid unit. The mass of the connecting links is negligible as compared to the disc. If the upper disc rolls without slipping, determine the time period of small oscillations. The length of the connecting rod is L.

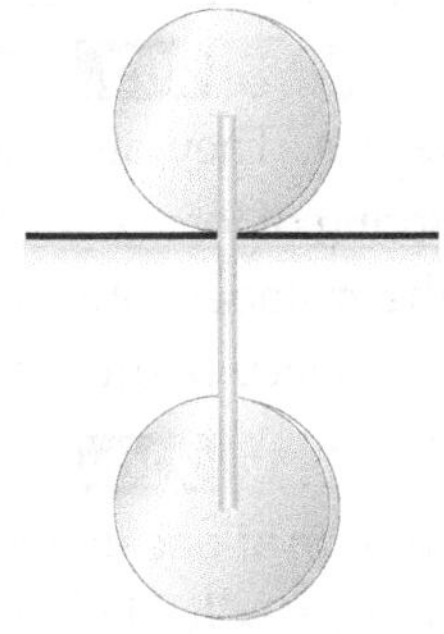

Problem 204. A solid sphere (radius = R) rolls without slipping in a cylindrical trough (radius = $5R$). Find the time period of small oscillations.

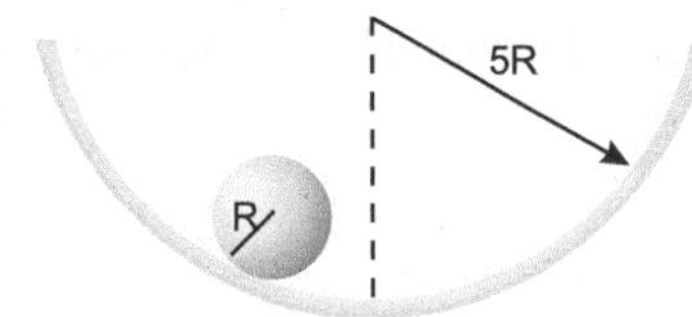

Problem 205. Find the time period of small oscillations of the system composed of two homogeneous circular cylinders, each of mass M and the connecting link AB. Assume no slipping on ground.

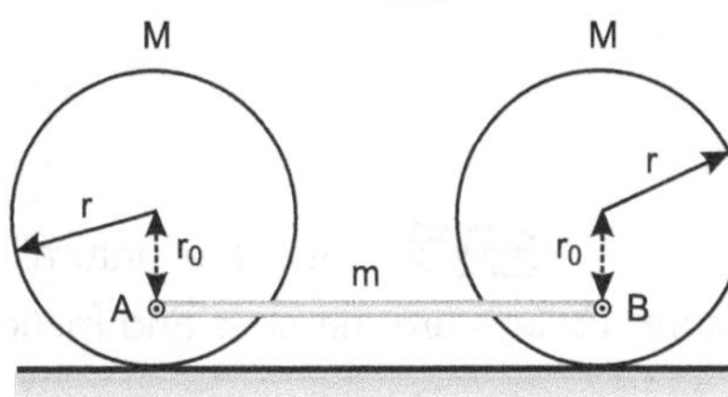

Problem 206. A solid cylinder of mass $m = 1$ kg is kept in equilibrium on a horizontal surface. Two unstretched springs of force constant $k_1 = 10$ N/m and $k_2 = 20$ N/m are attached to the cylinder shown in figure. Find the period of small oscillations. Cylinder rolls without sliding.

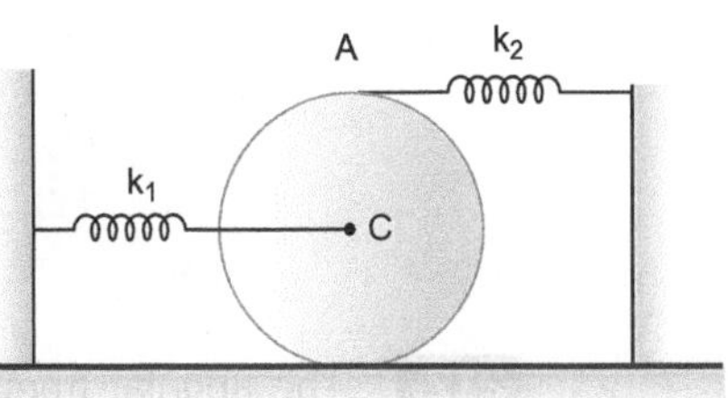

Problem 207. Two blocks A (2 kg) and B (3 kg) rest up on a smooth horizontal surface are connected by a spring of stiffness 120 N/m. Initially the spring is undeformed. A is imparted a velocity of 2 m/s along the line of the spring away from B. Find the displacement of A t seconds later.

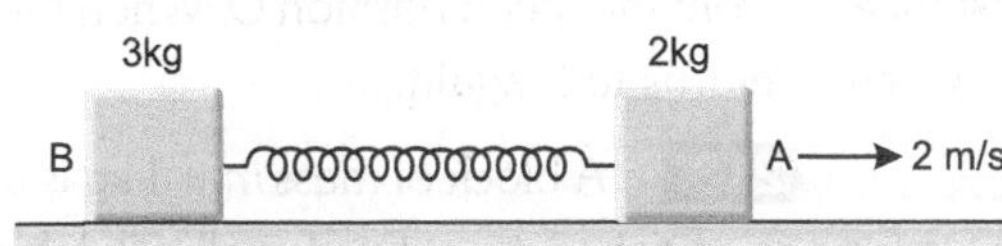

Problem 208. A block of mass m is tied to one end of a string which passes over a smooth fixed pulley A and under a light smooth movable pulley B as shown in figure. The other end of the string is attached to the lower end of a spring of spring constant k_2. Find the period of small oscillations of mass m about its equilibrium position.

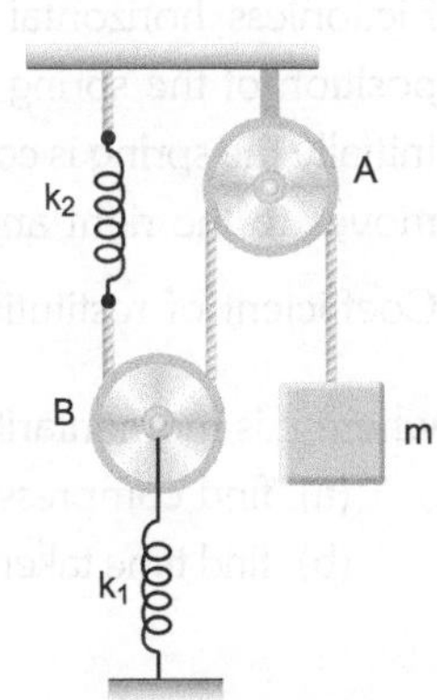

Problem 209. Two cylinders of masses m_1 and m_2 have the same radius. They are connected from the centre with the help of an ideal spring of force constant k. Two equal and opposite forces are applied on the cylinders and then released on rough horizontal surface. Find the frequency of oscillation. Slipping is absent.

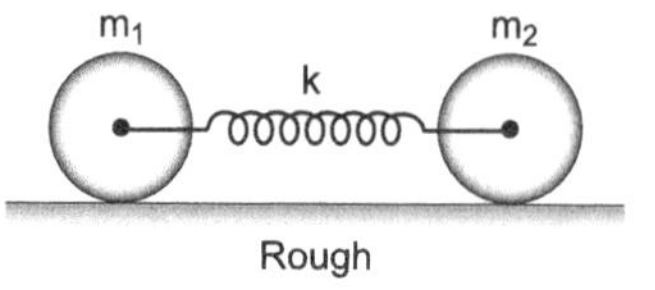

Problem 210. A massless rod is hinged at O. A string carrying a mass m at one end is attached to point A on the rod so that $OA = a$ and $OB = b$. Find the period of small vertical oscillations of mass m around its equilibrium position.

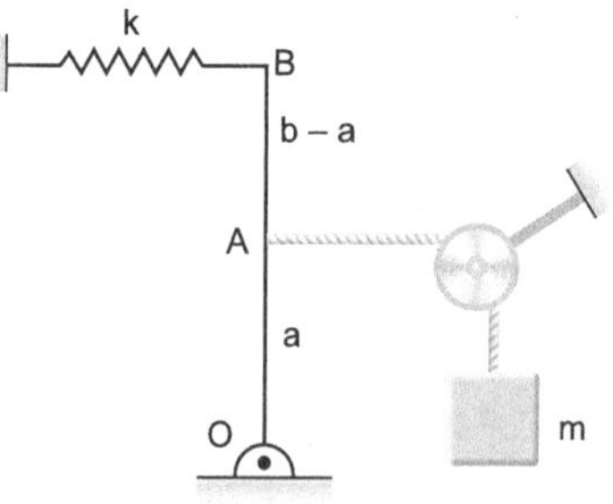

Problem 211. Find the natural frequency of oscillation of the system as shown in figure. Pulleys are massless and frictionless. Spring and string are also massless.

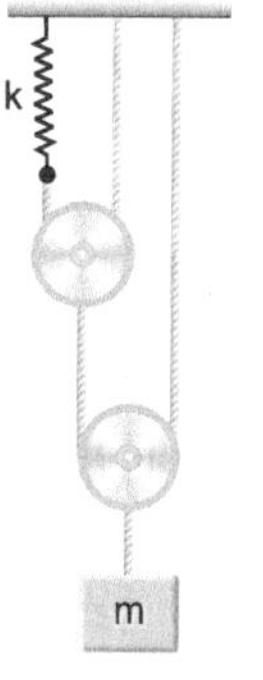

Problem 212. The angular frequency of a particle executing SHM is ω. There is a point P, at a distance 'x' from the mean position O. When the particle passes P along OP, it has speed v. Find the time in which it returns to P again.

Problem 213. A block of mass $m = 1$ kg is attached to a spring of constant $k = 64$ N/m, other end of the spring is attached to a fixed wall. The block can move on a frictionless horizontal surface. In the given figure O is the mean position of the spring block system, when there is no compression. Initially the spring is compressed by $OA = 4$ cm. When it is released it moves to the right and makes inelastic collision with the wall at B. Coefficient of restitution being $e = \frac{1}{\sqrt{2}}$. Distance $OB = 2\sqrt{2}$ cm. After the collision block return to A' where it is momentarily at rest.

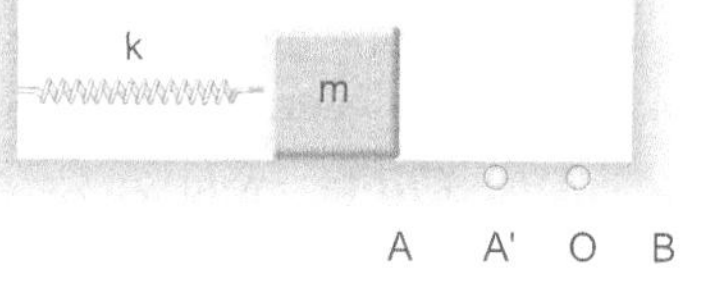

(a) find compression in the spring OA'.

(b) find time taken by the block to return to A'. [Take $t = 0$, at the moment block was released from A]

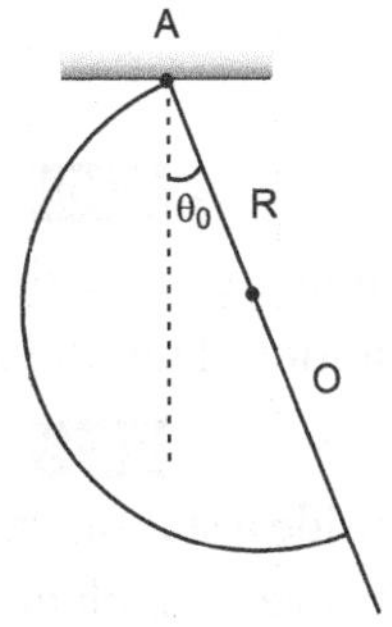

Problem 214. A semicircular plate of mass M is pivoted at one of its end so that it can freely rotate in a vertical plane, as shown in the figure.

(a) Find the value of θ_0 at equilibrium.

(b) If it is slightly displaced from its equilibrium position, find the frequency of oscillation. Distance $OC = \dfrac{4R}{3\pi}$ where C is the centre of mass of the plate.

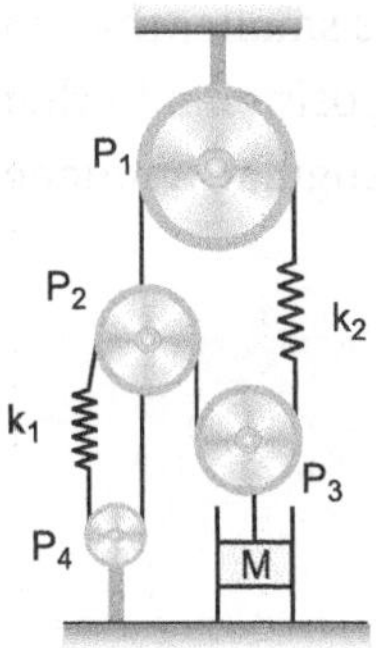

Problem 215. An ideal gas having adiabatic constant γ is enclosed in a adiabatic cylindrical container having area of cross-section A. A piston of mass M can move freely in the cylinder is attached with ideal springs k_1 and k_2 and light pulleys P_1, P_2 and P_3 as shown in figure. When the piston is in equilibrium the volume of the gas in the cylinder is V_0 and its pressure is P_0. The piston is slightly displaced from the equilibrium position and released. Find the time period of oscillation. Pulley P_2 is fixed.

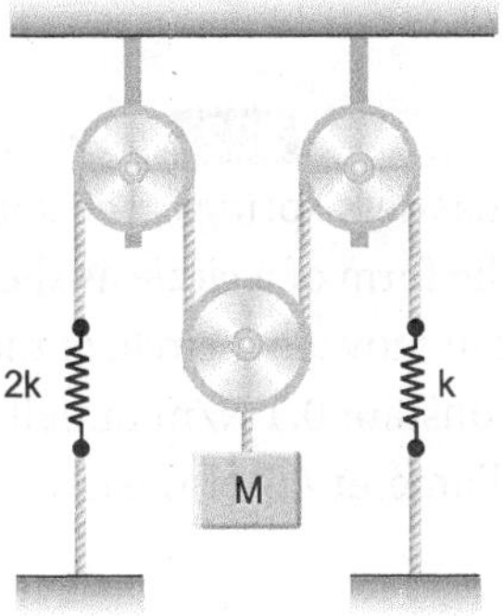

Problem 216. Find the time period of oscillation for the arrangement shown in the figure. The pulleys are smooth and massless.

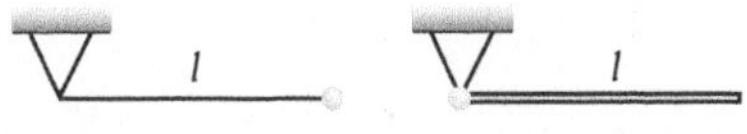

Problem 217. A simple pendulum and a homogeneous rod pivoted at its end are released from horizontal positions. What is the ratio of their periods of swing if their lengths are identical ?

Problem 218. An elastic string of natural length $(a + b)$ where $a > b$ and modulus of elasticity Y has a particle of mass m attached to it at a distance a from one end, which is fixed to a point A of a smooth horizontal plane. The other end of the string is fixed to a point B so that the string is just unstretched. If the particle be held at B (by stretching the length a of the string) and then released, show that it will oscillate to and fro with a time period of $\pi(\sqrt{a} + \sqrt{b})\sqrt{\dfrac{m}{Y}}$. Take the cross-sectional area of the string as unity.

Problem 219. In the above problem, show that the particle oscillate to and fro through a distance $\dfrac{b(\sqrt{a} + \sqrt{b})}{\sqrt{a}}$.

IIT JEE PROBLEMS

Problem 220. A solid sphere of radius R is floating in a liquid of density ρ with half of its volume submerged. If the sphere is slightly pushed and released, it starts performing simple harmonic motion. Find the frequency of these oscillations. **(JEE 2004)**

Problem 221. A thin rod of length L and uniform cross-section is pivoted at its lowest point P inside a stationary homogeneous and non-viscous liquid (as shown in figure). The rod is free to rotate in a vertical plane about a horizontal axis passing through P. The density d_1 of the material of the rod is smaller than the density d_2 of the liquid. The rod is displaced by small angle θ from its equilibrium position and then released. Show that the motion of the rod is simple harmonic and determine its angular frequency in terms of the given parameters. **(JEE 1996)**

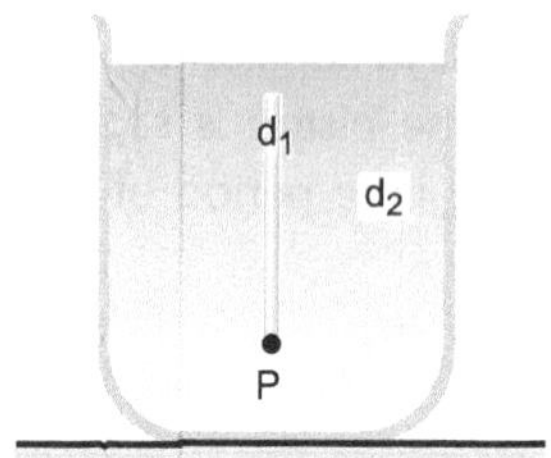

Problem 222. Two identical balls A and B, each of mass 0.1 kg, are attached to two identical massless springs. The spring-mass system is constrained to move inside a rigid smooth pipe bent in the form of a circle as shown in figure. The pipe is fixed in a horizontal plane. The centres of the balls can move in a circle of radius 0.06 metre. Each spring has a natural length of $0.06\,\pi$ metre and spring constant 0.1 N/m Initially, both the balls are displaced by an angle $\theta = \pi/6$ radian with respect to the diameter PQ of the circle (as shown in figure) and released from rest.

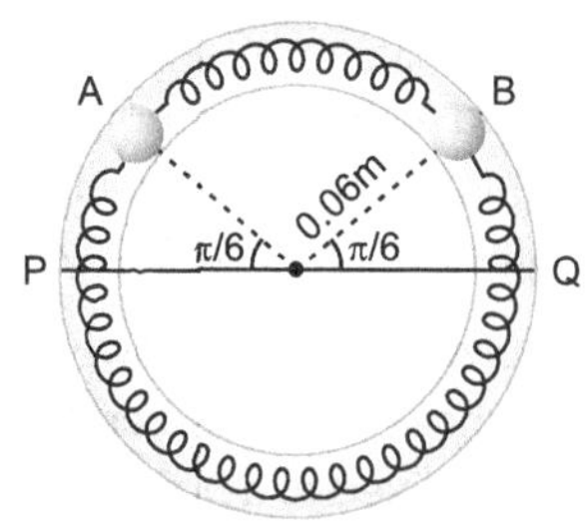

(i) calculate the frequency of oscillation of ball B.

(ii) find the speed of ball A when A and B are at the two ends of the diameter PQ.

(iii) what is the total energy of the system ? **(JEE 1993)**

CHAPTER 8

Solids And Fluids

Problem 223. A thin ring of radius R is made of a material of density ρ and Young's modulus Y. If the ring is rotated about its centre and its own plane with angular velocity ω. Find the small increase in its radius.

Problem 224. Find the volume density of the elastic deformation energy in fresh water at a depth of $h = 1$ km. Take density of water $= 10^3$ kg / m^3, $g = 9.8$ m/s^2 and bulk modulus of elasticity of water $B = 2 \times 10^9$ N/m^2.

Problem 225. Two opposite forces F_1 and F_2 ($< F_1$) act on an elastic plank of modulus of elasticity Y and length l placed over a smooth horizontal surface. The cross-sectional area of the plank is S. Find the change in length of the plank in the direction of the force.

Problem 226. Two rods of different metals having the same area of cross-section A are placed between the two massive walls as shown in figure. The first rod has a length l_1, coefficient of linear expansion α_1 and Young's modulus Y_1. The corresponding quantities for second rod are l_2, α_2 and Y_2. The temperature of both the rods is now raised by $t°$ C.

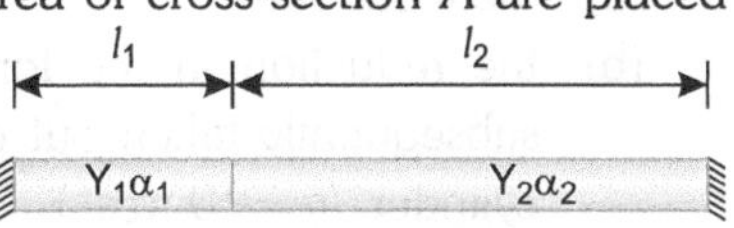

(a) find the force with which the rods act on each other at higher temperature.
(b) the lengths of the rods at higher temperature.

Problem 227. A solid sphere of mass $m = 2$ kg and density $\rho = 500$ kg/m^3 is held stationary relative to a tank filled with water. The tank is accelerating upward with acceleration 2 m/s^2. Calculate

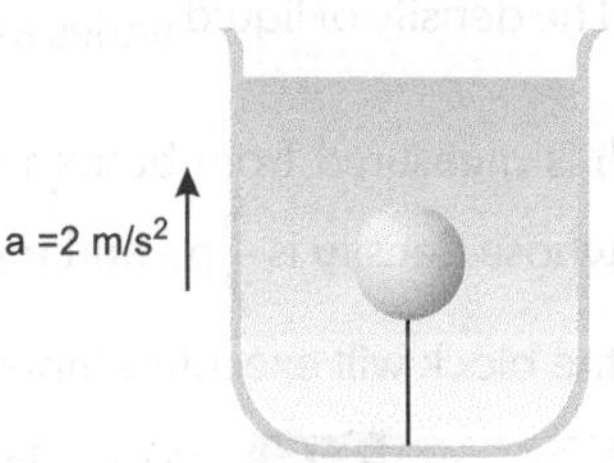

(a) tension in the thread connected between the sphere and the bottom of the tank
(b) if the thread snaps, calculate the acceleration of sphere with respect to the tank. [density of water $= 1000$ kg/m^3, $g = 10$ m/s^2]

Problem 228. A cylindrical tank of base area A has a small hole of area 'a' at the bottom. At time $t = 0$, a tap starts to supply water into the tank at a constant rate α m^3/s.
(a) what is the maximum level of water h_{max} in the tank ?
(b) find the time when level of water becomes h ($< h_{max}$).

Problem 229. Two cylindrical tanks having cross-sectional area A and $2A$ are kept on a horizontal floor. First tank is filled with water to a height H and the other is empty. If two tanks are connected by a pipe of cross-sectional area a ($<< A$) at the bottom at $t = 0$, calculate time t when level of water in two tanks becomes same.

Problem 230. An ice cube of size $l = 20$ cm is floating in a tank (base area $A = 50$ cm $\times$ 50 cm) partially filled with water. Density of water is $\rho_\omega = 1000$ kg/m^3 and that of ice is $\rho_i = 900$ kg/m^3. Calculate increase in gravitational potential energy when ice melts completely. ($g = 10$ m/s^2)

Problem 231. Two cylinders with a horizontal and a vertical axis respectively rest on a horizontal surface. The cylinders are connected at the lower parts through a thin tube. The horizontal cylinder of radius r has a piston at one end as shown.

The vertical cylinder is open at the top. The cylinders contain water which completely fills the part of the horizontal cylinder behind the piston and is at a certain level in the vertical cylinder.

Determine the level 'h' of water in the vertical cylinder at which the piston is in equilibrium. Neglect friction.

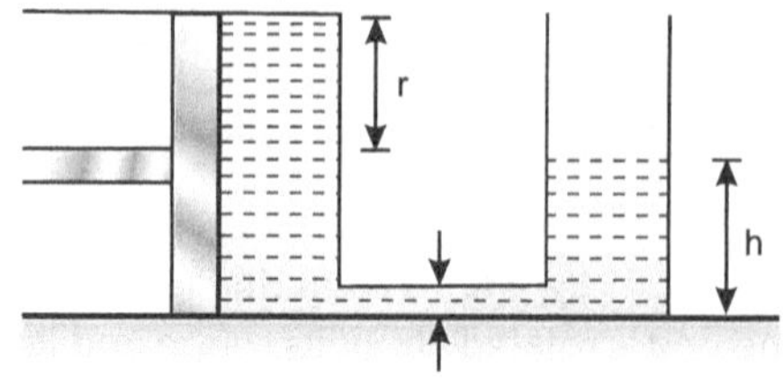

Problem 232. A cubical block of wood has density $\rho_1 = 500\ \text{kg/m}^3$ and side $l = 30$ cm. It is floating in a rectangular tank partially filled with water of density $\rho_2 = 1000\ \text{kg/m}^3$ and having base area, $A = 45\ \text{cm} \times 60$ cm. Calculate work done to press the block slowly so that it is just immersed in water. $(g = 10\ \text{m/s}^2)$

Problem 233. A cylinder of radius 1 cm, length 4 cm made of a material of specific gravity 0.75 floats in water with its axis vertical. It is then pushed vertically downwards, so as to be just immersed. Find

(a) the work done

(b) the reduction in the force on the bottom of the containing vessel when the cylinder is subsequently taken out of water. Assume that volume of water in the vessel >> volume of cylinder. $(g = 10\ \text{m/s}^2)$

Problem 234. Figure shows a container having liquid of variable density. The density of liquid varies as $\rho = \rho_0 \left(4 - \frac{3h}{h_0}\right)$. Here h_0 and ρ_0 are constants and h is measured from bottom of the container. A solid block of small dimensions whose density is $\frac{5}{2}\rho_0$ and mass m is released from bottom of the tank. Prove that the block will execute simple harmonic motion. Find the frequency of oscillation.

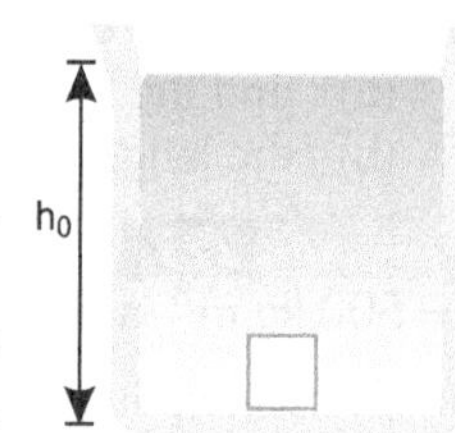

Problem 235. A wooden cylinder of length l floats vertically in a liquid of specific gravity ρ with part of its length submerged. Another liquid that is immiscible with the previous liquid is poured into it, to just completely submerge the cylinder. If the density of the liquid at the bottom, density of cylinder and density of the liquid at the top are in geometric progession (G.P.), calculate the fraction of the cylinder submerged in the lower liquid. If the cylinder is slightly depressed and released. Find its time period of small oscillations. Cross-sectional area of the container can be assumed large compared to cross-sectional area of the cylinder.

Problem 236. A 4.0 kg mass is hung from a vertical steel wire 2.0 m long and $5.0 \times 10^{-3}\ \text{cm}^2$ in cross-sectional area. The wire is securely fastened to the ceiling.

(a) Calculate the amount the wire is stretched by the hanging mass.

(b) Now assume that the mass is very slowly pulled downward 0.06 cm from its equilibrium position by an external force $\vec{F}$. Calculate

  (i) the work done by gravity

(ii) work done by the force $\vec{F}$

(iii) the work done by the force the wire exerts on the mass

(iv) the change in the elastic potential energy. Young's modulus of steel is $Y = 2.0 \times 10^{11}$ N/m^2 $g = 9.8$ m/s^2.

Problem 237. For the arrangement shown in figure, find the time interval after which the water jet ceases to cross the wall. Area of the tank is A and area of orifice is a.

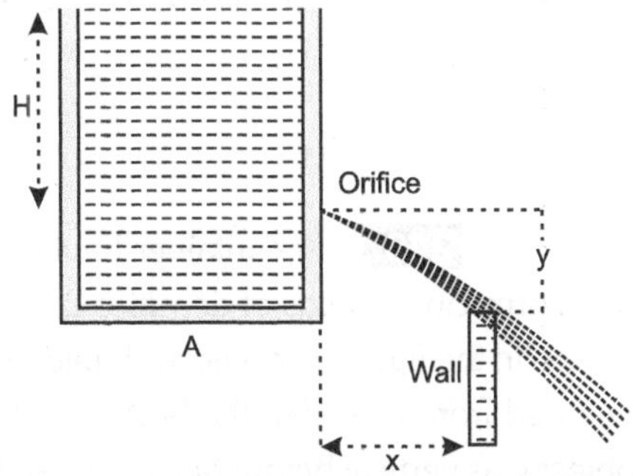

Problem 238. In a tank, having large base area, a liquid of density 1200 kg/m^3 is filled upto a height of 5 m, the tank is placed on a platform which is moving upward with an acceleration of 5 m/s^2. A very small hole is made in the tank at height of 2 m from the bottom. Find the distance of the point, where the liquid falls on the platform w.r.t. the edge of the tank. (Take $g = 10$ m/s^2)

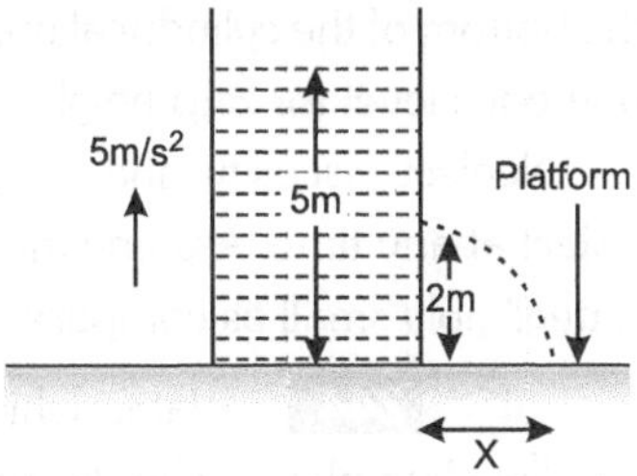

Problem 239. A wooden cone of mass 8.8 kg floats with its apex downward in a liquid of specific gravity 0.8. If the specific gravity of wood is 0.5,

(a) what mass of steel piece of specific gravity 7.8 suspended from the apex of the cone by a cord thats will just be able to submerge the cone?

(b) what will be tension in the cord.

Problem 240. An incompressible viscous liquid of density ρ and viscosity η is filled into a cylindrical vessel of cross-section A, up to a height h. The vessel has a horizontal capillary tube of length l and bore radius a attached to the bottom of the vessel as shown in the diagram. A piston of mass m is placed on top of the liquid in the tank and the liquid flows slowly and steadily out of the tank under pressure. Find the level of the liquid in the tank as a function of time t. Ignore the surface tension of the liquid.

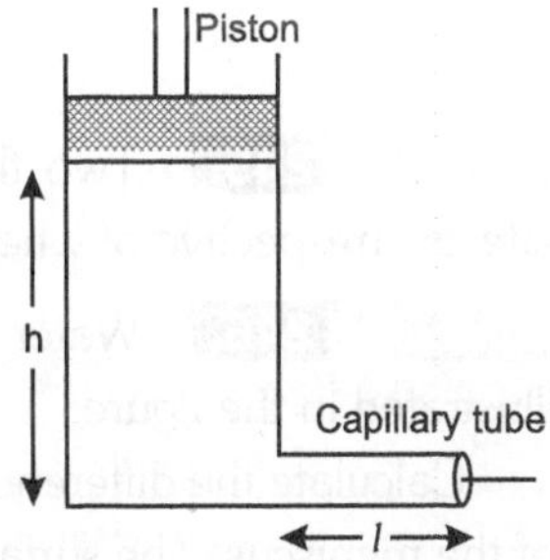

Problem 241. Figure shows two pipes P and Q connected with the help of three capillary tubes A, B and C. The radii and lengths of these tubes are shown in the figure. If pressure difference between points P and Q is p then find :

(a) the pressure difference across B.

(b) the ratio of flow rates in A and B.

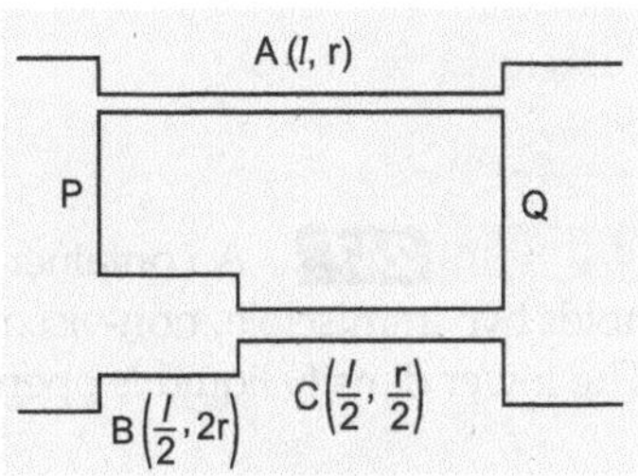

Problem 242. The pulley mass system is as shown in the figure. Mass M is a uniform cylinder ($M > m$). String is inextensible and there is no friction. Find the acceleration of the cylinder as a function of length of cylinder x inside the liquid. Find the velocity of system when cylinder is totally immersed. ρ_s and ρ_l are the densities of cylinder and the liquid respectively. Length of the cylinder is L.

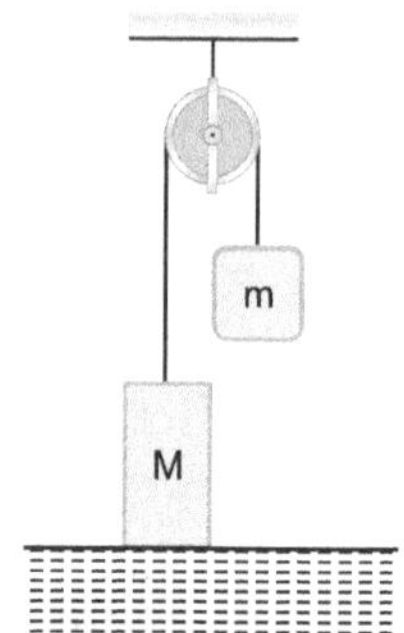

Problem 243. A container having very large cross-sectional area contains two immiscible liquids of densities 2ρ and ρ of heights H and $2H$ respectively as shown in the figure. A cylindrical solid object having density $\rho/2$ and height $H/2$ is released from rest from the bottom of the container. The axis of the cylindrical object was vertical before being released. Find the maximum height attained by the bottom of the cylindrical object, treating the bottom of the container to be the reference level for zero height.

Neglect viscosity and assume that there is no toppling of the cylindrical object at any time. Assume that the cross-sectional area of the cylindrical object is negligibly small as compared to the cross-sectional area of container.

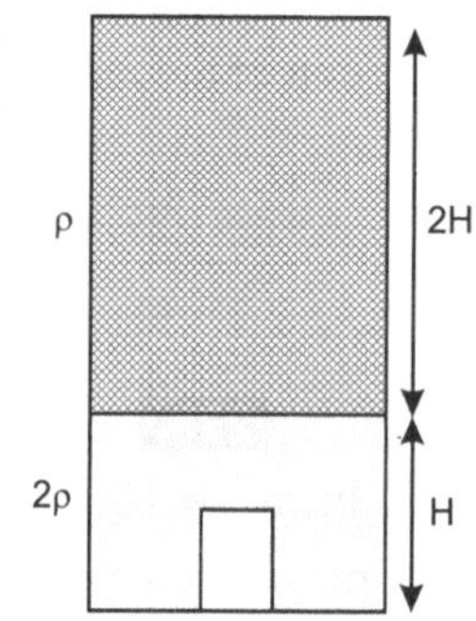

Problem 244. Water, which wets glass, is stuck between two parallel glass plates. The distance between the plates is d and the diameter of the trapped water 'disc' is $D >> d$. (S = surface tension of water)

What is the force acting between the plates?

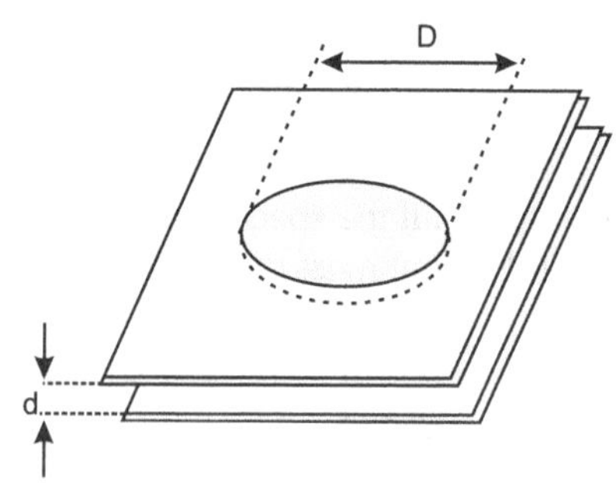

Problem 245. Two floating objects are attracted to each other as the result of surface tension effects, irrespective of whether they are floating on water or on mercury. Explain why this is so.

Problem 246. Water in a clean aquarium forms a meniscus, as illustrated in the figure.

Calculate the difference in height h between the centre and the edge of the meniscus. The surface tension of water is $\gamma = 0.073\ \text{Nm}^{-1}$

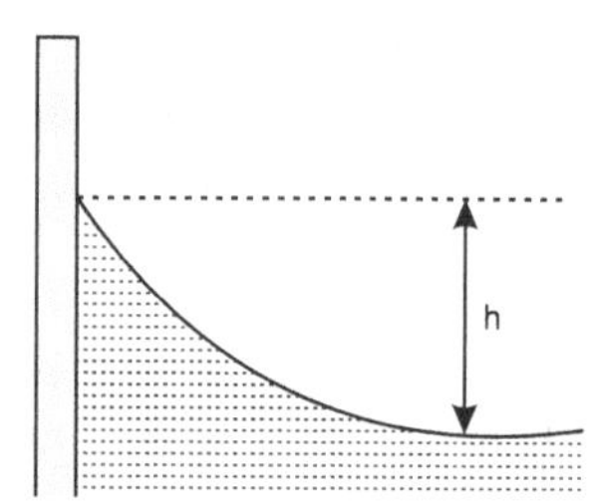

Problem 247. A container of large uniform cross-section area A resting on a horizontal surface, holds two immiscible, non-viscous and incompressible liquids of densities d and $2d$ each of height $(H/2)$. The lower density liquid is open to the atmosphere having pressure p_0.

(a) A homogeneous solid cylinder of length $L\left(L<\dfrac{H}{2}\right)$, cross-section area $\left(\dfrac{A}{5}\right)$ is immersed such that it floats with its axis vertical at the liquid-liquid interface with length $(L/4)$ in the denser liquid. Determine (i) The density of solid and (ii) The total pressure at the bottom of the container.

(b) The cylinder is removed and original arrangement is restored. A tiny hole of area s ($s << A$) is punched on the vertical side of the container at a height h ($h < H/2$). Determine (i) the initial speed of efflux of the liquid at the hole (ii) the horizontal distance x travelled by the liquid initially and (iii) the height h_m at which the hole should be punched so that the liquid travels the maximum distance x_m initially. Also calculate x_m.

Problem 248. A cylindrical tank 0.9 m in radius is filled with water ($\rho = 1\times10^3$ kg/m^3) up to a height $y = 3.0$ m. A pipe having cross-sectional area 6.3 cm^2 is used to siphon from the tank. End B of the pipe is open to atmosphere and $h = 5.0$ m.

(a) How should the value of x be adjusted so that whole of the water can be drained from the tank ?

(b) With maximum possible value of x, how much time will it take to empty the vessel?

(Atmospheric pressure $= 1.0\times10^5$ Nm^{-2})

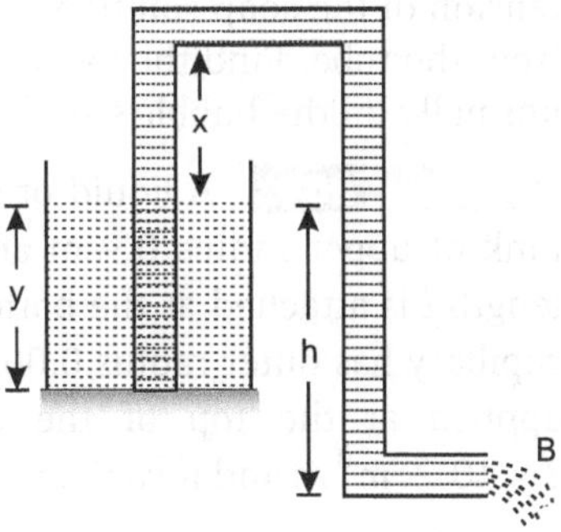

Problem 249. A vertical cylinder of radius 'R' is rotating with a constant angular velocity 'ω' along with holders A and B about the axis as shown. The vertical movement of cylinder is restricted by the holders. Oil is only in between the bottom of cylinder and surface as shown. The thickness of the oil layer is 't'. Assume that coefficient of viscosity is 'η' and that cylinder can only rotate if there is no friction elsewhere. Find the power required to overcome the viscous resistance.

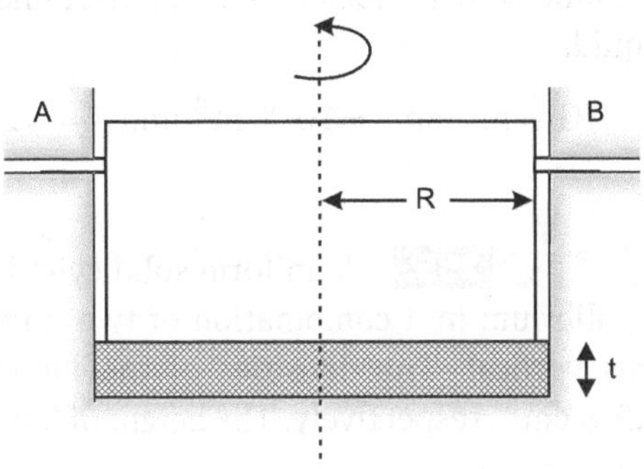

IIT JEE PROBLEMS

Problem 250. A container of width $2a$ is filled with a liquid. A thin wire weight of per unit length λ is gently placed over the liquid surface in the middle of the surface as shown in the figure. As a result, the liquid surface is depressed by a distance y ($y << a$). Determine the surface tension of the liquid.

(JEE 2004)

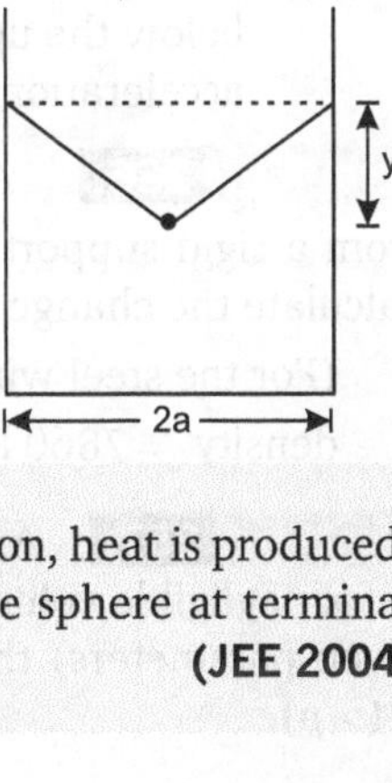

Problem 251. A small sphere falls from rest in a viscous liquid. Due to friction, heat is produced. Find the relation between the rate of production of heat and the radius of the sphere at terminal velocity. **(JEE 2004)**

Problem 252. Consider a horizontally oriented syringe containing water located at a height of 1.25 m above the ground. The diameter of the plunger is 8 mm and the diameter of the nozzle is 2 mm. The plunger is pushed with a constant speed of 0.25 m/s. Find the horizontal range of water stream on the ground. Take $g = 10\ \text{m/s}^2$. **(JEE 2004)**

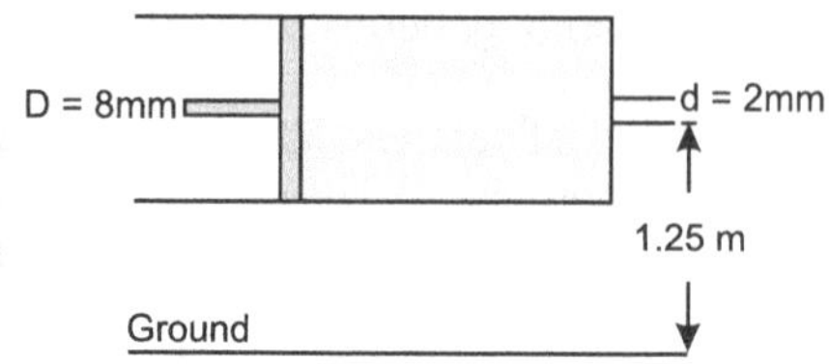

Problem 253. A soap bubble is being blown at the end of a very narrow tube of radius b. Air (density ρ) moves with a velocity v inside the tube and comes to rest inside the bubble. The surface tension of the soap solution is T. After some time the bubble, having grown to a radius r, separates from the tube. Find the value of r. Assume that $r >> b$ so that you can consider the air to be falling normally on the bubble's surface. **(JEE 2003)**

Problem 254. A liquid of density 900 kg/m^3 is filled in a cylindrical tank of upper radius 0.9 m and lower radius 0.3 m. A capillary tube of length l is attached at the bottom of the tank as shown in the figure. The capillary has outer radius 0.002 m and inner radius a. When pressure P is applied at the top of the tank volume flow rate of the liquid is $8 \times 10^{-6}\ \text{m}^3/\text{s}$ and if capillary tube is detached, the liquid comes out from the tank with a velocity 10 m/s. Determine the coefficient of viscosity of the liquid.

[Given: $\pi a^2 = 10^{-6}\ \text{m}^2$ and $\frac{a^2}{l} = 2 \times 10^{-6}\ \text{m}$] **(JEE 2003)**

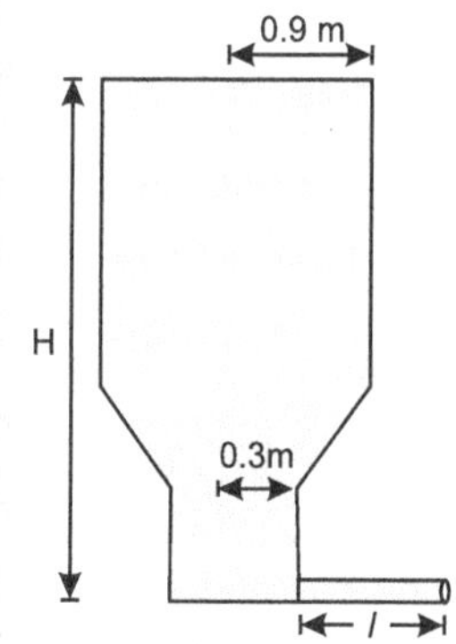

Problem 255. A uniform solid cylinder of density 0.8 g/cm^3 floats in equilibrium in a combination of two non-mixing liquids A and B with its axis vertical. The densities of the liquid A and B are 0.7 g/cm^3 and 1.2 g/cm^3, respectively. The height of liquid A is $h_A = 1.2$ cm. The length of the part of the cylinder immersed in liquid B is $h_B = 0.8$ cm.

(a) Find the total force exerted by liquid A on the cylinder.
(b) Find h, the length of the part of the cylinder in air.
(c) The cylinder is depressed in such a way that its top surface is just below the upper surface of liquid A and is then released. Find the acceleration of the cylinder immediately after it is released. **(JEE 2002)**

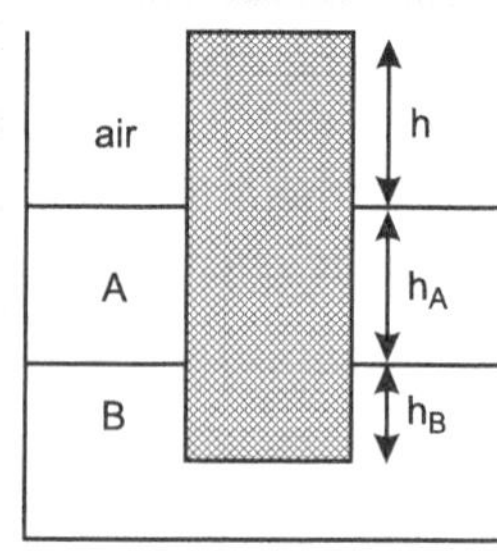

Problem 256. A 5 m long cylindrical steel wire with radius 2×10^{-3} m is suspended vertically from a rigid support and carries a bob of mass 100 kg at the other end. If the bob gets snapped, calculate the change in temperature of the wire ignoring radiation losses. Take $g = 10\ \text{m/s}^2$

(For the steel wire : Young's modulus $= 2.1 \times 10^{11}\ \text{N/m}^2$;
density $= 7860\ \text{kg/m}^3$; Specific heat $= 420$ J/kg-K) **(JEE 2001)**

Problem 257. A wooden stick of length L, radius R and density ρ has a small metal piece of mass m (of negligible volume) attached to its one end. Find the minimum value for the mass m (in terms of given parameters) that would make the stick float vertically in equilibrium in a liquid of density σ (> p). **(JEE 1999)**

Problem 258. A thin rod of negligible mass and area of cross-section $4 \times 10^{-6}\ \text{m}^2$, suspended vertically from one end, has a length of 0.5 m at 100° C. The rod is cooled to 0°C, but prevented from contracting by attaching a mass at the lower end. Find (i) this mass, and (ii) the energy stored in the rod. Given for the rod, Young's modulus $= 10^{11}\ \text{N/m}^2$, coefficient of linear expansion $= 10^{-5}\ \text{K}^{-1}$ and $g = 10\ \text{m/s}^2$. **(JEE 1997, Cancelled)**

Problem 259. A non-viscous liquid of constant density $1000\ \text{kg/m}^3$ flows in a streamline motion along a tube of variable cross-section. The tube is kept inclined in the vertical plane as shown in the figure. The area of cross-section of the tube at two points P and Q at heights of 2 metre and 5 metre are respectively $4 \times 10^{-3}\ \text{m}^2$ and $8 \times 10^{-3}\ \text{m}^2$. The velocity of the liquid at point P is 1 m/s. Find the work done per unit volume by the pressure and the gravity forces as the fluid flows from point P to Q. Take $g = 9.8\ \text{m/s}^2$ **(JEE 1997)**

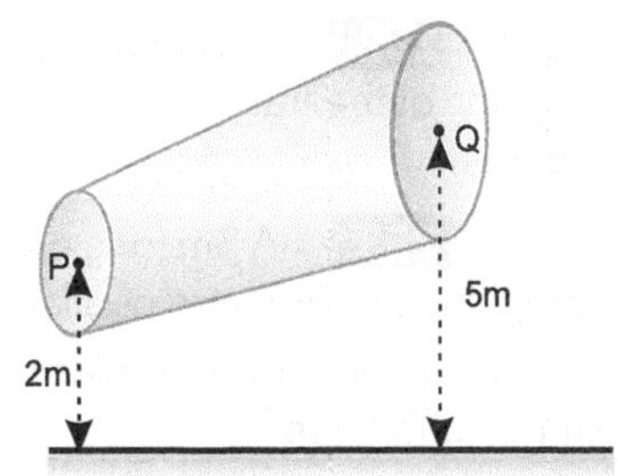

Problem 260. A container of large uniform cross-sectional area A resting on a horizontal surface, holds two immiscible, non-viscous and incompressible liquids of densities d and $2d$, each of height $H/2$ as shown in figure. The lower density liquid is open to the atmosphere having pressure P_0.

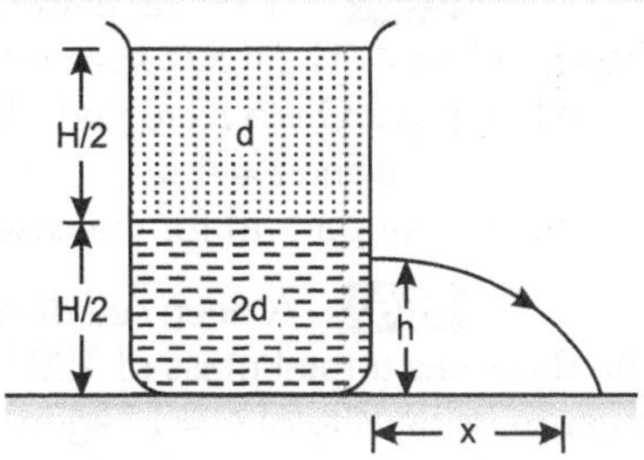

(a) A homogeneous solid cylinder of length L $(L < H/2)$, cross-sectional area $A/5$ is immersed such that if floats with its axis vertical at the liquid-liquid interface with length $L/4$ in the denser liquid. Determine :
  (i) the density D of the solid,
  (ii) the total pressure at the bottom of the container.

(b) The cylinder is removed and the original arrangement is restored. A tiny hole of area S $(S << A)$ is punched on the vertical side of the container at a height h $(h < H/2)$. Determine :
  (i) the initial speed of efflux of the liquid at the hole,
  (ii) the horizontal distance x travelled by the liquid initially, and
  (iii) the height h_m at which the hole should be punched so that the liquid travels the maximum distance x_m initially. Also calculate x_m.

(Neglect the air resistance in these calculations). **(JEE 1995)**

Problem 261. A ball of density d is dropped onto a horizontal solid surface. It bounces elastically from the surface and returns to its original position in a time t_1. Next, the ball is released and it falls through the same height before striking the surface of a liquid of density d_L.

(a) if $d < d_L$, obtain an expression (in terms of d, t_1 and d_L) for the time t_2 the ball takes to come back to the position from which it was released.

(b) is the motion of the ball simple harmonic ?

(c) if $d = d_L$, how does the speed of the ball depend on its depth inside the liquid ? Neglect all frictional and other dissipative forces. Assume the depth of the liquid to be large. **(JEE 1992)**

CHAPTER 9

WAVES

Problem 262. A train of length l is moving with a constant speed υ along a circular track of radius R. The engine of the train emits a sound of frequency f. Find the frequency heard by a guard at the rear end of the train.

Problem 263. A 3m long organ pipe open at both ends is driven to third harmonic standing wave. If the amplitude of pressure oscillations is 1 percent of mean atmospheric pressure ($P_0 = 10^5$ N/m^2). Find the amplitude of particle displacement and density oscillations. Speed of sound $\upsilon = 332$ m/s and density of air $\rho = 1.03$ kg/m^3.

Problem 264. A siren creates a sound level of 60 dB at a location 500 m from the speaker. The siren is powered by a battery that delivers a total energy of 1.0 kJ. Assuming that the efficiency of siren is 30%, determine the total time the siren can sound.

Problem 265. A wave pulse on a string has the dimensions shown in figure. The wave speed is $\upsilon = 1$ cm/s.

(a) if point O is a fixed end, draw the total wave on the string at $t = 3$ s and $t = 4$ s.

(b) repeat part (a) for the case in which O is a free end.

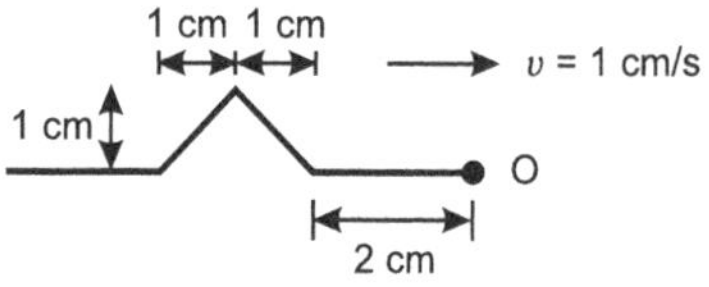

Problem 266. A deep sea diver is suspended beneath the surface of water by a 100 m long cable. The diver has a total mass of 120 kg and volume 0.08 m^3. The cable has a diameter of 2.0 cm and a linear mass density of $\mu = 1.1$ kg/m. The diver sends a transverse wave pulse from the bottom. Find the time required for the wave pulse to reach the surface. Density of water $= 10^3$ kg/m^3 and $g = 10$ m/s^2.

Problem 267. In a stationary wave that forms as a result of reflection of waves from an obstacle, the ratio of the amplitude at an antinode to the amplitude at node is 6. What percentage of energy is transmitted.

Problem 268. A conveyor belt moves to the right with speed $\upsilon = 300$ m/min. A very fast pieman puts pies on the belt at a rate of 20 per minute and they are received at the other end by a pieeater.

(a) If the pieman is stationary, find the spacing x between the pies and the frequency with which they are received by the stationary pieeater.

(b) The pieman now walks with speed 30m/min towards the receiver while continuing to put pies on the belt at 20 per minute. Find the spacing of the pies and the frequency with which they are received by the stationary pieeater.

Problem 269. A stationary observer receives a sound from a source of frequency 2000 Hz moving with a constant velocity. The apparent frequency varies with time as shown in figure. Find :

(a) speed of source (υ_s)

(b) maximum value of apparent frequency f_m. (speed of sound is $\upsilon = 300$ m/s)

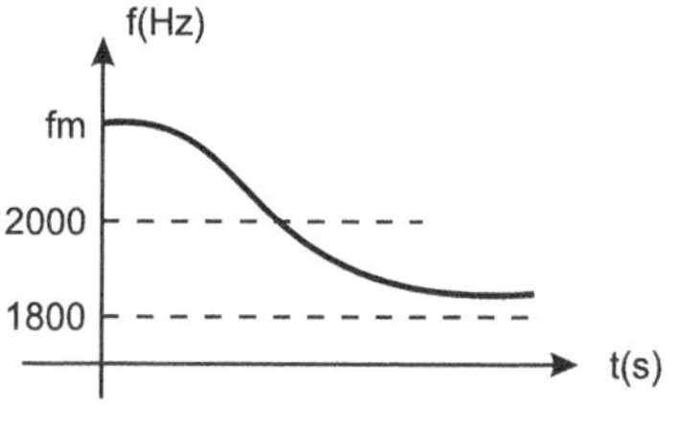

Problem 270. Consider the arrangement shown in given figure. $m_1 = 7.5$ kg, $m_2 = 10$ kg and mass per unit length of the string is 15×10^{-3} kg/m. At $t = 0$, the system is released from rest and at the same instant of time a transverse pulse is produced at A. The pulley B is at a distance $l = 10$ m from A at $t = 0$. Assuming all surfaces to be frictionless and all pulleys to be massless. Find the time taken by the pulse to reach to the pulley B.

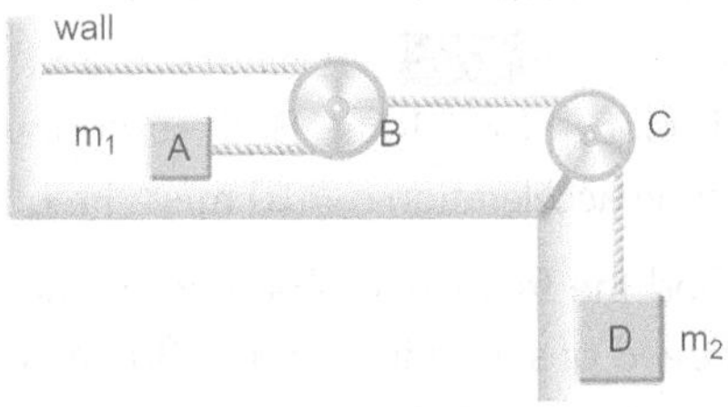

Problem 271. A point sound source is situated in a medium of bulk modulus 1.6×10^5 N/m^2. An observer standing at a distance 10 m from the source, writes down the equation for the wave as $y = A \sin(15\pi x - 6000\pi t)$; y and x are in metre and t is in second. The maximum pressure amplitude received to the observer's ear is 24π Pa, then find

(a) the density of the medium

(b) the displacement amplitude A of the waves received by the observer.

(c) the power of the sound source.

Problem 272. A source of sound with frequency 1000 Hz and a receiver are located at the same point. At the moment $t = 0$, the source and receiver starts receding from each other with acceleration 3.0 m/s^2 and 2 m/s^2 respectively. Find the oscillation frequency registered by the receiver at $t = 9$ sec after the start of motion.

Problem 273. A sinusoidal wave is propagating along a strechted string that lies along the x-axis. The displacement of the string as a function of time is graphed in figure for particles at $x = 0$ and $x = 0.09$ m. These two points are within one wavelength of each other. Determine the wavelength and the wave speed if the wave is moving in positive x-direction.

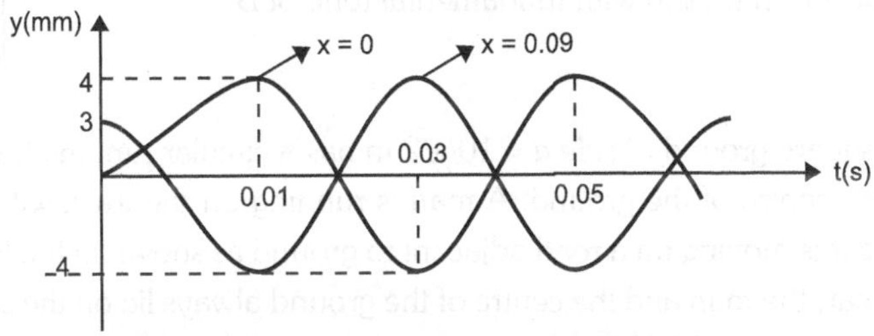

Problem 274. A standing wave $y = a \sin kx \cos \omega t$ is maintained in a homogeneous rod with cross-sectional area S and density ρ. Find the total mechanical energy confined between the sections corresponding to the adjacent nodes.

Problem 275. A uniform string of length l is fixed at both ends such that tension T is produced in it. The string is excited to vibrate with maximum displacement amplitude a_0. Calculate the total energy of the string for its first overtone.

Problem 276. A source of sonic oscillations with frequency f Hz and a receiver are located at the same point. At $t = 0$, the source starts receding from the receiver with constant acceleration a. Assuming the velocity of sound to be c, find the oscillation frequency registered by stationary receiver t_0 second after the start up of the motion.

Problem 277. A source S and a detector O are initially at a distance of $x = 1$ km. Both start moving towards one another with same acceleration $a = 10\ \text{m/s}^2$. Frequency of source is $f = 2000$ Hz. Find the frequency observed by the detector at time $t = 4$ second. Speed of sound in air is $v = 300$ m/s.

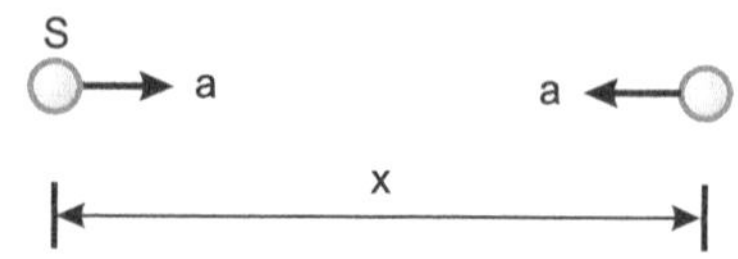

Problem 278. A string 120 cm in length sustains a standing wave with the points of the string at which the displacement amplitude is equal to 3.5 mm being separated by 15.0 cm. Find

(a) the maximum displacement amplitude.

(b) to which overtone do these oscillations correspond ?

Problem 279. An aluminium wire of cross-sectional area $10^{-6}\ \text{m}^2$ is joined to a steel wire of the same cross-sectional area. This compound wire is stretched on a sonometer pulled by a weight of 10kg. The total length of the compound wire between the bridges is 1.5 m of which the aluminium wire is 0.6 m and the rest is steel wire. Transverse vibrations are set up in the wire by using an external source of variable frequency. Find the lowest frequency of excitation for which the standing waves are formed such that the joint in the wire is a node. What is the total number of nodes at this frequency. The density of aluminium is $2.6 \times 10^3\ \text{kg/m}^3$ and that of steel is $1.04 \times 10^4 \text{kg/m}^3$. ($g = 10\ \text{m/s}^2$)

Problem 280. A U-tube having uniform cross-section but unequal arm lengths l_1 and $l_2\ (< l_1)$ has same liquid of density ρ_1 filled in it upto a height h as shown in figure. Another liquid of density $\rho_2\ (= \rho_1/2)$ is poured in arm A. Both liquids are immiscible. What length of the second liquid should be poured in A so that first overtone of A is in unison with fundamental tone of B.

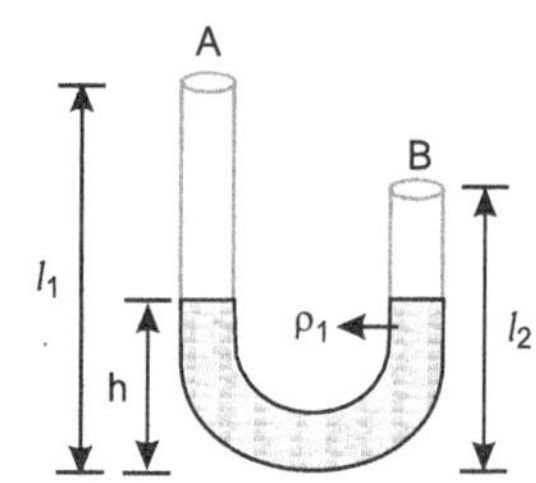

Problem 281. A square ground of side $a = 10/\sqrt{2}$ m has a circular running track of radius $a/2$ with its centre coinciding the centre of the ground. A man is running on the track with an angular velocity $\omega = 22\ \text{rad/s}$ while a car is moving on a road adjacent to ground as shown in the figure. The car moves in such a way that the car, the man and the centre of the ground always lie on the same straight line. If a source of sound of frequency $\nu = 300$ Hz is being placed at the centre of the ground find the minimum frequency received by the man in the car. Assume velocity of sound in air is $v = 330\ \text{m/s}$.

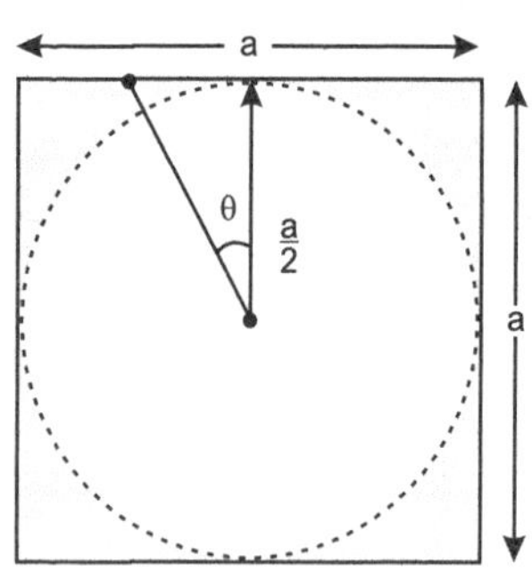

Problem 282. $y_1 = 8 \sin(\omega t - kx)$ and $y_2 = 6 \sin(\omega t + kx)$ are two waves travelling in a string of area of cross-section s and density ρ. These two waves are superimposed to produce a standing wave.

(a) Find the energy of the standing wave between two consecutive nodes.

(b) Find the total amount of energy crossing through a node per second.

IIT JEE PROBLEMS

Problem 283. In a resonance tube experiment to determine the speed of sound in air, a pipe of diameter 5 cm is used. The air column in pipe resonates with a tuning fork of frequency 480 Hz when the minimum length of the air column is 16 cm. Find the speed of sound in air at room temperature. **(JEE 2003)**

Problem 284. A string of mass per unit length μ is clamped at both ends such that one end of the string is at $x = 0$ and the other is at $x = l$. When string vibrates in fundamental mode amplitude of the mid-point of the string is a, and tension in the string is T. Find the total oscillation energy stored in the string. **(JEE 2003)**

Problem 285. Two narrow cylindrical pipes A and B have the same length. Pipe A is open at both ends and is filled with a monoatomic gas of molar mass M_A. Pipe B is open at one end and closed at the other end, and is filled with a diatomic gas of molar mass M_B. Both gases are at the same temperature.

(a) If the frequency to the second harmonic of the fundamental mode in pipe A is equal to the frequency of the third harmonic of the fundamental mode in pipe B, determine the value of $\frac{M_A}{M_B}$.

(b) Now the open end of pipe B is also closed (so that the pipe is closed at both ends). Find the ratio of the fundamental frequency in pipe A to that in pipe B. **(JEE 2002)**

Problem 286. A boat is travelling in a river with a speed 10 m/sec along the stream flowing with a speed 2 m/sec. From this boat, a sound transmitter is lowered into the river through a rigid support. The wavelength of the sound emitted from the transmitter inside the water is 14.45 mm. Assume that attenuation of sound in water and air is negligible.

(a) What will be the frequency detected by a receiver kept inside the river downstream ?

(b) The transmitter and the receiver are now pulled up into air. The air is blowing with a speed 5 m/sec in the direction opposite the river stream. Determine the frequency of the sound detected by the receiver.

(Temperature of the air and water = 20°C; Density of river water $= 10^3$ kg/m^3;

Bulk modulus of the water $= 2.088 \times 10^9$ Pa; Gas constant, R = 8.31 J/mol-K; Mean molecular mass of air $= 28.8 \times 10^{-3}$ kg/mol; C_p/C_V for air = 1.4) **(JEE 2001)**

Problem 287. A 3.6 m long pipe resonates with a source of frequency 212.5 Hz when water level is at certain heights in the pipe. Find the heights of water level (from the bottom of the pipe) at which resonances occur. Neglect end correction. Now the pipe is filled to a height H ($\approx$3.6 m). A small hole is drilled very close to its bottom and water is allowed to leak. Obtain an expression for the rate of fall of water level in the pipe as a function of H. If the radii of the pipe and the hole are 2×10^{-2} m and 1×10^{-3} m respectively, calculate the time interval between the occurrence of first two resonances. Speed of sound in air is 340 m/s and $g = 10$ m/s^2. **(JEE 2000)**

Problem 288. A long wire PQR is made by joining two wires PQ and QR of equal radii. PQ has a length 4.8 m and mass 0.06 kg. QR has length 2.56 m and mass 0.2 kg. The wire PQR is under a tension of 80 N. A sinusoidal wave pulse of amplitude 3.5 cm is sent along the wire PQ from the end P. No power is dissipated during the propagation of the wave pulse. Calculate :

(a) the time taken by the wave pulse to reach the other end R, and

(b) the amplitude of the reflected and transmitted wave pulse after the incident wave pulse crosses the joint Q. **(JEE 1999)**

Problem 289. The air column in a pipe closed at one end is made to vibrate in its second overtone by tuning fork of frequency 440 Hz. The speed of sound in air is 330 m/s. End corrections may be neglected. Let P_0 denote the mean pressure at any point in the pipe, and ΔP_0 the maximum amplitude of pressure variation.

(a) Find the length L of the air column.

(b) What is the amplitude of pressure variation at the middle of the column ?

(c) What are the maximum and minimum pressures at the open end of the pipe ?

(d) What are the maximum and minimum pressures at the closed end of the pipe ? **(JEE 1998)**

Problem 290. The first overtone of an open organ pipe beats with the first overtone of a closed organ pipe with a beat frequency of 2.2 Hz. The fundamental frequency of the closed organ pipe is 110 Hz. Find the lengths of the pipes. Speed of sound in air $v = 330$ m/s. **(JEE 1997, Cancelled)**

Problem 291. A metallic rod of length 1m is rigidly clamped at its mid-point. Longitudinal stationary waves are set up in the rod in such a way that there are two nodes on either side of the mid-point. The amplitude of an antinode is 2×10^{-6} m. Write (a) the equation of motion at a point 2 cm from the mid-point and (b) equation of the constituent waves in the rod. (Young's Modulus of the material of the rod $= 2 \times 10^{11}$ Nm^{-2}; density $= 8000$ kgm^{-3})

Problem 292. Two radio stations broadcast their programmes at the same amplitude, A and at slightly different frequencies ω_1 and ω_2 respectively, where $\omega_1 - \omega_2 = 10^3$ Hz. A detector receives the signals from the two stations simultaneously. It can only detect signals of intensity $\geq 2A^2$.

(a) find the time interval between successive maxima of the intensity of the signal received by the detector.

(b) find the time for which the detector remains idle in each cycle of the intensity of the signal. **(JEE 1993)**

CHAPTER 10 HEAT AND THERMODYNAMICS

Problem 293. Two moles of a monoatomic ideal gas undergo a cyclic process *ABCDA* as shown in figure. *BCD* is a semicircle. Find the efficiency of the cycle.

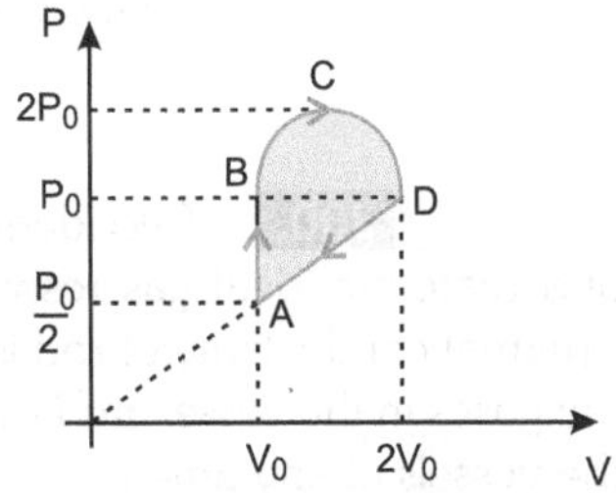

Problem 294. Three moles of an ideal gas initially at temperature $T = 273$ K is expanded isothermally so that its volume increases 5 times. It is then heated at constant volume till the final pressure becomes equal to its initial value. If the total amount of heat transferred is $Q = 80$ kJ. Find the ratio $\gamma = \frac{C_p}{C_v}$ of the gas.

Problem 295. Two glass spheres of equal volume are connected by a small tube containing a small amount of mercury as shown in figure. The spheres are sealed at 20° C with exactly 1 litre of air in each side. If the cross-sectional area of tube is 5 mm^2, how far will the mercury be displaced if the temperature of one sphere is raised by 0.1° C while the other is maintained at 20° C?

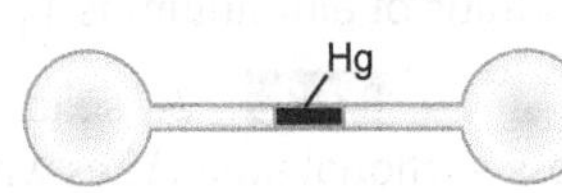

Problem 296. Pressure *P*, volume *V* and temperature *T* for a certain material are related by

$$P = \frac{\alpha T - \beta T^2}{V}$$

where α and β are constants. Find the work done by the material if the temperature changes from T_1 to T_2 while the pressure remains the constant.

Problem 297. A thin *U*-tube sealed at one end consists of three bends of length $l = 250$ mm each forming right angles. The vertical parts of the tube are filled with mercury to half the height. All of mercury is displaced from the tube by heating slowly the gas in the sealed end of the tube. Determine the work done by the gas if atmospheric pressure is $P_0 = 10^5$ N/m^2, the density of mercury in $\rho_{Hg} = 13.6 \times 10^3$ kg/m^3 and the cross-sectional area of tube is $S = 1$ cm^2. ($g = 9.8$ m/s^2)

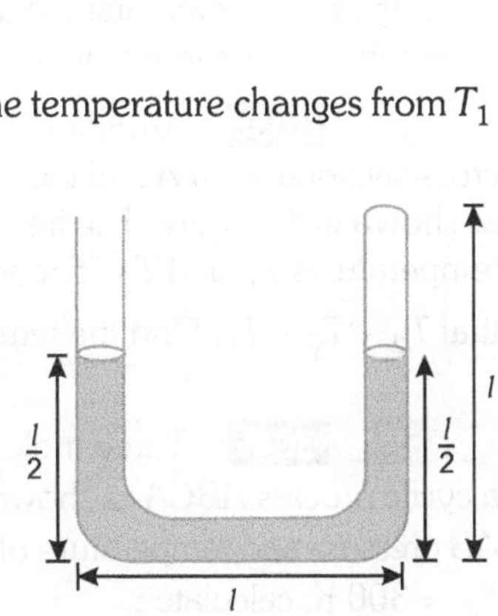

Problem 298. A pond of water at 0°C is covered with a layer of ice 4.00 cm thick. If the air temperature stays constant at – 10°C, how long does it take the ice's thickness to increase to 8.00 cm. Thermal conductivity of ice is 2 W/m °C and latent heat of fusion is 80 cal/g. Density of ice is 900 kg/m^3.

Problem 299. An aluminium rod of length $l = 0.5$ m is attached to the bottom of an iron rod of length $L = 1$ m as shown in figure. Find the decrease in height per unit change in temperature of the centre of mass of both the rods.

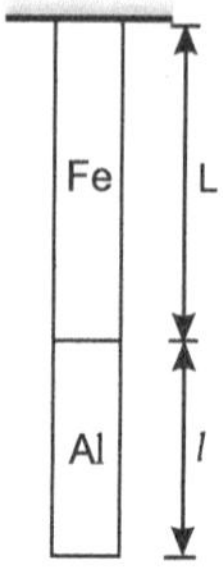

Assume that cross-sectional area of both the rods are equal.

Given : density of iron = 7.86 g/cc, density of aluminium = 4.4 g/cc

$\alpha_{iron} = 11 \times 10^{-6}$ per °C and $\alpha_{aluminium} = 25 \times 10^{-6}$ per °C.

Problem 300. Two identical adiabatic vessels A and B, each containing n moles of a monoatomic and diatomic ideal gas respectively, are connected by a rod of length l and cross-sectional area A. Thermal conductivity of rod is K and lateral surface of the rod is insulated. At time $t = 0$, temperature of the gases in the vessel are T_1 and T_2 respectively with $T_1 > T_2$. Find the temperature difference between the vessels at any time t.

Problem 301. Two steel rods and an aluminium rod of equal length l_0 and equal cross-section are joined rigidly at their ends as shown in the figure. All the rods are in a state of zero tension at 0° C. Find the length of the system when the temperature is raised to θ. Coefficient of linear expansion of aluminium and steel are α_1 and α_2 respectively. Young's modulus of aluminium is Y_1 and of steel is Y_2.

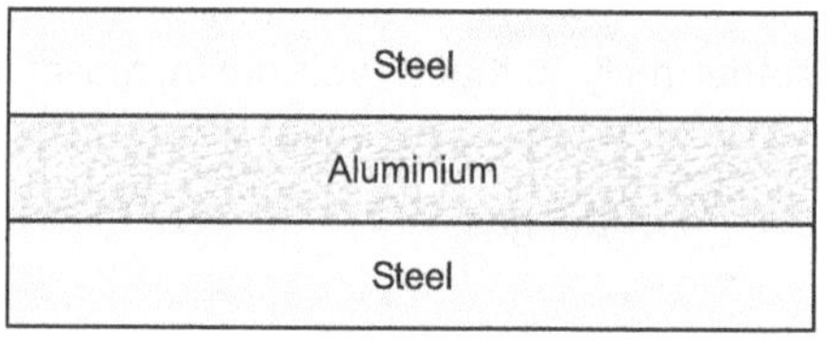

Problem 302. A solid cylindrical rod of length L_0 and cross-sectional area A lies with its axis along the x-axis and one of its ends at the origin O. The conductivity of the material of the cylinder varies with temperature T as,

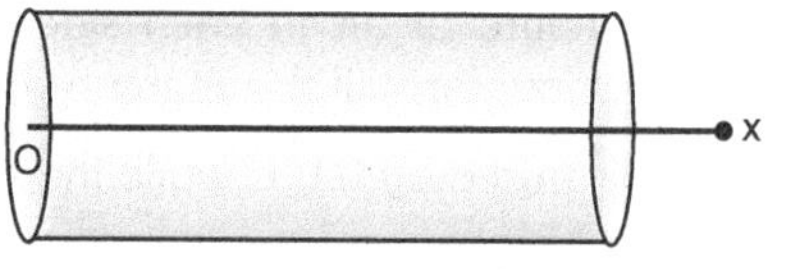

$$K = K_0 (1 + \alpha T)$$

If the end O is maintained at a temperature $2T_0$ and the other end is at T_0. Find the rate of heat flow across the rod assuming no loss of heat from the sides of the rods.

Problem 303. A body of heat capacity C is connected by 2 rods of same cross-sectional area A and length l but different thermal conductivities k_1 and k_2 as shown in the figure. The free ends of both the rods are maintained at constant temperatures T_1 and T_2. The temperature of the body is T_0 at time $t = 0$, such that $T_2 < T_0 < T_1$. Find the temperature of the body after time $t_0 = \dfrac{lC}{A(k_1 + k_2)}$.

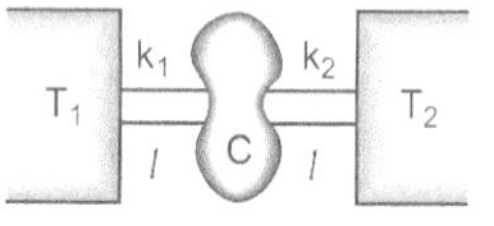

Problem 304. Three moles of an ideal diatomic gas is taken through a cyclic process $ABCA$ as shown in the $P-T$ diagram. During the process AB pressure and temperature of the gas varies such that PT^2 is constant. If $T_B = 300$ K, calculate :

(a) the work done on the gas in the process AB.

(b) the heat absorbed or released by the gas in each of the processes.

Give answer in terms of the gas constant R.

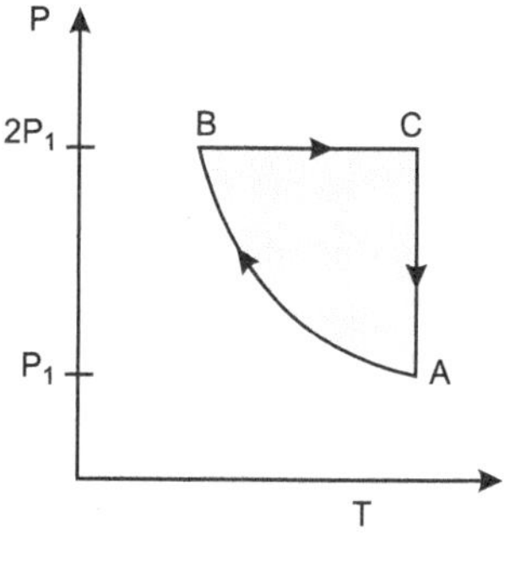

Problem 305. A gas is undergoing an adiabatic process. At a certain stage A, the values of volume and temperature $= (V_0, T_0)$ and the magnitude of the slope of VT curve is m. Find the value of C_p and C_v.

Problem 306. 2.8 grams of an ideal gas is taken in a container of volume 100 litres at temperature $T_1 = 288$ K and pressure $P = 1.5 \times 10^3$ Pa. The gas in the tube is heated isobarically to a temperature T_2. Standing waves of frequency 5 kHz is produced in tube. The separation between the nodes at temperatures T_1 and T_2 are 3 cm and 4 cm respectively find

(a) the final temperature T_2
(b) adiabatic constant γ
(c) amount of heat supplied

Prbolem 307. The minimum velocity of projection of a body to send it to infinity from the surface of a planet is $1/\sqrt{6}$ times that is required from the surface of the earth. The radius of the planet is 1/36 times the radius of the earth. The planet is surrounded by an atmosphere which contains monatomic inert gas ($\gamma = 5/3$) of constant density up to a height h(h << radius of the planet). Find the velocity of sound on the surface of the planet in terms of g_e the acceleration due to gravity on earth surface and h.

Problem 308. One mole of an ideal gas is taken from an initial state P_0, V_0 and temperature T_0 through the following cycle : (C_p/C_v for the gas is 1.4)

(a) Heating at constant volume to a temperature $3T_0$.
(b) Adiabatic expansion to a volume $2V_0$.
(c) Cooling at constant volume.
(d) Adiabatic compression so that it is returned to its initial state.
  (i) Show this cycle on a PV diagram.
  (ii) Calculate efficiency of the cycle.

Given : $(2)^{-1.4} = 0.38$

Problem 309. A composite bar is rigidly attached to the end supports. The temperature of the composite system is raised by 60°C. Find out the stress in three portions of the bar.
[$Y_s = 200$ GPa, $\alpha_s = 12 \times 10^{-6}/°C$, $Y_a = 90$ GPa,
$\alpha_a = 20 \times 10^{-6}/°C$, $Y_c = 100$ GPa, $\alpha_c = 16 \times 10^{-6}/°C$]

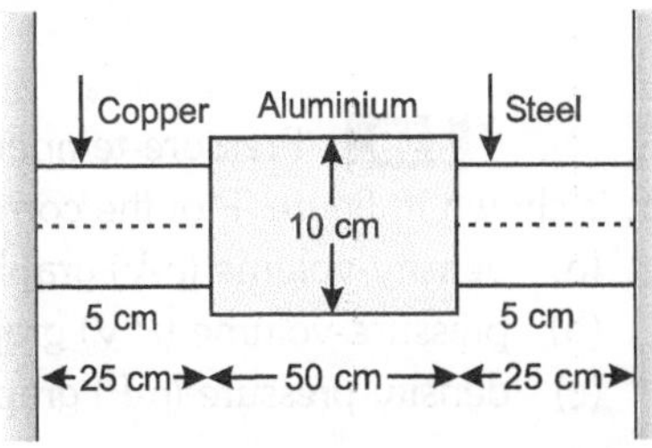

Problem 310. An ideal gas has a specific heat at constant pressure $C_P = \frac{5R}{2}$. The gas is kept in a closed vessel of volume V_0 m^3 at a temperature T_0 K and pressure P_0 N/m^2. An amount of $10\,P_0V_0$ J of heat is supplied to the gas.

(a) Calculate the final pressure and temperature of the gas.
(b) Show the process on $P-V$ diagram.

Problem 311. A vertical hollow cylinder of height 1.52 m is fitted with a movable piston of negligible mass and thickness. The lower half portion of the cylinder contains an ideal gas and the upper half is filled with mercury. The cylinder is initially at 300 K. When the temperature is raised, half of the mercury comes out of the cylinder. Find this temperature, assuming the thermal expansion of mercury to be negligible. (P_0 = atmospheric pressure = 76 cm of Hg)

Problem 312. In the given figure, an ideal gas changes its state from state A to state C by two paths ABC and AC.

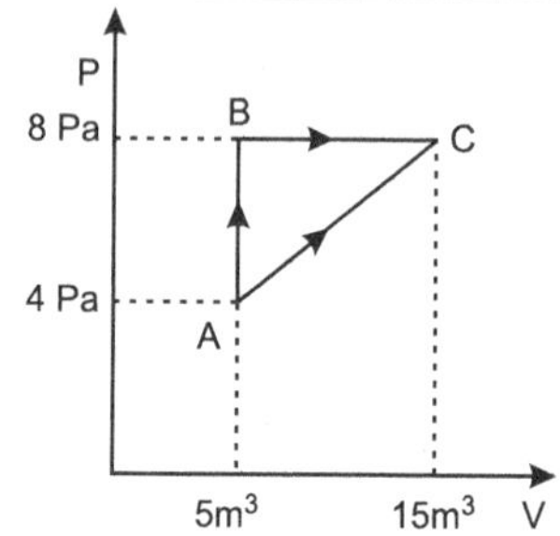

(a) Find the path along which work done is less.
(b) The internal energy of gas at A is 10 J and the amount of heat supplied to change its state to C through path AC is 200 J. Calculate the internal energy of gas at C.
(c) The internal energy of gas at state B is 20 J. Find the amount of heat supplied to the gas to go from A to B.

Problem 313. A gas undergoes cyclic process as shown in the figure, 5-1 and 3-4 are adiabatic process, 1-2 and 4-5 are isochoric process, 2-3 is isobaric process. Find efficiency of the cycle.

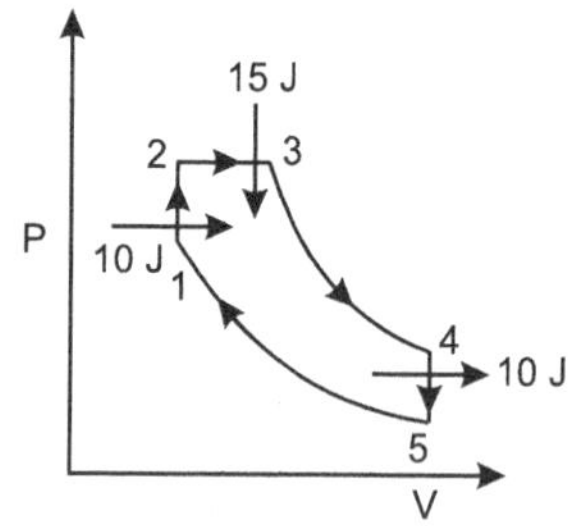

Problem 314. One mole of an ideal gas goes through a cycle consisting of alternate isothermal and adiabatic curves. AB, CD, EF are isothermal and BC, DE, FA are adiabatic curves. Temperature of each isothermal curve is given in the graph. The volume changes two fold in every isothermal expansion. Find

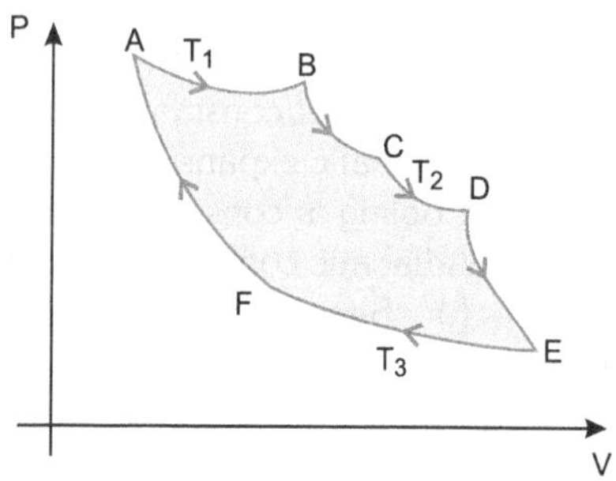

(a) work done by the gas in the cycle
(b) heat absorbed by the gas
(c) efficiency of the cycle

Problem 315. Pressure-temperature (P-T) graph of n moles of an ideal gas is shown in figure. Plot the corresponding

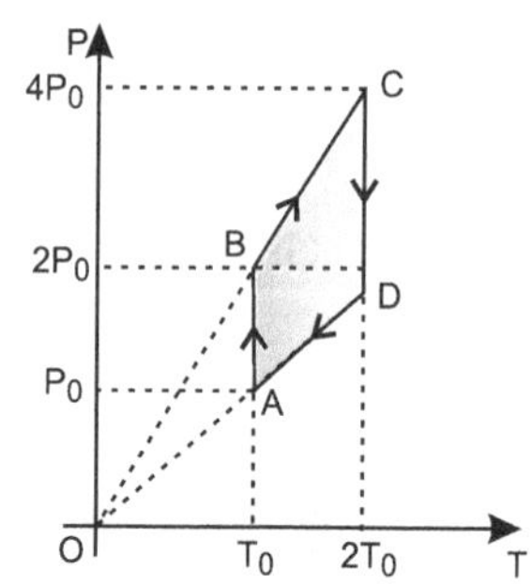

(a) density-volume (ρ-V) graph
(b) pressure-volume (P-V) graph and
(c) density-pressure (ρ-P) graph.

Problem 316. Helium is used as working substance in an engine working on the cycle as shown in figure. Process AB is isobaric, BC is adiabatic, CD is isochoric and DA is isothermal. The ratio of maximum to minimum volume of helium during the cycle is $8\sqrt{2}$ and that of maximum to minimum absolute temperature is 4. Calculate efficiency of the cycle in percentage.

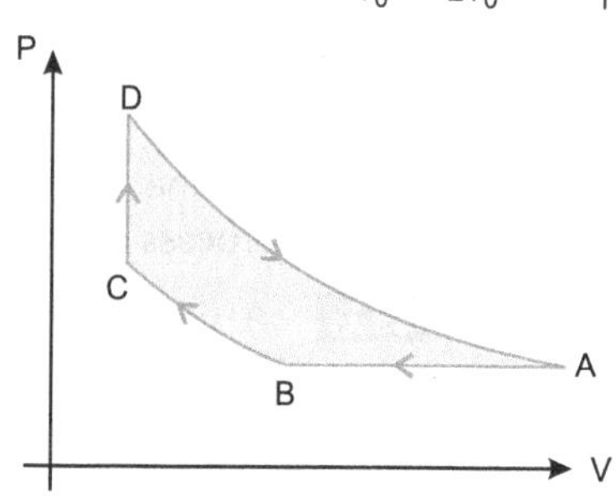

Problem 317. Two samples A and B of same gas have equal volumes and pressures. The gas in sample A is expanded isothermally to double its volume and the gas in sample B is expanded to double its volume adiabatically. If work done by the gas is same in the two processes. Show that γ $(= C_p/C_v)$ satisfies the equation

$$1 - 2^{1-\gamma} = (\gamma - 1) \ln 2$$

Problem 318. Consider a cylindrical column of air in uniform gravitational field. The temperature of air varies with height in such a way that $\frac{P}{\rho^{\gamma}}$ = constant. Here ρ is density and γ $(= C_p/C_v)$ does not vary with height. Considering air as an ideal diatomic gas, find the change in temperature for a height of 1 km. Molecular weight of air is 0.029 kg/mol, $R = 8.31$ J/mol-K and $g = 9.8$ m/s^2

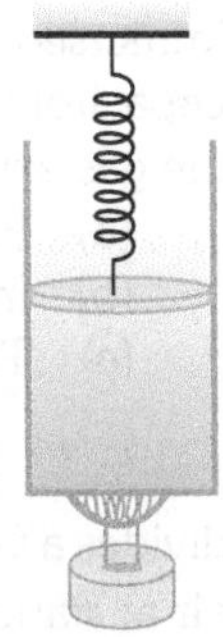

Problem 319. A gas is inside a cylinder closed by a piston. The piston is held from above by a spring whose elastic properties obey Hooke's law. Plot a rough P-V diagram. If the gas is heated then determine the work done by the gas in the process if volume of the gas varies from V_1 to V_2 and the pressure varies from P_1 to P_2.

Problem 320. A vertical thermally insulated cylinder of volume V contains n moles of an ideal monoatomic gas under a weightless piston. A load of weight W is placed on the piston as a result of which the piston is displaced by a distance h. Determine the final temperature of the gas. The area of the piston is A and atmospheric pressure is P_0.

Problem 321. A horizontal insulated cylindrical vessel of length $2l$ is separated by a thin insulating piston into two equal parts each of which contains n moles of an ideal monoatomic gas at temperature T. The piston is connected to the end faces of the vessel by undeformed springs of force constant k each. The left part is in contact with a thermostat (a device which maintains a constant temperature). When an amount of heat Q is supplied to the gas in the right part, the piston is displaced to the left by a distance $x = l/2$. Determine the heat Q' given away at the temperature T to the thermostat by the left part of the piston.

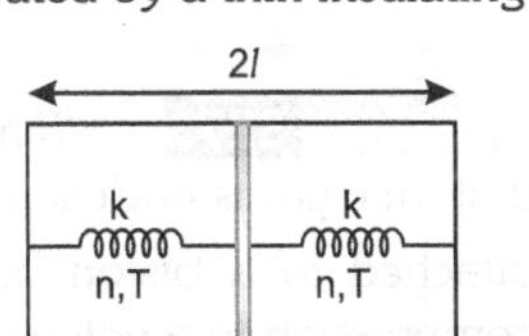

Problem 322. A solid material has a density ρ coefficient of linear expansion α and mass m.

(a) Find the difference of heat capacities C_p and C_v at pressure P.

(b) Find the difference (in joule per mole-Kelvin) in the molar heat capacities of aluminium at atmospheric pressure using $\alpha = 23 \times 10^{-6}\ K^{-1}$, $M = 27 \times 10^{-3}$ kg/mol, $\rho = 2700$ kg/m^3 and atmospheric pressure, $P_0 = 1.01 \times 10^5$ Pa.

Problem 323. A vessel is completely filled with 500 g of water and 1000 g of mercury. When 2.12×10^4 cal of heat is given to it, water of mass 3.52 g overflows. Calculate the coefficient of volume expansion of mercury. The expansion of vessel may be neglected. Coefficient of volume expansion of water is 1.5×10^{-4} per °C. Density of mercury is 13.6 g/cc and density of water is 1 g/cc before heating and specific heat of mercury and water are 0.03 and 1 cal/g-°C respectively.

Problem 324. An ideal gas has specific heat at constant pressure $C_p = \frac{5R}{2}$. The gas is kept in a cylindrical vessel fitted with a movable piston as shown in the figure. Mass of the frictionless piston is 9 kg. Initial volume of the gas is 0.027 m^3 and cross-section area of the piston is 0.09 m^2. The initial temperature of the gas is 300 K. An amount of 2.5×10^4 J of heat energy is supplied to the gas. Calculate the initial pressure, final pressure, final temperature and the work done by the gas. The walls of the cylinder and piston are thermally insulated. ($P_{atm} = 1.05 \times 10^5$ N/m^2)

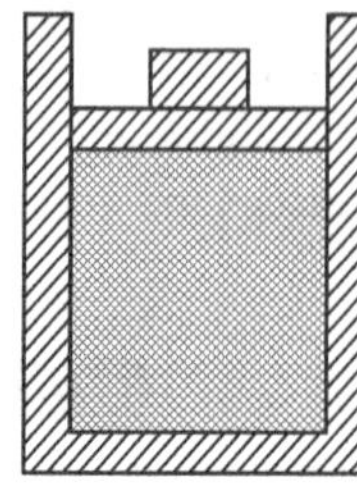

Problem 325. One mole of a gas is taken in a cylinder with a movable piston. A resistor R connected to a capacitor through a key is immersed in the gas. Initial potential difference across the plates of the capacitor is equal to 640/3 V. When the key is closed for (2.5 ln 4) minutes, the gas expands isobarically and its temperature changes by 22 K.

(a) Find the work done by the gas
(b) Increment in the internal energy of the gas
(c) The value of γ.

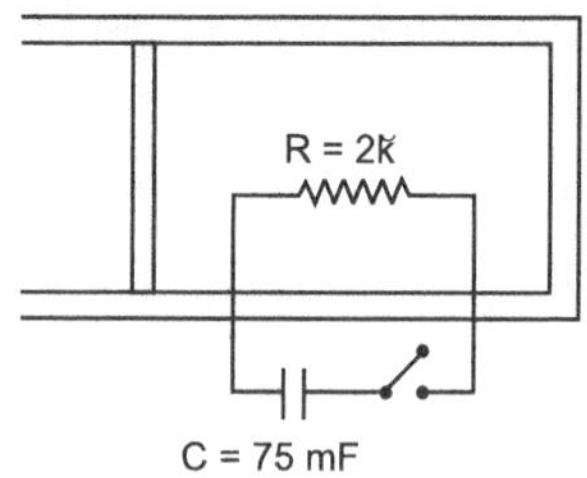

Problem 326. A horizontal frictionless piston, of negligible mass and heat capacity, divides a vertical insulated cylinder into two halves. Each half of the cylinder contains 1 mole of air at standard temperature and pressure p_0.

A load of weight W is now suspended from the piston, as shown in the figure. It pulls the piston down and comes to rest after a few oscillations. How large a volume does the compressed air in the lower part of the cylinder ultimately occupy if W is very large?

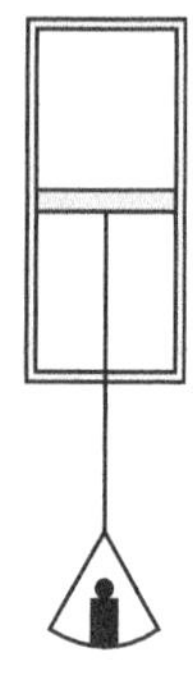

Problem 327. $1/R$ mole (where R is the magnitude of the gas constant in the SI system) of an ideal diatomic gas is enclosed in a container of volume 4 litres at a pressure of 2×10^5 Pa. The container is attached to a piston by which its volume can be changed. Initially the gas undergoes adiabatic compression to a volume of 2 litres. Then the gas is given 200 J of heat at constant pressure.

(a) Plot the complete process on a PT diagram. (Take $2^{1.4} = 2.63$)
(b) Find the total work done by the gas.

IIT JEE PROBLEMS

Problem 328. The piston cylinder arrangement shown contains a diatomic gas at temperature 300 K. The cross-sectional area of the cylinder is 1 m^2. Initially the height of the piston above the base of the cylinder is 1 m. The temperature is now raised to 400 K at constant pressure. Find the new height of the piston above the base of the cylinder. If the pistorn is now brough back to its original height without any heat loss, find the new equilibrium temperature of the gas. You can leave the answer in fraction. **(JEE 2004)**

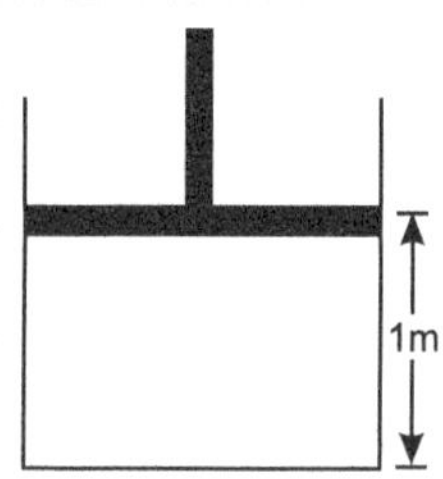

Problem 329. A cube of coefficient of linear expansion α_s is floating in a bath containing a liquid of coefficient of volume expansion γ_1. When the temperature is raised by ΔT, the depth upto which the cube is submerged in the liquid remains the same. Find the relation between α_s and γ_l, showing all the steps. **(JEE 2004)**

Problem 330. One end of a rod of length L and cross-sectional area A is kept in a furnace of temperature T_1. The other end of the rod is kept at a temperature T_2. The thermal conductivity of the material of the rod is K and emissivity of the rod is e. It is given that $T_2 = T_s + \Delta T$, where $\Delta T \le T_s$, T_s being the temperature of the surroundings. If $\Delta T \propto (T_1 - T_s)$, find the proportionality constant. Consider that heat is lost only by radiation at the end where the temperature of the rod is T_2. **(JEE 2004)**

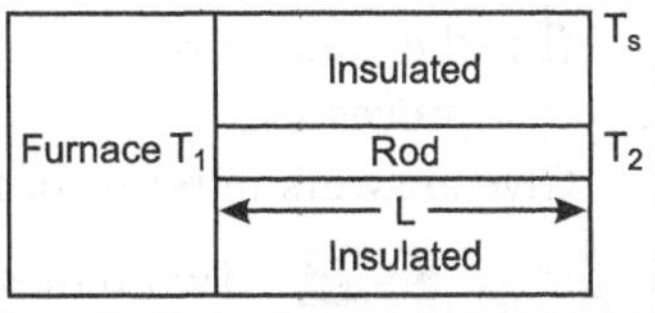

Problem 331. An insulated box containing a monoatomic gas of molar mass M moving with a speed v_0 is suddenly stopped. Find the increment in gas temperature as a result of stopping the box. **(JEE 2003)**

Problem 332. The top of an insulated cylindrical container is covered by a disc having emissivity 0.6 and conductivity 0.167 W/Km and thickness 1 cm. The temperature is maintained by circulating oil as shown :

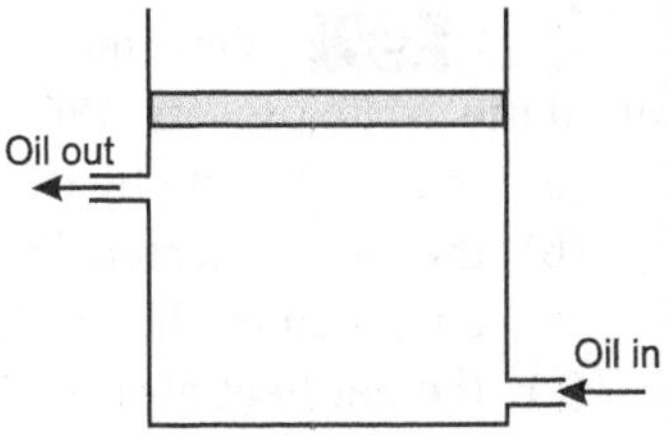

(a) Find the radiation loss to the surroundings in J/m^{2}s if temperature of the upper surface of disc is 127°C and temperature of surroundings is 27°C.

(b) Also find the temperature of the circulating oil. Neglect the heat loss due to convection.

$$\left[\text{Given: } \sigma = \frac{17}{3} \times 10^{-8} \text{ Wm}^{-2}\text{K}^{-4}\right]$$

(JEE 2003)

Problem 333. A cubical box of side 1 meter contains helium gas (atomic weight 4) at a pressure of 100 N/m^2. During an observation time of 1 second, an atom travelling with the root-mean-square speed parallel to one of the edges of the cube, was found to make 500 hits with a particular wall, without any collision with other atoms. Take $R = \frac{25}{3}$ J/mol-K and $k = 1.38 \times 10^{-23}$ J/K.

(a) Evaluate the temperature of the gas.
(b) Evaluate the average kinetic energy per atom.
(c) Evaluate the total mass of helium gas in the box. **(JEE 2002)**

Problem 334. A monoatomic ideal gas of two moles is taken through a cyclic process starting from A as shown in the figure. The volume ratios are $\frac{V_B}{V_A} = 2$ and $\frac{V_D}{V_A} = 4$. If the temperature T_A at A is 27°C, calculate

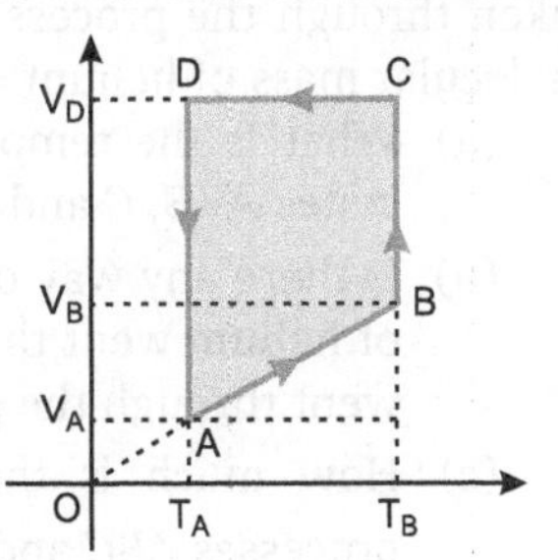

(a) the temperature of the gas at point B.
(b) heat absorbed or released by the gas in each process.
(c) the total work done by the gas during the complete cycle.

Express your answer in terms of the gas constant R. **(JEE 2001)**

Problem 335. Two moles of an ideal monoatomic gas is taken through a cycle $ABCA$ as shown in the P–T diagram. During the process AB, pressure and temperature of the gas vary such that PT = constant. If $T_1 = 300$ K, calculate

(a) the work done on the gas in the process AB and

(b) the heat absorbed or released by the gas in each of the processes.

Give answers in terms of the gas constant R. **(JEE 2000)**

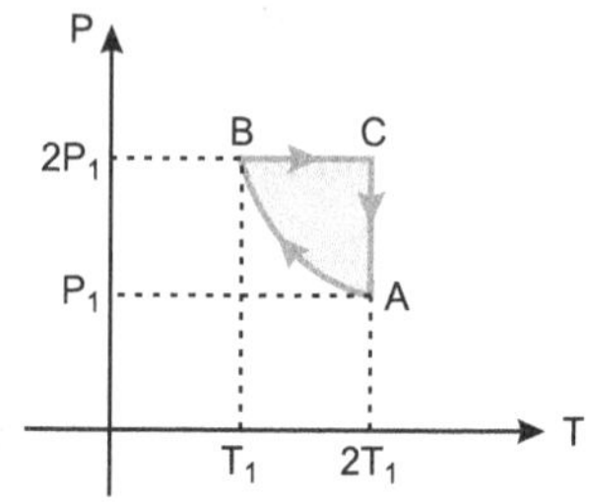

Problem 336. Two moles of an ideal monoatomic gas, initially at pressure P_1 and volume V_1, undergo an adiabatic compression until its volume is V_2. Then the gas is given heat Q at constant volume V_2.

(a) Sketch the complete process on a P–V diagram.

(b) Find the total work done by the gas, the total change in internal energy and the final temperature of the gas.

[Give your answer in terms of $P_1 : V_1 : V_2 : Q$ and R] **(JEE 1999)**

Problem 337. One mole of an ideal monoatomic gas is taken round the cyclic process $ABCA$ as shown in figure. Calculate :

(a) the work done by the gas

(b) the heat rejected by the gas in the path CA and the heat absorbed by the gas in the path AB.

(c) the net heat absorbed by the gas in the path BC

(d) the maximum temperature attained by the gas during the cycle. **(JEE 1998)**

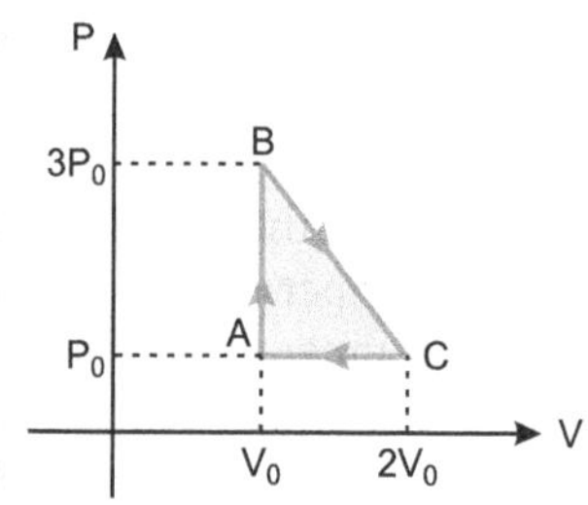

Problem 338. A solid body X of heat capacity C is kept in an atmosphere whose temperature is $T_A = 300$ K. At time $t = 0$, the temperature of X is $T_0 = 400$ K. It cools according to Newton's law of cooling. At time t_1 its temperature is found to be 350 K.

At this time (t_1) the body X is connected to a large body Y at atmospheric temperature T_A through a conducting rod of length L, cross-sectional area A and thermal conductivity K. The heat capacity of Y is so large that any variation in its temperature may be neglected. The cross-sectional area A of the connecting rod is small compared to the surface area of X. Find the temperature of X at time $t = 3t_1$.

(JEE 1998)

Problem 339. A sample of 2 kg of monoatomic helium gas (assumed ideal) is taken through the process ABC and another sample of 2 kg of the same gas is taken through the process ADC (see figure). Given, relative molecular mass of helium = 4.

(a) What is the temperature of helium in each of the states A, B, C and D?

(b) Is there any way of telling afterwards which sample of helium went through the process ABC and which went through the process ADC? Write Yes or No.

(c) How much is the heat involved in each of the processes ABC and ADC? **(JEE 1997, Cancelled)**

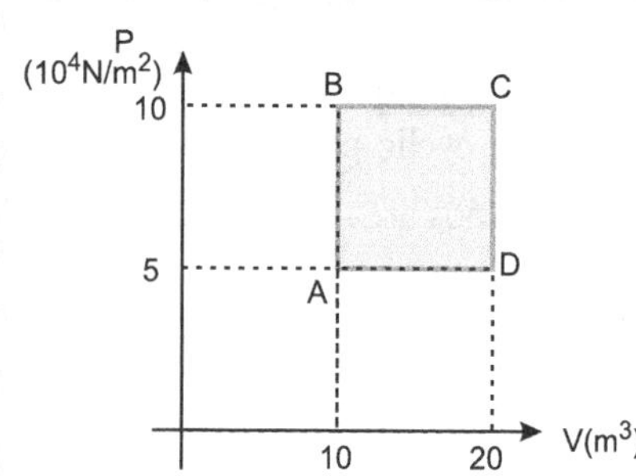

Problem 340. One mole of a diatomic ideal gas ($\gamma = 1.4$) is taken through a cyclic process starting from point A. The process $A \to B$ is an adiabatic compression. $B \to C$ is isobaric expansion, $C \to D$ an adiabatic expansion and $D \to A$ is isochoric.

The volume ratios are $V_A/V_B = 16$ and $V_C/V_B = 2$ and the temperature at A is $T_A = 300$ K. Calculate the temperature of the gas at the points B and D and find the efficiency of the cycle. **(JEE 1997)**

Problem 341. The apparatus shown in figure consists of four glass columns connected by horizontal sections. The height of two central columns B and C are 49 cm each. The two outer columns A and D are open to the atmosphere. A and C are maintained at a temperature of 95°C while the columns B and D are maintained at 5°C. The height of the liquid in A and D measured from the base line are 52.8 cm and 51 cm respectively. Determine the linear coefficient of thermal expansion of the liquid. **(JEE 1997)**

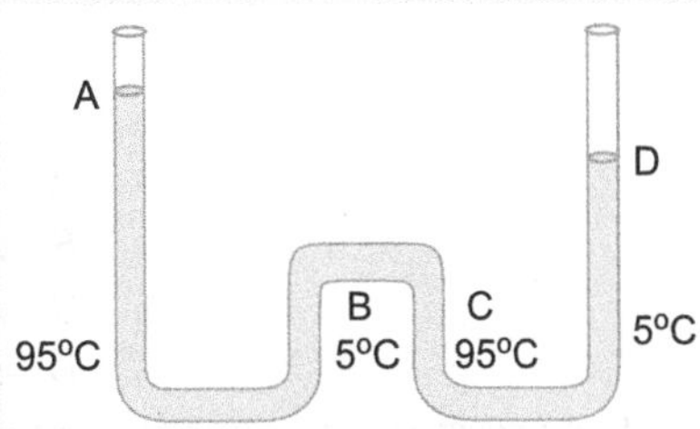

Problem 342. At 27°C two moles of an ideal monoatomic gas occupy a volume V. The gas expands adiabatically to a volume $2V$. Calculate :

(a) the final temperature of the gas
(b) change in its internal energy
(c) the work done by the gas during this process. **(JEE 1996)**

Problem 343. A gaseous mixture enclosed in a vessel of volume V consists of one gram mole of a gas A with $\gamma\ (= C_p/C_v = 5/3)$ and another gas B with $\gamma = 7/5$ at a certain temperature T. The gram molecular weights of the gases A and B are 4 and 32 respectively. The gases A and B do not react with each other and are assumed to be ideal. The gaseous mixture follows the equation $PV^{19/13}$ = constant, in adiabatic process.

(a) Find the number of gram moles of the gas B in the gaseous mixture.
(b) Compute the speed of sound in the gaseous mixture at 300 K.
(c) If T is raised by 1 K from 300 K, find the percentage change in the speed of sound in the gaseous mixture.
(d) The mixture is compressed adiabatically to 1/5 of its initial volume V. Find the change in its adiabatic compressibility in terms of the given quantities. **(JEE 1995)**

Problem 344. An ideal gas is taken through a cyclic thermodynamic process through four steps. The amounts of heat involved in these steps are $Q_1 = 5960$ J, $Q_2 = -5585$ J, $Q_3 = -2980$ J and $Q_4 = 3645$ J respectively. The corresponding quantities of work involved are $W_1 = 2200$ J, $W_2 = -825$ J, $W_3 = -1100$ J and W_4 respectively.

(a) Find the value of W_4.
(b) What is the efficiency of the cycle ? **(JEE 1994)**

Problem 345. A closed container of volume 0.02 m^3 contains a mixture of neon and argon gases, at a temperature of 27°C and pressure of $1 \times 10^5\ Nm^{-2}$. The total mass of the mixture is 28 g. If the molar masses of neon and argon are 20 and 40 g mol^{-1} respectively, find the masses of the individual gases in the container assuming them to be ideal (universal gas constant, $R = 8.314$ J/mol-K). **(JEE 1994)**

Problem 346. One mole of a monoatomic ideal gas is taken through the cycle shown in figure :

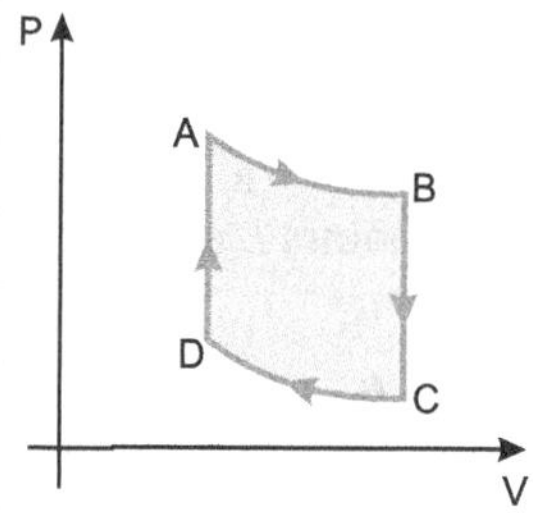

$A \to B$: adiabatic expansion
$B \to C$: cooling at constant volume
$C \to D$: adiabatic compression
$D \to A$: heating at constant volume.

The pressure and temperature at A, B etc. are denoted by P_A, T_A, P_B, T_B etc., respectively. Given that $T_A = 1000$ K, $P_B = (2/3)P_A$ and $P_C = (1/3)P_A$, calculate the following quantities :

(a) The work done by the gas in the process $A \to B$.
(b) The heat lost by the gas in the process $B \to C$.
(c) The temperature T_D. [Given : $(2/3)^{2/5} = 0.85$] **(JEE 1993)**

Problem 347. Two moles of helium gas undergo a cyclic process as shown in figure. Assuming the gas to be ideal, calculate the following quantities in this process :

(a) The net change in the heat energy.
(b) The net work done.
(c) The net change in internal energy. **(JEE 1992)**

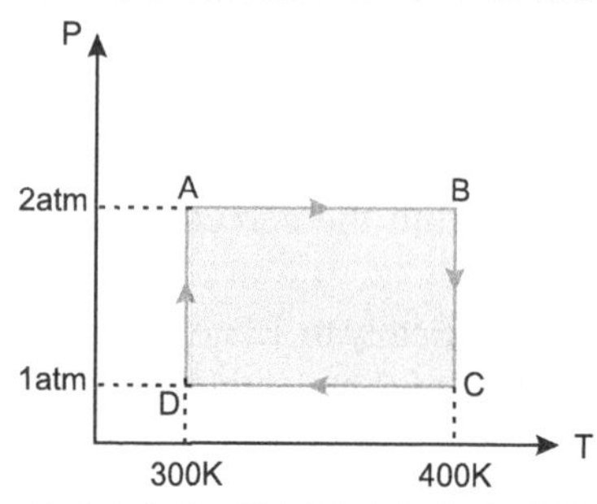

Problem 348. A cylindrical block of length 0.4 m and area of cross-section 0.04 m^2 is placed coaxially on a thin metal disc of mass 0.4 kg and of the same cross-section. The upper face of the cylinder is maintained at a constant temperature of 400 K and the initial temperature of the disc is 300 K. If the thermal conductivity of the material of the cylinder is 10 watt/m K and the specific heat capacity of the material of the disc is 600 J/kg-K, how long will it take for the temperature of the disc to increase to 350 K ? Assume, for purposes of calculation, the thermal conductivity of the disc to be very high and the system to be thermally insulated except for the upper face of the cylinder. **(JEE 1992)**

CHAPTER 11 OPTICS

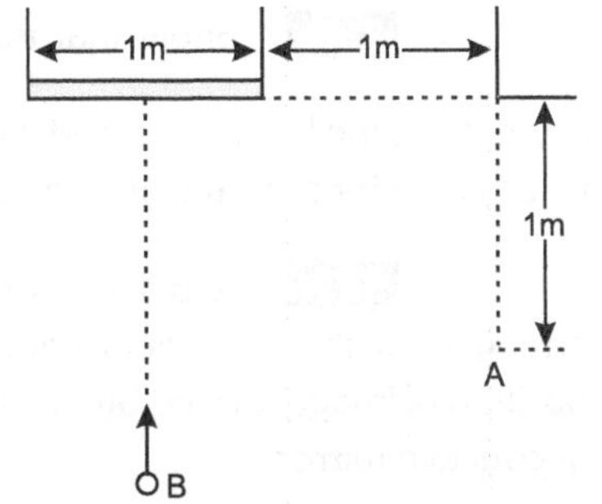

Problem 349. A man *A* stands to one side of a mirror; a second man *B* approaches the mirror along the line perpendicular to it which passes through its centre. At what distance from the mirror will *B* be at the moment when *A* and *B* see each other in the mirror?

Problem 350. A ray of light failing on a glass sphere of $\mu = \sqrt{3}$ such that the directions of the incident ray and emergent ray when produced meet the surface at the same point on the surface. Draw the ray diagram and find the value of angle of incidence.

Problem 351. A ray incident on a spherical drop of water at an angle of incidence *i* undergoes two reflections (not necessarily total internal reflections) and emerges from the drop. If the deviation suffered by the ray within the drop is minimum and the refractive index of the drop is μ, then show that $\cos i = \sqrt{\frac{\mu^2 - 1}{8}}$.

Problem 352. Two rays are incident on a spherical mirror of radius $R = 5$ cm parallel to its optical axis at a distance $h_1 = 0.5$ cm and $h_2 = 3$ cm. Determine the distance Δx between the points at which these rays intersect the optical axis after being reflected at the mirror.

Problem 353. You are given a parabolic mirror whose inner surface is silvered. The equation of the curve formed by its intersection with $x - y$ plane is given by $y = \frac{x^2}{4}$. A ray travelling in $x - y$ plane along line $y = x + 3$ hits the mirror in second quadrant and gets reflected. Find the unit vector in the direction of reflected ray.

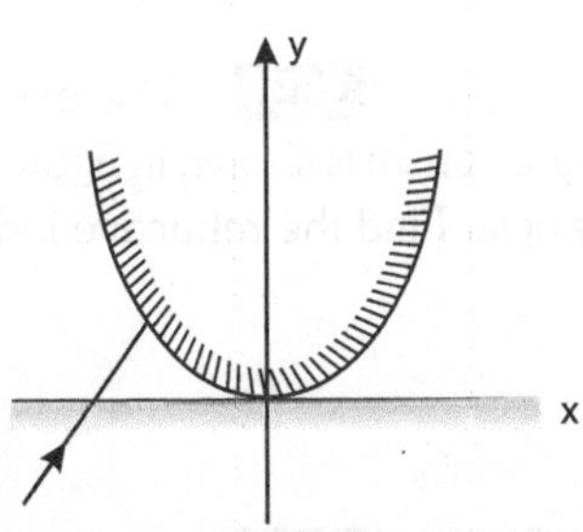

Problem 354. A point source of light is placed inside water and a thin converging lens ($\mu_{lens} = \mu_2$) is placed just outside the plane surface of water. The image of source is formed at a distance *x* from the surface of water. If the lens is now placed just inside water and the image is now formed at a distance *x'* from the surface of water, then show $\frac{1}{x} - \frac{1}{x'} = \frac{1}{f}\left(\frac{\mu_1 - 1}{\mu_2 - 1}\right)$, where *f* is focal length of the lens and μ_1 is the refractive index of water.

Problem 355. An object is moving with a velocity of 5 cm/s at an angle of 60° with respect to the optic axis of a concave mirror of focal length 100 cm. The object is at a distance of 50 cm from the mirror. Find the velocity of image at the given instant.

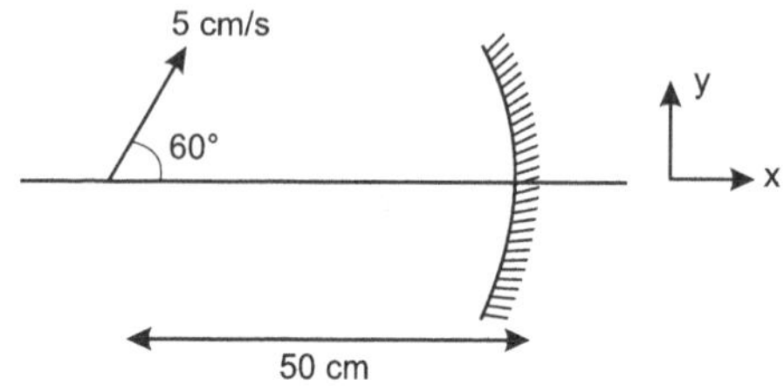

Problem 356. Show that the equation, $\frac{1}{v}+\frac{1}{u}=\frac{1}{f}$ is that of a rectangular hyperbola having the asymptotes the lines, $x = f$ and $y = f$. Plot a graph with object distance u as abscissa and image distance v as ordinate for a mirror of focal length 'f' and for object distances from 0 to ∞.

Problem 357. A plane mirror and a concave mirror are arranged as shown in figure and O is a point object. Find the position of image formed by two reflections, first one taking place at concave mirror.

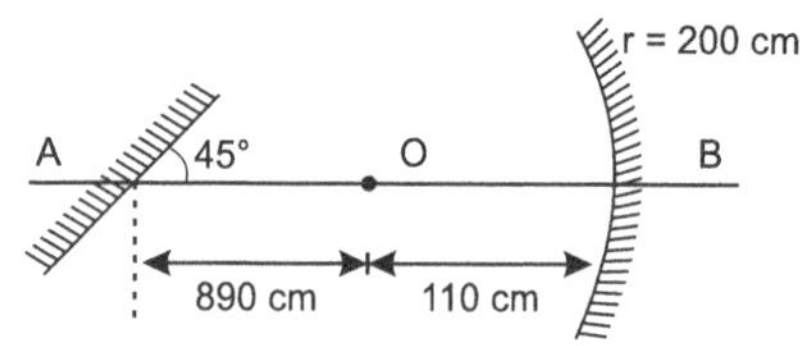

Problem 358. A concave mirror and a glass slab ($\mu = 1.5$) are arranged as shown in the figure. A converging bundle of paraxial rays is incident on the slab as shown. Find the position of the final image.

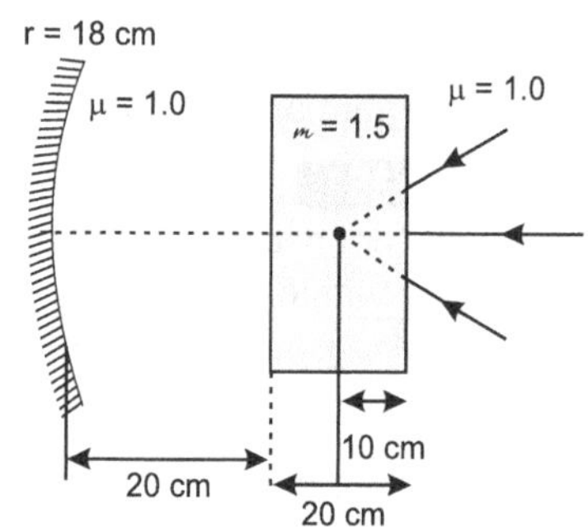

Problem 359. The path of a ray of light passing through an equilateral glass prism is shown in figure. The ray of light striking at AC is incident at critical angle. Find the refractive index of glass.

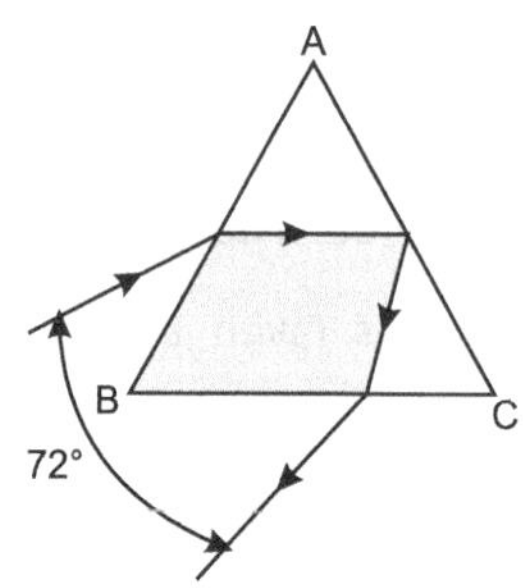

Problem 360. The diagonal AC of a right angled prism ABC is silvered. A ray of light falls on face AB perpendicular to AC and finally comes out of the face BC. What is the angle ϕ made by the outgoing ray with the normal to the face BC. Refractive index of prism is 1.5 and $\angle BAC$ is 60°.

Problem 361. The refracting angle of prism is 60°. A ray of light passing through the prism undergoes a deviation of 30°. If the prism is rotated keeping the incident ray fixed through an angle 30°, the same deviation is again obtained. Calculate the refractive index of the prism.

Problem 362. A 16 cm long pencil is placed at 45° angle with its centre 15.0 cm above the optic axis and 45.0 cm from a convex lens of focal length 20 cm as shown in figure. Assume that the diameter of the lens is large enough for the paraxial approximation to be valid :

(a) where is the image of the pencil formed. (give the location of the images of the points A, B and C)

(b) what is the length of the image (the distance between the images of points A and B).

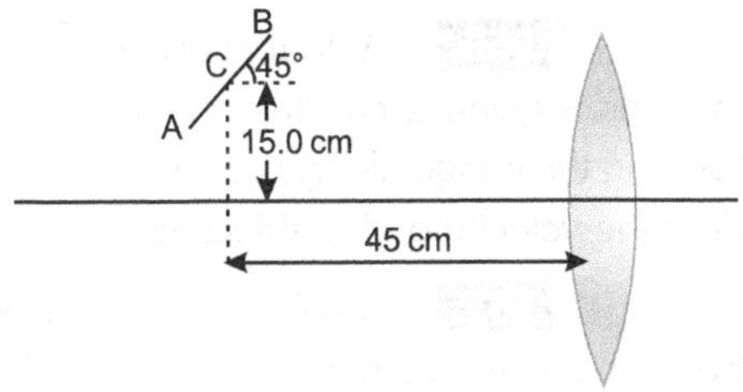

Problem 363. A fly F is sitting on a glass slab A, 45 cm thick and of refractive index 3/2. The slab covers the top of a container B containing water (Refractive index 4/3) upto a height of 20 cm. The bottom of container is closed by a concave mirror C of radius of curvature 40 cm. Locate the final image formed by all refractions and reflection assuming paraxial rays.

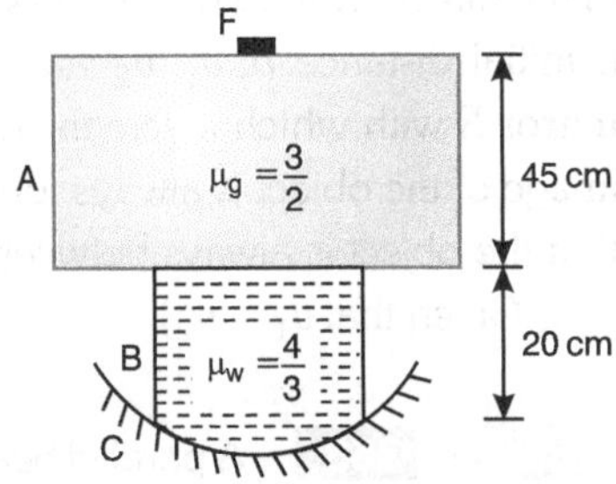

Problem 364. A composite slab consisting of different media is placed in front of a concave mirror of radius of curvature 150 cm. The whole arrangement is placed in water ($\mu_w = 4/3$). An object O is placed at a distance 20 cm from the slab. The refractive index of different media are given in the diagram. Find the position of final image formed by the system.

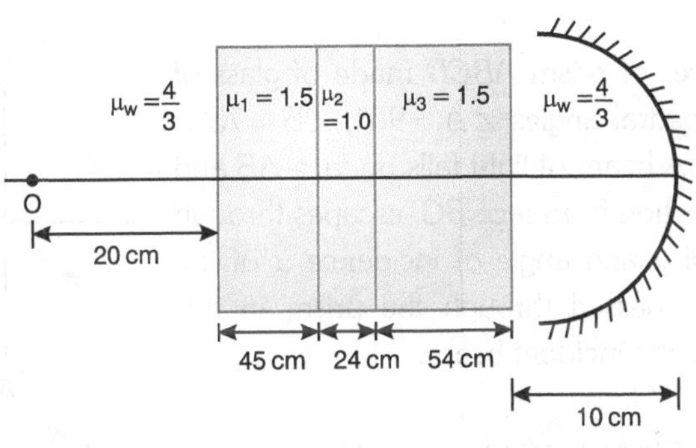

Problem 365. Two converging lenses of the same focal length f are separated by distance $2f$. The axis of the second lens is inclined at angle $\theta = 60°$ with respect to the axis of the first lens. A parallel paraxial beam of light is incident from left side of the lens. Find the coordinates of the final image with respect to the origin of the first lens.

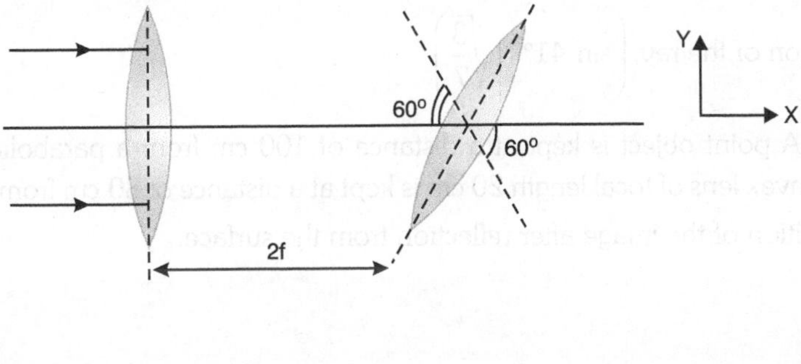

Problem 366. A double convex lens forms a real image of an object on a screen which is fixed. Now the lens is given a constant velocity v_0 along its axis and away from the screen. For the purpose of forming the image always on the screen, the object is also required to be given an appropriate velocity. Find the velocity of the object at the instant its size is n times the size of the image. $(n < 1)$

Problem 367. An object AB is placed at a distance d in front of a concave mirror at time $t = 0$. If the mirror and object move with constant velocities v_1 and v_2 respectively in the direction shown, find the velocity as a function of time t, initial distance d, v_1, v_2 and radius of curvature of the mirror R with which a screen should be moved so that the image of the object is always formed on the screen. Assume that the object is always between focus and infinity.

(Given that $v_1 > v_2$)

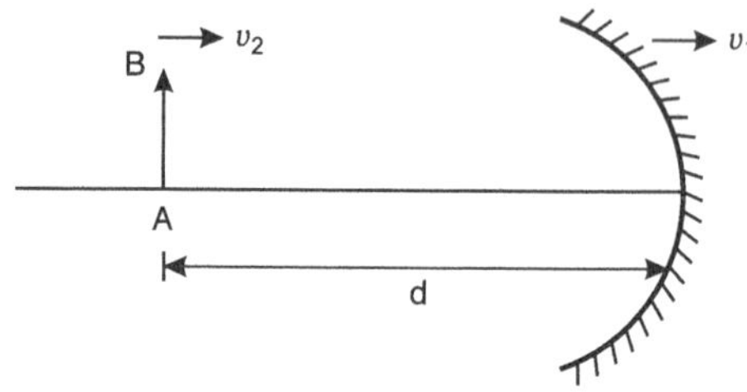

Problem 368. A parallel beam of light is incident on the system as shown in figure. Refractive indices of different parts are shown. Assuming α to be small, find the deviation in the emergent beam $(\mu > \mu')$.

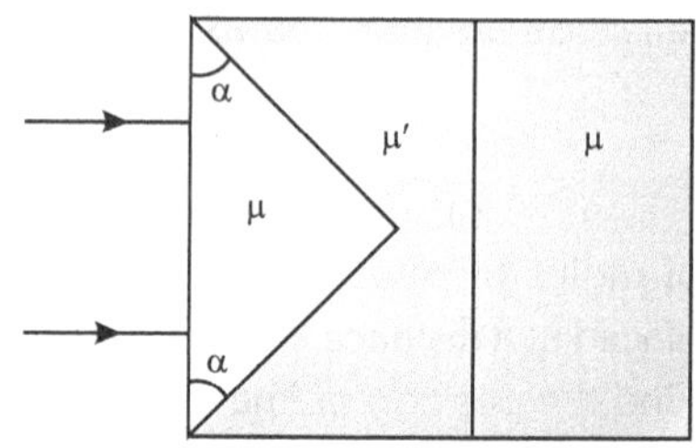

Problem 369. The faces of prism $ABCD$ made of glass of refractive index μ form dihedral angles $\angle A = 90°$, $\angle B = 75°$, $\angle C = 135°$ and $\angle D = 60°$. A beam of light falls on face AB and after complete internal reflection from face BC, escapes through face AD. Find the range of μ and angle of incidence α of the beam, if a beam that has passed through the prism in this manner, is perpendicular to the incident beam.

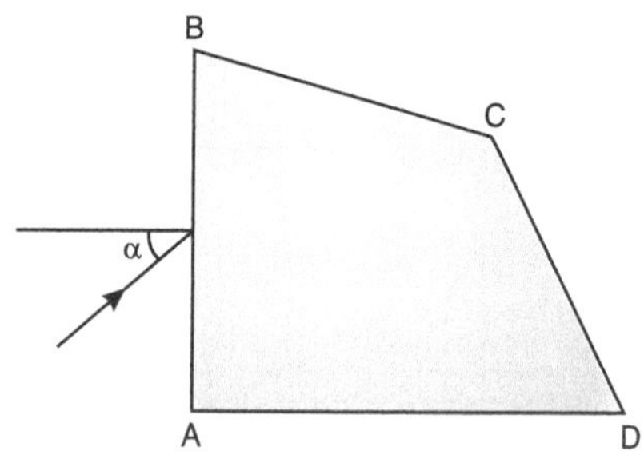

Problem 370. A ray of light is incident on the surface of a sphere of refractive index $\sqrt{7}/2$. Other half of the sphere is silvered. After refraction it is reflected and then refracted out of the sphere again such that the total deviation is minimum. Find :

(a) the angle of incidence of the ray

(b) total deviation of the ray. $\left(\sin 41° = \sqrt{\frac{3}{7}}\right)$

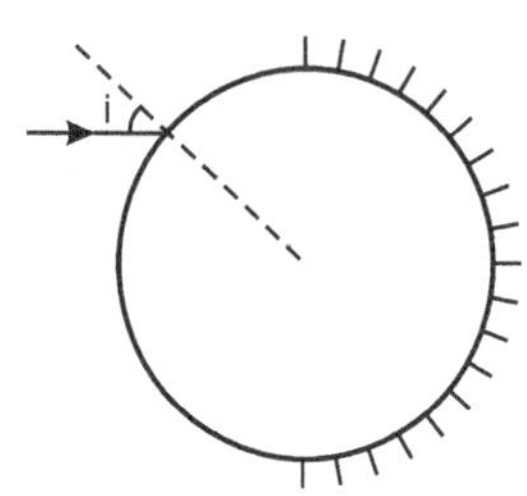

Problem 371. A point object is kept at a distance of 100 cm from a parabolic reflecting surface $x = 8y^2$. An equiconvex lens of focal length 20 cm is kept at a distance of 80 cm from origin as shown in figure. Find the position of the image after reflection from the surface.

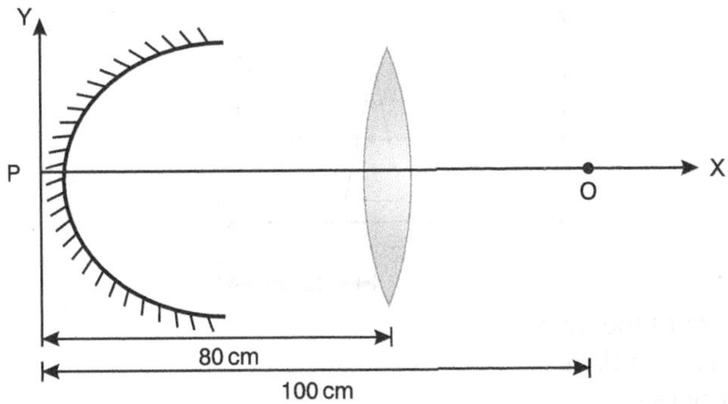

Problem 372. A transparent cylinder of radius $R = 2.00$ m has a mirrored surface on its right half as shown in figure. A light ray travelling in air is incident on the left side of the cylinder. The incident ray and the exiting ray are parallel and at a distance $d = 2.00$ m. Determine the refractive index of the material.

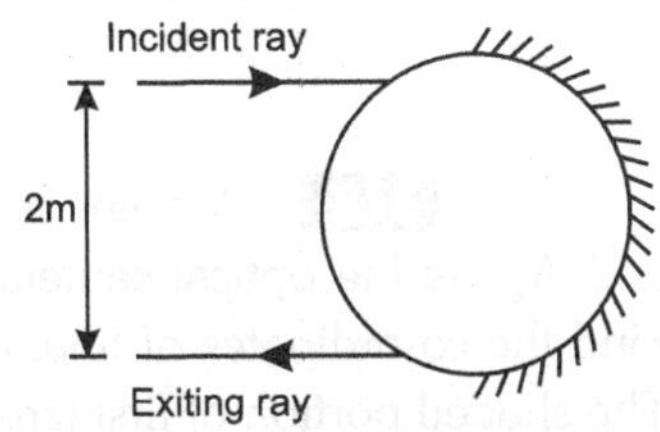

Problem 373. When the object is placed at 4 cm from the objective of a compound microscope, the final image formed coincides with the object and is at the least distance of distinct vision (25 cm). If the magnifying power of the microscope is 14, calculate the focal length of the objective and the eye piece.

Problem 374. Three right angled prisms of refractive indices μ_1, μ_2 and μ_3 are joined together so that the faces of the middle prism are in contact each with one of the outside prisms. If the ray passes through the composite block undeviated, show that

$$\mu_1^2 + \mu_3^2 - \mu_2^2 = 1$$

Problem 375. A lake is lit by an underwater isotropic lamp. If the surface of the lake is covered by a layer of oil of refractive index 1.2. Calculate the percentage of light

(a) escaping from the lake surface

(b) totally internally reflected in the water-oil layer.

Assume that the lake is absolutely calm and that both the oil and water are 100% transparent. The refractive index of water is 1.33.

Problem 376. A point source of light is placed on optical axis of a concave lens having focal length 30 cm at a distance of 90 cm from the lens. A thin prism having prism angle 5° and made of two materials having refractive indices 3/2 and 4/3 respectively, is placed at a distance of 44.5 cm from the lens. Find the co-ordinates of images formed assuming position of source as origin and optical axis of lens as x-axis, also state whether the images are real or virtual.

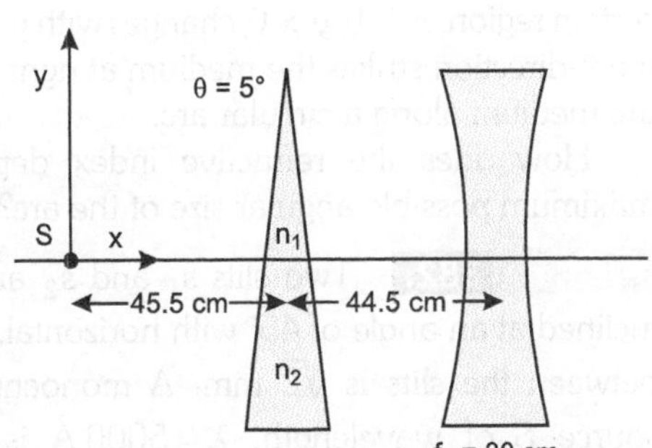

Problem 377. In the arrangement shown, the angle of thin prism is 1°. The image of the object kept at origin O is formed at a point $\left(10 \text{ cm}, -\dfrac{\pi}{24} \text{ cm}\right)$. Find

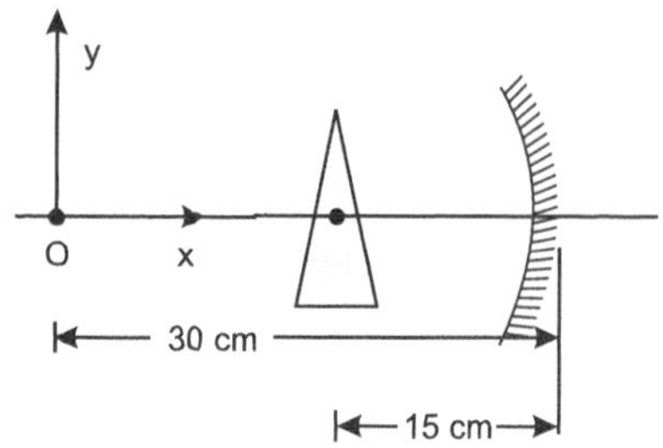

(a) radius of curvature of the mirror,
(b) refractive index of the prism.
Consider only paraxial rays.

Problem 378. A point object is kept at $O(0, 0)$. A_1 and A_2 are the optical centers of the two lenses shown. Find the co-ordinates of final image. Given : $d = 10$ cm. The shaded portion of first lens is opaque.

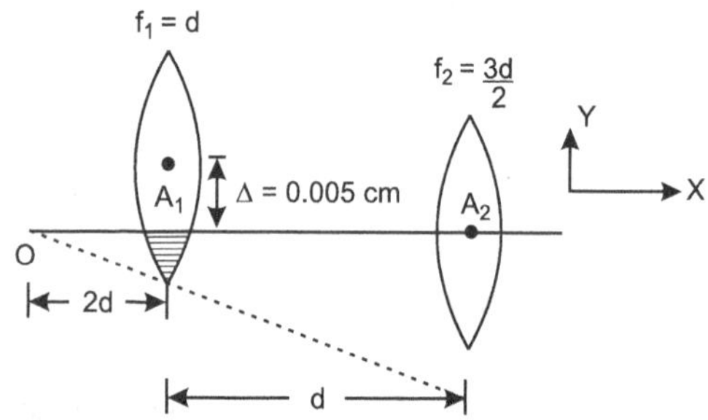

Problem 379. An equilateral prism has an angle of deviation 30° when the angle of incidence is 60°. Find the angle of deviation if a ray is incident normally on a surface.

Problem 380. A glass prism in the shape of a quarter-cylinder lies on a horizontal table. A uniform, horizontal light beam falls on its vertical plane surface, as shown in the figure.

If the radius of the cylinder is $R = 5$ cm and the refractive index of the glass is $n = 1.5$, where, on the table beyond the cylinder, will a patch of light be found?

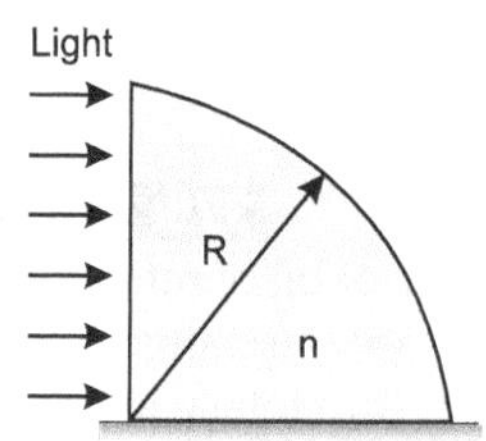

Problem 381. The refractive index of the medium within a certain region, $x > 0$, $y > 0$, changes with y. A thin light ray travelling in the x-direction strikes the medium at right angles and moves through the medium along a circular arc.

How does the refractive index depend on y? What is the maximum possible angular size of the arc?

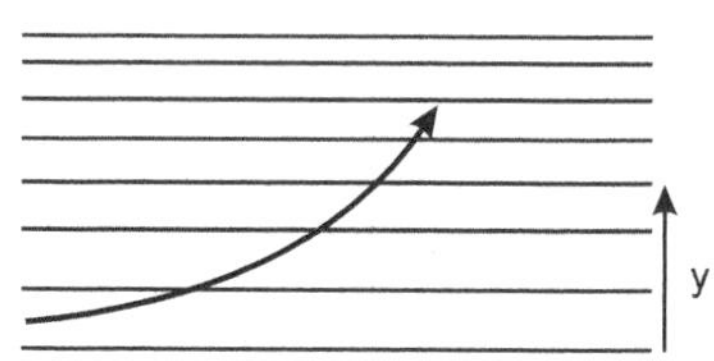

Problem 382. Two slits s_1 and s_2 are on a plane inclined at an angle of 45° with horizontal. The distance between the slits is $\sqrt{2}$ mm. A monochromatic point source S of wavelength, $\lambda = 5000$ Å is placed at a distance $1/\sqrt{2}$ mm from the midpoint of slits as shown in figure. The screen is placed at a distance of $\sqrt{2}$ m. Find the fringe width of interference pattern on the screen.

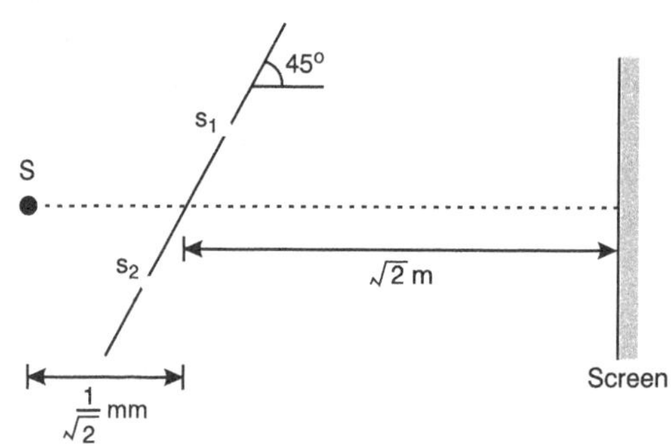

Problem 383. Parallel beam of a mono- chromatic light of wavelength λ is incident on a converging lens containing two identical parts L_1 and L_2 having refractive indices μ and $\mu - \Delta\mu$ ($\Delta\mu << \mu$). A screen is situated at a large distance D from the focus F of the combination. Find the distance y of the first maxima on the screen. Radius of curvature of each surface of the lens is R.

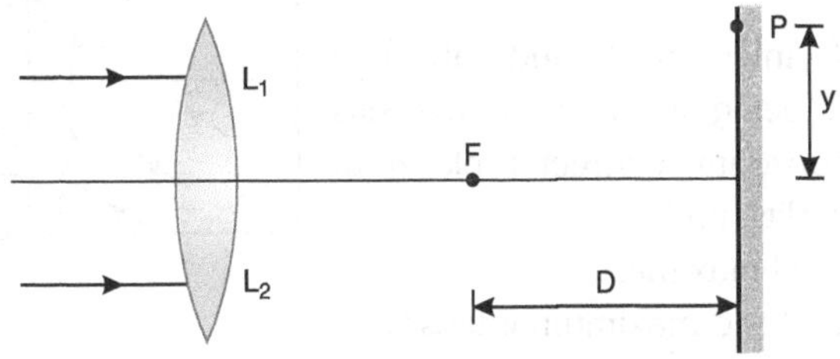

Problem 384. Monochromatic light passes through an orifice on a screen S and being reflected from a thin transparent plate P produces fringes of equal inclination on the screen. The thickness of the plate is equal to d, the distance between the plate and the screen is l ($>> d$), the radii of the ith and kth dark fringes are r_i and r_k. Find the wavelength of the light used in the experiment. Given that r_i and $r_k << l$.

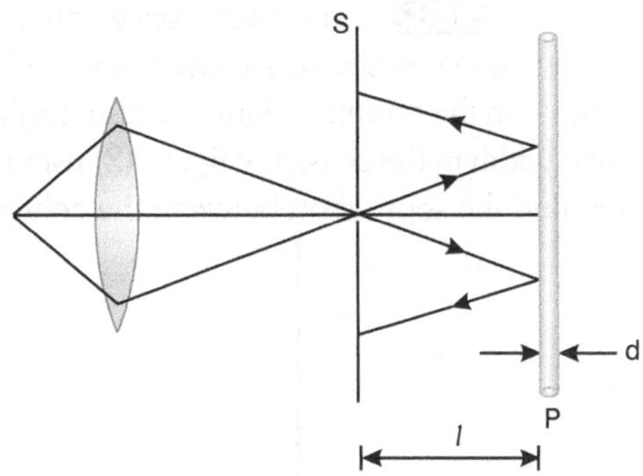

Problem 385. A system illustrated in figure consists of two coherent point sources 1 and 2 located in a certain plane. The sources are separated by a distance d, the radiation wavelength is equal to λ. Taking into account that the oscillations of source 2 lag in phase behind the oscillations of source 1 by ϕ ($\phi < \pi$), find :

(a) the angle θ at which the radiation intensity is maximum.

(b) the conditions under which the radiation intensity in the direction $\theta = \pi$ is maximum and in the opposite direction, minimum.

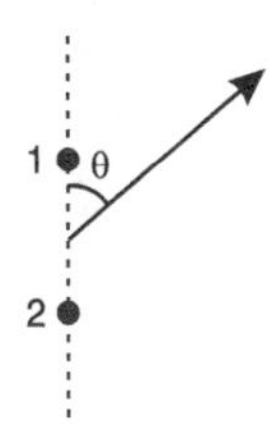

Problem 386. A point source emitting light having wavelength λ is placed at a distance $d/2$ below the principal axis of an equiconvex lens of refractive index 3/2 and radius 20 cm. The emergent light from lens fall on the slits s_1 and s_2 placed symmetrically with the principal axis. The resulting interference pattern is observed on the screen kept at a distance $D = 1$ m from the slit plane. Find

(a) the position of central maxima and its width

(b) ratio of intensity at O and maximum intensity. (give your answer in terms of d and λ) You can take the approximation $\sin\theta = \tan\theta$, where required.

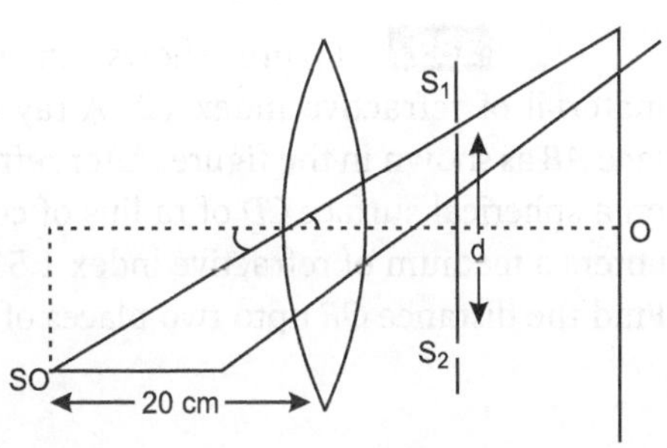

Problem 387. In a YDSE experiment the two slits are covered with a transparent membrane of negligible thickness which allows light to pass through it but does not allow water. A glass slab of thickness $t = 0.41$ mm and refractive index $\mu_g = 1.5$ is placed in front of one of the slits as shown in the

figure. The separation between the slits is $d = 0.30$ mm. The entire space to the left of the slits is filled with water of refractive index $\mu_w = \frac{4}{3}$.

A coherent light of intensity I and absolute wavelength $\lambda = 5000$ Å is being incident on the slits making an angle 30° with horizontal. If screen is placed at a distance $D = 1$ m from the slits, find

(a) the position of central maxima.

(b) ratio of intensity at O and maximum intensity.

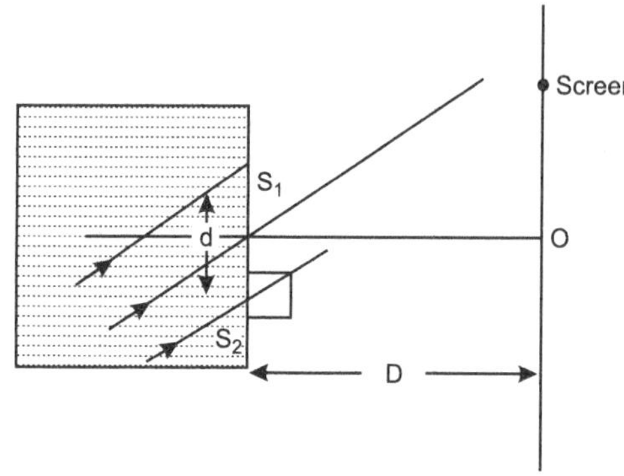

Problem 388. The interference pattern of a Young's double slit experiment is observed in two ways by placing the screen as shown in figure (1) and (2). The distance between two consecutive right most minima on the screen of figure (1) using light of wavelength $\lambda_1 = 4000$ Å is observed to be 600 times the fringe width in the screen of figure (2) using the wavelength $\lambda_2 = 6000$ Å. If D (as shown in figure) is 1 m then find the separation between the coherent sources S_1 and S_2.

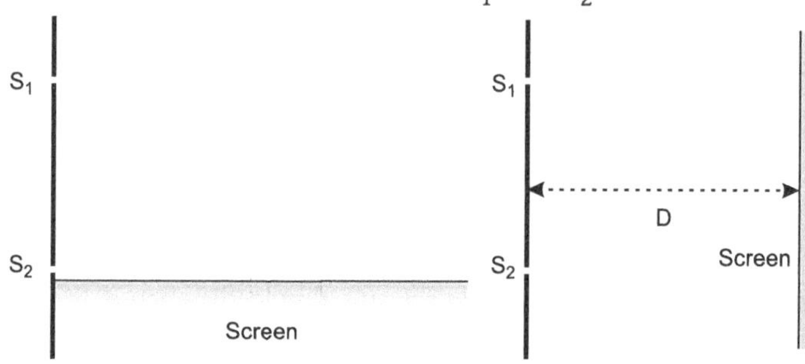

Fig. 1 Fig. 2

IIT JEE PROBLEMS

Problem 389. Figure shows an irregular block of material of refractive index $\sqrt{2}$. A ray of light strikes the face AB as shown in the figure. After refraction it is incident on a spherical surface CD of radius of curvature 0.4 m and enters a medium of refractive index 1.514 to meet PQ at E. Find the distance OE upto two places of decimal.

(JEE 2004)

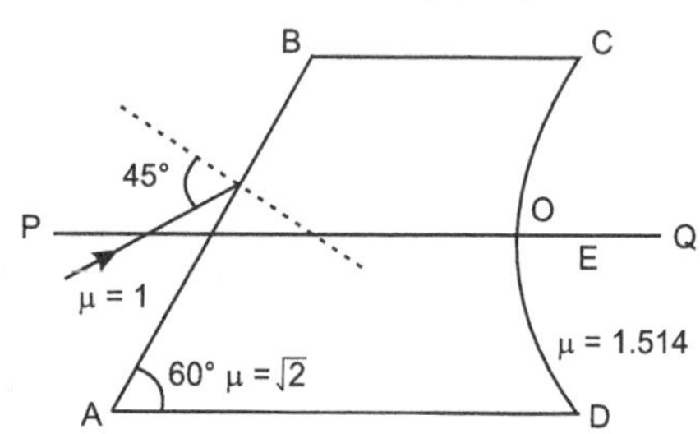

Problem 390. An object is approaching a thin convex lens of focal length 0.3 m with a speed of 0.01 m/s. Find the magnitudes of the rates of change of position and lateral magnification of image when the object is at a distance of 0.4 m from the lens. **(JEE 2004)**

Problem 391. In the figure, light is incident on the thin lens as shown. The radius of curvature for both the surface is R. Determine the focal length of this system. **(JEE 2003)**

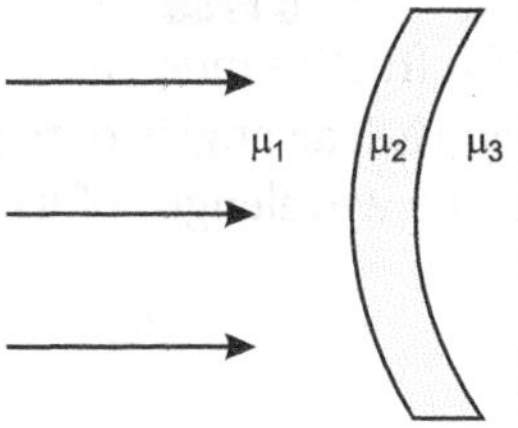

Problem 392. A prism of refracting angle 30° is coated with a thin film of transparent material of refractive index 2.2 on face AC of the prism. A light of wavelength 6600 Å is incident on face AB such that angle of incidence is 60°, find

(a) the angle of emergence, and

(b) the minimum value of thickness of the coated film on the face AC for which the light emerging from the face has maximum intensity.

[Given refractive index of the material of the prism is $\sqrt{3}$] **(JEE 2003)**

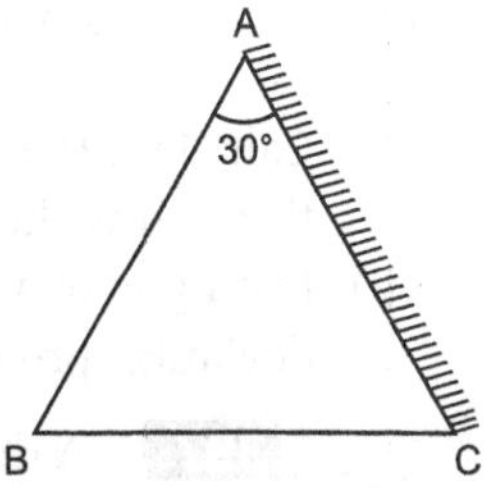

Problem 393. A thin biconvex lens of refractive index 3/2 is placed on a horizontal plane mirror as shown in the figure. The space between the lens and the mirror is then filled with water of refractive index 4/3. It is found that when a point object is placed 15 cm above the lens on its principal axis, the object coincides with its own image. On repeating with another liquid, the object and the image again coincide at a distance 25 cm from the lens. Calculate the refractive index of the liquid. **(JEE 2001)**

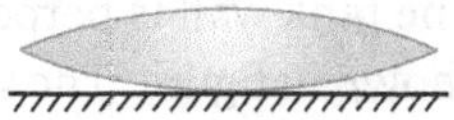

Problem 394. A convex lens of focal length 15 cm and a concave mirror of focal length 30 cm are kept with their optic axis PQ and RS parallel but separated in vertical direction by 0.6 cm as shown. The distance between the lens and mirror is 30 cm. An upright object AB of height 1.2 cm is placed on

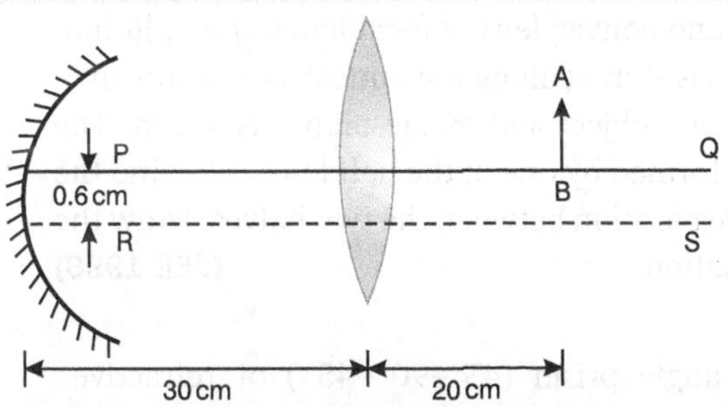

the optic axis PQ of the lens at a distance of 20 cm from the lens. If $A'B'$ is the image after refraction from the lens and the reflection from the mirror, find the distance of $A'B'$ from the pole of the mirror and obtain its magnification. Also locate positions of A' and B' with respect to the optic axis RS. **(JEE 2000)**

Problem 395. The X-Y plane is the boundary between two transparent media. Medium-1 with $z \geq 0$ has a refractive index $\sqrt{2}$ and medium-2 with $z \leq 0$ has a refractive index $\sqrt{3}$. A ray of light in medium-1 given by vector $\vec{A} = 6\sqrt{3}\hat{i} + 8\sqrt{3}\hat{j} - 10\hat{k}$ is incident on the plane of separation. Find the unit vector in the direction of the refracted ray in medium-2. **(JEE 1999)**

Problem 396. A prism of refractive index n_1 and another prism of refractive index n_2 are stuck together with a gap as shown in the figure. The angles of the prism are as shown. n_1 and n_2 depend on λ, the wavelength of light, according to :

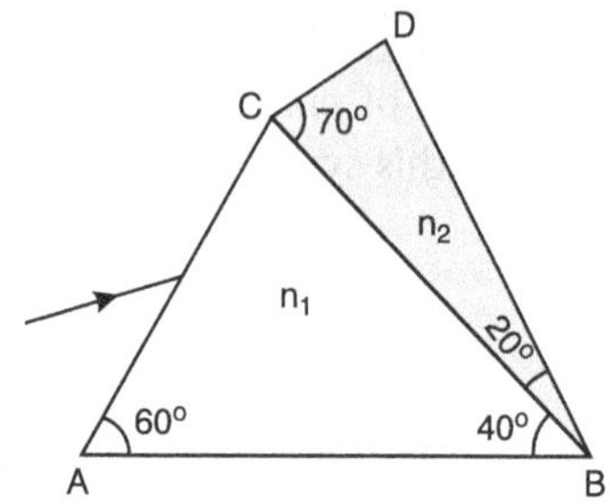

$$n_1 = 1.20 + \frac{10.8 \times 10^4}{\lambda^2}$$

and $$n_2 = 1.45 + \frac{1.80 \times 10^4}{\lambda^2}$$ where λ is in nm.

(a) Calculate the wavelength λ_0 for which rays incident at any angle on the interface BC pass through without bending at that interface.

(b) For light of wavelength λ_0, find the angle of incidence i on the face AC such that the deviation produced by the combination of prisms is minimum. **(JEE 1998)**

Problem 397. A thin equiconvex lens of glass of refractive index $\mu = 3/2$ and of focal length 0.3 m in air is sealed into an opening at one end of a tank filled with water ($\mu = 4/3$). On the opposite side of the lens, a mirror is placed inside the tank. The tank wall is perpendicular to the lens axis as shown in figure. The separation between the lens and the mirror is 0.8 m. A small object is placed outside the tank in front of the lens at a distance of 0.9 m from the lens along its axis. Find the position (relative to the lens) of the image of the object formed by the system. **(JEE 1997)**

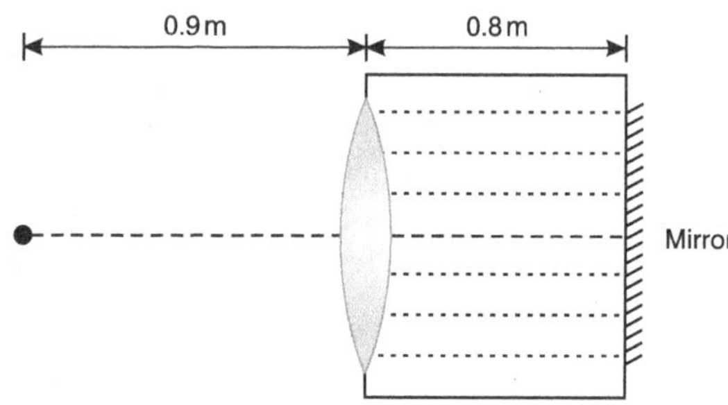

Problem 398. A thin plano-convex lens of focal length f is split into two halves. One of the halves is shifted along the optical axis as shown in figure. The separation between object and image planes is 1.8 m. The magnification of the image, formed by one of the half lens, is 2. Find the focal length of the lens and separation between the two halves. Draw the ray diagram for image formation. **(JEE 1996)**

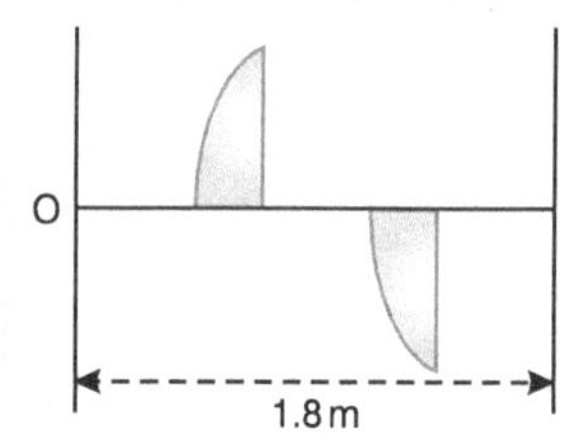

Problem 399. A right angle prism (45°–90°–45°) of refractive index n has a plane of refractive index $n_1 (n_1 < n)$ cemented to its diagonal face. The assembly is in air. The ray is incident on AB as shown in figure.

(i) Calculate the angle of incidence at AB for which the ray strikes the diagonal face at the critical angle.

(ii) Assuming $n = 1.352$, calculate the angle of incidence at AB for which the refracted ray passes through the diagonal face undeviated.

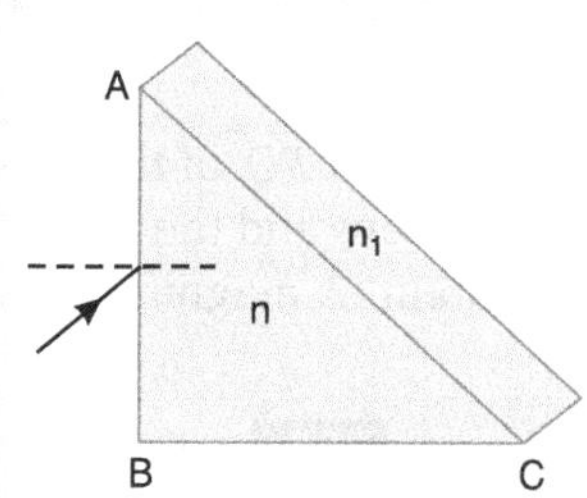

(JEE 1996)

Problem 400. A ray of light travelling in air is incident at grazing angle (incident angle = 90°) on a long rectangular slab of a transparent medium of thickness $t = 1.0$ m as shown in figure. The point of incidence is the origin $A\ (0, 0)$. The medium has a variable index of refraction $n\ (y)$ given by

$$n\ (y) = [ky^{3/2} + 1]^{1/2}$$

where $k = 1.0\ (\text{metre})^{-3/2}$. The refractive index of air is 1.0.

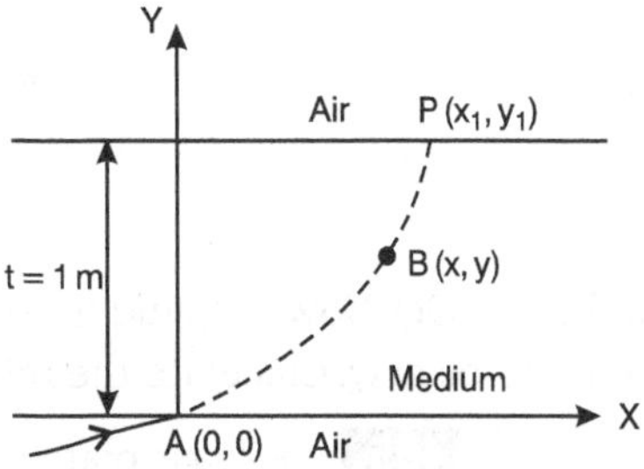

(a) Obtain a relation between the slope of the trajectory of the ray at a point $B\ (x, y)$ in the medium and the incident angle at that point.
(b) Obtain an equation for the trajectory $y\ (x)$ of the ray in the medium.
(c) Determine the coordinates (x_1, y_1) of the point P, where the ray intersects the upper surface of the slab-air boundary.
(d) Indicate the path of the ray subsequently. **(JEE 1995)**

Problem 401. An image Y is formed of point object X by a lens whose optic axis is AB as shown in figure. Draw a ray diagram to locate the lens and its focus. If the image Y of the object X is formed by a concave mirror (having the same optic axis as AB) instead of lens, draw another ray diagram to locate the mirror and its focus.

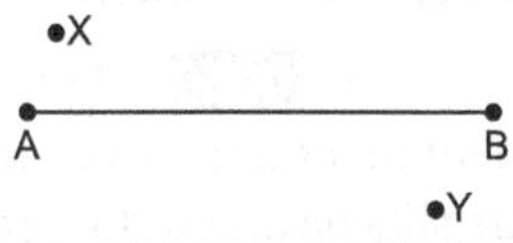

Write down the steps of construction of the ray diagrams. **(JEE 1994)**

Problem 402. In a Young's double slit experiment, two wavelengths of 500 nm and 700 nm were used. What is the minimum distance from the central maximum where their maximas concide again? Take $\frac{D}{d} = 10^3$. Symbols have their usual meanings. **(JEE 2004)**

Problem 403. A point source S emitting light of wavelength 600 nm is placed at a very small height h above a flat reflecting surface AB (see figure). The intensity of the reflected light is 36% of the incident intensity. Interference fringes are observed on a screen placed parallel to the reflecting surface at very large distance D from it.

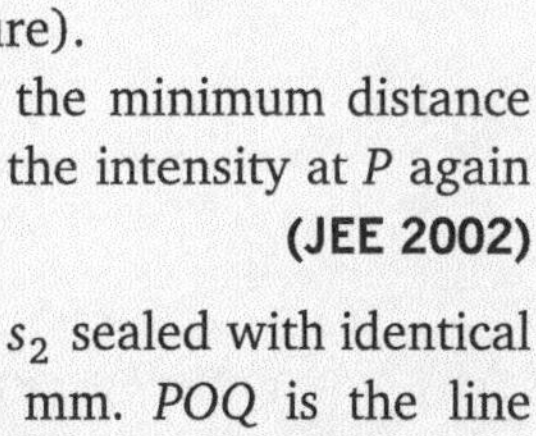

(a) What is the shape of the interference fringes on the screen ?
(b) Calculate the ratio of the minimum to the maximum intensities in the interference fringes formed near the point P (shown in the figure).
(c) If the intensity at point P corresponds to a maximum, calculate the minimum distance through which the reflecting surface AB should be shifted so that the intensity at P again becomes maximum. **(JEE 2002)**

Problem 404. A vessel $ABCD$ of 10 cm width has two small slits s_1 and s_2 sealed with identical glass plates of equal thickness. The distance between the slits is 0.8 mm. POQ is the line perpendicular to the plane AB and passing through O, the middle point of s_1 and s_2. A monochromatic light source is kept at S, 40 cm below P and 2 m from the vessel, to illuminate the slits as shown in the figure alongside. Calculate the position of the central bright fringe on the other wall CD with respect

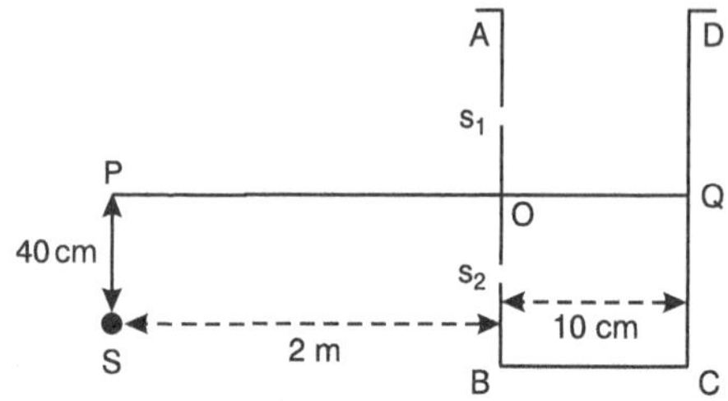

to the line OQ. Now, a liquid is poured into the vessel and filled upto OQ. The central bright fringe is found to be at Q. Calculate the refractive index of the liquid. **(JEE 2001)**

Problem 405. A glass plate of refractive index 1.5 is coated with a thin layer of thickness t and refractive index 1.8. Light of wavelength λ travelling in air is incident normally on the layer. It is partly reflected at the upper and the lower surfaces of the layer and the two reflected rays interfere. Write the condition for their constructive interference. If $\lambda = 648$ nm, obtain the least value of t for which the rays interfere constructively. **(JEE 2000)**

Problem 406. The Young's double slit experiment is done in a medium of refractive index 4/3. A light of 600 nm wavelength is falling on the slits having 0.45 mm separation. The lower slit S_2 is covered by a thin glass sheet of thickness 10.4 μm and refractive index 1.5. The interference pattern is observed on a screen placed 1.5 m from the slits as shown.

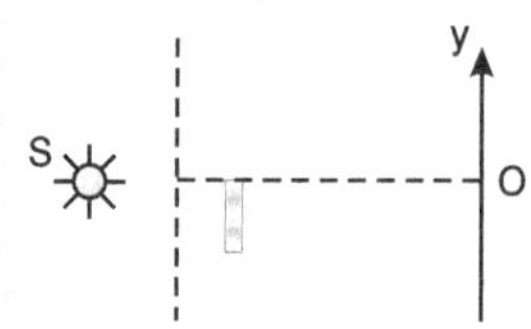

(a) Find the location of central maximum (bright fringe with zero path difference) on the Y-axis.
(b) Find the light intensity at point O relative to the maximum fringe intensity.
(c) Now if 600 nm light is replaced by white light of range 400 to 700 nm. Find the wavelength of the light that form maxima exactly at point O.

[All wavelengths in the problem are for the given medium of refractive index 4/3. Ignore dispersion] **(JEE 1999)**

Problem 407. A coherent parallel beam of microwaves of wavelength, $\lambda = 0.5$ mm falls on a Young's double slit apparatus. The separation between the slits is 1.0 mm. The intensity of microwaves is measured on a screen placed parallel to the plane of the slits at a distance of 1.0 m from it as shown in the figure.

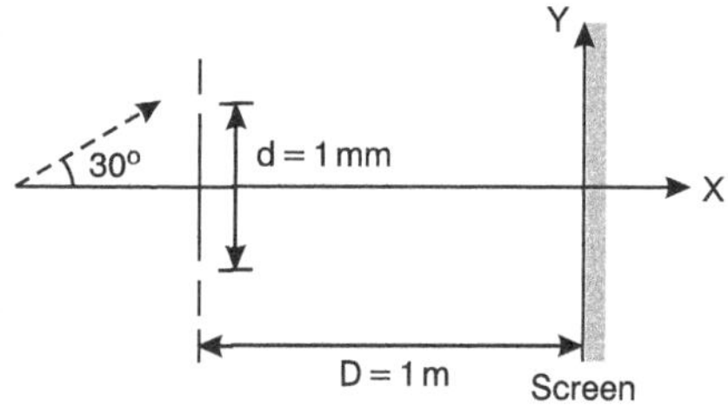

(a) If the incident beam falls normally on the double slit apparatus, find the y-coordinates of all the interference minima on the screen.

(b) If the incident beam makes an angle of 30° with the X-axis (as in the dotted arrow shown in figure), find the y-coordinates of the first minima on either side of the central maximum. **(JEE 1998)**

Problem 408. In a Young's experiment, the upper slit is covered by a thin glass plate of refractive index 1.4, while the lower slit is covered by another glass plate, having the same thickness as the first one but having refractive index 1.7. Interference pattern is observed using light of wavelength 5400 Å. It is found that the point P on the screen, where the central maximum ($n = 0$) fall before the glass plates were inserted, now has 3/4 the original intensity. It is further observed that what used to be the fifth maximum earlier, lies below the point P while the sixth minima lies above P. Calculate the thickness of glass plate. (absorption of light by glass plate may be neglected). **(JEE 1997)**

Problem 409. In Young's experiment, the source is red light of wavelength 7×10^{-7} m. When a thin glass plate of refractive index 1.5 at this wavelength is put in the path of one of the interfering beams, the central bright fringe shifts by 10^{-3} m to the position previously occupied by the 5th bright fringe. Find the thickness of the plate. When the source is now changed to green light of wavelength 5×10^{-7} m, the central fringe shifts to a position initially occupied by the 6th bright fringe due to red light. Find the refractive index of glass for the green light. Also estimate the change in fringe width due to the change in wavelength. **(JEE 1997, Cancelled)**

Problem 410. Angular width of central maximum in the Fraunhofer diffraction pattern of a slit is measured. The slit is illuminated by light of wavelength 6000 Å. When the slit is illuminated by light of another wavelength, the angular width decreases by 30%. Calculate the wavelength of this light. The same decrease in the angular width of central maximum is obtained when the original apparatus is immersed in a liquid. Find refractive index of the liquid. **(JEE 1996)**

Problem 411. In figure, S is a monochromatic point source emitting light of wavelength $\lambda = 500$ nm. A thin lens of circular shape and focal length 0.10 m is cut into two identical halves L_1 and L_2 by a plane passing through a diameter. The two halves are placed symmetrically about the central axis SO with a gap of 0.5 mm. The distance along the axis from S to L_1 and L_2 is 0.15 m while that from L_1 and L_2 to O is 1.30 m. The screen at O is normal to SO.

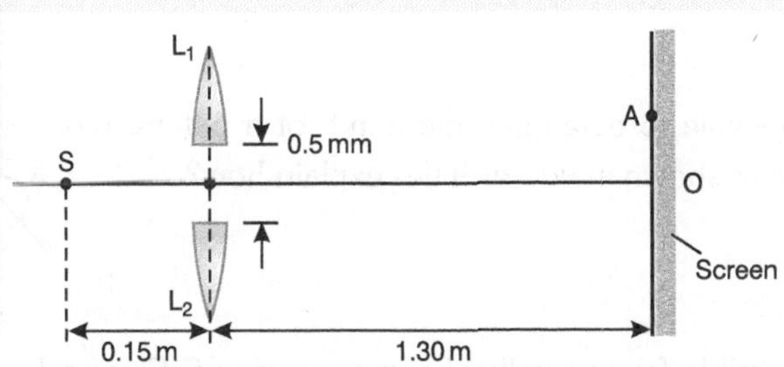

(i) If the third intensity maximum occurs at the point A on the screen, find the distance OA.

(ii) If the gap between L_1 and L_2 is reduced from its original value of 0.5 mm, will the distance OA increase, decrease, or remain the same ? **(JEE 1993)**

CHAPTER 12 CURRENT ELECTRICITY

Problem 412. Two wires *MACN* and *MBDN*, of the same length but of different resistances R_1 and R_2 are connected as shown in the figure. How should contacts *A*, *B*, *C* and *D* be arranged so that there should be no current passing through wires *AB* and *CD*? Will current pass through *AB* and *CD* with this arrangement of contacts if two points *E* and *F* on these wires are connected?

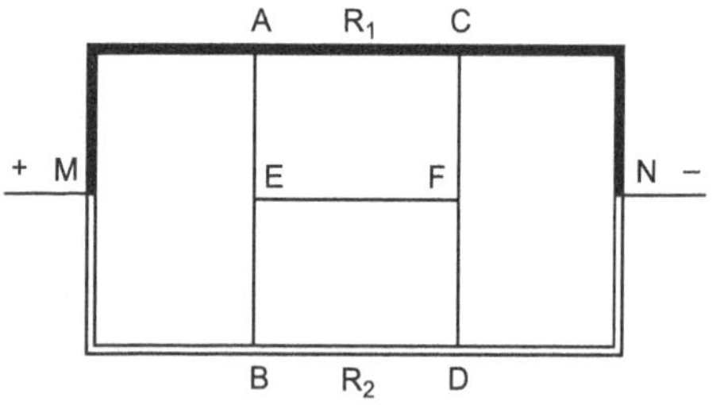

Problem 413. Two arbitrary points *A* and *B* are chosen on a direct current two-wire system, one point on each wire. How can we find in which direction is the source of current with the help of a voltmeter and a magnetic needle?

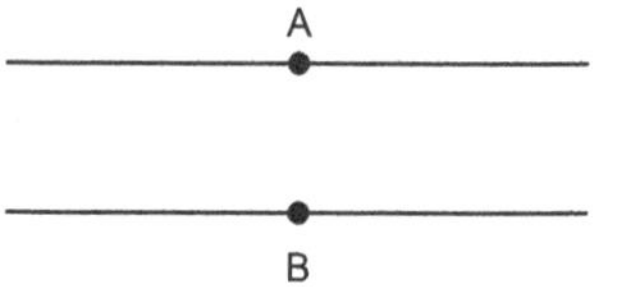

Problem 414. How can the resistance of a galvanometer which is normally connected on the diagonal of a Wheatstone bridge be measured on the bridge without using another galvanometer?

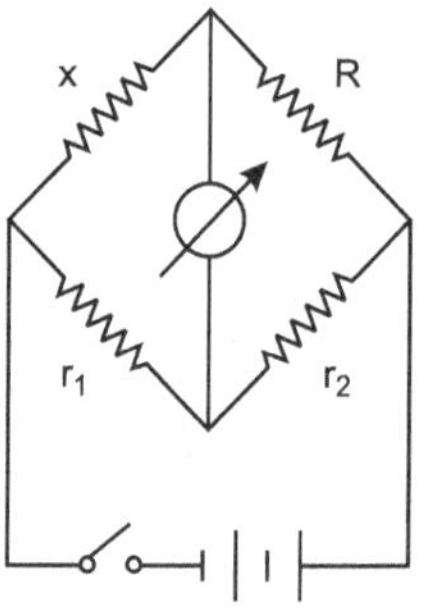

Problem 415. Is it possible to determine the e.m.f. of a battery by connecting up instruments as shown in figure? If yes explain how?

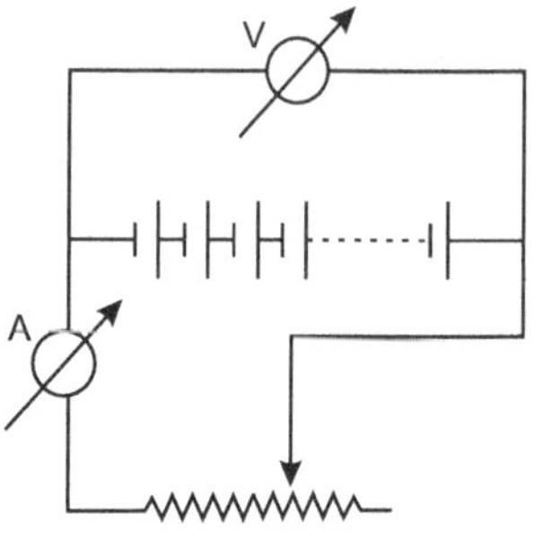

Problem 416. Is it possible for two cells of respective e.m.f. E_1 and E_2 and respective internal resistance r_1 and r_2 to produce a weaker current when connected in series to an external resistance R than one of the cells by itself, connected to the same resistance?

Problem 417. Find the electrical resistance of a homogeneous wire frame in the shape of a regular hexagon with two diagonals linked together at O. The current is led into the frame at the mid-points A and B of opposite sides of the hexagon. The resistance of one side of the hexagon is R.

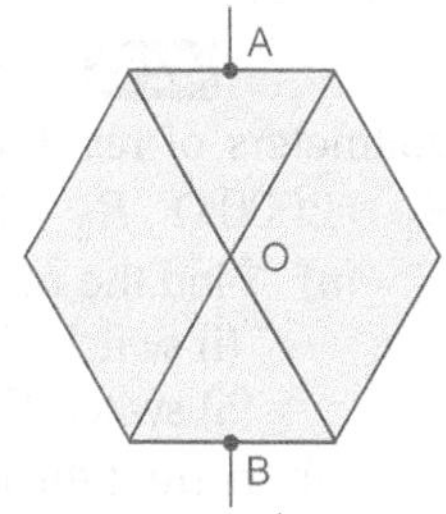

Problem 418. The plates of a flat capacitor are connected to a galvanometer. One of the plates is earthed. A positive charge is passed between the plates. What will be the reading of the galvanometer?

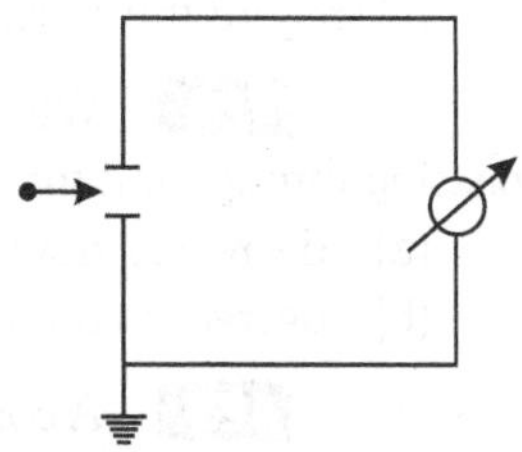

Problem 419. If $r = 1\,\Omega$ and $E = 10$ V in the network shown in figure. Find the current i and current in the branch PQ of the circuit.

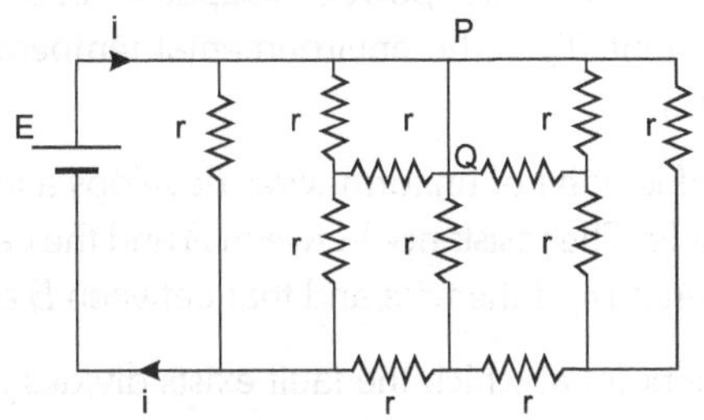

Problem 420. A resistor of resistance $r = 500\,\Omega$ connected to an external battery is placed inside a cylinder fitted with a frictionless piston of mass $m = 10$ kg, area of cross section $10^{-3}\ \text{m}^2$ and containing 2 moles of an ideal monoatomic gas. A current $i = 0.3$ A flows through the resistance.

(a) If cylinder is thermally insulated, at what speed v must the piston be moved upward in order that the temperature of the gas remains constant.

(b) If the piston is fixed and cylinder is insulated find the rise in temperature of the gas after 30 second. Take $g = 10\ \text{m/s}^2$, $R = 8.31$ J/mol - K and atmospheric pressure $P_0 = 10^5\ \text{N/m}^2$.

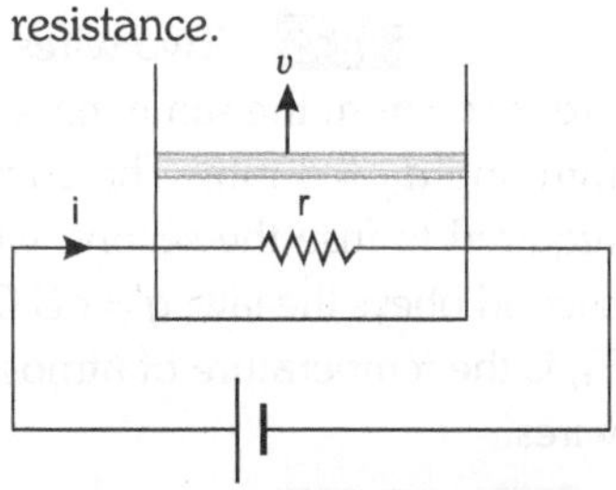

Problem 421. Suppose a voltmeter reads the voltage of a very old cell to be 1.40 V while a potentiometer reads its voltages to be 1.55 V. What is

(a) the internal resistance of the cell

(b) the current it would supply to a $5\,\Omega$ resistor. Assume the voltmeter resistance to be $280\,\Omega$.

Problem 422. The resistance of three wires BC, CA and AB of the same uniform cross-section and material are a, b and c respectively another wire from A of constant resistance d can make a sliding contact with BC. If a battery of constant emf E is connected between A and point of contact with BC. Find the minimum current drawn from the battery.

Problem 423. In the circuit shown in figure V_1 and V_2 are two voltmeters of resistances $3000\,\Omega$ and $2000\,\Omega$ respectively. In addition $R_1 = 2000\,\Omega$, $R_2 = 3000\,\Omega$ and $E = 200$ V.

(a) Find the reading of voltmeters V_1 and V_2 when
   (i) switch S is open
   (ii) switch S is closed

(b) current through S, when it is closed.

(Disregard the resistance of battery)

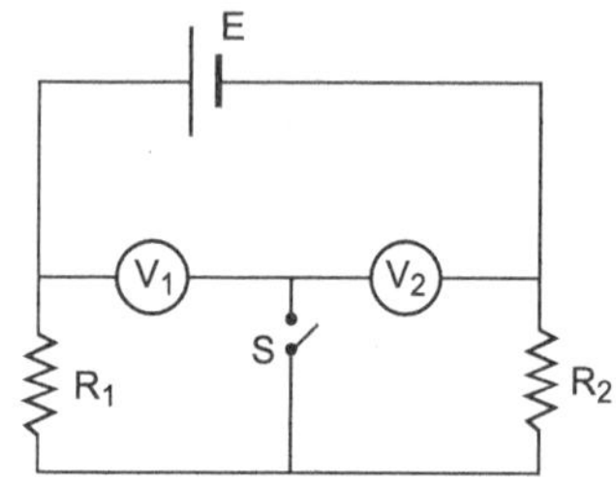

Problem 424. What amount of heat will be generated in a coil of resistance R due to a charge q passing through it if the current in the coil

(a) decreases down to zero uniformly during a time interval t_0

(b) decrease continuously down to zero halving its value in every t_0 second.

Problem 425. A conductor has a temperature independent resistance R and a total heat capacity C. At the moment $t = 0$, it is connected to a DC voltage V. Find the time dependence of conductor's temperature T assuming that the thermal power dissipated into surrounding space varies as $q = \alpha\,(T - T_0)$ where α is a constant, T_0 is the environmental temperature (equal to the conductor's temperature at the initial moment).

Problem 426. A cable AB of length l of uniform wires develops a leak at a certain point. To locate the fault two observations are made. The resistance between A and the earth through the cable when B is earthed is found to be that of a length l_1 of the wire and that between B and the earth when A is earthed that of a length l_2. Show that the point at which the fault exists divides AB in the ratio $\sqrt{\dfrac{l_1\,(l - l_2)}{l_2\,(l - l_1)}}$.

Hint: Consider some resistance of leak also

Problem 427. Two wires of same material and same length but of different square cross-sections are made from the same material. The sides of the cross-section of the first and second wires are $d_1 = 1$ mm and $d_2 = 4$ mm. The current required to fuse the first wire is $i_1 = 10$ A. Determine the current i_2 required to fuse the second wire assuming that the amount of heat dissipated to the atmosphere per second obeys the law, $q = \alpha S\,(T - T_0)$ where S is the lateral surface area of the wire, T is its temperature, T_0 is the temperature of atmosphere and α is the constant of proportionality which is same for the two wires.

Problem 428. In the circuit shown in figure $E_1 = 7$ V, $E_2 = 1$ V, $R_1 - 2\,\Omega$, $R_2 = 2\,\Omega$ and $R_3 = 3\,\Omega$ respectively. Find the power supplied by the two batteries.

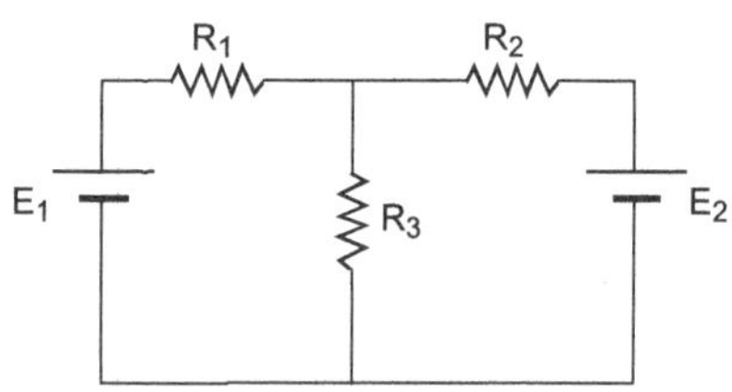

Problem 429. Find the net emf of the three batteries shown in figure.

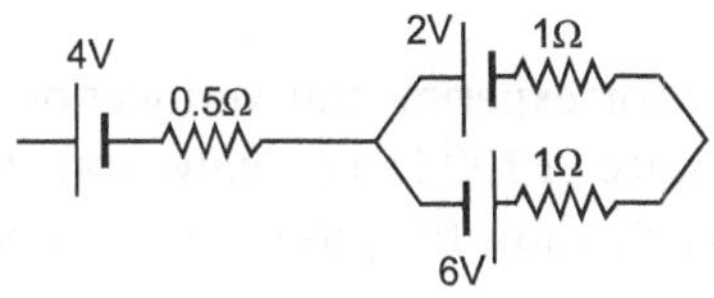

Problem 430. The figure shows part of certain circuit find:

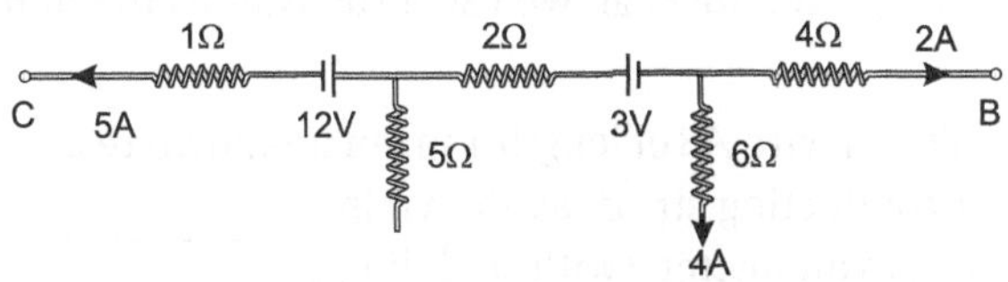

(a) Power dissipated in 5 Ω resistance.
(b) Potential difference $V_C - V_B$
(c) Which battery is being charged.

Problem 431. Find the equivalent resistance of the network shown in figure, across the points *O* and *A*. The resistance of each branch of the octagon is r_o.

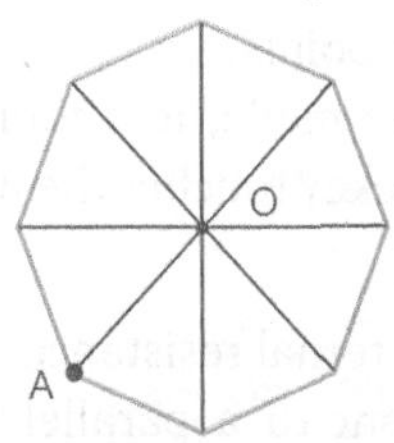

Problem 432. In the circuit shown the resistance *R* is kept in a chamber whose temperature is 20°C which remains constant. The initial temperature and resistance of *R* is 50°C and 15 Ω respectively. The rate of change of resistance *R* with temperature is $\frac{1}{2}\ \Omega/°C$ and the rate of decrease of temperature of *R* is (ln 3)/100 times the temperature difference from the surrounding (assume the resistance *R* loses heat only in accordance with Newton's law of cooling). *K* is closed at $t = 0$ then find the

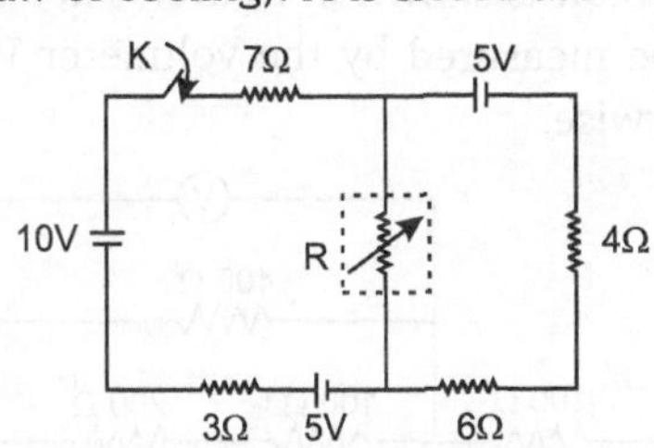

(a) value of *R* for which power dissipation in it is maximum.
(b) temperature of *R* when power dissipation is maximum.
(c) time after which the power dissipation will be maximum.

IIT JEE PROBLEMS

Problem 433. Draw the circuit for experimental verification of Ohm's law using a source of variable D.C. voltage, a main resistance of 100 Ω, two galvanometers and two resistances of values 10^6 Ω and 10^{-3} Ω respectively. Clearly show the positions of the voltmeter and the ammeter.

(JEE 2004)

Problem 434. Show by diagram, how can we use a rheostat as the potential divider.

(JEE 2003)

Problem 435. A thin uniform wire *AB* of length 1 m, an unknown resistance *X* and a resistance of 12 Ω are connected by thick conducting strips, as shown in the figure. A battery and a galvanometer (with a sliding jockey connected to it) are also available. Connections are to be made to measure the unknown resistance *X* using the principle of Wheatstone bridge. Answer the following questions.

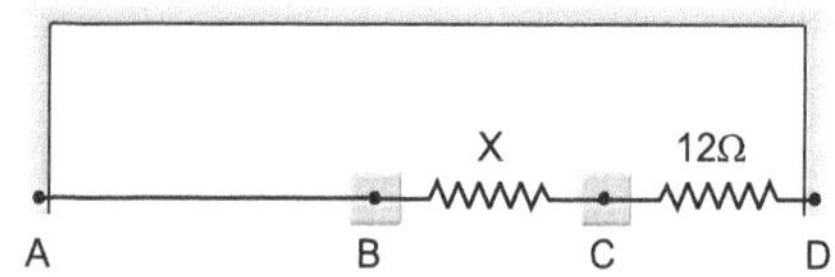

(a) Are there positive and negative terminals on the galvanometer?

(b) Copy the figure in your answer book and show the battery and the galvanometer (with jockey) connected at appropriate points.

(c) After appropriate connections are made, it is found that no deflection takes place in the galvanometer when the sliding jockey touches the wire at a distance of 60 cm from *A*. Obtain the value of the resistance *X*.

(JEE 2002)

Problem 436. Find the emf (V) and internal resistance (r) of a single battery which is equivalent to a parallel combination of two batteries of emfs V_1 and V_2 and internal resistances r_1 and r_2 respectively, with polarities as shown in figure. **(JEE 1997, Cancelled)**

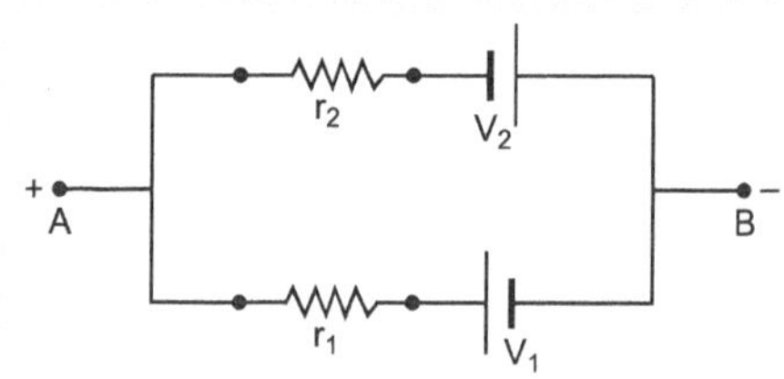

Problem 437. An electrical circuit is shown in figure. Calculate the potential difference across the resistor of 400 ohm as will be measured by the voltmeter *V* of resistance 400 ohm either by applying Kirchhoff's rules or otherwise.

(JEE 1996)

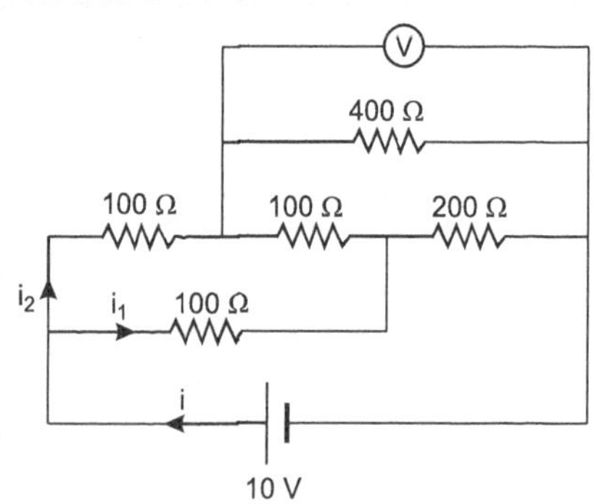

CHAPTER 13

Electrostatics

Problem 438. A capacitor is connected up with an accumulator. If we move the plates of the capacitor apart, we overcome the force of electrostatic attraction between the plates and consequently we do positive work. On what does this work go? What happens to the energy of the capacitor?

Problem 439. A conducting disc of radius R is rotating with an angular velocity ω. Allowing for the fact that electrons are the current carriers in a conductor, determine the potential difference between the centre of the disc and the edge. Mass of the electron is m and charge is e.

Problem 440. Two balls having like charges q_1 and q_2 initially have a velocity of the same magnitude and direction.After a uniform electric field has been applied during a certain time, the direction of the first ball changes by 60° and the speed becomes two times. The direction of the velocity of the second ball changes by 30°. In what proportion will the velocity of the second ball change ? Determine the magnitude of the charge to mass ratio for the second ball, if it is equal to α_1 for the first ball.

Problem 441. A point charge q is fixed at the centre of an insultated disc of mass M and radius R. The disc is resting on a rough horizontal plane. Another charge $-Q$ is fixed vertically above the centre of the disc at a height h. If the disc is displaced slightly in the horizontal direction, find the period of oscillation of the disc. Friction is sufficient to prevent slipping.

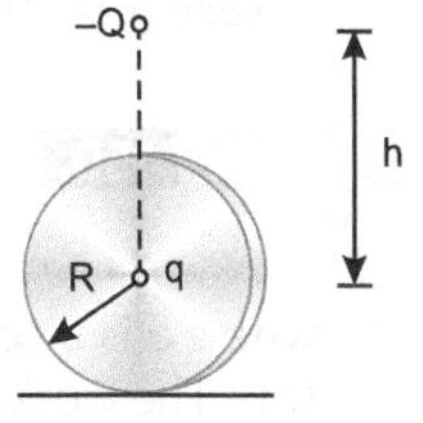

Problem 442. A parallel plate capacitor is made up of two square plates of sides 'a'. The distance between the plates is d. A dielectric slab of dielectric constant 4 completely fills the gap between the plates. One end of the slab is connected to a massless spring of force constant k, other end of which is fixed. Mass of the slab is M. Initially spring is unstretched. A constant force F_0 is applied to the slab as shown in figure. Prove that the slab will execute simple harmonic motion. Find its amplitude and time period.

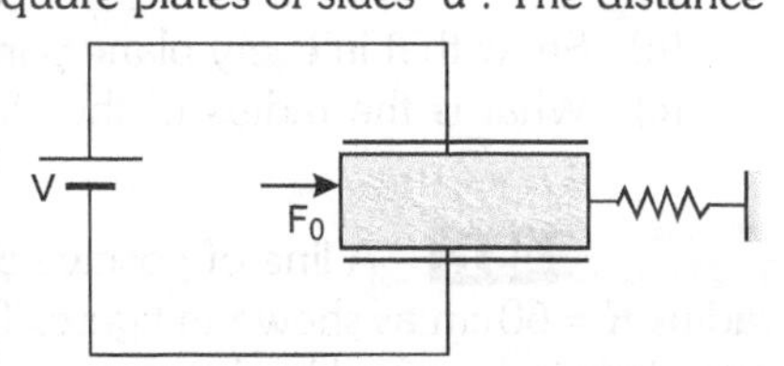

Problem 443. Two small identical balls lying on a horizontal plane are connected by a massless spring. One ball is fixed and the other is free. The balls are charged identically as a result of which the spring length increases two fold. Determine the factor by which frequency of small harmonic vibrations of the system will change. Assume that force constant of spring is constant. It does not change with length of spring.

Problem 444. Two small balls having the same mass and charge are located on the same vertical line at heights h_1 and h_2. They are thrown simultaneously with the same velocity in the same direction along the horizontal.

The first ball hits the ground at a distance l from the initial vertical line. At what height will the second ball be at this instant ($h_1 < h_2$).

Problem 445. The diameter of the outer conductor of a cylindrical capacitor is D_2. What should be the diameter of the core (inner cylinder) D_1 of this capacitor be, so that for a given potential difference between the outer conductor and the core, the electric field strength at the core is minimum.

Problem 446. Find the electrostatic potential energy of two small coplanar dipoles p_1 and p_2 at a distance r apart. The two dipoles make angles θ_1 and θ_2 with the line joining their centres.

Problem 447. Three concentric thin spherical shells are of radii a, b and c $(a < b < c)$. The first and third are connected by a fine wire through a small hole in the second. The second is connected to earth. Find the capacity of the condenser so formed.

Problem 448. (a) Two similar point charges q_1 and q_2 are placed at a distance r apart in air. If a dielectric slab of thickness t and dielectric constant k is put between the charges, calculate the coulomb force of repulsion.

(b) If the thickness of the slab covers half the distance between the charges, the coulomb repulsive force is reduced in the ratio 4 : 9. Calculate the dielectric constant of slab.

Problem 449. Three particles each of mass 1 g and carrying a charge q are suspended from a common point by insulated massless string each 100 cm long. If the particles are in equilibrium and are located at the corners of an equilateral triangle of side length 3 cm, calculate the charge q on each particle.

$$\left(\text{Take } g = 10 \text{ m/s}^2 \text{ and } \frac{1}{4\pi\varepsilon_0} = 9 \times 10^9 \frac{\text{N-m}^2}{\text{kg}^2}\right)$$

Problem 450. The electric potential V in volt in a region of space in given by

$$V = ax^2 + ay^2 + 2az^2$$

where a is a constant of proper dimensions.

(a) The work done by the field when a 2 μC test charge moves from point (0, 0, 0.1 m) to origin is 5×10^{-5} J. Determine a.

(b) Show that in every plane parallel to X-Y plane the equipotential lines are circles.

(c) What is the radius of the circle of the equipotential line corresponding to $V = 6250$ V and $z = \sqrt{2}$ m.

Problem 451. A line of positive charge is formed into a semicircle of radius $R = 60$ cm as shown in figure. The charge per unit length along the semicircle is described by the expression $\lambda = \lambda_0 \cos\theta$. The total charge on the semicircle is 12.0 μC. Calculate the total force on a charge of 3.00 μC placed at the centre of curvature.

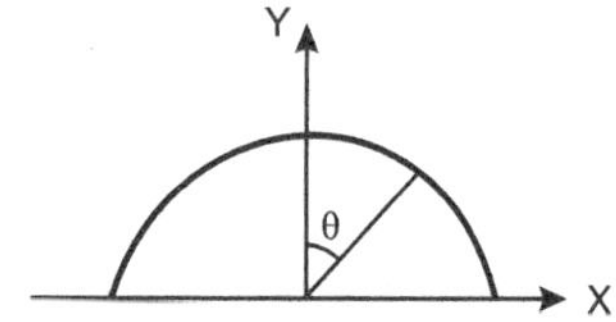

Problem 452. The plates of a capacitor of capacitance $C = 2\,\mu\text{F}$ carry opposite charges $q_0 = 10$ mC. This is now discharged through a resistance $R = 5\,\Omega$ and an identical capacitor. Find

(a) the charge flowing through the resistance during a time interval $\Delta t = 1\,\mu\text{s}$

(b) amount of heat generated in the resistance during the same interval.

Problem 453. The radii of a spherical capacitor electrodes are equal to a and b $(a < b)$. The space between them is filled with a dielectric of dielectric constant K and resistivity ρ. Initially at time $t = 0$, the inner electrode gets a charge q_0. Find

(a) time dependence of charge on the inner electrode

(b) the amount of heat generated during the spreading of charge.

Problem 454. In the circuit shown in figure S_1 and S_2 are simultaneously closed at time $t = 0$. Power dissipated in the resistor R_1 is 0.2 watt and the initial current through R_2 is 10^{-2} A.

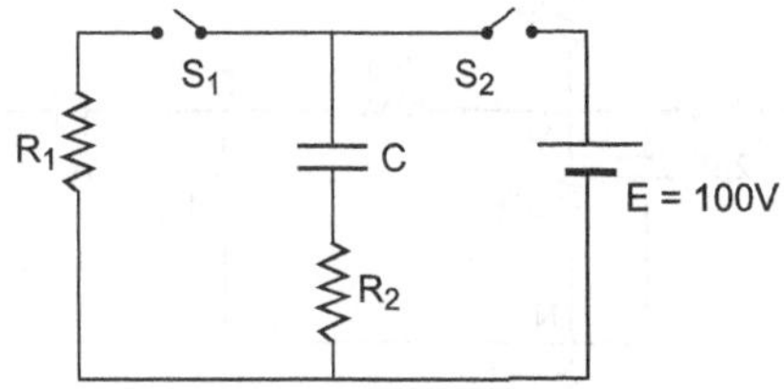

(a) Find R_1 and R_2

(b) When capacitor gets fully charged switch S_2 is opened. It is observed after 5 s that the current in R_1 is 0.74 mA. Find the value of C.

(c) At this moment switch S_1 is opened and S_2 is closed simultaneously. Find the instantaneous charge on the capacitor after 't' seconds.

Problem 455. Two positive charges q_1 and q_2 are located at the points with radius vectors $\vec{r}_1$ and $\vec{r}_2$. Find a negative charges q_3 and a radius vector $\vec{r}_3$ of the point at which it has to be placed for the force acting on each of the three charges to be equal to zero.

Problem 456. The electric field strength in a region is given as $\vec{E} = \dfrac{x\hat{i} + y\hat{j}}{x^2 + y^2}$. Find the net charge inside a sphere of radius 'a' with its centre at origin.

Problem 457. The electric potential in a certain region of space depends only on the x co-ordinate as

$$V = -\alpha x^3 + \beta$$

where α and β are constants. Find the volume distribution of space charge $\rho(x)$.

Problem 458. A point electric dipole with a moment $\vec{P}$ is placed in the external uniform electric field $\vec{E}_0$ with $\vec{P} || \vec{E}_0$. In this case one of the equipotential surfaces enclosing the dipole forms a sphere. Find the radius of this sphere.

Problem 459. The given C-R circuit has two switches S_1 and S_2. Switch S_2 is closed and S_1 is opened till capacitor is fully charged to q_0. Then S_2 is opened and S_1 is closed simultaneously till charge on capacitor remains $q_0/2$. It takes time t_1. Now S_1 is again opened and S_2 is closed till charge on capacitor becomes $3q_0/4$. It takes time t_2. Find the ratio t_1/t_2.

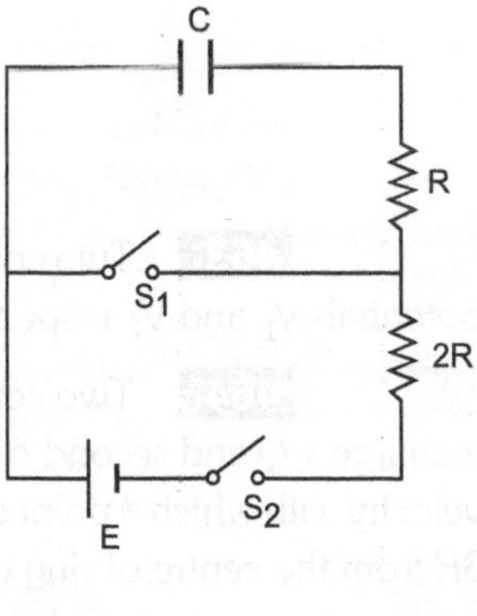

Problem 460. Figure shows a network of capacitors and resistances. Potentials of some of the points are given. Find

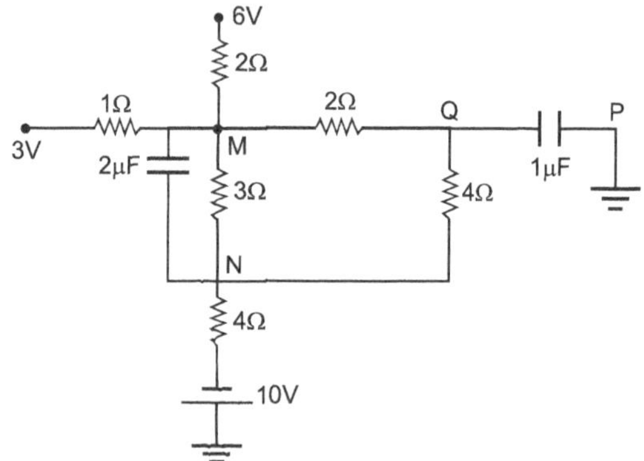

(a) the potentials of points M and N
(b) the charge stored in both the capacitors in steady state.

Problem 461. Three charged capacitors C_1, C_2 and C_3 of capacitances $3\,\mu F$, $6\,\mu F$ and $6\,\mu F$ respectively and having charge $30\,\mu C$ each are connected with a resistor of resistance $R = 10\,\Omega$ through a switch S as shown. The switch S is closed at time $t = 0$. Find :

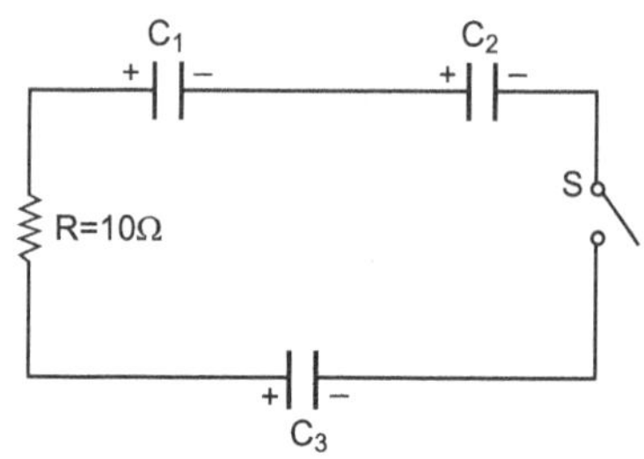

(a) the initial current in the circuit
(b) the final charge deposited on the capacitors
(c) heat dissipated in the resistor.

Problem 462. Consider the circuit shown in the figure :

(a) Find the current i flowing through the circuit when the key is open.
(b) Find the net charge on the capacitor when K is open and also when K is closed.
(c) What is the change in the current i, when K is closed?

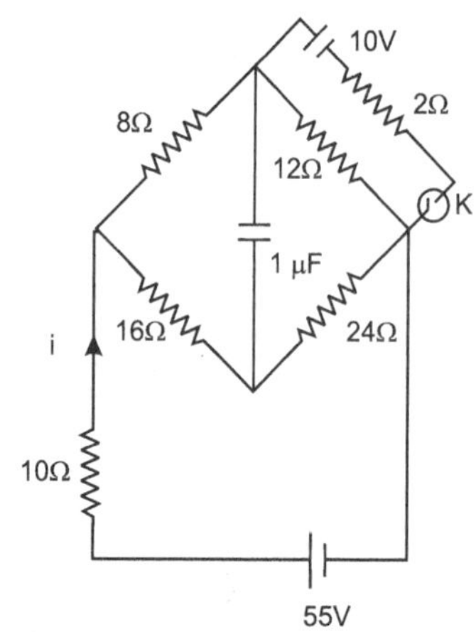

Problem 463. Two concentric conducting spherical shells of radii a_1 and a_2 $(a_2 > a_1)$ are charged to potentials V_1 and V_2 respectively. Find the charge on the inner shell.

Problem 464. Two concentric rings placed in a gravity free region in yz plane one of radius R carries a charge $+Q$ and second of radius $4R$ and charge $-8Q$ distributed uniformly over it. Find the minimum velocity with which a point charge of mass m and charge $-q$ should be projected from a point at a distance $3R$ from the centre of ring on its axis so that it will reach to the centre of the rings.

Problem 465. In the circuit diagram shown in figure, the capacitor of capacitance C is uncharged when the key k is open. The key is closed for time during which the capacitor becomes charged to a voltage V. Determine the amount of heat Q_2 liberated during this time in the resistor of resistance R_2 if the emf of the source is E, and its internal resistance can be neglected.

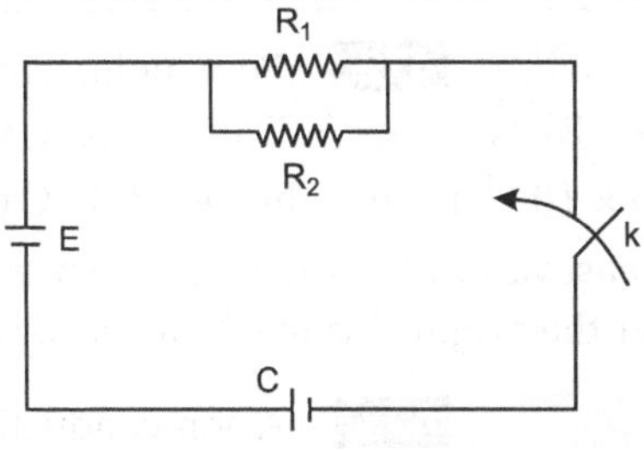

Problem 466. A metal sphere, of radius R and cut in two along a plane whose minimum distance from the sphere's centre is h, is uniformly charged by a total electric charge Q. What force is necessary to hold the two parts of the sphere together?

Problem 467. A small positively charged ball of mass m is suspended by an insulating thread of negligible mass. Another positively charged small ball is moved very slowly from a large distance until it is in the original position of the first ball. As a result, the first ball rises by h. How much work has been done?

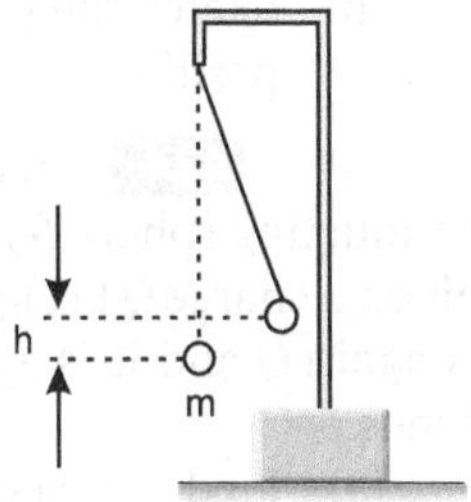

IIT JEE PROBLEMS

Problem 468. There are two large parallel metallic plates S_1 and S_2 carrying surface charge densities σ_1 and σ_2 respectively ($\sigma_1 > \sigma_2$) placed at a distance d apart in vacuum. Find the work done by the electric field in moving a point charge q a distance $a(a < d)$ from S_1 towards S_2 along a line making an angle $\pi/4$ with the normal to the plates. **(JEE 2004)**

Problem 469. Eight point charges are placed at the corners of a cube of edge a as shown in figure. Find the work done in disassembling this system of charges. **(JEE 2003)**

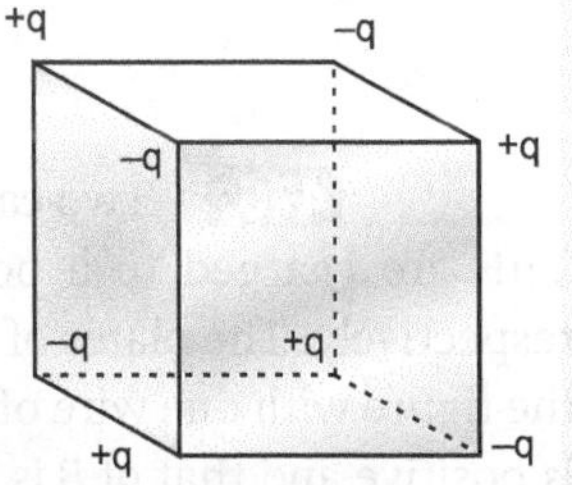

Problem 470. A positive point charge q is fixed at origin. A dipole with a dipole moment $\vec{p}$ is placed along the x-axis far away from the origin with $\vec{p}$ pointing along positive x-axis. Find:

(a) the kinetic energy of the dipole when it reaches a distance d from the origin, and
(b) the force experienced by the charge q at this moment. **(JEE 2003)**

Problem 471. A small ball of mass 2×10^{-3} kg having a charge of 1 μC is suspended by a string of length 0.8 m. Another identical ball having the same charge is kept at the point of suspension. Determine the minimum horizontal velocity which should be imparted to the lower ball so that it can make complete revolution. Take $g = 10\ m/s^2$. **(JEE 2001)**

Problem 472. Four point charges $+8\ \mu C$, $-1\ \mu C$, $-1\ \mu C$ and $+8\ \mu C$ are fixed at the points $-\sqrt{27/2}$ m $-\sqrt{3/2}$ m, $+\sqrt{3/2}$ m and $+\sqrt{27/2}$ m respectively on the Y-axis. A particle of mass 6×10^{-4} kg and charge $+0.1\ \mu C$ moves along the $-X$ direction. Its speed at $x = +\infty$ is v_0. Find the least value of v_0 for which the particle will cross the origin. Find also the kinetic energy of the particle at the origin. Assume that space is gravity free. $1/4\pi\varepsilon_0 = 9 \times 10^9\ Nm^2/C^2$. **(JEE 2000)**

Problem 473. A non-conducting disc of radius a and uniform positive surface charge density σ is placed on the ground with its axis vertical. A particle of mass m and positive charge q is dropped, along the axis of the disc from a height H with zero initial velocity. The particle has $q/m = 4\varepsilon_0 g/\sigma$.

(a) Find the value of H if the particle just reaches the disc.

(b) Sketch the potential energy of the particle as a function of its height and find its equilibrium position. **(JEE 1999)**

Problem 474. A conducting sphere S_1 of radius r is attached to an insulating handle. Another conducting sphere S_2 of radius R is mounted on an insulating stand, S_2 is initially uncharged. S_1 is given a charge Q brought into contact with S_2 and removed. S_1 is recharged such that the charge on it is again Q and it is again brought into contact with S_2 and removed. This procedure is repeated n times.

(a) Find the electrostatic energy of S_2 after n such contacts with S_1.

(b) What is the limiting value of this energy as $n \to \infty$? **(JEE 1998)**

Problem 475. In the circuit shown in figure, the battery is an ideal one with emf V. The capacitor is initially uncharged. The switch S is closed at time $t = 0$.

(a) Find the charge Q on the capacitor at time t.

(b) Find the current in AB at time t. What is its limiting value as $t \to \infty$? **(JEE 1998)**

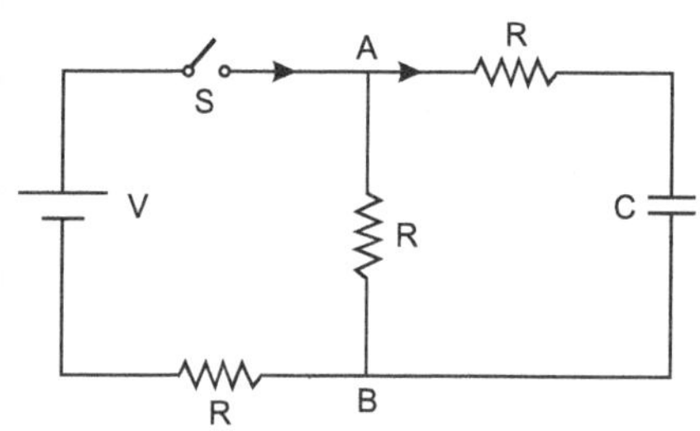

Problem 476. Two capacitors A and B with capacities $3\ \mu F$ and $2\ \mu F$ are charged to a potential difference of 100 V and 180 V respectively. The plates of the capacitors are connected as shown in the figure with one wire of each capacitor free. The upper plate of A is positive and that of B is negative. An uncharged $2\ \mu F$ capacitor C with lead wires falls on the free ends to complete the circuit. Calculate :

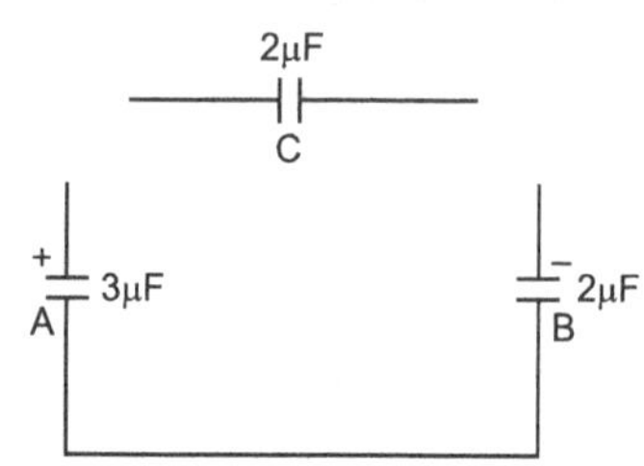

(i) the final charge on the three capacitors, and

(ii) the amount of electrostatic energy stored in the system before and after completion of the circuit. **(JEE 1997)**

Problem 477. A leaky parallel plate capacitor is filled completely with a material having dielectric constant $K = 5$ and electrical conductivity $\sigma = 7.4 \times 10^{-12}\ \Omega^{-1}m^{-1}$. If the charge on the capacitor at the instant $t = 0$ is $q = 8.85\ \mu C$, then calculate the leakage current at the instant $t = 12$ s. **(JEE 1997, Cancelled)**

Problem 478. (a) The capacitance of a parallel plate capacitor with plate area A and separation d, is C. The space between the plates is filled with two wedges of dielectric constants K_1 and K_2 respectively (fig.). Find the capacitance of the resulting capacitor.

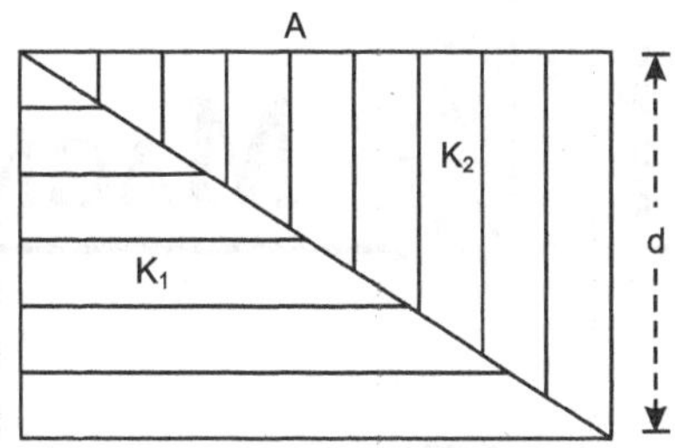

(b) Two isolated metallic solid spheres of radii R and $2R$ are charged such that both of these have same charge density σ. The spheres are located far away from each other and connected by a thin conducting wire. Find the new charge density on the bigger sphere. **(JEE 1996)**

Problem 479. Two square metal plates of side 1 m are kept 0.01 m apart like a parallel plate capacitor in air in such a way that one of their edges is perpendicular to an oil surface in a tank filled with an insulating oil. The plates are connected to a battery of emf 500 V. The plates are then lowered vertically into the oil at a speed of 0.001 ms^{-1}. Calculate the current drawn from the battery during the process. (Dielectric constant of oil $= 11$, $\varepsilon_0 = 8.85 \times 10^{-12}\ C^2N^{-1}m^{-2}$) **(JEE 1994)**

Problem 480. **(a)** Two parallel plate capacitors A and B have the same separation $d = 8.885 \times 10^{-4}$ m between the plates. The plate areas of A and B are 0.04 m^2 and 0.02 m^2 respectively. A slab of dielectric constant (relative permittivity) $K = 9$ has dimensions such that it can exactly fill the space between the plates of capacitor B.

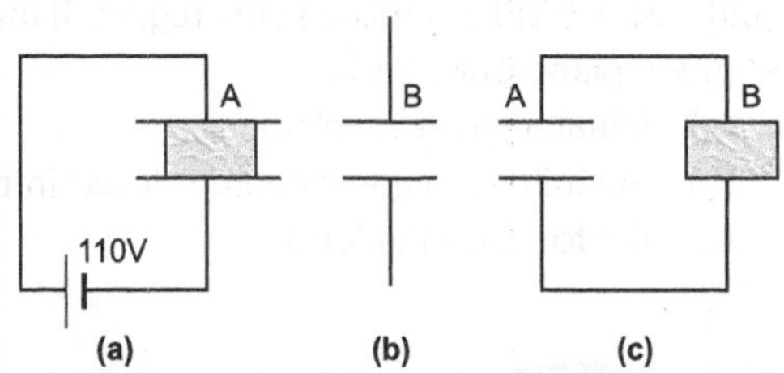

(i) The dielectric slab is placed inside A as shown in figure (a). A is then charged to a potential difference of 110 V. Calculate the capacitance of A and the energy stored in it.

(ii) The battery is disconnected and then the dielectric slab is removed from A. Find the work done by the external agency in removing the slab from A.

(iii) The same dielectric slab is now placed inside B, filling it completely. The two capacitors A and B are then connected as shown in figure (c). Calculate the energy stored in the system.

(b) A circular ring of radius R with uniform positive charge density λ per unit length is located in the Y-Z plane with its centre at the origin O. A particle of mass m and positive charge q is projected from the point P $(R\sqrt{3}, 0, 0)$ on the positive X-axis directly towards O, with an initial speed v. Find the smallest (non-zero) value of the speed v such that the particle does not return to P. **(JEE 1993)**

Problem 481. (a) A charge Q is uniformly distributed over a spherical volume of radius R. Obtain an expression for the energy of the system.

(b) What will be the corresponding expression for the energy needed to completely disassemble the planet earth against the gravitational pull amongst its constituent particles ?
Assume the earth to be a sphere of uniform mass density. Calculate this energy, given the product of the mass and the radius of the earth to be 2.5×10^{31} kg-m.

(c) If the same charge of Q as in part (a) above is given to a spherical conductor of the same radius R, what will be the energy of the system ? Take $g = 10\ m/s^2$ **(JEE 1992)**

CHAPTER 14

MAGNETICS

Problem 482. A particle of mass m having a charge q enters into a circular region of radius R with velocity v directed towards the centre. The strength of magnetic field is B. Find the deviation in the path of the particle.

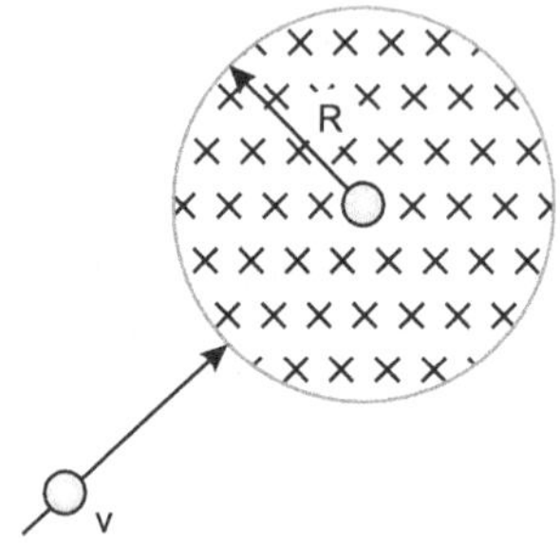

Problem 483. A ring of mass m and radius r carrying current I_0 is lying in the x-y plane with centre at the origin. A uniform magnetic field of strength $B_0(2\hat{i} - 3\hat{j} + 5\hat{k})$ T is applied in the region. If the ring can rotate about the axis AA' in x-y plane only, find :

(a) initial angular acceleration.

(b) the initial magnetic energy stored in the ring.

(c) the force on the loop.

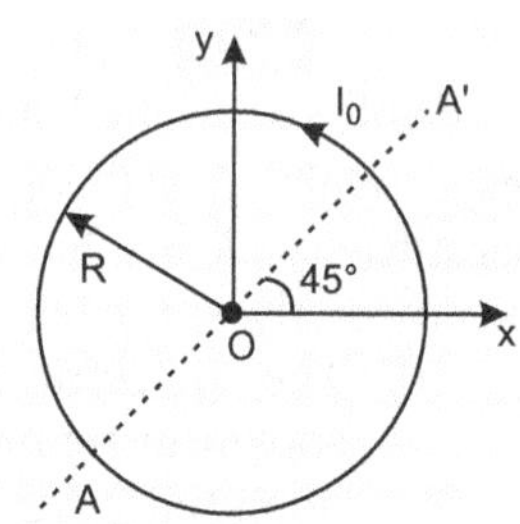

Problem 484. A particle of specific charge α (charge per unit mass) is released at time $t = 0$ from origin with an initial velocity of $\vec{v} = v_0\hat{i}$ in a uniform magnetic field $\vec{B} = -B_0\hat{k}$. Find the velocity and position of particle at any time t.

Problem 485. A positively charged particle having charge $q_1 = 1$ C and mass $m_1 = 40$ g is revolving along a circle of radius $R = 40$ cm with speed $v_1 = 5$ m/s in a uniform magnetic field with centre of circle at origin. At time $t = 0$, the particle is at (0, 0.4 m, 0) and velocity is directed along positive x-direction. Another particle having charge $q_2 = 1$ C and mass $m_2 = 10$ g moving uniformly parallel to z-direction with velocity $v_2 = 40$ m/s collides with revolving particle at $t = 0$ and gets stuck to it. Neglecting gravitational force and coulomb force calculate x, y and z co-ordinates of the combined particle at $t = \frac{\pi}{40}$ second.

Problem 486. A particle with specific charge α (charge per unit mass) moves in the region of space where mutually perpendicular electric field $\vec{E} = E_0\hat{j}$ and magnetic field $\vec{B} = B_0\hat{k}$ are present. At time $t = 0$ particle is located at origin with zero initial velocity. Find x and y co-ordinates of particle at any time t.

Problem 487. A hypothetical magnetic field existing in a region is given by $\vec{B} = B_0 \vec{e_r}$, where $\vec{e_r}$ denotes the unit vector along the radial direction. A circular loop of radius 'a' carrying a current i, is placed with its plane parallel to the x-y plane and the centre at $(0, 0, d)$. Find the magnitude of the magnetic force acting on the loop.

Problem 488. A regular polygon of n side is formed by bending a wire of total length $2\pi r$ which carries a current i.

(a) Find the magnetic field B at the centre of the polygon.
(b) By letting $n \to \infty$, deduce the expression for the magnetic field at the centre of a circular coil.

Problem 489. Figure shows a circular wire loop of radius a carrying a current i, placed in a perpendicular magnetic field B

(a) consider a small part dl of the wire, find the force on this part of the wire exerted by the magnetic field.
(b) find the force of compression in the wire
(c) suppose that the radius of cross-section of the wire used is r. Find the increase in the radius of the loop if the magnetic field is switched off. The Young's modulus of the material of the wire is Y.

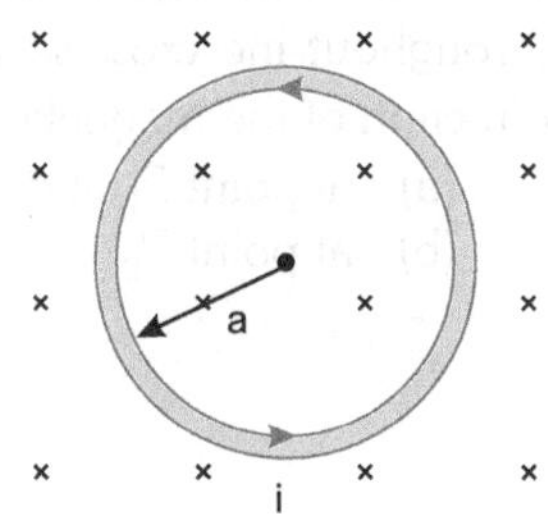

Problem 490. In the configuration shown in figure wires 1 and 2 are infinitely long and carry currents i_1 and i_2 in a direction perpendicular to plane of paper and inwards. Wire 3 of length l carries a current i_3 from left to right. Find the net magnetic force on wire 3.

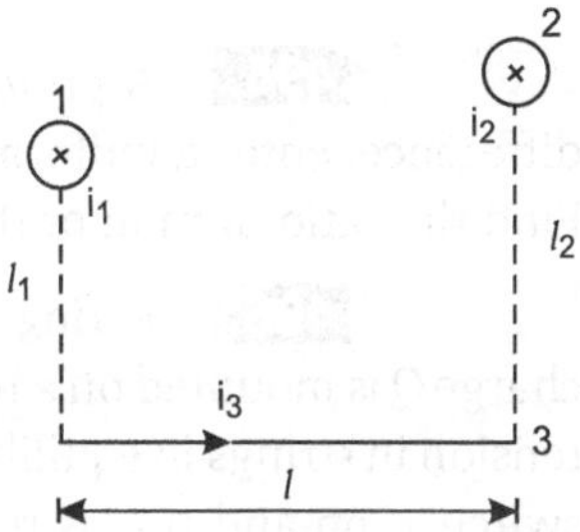

Problem 491. A rod of mass 0.72 kg and radius 6 cm rests on two parallel rails, that are $d = 12$ cm apart and $L = 45$ cm long. The rod carries a current $i = 48$ A (in the direction shown) and rolls along the rails without slipping. If it starts from rest, what is the speed of the rod as it leaves the rails if a uniform magnetic field of magnitude 0.24 T is directed perpendicular to the rod and the rails.

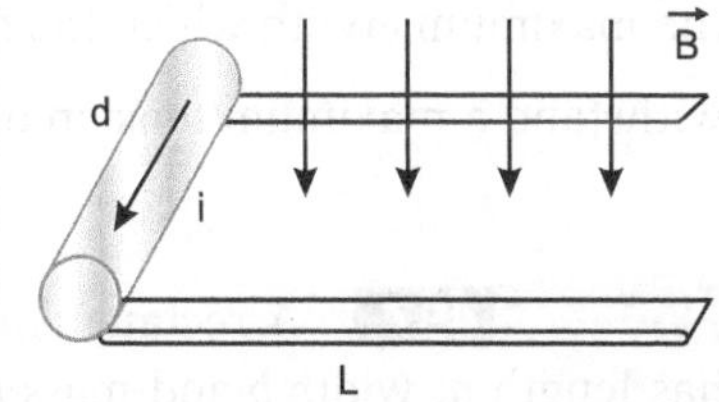

Problem 492. Magnetic field $\vec{B} = -B_0 x \hat{k}$ exists in a region of space. A particle of specific charge α (charge per unit mass) enters this region of space. Its velocity and position at time $t = 0$, are $\vec{v} = v_0 \hat{i}$ and $(0, 0, 0)$. Find the maximum x-displacement of the particle.

Problem 493. Electric field and magnetic field in a region of space are given by and $\vec{E} = E_0 \hat{j}$ and $\vec{B} = B_0 \hat{j}$. A particle of specific charge α (charge per unit mass) is released from origin with velocity $\vec{v} = v_0 \hat{i}$. Determine

(a) the path of the particle
(b) velocity of particle at any time t
(c) position of particle at time t
(d) the y coordinate of particle when it crosses the y-axis for the nth time and
(e) angle ϕ between particle's velocity and y-axis at that moment

Problem 494. A long cylindrical conductor of radius a has two cylindrical cavities of diameter a through its entire length as shown in cross-section in figure. A current I is directed out of the page and is uniform throughout the cross-section of the conductor. Find the magnitude and direction of the magnetic field in terms of μ_0, I, r and a.
(a) at point P_1 and
(b) at point P_2

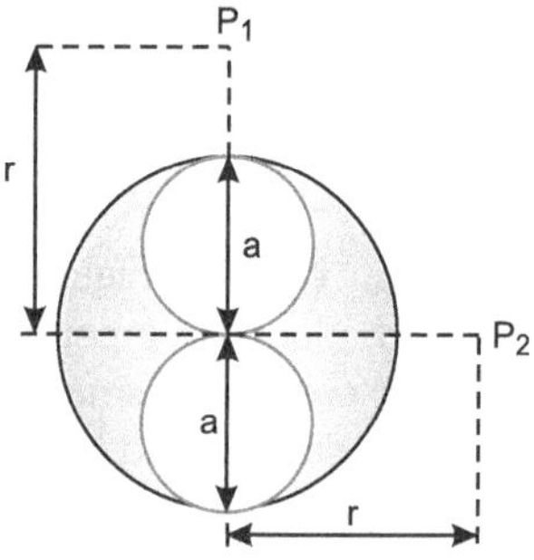

IIT JEE PROBLEMS

Problem 495. A proton and an alpha particle, after being accelerated through same potential difference, enter a uniform magnetic field the direction of which is perpendicular to their velocities. Find the ratio of radii of the circular paths of the two particles. **(JEE 2004)**

Problem 496. A ring of radius R having uniformly distributed charge Q is mounted on a rod suspended by two identical strings. The tension in strings in equilibrium is T_0. Now a vertical magnetic field is switched on and ring is rotated at constant angular velocity ω. Find the maximum ω with which the ring can be rotated if the strings can withstand a maximum tension of $\frac{3T_0}{2}$. **(JEE 2003)**

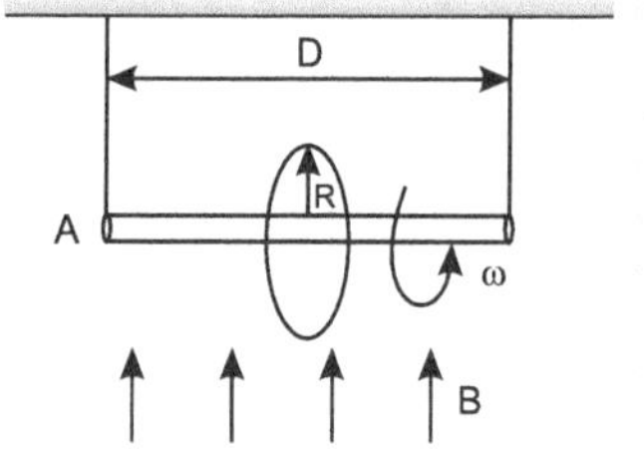

Problem 497. A rectangular loop $PQRS$ made from a uniform wire has length a, width b and mass m. It is free to rotate about the arm PQ, which remains hinged along a horizontal line taken as the y-axis (see figure). Take the vertically upward direction as the z-axis. A uniform magnetic field $\vec{B} = (3\hat{i} + 4\hat{k})B_0$ exists in the region. The loop is held in the x-y plane and a current I is passed through it. The loop is now released and is found to stay in the horizontal position in equilibrium.
(a) What is the direction of the current I in PQ?
(b) Find the magnetic force on the arm RS.
(c) Find the expression for I in terms of B_0, a, b and m. **(JEE 2002)**

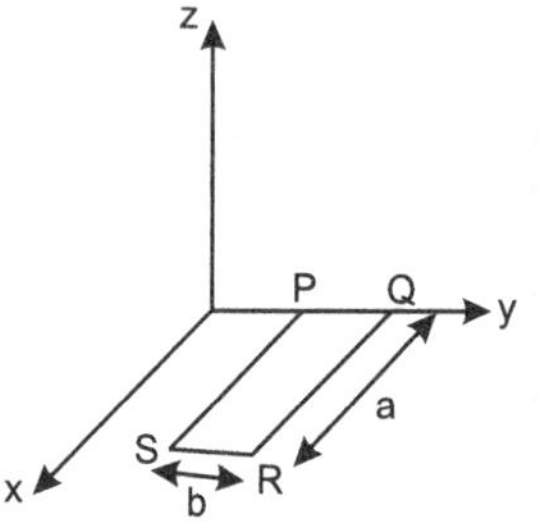

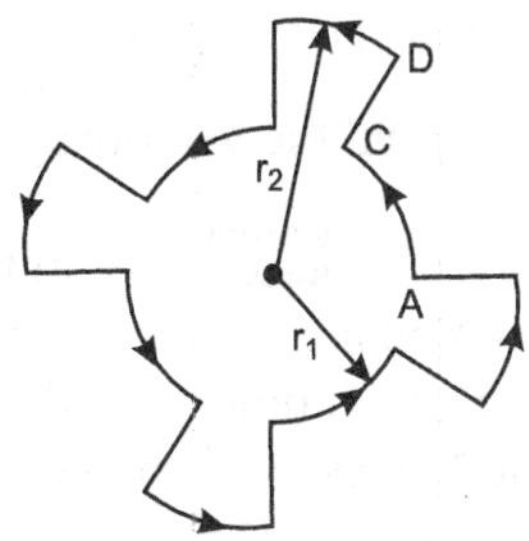

Problem 498. A current of 10 A flows around a closed path in a circuit which is in the horizontal plane as shown in the figure. The circuit consists of eight alternating arcs of radii $r_1 = 0.08$ m and $r_2 = 0.12$ m. Each subtends the same angle at the centre.

(a) Find the magnetic field produced by this circuit at the centre.

(b) An infinitely long straight wire carrying a current of 10 A is passing through the centre of the above circuit vertically with the direction of the current being into the plane of the circuit. What is the force acting on the wire at the centre due to the current in the circuit ? What is the force acting on the arc *AC* and the straight segment *CD* due to the current at the centre ? **(JEE 2001)**

Problem 499. A circular loop of radius *R* is bent along a diameter and given a shape as shown in figure. One of the semicircles (*KNM*) lies in the *X-Z* plane and the other one (*KLM*) in the *Y-Z* plane with their centres at origin. Current *I* is flowing through each of the semicircles as shown in figure.

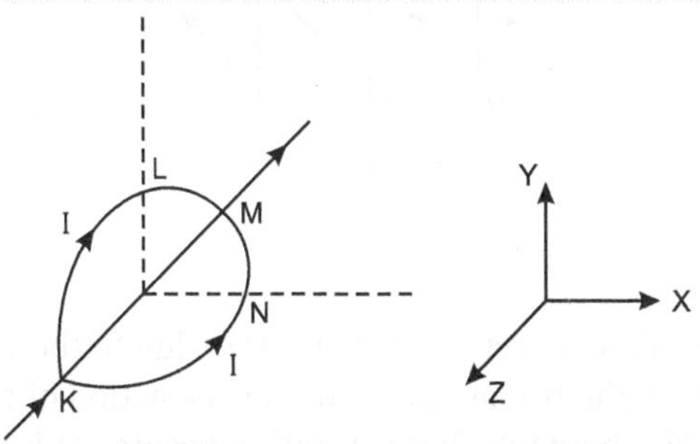

(a) A particle of charge q is released at the origin with a velocity $\vec{v} = -v_0\hat{i}$. Find the instantaneous force $\vec{F}$ on the particle. Assume that space is gravity free.

(b) If an external uniform magnetic field $B_0\hat{j}$ is applied, determine the force $\vec{F_1}$ and $\vec{F_2}$ on the semicircles *KLM* and *KNM* due to the field and the net force $\vec{F}$ on the loop. **(JEE 2000)**

Problem 500. The region between $x = 0$ and $x = L$ is filled with uniform steady magnetic field $B_0\,\hat{k}$. A particle of mass m, positive charge q and velocity $v_0\hat{i}$ travels along x-axis and enters the region of the magnetic field.

Neglect the gravity throughout the question

(a) Find the value of L if the particle emerges from the region of magnetic field with its final velocity at an angle $30°$ to its initial velocity.

(b) Find the final velocity of the particle and the time spent by it in the magnetic field, if the magnetic field now expands upto $2.1\,L$. **(JEE 1999)**

Problem 501. A particle of mass m and charge q is moving in a region where uniform, constant electric and magnetic fields $\vec{E}$ and $\vec{B}$ are present. $\vec{E}$ and $\vec{B}$ are parallel to each other. At time $t = 0$, the velocity $\vec{v_0}$ of the particle is perpendicular to $\vec{E}$ (Assume that its speed is always << c, the speed of light in vacuum). Find the velocity $\vec{v}$ of the particle at time t. You must express your answer in terms of t, q, m, the vectors $\vec{v}_0$, $\vec{E}$ and $\vec{B}$ and their magnitudes v_0, E and B. **(JEE 1998)**

Problem 502. A uniform constant magnetic field $\vec{B}$ is directed at an angle of 45° to the X-axis in $X-Y$ plane. $PQRS$ is a rigid square wire frame carrying a steady current I_0, with its centre at the origin O. At time $t=0$, the frame is at rest in the position shown in the figure with its sides parallel to X and Y axes. Each side of the frame is of mass M and length L.

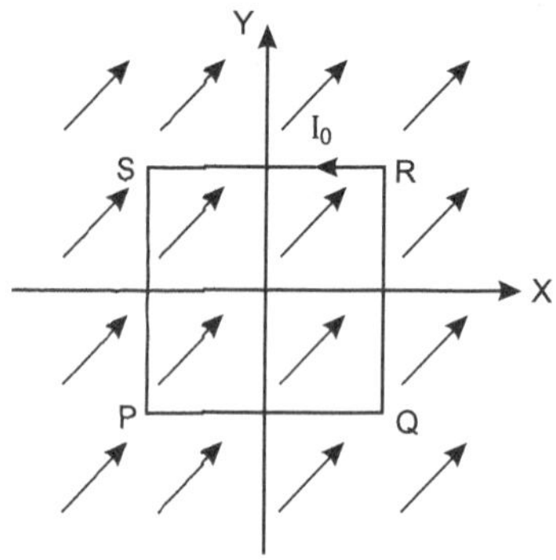

(a) What is the torque $\vec{\tau}$ about O acting on the frame due to the magnetic field ?

(b) Find the angle by which the frame rotates under the action of this torque in a short interval of time Δt, and the axis about which this rotation occurs (Δt is so short that any variation in the torque during this interval may be neglected). Given : the moment of inertia of the frame about an axis through its centre perpendicular to its plane is $\frac{4}{3}ML^2$. **(JEE 1998)**

Problem 503. Three infinitely long thin wires, each carrying current i in the same direction, are in the X-Y plane of a gravity free space. The central wire is along the y-axis while the other two are along $x = \pm d$.

(i) Find the locus of the points for which the magnetic field B is zero.

(ii) If the central wire is displaced along the z-direction by a small amount and released, show that it will execute simple harmonic motion. If the linear density of the wires is λ, find the frequency of oscillation. **(JEE 1997)**

Problem 504. An infinitesimally small bar magnet of dipole moment $\vec{M}$ pointing and moving with the speed v in the X-direction. A small closed circular conducting loop of radius a and negligible self-inductance lies in the Y-Z plane with its centre at $x=0$, and its axis coinciding with the X-axis. Find the force opposing the motion of the magnet, if the resistance of the loop is R. Assume that the distance x of the magnet from the centre of the loop is much greater than a. **(JEE 1997, Cancelled)**

Problem 505. An electron in the ground state of hydrogen atom is revolving in anticlockwise direction in a circular orbit of radius R (figure).

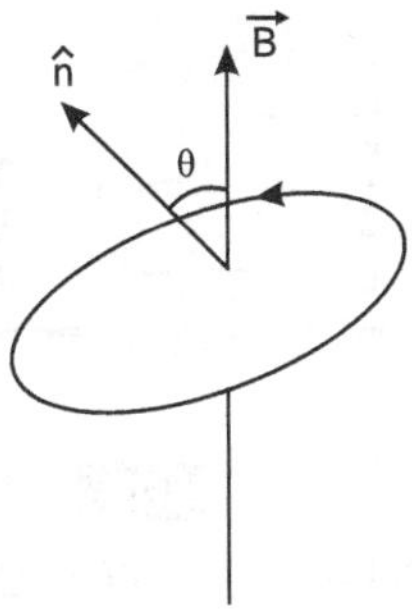

(i) Obtain an expression for the orbital magnetic moment of the electron.

(ii) The atom is placed in a uniform magnetic induction $\vec{B}$ such that the plane normal of the electron orbit makes an angle of 30° with the magnetic induction. Find the torque experienced by the orbiting electron. **(JEE 1996)**

Problem 506. A long horizontal wire AB, which is free to move in a vertical plane and carries a steady current of 20 A, is in equilibrium at a height of 0.01 m over another parallel long wire CD which is fixed in a horizontal plane and carries a steady current of 30 A, as shown in figure. Show that when AB is slightly depressed, it executes simple harmonic motion. Find the period of oscillations. **(JEE 1994)**

Problem 507. An electron gun G emits electrons of energy 2 keV travelling in the positive X-direction. The electrons are required to hit the spot S where $GS = 0.1$ m, and the line GS make an angle of 60° with the x-axis as shown in figure. A uniform magnetic field $\vec{B}$ parallel to GS exists in the region outside the electron gun. Find the minimum value of B needed to make the electrons hit S.

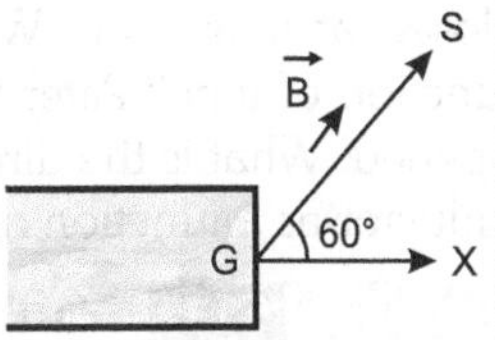

(JEE 1993)

Problem 508. A straight segment OC (of length L) of a circuit carrying a current i is placed along the X-axis as shown in figure. Two infinitely long straight wires A and B, each extending from $z = -\infty$ to $+\infty$, are fixed at $y = -a$ and $y = +a$ respectively, as shown in the figure. If the wires A and B each carry a current i into the plane of the paper, obtain the expression for the force acting on the segment OC. What will be force on OC if the current in the wire B is reversed ? **(JEE 1992)**

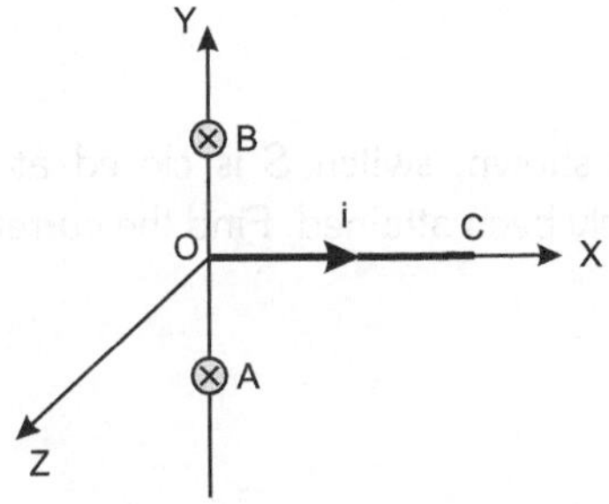

CHAPTER 15 ELECTROMAGNETIC INDUCTION

Problem 509. In circuit power factor of box is given 0.5 and power factor of circuit is given $\sqrt{3}/2$. Current leading the voltage. Find the effective resistance of the box.

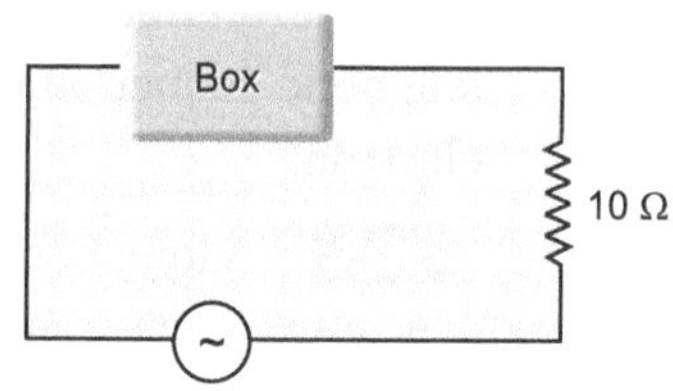

Problem 510. An inductor of inductance $L = 400$ mH and three resistors of resistances $R_1 = 4\,\Omega$ and $R_2 = 2\,\Omega$ are connected to a battery of e.m.f $E = 10$ V as shown in the figure. The internal resistance of the battery is negligible. The switch S is closed at time $t = 0$. What is the potential drop across L as a function of time? After the steady state is reached, the switch is opened. What is the direction and magnitude of current through inductor as a function of time?

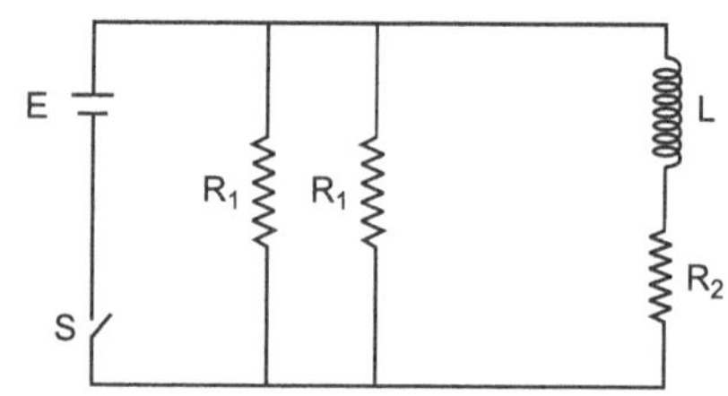

Problem 511. In the figure shown the reading of voltmeters are $V_1 = 40$ V, $V_2 = 40$ V and $V_3 = 10$ V.

Find :

(a) the peak value of current.

(b) the peak value of emf.

(c) the value of L and C.

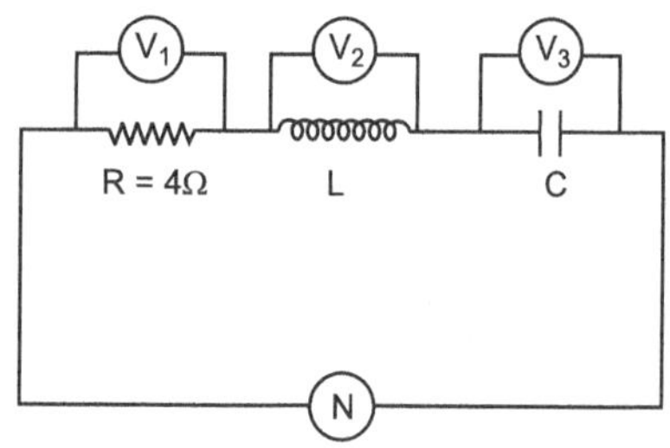

$E = E_0 \sin\left(100\,\pi t + \frac{\pi}{6}\right)$

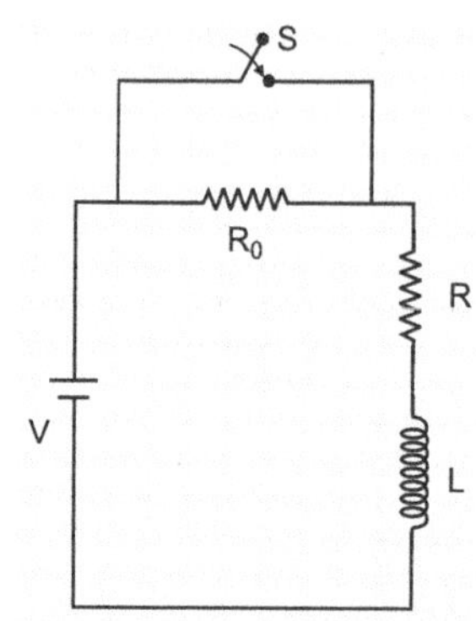

Problem 512. In the network shown, switch S is closed at $t = 0$, a steady state current having previously been attained. Find the current in the circuit at time t.

Problem 513. In the circuit shown in the figure $R = 50\,\Omega$, $E_1 = 25\sqrt{3}$ volt and $E_2 = 25\sqrt{6} \sin \omega t$ volt where $\omega = 100\pi\ s^{-1}$. The switch s is closed at time $t = 0$, and remains closed for 14 minutes, then it is opened.

(a) Find the amount of heat produced in the resistor.

(b) If total heat produced is used to raise the temperature of 3 kg of water at 20°C, what would be the final temperature of water?

(c) Find the value of the direct current that will produce same amount of heat in the resistor in same time as combination of DC source and AC source has produced.

Specific heat of water = 4200 J/kg-°C.

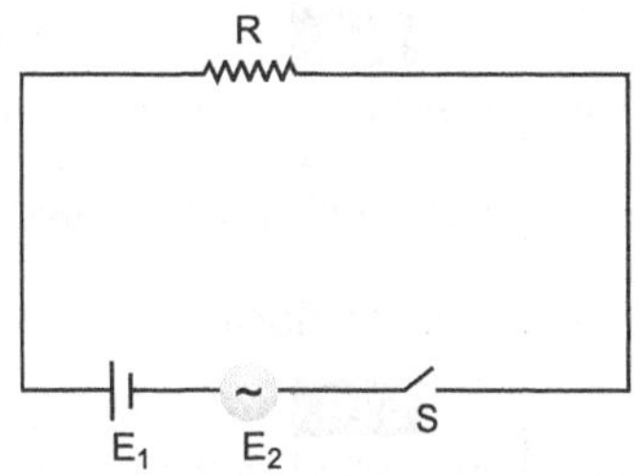

Problem 514. A circuit element shown in the figure as a box is having either a capacitor or, an inductor. The power factor of above circuit is 0.8, while current lags behind the voltage. Find :

(a) the source voltage V.

(b) the nature of the element in box and find its value.

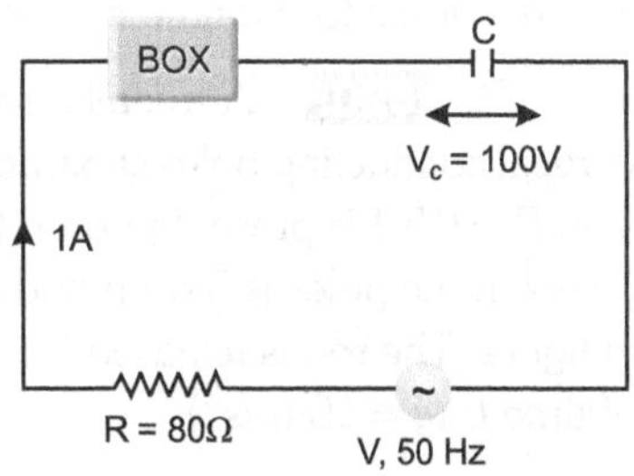

Problem 515. Initially the capacitor is charged to a potential of 5 V and then connected to position 1 with the shown polarity for 1 second. After 1 sec it is connected across the inductor at position 2.

(a) Find the potential across the capacitor after 1 second of its connection to position 1.

(b) Find the maximum current flowing in the LC circuit when capacitor is connected across the inductor. Also find the frequency of LC oscillations.

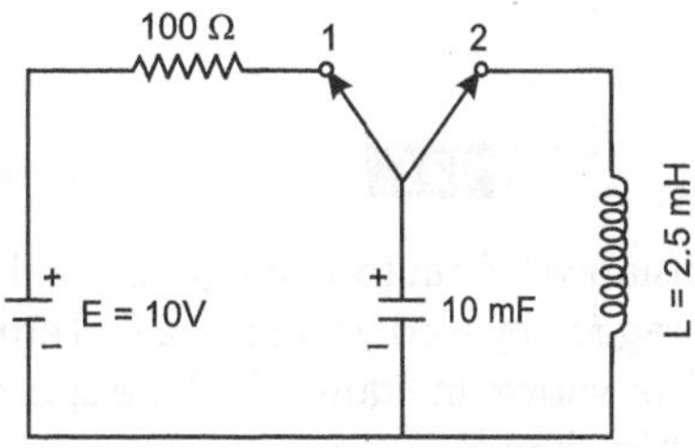

Problem 516. A metal rod of mass m can slide without friction along two parallel metal guides inclined at an angle α to the horizontal and separated by a distance b. The guides are connected at the bottom through an uncharged capacitor of capacitance C, and the entire system is placed in an upward magnetic field of induction $\vec{B}$. At time $t = 0$, the rod is at a distance l from the bottom.

(a) find the time when the rod reaches the bottom and

(b) velocity of rod at that moment.

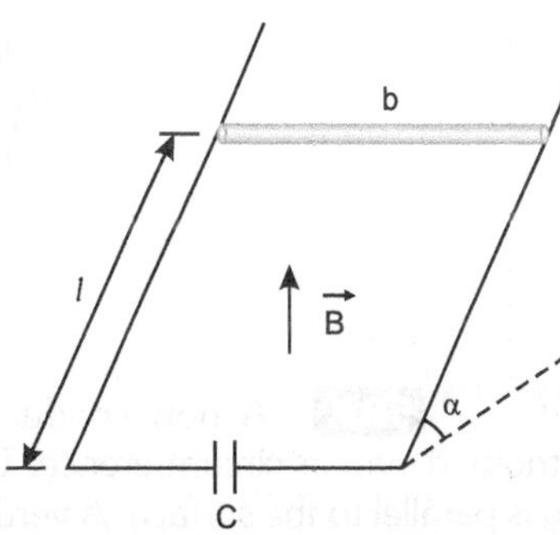

Problem 517. A square loop of side 'l' containing a resistance R, and a capacitance C is placed near an infinitely long current carrying wire as shown in figure. The current in the wire varies with time as

$$i = \left(\frac{i_0}{t_0}\right)t$$

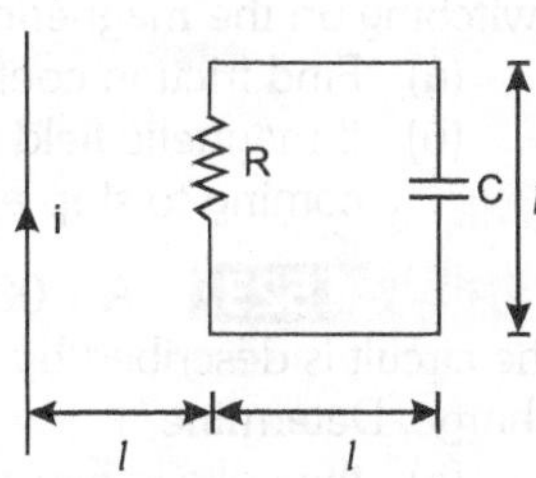

(a) find the charge stored in the capacitor at any time t.

(b) find the ratio of heat geneated in the resistor and the energy stored in the capacitor as a function of time t.

(c) if the breakdown potential of capacitor is V_0, find the time taken for breakdown to take place.

Problem 518. A conducting rod of zero resistance slides on a smooth wire of resistance per unit length ρ, bent at an angle $\theta = 60°$ as shown in figure. Find the force $\vec{F}$ to be applied on the rod as a function of x to make it move with constant velocity $\vec{v}$.

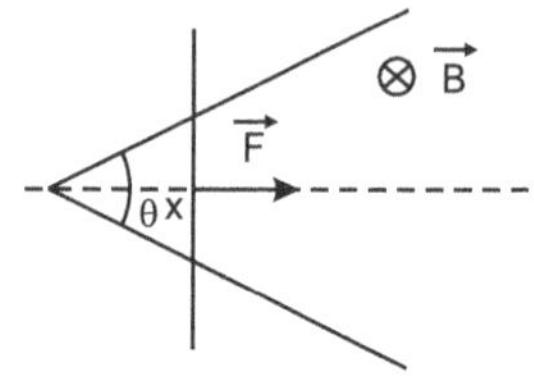

Problem 519. A 5 henry inductor is placed in series with a 10 ohm resistor. An emf of 5 volt is suddenly applied to the combination. Using these values prove the principle of conservation of energy for time equal to the time constant.

Problem 520. A metallic rod AC of mass 1 kg can slide freely on two vertical conducting poles separated by a distance of 10 m. A uniform magnetic field $B = 0.5$ T is present perpendicular to poles inwards. Resistance of the rod between the poles is 5 ohm and a constant emf $E = 20$ V is applied as shown in figure. The rod is released from rest. Find the speed of the rod as a function of time t. $(g = 10 \text{ m/s}^2)$

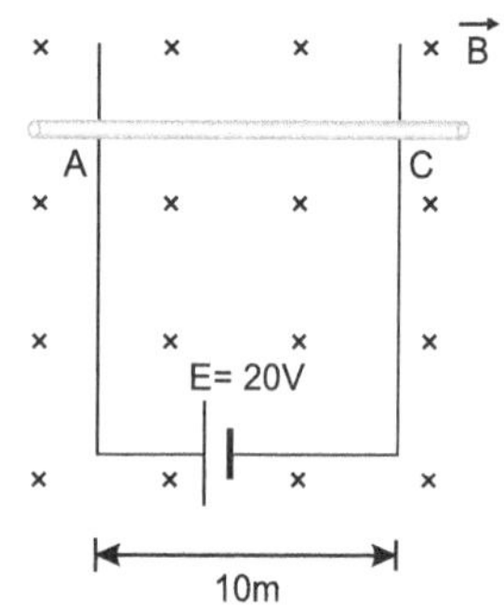

Problem 521. A conducting rod PQ of mass m is placed over two conducting rails separated by a distance l. A uniform magnetic field $\vec{B}$ perpendicular to the plane of the rails is switched on and the rod is dragged by a constant force F. Terminals of the rails are connected across a resistance R and a capacitor C as shown in figure. Find the speed of the rod at any time t. Neglect friction everywhere and resistance other than R.

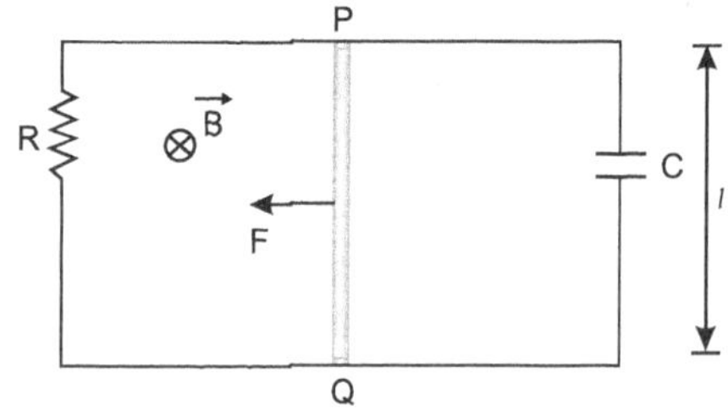

Problem 522. A non conducting ring of mass 'm' and radius 'R' has a charge Q uniformly distributed over its circumference. The ring is placed on a rough horizontal surface such that plane of the ring is parallel to the surface. A vertical magnetic field $B = B_0 t^2$ tesla is switched on. After 2 second from switching on the magnetic field the ring is just about to rotate about vertical axis through its centre.

(a) Find friction coefficient 'μ' between the ring and the surface.

(b) If magnetic field is switched off after 4 second, then find the angle rotated by the ring before coming to stop after switching off the magnetic field.

Problem 523. A 1.00 mH inductor and a 1.00 μF capacitor are connected in series. The current in the circuit is described by $i = 20t$, where t is in second and i is in ampere. The capacitor initially has no charge. Determine,

(a) the voltage across the inductor as a function of time.

(b) the voltage across the capacitor as a function of time.

(c) the time when the energy stored in the capacitor first exceeds that in the inductor.

Problem 524. Two long parallel horizontal rails, a distance l apart and each having a resistance λ per unit length are joined at one end by a resistance R. A perfectly conducting rod MN of mass m is free to slide along the rails without friction. There is a uniform magnetic field of induction B normal to the plane of paper and directed into the paper. A variable force F is applied to the rod MN such that, as the rod moves, a constant current i flows through R.

(a) Find the velocity of rod and the applied force F as function of distance x of the rod from R and

(b) What fraction of work per second by F is converted into heat.

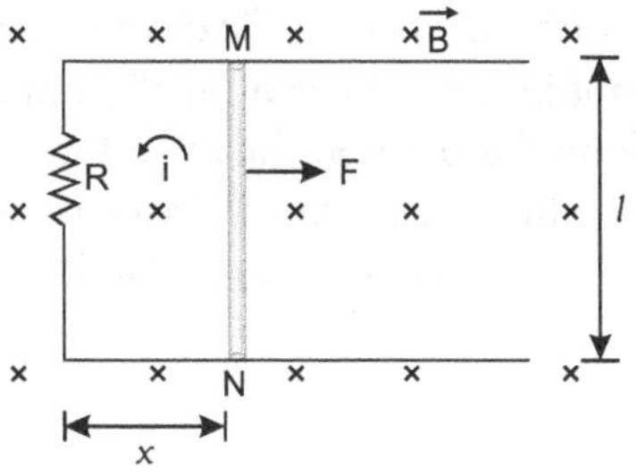

Problem 525. A conducting rod AC of mass m free to rotate in a horizontal plane about one end A over a semicircular conducting ring of radius 'a' is joined with an external resistance R as shown in figure. The resistance of the rod is r. The rod is given an initial angular velocity ω_0. A uniform magnetic field of magnitude B exists perpendicular to the plane of semicircular loop. Find the current in the circuit at angle θ.

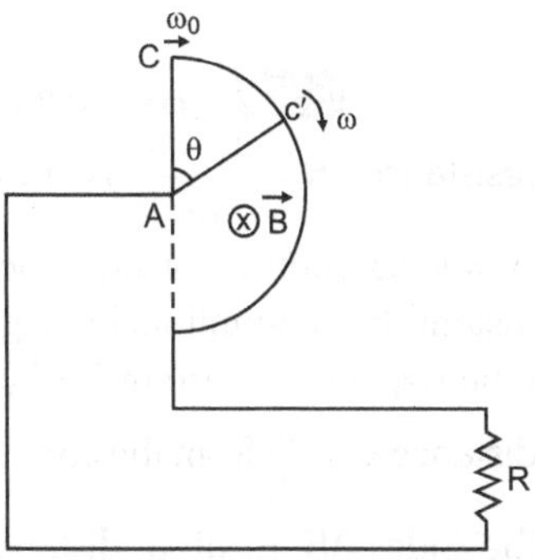

Problem 526. Switch S is closed in the circuit at time $t = 0$. Find the current through capacitor and inductor at any time t.

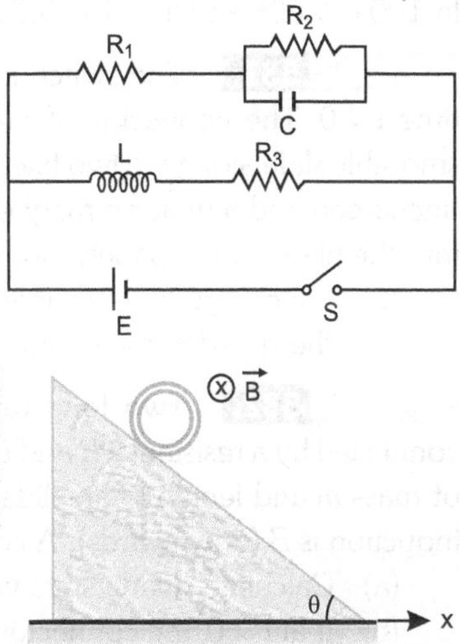

Problem 527. A small conducting loop of mass m, radius r and resistance R is released from rest along a smooth inclined plane such that the plane of the loop is perpendicular to a magnetic field which varies along x-axis as $B = B_0(1 + ax)$. Here B_0 and a are constants. Find the speed of the ring at time t.

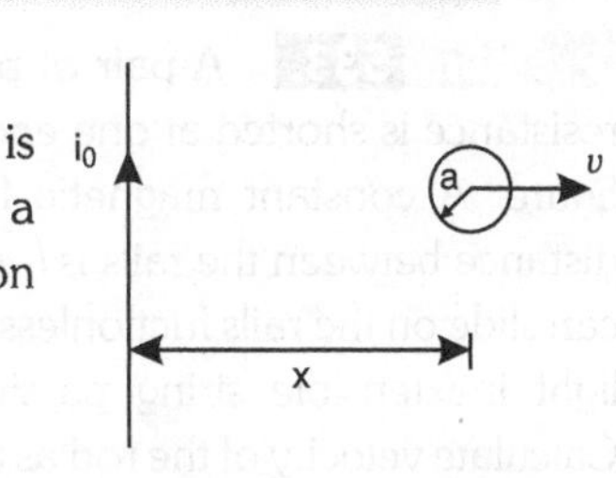

Problem 528. A small conducting loop of radius a and resistance r is pulled with velocity v perpendicular to a long straight conductor carrying a current i_0. If a constant power P is dissipated in the loop, find the variation of velocity of the loop as a function of x. Given that $x >> a$.

Problem 529. A rectangular loop is moved through a region in which the magnetic field is given by $B_y = B_z = 0, B_x = (6 - y)$ T. Find the emf in the loop as a function of time, if at $t = 0$, the loop starts moving from the position shown in the figure,

(a) at constant velocity $v = 6.5$ m/s.

(b) at constant acceleration of 2 m/s^2 from state of rest.

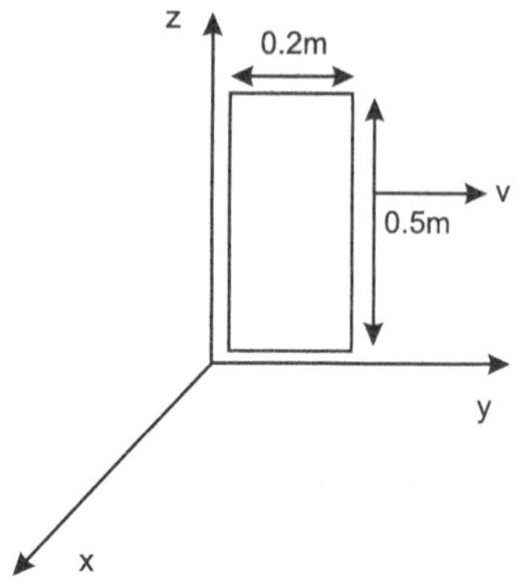

Problem 530. As shown in the figure a square shaped conducting loop having side length a and resistance $R = \frac{\mu_0 I_0 a}{2\pi}$ is moving away with a velocity $v = 5x$ (x = separation between conductor and side AB of the loop at any instant) from an infinitely long current carrying conductor in which a time dependent current $I = I_0 t$ is flowing. At $t = 0$ the side AB is at a distance $x = \frac{a}{10}$ from the conductor. Find the current in the loop when the side AB is at a distance $10a$ from the conductor. (Given: ln 100 = 4.605 and ln 1.1 = 0.095)

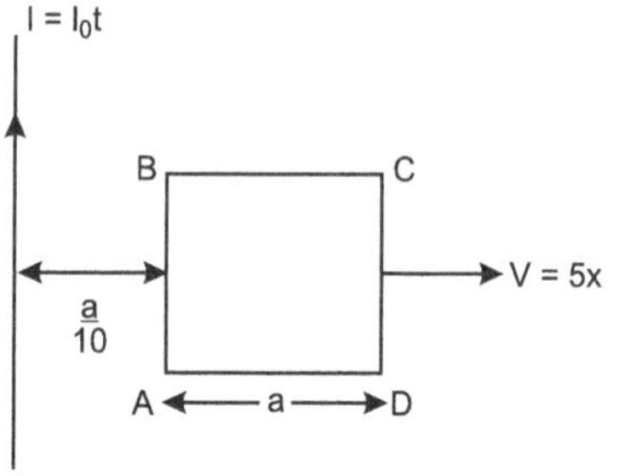

Problem 531. The switch s of the adjacent network is closed at time $t = 0$. The connector of mass m, length l and resistance R can smoothly slide along the two fixed parallel rails. The capacitor is initially uncharged and a uniform magnetic field B exists in the region directed into the plane of the paper. Find :

(a) the charge on the capacitor after a long time.

(b) the speed of the connector after a long time.

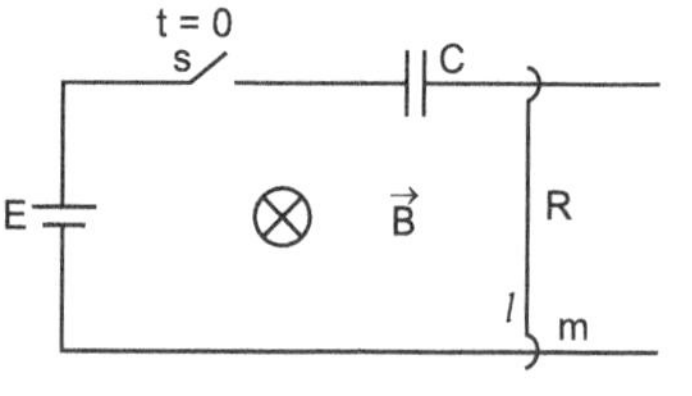

Problem 532. Two long rails are horizontal parallel to each other. On one end, the rails are connected by a resistance R and on the other end a capacitor of capacitance C is connected. A connector of mass m and length l can slide on the rails without friction. Vertical component of earth's magnetic induction is B (downwards). A constant horizontal force F starts acting on the connector horizontally.

(a) Calculate steady state velocity of the connector and steady charge on capacitor.

(b) If at $t = 0$, the connector was at rest, calculate its velocity as a function on time t.

Problem 533. A pair of parallel horizontal conducting rails of negligible resistance is shorted at one end of it and is fixed on the table as shown in the figure. A constant magnetic field B exists perpendicular to the table. The distance between the rails is l. An inductor of negligible resistance and mass m can slide on the rails frictionlessly. The inductor is connected to the mass m by a light inextensible string passing over a light frictionless pulley as shown. Calculate velocity of the rod as a function of displacement from initial position.

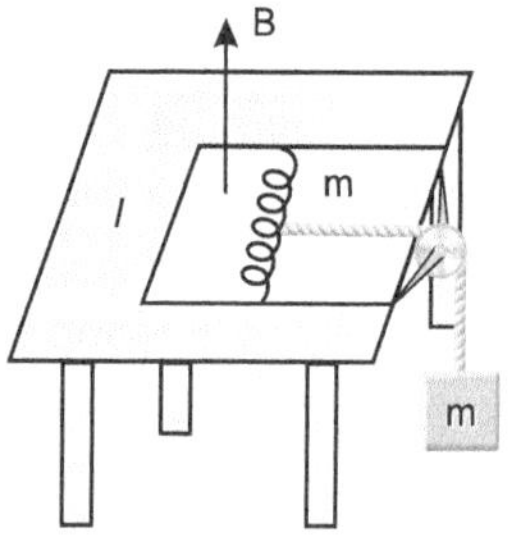

Problem 534. Two infinitely long parallel wires, having resistance per unit length λ are connected as shown in the figure. A slide wire of negligible resistance and having mass 'm' and length 'l' can slide between the parallel wires, without any frictional resistance. If the system of wires is introduced to a magnetic field of intensity 'B' (directed into the plane of paper) and the slide wire is pulled with a force which varies with the velocity of the slide wire as $F = F_0V$, then find the velocity of the slide wire as a function of the distance x travelled. (The slide wire is initially at origin and has a velocity v_o)

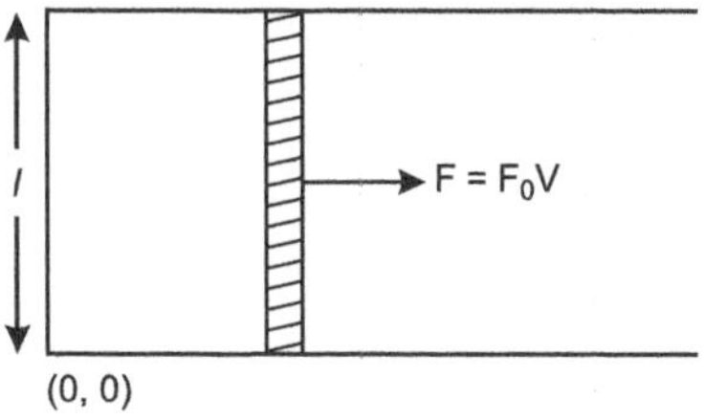

Problem 535. A homogeneous field of magnetic induction B is perpendicular to a track of gauge l which is inclined at an angle α to the horizontal. A frictionless conducting rod of mass m straddles the two rails of the track as shown in the figure.

How does the rod move, after being released from rest, if the circuit formed by the rod and the track is closed by :

(i) a resistor of resistance R,

(ii) a capacitor of capacitance C, or

(iii) a coil of inductance L?

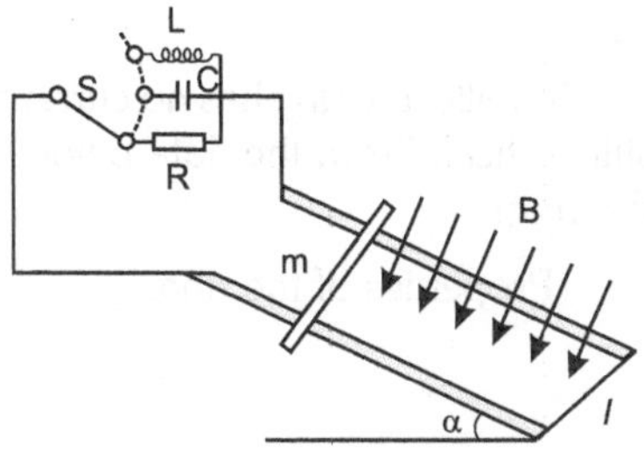

Problem 536. One end of a horizontal track of gauge l and negligible resistance, is connected to a capacitor of capacitance C charged to voltage V_0. The inductance of the assembly is negligible. The system is placed in a homogeneous, vertical magnetic field of induction B, as shown in the figure.

A frictionless conducting rod of mass m and resistance R is placed perpendicularly onto the track. The polarity of the capacitor is such that the rod is repelled from the capacitor when the switch is turned over.

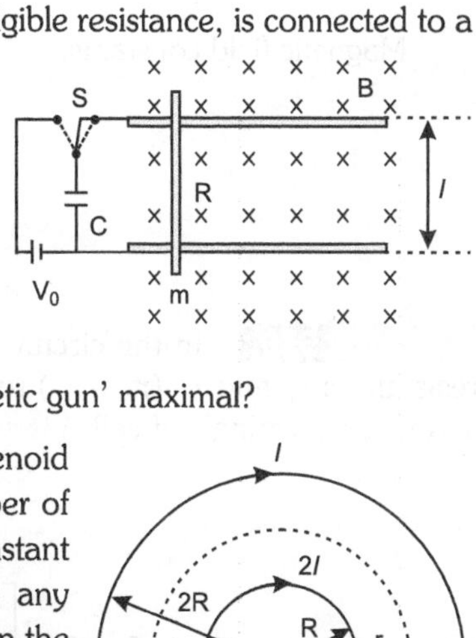

(i) What is the maximum velocity of the rod ?

(ii) Under what conditions is the efficiency of this 'electromagnetic gun' maximal?

Problem 537. A long solenoid contains another coaxial solenoid (whose radius R is half of its own). Their coils have the same number of turns per unit length and initially both carry no current. At the same instant currents start increasing linearly with time in both solenoids. At any moment the current flowing in the inner coil is twice as large as that in the outer one and their directions are the same. As a result of the increasing currents a charged particle, initially at rest between the solenoids, starts moving along a circular trajectory (see figure). What is the radius r of the circle?

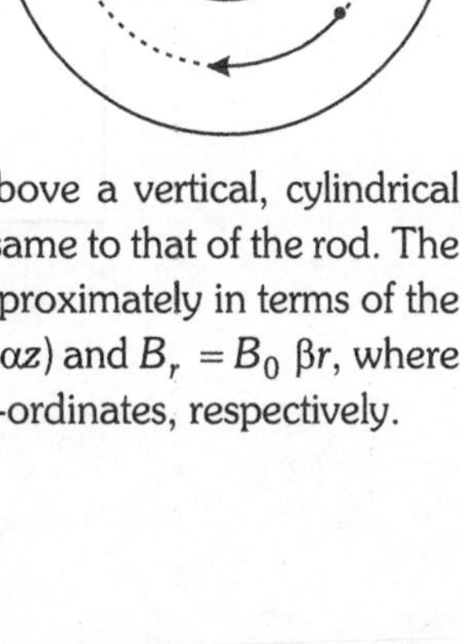

Problem 538. A thin superconducting (zero resistance) ring is held above a vertical, cylindrical magnetic rod, as shown in the figure. The axis of symmetry of the ring is the same to that of the rod. The cylindrically symmetrical magnetic field around the ring can be described approximately in terms of the vertical and radial components of the magnetic field vector as $B_z = B_0 (1 - \alpha z)$ and $B_r = B_0 \beta r$, where B_0, α and β are constants and z and r are the vertical and radial position co-ordinates, respectively.

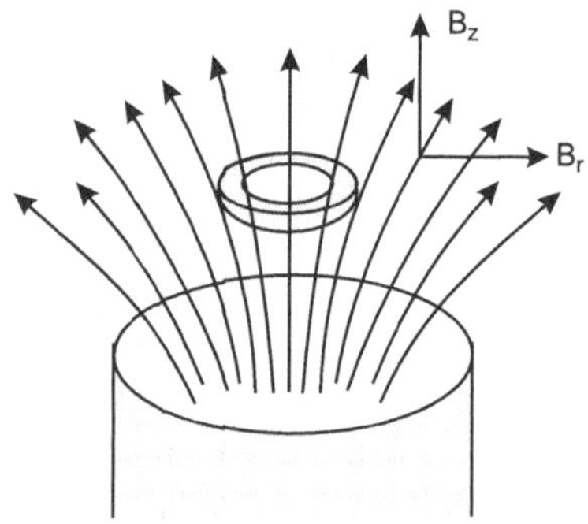

Initially, the ring has no current flowing in it. When released, it starts to move downwards with its axis still vertical. From the data below, determine how the ring moves subsequently? What current flows in the ring?

Properties of the ring:	mass	$m = 50$ mg
	radius	$r_0 = 0.5$ cm
	inductance	$L = 1.3 \times 10^{-8}$ H
Initial co-ordinates of the centre of the ring:		$z = 0$
		$r = 0$
Magnetic field constants:		$B_0 = 0.01$ T
		$\alpha = 2\ \text{m}^{-1}$
		$\beta = 32\ \text{m}^{-1}$

IIT JEE PROBLEMS

Problem 539. In the circuit shown A and B are two cells of same emf E but different internal resistances r_1 and r_2 $(r_1 > r_2)$ respectively. Find the value of R such that the potential difference across the terminals of cell A is zero a long time after the key K is closed. **(JEE 2004)**

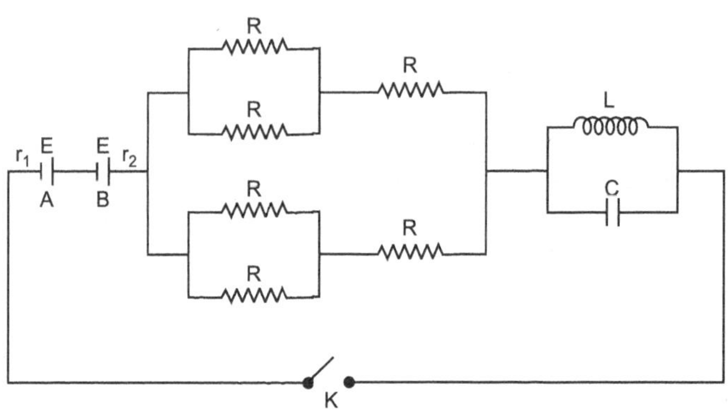

Problem 540. In an LR series circuit, a sinusoidal voltage $V = V_0 \sin \omega t$ is applied. It is given that $L = 35$ mH, $R = 11\,\Omega$, $V_{rms} = 220$ V, $\frac{\omega}{2\pi} = 50$ Hz and $\pi = \frac{22}{7}$. Find the amplitude of current in the steady state and obtain the phase difference between the current and the voltage. Also plot the variation of current for one cycle on the given graph. **(JEE 2004)**

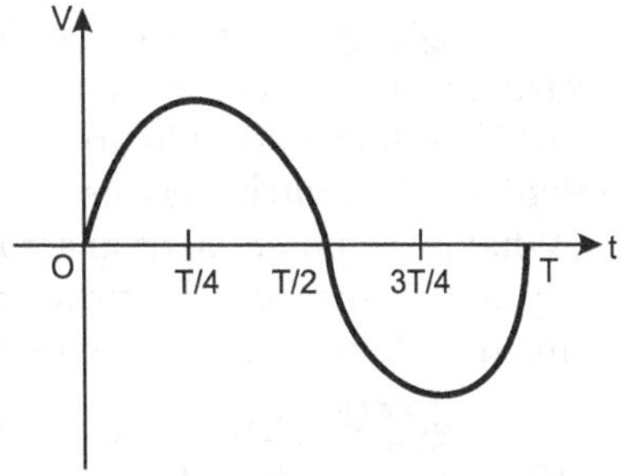

Problem 541. Two infinitely long parallel wires carrying current $I = I_0 \sin \omega t$ in opposite directions are placed a distance $3a$ apart. A square loop of side a of negligible resistance with a capacitor of capacitance C is placed in the plane of wires as shown. Find the maximum current in the square loop. Also sketch the graph showing the variation of charge on the upper plate of the capacitor as a function of time for one complete cycle taking anticlockwise direction for the current in the loop as positive. **(JEE 2003)**

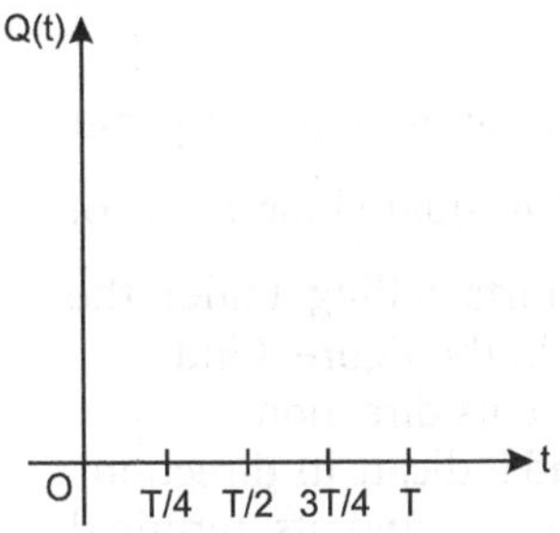

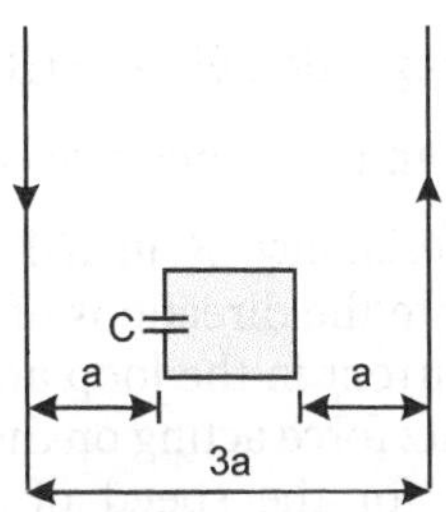

Problem 542. A metal bar AB can slide on two parallel thick metallic rails separated by a distance l. A resistance R and an inductance L are connected to the rails as shown in the figure. A long straight wire carrying a constant current I_0 is placed in the plane of the rails and perpendicular to them as shown. The bar AB is held at rest at a distance x_0 from the long wire. At $t = 0$, it is made to slide on the rails away from the wire. Answer the following questions.

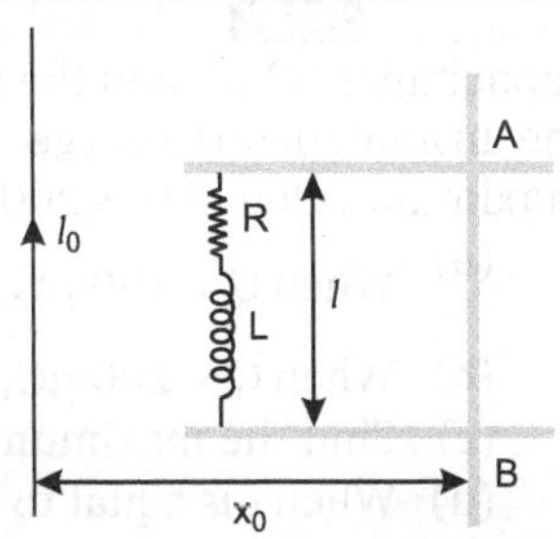

(a) Find a relation among i, $\frac{di}{dt}$ and $\frac{d\phi}{dt}$, where i is the current in the circuit and ϕ is the flux of the magnetic field due to the long wire through the circuit.

(b) It is observed that at time $t = T$, the metal bar AB is at a distance of $2x_0$ from the long wire and the resistance R carries a current i_1. Obtain an expression for the net charge that has flown through resistance R from $t = 0$ to $t = T$.

(c) The bar is suddenly stopped at time T. The current through resistance R is found to be $\frac{i_1}{4}$ at time $2T$. Find the value of L/R in terms of the other given quantities. **(JEE 2002)**

Problem 543. An inductor of inductance $L = 400$ mH and resistors of reistances $R_1 = 2\ \Omega$ and $R_2 = 2\ \Omega$ are connected to a battery of e.m.f. $E = 12$ V as shown in the figure. The internal resistance of the battery is negligible. The switch S is closed at time $t = 0$.

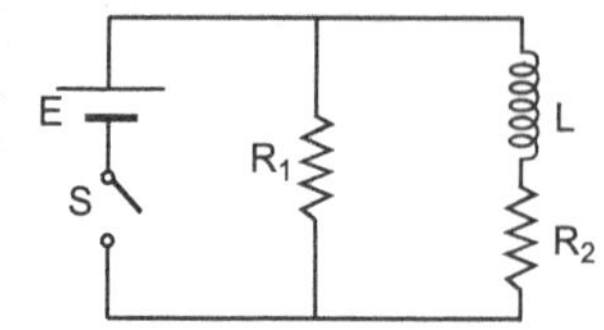

What is the potential drop across L as a function of time ? After the steady state is reached, the switch is opened. What is the direction and the magnitude of current through R_1 as a function of time ? **(JEE 2001)**

Problem 544. A thermocole vessel contains 0.5 kg of distilled water at 30°C. A metal coil of area $5 \times 10^{-3}\ m^2$, number of turns 100, mass 0.06 kg and resistance 1.6 Ω is lying horizontally at the bottom of the vessel. A uniform time varying magnetic field is set up to pass vertically through the coil at time $t = 0$. The field is first increased from zero to 0.8 T at a constant rate between 0 and 0.2 s and then decreased to zero at the same rate between 0.2 and 0.4 s. The cycle is repeated 12000 times. Make sketches of the current through the coil and the power dissipated in the coil as a function of time for the first two cycles. Clearly indicate the magnitudes of the quantities on the axes. Assume that no heat is lost to the vessel or the surroundings. Determine the final temperature of the water under thermal equilibrium. Specific heat of metal = 500 J/kg-K and the specific heat of water = 4200 J/kg-K. Neglect the inductance of coil. **(JEE 2000)**

Problem 545. A magnetic field $B = (B_0 y/a)\ \hat{k}$ is acting into the paper in the $+Z$ direction. B_0 and a are positive constants. A square loop $EFGH$ of side a, mass m and resistance R in X-Y plane starts falling under the influence of gravity. Note the directions of X and Y in the figure. Find :

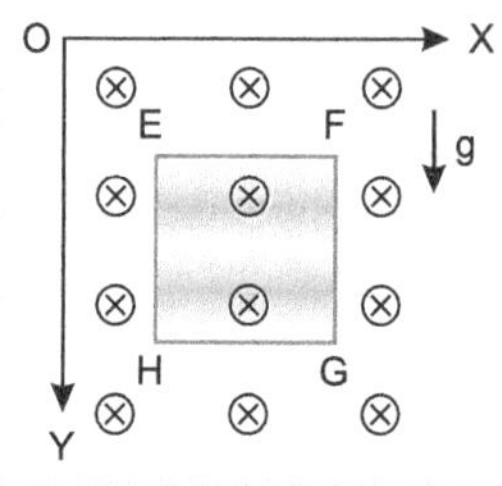

(a) the induced current in the loop and indicate its direction
(b) the total Lorentz force acting on the loop and indicate its direction
(c) an expression for the speed of the loop $v_{(t)}$ and its terminal velocity. **(JEE 1999)**

Problem 546. An inductor of inductance 2.0 mH is connected across a charged capacitor of capacitance 5.0 μF and the resulting LC circuit is set oscillating at its natural frequency. Let Q denote the instantaneous charge on the capacitor and i the current in the circuit. It is found that the maximum value of Q is 200 μC.

(a) When $Q = 100$ μC, what is the value of $\left|\frac{di}{dt}\right|$?
(b) When $Q = 200\ \mu C$, what is the value of i ?
(c) Find the maximum value of i.
(d) When i is equal to one-half its maximum value, what is the value of $|Q|$? **(JEE 1998)**

Problem 547. A pair of parallel horizontal conducting rails of negligible resistance shorted at one end is fixed on a table. The distance between the rails is L. A conducting massless rod of resistance R can slide on the rails frictionlessly. The rod is tied to a massless string which passes over a pulley fixed to the edge of the table. A mass m tied to the other end of the string hangs vertically. A constant magnetic field B exists perpendicular to the table. If the system is released from rest, Calculate :

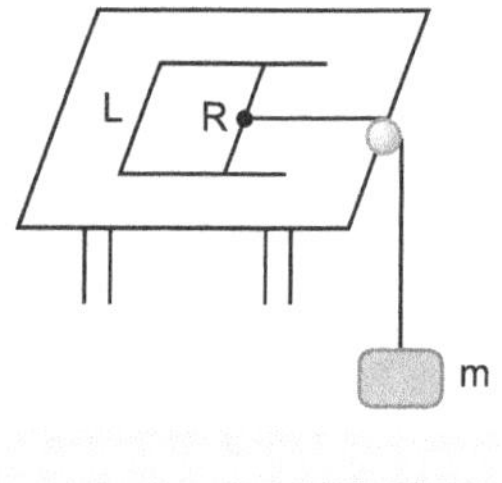

(i) the terminal velocity achieved by the rod, and
(ii) the acceleration of the mass at the instant when the velocity of the rod is half the terminal velocity. **(JEE 1997)**

Problem 548. A metal rod OA of mass m and length r kept rotating with a constant angular speed ω in a vertical plane about a horizontal axis at the end O. The free end A is arranged to slide without friction along a fixed conducting circular ring in the same plane as that of rotation. A uniform and constant magnetic induction $\vec{B}$ is applied perpendicular and into the plane of rotation as shown in figure. An inductor L and an external resistance R are connected through a switch S between the point O and a point C on the ring to form an electrical circuit. Neglect the resistance of the ring and the rod. Initially, the switch is open.

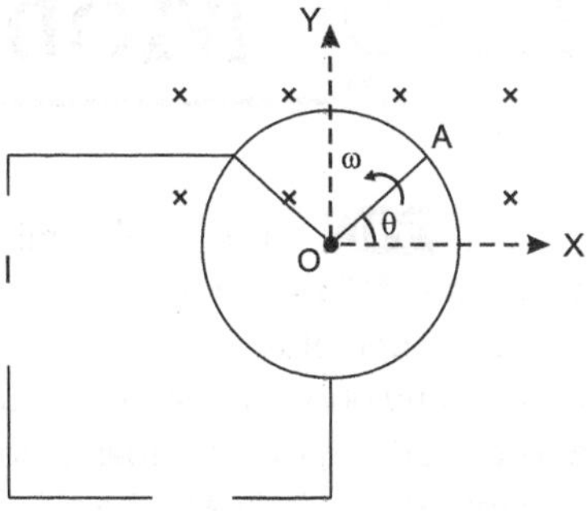

(a) What is the induced emf across the terminals of the switch ?

(b) The switch S is closed at time $t = 0$.

(i) Obtain an expression for the current as a function of time.

(ii) In the steady state, obtain the time dependence of the torque required to maintain the constant angular speed. Given that the rod OA was along the positive X-axis at $t = 0$.

(JEE 1995)

Problem 549. Two parallel vertical metallic rails AB and CD are separated by 1 m. They are connected at two ends by resistances R_1 and R_2 as shown in Figure. A horizontal metallic bar of mass 0.2 kg slides without friction vertically down the rails under the action of gravity. There is a uniform horizontal magnetic field of 0.6 tesla perpendicular to the plane of the rails. It is observed that when the terminal velocity is attained, the powers dissipated in R_1 and R_2 are 0.76 watt and 1.2 watt respectively. Find the terminal velocity of the bar L and the values of R_1 and R_2.

(JEE 1994)

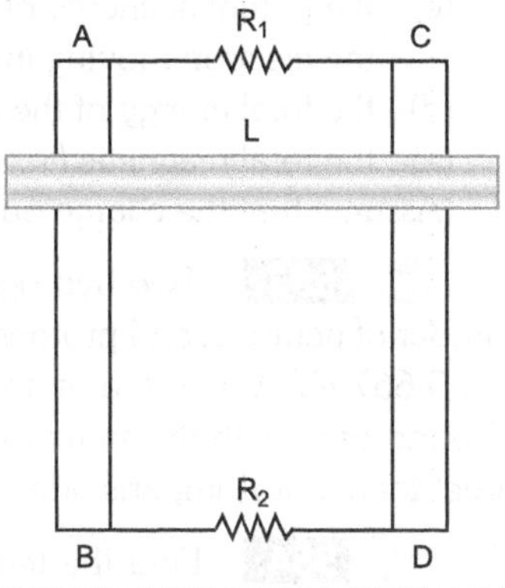

Problem 550. A rectangular frame $ABCD$, made of a uniform metal wire, has a straight connection between E and F made of the same wire, as shown in figure. $AEFD$ is a square of side 1 m, and $EB = FC = 0.5$ m The entire circuit is placed in a steadily increasing, uniform magnetic field directed into the plane of the paper and normal to it. The rate of change of the magnetic field is 1 T/s. The resistance per unit length of the wire is 1 Ω/m. Find the magnitudes and directions of the currents in the segments AE, BE and EF.

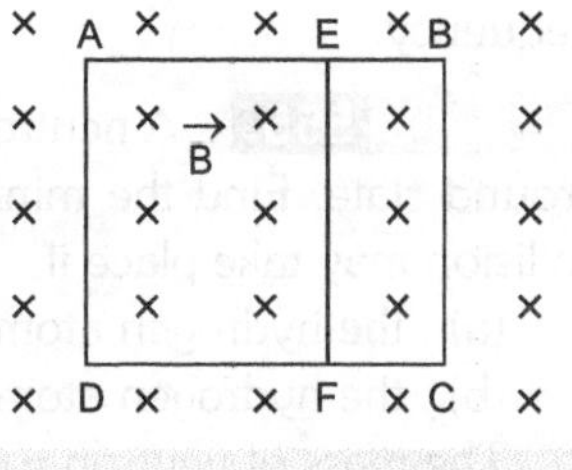

(JEE 1993)

CHAPTER 16

MODERN PHYSICS

Problem 551. The peak emission from a black body at a certain temperature occurs at a wavelength of 9000 Å. On increasing its temperature the total radiation emitted is increased 81 times. At the initial temperature when the peak radiation from the black body is incident on a metal surface, it does not cause any photoemission from the surface. After the increase of temperature the peak radiation from the black body caused photoemission. To bring these photoelectrons to rest a potential equivalent to the excitation energy between the $n = 2$ to $n = 3$. Bohr levels of hydrogen atom is required. Find the work function of the metal.

Problem 552. An electron is orbiting in a circular orbit of radius r under the influence of a constant magnetic field of strength B. Assuming that Bohr's postulate regarding the quantisation of angular momentum holds good for this electron, find:

(a) the allowed values of the radius 'r' of the orbit.
(b) the kinetic energy of the electron in orbit.
(c) the potential energy of interaction between the magnetic moment of the orbital current due to the electron moving in its orbit and the magnetic field B.
(d) the total energy of the allowed energy levels.
(e) the total magnetic flux due to the magnetic field B passing through the nth orbit.

(Assume that the charge on the electron is $-e$ and the mass of the electron is m).

Problem 553. Two hydrogen like atoms A and B are of different masses and each contains equal number of neutrons and protons. The energy difference between the first Balmer lines emitted by A and B is 5.667 eV. When the atoms A and B moving with the same velocity, strike a heavy target, they rebound back with the same velocity. In this process the atom B imparts twice the momentum to the target than that A imparts. Identify the atoms.

Problem 554. Find the frequency of a photon of frequency ν after falling through a height H in earth's gravitational field. Assume that a photon manifest any increase in its energy by changing its frequency.

Problem 555. A neutron moving with speed υ makes a head on collision with a hydrogen atom in ground state. Find the minimum velocity of the neutron for which inelastic (completely or partially) collision may take place if

(a) the hydrogen atom is initially at rest
(b) the hydrogen atom is also moving with same velocity υ towards the neutron.

The mass of neutron $\approx$ mass of hydrogen $= 1.67 \times 10^{-27}$ kg.

Problem 556. A photon with an energy of 4.9 eV ejects photoelectrons from tungsten. When the ejected electron enters a constant magnetic field of strength $B = 2.5$ mT at an angle of 60° with the field direction, the maximum pitch of the helix described by the electron is found to be 2.7 mm. Find the work function of the metal in electron volts. Given that specific charge of electron is 1.76×10^{11} C/kg.

Problem 557. Photons of energy 3 eV fall on a photosensitive metal of work function 1 eV. The de-Broglie wavelength of the most energetic ejected electron is found to be $(1.09)^2 \sqrt{10}$ times the wavelength of K_α, X-ray coming from a certain element A when it is bombarded by fast moving electron. Find the atomic number (Z) of the element A.

Take h (Planck constant) $= 6.54 \times 10^{-34}$ J-s, mass of the electron $= 9 \times 10^{-31}$ kg, Rydberg's constant $= 1.09 \times 10^7 \text{ m}^{-1}$.

Problem 558. A stable nuclei C is formed from two radioactive nuclei A and B with decay constant of λ_1 and λ_2 respectively. Initially the number of nuclei of A is N_0 and that of B is zero. Nuclei B are produced at a constant rate of P. Find the number of nuclei of C after time t.

Problem 559. Demonstrate that the frequency ν of a photon emerging when an electron jumps between neighbouring circular orbits of a hydrogen like atom satisfies the inequality $\nu_{n+1} < \nu < \nu_n$ where ν_n and ν_{n+1} are the frequencies of revolution of that electron around the nucleus along the circular orbits. Also show that for large values of n all these three are almost equal.

Problem 560. A particle of mass m moves along a circular orbit in a centrosymmetrical potential field $U(r) = \dfrac{kr^2}{2}$. Using the Bohr's quantization condition, find the permissible orbital radii and energy levels of that particle.

Problem 561. Taking into account the motion of the nucleus of a hydrogen atom, find the expressions for the electron's binding energy in the ground state. How much (in per cent) do the binding energy obtained without taking into account the motion of the nucleus differ from the more accurate corresponding value of this quantity. Given $\dfrac{m}{M} = 0.00055$, where m and M are the masses of an electron and a proton.

Problem 562. Consider the β-decay $\;^{198}\text{Au} \to \;^{198}\text{Hg} + \beta^- + \bar{\nu}$

where ^{198}Hg represents a mercury nucleus in an excited state of energy 1.088 MeV above the ground state. What can be the maximum kinetic energy of the electron emitted. The atomic mass of ^{198}Au is 197.968233 u and that of ^{198}Hg is 197.966760 u. 1 u = 931 MeV/c^2.

Problem 563. A sample of Uranium is a mixture of two isotopes $_{92}\text{U}^{234}$ and $_{92}\text{U}^{238}$ present in the ratio 10% and 90% by weight. The half lives of these isotopes are 2.5×10^5 years and 4.5×10^5 years respectively. Calculate the contribution to activity in percentage of each isotope in this sample.

Problem 564. A number N_0 of atoms of a radio active element are placed inside a closed volume. The radioactive decay constant for the nucleus of this element is λ_1. The daughter nucleus that form as a result of the decay process are assumed to be radioactive too with a radioactive decay constant λ_2. Determine the time variation of the number of such nucleus. Consider two limiting cases $\lambda_1 >> \lambda_2$ and $\lambda_1 << \lambda_2$.

Problem 565. A nucleus at rest undergoes α-decay according to the equation.

$$^{225}_{92}X \to Y + \alpha$$

At time $t = 0$, the emitted α-particle enters in a region of space where a uniform magnetic field $\vec{B} = B_0 \hat{i}$ and electric field $\vec{E} = E_0 \hat{i}$ exist. The α-particle enters in the region with velocity $\vec{v} = v_0 \hat{j}$ from

$x = 0$. At time $t = \sqrt{3} \times 10^7 \dfrac{m_\alpha}{q_\alpha E_0}$ sec, the particle was observed to have speed twice the initial velocity v_0 then find

(a) the velocity of α-particle at time t.
(b) the initial velocity v_0 of the α-particle.
(c) the binding energy per nucleon of α-particle.

Given that : $m(Y) = 221.03\,u$, $m(\alpha) = 4.003\,u$, $m(n) = 1.009\,u$, $m(p) = 1.0084\,u$

charge on α-particle $q_\alpha = 6.4 \times 10^{-19}$ C

and $1\,u = 1.67 \times 10^{-27}$ kg $= 931$ MeV/c^2

Problem 566. A solution contains a mixture of two isotopes A (half life = 10 days) and B (half life = 5 days). Total activity of the mixture is 10^{10} disintegrations per second at time $t = 0$, the activity reduces to 20% in 20 days. Find

(a) the initial activities of A and B.
(b) the ratio of initial number of their nuclei.

Problem 567. Consider an excited hydrogen atom in state n moving with a velocity v ($v << c$). It emits a photon in the direction of its motion and changes its state to a lower state m. Apply momentum and energy conservation principle to calculate the frequency ν of the emitted radiation. Compare this with the frequency ν_0 emitted if the atom were at rest.

Problem 568. Polonium ($_{84}Po^{210}$) emits α-particles and is converted into lead ($_{82}Pb^{206}$). This reaction is used for producing electric power. Polonium has half life 138.6 days. Assuming an efficiency of 10% for the thermoelectric machine. How much polonium is required to produce 1.2×10^7 J of electric energy per day at the end of 693 days. Also find the initial activity of the material. Masses of nuclei are

$$Po^{210} = 209.98264 \text{ amu}$$
$$Pb^{206} = 205.97440 \text{ amu}$$
$$He^4 = 4.00260 \text{ amu}$$
$$1 \text{ amu} = 931 \text{ MeV/c}^2$$
$$\text{Avogadro's number} = 6 \times 10^{23} \text{ per mol.}$$

Problem 569. A radio nuclide with half life $T = 14.3$ days is produced in a reactor at a constant rate $q = 10^9$ nuclei per second. How soon after the beginning of production of that radionuclide will, its activity be equal to $A = 10^8$ disintegrations per second. Plot a rough graph of its activity with time.

Problem 570. A body of mass m_0 is placed on a smooth horizontal surface. The mass of the body is decreasing exponentially with disintegration constant λ. Assuming that the mass is ejected backwards with a relative velocity u. Initially the body was at rest. Find the velocity of it after time t.

Problem 571. A radionuclide with disintegration constant λ is produced in a reactor at a constant rate α nuclei per second. During each decay energy E_0 is released. 20% of this energy is utilised in increasing the temperature of water. Find the increase in temperature of m mass of water in time t. Specific heat of water is s. Assume that there is no loss of energy through water surface.

Problem 572. A parallel plate capacitor of capacitance 100 μF and a separation of 1 cm is charged with a battery to a potential difference of 10 V. The battery is then disconnected. Electromagnetic wave is now incident on negatively charged plate which emits electrons with kinetic energies ranging from 0 to 1.5 eV.

The electrons are attracted to the positive plate. The current which flows between the two plates varies with time t as shown in figure.

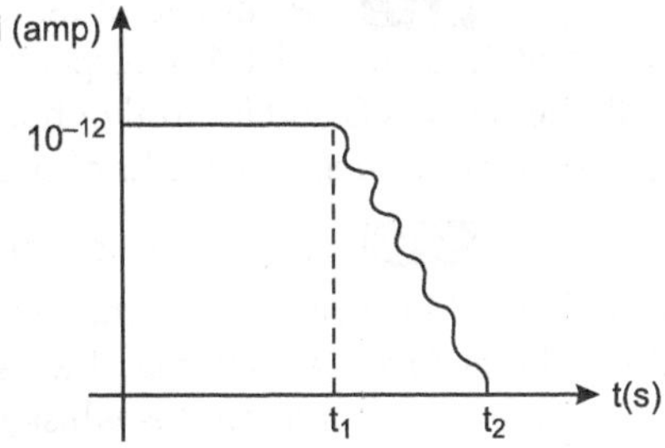

(a) What is the potential difference between the plates at time t_1

(b) What is the numerical value of t_1

(c) What is the potential difference between the plates for $t > t_2$

Problem 573. A photon of momentum p_1 is absorbed by an electron initially at rest, which instantly recoils and emits a second photon of momentum p_2 in a direction making an angle θ with the direction of p_1. The electron at rest has an energy of m_0c^2 where m_0 is its rest mass and an energy of $c\sqrt{p^2 + m_0^2c^2}$ when momentum p. Here c is speed of light. Show that

$$\frac{1}{p_2} - \frac{1}{p_1} = \frac{1}{m_0c}(1 - \cos\theta)$$

Problem 574. A NPN transistor is connected in common emitter configuration in which collector supply is 8 V and the voltage drop across the load resistance of 800 Ω connected in collector circuit is 0.8 V. If current amplification factor β = 25/26, determine collector emitter voltage and base current. If the internal resistance of the transistor is 200 Ω, calculate the voltage gain and power gain.

Problem 575. A NPN transistor in a common emitter mode is used as a simple voltage amplifier with a collector current of 4 mA. The terminal of a 8V battery is connected to the collector through a load resistance R_L and to the base through a resistance R_B. The collector emitter voltage $V_{CE} = 4$ V, base emitter voltage $V_{BE} = 0.6$ V and current amplification factor is β = 100. Calculate the values of R_L and R_B.

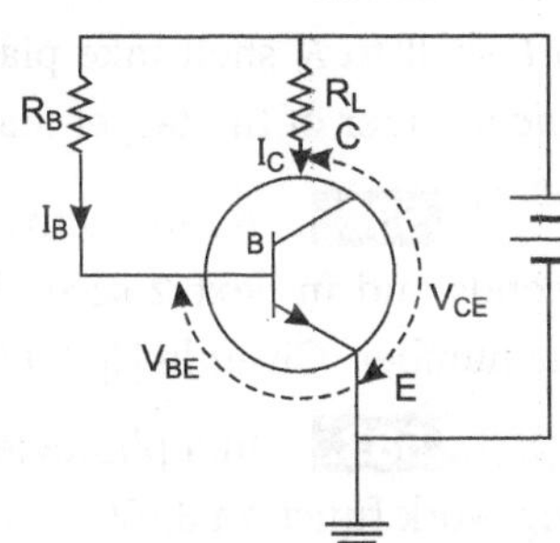

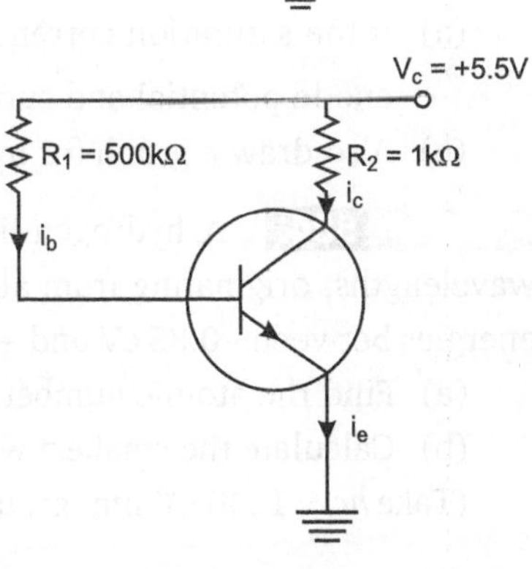

Problem 576. In the circuit shown in figure, the base current $i_b = 10$ μA and the collector current $i_c = 5.2$ mA. Can this transistor be used as an amplifier ? Explain why or why not ?

Problem 577. A sample of radioactive material decays simultaneously by two processes A and B with half lives 1/2 and 1/4 hr respectively. For first half hour it decays with the process A, next one hour with the process B and for further half an hour with both A and B. If originally there were N_0 nuclei, find the number of nuclei after 2 hr of such decay.

Problem 578. Natural water contains a small amount of tritium (^3_1H). This isotope beta decays with a half life of 20 years. A mountaineer find debris of some earlier unsuccessful attempt. Among other things he finds a sealed bottle of whisky.

On return he analyses the whisky and finds that it contains only 2.5 per cent of the ^3_1H radioactivity as compared to a recently purchased bottle marked 10 years old. Estimate the time of that unsuccessful attempt.

IIT JEE PROBLEMS

Problem 579. A rock is 1.5×10^9 years old. The rock contains ^{238}U, which disintegrates to form ^{206}Pb. Assume that there was no ^{206}Pb in the rock initially and it is the only stable product formed by the decay. Calculate the ratio of number of nuclei of ^{238}U to that of ^{206}Pb in the rock. Half-life of ^{238}U is 4.5×10^9 years. ($2^{1/3} = 1.259$) **(JEE 2004)**

Problem 580. Wavelengths belonging to Balmer series lying in the range of 450 nm to 750 nm were used to eject photoelectrons from a metal surface whose work function is 2.0 eV. Find (in eV) the maximum kinetic energy of the emitted photoelectrons. Take $hc = 1242$ eV nm. **(JEE 2004)**

Problem 581. Characteristic X-rays of frequency 4.2×10^{18} Hz are produced when transitions from L shell to K shell take place in a certain target material. Use Mosley's law to determine the atomic number of the target material. Given Rydberg constant $R = 1.1 \times 10^7\ m^{-1}$. **(JEE 2003)**

Problem 582. A radioactive element decays by β-emission. A detector records n beta particles in 2 seconds and in next 2 seconds it records $0.75n$ beta particles. Find mean life correct to nearest whole number. Given $\ln|2| = 0.6931$, $\ln|3| = 1.0986$. **(JEE 2003)**

Problem 583. In a photoelectric experiment setup, photons of energy 5 eV falls on the cathode having work function 3 eV.

(a) If the saturation current is $i_A = 4\ \mu A$ for intensity $10^{-5}\ W/m^2$, then plot a graph between anode potential and current.

(b) Also draw a graph for intensity of incident radiation $2 \times 10^{-5}\ W/m^2$. **(JEE 2003)**

Problem 584. A hydrogen-like atom (described by the Bohr model) is observed to emit six wavelengths, originating from all possible transitions between a group of levels. These levels have energies between –0.85 eV and –0.544 eV (including both these values).

(a) Find the atomic number of the atom.

(b) Calculate the smallest wavelength emitted in these transitions.

(Take $hc = 1240$ eV-nm, ground state energy of hydrogen atom $= -13.6$ eV). **(JEE 2002)**

Problem 585. Two metallic plates A and B, each of area 5×10^{-4} m^2, are placed parallel to each other at a separation of 1 cm. Plate B carries a positive charge of 33.7×10^{-12} C. A monochromatic beam of light, with photons of energy 5 eV each, starts falling on plate A at $t = 0$ so that 10^6 photons fall on it per square metre per second. Assume that one photoelectron is emitted for every 10^6 incident photons. Also assume that all the emitted photoelectrons are collected by plate B and the work function of plate A remains constant at the value 2 eV. Determine :

(a) the number of photoelectrons emitted up to $t = 10$ s.

(b) the magnitude of the electric field between the plates A and B at $t = 10$ s.

(c) the kinetic energy of the most energetic photo-electron emitted at $t = 10$ s when it reaches plate B. Neglect the time taken by the photoelectron to reach plate B.
Take $\varepsilon_0 = 8.85 \times 10^{-12}$ $C^2/N\text{-}m^2$ **(JEE 2002)**

Problem 586. A radioactive nucleus X decays to a nucleus Y with a decay constant $\lambda_X = 0.1$ sec^{-1}. Y further decays to a stable nucleus Z with a decay constant $\lambda_Y = 1/30$ sec^{-1}. Initially, there are only X nuclei and their number is $N_0 = 10^{20}$. Set up the rate equations for the populations of X, Y and Z. The population of the Y nucleus as a function of time is given by $N_Y(t) = \{N_0 \lambda_X / (\lambda_X - \lambda_Y)\} \{\exp(-\lambda_Y t) - \exp(-\lambda_X t)\}$. Find the time at which N_Y is maximum and determine the population of X and Z at that instant. **(JEE 2001)**

Problem 587. In a nuclear reactor ^{235}U undergoes fission liberating 200 MeV of energy. The reactor has a 10% efficiency and produces 1000 MW power. If the reactor is to function for 10 years, find the total mass of uranium required. **(JEE 2001)**

Problem 588. A nucleus at rest undergoes a decay emitting an α-particle of de-Broglie wavelength, $\lambda = 5.76 \times 10^{-15}$ m. If the mass of the daughter nucleus is 223.610 amu and that of the α-particle is 4.002 amu, determine the total kinetic energy in the final state. Hence, obtain the mass of the parent nucleus in amu.
(1 amu = 931.470 MeV/c^2) **(JEE 2001)**

Problem 589. (a) A hydrogen like atom of atomic number Z is in an excited state of quantum number $2n$. It can emit a maximum energy photon of 204 eV. If it makes a transition to quantum state n, a photon of energy 40.8 eV is emitted. Find n, Z and the ground state energy (in eV) of this atom. Also calculate the minimum energy (in eV) that can be emitted by this atom during de-excitation. Ground state energy of hydrogen atom is −13.6 eV.

(b) When a beam of 10.6 eV photons of intensity 2.0 W/m^2 falls on a platinum surface of area 1.0×10^{-4} m^2 and work function 5.6 eV, 0.53% of the incident photons eject photoelectrons. Find the number of photoelectrons emitted per second and their minimum and maximum energies (in eV). Take 1 eV = 1.6×10^{-19} J. **(JEE 2000)**

Problem 590. Photoelectrons are emitted when 400 nm radiation is incident on a surface of work function 1.9 eV. These photoelectrons pass through a region containing α-particle. A maximum energy electron combines with an α-particle to form a He^+ ion, emitting a single photon in this process. He^+ ions thus formed are in their fourth excited state. Find the energies in eV of the photons lying in the 2 to 4 eV range, that are likely to be emitted during and after the combination.
[Take $h = 4.14 \times 10^{-15}$ eV-s] **(JEE 1999)**

Problem 591. Nuclei of a radioactive element A are being produced at a constant rate α. The element has a decay constant λ. At time $t = 0$, there are N_0 nuclei of the element.

(a) Calculate the number N of nuclei of A at time t

(b) If $\alpha = 2N_0\lambda$, calculate the number of nuclei of A after one half life of A, and also the limiting value of N as $t \to \infty$. **(JEE 1998)**

Problem 592. Assume that the de-Broglie wave associated with an electron can form a standing wave between the atoms arranged in a one dimensional array with nodes at each of the atomic sites. It is found that one such standing wave is formed if the distance d between the atoms of the array is 2 Å. A similar standing wave is again formed if d is increased to 2.5 Å but not for any intermediate value of d. Find the energy of the electron in eV and the least value of d for which the standing wave of the type described above can form. **(JEE 1997)**

Problem 593. The element Curium $^{248}_{96}Cm$ has a mean life of 10^{13} second. Its primary decay modes are spontaneous fission and α-decay, the former with a probability of 8% and the latter with a probability of 92%. Each fission releases 200 MeV of energy. The masses involved in decay are as follows :

$$^{248}_{96}Cm = 248.072220\,u, \quad ^{244}_{94}Pu = 244.064100\text{ u and } ^{4}_{2}He = 4.002603\text{ u.}$$

Calculate the power output from a sample of 10^{20} Cm atoms. ($1\text{ u} = 931\text{ meV/C}^2$) **(JEE 1997)**

Problem 594. In an ore containing Uranium, the ratio of ^{238}U to ^{206}Pb nuclei is 3. Calculate the age of the ore, assuming that all the lead present in the ore is the final stable product of ^{238}U. Take the half life of ^{238}U to be 4.5×10^9 years. **(JEE 1997, Cancelled)**

Problem 595. (a) An electron in a hydrogen like atom is in an excited state. It has a total energy of – 3.4 eV. Calculate :

(i) the kinetic energy

(ii) the de-Broglie wavelength of the electron

(b) At a given instant there are 25% undecayed radioactive nuclei in a sample. After 10 second the number of undecayed nuclei reduces to 12.5%. Calculate :

(i) mean life of the nuclei

(ii) the time in which the number of undecayed nuclei will further reduce to 6.25% of the reduced number. **(JEE 1996)**

Problem 596. In a photoelectric effect set-up a point source of light of power 3.2×10^{-3} W emits monoenergetic photons of energy 5.0 eV. The source is located at a distance of 0.8 m from the centre of a stationary metallic sphere of work function 3.0 eV and of radius 8.0×10^{-3} m. The efficiency of photoelectron emission is one for every 10^6 incident photons. Assume that the sphere is isolated and initially neutral and that photoelectrons are instantly swept away after emission.

(a) Calculate the number of photoelectrons emitted per second.

(b) Find the ratio of the wavelength of incident light to the de-Broglie wavelength of the fastest photoelectrons emitted.

(c) It is observed that the photoelectron emission stops at a certain time t after the light source is switched on why ?

(d) Evaluate the time t. **(JEE 1995)**

Problem 597. A hydrogen like atom (atomic number Z) is in a higher excited state of quantum number n. The excited atom can make a transition to the first excited state by successively emitting two photons of energy 10.2 and 17.0 eV respectively. Alternately, the atom from the same excited state can make a transition to the second excited state by successively emitting two photons of energies 4.25 eV and 5.95 eV respectively. Determine the values of n and Z. (Ionization energy of H-atom = 13.6 eV)
(JEE 1994)

Problem 598. A small quantity of solution containing Na^{24} radio nuclide (half life = 15 hour) of activity 1.0 microcurie is injected into the blood of a person. A sample of the blood of volume 1 cm^3 taken after 5 hour shows an activity of 296 disintegrations per minute. Determine the total volume of the blood in the body of the person. Assume that the radioactive solution mixes uniformly in the blood of the person.

(1 curie = 3.7×10^{10} disintegrations per second) **(JEE 1994)**

Problem 599. A neutron of kinetic energy 65 eV collides inelastically with a singly ionized helium atom at rest. It is scattered at an angle of 90° with respect of its original direction.

(i) Find the allowed values of the energy of the neutron and that of the atom after the collision
(ii) If the atom gets de-excited subsequently by emitting radiation, find the given frequencies of the emitted radiation.

[Given : Mass of He atom = 4 × (mass of neutron) Ionization energy of H atom = 13.6 eV]
(JEE 1993)

Problem 600. Light from a discharge tube containing hydrogen atoms falls on the surface of a piece of sodium. The kinetic energy of the fastest photoelectrons emitted from sodium is 0.73 eV. The work function for sodium is 1.82 eV. Find :

(a) the energy of the photons causing the photoelectric emission.
(b) the quantum numbers of the two levels involved in the emission of these photons.
(c) the change in the angular momentum of the electron in the hydrogen atom in the above transition, and
(d) the recoil speed of the emitting atom assuming it to be at rest before the transition. (Ionization potential of hydrogen is 13.6 eV). **(JEE 1992)**

ANSWERS

ANSWERS

1. (a) $\frac{x^2}{16} + \frac{y^2}{36} = 1$ (b) 90° **2.** $\frac{\pi u}{4g}, \frac{u^2}{2g} \ln(2)$ **3.** 16.67 s

4. $\frac{2E_0}{\alpha}, \frac{-qE_0\hat{i}}{m}$ oscillatory between $x = 0$ and $x = \frac{2E_0}{\alpha}$ **5.** $\frac{1}{2}$ **6.** $\sqrt{\frac{2nh}{(n-1)g}}$ **9.** $4\sqrt{3}$ m/s

10. 64.2 min **11.** 3 km/hour **12.** $\sin^{-1}\left(\frac{3}{5}\right)$, 1.09 sec, 62.64 m

14. (a) 2.67 km (b) 0.9 km/s **15.** $\sqrt{7gR}$ **16.** 34.8 m/s^2

17. 0.6s, 23.43 m/s, 6.59 N, zero **18.** (a) 0.833 m/s (b) 2.63 m/s

19. *B* should move up with initial velocity $\frac{v}{2}$ and downward acceleration $\frac{a}{2}$

21. $t = \frac{\sqrt{x^2 + y^2}}{\{v \cos\beta - u \cos\alpha\}}$, Here $\alpha = \sin^{-1}\left[\frac{y}{\sqrt{x^2 + y^2}}\right]$ and $\beta = \sin^{-1}\left[\frac{u}{v}\sin\alpha\right]$

22. 2473 m **23.** 2.76 m/s at an angle of 72.8° with *OA*

24. 2.88 cm/s and 4.5 cm/s^2 both upwards **25.** $v(2.41\hat{i} + \hat{j})$ **26.** $x = t - \sin t$

27. $S = A\left[n + 1 - \cos\left(\omega t - \frac{n\pi}{2}\right)\right]$ when n is even

$= A\left[n + \sin\left(\omega t - \frac{n\pi}{2}\right)\right]$ when n is odd where $t = t_1 + t_2$, $t_2 < \frac{T}{4}$ and $t_1 = \frac{nT}{4}$, $n = 0, 1, 2, \ldots$

28. (a) $\frac{u}{\sqrt{2}}$ (b) $u\left(1 - \frac{1}{\sqrt{2}}\right)$ **29.** (a) $v = \frac{x^2}{2a} - \frac{x^2}{3a^2}$ (b) $\frac{a}{v}$ (c) v (due east) (d) $a\hat{i} + \frac{a\hat{j}}{6}$

30. (a) $\frac{\omega\pi}{6(\sqrt{3} - 1)v_0}$ (b) v_0 **31.** $\cot^{-1}(2)$ **34.** (a) 45° (b) 12.8 m/s

35. (a) circle (b) $\sqrt{3}\,\omega$ (c) $\frac{1.317\omega}{v}$ **36.** 10 m **37.** $R = \sqrt{7}a$ **38.** 6th step

39. $\sqrt{26}$ m/s at angle $\theta = \tan^{-1}(5)$ with *x*-axis **40.** $v_0 = \sqrt{\frac{8gd}{(1 + 2\sin\theta)[8t\cos\theta + \tan\theta(1 + 2\sin\theta)]}}$

41. $x = 373$ m, $y = 18.75$ m **42.** 1.09×10^{10} N/m^2 **43.** 2.6 cm^2 **44.** $\frac{a}{N+1}$

45. (a) 45° (b) 2 m/s **46.** 1 s, $(3.75\hat{i} + 6.25\hat{j})$ m/s **47.** (a) 1 m/s along negative *x*-direction (b) 1.48 s

48. (a) $\frac{u^2 \sin 2\alpha}{g\cos\theta}$ (b) $\frac{u\cos(\alpha + \theta)}{\cos\theta}$ **49.** (i) 2.5 m/s (ii) 0.32 m

50. (a) 1 s (b) $(5\sqrt{3}$ m, 5 m) **51.** (i) 17.32 m (ii) 11.55 m from *B* and 5.77 m from *D*

52. (a) 30° (b) (108.25 m, 31.25 m)

53. (a) The lighter man will reach the pulley first, (b) $\frac{m}{M + m}\left[\frac{gt^2}{2} + h\right]$

54. 2 m/s^2 (upwards), 1 m/s^2 (upwards), 3 m/s^2 (downwards), 3 m/s^2 (downwards)

55. 2500 N, 1294 N **56.** $2a_A + 4a_B + a_C = 0$

57. 0.5 m/s (towards right), 0.75 m/s^2 (towards right), 1 m/s (towards right)

58. $N = 3mg\left(\cos\theta - \frac{1}{2}\right), a_t = g\sin\theta$ **59.** 275 N **60.** 20 m/s **61.** 0.317 m

62. 0.18 **63.** 1.82 sec **64.** (a) $\frac{20}{3}$ m/s^2 (↓), $\frac{10}{3}$ m/s^2 (↑) (b) $\frac{10}{11}$ m/s^2 (↑), $\frac{30}{11}$ m/s^2 (↓)

65. $\frac{4}{13}$ m/s^2, $\frac{6}{13}$ m/s^2 **66.** 20 rad/s **67.** 8.40 kN **68.** 1.098, $\sqrt{3}\,mg$

69. 6.28 m/s^2 down the wedge, 8.14 m/s^2

70. 1.272 m/s^2 (down the plane), 1.8 m/s^2 (at 15° with horizontal), 149 N

71. (a) 6.36 m/s^2 (down the plane) (b) 5.5 m/s^2 in horizontal direction (towards right)

72. (a) 24.5 N (b) 109 N **73.** (a) $\left(\frac{5\mu}{4-\mu}\right)m$ (b) $\sqrt{\frac{28(9\mu+4)l}{(20\mu-5\mu^2)g}}$ **74.** 0.474

75. $\sqrt{\frac{2(L-l)}{(\mu_2-\mu_1)g}}$ **76.** 10 m, 5 m **77.** 12.5 rad **78.** $\frac{1}{5}$ sec

79. (a) 1.30 kg (b) 0.065 J **80.** 2 sec, after A has travelled a distance $8\sqrt{2}$ m

81. (b) (i) $f_1 = 30$ N, $f_2 = 15$ N

(ii) Equations of motion for m_1, m_2 and M are $30 - T = 20\,a$, $T - 15 = 5a$ and $F - 30 = 50\,a$

(iii) $F = 60$ N, $T = 18$ N, $a = \frac{3}{5}$ m/s^2 of all the masses towards right

82. (i) 36 N, towards centre (ii) 11.67 rad/s; (iii) $r_1 = 0.1$ m, $r_2 = 0.2$ m **83.** (ii) $\frac{5\sqrt{3}}{8}g, \frac{3mg}{8}$

84. (i) $\omega^2 = \frac{g}{R-h}$, 9.89 rad/s (ii) 9.78×10^{-3} m/s^2 **85.** (a) $3\theta^2 = 1 + \cos\theta$ (b) 2.45 m/s

86. $\mu\, mg^2\cos\alpha\,(\sin\alpha - \mu\cos\alpha)t$ **87.** (a) $0.864\sqrt{gr}$ (b) $0.908\sqrt{gr}$ (c) 126.9°

88. $\frac{5}{2}\sqrt{gR}, 2R$ **89.** 1.825 m/s, 2.74 m/s **90.** $\sqrt{2gy\left(\frac{L-\frac{y}{2}}{L-y}\right)}$

91. $\cos^{-1}\sqrt{\frac{v_1^2\cos^2\alpha_1 - 2gh}{v_1^2 - 2gh}}, v_1\cos\alpha_1 < \sqrt{2gh}$ **92.** $2.414R \le H \le 2.5R$, $45° \le \alpha \le 60°$

93. (a) $\frac{v_0}{6}$ (b) $\sqrt{\frac{m}{3k}}\,v_0$ **94.** 0.124 sec **95.** (a) 90° (b) $mg\,[3\sqrt{2}-2]$ **96.** $\sqrt{5gl}$, $\tan^{-1}(2)$, $2l$

97. (a) $N = mg\,(3\cos\theta - 2)$, (b) For $\theta \le \cos^{-1}\left(\frac{2}{3}\right)$

$N_B = 0$, $N_A = mg\,(3\cos\theta - 2)$ and for $\theta \ge \cos^{-1}\left(\frac{2}{3}\right)$, $N_A = 0$, $N_B = mg\,(2 - 3\cos\theta)$

98. $2.14\sqrt{gL}$ **99.** $\frac{11x_0 - 30}{13}$ **100.** $5m$ **101.** $\frac{\vec{p}}{\Delta t}\sin\alpha + |m\vec{g}|\cos\alpha$

102. (i) 25.981 m/s (ii) 25.918 m/s **103.** $\cos^{-1}\left\{1 - \frac{v_0^2}{8gl}\right\}$ **104.** 4.33 J

105. (a) $\vec{J}_m = -\vec{J}_M = -m\,(5\hat{i} - \hat{j})$ (b) $\frac{1}{13}(5\hat{i} + \hat{j})$ m/s (c) $\frac{11}{17}$ **106.** $\frac{(2-e)}{3}v_0, \frac{2(1+e)v_0}{3}$

107. yes **108.** $\frac{m}{2M}\Delta\vec{r}$ **109.** $\frac{2}{3}$ **111.** 1 : 2 : 3 **112.** $\frac{1}{2}\cot\alpha\cot\theta - 1$

113. (a) 37.5 m/s (b) (50 m, 17 m) **114.** 3.43 m, 13.12 m

115. $\frac{(1+e)mu\cos\theta}{M+m}, \tan^{-1}\sqrt{\frac{eM-m}{M+m}}$ **116.** 0.5 **117.** (a) $3d\sqrt{\frac{k}{8m}}$ (b) $\frac{d}{2}$

118. $\frac{\sqrt{2}J}{7m}, \frac{\sqrt{10}J}{7m}, \frac{3J}{7m}$ **119.** $\frac{\sqrt{2gh}}{3}$ **120.** $\frac{l}{\sqrt{2gh}}$ **121.** (a) 1.65 m/s (b) 2.25 J

122. 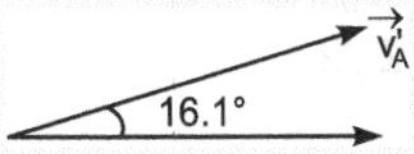

$|\vec{v}_A'| = 0.721\, v_0$ **123.** 4.74 m **124.** (a) $v_0 e^{-\frac{qL}{m_0 v_0}}$ (b) $m_0 e^{\frac{qL}{m_0 v_0}}$

125. $T \ln(2), 2m_0$ **126.** $\frac{g}{2}\left[\frac{t^2}{2} + \frac{m_0 t}{\mu} - \frac{m_0^2}{\mu^2}\ln\left(1 + \frac{\mu t}{m_0}\right)\right]$ **127.** $u \ln\left(\frac{M_0}{M}\right) - ugt$

128. (a) $\frac{3}{4}$ m/s (b) $\frac{1}{2}$ N **129.** (a) $\frac{1}{2}$ m (b) 2.5 sec **130.** The particle will strike at $(\sqrt{20}$ m, 1 m), No

131. $m\left[\left(-v_2 \sin\frac{v_2}{R}t\right)\hat{i} + \left(v_2\cos\frac{v_2}{R}t - v_1\right)\hat{j}\right]$ **132.** $x_2 = v_2 t + \frac{m_1}{m_2}A(1-\cos\omega t), l_0 = \left(\frac{m_1}{m_2}+1\right)A$

133. 12 s, 15.75 m/s **134.** 0.84, 15.12 kg **135.** (i) $\frac{g}{50}$ (ii) $\sqrt{\frac{m_0 g}{2A\rho}}$ **136.** $(L + 2R, 0)$ **137.** 10^5 m

138. $\vec{r}_p = R(\omega t - \sin\omega t)\,\hat{i} + R(1-\cos\omega t)\,\hat{j}, \quad \vec{v}_p = R\omega(1-\cos\omega t)\,\hat{i} + R\omega\sin\omega t\,\hat{j},$

$\vec{a}_p = R\omega^2(\sin\omega t\,\hat{i} + \cos\omega t\,\hat{j})$

139. – 0.732 rad/s, 0.771 rad/s^2 **140.** 71.4 N **141.** $\mu \ge \sqrt{\frac{d}{D}}$

142. $\sqrt{\left(a_0\cos\theta + \frac{r_0 v_0^2}{r^2}\right)^2 + \left(a_0\sin\theta + \frac{r_0 a_0}{r}\right)^2}, \frac{v_0^2}{r}$

143. (a) $\frac{2}{5}v_1$ (b) $\frac{2v_1}{5\mu_k g}$ (c) $\frac{2v_1^2}{25\mu_k g}$ **144.** $0.423\frac{g}{l}$ **145.** $\sqrt{2}\,v_0$ at 45° with positive x-axis

146. (a) No **147.** $\sqrt{\frac{3g}{L}(1-\sin\theta)}, \frac{\sqrt{5}}{3}L$

148. $\frac{M\omega}{12}(b^2\cos\theta\,\hat{i} + l^2\sin\theta\,\hat{j}), I_{ACB}\,\omega$ where $I_{ACB} = I_x\cos^2\theta + I_y\sin^2\theta$ or $I_{ACB} = \frac{Mb^2}{12}\cos^2\theta + \frac{Ml^2}{12}\sin^2\theta$

149. $\frac{2mg}{11}, \frac{3\sqrt{3}}{11}\frac{g}{L}$

150. (i) $\frac{\sqrt{37}}{4}mg$ at an angle $\alpha = \tan^{-1}\left(\frac{1}{6}\right)$ with horizontal, (ii) $\omega(t) = \sqrt{\frac{3g}{L}}$, (iii) $v_c(t) = \frac{\sqrt{3gL}}{2} + gt$

151. (40.0 cm, 33.75 cm) **152.** (a) 4.9 m/s^2 (b) 6.53 m/s^2 (c) 40.82 m

153. (a) $\frac{1}{9}$ (b) $\frac{4}{9}$ **154.** 821.84 N **155.** 3.45 N **156.** $\frac{1}{6}$ m

157. (a) $\frac{29}{19}u$ (downwards) (b) $\frac{12u}{19a}$ (c) $\frac{27}{19}u$ (downwards) **158.** $4\,v$ **159.** $\frac{M}{m}[3ag(\sqrt{2}-1)]^{\frac{1}{2}}$

160. 0.398 **161.** (a) $\frac{g}{8r}$ (b) $\frac{g}{8}\rightarrow$ and $\frac{g}{8}\downarrow$ (c) 3.88 $mg\uparrow$ and 0.5 $mg\rightarrow$

162. (a) 7.1 m/s^2 (b) 53.9 rad/s^2 **163.** $\frac{4}{5}\left(\frac{M+m}{M}\right)\frac{v_0 r\,\omega_0}{g}$

164. (a) 6.84 rad/s^2 (b) 47.27 N (c) 9.45 N

165. (a) $\frac{2}{3} m_1 g \sin\alpha$ (b) $\frac{2}{3}\frac{m_1}{m_A}\sin\alpha$ (c) $\frac{2m_1 g \sin\alpha}{3k}$ (d) 3

166. $12P_0$ **167.** (a) 79 N (b) 0.55 sec **168.** (a) $\sqrt{3}\, ml\,\omega^2$ (b) $F_x = \frac{-F}{4}, F_y = \sqrt{3}\, ml\,\omega^2$

169. (a) 0.1 m (b) 1 rad/s (c) will never come to rest **170.** (a) $\frac{1}{4}$ (b) $\frac{2}{3}L$ (c) $\frac{v_0}{2\sqrt{2}}$

171. (a) $\frac{8F}{3m_1 + 8m_2}, \frac{4F}{3m_1 + 8m_2}$

(b) $\frac{3m_1 F}{3m_1 + 8m_2}$ (between plank and cylinder), $\frac{m_1 F}{3m_1 + 8m_2}$ (between cylinder and ground)

172. $\sqrt{5gR}$ **173.** (i) $(6\hat{i})$ N (ii) $\vec{\tau}_1 = 0.6(\hat{k} - \hat{j})$ N-m, $\vec{\tau}_2 = 0.6(-\hat{j} - \hat{k})$ N-m, $|\vec{\tau}_1| = |\vec{\tau}_2| = 0.85$ N-m

174. (i) $\frac{2}{3}v_0$ (ii) $\frac{m\mu g t}{2}[3\mu g t - 2v_0], -\frac{m v_0^2}{6}$ **175.** (a) $\cos^{-1}\left(\frac{4}{7}\right)$ (b) $\sqrt{\frac{4}{7}gR}$ (c) 6

176. (i) 1.63 N (ii) 1.22 m **177.** 6.3 m/s **178.** $\cos^{-1}\left[\frac{\sqrt{r_1 r_2}(\sqrt{r_1} + \sqrt{r_2})}{r_2\sqrt{r_2} + r_1\sqrt{r_1}}\right]$

180. $2\sqrt{\frac{R^3}{GM}}$ **181.** $\frac{5r}{3}, r$ **182.** 0.8 hrs, 4.5 days **183.** 2.323 R **184.** $2a, \frac{2a}{3}$

185. 3×10^{30} kg, 3.273×10^6 km **186.** $\sqrt{2G(m_1 + m_2)\left[\frac{1}{R_1 + R_2} - \left\{\frac{4\pi^2}{G(m_1 + m_2)T^2}\right\}^{1/3}\right]}$

187. $\sqrt{\frac{2GM}{a}}\left(3\sqrt{3} - \sqrt{\frac{3}{28}}\right)$ **189.** $T_0\left(\frac{1}{2} + \frac{1}{\pi}\cos\frac{\theta}{2}\right), R\cos\frac{\theta}{2}$ **190.** $h = 99.5R$

191. $\frac{3\sqrt{5}}{2}\sqrt{\frac{GM}{a}}$ **192.** 0.14% **193.** (a) $5.36\sqrt{\frac{m}{k}}$ (b) $-\frac{A}{\sqrt{2}}$ **194.** $\sqrt{\frac{M}{M+m}}$

195. 0.2 m, 1.4 s **196.** (a) 0.352 s (b) 0.352 s **197.** $\sqrt{\frac{k}{5m}}$ **198.** (a) 0.72 sec (b) 87.3 mm/s

199. 82.6 mm/s **200.** $2\pi\sqrt{\frac{R(\pi - 2)}{g}}$ **201.** 3.2 Hz **202.** $\{9.2 - 60\sin(10.16t + 2.99)\}$ mm

203. $2\pi\sqrt{\frac{R^2 + L^2 - 2RL}{gL}}$ **204.** $2\pi\sqrt{\frac{28R}{5g}}$ **205.** $2\pi\sqrt{\frac{3Mr^2 + m(r - r_0)^2}{mgr_0}}$ **206.** 0.81 sec

207. $0.8t + 1.2\sin 10t$ **208.** $2\pi\sqrt{\frac{m(4k_2 + k_1)}{k_1 k_2}}$ **209.** $\frac{1}{2\pi}\sqrt{\frac{2k(m_1 + m_2)}{3m_1 m_2}}$ **210.** $\frac{2\pi a}{b}\sqrt{\frac{m}{k}}$

211. $\frac{2}{\pi}\sqrt{\frac{k}{m}}$ **212.** $\frac{1}{\omega}\tan^{-1}\left[\frac{2vx\omega}{\omega^2 x^2 - v^2}\right]$ **213.** (a) $2\sqrt{3}$ cm (b) $\frac{5\pi}{4}\sqrt{\frac{m}{k}} + \sqrt{\frac{m}{k}}\sin^{-1}\left(\sqrt{\frac{2}{3}}\right)$

214. (a) $\theta_0 = \tan^{-1}\left(\frac{4}{3\pi}\right)$ (b) $\frac{1}{2\pi}\sqrt{\frac{g\sqrt{9\pi^2 + 16}}{18\pi^3 R}}$ **215.** $T = 2\pi\sqrt{\frac{M}{\left(\frac{4k_1 k_2}{k_1 + k_2} + \frac{\gamma P_0 A^2}{V_0}\right)}}$

216. $T = 2\pi\sqrt{\frac{3M}{8k}}$ **217.** $\sqrt{\frac{2}{3}}$ **220.** $f = \frac{1}{2\pi}\sqrt{\frac{3g}{2R}}$ **221.** $\sqrt{\frac{3g(d_2 - d_1)}{2d_1 L}}$

222. (i) $\frac{1}{\pi}$ Hz (ii) 0.0628 m/s (iii) 3.9×10^{-4} J **223.** $\frac{\rho\omega^2 R^3}{Y}$ **224.** 2.4×10^4 J/m^3

225. $\frac{(F_1 + F_2)l}{2SY}$ **226.** (a) $\frac{At\,(l_1\alpha_1 + l_2\alpha_2)}{\left(\frac{l_1}{Y_1} + \frac{l_2}{Y_2}\right)}$ (b) $l'_1 = l_1 + \Delta l, l'_2 = l_2 - \Delta l$, where $\Delta l = \frac{l_1 l_2 t\,(Y_1\alpha_1 - Y_2\alpha_2)}{Y_1 l_2 + Y_2 l_1}$

227. (a) 24 N (b) 12 m/s^2 **228.** (a) $\frac{\alpha^2}{2ga^2}$ (b) $\frac{A}{ag}\left[\frac{\alpha}{a}\ln\left\{\frac{\alpha - a\sqrt{2gh}}{\alpha}\right\} - \sqrt{2gh}\right]$

229. $\frac{2A}{3a}\sqrt{\frac{2H}{g}}$ **230.** – 0.72 J **231.** r **232.** 6.75 J

233. (a) 1.57×10^{-4} J (b) 9.42×10^{-2} N **234.** $\frac{1}{2\pi}\sqrt{\frac{6g}{5h_0}}$ **235.** $\frac{1}{n+1}, \pi\left[\sqrt{\frac{l}{ng}} + \sqrt{\frac{nl}{g\,(n^2 - 1)}}\right]$

236. (a) 7.84×10^{-4} m (b) (i) 0.024 J (ii) 0.009 J (iii) – 0.033 J (iv) 0.033 J

237. $t = \frac{A}{a}\sqrt{\frac{2}{g}}\left[\sqrt{H} - \frac{x}{\sqrt{4y}}\right]$ **238.** 4.9 m **239.** (a) 5.88 kg (b) 54.88 N

240. $x = \left(h + \frac{m}{\rho A}\right)e^{-\lambda t} - \frac{m}{\rho A}$ where $\lambda = \frac{\pi\rho\, g a^4}{8\eta\, l\, A}$ **241.** (a) $\frac{p}{257}$ (b) $\frac{257}{32}$

242. $v = \sqrt{\frac{2g}{m + M}\left(ML - \frac{M\rho_l\, L}{2\rho_S} - mL\right)}$ **243.** $4H$ **244.** $F = \frac{\pi D^2}{4}\left(\frac{2s}{d}\right)$, attraction

246. Approximately 4mm

247. (a) (i) $\frac{5}{4}d$ (ii) $p = p_0 + dg\left(\frac{3}{2}H + \frac{L}{4}\right)$ (b) (i) $\sqrt{\frac{g}{2}(3H - 4h)}$ (ii) $\sqrt{h(3H - 4h)}$ (iii) $\frac{3}{8}H, \frac{3}{4}H$

248. (a) $x_{max} = 5.2$ m (b) 68 minute **249.** Power $= \left[\frac{\pi\eta\,\omega^2 R^4}{2t}\right]$ **250.** $T = \frac{\lambda a}{2y}$

251. $\frac{dQ}{Qt} \propto r^5$ **252.** 2m **253.** $\frac{4T}{\rho v^2}$ **254.** $\frac{1}{720}$ N-S/m^2

255. (a) zero (b) 0.25 cm (c) $\frac{g}{6}$ (upward) **256.** 4.568×10^{-3}°C

257. $\pi R^2 L\,(\sqrt{\rho\sigma} - \rho)$ **258.** (i) 40 kg (ii) 0.1 J **259.** 29025 J/m^3, – 29400 J/m^3

260. (a) (i) $\frac{5}{4}d$ (ii) $P_0 + \frac{dg\,(6H + L)}{4}$ (b) (i) $\sqrt{(3H - 4h)\frac{g}{2}}$ (ii) $\sqrt{h(3H - 4h)}$ (iii) $\frac{3H}{8}, \frac{3}{4}H$

261. (a) $\frac{t_1 d_L}{d_L - d}$ (b) no (c) The ball will continue to move with constant velocity $v = \frac{gt_1}{2}$ inside the liquid.

262. f **263.** 0.28 cm, 9×10^{-3} kg /m^3 **264.** 95.5 s

265. (a) (i)  (ii) 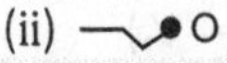 (b) (i) 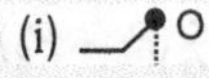 (ii)  **266.** 3.85 s **267.** 49%

268. (a) 15 m, 20 min^{-1} (b) 13.5 m, 22.22 min^{-1} **269.** (a) 33.33 m/s (b) 2250 Hz

270. 0.2 sec **271.** (a) 1 kg/m^2 (b) 10 μm (c) 288 π^2 watt **272.** 878.2 Hz **273.** 0.142 m, 3.55 m/s

274. $\frac{\pi S\rho\omega^2 a^2}{4k}$ **275.** $\frac{a_0^2\,\pi^2 T}{l}$ **276.** $\frac{f}{\sqrt{1 + \frac{2at_0}{c}}}$ **277.** 2340.79 Hz

278. (a) 4.95 mm (b) Third overtone **279.** 163.4 Hz, Five **280.** $\frac{2(l_1 - 3l_2 + 2h)}{3}$

281. 200 Hz **282.** (a) $\frac{50\pi}{k}\rho\omega^2 s$ (b) $\frac{2\rho\omega^3 s}{k}$ **283.** 336 m/s **284.** $\frac{\pi^2 a^2 T}{4l}$

285. (a) $\frac{400}{189}$ (b) $\frac{3}{4}$ **286.** (a) 1.0069×10^5 Hz (b) 1.0304×10^5 Hz

287. (i) 3.2 m, 2.4 m, 1.6 m and 0.8 m (ii) $\left(-\frac{dH}{dt}\right)=(1.11\times10^{-2})\sqrt{H}$ (iii) 43 s

288. (a) 0.14 s (b) – 1.5 cm, 2.0 cm

289. (a) $\frac{15}{16}$ m (b) $\pm\frac{\Delta P_0}{\sqrt{2}}$ (c) $P_{max}=P_{min}=P_0$ (d) $P_{max}=P_0+\Delta P_0, P_{min}=P_0-\Delta P_0$

290. Length of closed organ pipe is $l_1 = 0.75$ m while length of open pipe in either $l_2 = 0.99$ m or 1.0067 m

291. (a) $y=10^{-6}\sin(0.1\,\pi)\sin(25000\,\pi t)$ m (b) $y_1=10^{-6}\sin(25000\,\pi t-5\pi x)$ m,
$y_2=10^{-6}\sin(25000\,\pi t+5\pi x)$ m, Here x is in metres and t in seconds.

292. (i) 6.28×10^{-3} s (ii) 1.57×10^{-3} s **293.** 25.8% **294.** 1.4 **295.** 3.4 cm

296. $\alpha(T_2-T_1)-\beta(T_2^2-T_1^2)$ **297.** 7.7 J **298.** 10.03 hr **299.** 8.06×10^{-6} m/°C

300. $(T_1-T_2)\,e^{-\frac{16KA}{15nRl}t}$ **301.** $l_0\left[1+\frac{Y_1\alpha_1+2Y_2\alpha_2}{Y_1+2Y_2}\cdot\theta\right]$ **302.** $H=\frac{K_0AT_0}{L_0}\left[1+\frac{3\,\alpha\,T_0}{2}\right]$

303. $\frac{T_0}{e}+\left(1-\frac{1}{e}\right)\left(\frac{k_1T_1+k_2T_2}{k_1+k_2}\right)$ **304.** (a) $-2700R$ (b) $-4950R, 3150R, 1247.4R$

305. $C_V=\frac{RT_0m}{V_0}, C_P=\left(1+\frac{T_0m}{V_0}\right)R$ **306.** (a) 512 K (b) 1.68 (c) 288.34 J

307. $\sqrt{10g_eh}$ **308.** (b) 25% **309.** $\sigma_c=\sigma_s=19.84\times10^7$ N/m$^2=4\sigma_a$ **310.** (a) $\frac{23}{3}P_0, \frac{23}{3}T_0$

311. 337.5 K **312.** (a) AC (b) 150 J (c) 10 J **313.** 60%

314. (a) $R(T_1+T_2-2T_3)\ln(2)$ (b) $R(T_1+T_2)\ln(2)$ (c) $1-\frac{2T_3}{T_1+T_2}$

315. (a) $\rho - V$

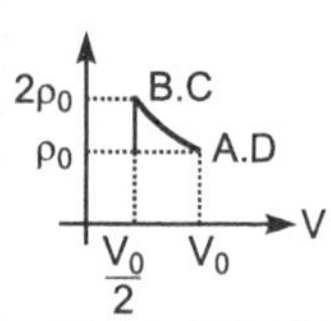

(b) $P - V$

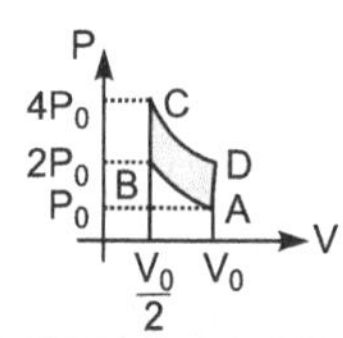

(c) $\rho - P$

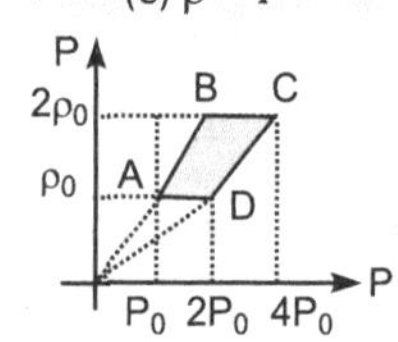

Here $V_0=\frac{nRT_0}{P_0}$ and $\rho_0=\frac{P_0M}{RT_0}$

316. 41% **318.** -9.8 K **319.**

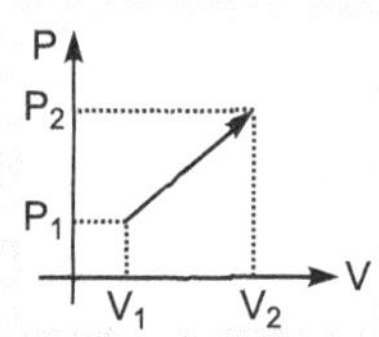

$W=\frac{1}{2}(P_1+P_2)(V_2-V_1)$

320. $\frac{(P_0A+W)(3P_0AV+2WV)}{nAR\,(3P_0A+5W)}$ **321.** $Q-\frac{5}{2}kl^2-3nRT$ **322.** (a) $\frac{3P\alpha m}{\rho}$ (b) 69.69 μJ/mol-K

323. 1.84×10^{-4} per°C **324.** 1.06×10^5 N/m^2, 1.06×10^5 N/m^2, 1346 K, 10^4 J

325. (a) 0.182 kJ (b) 1.418 kJ (c) 1.13 **326.** 15% of original volume

327. (b) – 583 J **328.** $\frac{4}{3}$ m, 448.8 K **329.** $\gamma_l = 2\alpha_s$ **330.** Proportionality constant $= \frac{K}{4e\sigma LT_s^3 + K}$

331. $\frac{Mv_0^2}{3R}$ **332.** (a) 595 watt/m^2 (b) 162.6°C **333.** (a) 160 K (b) 3.312×10^{-21} J (c) 0.3 g

334. (a) 600 K (b) $Q_{AB} = 1500R$ (absorbed), $Q_{BC} = 831.6R$ (absorbed)
$Q_{CD} = -900R$ (released), $Q_{DA} = -831.6R$ (released) (c) $600R$

335. (a) $-1200R$ (b) $Q_{AB} = -2100R$ (released), $Q_{BC} = 1500R$ (absorbed), $Q_{CA} = 831.6R$ (absorbed)

336. (a) 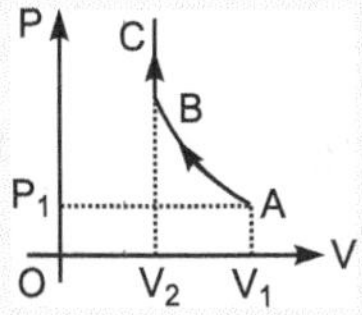

(b) (i) $-\frac{3}{2}P_1V_1\left[\left(\frac{V_1}{V_2}\right)^{2/3} - 1\right]$ (ii) $\frac{3}{2}P_1V_1\left[\left(\frac{V_1}{V_2}\right)^{2/3} - 1\right] + Q$ (iii) $\frac{Q}{3R} + \frac{P_1V_1}{2R}\left(\frac{V_1}{V_2}\right)^{2/3}$

337. (a) P_0V_0 (b) $\frac{5}{2}P_0V_0, 3P_0V_0$ (c) $\frac{P_0V_0}{2}$ (d) $\frac{25}{8}\frac{P_0V_0}{R}$ **338.** $\left(300 + 12.5e^{-\frac{2KAt_1}{CL}}\right)$ kelvin

339. (i) 120.34 K, 240.68 K, 481.36 K, 240.68 K (ii) No (iii) 3.25×10^6 J, 2.75×10^6 J

340. 909 K, 791.4 K, 61.4% **341.** 6.7×10^{-5} per°C **342.** (i) 189 K (ii) –2767 J (iii) 2767 J

343. (a) 2 (b) 401 m/s (c) 0.167% (d) 8.27×10^{-5} V **344.** (i) 765 J (ii) 10.82%

345. 4.074 g Neon and 23.926 g Argon **346.** (i) 1869.75 J (ii) 5297.6 J (iii) 500 K

347. (a) 1152 J (b) 1152 J (c) zero **348.** 166.32 s **349.** $\frac{1}{2}$ m **350.** 60°

352. 0.63 cm **353.** $\frac{1}{\sqrt{2}}(-\hat{i} - \hat{j})$ **355.** $(-10\hat{i} + 5\sqrt{3}\,\hat{j})$ cm/s **357.** 100 cm vertically below A

358. 21.3 cm from the mirror (towards right) **359.** 1.45 **360.** 7.96° **361.** 1.382

362. (a) co-ordinates of images of points A, C and B in centimeters with respect to optical centre of lens are :
$A \equiv (33.0, -6.1)$, $C \equiv (36, -12)$, $B \equiv (40.7, -21.36)$ (b) 17.1 cm **363.** At the middle of the glass slab

364. final image will coincide with the object **365.** (0, 0) **366.** $\left(\frac{1-n^2}{n^2}\right)v_0$ towards the screen

367. $v_1 + \left(\frac{R}{2x - R}\right)^2(v_1 - v_2)$ in the direction of v_1 and v_2, Here $x = d + (v_1 - v_2)t$

368. $(\mu - \mu')\alpha$ **369.** $\mu \geq \sqrt{2}, \alpha \geq 45°$ **370.** (a) 60° (b) 136°

371. Image is formed at $x = 3.125$ cm **372.** 1.932 **373.** 3.136 cm, 8.75 cm

375. (a) 17% (b) 71.5% **376.** (67.5 cm, 0.5 cm), (67.5 cm, 0.33 cm), virtual

377. (a) 24 cm (b) 1.5 **378.** (36 cm, 0.006 cm) **379.** 60° **380.** 1.71 cm

381. $n(y) = \frac{R}{R - y}n_0$, 66.4° **382.** 0.7 mm **383.** $D\sqrt{2\left\{1 - \frac{2(\mu - 1)^2\lambda}{\Delta\mu R}\right\}}$

384. $\frac{d(r_k^2 - r_i^2)}{4l^2(i - k)}$ **385.** (a) $\cos^{-1}\left(n - \frac{\phi}{2\pi}\right)\frac{\lambda}{d}, n = 0, \pm 1, \pm 2, \ldots$ (b) $\phi = \frac{\pi}{2}$ and $\frac{d}{\lambda} = n + \frac{1}{4}, n = 0, \pm 1, \pm 2, \ldots$

386. (a) central maxima is obtained at a distance $y = \frac{d}{40}$ above O. Fringe width $= \frac{\lambda}{d}$ (b) $\frac{I_0}{I_{max}} = \cos^2\left(\frac{\pi d^2}{40\lambda}\right)$

387. (a) 1.66 cm below point O (b) 1 **388.** 0.6 mm **389.** $OE = 6.06$ m **390.** 0.09 m/s, 0.3 per second

391. $f=\dfrac{\mu_3 R}{\mu_3-\mu_1}$ **392.** (a) 0° (b) 1500 Å **393.** 1.6 **394.** 15 cm, $-\dfrac{3}{2}$ **395.** $\dfrac{1}{5\sqrt{2}}(3\hat{i}+4\hat{j}-5\hat{k})$

396. (a) 600 nm (b) $\sin^{-1}\left(\dfrac{3}{4}\right)$ **397.** 0.9 m (rightwards) **398.** 0.4 m, 0.6 m

399. (i) $\sin^{-1}\left\{\dfrac{1}{\sqrt{2}}(\sqrt{n^2-n_1^2}-n_1)\right\}$ (ii) 73°

400. (a) slope = $\cot i$ (b) $4y^{1/4}=x$ (c) (4.0 m, 1.0 m)

(d) The ray will emerge parallel to the boundary at P *i.e.*, at grazing emergence

402. 3.5 mm **403.** (a) circular (b) $\dfrac{1}{16}$ (c) 300 nm **404.** 2 cm above Q, 1.0016

405. (i) $3.6\,t=\left(n-\dfrac{1}{2}\right)\lambda, n=1,2,3,\ldots$ (ii) 90 nm

406. (a) $y=-4.33$ mm (b) $\dfrac{3}{4}I_{max}$ (c) 650 nm and 433.33 nm

407. (a) ± 0.26 m, ± 1.13 m (b) 0.26 m, 1.13 m **408.** 9.3 μm

409. 7×10^{-6} m, 1.6; -5.71×10^{-5} m **410.** 4200 Å, 1.429 **411.** 1 mm, increase

415. yes **416.** It is possible if $\dfrac{E_2}{E_1}<\dfrac{r_2}{r_1+R}$ **417.** R **419.** 22.85 A, 7.62 A from P to Q

420. (a) 0.225 m/s (b) 54.15° C **421.** (a) 30 Ω (b) 0.044 A **422.** $\dfrac{E(a+b+c+4d)}{(a+b+c)d}$

423. (a) (i) 120 V, 80 V (ii) 100 V, 100 V (b) $\dfrac{1}{60}$ A from D to C

424. (a) $\dfrac{4}{3}\dfrac{q^2R}{t_0}$ (b) $\left\{\dfrac{1}{2}\ln(2)\right\}\dfrac{q^2R}{t_0}$ **425.** $T_0+\left(1-e^{-\frac{\alpha t}{C}}\right)\dfrac{V^2}{\alpha R}$ **427.** 80 A

428. + 14 watt, − 1 watt **429.** 2V **430.** (a) 605 watt (b) 6 volt (c) both

431. $\dfrac{47}{105}r_0$ **432.** (a) 5 Ω (b) 30°C (c) 100 sec

435. (a) No (c) 8 Ω **436.** $\dfrac{V_1r_2-V_2r_1}{r_1+r_2};\ \dfrac{r_1r_2}{r_1+r_2}$ **437.** $\dfrac{20}{3}$ V **439.** $\dfrac{m\omega^2R^2}{2e}$

440. $\dfrac{2}{\sqrt{3}},\dfrac{\alpha_1}{3}$ **441.** $2\pi h\sqrt{\left(\dfrac{6\pi\varepsilon_0 MR}{Qq}\right)}$ **442.** $\dfrac{1}{k}\left[F_0-\dfrac{3}{2}\dfrac{\varepsilon_0 aV^2}{d}\right], 2\pi\sqrt{\dfrac{M}{k}}$

443. $\sqrt{2}$ **444.** $h_1+h_2-g\left(\dfrac{l}{v}\right)^2$ **445.** $\dfrac{D_2}{e}$ **446.** $\dfrac{p_1p_2}{4\pi\varepsilon_0 r^3}(\sin\theta_1\sin\theta_2-2\cos\theta_1\cos\theta_2)$

447. $4\pi\varepsilon_0\left[\dfrac{ab}{b-a}+\dfrac{c^2}{c-b}\right]$ **448.** (a) $\dfrac{1}{4\pi\varepsilon_0}\dfrac{q_1q_2}{(r-t+t\sqrt{k})^2}$ (b) 4 **449.** 3.174×10^{-9} C

450. (a) 1.25×10^3 V/m² (c) 1 m **451.** $-0.45\hat{i}$ N **452.** (a) 0.9 mC (b) 4.12 J

453. (a) $q_0e^{-t/\rho\varepsilon_0 K}$ (b) $\left(\dfrac{1}{a}-\dfrac{1}{b}\right)\dfrac{q_0^2}{8\pi\varepsilon_0 K}$ **454.** (a) 50 kΩ, 10 kΩ (b) 100 μF (c) $(10-5.66\,e^{-t})$ mC

455. $q_3=\dfrac{-q_1q_2}{(\sqrt{q_1}+\sqrt{q_2})^2}, \vec{r_3}=\dfrac{\sqrt{q_1}\,\vec{r_2}+\sqrt{q_2}\,\vec{r_1}}{\sqrt{q_1}+\sqrt{q_2}}$ **456.** $4\pi\varepsilon_0 a$ **457.** $6\,\alpha\varepsilon_0 x$

458. $\left(\frac{P}{4\pi\varepsilon_0 E_0}\right)^{1/3}$ **459.** $\frac{1}{3}$ **460.** (a) + 2.6 V, – 1.6 V (b) 1.2 μC, 8.4 μC

461. (a) 1 A (b) 15 μC, 15 μC, 45 μC (c) 75 μJ **462.** (a) 2.36 A (b) zero, 4.8 μC (c) 0.34 A

463. $4\pi\varepsilon_0\left(\frac{V_1 - V_2}{a_2 - a_1}\right) a_1 a_2$ **464.** $\left[\frac{2}{m}\cdot\frac{1}{4\pi\varepsilon_0}\cdot\frac{Qq}{R}\cdot\frac{(6\sqrt{5}+\sqrt{10}-16)}{10}\right]^{1/2}$ **465.** $\frac{R_1 CV}{2(R_1 + R_2)}(E - V)$

466. $\frac{Q^2(R^2 - h^2)}{32\,\pi\varepsilon_0 R^4}$ **467.** $3mgh$ **468.** $\frac{(\sigma_1 - \sigma_2)qa}{\sqrt{2}\,\varepsilon_0}$ **469.** $W = 5.824\left(\frac{1}{4\pi\varepsilon_0}\cdot\frac{q^2}{a}\right)$

470. (a) $\frac{qp}{4\pi\varepsilon_0 d^2}$ (b) $\frac{pq}{2\pi\varepsilon_0 d^3}\hat{i}$ **471.** 5.86 m/s **472.** 3 m/s, 3×10^{-4} J

473. (a) $\frac{4}{3}a$ (b) Equilibrium position is $H = \frac{a}{\sqrt{3}}$

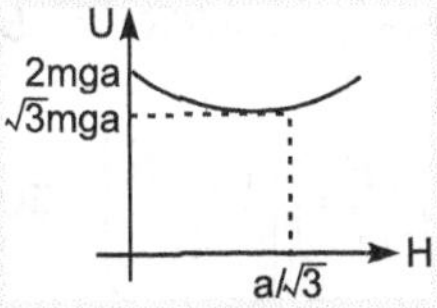

474. (a) $\frac{q_n^2}{8\pi\varepsilon_0 R}$ where $q_n = \frac{QR}{r}\left[1 - \left(\frac{R}{R+r}\right)^n\right]$, (b) $\frac{Q^2 R}{8\pi\varepsilon_0 r^2}$

475. (a) $\frac{CV}{2}\left(1 - e^{-\frac{2t}{3RC}}\right)$ (b) $\frac{V}{2R} - \frac{V}{6R}e^{-\frac{2t}{3RC}}, \frac{V}{2R}$

476. (i) 90 μC, 150 μC, 210 μC (ii) (a) 47.4 mJ (b) 18 mJ

477. 0.198 μA **478.** (a) $\frac{CK_1K_2}{K_2 - K_1}\ln\left(\frac{K_2}{K_1}\right)$ (b) $\frac{5}{6}\sigma$

479. 4.43×10^{-9} A **480.** (a) (i) 2.0×10^{-9} F, 1.21×10^{-5} J (ii) 4.84×10^{-5} J (iii) 1.1×10^{-5} J (b) $\sqrt{\frac{q\lambda}{2\varepsilon_0 m}}$

481. (a) $\frac{3}{20}\frac{Q^2}{\pi\,\varepsilon_0 R}$ (b) 1.5×10^{32} J (c) $\frac{Q^2}{8\pi\,\varepsilon_0 R}$ **482.** $\pi - 2\tan^{-1}\left(\frac{mv}{BqR}\right)$

483. (a) $\frac{5\sqrt{2}\,I_0 B_0 \pi}{m}$ rad/s^2 (b) $-5\,\pi R^2 I_0 B_0$ J (c) zero

484. $v_0\cos(B_0\,\alpha\,t)\hat{i} + v_0\sin(B_0\,\alpha t)\hat{j}, \frac{v_0}{B_0\alpha}[\sin(B_0\alpha t)\hat{i} + \{1 - \cos(B_0\alpha t)\}\hat{j}]$

485. (0.2 m, 0, 0.628 m) **486.** $x = a(\omega t - \sin\omega t)$, $y = a(1 - \cos\omega t)$, where $\omega = \alpha B_0$ and $a = \frac{E_0}{\alpha B_0^2}$

487. $\frac{2\pi a^2 iB_0}{\sqrt{a^2 + d^2}}$ **488.** (a) $\frac{\mu_0 in^2\sin\left(\frac{\pi}{n}\right)\cdot\tan\left(\frac{\pi}{n}\right)}{2\pi^2 r}$ (b) $\frac{\mu_0 i}{2r}$

489. (a) $iBdl$ (towards centre) (b) iBa (c) $\frac{ia^2B}{\pi r^2 Y}$

490. $\frac{\mu_0 i_3}{2\pi}\left[i_1\ln\sqrt{1+\left(\frac{l}{l_1}\right)^2} - i_2\ln\sqrt{1+\left(\frac{l}{l_2}\right)^2}\right]$ inwards **491.** 1.07 m/s **492.** $\sqrt{\frac{2v_0}{B_0\,\alpha}}$

493. (a) Helix with increasing pitch (b) $v_0 \cos(B_0\,\alpha t)\,\hat{i} + (E_0\,\alpha t)\,\hat{j} + v_0 \sin(B_0 \alpha t)\,\hat{k}$

(c) $\dfrac{v_0}{B_0\,\alpha}\sin(B_0\,\alpha t)\,\hat{i} + \dfrac{1}{2}E_0\,\alpha t^2\,\hat{j} + \dfrac{v_0}{B_0\,\alpha}(1-\cos B_0\,\alpha t)\,\hat{k}$ (d) $\dfrac{2n^2\pi^2 E_0}{B_0^2\,\alpha}$ (e) $\tan^{-1}\left(\dfrac{v_0 B_0}{2n\pi E_0}\right)$

494. (a) $\dfrac{\mu_0 I}{\pi r}\left(\dfrac{2r^2-a^2}{4r^2-a^2}\right)$ to the left (b) $\dfrac{\mu_0 I}{\pi r}\left(\dfrac{2r^2+a^2}{4r^2+a^2}\right)$ towards the top of the page

495. $\dfrac{1}{\sqrt{2}}$ **496.** $\omega_{max} = \dfrac{DT_0}{BQR^2}$ **497.** (a) P to Q (b) $Ib\,B_0\,(3\hat{k}-4\hat{i})$ (c) $\dfrac{mg}{6b\,B_0}$

498. (a) 6.54×10^{-5} T (vertically upward or outward normal to the paper)

(b) zero, zero, 8.1×10^{-6} N (inwards)

499. (a) $-\dfrac{\mu_0 q v_0 I}{4R}\hat{k}$ (b) $\vec{F_1} = \vec{F_2} = 2BIR\hat{i}, \vec{F} = 4BIR\hat{i}$ **500.** (a) $\dfrac{mv_0}{2B_0 q}$ (b) $-v_0\,\hat{i}, \dfrac{\pi m}{B_0 q}$

501. $\cos\left(\dfrac{qB}{m}t\right)\vec{v_0} + \left(\dfrac{q}{m}t\right)(\vec{E}) + \sin\left(\dfrac{qB}{m}t\right)\left(\dfrac{\vec{v_0}\times\vec{B}}{B}\right)$ **502.** (a) $\dfrac{I_0 L^2 B}{\sqrt{2}}(\hat{j}-\hat{i})$ (b) $\dfrac{3}{4}\dfrac{I_0 B}{M}(\Delta t)^2$

503. (i) $x = 0 = z, x = \dfrac{d}{\sqrt{3}}, x = -\dfrac{d}{\sqrt{3}}$ $(z=0)$ (ii) $\dfrac{i}{2\pi d}\sqrt{\dfrac{\mu_0}{\pi\lambda}}$ **504.** $\dfrac{21}{4}\dfrac{\mu_0^2 M^2 a^4 v^2}{Rx^8}$

505. (i) $M = \dfrac{eh}{4\pi m}$ (ii) $\tau = \dfrac{ehB}{8\pi m}$ (perpendicular to both $\vec{M}$ and $\vec{B}$) **506.** 0.2 s

507. 4.73×10^{-3} T **508.** (a) $-\dfrac{\mu_0 i^2}{2\pi}\ln\left(\dfrac{L^2+a^2}{a^2}\right)\hat{k}$ (b) zero

509. 5 ohm **510.** $10\,e^{-5t}$ volts, $5e^{-10t}$ (downwards)

511. (a) $10\sqrt{2}$ A (b) $50\sqrt{2}$ volts (c) $\dfrac{1}{25\pi}$ henry, $\dfrac{1}{100\pi}$ Farad

512. $\dfrac{V}{R} - \dfrac{VR_0}{(R+R_0)R}e^{-Rt/L}$ **513.** (a) 63000 J (b) 25° C (c) 1.22 A

514. (a) 100 volt (b) Inductor of $L = \dfrac{0.4}{\pi}$ henry or $\dfrac{1.6}{\pi}$ henry

515. (a) 8.15 volt (b) 16.3 A, $\dfrac{20}{\pi}$ Hz **516.** (a) $\sqrt{\dfrac{2l\,(m + Cb^2B^2\cos^2\alpha)}{mg\sin\alpha}}$ (b) $\sqrt{\dfrac{2mgl\sin\alpha}{m + Cb^2B^2\cos^2\alpha}}$

517. (a) $\dfrac{\mu_0\, l i_0 C \ln(2)}{2\pi t_0}(1-e^{-t/CR})$ (b) $\dfrac{1-e^{-2t/CR}}{(1-e^{-t/CR})^2}$ (c) $CR\ln\left(\dfrac{e_t}{e_t - V_0}\right)$ where $e_t = \dfrac{\mu_0\, l i_0 \ln(2)}{2\pi\, t_0}$

518. $\left(\dfrac{B^2 v}{\sqrt{3}\rho}\right)x$ **520.** $2(1-e^{-5t})$ m/s **521.** $\dfrac{FR}{B^2l^2}\left[1 - e^{-\frac{B^2l^2t}{R(m+B^2l^2C)}}\right]$ **522.** (a) $\dfrac{2B_0RQ}{mg}$ (b) $\dfrac{B_0Q}{m}$

523. (a) -20 mV (b) $10^7\,t^2$ volt (c) 63.2 μs

524. (a) $\dfrac{i\,(R+2\lambda x)}{Bl}, ilB + \dfrac{2m\lambda i^2}{B^2l^2}(R+2\lambda x)$ (b) $\left\{1 + \dfrac{2m\lambda i\,(R+2\lambda x)}{B^3l^3}\right\}^{-1}$

525. $\dfrac{Ba^2}{2(R+r)}\left[\omega_0 - \dfrac{3}{4}\dfrac{B^2a^2\theta}{m\,(R+r)}\right]$ **526.** $\dfrac{E}{R_1}e^{-\left(\frac{R_1+R_2}{CR_1R_2}\right)t}, \dfrac{E}{R_3}\left(1 - e^{-\frac{R_3}{L}t}\right)$

527. $\dfrac{g\sin\theta}{K}(1-e^{-Kt})$ where $K=\dfrac{\pi^2r^4B_0^2a^2\cos^2\theta}{mR}$ **528.** $\dfrac{2x^2}{\mu_0 i_0 a^2}\sqrt{Pr}$

529. (a) 0.65 volt (b) $0.2t$ volt **530.** 0.32 A **531.** (a) $\dfrac{EmC}{CB^2l^2+m}$ (b) $\dfrac{BlCE}{(m+CB^2l^2)}$

532. (a) $\dfrac{FR}{B^2l^2},\dfrac{CFR}{Bl}$ (b) $\dfrac{FR}{B^2l^2}(1-e^{-\lambda t})$ where $\lambda=\dfrac{B^2l^2}{R(m+CB^2l^2)}$

533. $v=\left(gx-\dfrac{B^2l^2x^2}{2mL}\right)^{1/2}$ **534.** $v=v_0+\dfrac{F_0x}{m}-\dfrac{B^2l^2}{2\lambda m}\ln\left(\dfrac{2x+l}{l}\right)$

535. (i) Rod moved with terminal velocity $v_T=\dfrac{mgR\sin\alpha}{B^2l^2}$

(ii) Rod moves with uniform acceleration $a=\dfrac{mg\sin\alpha}{m+B^2l^2C}$

(iii) Rod oscillates $x(t)=A(1-\cos\omega t)$ where, $A=\dfrac{mgL\sin\alpha}{B^2l^2}$ and $\omega^2=\dfrac{B^2l^2}{mL}$

536. (i) $v_{max}=\dfrac{BlCV_0}{m+B^2l^2C}$ (ii) $m=CB^2l^2$ for maximum efficiency **537.** $r=\sqrt{2}R$

538. $z(t)=\dfrac{g}{\omega^2}(\cos\omega t-1)$ where $\omega^2=\dfrac{k}{m}, k=\dfrac{2\pi^2r_0^4B_0^2\alpha\beta}{L}, i_{max}=39$ A

539. $R=\dfrac{4}{3}(r_1-r_2)$ **540.** 20 A, $\dfrac{\pi}{4}$ **541.** $i_{max}=\dfrac{\mu_0 aCI_0\omega^2\ln(2)}{\pi}$

542. (a) $\dfrac{d\phi}{dt}=iR+L\dfrac{di}{dt}$ (b) $\dfrac{1}{R}\left[\dfrac{\mu_0 I_0 l}{2\pi}\ln(2)-Li_1\right]$ (c) $\dfrac{T}{\ln 4}$

543. $12e^{-5t}$ volt, $6e^{-10t}$ ampere (clockwise) **544.** 35.6°C

545. (a) $\dfrac{B_0av}{R}$ (counter clockwise) (b) $-\dfrac{B_0^2a^2v}{R}\hat{j}$ (c) (i) $v(t)=\dfrac{g}{K}(1-e^{-Kt})$, where $K=\dfrac{B_0^2a^2}{mR}$ (ii) $v_t=\dfrac{g}{K}$

546. (a) 10^4 A/s (b) zero (c) 2.0 A (d) 1.732×10^{-4} C

547. (i) $\dfrac{mgR}{B^2L^2}$ (ii) $\dfrac{g}{2}$ **548.** (a) $\dfrac{B\omega r^2}{2}$ (b) (i) $i=\dfrac{B\omega r^2}{2R}\left[1-e^{-\left(\frac{R}{L}\right)t}\right]$

(ii) $\tau_{ext}=\dfrac{B^2\omega r^4}{4R}+\dfrac{mgr}{2}\cos\omega t$ (anticlockwise) **549.** 1.0 m/s, 0.47 Ω, 0.3 Ω,

550. Current in segment AE is $\dfrac{7}{22}$ A from E to A

Current in segment BE is $\dfrac{6}{22}$ A from B to E and current in segment EF is $\dfrac{1}{22}$ A from F to E.

551. 2.225 eV **552.** (a) $r_n=\sqrt{\dfrac{nh}{2\pi Be}}$ (b) $K_n=\dfrac{Benh}{4\pi m}$ (c) $U_n=\dfrac{Benh}{2m}$

553. A is ${}_1H^2$ and B is ${}_2He^4$ **554.** $\nu\left(1+\dfrac{gH}{c^2}\right)$ **555.** (a) 6.25×10^4 m/s (b) 3.13×10^4 m/s

556. 4.5 eV **557.** $Z=24$ **558.** $N_c=N_0(1-e^{-\lambda_1 t})+P\left(t+\dfrac{e^{-\lambda_2 t}-1}{\lambda_2}\right)$

560. $r_n = \sqrt{\frac{n\hbar}{m\omega}}$ and $E_n = n\,\hbar\,\omega$ where $\hbar = \frac{h}{2\pi}$ and $\omega = \sqrt{\frac{k}{m}}$

561. $\frac{\mu e^4}{32\pi^2\varepsilon_0^2 h^2}$, 0.055%. Here $\hbar = \frac{h}{2\pi}$ and $\mu = \frac{mM}{m+M}$ **562.** 0.2834 MeV **563.** 16.67% and 83.33%

564. $N_0 e^{-\lambda_2 t}, \frac{\lambda_1 N_0}{\lambda_2}(1 - e^{-\lambda_2 t})$

565. (a) $\left(\frac{q_\alpha E_0}{m_\alpha} t\right)\hat{i} + v_0 \cos\theta\,\hat{j} - v_0 \sin\theta\,\hat{k}$ where $\theta = \omega t$ and $\omega = \frac{q_\alpha B}{m_\alpha}$ (b) 10^7 m/s (c) 8.00 MeV

566. (a) 0.73×10^{10} dps, 0.27×10^{10} dps (b) 5.4 **567.** $\nu = \nu_0\left(1 + \frac{v}{c}\right)$

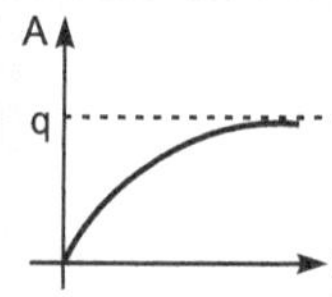

568. 319.984 g, 4.5712×10^{21} disintegrations per day **569.** 2.167 days

570. $u\lambda t$ **571.** $\frac{0.2 E_0\left[\left[\alpha t - \frac{\alpha}{\lambda}(1 - e^{-\lambda t})\right]\right]}{ms}$ **572.** (a) 0 V (b) 10^9 s (c) 1.5 V

574. 7.2 V, 1.04×10^{-3} A, 3.846, 3.698 **575.** $10^3\ \Omega$, $1.85 \times 10^5\ \Omega$

576. The circuit cannot be used as an amplifier. **577.** $N_0\left(\frac{1}{2}\right)^8$ **578.** 116.8 years

579. 3.846 **580.** 0.55 eV **581.** $Z = 42$ **582.** 6.947 sec **584.** (a) $Z = 3$ (b) 4052.3 nm

585. (a) 5.0×10^7 (b) 2×10^3 N/C (c) 23 eV

586. (i) $\frac{dN_X}{dt} = -\lambda_X N_X, \frac{dN_Y}{dt} = \lambda_X N_X - \lambda_Y N_Y, \frac{dN_Z}{dt} = \lambda_Y N_Y$

(ii) 16.48 s (iii) $N_X = 1.92 \times 10^{19}$, $N_Y = 5.76 \times 10^{19}$, $N_Z = 2.32 \times 10^{19}$

587. 3.847×10^4 kg **588.** 6.25 MeV, 227.62 a.m.u.

589. (a) 2, 4, – 217.6 eV, 10.58 eV (b) 6.25×10^{11}, 0, 5.0 eV

590. 3.376 eV during combination and 3.94 and 2.64 eV after combination.

591. (a) $\frac{1}{\lambda}[\alpha - (\alpha - \lambda N_0)\,e^{-\lambda t}]$ (b) $\frac{3}{2}N_0, 2N_0$ **592.** 150.8 eV, 0.5 Å **593.** 3.32×10^{-5} watt

594. 1.88×10^9 years **595.** (a) (i) 3.4 eV (ii) 6.63 Å (b) (i) 14.43 s (ii) 40 s

596. (a) 10^5 (b) 285.1 (d) 111 s **597.** 6, 3 **598.** 5.95 litre

599. (i) Allowed values of K.E. of neutron are 6.36 eV and 0.312 eV and that of atom are 17.84 eV and 16.328 eV

(ii) 1.82×10^{15} Hz, 11.67×10^{15} Hz, 9.84×10^{15} Hz

600. (a) 2.55 eV (b) 4 and 2 (c) $-\frac{h}{\pi}$ (d) 0.814 m/s

SOLUTIONS

CHAPTER 1

KINEMATICS

1. (a) Given : $x = 4\cos 6t$

or $\cos 6t = \frac{x}{4}$ ...(1)

$y = 6\sin 6t$

or $\sin 6t = \frac{y}{6}$ ...(2)

Squaring and adding (1) and (2), we get :

$$\mathbf{\frac{x^2}{16} + \frac{y^2}{36} = 1}$$

This is the desired equation of the path of the particle which is an ellipse.

(b) The position vector and velocity of particle at any time t are as follows :

$\vec{r} = x\hat{i} + y\hat{j} = 4\cos 6t\,\hat{i} + 6\sin 6t\,\hat{j}$ ...(3)

$\vec{v} = \frac{dx}{dt}\hat{i} + \frac{dy}{dt}\hat{j}$

$= -24\sin 6t\,\hat{i} + 36\cos 6t\,\hat{j}$ ...(4)

at $t = \pi/12$ or $6t = \pi/2$.

$\vec{r} = 6\hat{j}$ and $\vec{v} = -24\hat{i}$

i.e. **angle between $\vec{r}$ and $\vec{v}$ is 90°.**

(c) Path of the particle is an ellipse with its centre at origin.

Radius vector and velocity of particle at any time t are :

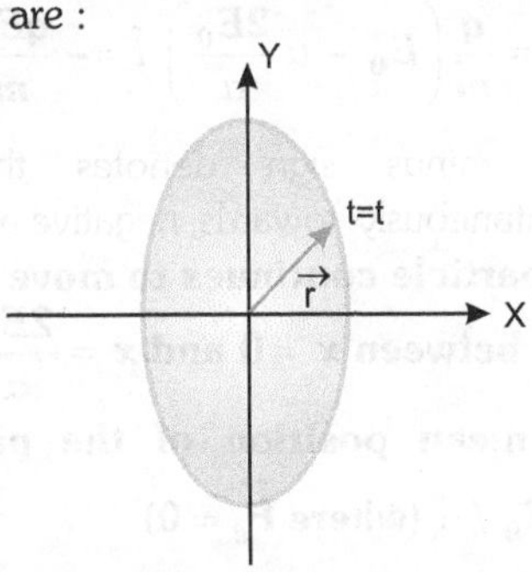

$\vec{r} = 4\cos 6t\,\hat{i} + 6\sin 6t\,\hat{j}$

$\vec{v} = -24\sin 6t\hat{i} + 36\cos 6t\hat{j}$

and its acceleration would be

$$\vec{a} = \frac{d\vec{v}}{dt} = -144\cos 6t\,\hat{i} - 216\sin 6t\,\hat{j}$$

or $\vec{a} = -36(4\cos 6t\,\hat{i} + 6\sin 6t\,\hat{j})$

or $\vec{a} = -36\,\vec{r}$

i.e. $\vec{a}$ is directed towards origin.

2. Initially net force is $2mg$ (downwards) or resistive force at speed u is mg.

so $F_r \propto v^2$ (F_r = resistive force)

or $F_r = kv^2$ ...(1)

then $mg = ku^2$ (at $v = u$, $F_r = mg$)

or $k = \frac{mg}{u^2}$

hence from equation (1) we get

$$F_r = \left(\frac{mg}{u^2}\right)v^2$$

now when the speed is v

$$m\cdot\frac{dv}{dt} = -(\text{weight} + F_r)$$

or $m\frac{dv}{dt} = -\left(mg + \frac{mg}{u^2}v^2\right)$ ...(2)

or $\frac{dv}{dt} = -\left(g + \frac{g}{u^2}\cdot v^2\right) = -g\left(1 + \frac{v^2}{u^2}\right)$

or $\frac{dv}{-g\left(1 + \frac{v^2}{u^2}\right)} = dt \quad \int_0^t dt = -\int_u^0 \frac{dv}{g\left(1 + \frac{v^2}{u^2}\right)}$

solving this, we get

$$\boldsymbol{t = \frac{\pi u}{4g}}$$

equation (2) can also be written as

$$m\left(v\frac{dv}{ds}\right) = -\left(mg + \frac{mg}{u^2}\cdot v^2\right)$$

or $$\frac{v\,dv}{-g\left(1+\frac{v^2}{u^2}\right)} = ds$$

or $$\int_0^s ds = -\int_u^0 \frac{v\,dv}{g\left(1+\frac{v^2}{u^2}\right)}$$

solving this, we get

$$\boldsymbol{s = \frac{u^2}{2g}\ln(2)}$$

3. Acceleration-time graph of the car for $t \le 10\,\text{s}$ will be as shown in figure.

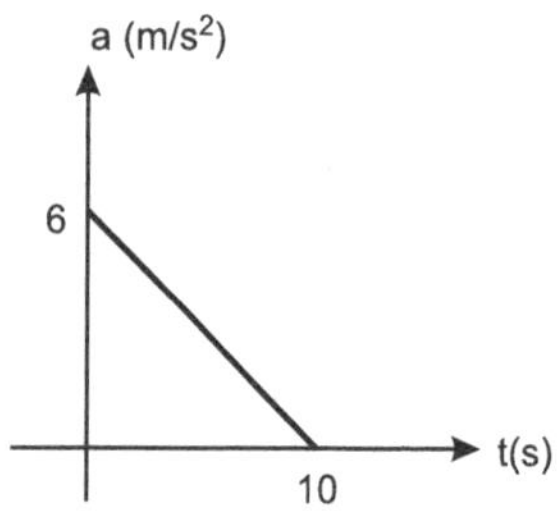

Acceleration at time t is,

$$a = (6 - 0.6t)\ \text{m/s}^2$$

or $$\frac{dv}{dt} = (6 - 0.6t)$$

or $$dv = (6 - 0.6t)\,dt$$

or $$\int_0^v dv = \int_0^t (6 - 0.6t)\,dt$$

or $$v = (6t - 0.3t^2) \quad \ldots(1)$$

or $$\frac{ds}{dt} = (6t - 0.3t^2)$$

or $$\int_0^s ds = \int_0^t (6t - 0.3t^2)\,dt$$

or $$s = 3t^2 - 0.1t^3 \quad \ldots(2)$$

from equations (1) and (2)

at $t = 10\,\text{s}$

$$v = 6\times 10 - 0.3(10)^2 = 30\ \text{m/s}$$

and $s = 3(10)^2 - 0.1(10)^3 = 200\ \text{m}$

The remaining distance $(400 - 200)$ m or 200 m is covered at a constant speed of 30 m/s. Hence time taken,

$$t = \frac{200}{30} = 6.67\ \text{s}$$

$\therefore$ total time taken $= (10 + 6.67) =$ **16.67 s**

4. Electrostatic force (F_e) on the particle is

$$F_e = qE = q(E_0 - \alpha x)$$

or $$m\cdot v\cdot\frac{dv}{dx} = q(E_0 - \alpha x)$$

or $$\int mv\,dv = q\int (E_0 - \alpha x)\,dx$$

or $$\frac{1}{2}mv^2 = q\left(E_0 x - \frac{1}{2}\alpha x^2\right) + C$$

Here C is constant of integration.

at $x = 0,\ v = 0 \quad \therefore\ C = 0$

We therefore, get

$$v^2 = \frac{2qx}{m}\left(E_0 - \frac{1}{2}\alpha x\right)$$

Velocity of particle becomes zero at $x = 0$ and

$$x = \frac{2E_0}{\alpha}$$

Therefore, the desired distance is $\boldsymbol{\frac{2E_0}{\alpha}}$

The acceleration of the particle at $x = \frac{2E_0}{\alpha}$ is $\frac{\vec{F}_e}{m}$

or $\frac{q\vec{E}}{m}$

$$\therefore\ \boldsymbol{\vec{a} = \frac{q}{m}\left(E_0 - \alpha\frac{2E_0}{\alpha}\right)\hat{i} = -\frac{qE_0}{m}\hat{i}}$$

Here minus sign denotes that it is instantaneously towards negative x-direction. **The particle continues to move back and forth between $x = 0$ and $x = \frac{2E_0}{\alpha}$.**

The mean position of the particle is $x = E_0/\alpha$ (where $\vec{F}_e = 0$)

5. Let s_0 be the total displacement of the particle till it stops in time t_0. Then average velocity

$$<v> = \frac{s_0}{t_0} \qquad ...(1)$$

Given that $F = -kv^n$

or $$m\left(v \cdot \frac{dv}{ds}\right) = -kv^n$$

or $$v^{1-n}\, dv = -\frac{k}{m} \cdot ds$$

or $$\int_{v_0}^{0} v^{1-n} dv = -\frac{k}{m}\int_0^{s_0} ds$$

v_0 = initial velocity of particle

or $$s_0 = \frac{mv_0^{2-n}}{k\,(2-n)} \qquad ...(2)$$

We can also write

$$m \cdot \left(\frac{dv}{dt}\right) = -kv^n$$

or $$v^{-n} dv = -\frac{k}{m}\, dt$$

or $$\int_{v_0}^{0} v^{-n} \cdot dv = -\frac{k}{m}\int_0^{t_0} dt$$

or $$t_0 = \frac{mv_0^{1-n}}{k\,(1-n)} \qquad ...(3)$$

From equations (1), (2) and (3),

$$<v> = \left(\frac{1-n}{2-n}\right) v_0$$

Given; $$<v> = \frac{v_0}{3}$$

or $$\frac{1-n}{2-n} = \frac{1}{3}$$

or $$\boldsymbol{n = \frac{1}{2}}$$

6. Weight of the lift $= Mg$ newton

Maximum tension $= nMg$ newton

$\therefore$ Net force on lift $= nMg - Mg = (n-1)\,Mg$

$\therefore$ Maximum acceleration

$$= \frac{(n-1)\,Mg}{M} = (n-1)\,g$$

Let the maximum velocity be v_m. Figure shows

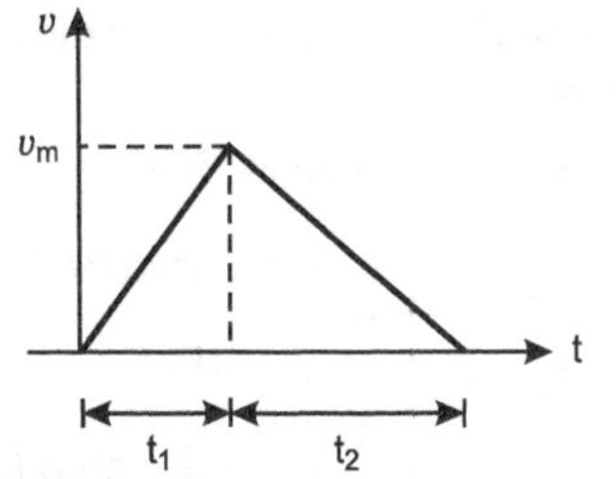

velocity-time graph

Maximum retardation can be g.

Let t_1 is the time of acceleration and t_2 the time of retardation. Then :

$$\frac{v_m}{t_1} = (n-1)g \quad i.e. \; t_1 = \frac{v_m}{(n-1)\,g}$$

$$\frac{v_m}{t_2} = g \quad i.e. \quad t_2 = \frac{v_m}{g}$$

Now area under $v - t$ graph gives, the displacement. Hence

$$\frac{1}{2} v_m\,(t_1 + t_2) = h$$

Substituting values of t_1 and t_2 we get

$$v_m^2\left[\frac{1}{(n-1)\,g} + \frac{1}{g}\right] = 2h$$

or $$\frac{v_m^2}{g}\left(\frac{n}{n-1}\right) = 2h$$

or $$v_m^2 = 2gh\left(\frac{n-1}{n}\right) \qquad ...(1)$$

Total time $t = t_1 + t_2$

or $$t = \frac{v_m}{(n-1)\,g} + \frac{v_m}{g} = \frac{v_m}{g}\left(\frac{n}{n-1}\right)$$

Substituting the value of v_m from equation (1), we get

$$t = \frac{n}{g\,(n-1)}\sqrt{\frac{2gh\,(n-1)}{n}} \Rightarrow \boldsymbol{t = \sqrt{\frac{2nh}{(n-1)\,g}}}$$

7. Velocity-time graph of the particle is as shown below :

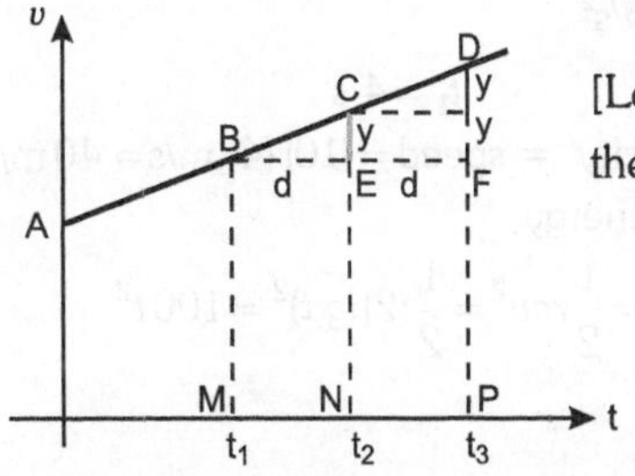

[Let $CE = y$ then $DF = 2y$]

Area under $v - t$ graph gives the displacement. hence

$$s_2 - s_1 = \text{area } BCNM$$
$$= \frac{1}{2} y.d + \text{area } BENM \quad \dots(1)$$
$$s_3 - s_2 = \text{area } CDPN$$
$$= \frac{1}{2} yd + yd + \text{area } EFPN \quad \dots(2)$$

(1) — (2) gives :

$$2s_2 - s_1 - s_3 = -y.d$$

or $$y \cdot d = s_1 + s_3 - 2s_2$$

or $$y = \frac{s_1 + s_3 - 2\sqrt{s_1 s_3}}{d} \qquad (s_2 = \sqrt{s_1 s_3})$$

or $$\frac{y}{d} = \frac{(\sqrt{s_1} - \sqrt{s_3})^2}{d^2} = \text{slope of line } AD$$
$$= \text{acceleration}$$

$\therefore$ acceleration of particle

$$= \frac{(\sqrt{s_1} - \sqrt{s_3})^2}{d^2}$$ **Proved**

8. Let t_1 be the time of first collision. Applying $h = \frac{1}{2} g t^2$, we get

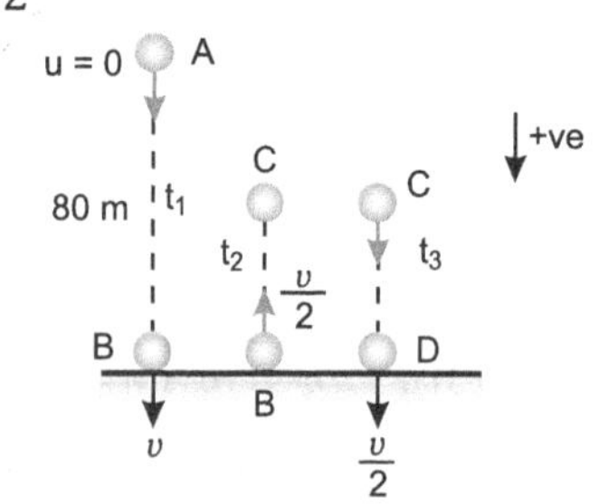

$$80 = \frac{1}{2} g t_1^2 \qquad \therefore \; t_1 = 4\text{s}$$

A* to *B velocity = speed $= g\,t = 10t$

hence velocity-time and speed time graphs are straight lines passing through origin with slope $g = 10 \text{ m/s}^2$.

At B $\qquad t_1 = 4$ s

$\therefore$ velocity = speed = (10) (4) m/s = 40 m/s

kinetic energy,

$$K = \frac{1}{2} mv^2 = \frac{1}{2}(2)(g\,t)^2 = 100\,t^2$$

therefore, kinetic energy-time graph is a parabola passing through origin.

At B $\qquad t_1 = 4$s

$\therefore \qquad K = (100)(4)^2 = 1600$ J

B* to *C velocity $= -\left(\frac{v}{2} - gt\right) = -20 + 10t$

$(v = 40 \text{ m/s})$

speed = | velocity | = $20 - 10t$

At C : velocity or speed = 0 $\quad \therefore \; t_2 = 2$s

kinetic energy, $K = \frac{1}{2} mv^2 = \frac{1}{2}(2)(20 - 10t)^2$
$$= (20 - 10t)^2$$

C* to *D velocity = speed $= gt = 10t$; $\quad t_3 = 2$ s

kinetic energy $= \frac{1}{2} mv^2$
$$= \frac{1}{2}(2)(10\,t)^2 = 100\,t^2$$

At D : speed = velocity = (10) (2) = 20 m/s

and kinetic energy $K = (100)\,(2)^2 = 400$ J.

The corresponding graphs are as follows :

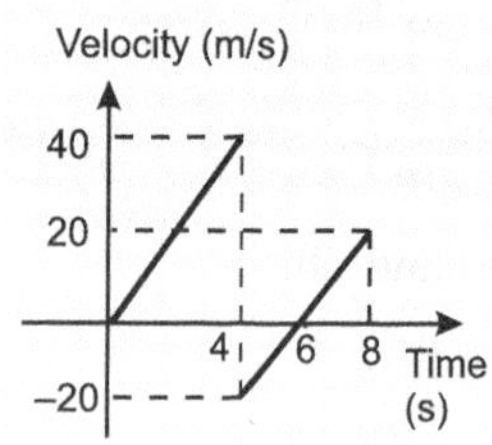

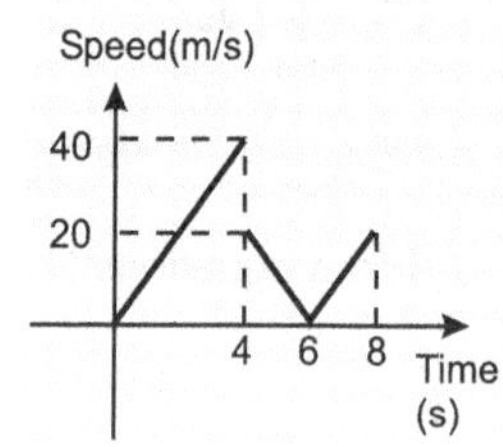

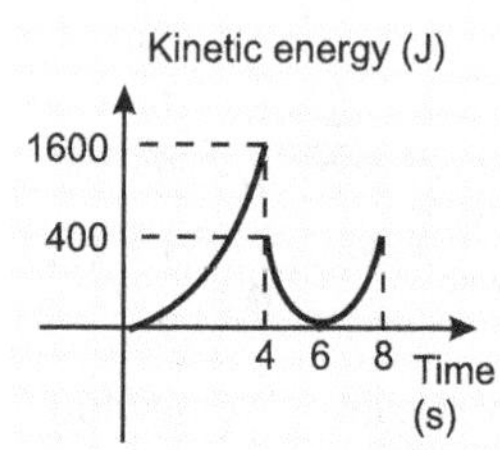

9. We know that

$$v \cdot \frac{dv}{ds} = a \quad \text{or} \quad v \cdot dv = ads$$

or $$\int_0^v v \cdot dv = \int ads$$

or $\dfrac{v^2}{2}$ = area under a-s graph

or $v = \sqrt{2(\text{area under } a - s \text{ graph})}$

$$\text{area} = \frac{1}{2}(2)(2) + 6 \times 2 + \frac{1}{2}(2+4)(2) + \frac{1}{2} \times 2 \times 4$$

$$= 2 + 12 + 6 + 4 = 24 \text{ m}^2/\text{s}^2$$

$\therefore$ $\boldsymbol{v = \sqrt{2 \times 24} = 4\sqrt{3}}$ **m/s**

10. ***A* to *B***

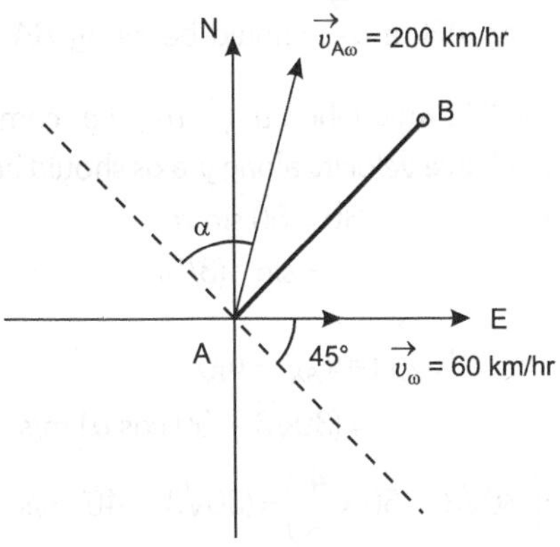

$$AB = 100 \text{ km}$$

$\vec{v}_w$ = 60 km/hr = velocity of wind

$\vec{v}_{Aw}$ = 200 km/hr = velocity of aircraft with respect to wind

$\vec{v}_A = \vec{v}_w + \vec{v}_{Aw}$ = velocity of aircraft with respect to ground

Let $\vec{v}_{Aw}$ makes an angle α with the perpendicular to *AB* (shown as dotted line in figure).

Aircraft has to reach from *A* to *B*, hence $\vec{v}_A$ should be along *AB* or components of $\vec{v}_{Aw}$ and $\vec{v}_w$ perpendicular to *AB* should be zero. Therefore,

$200 \cos \alpha = 60 \cos 45°$ or $\alpha = 77.7°$

and resultant velocity of aircraft along *AB* would be

$$|\vec{v}_A| = 60 \sin 45° + 200 \sin \alpha$$

$$= 237.87 \text{ km/hr}$$

$$\therefore \quad t_{AB} = \frac{AB}{|\vec{v}_A|} = \frac{100}{237.87} \text{ hr} = 0.42 \text{ hr}$$

B* to *A

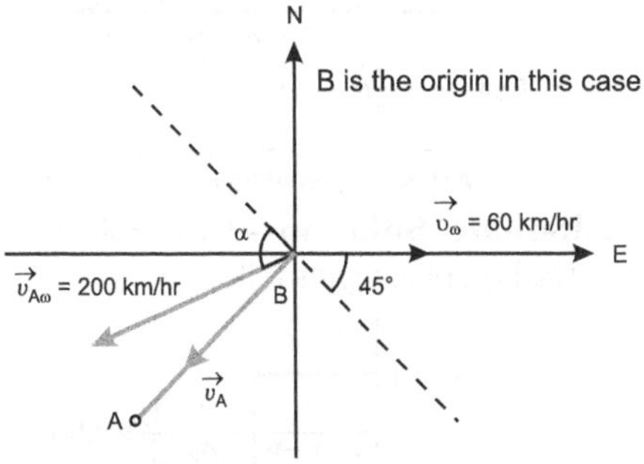

Similarly, components of $\vec{v}_{A\omega}$ and $\vec{v}_{\omega}$ perpendicular to *BA* should be zero. For this $\alpha = 77.7°$

Now resultant velocity of aircraft along *BA* is

$$|\vec{v}_A| = 200 \sin \alpha - 60 \sin 45° = 153 \text{ km/hr}$$

$$\therefore \quad t_{BA} = \frac{BA}{|\vec{v}_A|} = \frac{100}{153} \text{ hr}$$

or $t_{BA} = 0.65$ hr

$\therefore$ Total time of flight $= t_{AB} + t_{BA}$

$= (0.42 + 0.65)$ hr

$= 1.07$ **hr**

$= 64.2$ **min**

11. Let v_{br} be the velocity of boatman with respect to river or velocity of boatman in still water and v_r the river velocity. Velocity of raft is also v_r. Assuming river to be at rest or raft to be at rest. The boatman will move with v_{br} both during its downstream and upstream motion. Therefore Time of upstream motion = time of downstream motion = 1 hr.

$\therefore$ Total time of motion of raft = 2 hrs. In this 2 hrs, the raft moves 6 km. Hence the raft velocity or river velocity is 3 km/hr. This can be shown on next page :

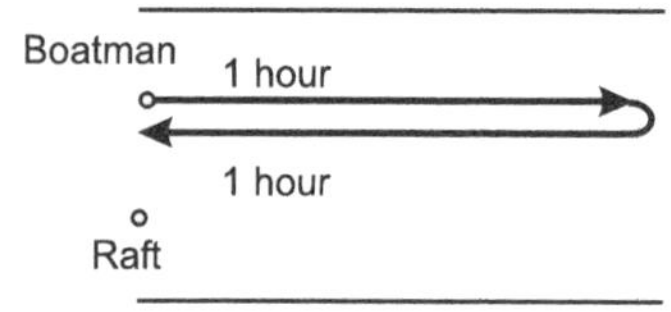

Assuming river to be at rest

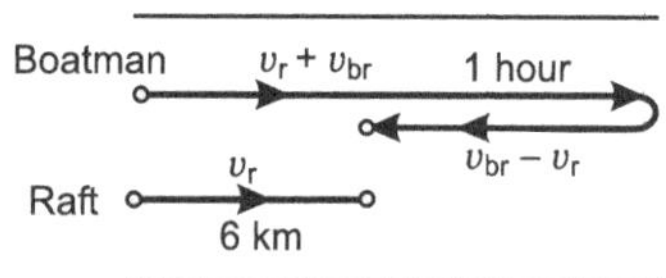

Actual motion of boatman and raft

Alternate Solution—This problem can be solved graphically as follows :

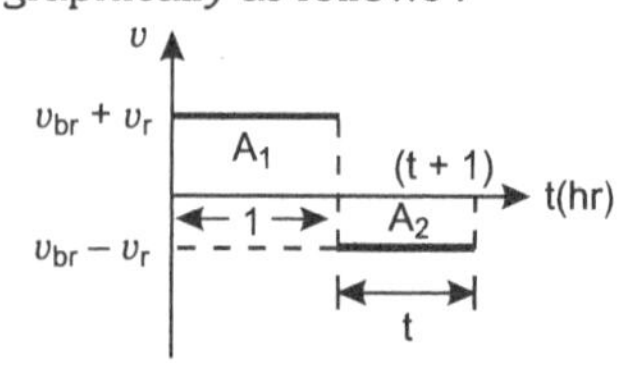

v–t graph of boatman

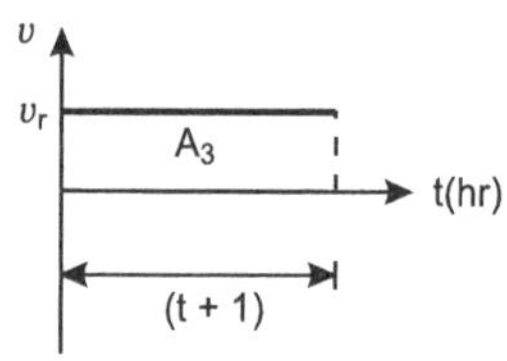

v–t graph of raft

Displacement of boatman = Displacement of raft

$$\therefore \qquad A_1 - A_2 = A_3$$

$$\therefore \quad (v_{br} + v_r)(1) - (v_{br} - v_r)(t) = (v_r)(t+1)$$

This gives $\quad t = 1$ hour

Now displacement of raft *i.e.* A_3 is 6 km

Hence $(t+1)v_r = 6$ or $2v_r = 6$

or $\quad \boldsymbol{v_r = 3}$ **km/hr**

12. Taking x and y directions as shown in figure :

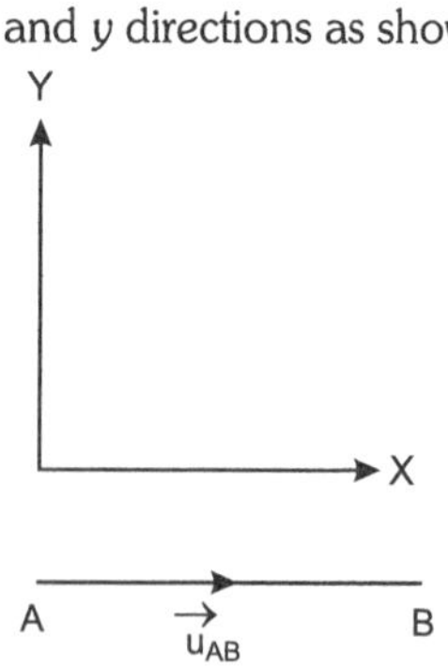

$$\vec{a}_A = -g\,\hat{j}, \qquad \vec{a}_B = -g\,\hat{j}$$

$$u_{AX} = 60\cos 30° = 30\sqrt{3} \text{ m/s}$$

$$u_{AY} = 60\sin 30° = 30 \text{ m/s}$$

$$u_{BX} = -\ 50\cos\alpha \quad \text{and} \quad u_{BY} = 50\sin\alpha$$

Relative acceleration between the two is zero ($\vec{a}_A = \vec{a}_B$). Hence the relative motion between the two is uniform.

Condition of collision is that $\vec{u}_{AB}$ should be along AB or $\vec{u}_{BA}$ should be along BA. This is possible only when $u_{AY} = u_{BY}$ *i.e.* component of relative velocity along y-axis should be zero.

or $\qquad 30 = 50\sin\alpha$

or $\qquad \alpha = \sin^{-1}(3/5)$

Now $\quad |\vec{u}_{AB}| = u_{AX} - u_{BX}$

$$= (30\sqrt{3} + 50\cos\alpha) \text{ m/s}$$

$$= \left(30\sqrt{3} + 50 \times \frac{4}{5}\right) = (30\sqrt{3} + 40) \text{ m/s}$$

$$\therefore \quad t = \frac{100}{|\vec{u}_{AB}|} = \frac{100}{30\sqrt{3} + 40} = \mathbf{1.09\ s}$$

therefore the particles collide after a time t = 1.09 s

Distance of point P from A where collision takes place is

$$S = \sqrt{(u_{AX}\,t)^2 + (u_{AY}t - \frac{1}{2}gt^2)^2}$$

$$= \sqrt{(30\sqrt{3} \times 1.09)^2 + (30 \times 1.09 - \frac{1}{2} \times 10 \times 1.09 \times 1.09)^2}$$

or $\qquad$ **S = 62.64m**

13. Point P is the point of intersection of two curves

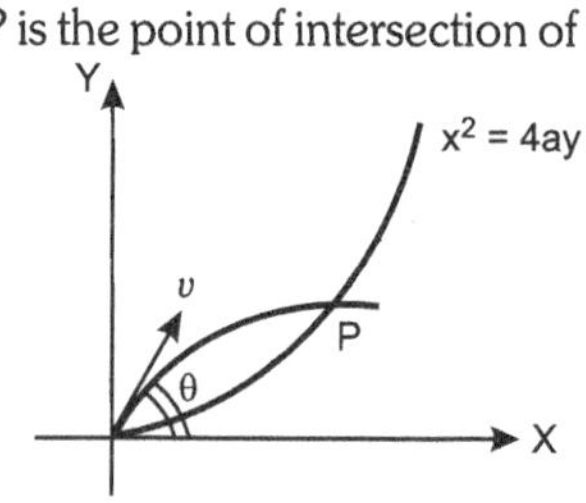

$$x^2 = 4ay \text{ or } y = \frac{x^2}{4a} \quad \dots (1)$$

and $$y = x \tan\theta - \frac{gx^2}{2v^2}(1+\tan^2\theta) \quad \dots (2)$$

(equation of projectile)

equating (1) and (2), we get

$$\frac{x^2}{4a} = x\tan\theta - \frac{gx^2}{2v^2}(1+\tan^2\theta)$$

$$x^2\left[\frac{1}{4a} + \frac{g}{2v^2}(1+\tan^2\theta)\right] = x\tan\theta$$

so either $x = 0$ or

$$\boldsymbol{x = \frac{4av^2\tan\theta}{v^2 + 2ag + 2ag\tan^2\theta}}$$

14. (a) $$v_y = \frac{dy}{dt} = \sqrt{2gy} \quad \dots (1)$$

$$v_x = \frac{dx}{dt} = ay \quad \dots (2)$$

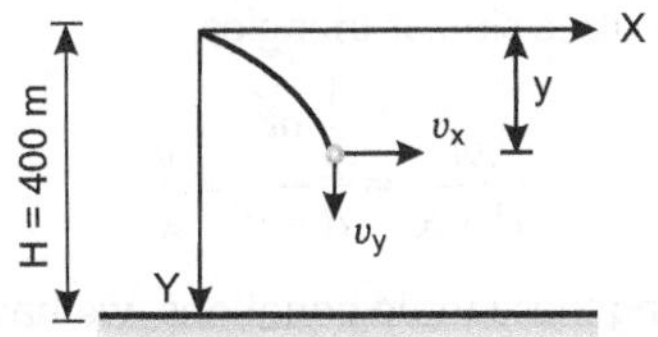

Dividing (1) by (2), we get

$$\frac{dy}{dx} = \frac{\sqrt{2gy}}{ay} = \frac{2}{\sqrt{y}}$$

($g = 10$ m/s^2 and $a = \sqrt{5}$ s^{-1})

or $$\sqrt{y}\cdot dy = 2\,dx$$

or $$\int_0^{400\text{m}} \sqrt{y}\,dy = 2\int_0^x dx$$

or $$\frac{2}{3}(y^{3/2})_0^{400\text{m}} = 2x$$

or $$x = \frac{1}{3}(400)^{3/2}$$

or $$x = \frac{8}{3}\times 10^3 \text{ m}$$

or $$\boldsymbol{x = 2.67 \textbf{ km}}$$

(b) Speed of particle when it strikes the ground will be

$$v = \sqrt{v_x^2 + v_y^2} = \sqrt{(2gy) + (a^2y^2)}$$

$$= \sqrt{(2)(10)(400) + (5)(400)^2}$$

or $v = 898.8$ m/s

or $\boldsymbol{v \approx 0.9}$ **km/s**

15. There is no loss of mechanical energy when the string hits the nail C, because the sudden change in tension is perpendicular to the direction of motion of the particle and therefore, has no effect on its speed.

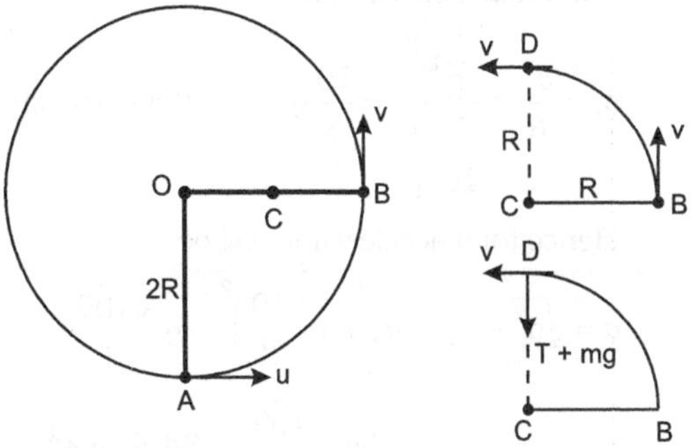

Let v be the minimum velocity at point D. Applying conservation of mechanical energy between points A and D

$$mg(3R) = \frac{1}{2}m(u^2 - v^2)$$

or $$v^2 = u^2 - 6gR \quad \dots (1)$$

Applying Newton's law radially at point D

$$T + mg = \frac{mv^2}{R} = \frac{m}{R}(u^2 - 6gR)$$

or $$T = \frac{m}{R}(u^2 - 7gR)$$

Now $T \geq 0$ or $u^2 \geq 7gR$

or $\boldsymbol{u_{\min} = \sqrt{7gR}}$

16. Let R be the radius of the circle, u be its velocity at bottommost point and $\frac{u}{2}$ at topmost point.

Then

$$\left(\frac{u}{2}\right)^2 = u^2 - 2g(2R)$$

or $$4gR = \frac{3}{4}u^2$$

or $$R = \frac{3}{16}\frac{u^2}{g}$$

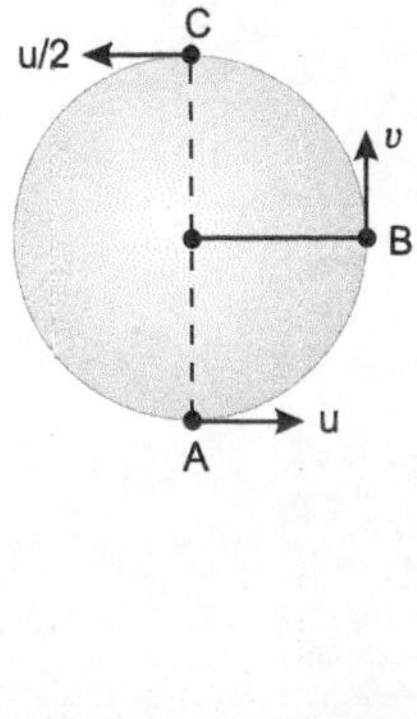

If υ is its velocity at point B, then

$$\upsilon^2 = u^2 - 2gR = u^2 - 2g\left(\frac{3}{16}\cdot\frac{u^2}{g}\right)$$

$$\upsilon^2 = \frac{5}{8}u^2$$

At point B its tangential acceleration is

$$a_t = g \qquad \text{(downwards)}$$

and radial acceleration,

$$a_r = \frac{\upsilon^2}{R} = \frac{\frac{5}{8}u^2}{\frac{3}{16}\frac{u^2}{g}} = \frac{10}{3}g \qquad \text{(towards centre)}$$

Hence total acceleration will be

$$a = \sqrt{a_t^2 + a_r^2} - g\sqrt{1 + \left(\frac{10}{3}\right)^2} = g\frac{\sqrt{109}}{3}$$

$$= (10)\left(\frac{\sqrt{109}}{3}\right) \approx \mathbf{34.8\ m/s^2}$$

17. (i) $g' = g + \frac{qE}{m}$

$$= 9.8 + \frac{10^{-6} \times 10^{6}}{0.01} = 109.8\ \text{m/s}^2$$

$$\therefore\ T = 2\pi\sqrt{\frac{l}{g'}} = 2\pi\sqrt{\frac{1}{109.8}} = \mathbf{0.6\ s}$$

(ii) $\upsilon_{\min} = \sqrt{5g'l}$

$$= \sqrt{5 \times 109.8 \times 1} = \mathbf{23.43\ m/s}$$

(iii) $T_{\mathbf{min}} = 0$ at topmost point

and $T_{\mathbf{max}} = 6mg'$ at bottommost point

$$= 6 \times 0.01 \times 109.8 = \mathbf{6.59\ N}$$

18. (a) Average velocity $= \dfrac{\int_0^5 \upsilon\,.\,dt}{\int_0^5 dt}$

$$= \frac{\int_0^5 (3t - t^2)\,dt}{5}$$

$$= \frac{\frac{3}{2}(25) - \frac{1}{3}(125)}{5}$$

$$= \mathbf{0.833\ m/s}$$

(b) For average speed let us put, $\upsilon = 0$, which gives,

$$t = 0 \quad \text{and} \quad t = 3\ \text{sec}$$

$$\therefore\ \text{Average speed} = \frac{\left|\int_0^3 \upsilon dt\right| + \left|\int_3^5 \upsilon dt\right|}{5}$$

Substituting the values and solving we get,

Average speed = **2.63 m / s**

19. After time 't' position of different particles will be as under:

Let $AB = BC = d$

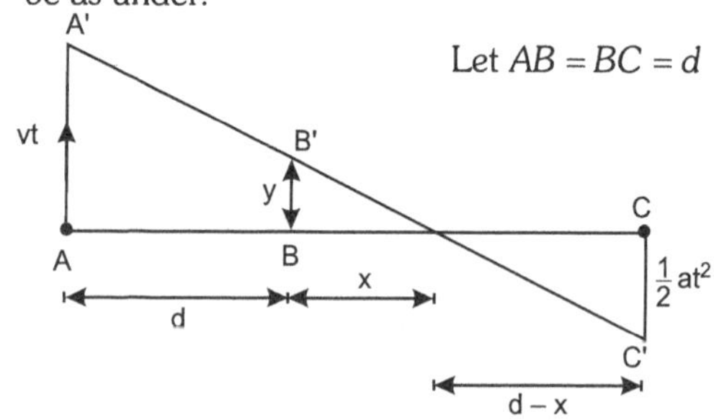

From three similar triangles,

$$\frac{\upsilon t}{d + x} = \frac{\frac{1}{2}at^2}{d - x} = \frac{y}{x}$$

Solving these three equations, we have,

$$y = \frac{\upsilon}{2}t - \frac{1}{4}at^2$$

Comparing with $s = ut + \frac{1}{2}at^2$ we get,

$$u = \frac{\upsilon}{2} \quad \text{and} \quad \text{acceleration} = -\frac{1}{2}a$$

Thus point B should move up with initial velocity $\upsilon/2$ and downward acceleration of $a/2$.

20. At the given instant component of $\vec{\upsilon}_{BA}$ along BA is $u - \upsilon\cos\theta$. This is basically the rate by which distance 'r' between A and B is decreasing. Thus,

$$(u - \upsilon\cos\theta) = -\frac{dr}{dt} \qquad \ldots(1)$$

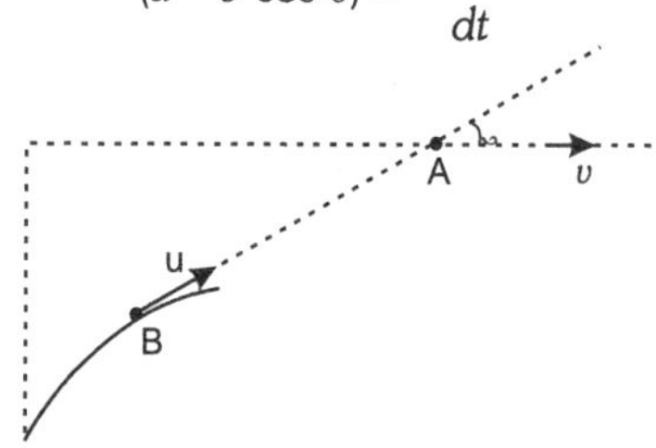

Further,

$$\frac{d\theta}{dt} = \frac{\text{Component of relative velocity perpendicular to } AB}{\text{distance } AB}$$

or $$\frac{d\theta}{dt} = \frac{v \sin\theta}{r} \quad \ldots(2)$$

Dividing (1) by (2), we have

$$-\frac{dr}{d\theta} = \frac{r\,(u - v\cos\theta)}{v\sin\theta}$$

$$\therefore \quad -\frac{dr}{r} = \frac{(u - v\cos\theta)}{v\sin\theta}\,d\theta$$

or $$\int_d^r -\frac{dr}{r} = \int_{\pi/2}^{\theta}\left(\frac{u - v\cos\theta}{v\sin\theta}\right).\,d\theta$$

Solving this equation we get the desired relation.

21. In the time when the boat moves from O to P, the displacement perpendicular to OP will zero. Let the time taken by the boat to move from O to $P = t$.

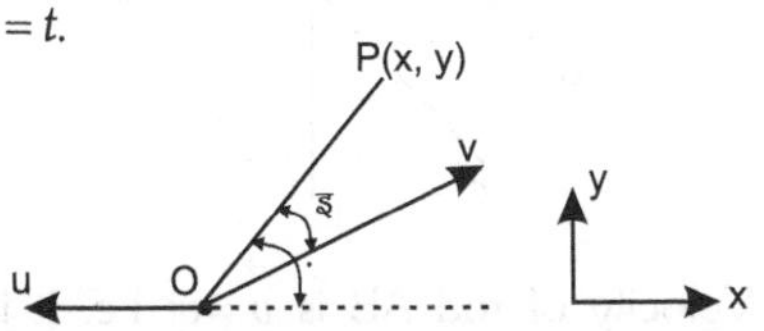

Resultant of u and v should be along OP. Therefore, their components perpendicular to OP should cancel each other. Hence,

$$u\sin\alpha = v\sin\beta$$

$$\therefore \quad \sin\beta = \frac{u}{v}\sin\alpha = \frac{u}{v}\cdot\frac{y}{\sqrt{x^2+y^2}}$$

$$\therefore \quad \beta = \sin^{-1}\left[\frac{u}{v}\cdot\frac{y}{\sqrt{x^2+y^2}}\right]$$

Further, $$t = \frac{OP}{v\cos\beta - u\cos\alpha}$$

Here, $$\alpha = \sin^{-1}\left(\frac{y}{\sqrt{x^2+y^2}}\right)$$

and $$\beta = \sin^{-1}\left[\frac{u}{v}\sin\alpha\right]$$

22. If it is being hit then

$$d = v_o t + \frac{1}{2}at^2 = (u\cos\theta)\,t$$

or $$t = \frac{u\cos\theta - v_o}{a/2}$$

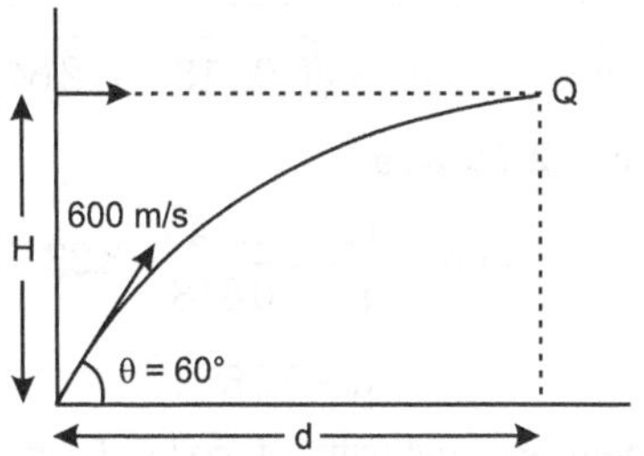

$$\therefore \quad t = \frac{600\times\frac{1}{2} - 250}{10} = 5\text{ s}$$

$$H = (u\sin\theta)\,t - \frac{1}{2}gt^2$$

$$= 600\times\frac{\sqrt{3}}{2}\times 5 - \frac{1}{2}\times 10\times 25$$

$$H = 1500\sqrt{3} - 125$$

$$\Rightarrow \quad H = \mathbf{2473\ m}$$

23. For $\theta = 30° = \frac{\pi}{6}$ radian

$$\frac{\pi}{6} = 0.5\,t^2 \quad \Rightarrow \quad t = 1.023\text{ s}$$

$$r = 3 - 0.4\,(1.023)^2 = 2.58\text{ m}$$

$$\frac{dr}{dt} = \frac{d}{dt}(3 - 0.4\,t^2) = -\,0.8\,t$$

or $$\frac{dr}{dt} = -\,(0.8)\,(1.023) = -\,0.818\text{ m/s}$$

(at $t = 1.023$ s)

$$\frac{d\theta}{dt} = t \qquad (\theta = 0.5\,t^2)$$

or $$\frac{d\theta}{dt} = 1.023\text{ rad/s} \qquad (\text{at } t = 1.023\text{ s})$$

Now two components of velocity of B at $\theta = \pi/6$ or $t = 1.023$ s are

$$v_r = \frac{dr}{dt} = -\,0.818\text{ m/s}$$

and $$v_\theta = r\left(\frac{d\theta}{dt}\right) = (2.58)\,(1.023) = 2.64\text{ m/s}$$

$\therefore$ Resultant velocity of B is

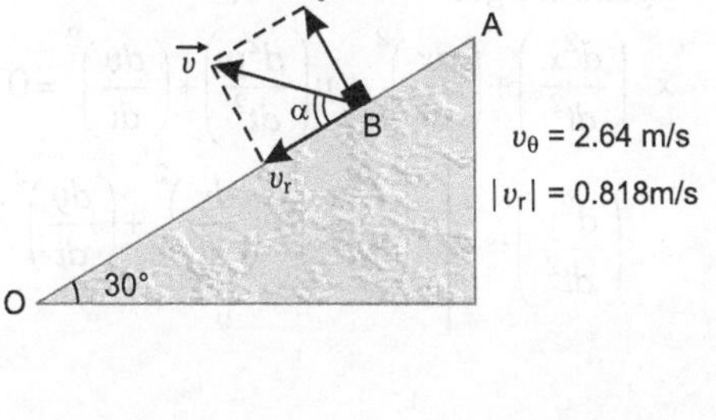

$$v = \sqrt{v_r^2 + v_\theta^2} = \sqrt{(-0.818)^2 + (2.64)^2}$$

$\mathbf{v = 2.76\ m/s}$

$$\tan\alpha = \frac{v_\theta}{|v_r|} = \frac{2.64}{0.818} = 3.23$$

or $\alpha = 72.8°$

Therefore, velocity of collar B at the given instant is 2.76 m/s at an angle $\alpha = 72.8°$ with OA in the direction shown in figure.

24. Point B moves with the same velocity and acceleration as A.

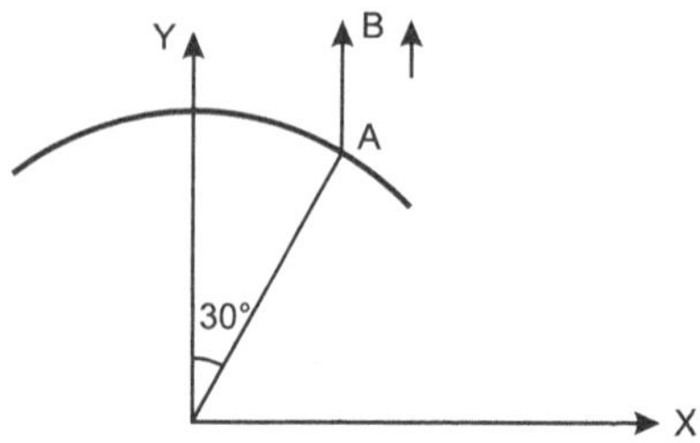

Assuming the cam to be at rest and the point A of the rod follower to slide up the circular path with the x-component of velocity – 5 cm/s and x-component of acceleration – 10 cm/s^2, the equation of the path, then being

$$x^2 + y^2 = (30)^2$$

Differentiating it with respect to time t

$$x \cdot \left(\frac{dx}{dt}\right) + y\left(\frac{dy}{dt}\right) = 0 \quad \ldots(1)$$

Here $x = 30 \sin 30° = 15$ cm

and $y = 30 \cos 30° = 26$ cm

From equation (1)

$$\frac{dy}{dt} = -\frac{x}{y}\cdot\frac{dx}{dt}$$

or $$\mathbf{\frac{dy}{dt} = -\left(\frac{15}{26}\right)(-5) = 2.88\ cm/s}$$

Which must be velocity of A and B (upwards)

Differentiating (1) with respect to time t once again, we get

$$x \cdot \left(\frac{d^2x}{dt^2}\right) + \left(\frac{dx}{dt}\right)^2 + y\left(\frac{d^2y}{dt^2}\right) + \left(\frac{dy}{dt}\right)^2 = 0$$

$$\therefore \left(\frac{d^2y}{dt^2}\right) = -\left[\frac{x\left(\frac{d^2x}{dt^2}\right) + \left(\frac{dx}{dt}\right)^2 + \left(\frac{dy}{dt}\right)^2}{y}\right]$$

Substituting the value we get

$$\left(\frac{d^2y}{dt^2}\right) = -\left[\frac{(15)(-10) + (-5)^2 + (2.88)^2}{26}\right]$$

$$= \mathbf{4.5\ cm/s^2}$$

This is also the acceleration of A and B (upwards).

25. Rod AB is lying along x-axis and rod CD along line $x = y$.

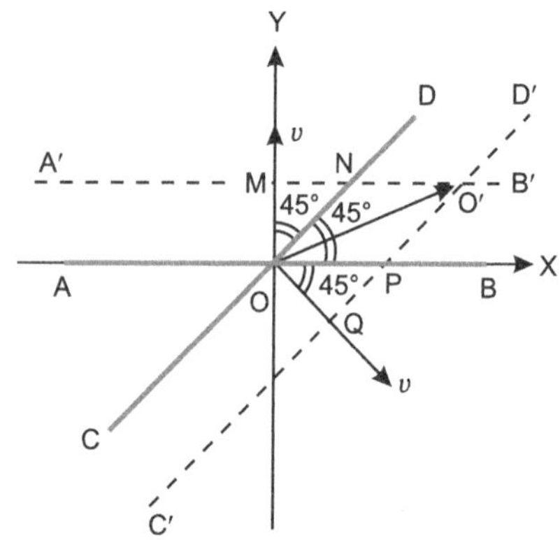

Velocity of rod AB is $\vec{v} = v\hat{j}$ i.e. it is moving along positive y-direction. Similarly velocity of rod CD is $\frac{v}{\sqrt{2}}\hat{i} - \frac{v}{\sqrt{2}}\hat{j}$. *i.e.* it is moving in the direction shown in figure with speed v. After time t,

$$OQ = vt$$

$$\therefore \quad OP = OQ \sec 45° = \sqrt{2}\,vt$$

similarly $OM = vt$

and $ON = OM \sec 45° = \sqrt{2}\,vt$

hence

$$OO' = \sqrt{(OP)^2 + (ON)^2 + 2(OP)(ON)\cos 45°}$$

$$= \sqrt{2v^2t^2 + 2v^2t^2 + 2(\sqrt{2}vt)(\sqrt{2}vt)\left(\frac{1}{\sqrt{2}}\right)}$$

$$OO' = vt\sqrt{4 + 2\sqrt{2}}$$

$\therefore$ velocity of point of intersection is

$$v_o = \frac{OO'}{t} = v\sqrt{4 + 2\sqrt{2}} = 2.61\,v$$

at bisector of ON and OP or at an angle of 22.5° with positive x-axis.

$\therefore \quad \vec{v}_o = v_o \cos 22.5° \hat{i} + v_o \sin 22.5° \hat{j}$

or $\quad \vec{\boldsymbol{v}}_o = \boldsymbol{v}\,(\mathbf{2.41}\,\hat{\boldsymbol{i}} + \hat{\boldsymbol{j}})$

26. Accelerating force

$$F_1 \propto t \quad \text{or} \quad F_1 = t$$

and retarding force

$F_2 \propto x \quad \text{or} \quad F_2 = x$

constant of proportionality being 1 in both the cases.

hence $\quad F_{net} = F_1 - F_2 = t - x$

or $\quad a = t - x \quad \ldots(1) \quad (\text{mass} = 1 \text{ kg})$

differentiating equation (1) w.r.t. time, we get

$$\frac{da}{dt} = 1 - \frac{dx}{dt}$$

or $$\frac{d^2v}{dt^2} = 1 - v \qquad \ldots(2)$$

Now, let us assume

$$V = 1 - v \qquad \ldots(3)$$

or $$-\frac{d^2v}{dt^2} = \frac{d^2V}{dt^2}$$

$\therefore$ Equation (2) can be written as

$$\frac{d^2V}{dt^2} = -V \qquad \ldots(4)$$

comparing this with standard equation of S.H.M. *i.e.* $\frac{d^2V}{dt^2} = -\omega^2 V$

$\therefore$ V will oscillate simple harmonically with angular frequency $\omega = 1$ rad/s

also at time $t = 0$, $v = 0$, hence $V = 1$ [from equation (3)]

$\therefore \quad V = \cos t \quad \text{or} \quad 1 - v = \cos t$

or $\quad v = 1 - \cos t \quad \text{or} \quad \frac{dx}{dt} = 1 - \cos t$

or $\quad \int_0^x dx = \int_0^t (1 - \cos t)\, dt$

or $\quad \boldsymbol{x = t - \sin t}$

27. The given equation *i.e.* $x = A \cos \omega t$ is an equation of SHM. Let T be the time period of SHM. Then distance travelled by the particle is 'A' during the time $T/4$ starting from extreme position or mean position. therefore, we may write

$$t = t_1 + t_2 \qquad \text{where } t_1 = nT/4$$

here $\quad n = 0, 1, 2, 3 \ldots$

and $\quad t_2 < \frac{T}{4} \quad \text{or} \quad t_2 = (t - nT/4)$

now two cases arise :

Case 1 : When *n* is even– Say $n = 0, 2, 4\ldots$

In this case, particle will be in its extreme position in time t_1. Hence distance travelled in time t would be

$$s = s_1 + s_2 \qquad \text{where } s_1 = nA$$

and s_2 = distance travelled in time t_2 from extreme position.

$$= A - A \cos \omega t_2$$

$$= A - A \cos\left(\omega t - \frac{n\pi}{2}\right)$$

$$\left\{t_2 = t - \frac{nT}{4} \text{ and } T = \frac{2\pi}{\omega}\right\}$$

$$\therefore \quad s = nA + A - A \cos\left(\omega t - \frac{n\pi}{2}\right)$$

$$= A\left[n + 1 - \cos\left(\omega t - \frac{n\pi}{2}\right)\right]$$

Case 2 : When *n* is odd – Say $n = 1, 3, 5\ldots.$

In this case, particle will be in its mean position in time t_1. Hence

$$s = s_1 + s_2$$

where $s_1 = nA$ and s_2 = distance travelled in time t_2 from mean position.

$$= A \sin \omega t_2 = A \sin\left(\omega t - \frac{n\pi}{2}\right)$$

$$\therefore \quad s = nA + A \sin\left(\omega t - \frac{n\pi}{2}\right)$$

or $$s = A\left[n + \sin\left(\omega t - \frac{n\pi}{2}\right)\right]$$

Hence $\boldsymbol{s = A\left[n + 1 - \cos\left(\omega t - \frac{n\pi}{2}\right)\right]}$

when *n* is even

and $\quad s = A\left[n + \sin\left(\omega t - \frac{n\pi}{2}\right)\right]$

when n is odd.

28. (i) $\vec{v}_A = u\hat{i}$ and $\vec{v}_B = u\hat{j}$

Let at any time t, coordinates of point A and B be $(X, 0)$ and $(0, Y)$ respectively.

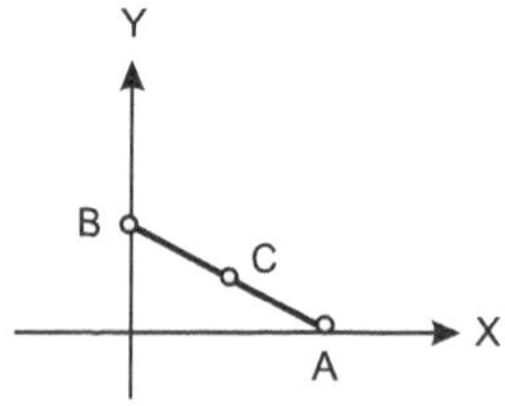

Then coordinates of point $C\,(x, y)$ will be

$$x = \frac{X}{2} \quad \text{or} \quad \frac{dx}{dt} = \frac{1}{2}\frac{dX}{dt} \quad \text{or} \quad v_x = \frac{u}{2}$$

$$\text{and } y = \frac{Y}{2} \quad \text{or} \quad \frac{dy}{dt} = \frac{1}{2}\frac{dY}{dt} \quad \text{or} \quad v_y = \frac{u}{2}$$

Hence velocity of C would be

$$\vec{v}_C = \frac{u}{2}\hat{i} + \frac{u}{2}\hat{j}$$

$\therefore$ Velocity of C with respect to A is

$$\vec{v}_{CA} = \vec{v}_C - \vec{v}_A = -\frac{u}{2}\hat{i} + \frac{u}{2}\hat{j} \qquad \ldots(1)$$

or $$|\vec{v}_{CA}| = \frac{u}{\sqrt{2}}$$

Velocity of C with respect to B will be

$$\vec{v}_{CB} = \frac{u}{2}\hat{i} - \frac{u}{2}\hat{j} \qquad \ldots(2)$$

From (1) and (2) we see that

$$\vec{v}_{CA} = -\vec{v}_{CB}$$

(ii) At points M and N, acceleration of the particle moving on the circle $x^2 + y^2 = 1$ with constant speed v is parallel to the line $x + y = 0$. This is its centripetal acceleration which acts towards centre. Velocity of particle at M and N is

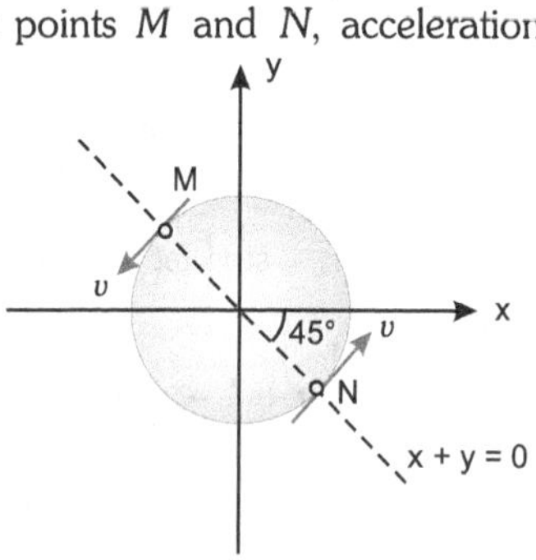

$$\vec{v}_M = -\frac{v}{\sqrt{2}}\hat{i} - \frac{v}{\sqrt{2}}\hat{j}$$

and $$\vec{v}_N = \frac{v}{\sqrt{2}}\hat{i} + \frac{v}{\sqrt{2}}\hat{j}$$

$$\vec{v}_{MA} = \vec{v}_M - \vec{v}_A = \left(-\frac{v}{\sqrt{2}} - u\right)\hat{i} - \frac{v}{\sqrt{2}}\hat{j}$$

and $$\vec{v}_{MB} = \vec{v}_M - \vec{v}_B = -\frac{v}{\sqrt{2}}\hat{i} + \left(-\frac{v}{\sqrt{2}} - u\right)\hat{j}$$

We can show that $|\vec{v}_{MA}| = |\vec{v}_{MB}|$

Similarly we can prove that

$$|\vec{v}_{NA}| = |\vec{v}_{NB}|$$

(iii) Magnitude of velocity of C is $\frac{u}{\sqrt{2}}$.

Therefore, the maximum value of relative velocity of P with respect to C will be $v + \frac{u}{\sqrt{2}}$, when P and C are moving in opposite direction. But this value is given to be equal to u. Hence

$$v + \frac{u}{\sqrt{2}} = u \quad \text{or} \quad v = u\left(1 - \frac{1}{\sqrt{2}}\right)$$

29. (a) Let $\vec{v}_{br}$ be the velocity of boatman relative to river, $\vec{v}_r$ the velocity of river and $\vec{v}_b$ is the absolute velocity of boatman. Then

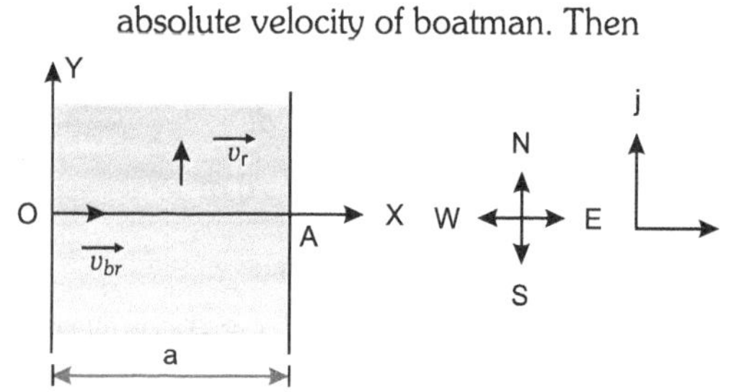

$$\vec{v}_b = \vec{v}_{br} + \vec{v}_r$$

Given; $|\vec{v}_{br}| = v$ and $|\vec{v}_r| = u$

Now $u = v_y = \frac{dy}{dt} = x(a-x)\frac{v}{a^2}$ ...(1)

and $v = v_x = \frac{dx}{dt} = v$ ...(2)

Dividing (1) by (2), we get

$$\frac{dy}{dx} = \frac{x(a-x)}{a^2} \quad \text{or} \quad dy = \frac{x(a-x)}{a^2}dx$$

or $\int_0^y dy = \int_0^x \frac{x(a-x)}{a^2}dx$

or $\boldsymbol{y = \frac{x^2}{2a} - \frac{x^3}{3a^2}}$ ...(3)

This is the desired equation of trajectory.

(b) Time taken to cross the river is

$$\boldsymbol{t = \frac{a}{v_x} = \frac{a}{v}}$$

(c) When the boatman reaches the opposite side, $x = a$ or $v_y = 0$ (from equation 1)

Hence resultant velocity of boatman is v along positive x-axis or due east.

(d) From equation (3)

$$y = \frac{a^2}{2a} - \frac{a^3}{3a^2} = \frac{a}{6}$$

at $x = a$ (at opposite bank)

Hence displacement of boatman will be

$$\vec{s} = x\hat{i} + y\hat{j} \quad \text{or} \quad \boldsymbol{\vec{s} = a\hat{i} + \frac{a}{6}\hat{j}}$$

30. (a) Given $|\vec{v}_R| = v_0\left[1 + \frac{\sqrt{3}-1}{w}y\right]$

and $|\vec{v}| = 2v_0$

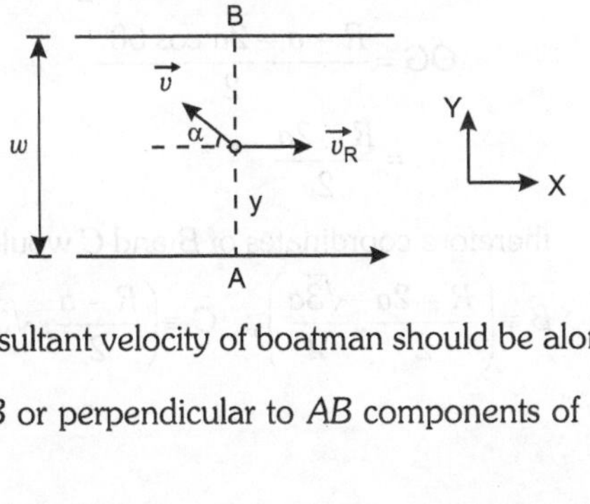

Resultant velocity of boatman should be along AB or perpendicular to AB components of $\vec{v}$ and $\vec{v}_R$ should be zero. Hence

$$v\cos\alpha = v_R$$

or $(2v_0)\cos\alpha = v_0\left[1 + \frac{\sqrt{3}-1}{w}y\right]$

or $\cos\alpha = \frac{1 + \frac{\sqrt{3}-1}{w}y}{2}$

Therefore, resultant velocity along AB is

$$v_y = v\sin\alpha$$

or $\frac{dy}{dt} = (2v_0)\sin\alpha$

$$= \frac{(2v_0)\sqrt{4w^2 - \{w + (\sqrt{3}-1)y\}^2}}{2w}$$

$$= \frac{v_0}{w}\sqrt{4w^2 - \{w + (\sqrt{3}-1)y\}^2}$$

or $\int_0^w \frac{dy}{\sqrt{4w^2 - \{w + (\sqrt{3}-1)y\}^2}} = \frac{v_0}{w}\int_0^t dt$

solving this, we get

$$\boldsymbol{t = \frac{w\pi}{6(\sqrt{3}-1)v_0}}$$

for integration, apply

$$\int \frac{dx}{\sqrt{a^2 - x^2}} = \sin^{-1}\left(\frac{x}{a}\right)$$

(b) When the boatman reaches the opposite bank :

$y = w$ or $v_R = \sqrt{3}v_0$ or $v\cos\alpha = v_R$

Hence, $(2v_0)\cos\alpha = \sqrt{3}v_0$

or $\cos\alpha = \frac{\sqrt{3}}{2}$ or $\alpha = 30°$

Hence resultant velocity will be

$$v_y = v\sin\alpha = (2v_0)\sin 30°$$

or $\boldsymbol{v_y = v_0}$

31. Let $-\vec{v}_B$ makes an angle θ with the line CD (a line perpendicular to AB).

$\vec{v}_{AB}$ or $\vec{v}_A - \vec{v}_B$ is along AB i.e. components of $\vec{v}_{AB}$ perpendicular to AB or along CD is zero.

Hence $\quad v \cos \alpha = 2v \cos \theta$

or $\quad \cos \alpha = 2 \cos \theta \quad \quad ...(1)$

In triangle OAB

$$\frac{\sin \alpha}{r} = \frac{\sin (180° - \theta)}{2r}$$

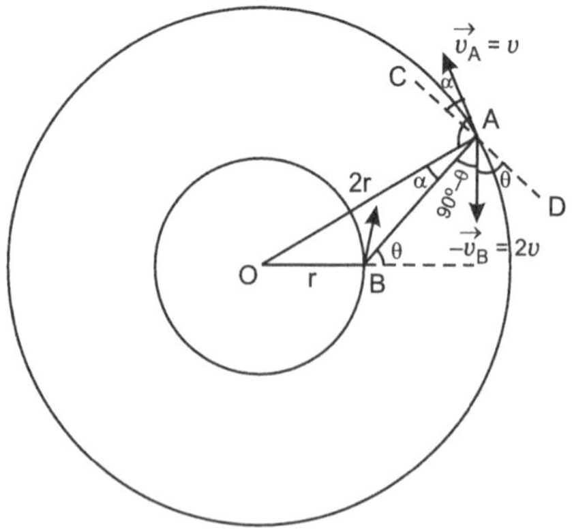

CD ⊥ AB

or $\quad \sin \alpha = \dfrac{\sin \theta}{2} \quad \quad ...(2)$

Squaring and adding (1) and (2) we get

$$1 = 4 \cos^2 \theta + \frac{1}{4} \sin^2 \theta$$

or $\quad 4 = 16 \cos^2 \theta + (1 - \cos^2 \theta)$

or $\quad 15 \cos^2 \theta = 3$ or $\cos \theta = \dfrac{1}{\sqrt{5}}$

or $\quad \cos \alpha = \dfrac{2}{\sqrt{5}} \quad$ [from Eq. (1)]

or $\quad \cot \alpha = 2$ or $\alpha = \mathbf{cot^{-1}(2)}$

32. $u_x = u \cos \theta, \quad u_y = u \sin \theta$

$a_x = - g \sin \alpha, \quad a_y = - g \cos \alpha$

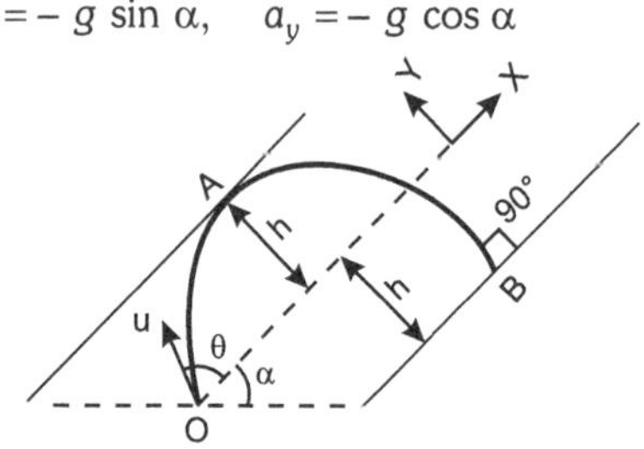

At point A: $s_y = h$ and $v_y = 0$

hence $\quad v_y^2 = u_y^2 + 2a_y s_y$

or $\quad 0 = (u \sin \theta)^2 - 2g \cos \alpha \cdot h$

or $\quad h = \dfrac{u^2 \sin^2 \theta}{2g \cos \alpha} \quad \quad ...(1)$

At point B

(i) $v_x = 0 = u_x + a_x t = (u \cos \theta) - (g \sin \alpha) t$

or $\quad t = \dfrac{u \cos \theta}{g \sin \alpha}$

(ii) $s_y = - h = u_y t + \dfrac{1}{2} a_y t^2$

or $\quad h = - (u_Y t + \dfrac{1}{2} a_Y t^2)$.

or $\quad h = - \left[u \sin \theta \cdot \dfrac{u \cos \theta}{g \sin \alpha} - \dfrac{1}{2} g \cos \alpha \left(\dfrac{u \cos \theta}{g \sin \alpha} \right)^2 \right] \quad ...(2)$

equating (1) and (2), we get the desired result *i.e.*,

$$\mathbf{\tan \theta = (\sqrt{2} - 1) \cot \alpha}$$

33. Let a be the side of hexagon and the particle is projected from origin O with an initial velocity u at an angle θ with horizontal as shown in figure.

$OJ = R \quad$ (range of the projectile)

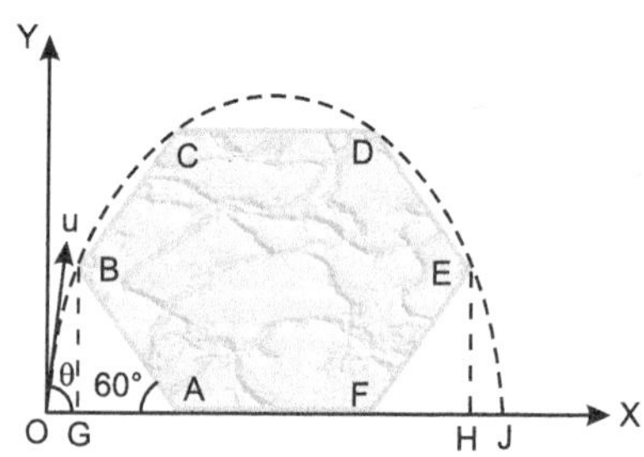

$$BG = EH = a \sin 60° = \frac{\sqrt{3}a}{2}$$

$$OG = \frac{R - a - 2a \cos 60°}{2} = \frac{R - 2a}{2}$$

therefore coordinates of B and C would be

$$B \equiv \left(\frac{R - 2a}{2}, \frac{\sqrt{3}a}{2} \right), \quad C \equiv \left(\frac{R - a}{2}, \sqrt{3}a \right)$$

equation of projectile in terms of R can be written as

$$y = x\left(1 - \frac{x}{R}\right)\tan\theta \qquad \ldots(1)$$

Satisfying co-ordinates of B and C with equation (1) we can find that

$$\cos\theta = \sqrt{\frac{3}{31}}$$

Now, $\dfrac{v_{max}}{v_{min}} = \dfrac{\boldsymbol{u}}{\boldsymbol{u}\cos\theta} = \dfrac{\mathbf{1}}{\cos\theta} = \sqrt{\dfrac{\mathbf{31}}{\mathbf{3}}}$

34. (a) The maximum range up the plane is

$$R_1 = \frac{u^2}{g\,(1 + \sin\beta)} \qquad \ldots(1)$$

where u = initial speed of particle and
β = angle of inclination of plane.
Similarly the maximum range down the plane is

$$R_2 = \frac{u^2}{g\,(1 - \sin\beta)} \qquad \ldots(2)$$

and maximum range on ground is

$$R = \frac{u^2}{g} \qquad \ldots(3)$$

Given that

$$R_1 + R_2 = 2\,(R + R) = 4R$$

or $$\frac{u^2}{g\,(1+\sin\beta)} + \frac{u^2}{g\,(1-\sin\beta)} = \frac{4u^2}{g}$$

or $$\frac{1}{1+\sin\beta} + \frac{1}{1-\sin\beta} = 4$$

or $$\frac{2}{1-\sin^2\beta} = 4 \quad \text{or} \quad \frac{1}{\cos^2\beta} = 2$$

or $$\cos\beta = \frac{1}{\sqrt{2}} \quad \therefore \ \beta = \mathbf{45°}$$

(b) Time period of particle projected up the plane is

$$T_1 = \frac{2u\sin(\alpha_1 - \beta)}{g\cos\beta}$$

where α_1 = angle of initial velocity with horizontal
For maximum range

$$\alpha_1 = \frac{\beta}{2} + \frac{\pi}{4}$$

$$\therefore \quad T_1 = \frac{2u\sin\left(\frac{\beta}{2} + \frac{\pi}{4} - \beta\right)}{g\cos\beta}$$

$$= \frac{2u\sin\left(\frac{\pi}{4} - \frac{\beta}{2}\right)}{g\cos\beta} \qquad \ldots(4)$$

Similarly, time period of particle projected down the plane is

$$T_2 = \frac{2u\sin(\alpha_2 + \beta)}{g\cos\beta}$$

where α_2 = angle of initial velocity with horizontal.
For maximum range

$$\alpha_2 = \frac{\pi}{4} - \frac{\beta}{2}$$

Hence $$T_2 = \frac{2u\sin\left(\frac{\pi}{4} - \frac{\beta}{2} + \beta\right)}{g\cos\beta}$$

$$T_2 = \frac{2u\sin\left(\frac{\pi}{4} + \frac{\beta}{2}\right)}{g\cos\beta} \qquad \ldots(5)$$

Given that,

$$T_2 - T_1 = 2 \text{ second}$$

or $$\frac{2u}{g\cos\beta}[\sin(\pi/4 + \beta/2) - \sin(\pi/4 - \beta/2)] = 2$$

Substituting $\beta = 45°$ or $\dfrac{\pi}{4}$

We get

$$u = \frac{g\cos(\pi/4)}{\left(\sin\frac{3\pi}{8} - \sin\frac{\pi}{8}\right)}$$

or $\boldsymbol{u} = \mathbf{12.8\ m/s}$

35. (a) Since the resultant velocity is always perpendicular to the line joining boat and R, the boat is moving in a circle of radius 2ω and centre at R.

(b) Drifting $= Qs = \sqrt{4\omega^2 - \omega^2} = \sqrt{3}\omega$

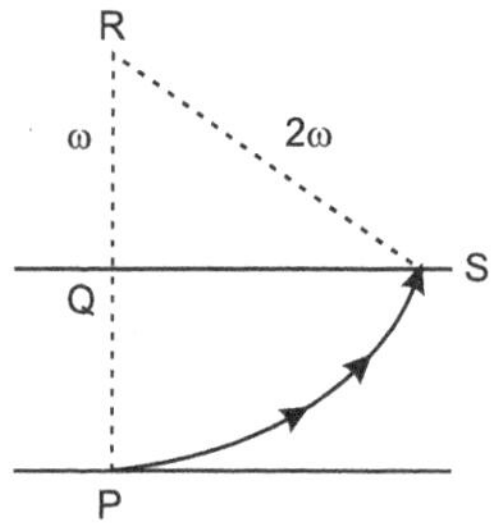

(c) Suppose at any arbitrary time, the boat is at point B.

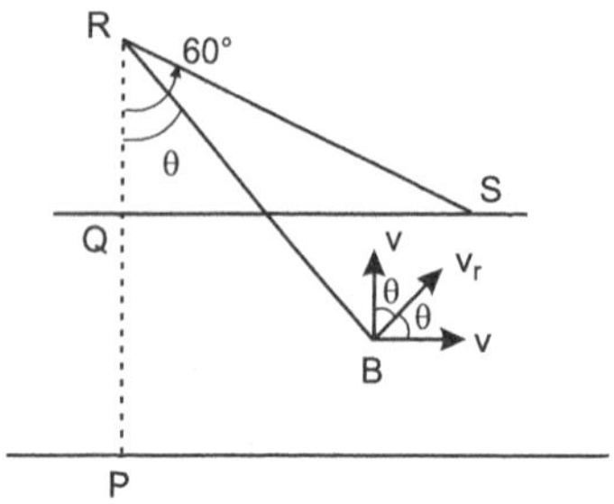

$$v_r = 2v \cos\theta$$

$$\frac{d\theta}{dt} = \frac{v_r}{2\omega} = \frac{v\cos\theta}{\omega}$$

or $$\frac{\omega}{v}\sec\theta\, d\theta = dt$$

$\therefore$ $$\int_0^t dt = \frac{\omega}{v}\int_0^{60^\circ} \sec\theta \,.\, d\theta$$

$\therefore$ $$t = \frac{\omega}{v}\left[\ln(\sec\theta + \tan\theta)\right]_0^{60^\circ}$$

or $$t = \frac{\mathbf{1.317}\,\omega}{\mathbf{v}}$$

36. Equation of trajectory of the body as a projectile is,

$$y = x\tan 45^\circ - \frac{10}{2\times(10)^2\cos^2 45^\circ}x^2$$

or $$y = x - \frac{x^2}{10} \qquad \ldots(1)$$

Equation of the inclined roof as a straight line is,

$$y = -(\tan 45^\circ)\,x + h$$

or $$y = h - x \qquad \ldots(2)$$

Solving (1) and (2) for the point of intersection, we get :

$$x - \frac{x^2}{10} = h - x$$

or $$\frac{x^2}{10} - 2x + h = 0 \qquad \ldots(3)$$

This is in the form,

$$ax^2 + bx + c = 0$$

Body just touches the roof only when there is only one point of intersection and this is possible only when equation (3) has only one root.

$\therefore$ $$b^2 - 4ac = 0$$

or $$4 - \frac{4h}{10} = 0$$

$\therefore$ $$h = 10\text{ m}$$

37. Let us take the origin as shown in figure. Mid point of CD is the origin O. Since the path is parabolic, let the equation be,

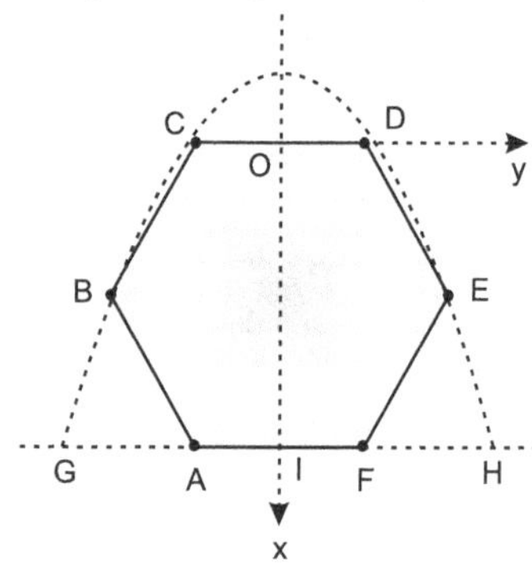

$$y^2 = m(x + b) \qquad \ldots(1)$$

Here m and b are unknowns. From geometry we can see that

$$D \equiv \left(0, \frac{a}{2}\right) \text{ and } E \equiv \left(\frac{\sqrt{3}a}{2}, a\right)$$

These co-ordinates should satisfy equation (1). So we can find,

$$m = \frac{\sqrt{3}a}{2} \text{ and } b = \frac{a}{2\sqrt{3}}$$

$\therefore$ Equation of parabola should be

$$y^2 = \frac{\sqrt{3}a}{2}\left(x + \frac{a}{2\sqrt{3}}\right) \qquad \ldots(2)$$

Now if R be the range (= GH), the co-ordinates of $H = \left(OI, \frac{R}{2}\right)$ or $\left(\sqrt{3}a, \frac{R}{2}\right)$ will satisfy the above equation. Putting $x = \sqrt{3}a$ and $y = \frac{R}{2}$ in above equation we get,

$$R = \sqrt{7}a$$

38. Equation of ball,

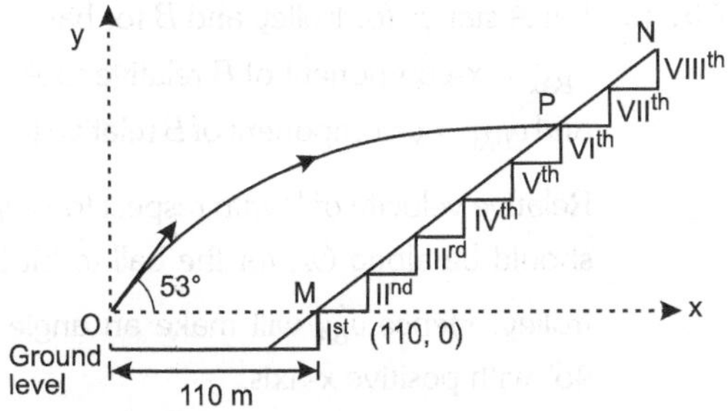

$$y = x \tan\theta - \frac{gx^2}{2u^2\cos^2\theta}$$

Substituting the values,

$$y = 1.33x - 0.0113x^2 \qquad \ldots(1)$$

Slope of line MN is 1 and it passes through point (110 m, 0). Hence the equation of this line can be written as,

$$y = x - 110 \qquad \ldots(2)$$

Point of intersection of two curves is say P. Solving (1) and (2) we get positive value of y equal to 4.5 m.

i.e., $\quad y_P = 4.5$

Height of one step is 1 m. Hence, the ball will collide somewhere between $y = 4$ m and $y = 5$ m. Which comes out to be 6th step.

39. $\quad v_0 \cos\theta \times t = 2 \qquad \ldots(1)$

$$v_0 \sin\theta \times t - \frac{1}{2}gt^2 + \frac{1}{2}gt^2 = 10$$

$$\Rightarrow \quad v_0 \sin\theta \times t = 10 \qquad \ldots(2)$$

From equation (1) and (2)

$$v_0 \cos\theta = 1$$

$$v_0 \sin\theta = 5$$

$$\tan\theta = 5$$

$$\therefore \quad v_0 = \sqrt{26}\ \text{ms}^{-1}$$

and $\quad \theta = \tan^{-1} 5$ with x-axis.

40. If they meet a distance 'x' from 'C', we have

$$(2v_0 \sin\theta)\, t = x \qquad \ldots(1)$$

and $\quad (4d - x) = v_0 t \qquad \ldots(2)$

or $\quad t = \frac{4d}{v_0(1 + 2\sin\theta)}$

from (1) and (2)

Also we have,

$$d\tan\theta + (2v_0\cos\theta)\, t - \frac{1}{2}gt^2 = 0$$

Substituting value of 't', we get

$$d\tan\theta + \frac{8d\cos\theta}{1 + 2\sin\theta} = \frac{8gd^2}{v_0^2(1 + 2\sin\theta)^2}$$

or $\quad v_0^2 = \frac{8gd}{(1 + 2\sin\theta)[8\cos\theta t + \tan\theta(1 + 2\sin\theta)]}$

$$\Rightarrow v_0 = \sqrt{\frac{8gd}{(1 + 2\sin\theta)[8\cos\theta t + \tan\theta(1 + 2\sin\theta)]}}$$

41. Let the projectile strikes the circular path at (x, y) and 'A' to be taken as origin. From the figure co-ordinates of the centre of the circular path is (300, 150). Then the equation of the circular path is

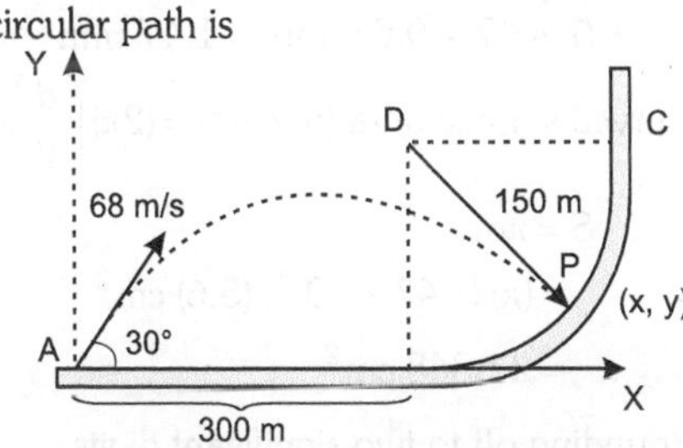

$$(x - 300)^2 + (y - 150)^2 = (150)^2 \qquad \ldots(1)$$

and the equation of the trajectory is

$$y = x\tan 30° - \frac{1}{2}\frac{gx^2}{(68)^2\cos^2 30°}$$

$$y = \frac{x}{\sqrt{3}} - \frac{2x^2 g}{9248} \qquad \ldots(2)$$

From Eqs. (1) and (2) we get

$$x = 373\text{ m}; \quad y \approx 18.75\text{ m}$$

Note: Had there been no solution, the projectile would not strike the circular part.

42. Young's modulus of elasticity is given by,

$$Y = \frac{\text{stress}}{\text{strain}} = \frac{F/A}{l/L} = \frac{FL}{lA} = \frac{FL}{l\left(\frac{\pi d^2}{4}\right)}$$

Substituting the values, we get

$$Y = \frac{50 \times 1.1 \times 4}{(1.25 \times 10^{-3}) \times \pi \times (5.0 \times 10^{-4})^2}$$

$$= 2.24 \times 10^{11}\text{ N/m}^2$$

Now, $$\frac{\Delta Y}{Y} = \frac{\Delta l}{L} + \frac{\Delta l}{l} + 2\frac{\Delta d}{d}$$

$$= \left(\frac{0.1}{110}\right) + \left(\frac{0.001}{0.125}\right) + 2\left(\frac{0.001}{0.05}\right)$$

$$= 0.0489$$

$$\Delta Y = (0.0489) Y$$

$$= (0.0489) \times (2.24 \times 10^{11})\text{ N/m}^2$$

$$= 1.09 \times 10^{10}\text{ N/m}^2$$

43. Least count of screw gauge

$$= \frac{1\text{ mm}}{100} = 0.01\text{ mm}$$

Diameter of wire

$$= (1 + 47 \times 0.01)\text{ mm} = 1.47\text{ mm}$$

Curved surface area (in cm^2) $= (2\pi)\left(\frac{d}{2}\right)(L)$

$$S = \pi d L$$

$$= (\pi)(1.47 \times 10^{-1})(5.6)\text{ cm}^2$$

$$= 2.5848\text{ cm}^2$$

Rounding off to two significant digits

$$S = 2.6\text{ cm}^2$$

44. $(N + 1)$ divisions on the vernier scale

$= N$ divisions on main scale

$\therefore$ 1 division on vernier scale

$$= \frac{N}{N+1}\text{ divisions on main scale}$$

Each division on the main scale is of a units.

$\therefore$ 1 division on vernier scale $= \left(\frac{N}{N+1}\right) a$ units

$= a'$ (say)

Least count = 1 main scale division − 1 vernier scale division

$$= a - a' = a - \left(\frac{N}{N+1}\right) a = \frac{a}{N+1}$$

45. (a) Let A stands for trolley and B for ball.

v_{BAx} = x-component of B relative to A

and v_{BAy} = y-component of B relative to A

Relative velocity of B with respect to $A(\vec{v}_{BA})$ should be along OA for the ball to hit the trolley. Hence $\vec{v}_{BA}$ will make an angle of 45° with positive x-axis.

or $\theta = \mathbf{45°}$

(b) $$\tan\theta = \frac{v_{BAy}}{v_{BAx}} = \tan 45°$$

or $$\frac{v_{BAy}}{v_{BAx}} = 1$$

or $$v_{BAy} = v_{BAx} \quad \text{...(1)}$$

Further $v_{BAx} = v_{Bx} - v_{Ax}$

or $v_{BAx} = v_{Bx} - 0$

or $$v_{BAx} = v_{Bx} \quad \text{...(2)}$$

Also $v_{BAy} = v_{By} - v_{Ay}$

$$v_{BAy} = v_{By} - (\sqrt{3} - 1) \quad \text{...(3)}$$

$$\tan\phi = \frac{v_{By}}{v_{Bx}}$$

or $$v_{By} = v_{Bx}\tan\phi \quad \text{...(4)}$$

From Eqs. (1), (2), (3) and (4) we get

or $$v_{Bx} = \frac{(\sqrt{3} - 1)}{\tan\phi - 1} \quad \text{...(5)}$$

and $$v_{By} = \frac{(\sqrt{3} - 1)}{\tan\phi - 1} \cdot \tan\phi \quad \text{...(6)}$$

$$\phi = \frac{4\theta}{3} = \frac{4}{3}(45°) = 60°$$

speed of ball with respect to surface

$$v_B = \sqrt{v_{Bx}^2 + v_{By}^2}$$

$$= \frac{\sqrt{3}-1}{\tan\phi - 1}\sqrt{1+\tan^2\phi}$$

$$= \frac{\sqrt{3}-1}{\tan\phi - 1}\sec\phi$$

substituting $\phi = 60°$, we get

$$v_B = \frac{\sqrt{3}-1}{\tan 60° - 1}\sec 60°$$

$$= \frac{\sqrt{3}-1}{\sqrt{3}-1}\cdot 2$$

$$\boldsymbol{v_B = 2\text{ m/s}}$$

46. Let 't' be the time after which the stone hits the object and θ be the angle which the velocity vector $\vec{u}$ makes with horizontal. According to question, we have following three conditions

(i) Vertical displacement of stone is 1.25 m

$$\therefore \quad 1.25 = (u\sin\theta)t - \frac{1}{2}gt^2$$

where $g = 10\text{ m/s}^2$

$$\text{or} \quad (u\sin\theta)\,t = 1.25 + 5t^2 \quad \ldots(1)$$

(ii) Horizontal displacement of stone $= 3 +$ displacement of object A

$$\therefore \quad (u\cos\theta)\,t = 3 + \frac{1}{2}at^2 \quad \text{where } a = 1.5\text{ m/s}^2$$

$$\text{or} \quad (u\cos\theta)\,t = 3 + 0.75\,t^2 \quad \ldots(2)$$

(iii) Horizontal component of velocity (of stone) = vertical component (because velocity vector is inclined at 45° with horizontal)

$$\therefore \quad (u\cos\theta) = gt - (u\sin\theta) \quad \ldots(3)$$

(The right hand side is written $gt - u\sin\theta$ because the stone is in its downward motion. Therefore, $gt > u\sin\theta$. In upward motion $u\sin\theta > gt$). Multiplying equation (3) with t, we can write

$$(u\cos\theta)\,t + (u\sin\theta)\,t = 10t^2 \quad \ldots(4)$$

Now (4)–(2)–(1) gives [Equation (4) minus (2) minus (1)]

$$4.25t^2 - 4.25 = 0$$

$$\text{or} \quad \boldsymbol{t = 1\text{ s}}$$

Substituting $t = 1s$ in (1) and (2), we get

$$u\sin\theta = 6.25\text{ m/s} \quad \text{or} \quad u_y = 6.25\text{ m/s}$$

$$\text{and} \quad u\cos\theta = 3.75\text{ m/s} \quad \text{or} \quad u_x = 3.75\text{ m/s}$$

$$\text{therefore,} \quad \vec{u} = u_x\hat{i} + u_y\hat{j}$$

$$\text{or} \quad \boldsymbol{\vec{u} = (3.75\,\hat{i} + 6.25\,\hat{j})\text{ m/s}}$$

Most of the problems of projectile motion are easily solved by breaking the motion of the particle in two suitable mutually perpendicular directions, say x and y. Find u_x, u_y, a_x and a_y and then apply

$$v_x = u_x + a_x t;$$

$$s_y = u_y t + \frac{1}{2}a_y t^2 \text{ etc.}$$

47. (a) Given $F(x) = -\dfrac{k}{2x^2}$

here k and x^2 are always positive hence F is always negative (whether x is positive or negative)

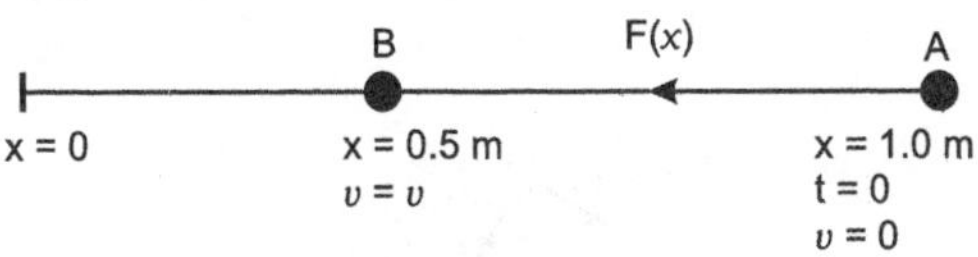

$$a(x) = \frac{F(x)}{m}$$

substituting the values, we have

$$a = \frac{-k}{2mx^2} = -\frac{10^{-2}}{2\times 10^{-2}\times x^2} = -\frac{1}{2x^2}$$

$$\text{or} \quad v\cdot\frac{dv}{dx} = -\frac{1}{2x^2} \quad \text{or} \quad v\,dv = -\frac{1}{2}\frac{dx}{x^2}$$

$$\text{or} \quad \int_0^v v\,dv = -\frac{1}{2}\int_{x=1.0}^{x=0.5}\frac{dx}{x^2}$$

$$\text{or} \quad \frac{v^2}{2} = \frac{1}{2}\left(\frac{1}{x}\right)_{x=1.0}^{x=0.5} \quad \ldots(1)$$

$$\text{or} \quad v^2 = \left[\frac{1}{0.5} - \frac{1}{1.0}\right]$$

$$\text{or} \quad v^2 = 1.0 \text{ or } v = \pm 1.0\text{ m/s}$$

so $\boldsymbol{v = -1.0}$ **m/s (because velocity is along negative X-direction)**

(b) To find velocity of particle at $x = x$ equation (1) can be written as

$$\frac{v^2}{2} = \frac{1}{2}\left(\frac{1}{x}\right)_{x=1.0}^{x=x}$$

or $$v^2 = \left(\frac{1}{x} - \frac{1}{1.0}\right) = \frac{1-x}{x}$$

or $$v = \left(-\frac{dx}{dt}\right) = \sqrt{\frac{1-x}{x}}$$

or $$\int \sqrt{\frac{x}{1-x}} dx = -\int dt$$

or $$\int_1^{0.25} \sqrt{\frac{x}{1-x}}\, dx = -\int_0^t dt$$

solving this, we get t = **1.48 s**

Note: For integration make the substitution $x = \sin^2 \theta$

48. (a) Accelerations of particle and box are shown in figures (a) and (b).

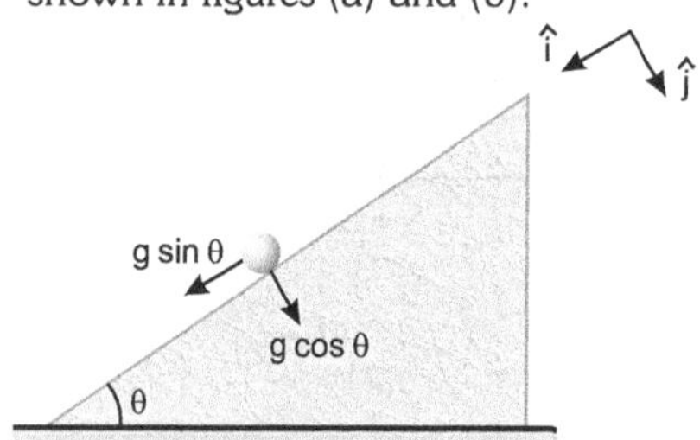

(a)

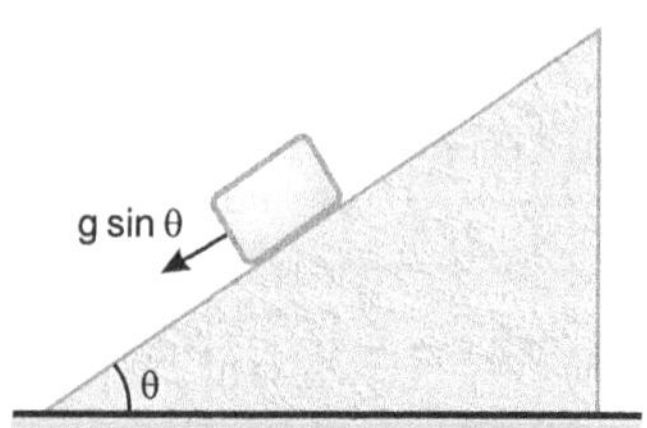

(b)

Acceleration of particle with respect to box
= acceleration of particle – acceleration of block

$= (g \sin\theta\, \hat{i} + g \cos\theta\, \hat{j}) - (g \sin\theta)\, \hat{i}$

$= g \cos\theta\, \hat{j}$

Now motion of particle with respect to box will be a projectile as shown in figure (c).

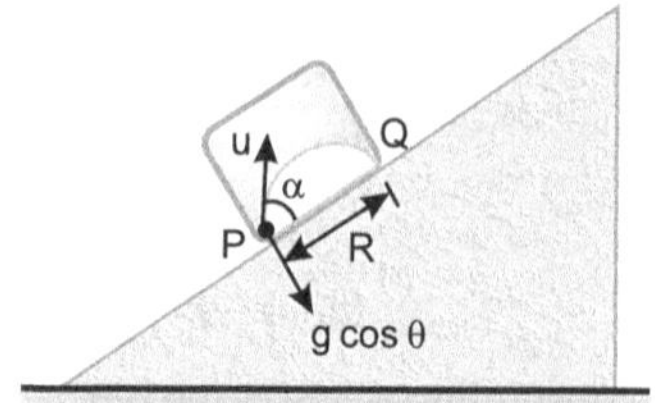

(c)

The only difference is that g will be replaced by $g \cos\theta$

$\therefore$ $$PQ = \text{range}\,(R) = \frac{u^2 \sin 2\alpha}{g \cos\theta}$$

$$\boldsymbol{PQ = \frac{u^2 \sin 2\alpha}{g \cos\theta}}$$

(b) Horizontal displacement of particle with respect to ground is zero. This implies that initial velocity of particle with respect to ground is in vertical direction or there is no horizontal component of the absolute velocity of the particle.

Velocity of particle with respect to box (figure d)

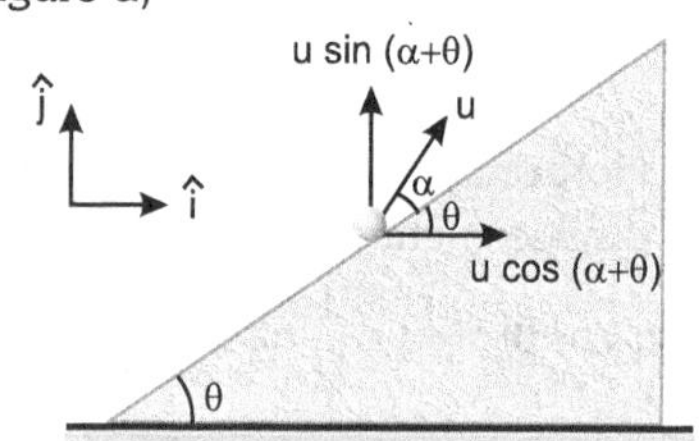

(d)

$= u \cos(\alpha + \theta)\, \hat{i} + u \sin(\alpha + \theta)\, \hat{j}$

velocity of box $= -v \cos\theta\, \hat{i} - v \sin\theta\, \hat{j}$

Here v is the velocity of the box down the plane (figure e). Then

velocity of particle with respect to ground

$= \{u \cos(\alpha + \theta) - v \cos\theta\}\, \hat{i}$

$+ \{u \sin(\alpha + \theta) - v \sin\theta\}\, \hat{j}$

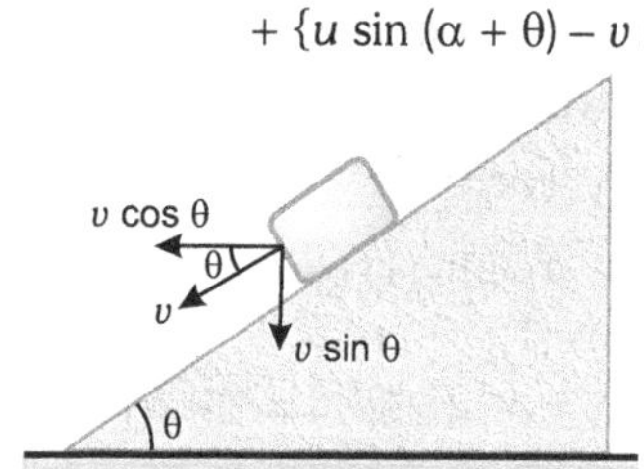

(e)

Now as we said earlier that horizontal component of absolute velocity should be zero.

Therefore,

$$u \cos(\alpha + \theta) - v \cos\theta = 0$$

or $$v = \frac{u \cos(\alpha + \theta)}{\cos\theta} \quad \textbf{(down the plane)}$$

49. (i) Let $\hat{i}$, $\hat{j}$ and $\hat{k}$ be the unit vectors along x, y and z directions respectively. Given

$$\vec{v}_{\text{cart}} = 4\hat{i} \text{ m/s}$$

$$\therefore \quad \vec{v}_{\text{stone, cart}} = (6 \sin 30°)\,\hat{j} + (6 \cos 30°)\,\hat{k}$$

$$= (3\,\hat{j} + 3\sqrt{3}\,\hat{k}) \text{ m/s}$$

$$\vec{v}_{\text{stone}} = \vec{v}_{\text{stone, cart}} + \vec{v}_{\text{cart}}$$

$$= (4\hat{i} + 3\hat{j} + 3\sqrt{3}\hat{k}) \text{ m/s}$$

This is the absolute velocity of stone (with respect to ground). At highest point of its trajectory, the vertical component of its velocity (v_z) will become zero, whereas the x and y components will remain unchanged. Therefore, velocity of stone at highest point will be

$$\vec{v} = (4\hat{i} + 3\hat{j}) \text{ m/s}$$

or speed at highest point,

$$v = |\vec{v}| = \sqrt{(4)^2 + (3)^2} \text{ m/s} = 5 \text{ m/s}$$

Now applying law of conservation of linear momentum. Let v_0 be the velocity of combined mass after collision. Then

$$mv = (2\text{ m})\,v_0$$

$$\therefore \quad v_0 = \frac{v}{2} = \frac{5}{2} \text{ m/s} = 2.5 \text{ m/s}$$

∴ Speed of combined mass just after collision is 2.5 m/s.

(ii) Tension in the string becomes zero at horizontal position. It implies that velocity of combined mass also becomes zero in horizontal position. Applying

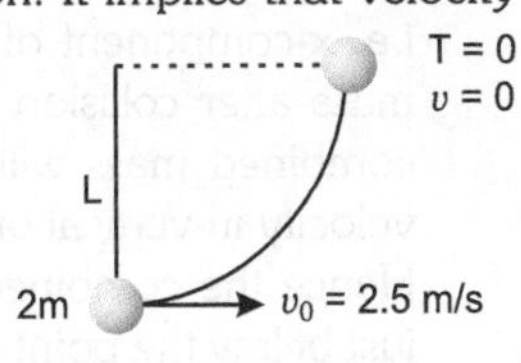

conservation of energy, we have

$$0 = v_0^{\,2} - 2gL$$

$$\therefore \quad L = \frac{v_0^{\,2}}{2g} = \frac{(2.5)^2}{2(9.8)} = 0.32 \text{ m}$$

Hence length of the string is 0.32 m.

50. $$u = 5\sqrt{3} \text{ m/s}$$

$$\therefore \quad u \cos 60° = (5\sqrt{3})\left(\frac{1}{2}\right) \text{m/s}$$

$$= 2.5\sqrt{3} \text{ m/s}$$

and $$u \sin 60° = (5\sqrt{3})\left(\frac{\sqrt{3}}{2}\right) \text{m/s} = 7.5 \text{ m/s}$$

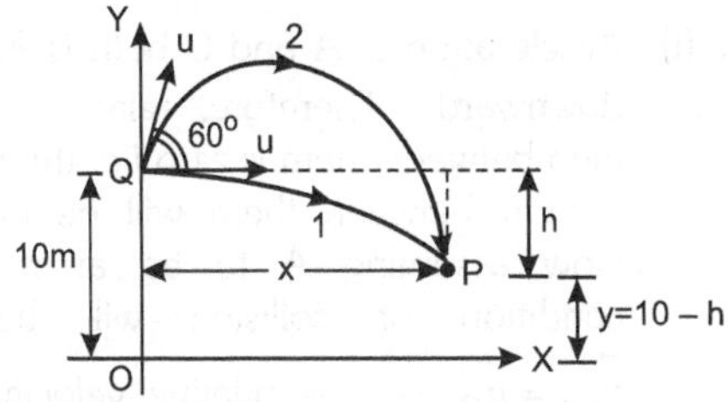

Since the horizontal displacement of both the shots are equal, the second should be fired early because its horizontal component of velocity $u \cos 60°$ or $2.5\sqrt{3}$ m/s is less than the other's which is u or $5\sqrt{3}$ m/s.

Now let first shot takes t_1 time to reach the point P and the second shot takes time t_2. Then

$$x = (u \cos 60°).\,t_2 = u.\,t_1$$

or $$x = 2.5\sqrt{3}.\,t_2 = 5\sqrt{3}\,t_1 \quad \ldots(1)$$

or $$t_2 = 2t_1 \quad \ldots(2)$$

and $$h = \left|(u \sin 60°)\,t_2 - \frac{1}{2}gt_2^2\right| = \frac{1}{2}gt_1^2$$

or $$h = \frac{1}{2}gt_2^2 - (u \sin 60°)\,t_2 = \frac{1}{2}gt_1^2$$

Taking $$g = 10 \text{ m/s}^2$$

$$h = 5t_2^2 - 7.5\,t_2 = 5t_1^2 \quad \ldots(3)$$

Substituting $t_2 = 2t_1$ in equation (3), we get

$$5(2t_1)^2 - 7.5(2t_1) = 5t_1^2$$

or $$15t_1^{\,2} = 15\,t_1 \Rightarrow t_1 = 0 \text{ and } 1\text{s}$$

Hence $t_1 = 1\text{s}$ and $t_2 = 2t_1 = 2\text{s}$

$\therefore \quad x = 5\sqrt{3}\, t_1 = 5\sqrt{3}$ m

(From equation 1)

and $\quad h = 5t_1^{\,2} = 5(1)^2 = 5$ m

(From equation 3)

$\therefore \quad y = 10 - h = (10 - 5) = 5$ m

Hence

(a) Time interval between the firings

$$= t_2 - t_1 = (2 - 1)\text{ s}$$
$$\boldsymbol{\Delta t = 1\text{ s}}$$

(b) Coordinates of point

$$P = (x, y) = \mathbf{(5\sqrt{3}\text{ m}, 5\text{ m})}$$

51. (i) Acceleration of A and C both is 9.8 m/s^2 downwards. Therefore, relative acceleraton between them is zero *i.e.* the relative motion between them will be uniform. Now assuming A to be at rest, the condition of collision will be that $\vec{v}_{CA} = \vec{v}_C - \vec{v}_A$ = relative velocity of C w.r.t. A, should be along CA

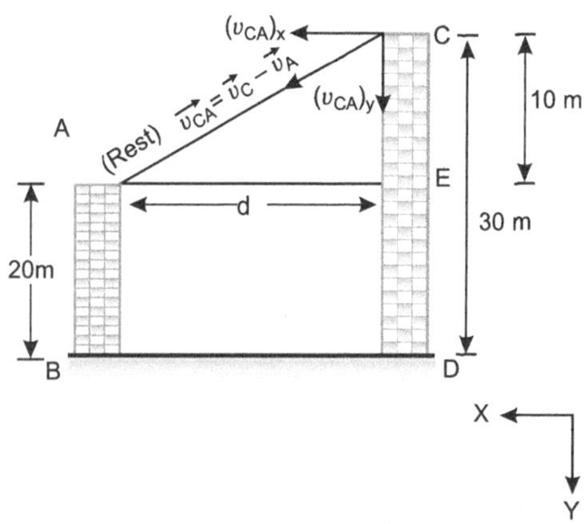

$$\therefore \quad \frac{(v_{CA})_y}{(v_{CA})_x} = \frac{CE}{AE} = \frac{10}{d}$$

$$\text{or} \quad \frac{v_{Cy} - v_{Ay}}{v_{Cx} - v_{Ax}} = \frac{10}{d}$$

$$\text{or} \quad \frac{5\sqrt{3} - 0}{5 - (-10)} = \frac{10}{d}$$

$$\therefore \quad \boldsymbol{d = 10\sqrt{3}\text{ m} \approx 17.32\text{ m}}$$

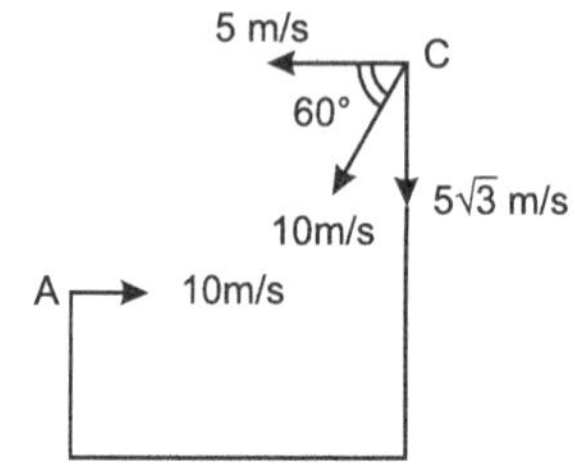

$v_{Ax} = -10$ m/s
$v_{Ay} = 0$
$v_{Cx} = 5$ m/s
$v_{Cy} = 5\sqrt{3}$ m/s

(ii) Time of collision,

$$t = \frac{AC}{|\vec{v}_{CA}|}$$

$$|\vec{v}_{CA}| = \sqrt{(v_{CAx})^2 + (v_{CAy})^2}$$
$$= \sqrt{\{5 - (-10)\}^2 + \{5\sqrt{3} - 0\}^2}$$
$$= 10\sqrt{3}\text{ m/s}$$

$$CA = \sqrt{(10)^2 + (10\sqrt{3})^2} = 20\text{ m}$$

$$\therefore \quad t = \frac{20}{10\sqrt{3}} = \frac{2}{\sqrt{3}}\text{ s}$$

Horizontal (or x) component of momentum of A, *i.e.*, $p_{Ax} = mv_{Ax} = -10$ m. Similarly, x component of momentum of C *i.e.*

$$p_{Cx} = (2m)v_{Cx}$$
$$= (2m)(5) = +10m$$

Since $\quad p_{Ax} + p_{Cx} = 0$

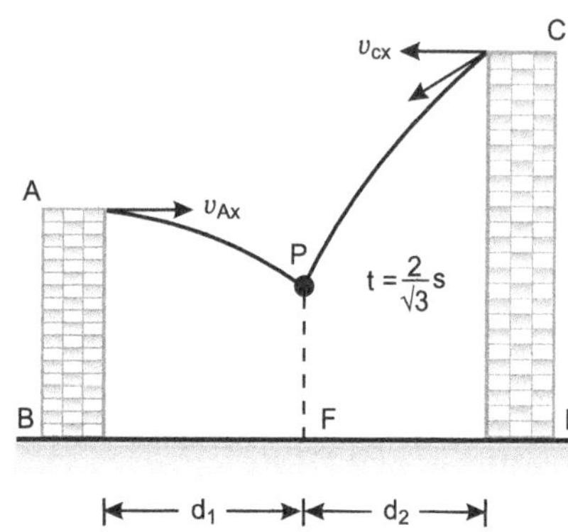

i.e. x-component of momentum of combined mass after collision will also be zero, *i.e.* the combined mass will have the momentum or velocity in vertical or y-direction only.

Hence the combined mass will fall at point F just below the point of collision P.

Here $d_1 = |(v_{Ax})|t$

$$= (10)\frac{2}{\sqrt{3}} = 11.55 \text{ m}$$

$\therefore$ $d_2 = (d - d_1)$

$$= (17.32 - 11.55) \text{ m} = 5.77 \text{ m}$$

d_2 should also be equal to

$$|v_{Cx}|t = (5)\left(\frac{2}{\sqrt{3}}\right) = 5.77 \text{ m}$$

Therefore, position where the objects hit the ground from B is d_1 *i.e.* 11.55 m and from D is d_2 or 5.77 m.

52. (a) At the highest point, velocity of bullet is $50 \cos \theta$. So by conservation of linear momentum

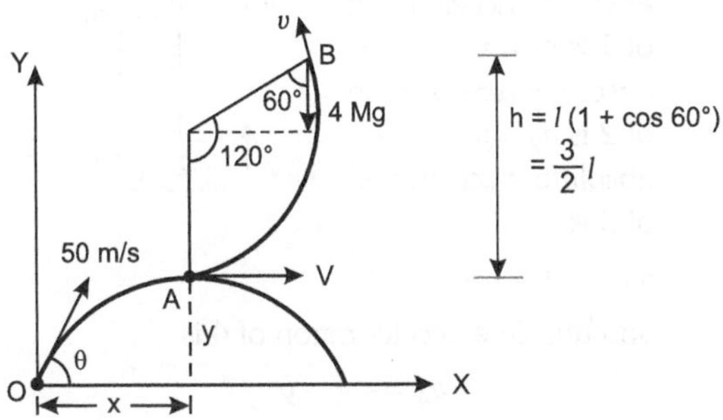

$$M(50 \cos \theta) = 4MV$$

$\therefore$ $$V = \left(\frac{50}{4}\right)\cos\theta \quad \ldots(1)$$

At point B, $T = 0$ but $v \neq 0$

Hence $$4Mg \cos 60° = \frac{(4M)v^2}{l}$$

or $$v^2 = \frac{g}{2}l = \frac{50}{3}$$

(as $l = \frac{10}{3}$ m and $g = 10 \text{ m/s}^2$) ...(2)

Also $$v^2 = V^2 - 2gh$$

or $$v^2 = V^2 - 2g\left(\frac{3}{2}l\right)$$

or $$v^2 = V^2 - 3(10)\left(\frac{10}{3}\right)$$

or $$v^2 = V^2 - 100 \quad \ldots(3)$$

Solving (1), (2) and (3) we get

$$\cos\theta = 0.86 \quad \text{or} \quad \theta \approx \mathbf{30°}$$

(b) $$x = \frac{\text{Range}}{2} = \frac{1}{2}\left(\frac{u^2 \sin 2\theta}{g}\right)$$

$$= \frac{50 \times 50 \times \sqrt{3}}{2 \times 10 \times 2}$$

$$= 108.25 \text{ m}$$

$$y = H = \frac{u^2 \sin^2\theta}{2g}$$

$$= \frac{50 \times 50 \times 1}{2 \times 10 \times 4}$$

$$= 31.25 \text{ m}$$

Hence the desired coordinates are **(108.25 m, 31.25 m)**

CHAPTER 2

LAWS OF MOTION

53. (a) Let a_1 and a_2 be the accelerations of the two men in upward direction, and T the tension in the rope. Then

$$T - Mg = Ma_1 \qquad \ldots(1)$$

and $$T - (M+m)g = (M+m)a_2 \qquad \ldots(2)$$

Subtracting (2) from (1) we get

or $$a_2 = \left(\frac{M}{M+m}\right)a_1 - \frac{mg}{M+m} \qquad a_2 < a_1$$

hence the lighter man will reach the pulley first.

(b) The lighter man ascends a distance h in time t with acceleration a_1. Hence

$$h = \frac{1}{2}a_1t^2 \qquad \ldots(3)$$

Let s be the distance travelled by the heavier man in this time t, then

$$s = \frac{1}{2}a_2t^2 = \frac{t^2}{2}\left[\frac{M}{M+m}a_1 - \frac{mg}{M+m}\right]$$

$$s = \frac{t^2}{2(M+m)}\left[M\left(\frac{2h}{t^2}\right) - mg\right]$$

$$= \frac{1}{2(M+m)}[2Mh - mgt^2]$$

The distance of the second man from the pulley $= h - s$

$$= h - \frac{1}{2(M+m)}[2Mh - mgt^2]$$

$$= \frac{1}{(M+m)}\left[Mh + mh - Mh + \frac{mgt^2}{2}\right]$$

$$= \mathbf{\frac{m}{(M+m)}\left[\frac{gt^2}{2} + h\right]}$$

54. Let acceleration of block 1 and pulley P_2 relative to ground is x in the directions shown in figure. Similarly, acceleration of block 4 and pulley P_3 relative to pulley P_2 is y and acceleration of block 2 and block 3 relative to pulley P_3 is z.

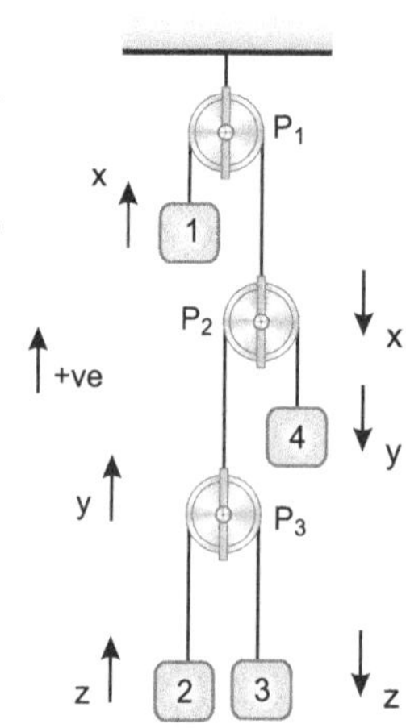

Then

absolute acceleration of 1 is $a_1 = x$

absolute acceleration of 2 is $a_2 = y + z - x$

absolute acceleration of 3 is

$$a_3 = -z + y - x$$

and absolute accele- ration of 4 is

$$a_4 = -x - y$$

Now given that

$$a_{21} = a_2 - a_1 = -1 \text{ m/s}^2$$

or $$y + z - 2x = -1 \qquad \ldots(1)$$

$$a_{31} = a_3 - a_1 = -5 \text{ m/s}^2$$

or $$y - z - 2x = -5 \qquad \ldots(2)$$

$$a_{34} = a_3 - a_4 = 0$$

or $$2y - z = 0 \qquad \ldots(3)$$

Solving these three equations we find

$$x = 2 \text{ m/s}^2, \quad y = 1 \text{ m/s}^2$$

and $$z = 2 \text{ m/s}^2$$

or $\mathbf{a_1 = 2 \text{ m/s}^2}$ **(upwards)**

$\mathbf{a_2 = 1 \text{ m/s}^2}$ **(upwards)**

$\mathbf{a_3 = -3 \text{ m/s}^2}$ **(downwards)**

$\mathbf{a_4 = -3 \text{ m/s}^2}$ **(downwards)**

55. Free body diagram of pulley A is shown in figure (a)

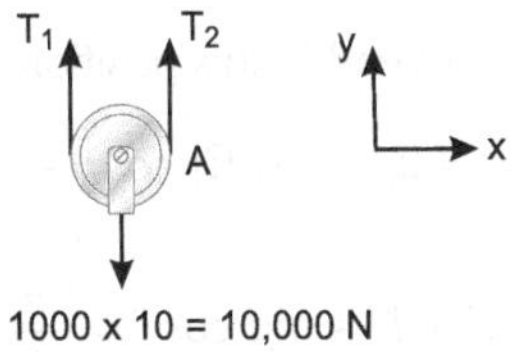

(a)

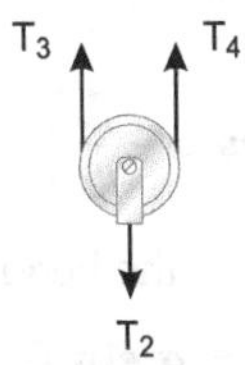

(b)

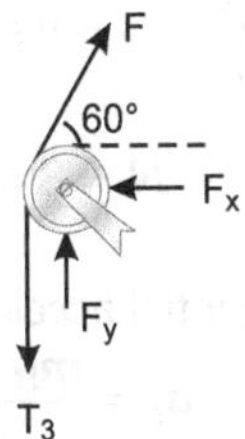

(c)

Let r_A = radius of pulley A

Equilibrium of moments about centre of pulley gives

$$T_1 r_A - T_2 r_A = 0$$

or $$T_1 = T_2 \quad \text{...(1)}$$

$$\Sigma F_y = 0$$

Hence $$T_1 + T_2 = 10{,}000 \text{ N} \quad \text{...(2)}$$

Solving (1) and (2), we get

$$T_1 = T_2 = 5000 \text{ N} \quad \text{...(3)}$$

Free body diagram of pulley B is shown in figure (b)

In the similar manner we can find that

$$T_3 = T_4 = \frac{T_2}{2} = 2500 \text{ N} \quad \text{...(4)}$$

Free body diagram of pulley C is shown in figure (c).

Moment equilibrium requires that

$$\boldsymbol{F = T_3 = 2500 \text{ N}}$$

Now $$F_x = F \cos 60° = 1250 \text{ N}$$

and $$F_y = T_3 - F \sin 60°$$

$$= 2500 - 2500\frac{\sqrt{3}}{2} = 335 \text{ N}$$

$\therefore$ magnitude of total force on the bearing of pulley C is $\sqrt{F_x^2 + F_y^2}$ or **1294 N**.

56.

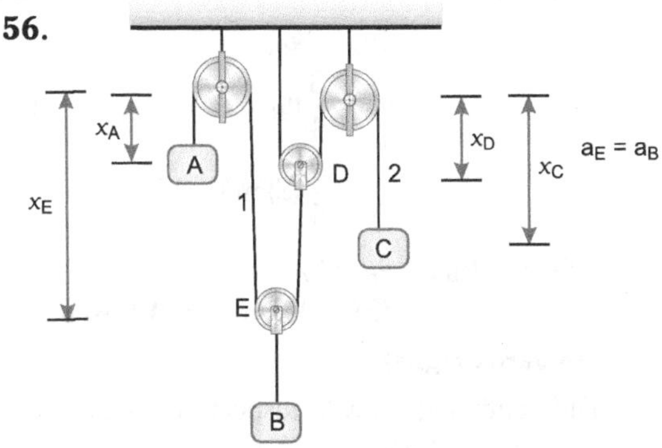

$$L_1 = x_A + x_E + (x_E - x_D) + C_1 \quad \text{...(1)}$$

Here L_1 = length of string 1

and C_1 = length of string 1 over pulleys.

Differentiating equation (1) twice with respect to time, we get

$$0 = a_A + 2a_E - a_D \quad \text{...(2)}$$

Similarly

$$L_2 = 2x_D + x_C + C_2$$

or $$0 = 2a_D + a_C \quad \text{...(3)}$$

Multiplying equation (2) by 2 and then adding with equation (3), we get

$$4a_E + 2a_A + a_C = 0$$

but $$a_E = a_B$$

Hence $$\boldsymbol{2a_A + 4a_B + a_C = 0}$$

57. In the figure we can see that

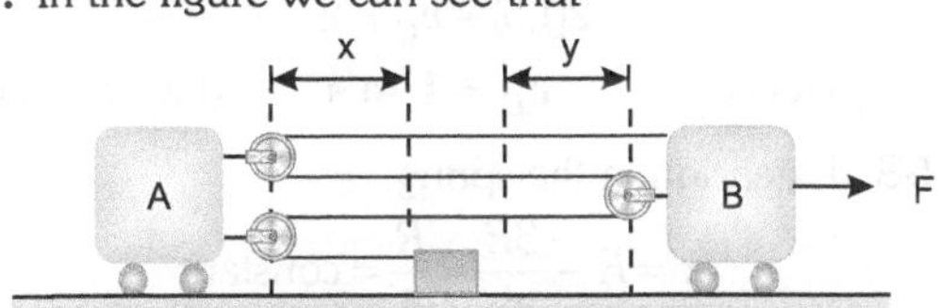

$$4x + 3y = \text{constant}$$

Differentiating this with respect to time, we have

$$4\frac{dx}{dt} + 3\frac{dy}{dt} = 0$$

or $$4\left(\frac{dx}{dt}\right) = -3\left(\frac{dy}{dt}\right)$$

Here negative sign implies that as x decreases y increases.

Hence $$4v_A = 3v_B$$

or $$v_A = \frac{3}{4}v_B \qquad \ldots(1)$$

$$= \frac{3}{4}(2)\text{ m/s} = 1.5\text{ m/s}$$

Hence $\boldsymbol{v_{BA}} = v_B - v_A$

$$= (2.0 - 1.5)\text{ m/s} = \mathbf{0.5\ m/s}$$

(towards right)

Differentiating (1) with respect to time, we get

$$a_A = \frac{3}{4}a_B = \left(\frac{3}{4}\right)(3)\text{ m/s}^2 = 2.25\text{ m/s}^2$$

Hence $a_{BA} = a_B - a_A$

$$= (3.0 - 2.25)\text{ m/s}^2$$

or $\boldsymbol{a_{BA}} = \mathbf{0.75\ m/s^2}$ **(towards right)**

Velocity of point C :

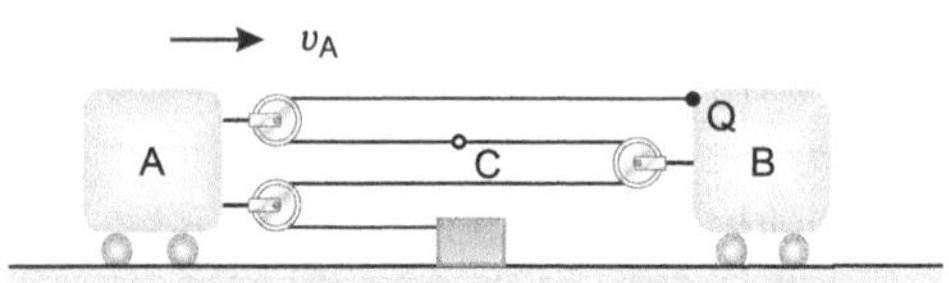

$$v_Q = v_B = 2\text{ m/s} \quad \text{(towards right)}$$

Let velocity of C is v_C (towards right)

Then $$2v_A - v_C = v_Q$$

or $$2(1.5) - v_C = 2$$

Hence $\boldsymbol{v_C} = \mathbf{1\ m/s}$ (towards right)

58. Extension in the spring,

$$x = R - \frac{3R}{4} = \frac{R}{4} = \text{constant}$$

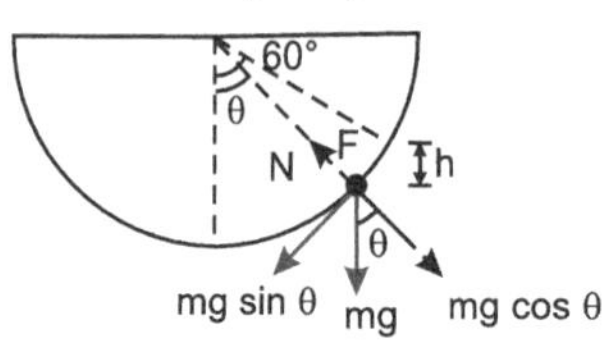

Hence spring force,

$$F = kx = \left(\frac{2mg}{R}\right)\left(\frac{R}{4}\right) = \frac{mg}{2}$$

Free body diagram of the bead at angle θ is shown in figure.

Here $h = R(\cos\theta - \cos 60°)$

or $$h = R\left(\cos\theta - \frac{1}{2}\right) \qquad \ldots(1)$$

Velocity of bead in this position would be

$$v^2 = 2gh = 2gR\left(\cos\theta - \frac{1}{2}\right) \qquad \ldots(2)$$

Further $F + N - mg\cos\theta = \dfrac{mv^2}{R}$

$$\therefore \quad \frac{mg}{2} + N - mg\cos\theta = 2mg\left(\cos\theta - \frac{1}{2}\right)$$

$$\boldsymbol{N = 3\,mg\left(\cos\theta - \frac{1}{2}\right)}$$

Tangential acceleration of the bead will be

$$\boldsymbol{a_t = \frac{mg\sin\theta}{m} = g\sin\theta}$$

At $\theta = 60°$, $N = 0$ and $a_t = \dfrac{\sqrt{3}g}{2}$

and at $\theta = 0°$, $N = \dfrac{3mg}{2}$ and $a_t = 0$

Hence N and a_t varies with θ as :

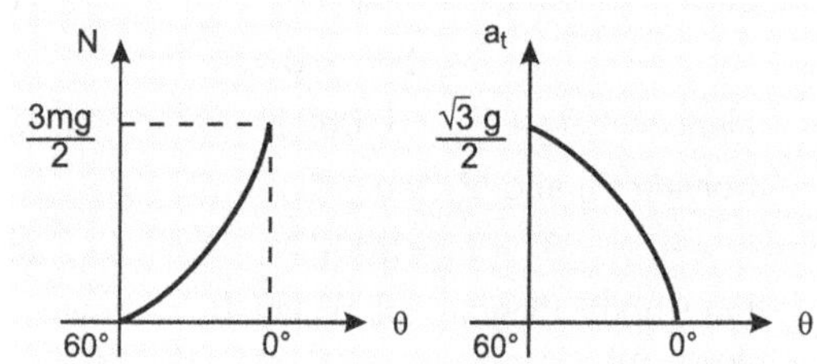

59. Let *R* be the radius of the disc.

When F is maximum, frictional force at A and B are $\frac{1}{2}N_A$ and $\frac{1}{2}N_B$.

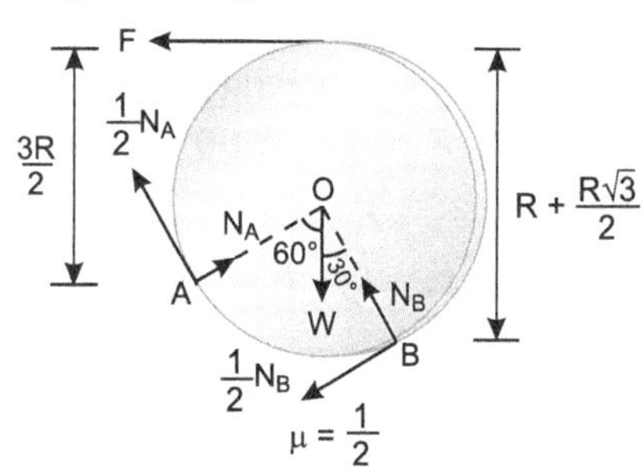

Weight W = 50 x 10 = 500 N

Taking moments about axes through A, B and O (which are not collinear) we have

$$\Sigma M_A = 0$$

$$\text{or } F\left(\frac{3R}{2}\right) + N_B R - \frac{1}{2} N_B R - \frac{500\sqrt{3}}{2} R = 0 \quad ...(1)$$

$$\Sigma M_B = 0$$

$$\text{or } F\left(R + R\frac{\sqrt{3}}{2}\right) + 500\frac{R}{2} - N_A R - \frac{1}{2} N_A R = 0 \quad ...(2)$$

$$\Sigma M_O = 0$$

$$\text{or } FR - \frac{1}{2} N_A R - \frac{1}{2} N_B R = 0 \quad ...(3)$$

From equation (1)

$$N_B = 500\sqrt{3} - 3F$$

From equation (2)

$$3N_A = 500 + F(\sqrt{3} + 2)$$

and from equation (3)

$$2F = N_A + N_B$$

Hence $2F = \frac{1}{3}(500 + 2F + \sqrt{3}F) + 500\sqrt{3} - 3F$

or $F = \left(\frac{11 + 20\sqrt{3}}{83}\right) 500$

or $\mathbf{F = 275\ N}$

60. Let T_1 and T_2 be the tensions in the strings as shown in figure. Then

$2T_1 = F$ or $2T_1 = 20t$

or $T_1 = 10t$ ...(1)

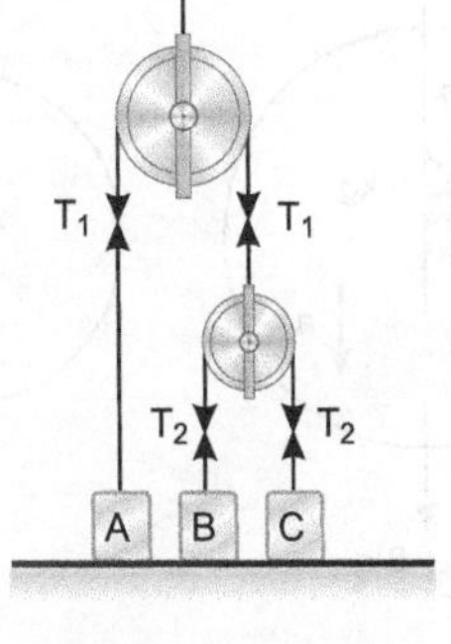

and $2T_2 = T_1$

or $T_2 = 5t$ ...(2)

Block A starts moving up at

$$T_1 = m_A g$$

or $10t = 10$ or $t = 1$ s

Block B starts moving up at

$$T_2 = m_B g$$

or $5t = 20$ or $t = 4$ s

and block C starts moving up at

$$T_2 = m_C g$$

or $5t = 10$ or $t = 2$ s

Acceleration of block C at any time $t \geq 2$ s is

$$a_c = \frac{T_2 - m_C g}{1} = (5t - 10)$$

or $\frac{dv_c}{dt} = (5t - 10)$

or $\int_0^{v_c} dv_c = \int_2^t (5t - 10)\, dt$

or $v_C = \left[\frac{5t^2}{2} - 10t\right]_2^t$

or $v_C = 2.5t^2 - 10t + 10$ ...(3)

Substituting $v_C = 2.5$ m/s in the above equation (3)

we get $t = 3$ s.

For time $t \geq 1$ s velocity of block A can be found as follows :

$$a_A = \frac{T_1 - m_A g}{m_A} = (10t - 10)$$

or $\frac{dv_A}{dt} = (10t - 10)$

or $dv_A = (10t - 10)\, dt$

velocity at time $t = 3$ s will be

$$\int_0^{v_A} dv_A = \int_1^3 (10t - 10)\, dt$$

or $v_A = [5t^2 - 10t]_1^3$ or $v_A = 20$ m/s

Velocity of block B at time $t = 3$ s is zero because it starts moving up at $t = 4$ s. **Hence relative velocity between blocks B and A at the desired instant is 20 m/s.**

61. Maximum velocity of belt

$$= (1.3)(2) \text{ m/s} = 2.6 \text{ m/s.}$$

The velocity-time graph of the belt till it comes to rest is shown in figure.

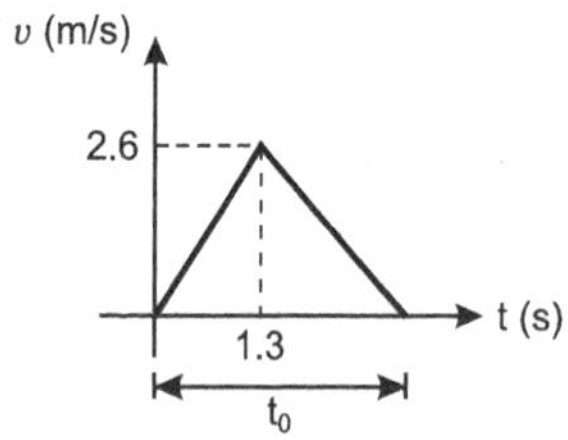

Total displacement of belt is 2.2 m.
Hence area under velocity-time graph should be 2.2 m

or $$\frac{1}{2}(t_0)(2.6) = 2.2$$

or $$t_0 = 1.692 \text{ s}$$

Hence the time of retardation will be

$$t = t_0 - 1.3 = 1.692 - 1.3$$
$$t = 0.392 \text{ s}$$

and retardation of belt,

$$a_2 = \frac{2.6}{0.392} \text{ m/s}^2 = 6.63 \text{ m/s}^2$$

Now during acceleration, there will be no relative motion between the block and the belt because

$$\mu_s g = (0.35) \times 10 \text{ m/s}^2$$
$$= 3.5 \text{ m/s}^2 > 2.0 \text{ m/s}^2$$

But during retardation, there will be relative motion between the two, because

$$\mu_s g = 3.5 \text{ m/s}^2 < 6.63 \text{ m/s}^2$$

Hence relative acceleration between the two will be

$$a_r = (6.63 - \mu_k g) \text{ m/s}^2$$
$$a_r = (6.63 - 2.5) \text{ m/s}^2 = 4.13 \text{ m/s}^2$$

Hence relative displacement between the two is

$$s = \frac{1}{2}a_r t^2 = \frac{1}{2}(4.13)(0.392)^2$$

$\mathbf{s = 0.317 \text{ m}}$

62. When there is no tendency to slip, no frictional force acts. Hence

$$N \sin\theta = \frac{m(28)^2}{200} \quad \ldots(1)$$

and $$N \cos\theta = mg \quad \ldots(2)$$

dividing (1) by (2), we get

$$\tan\theta = 0.4 \quad \ldots(3)$$

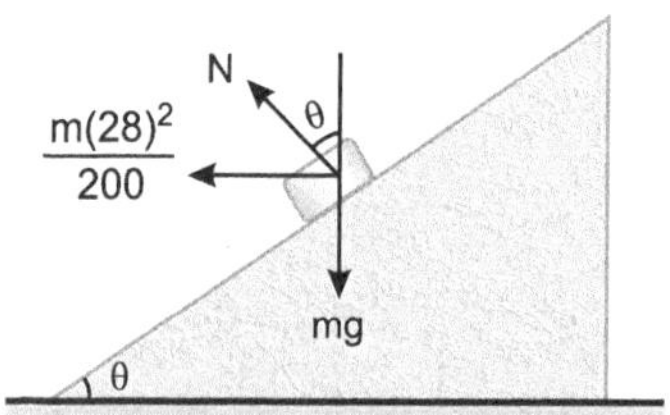

At speed 35 m/s, car has a tendency to slip up the track. Hence frictional force will act down the plane.
Therefore,

$$N' \sin\theta + \mu N' \cos\theta = \frac{m(35)^2}{200} \quad \ldots(4)$$

and $$N' \cos\theta - \mu N' \sin\theta = mg \quad \ldots(5)$$

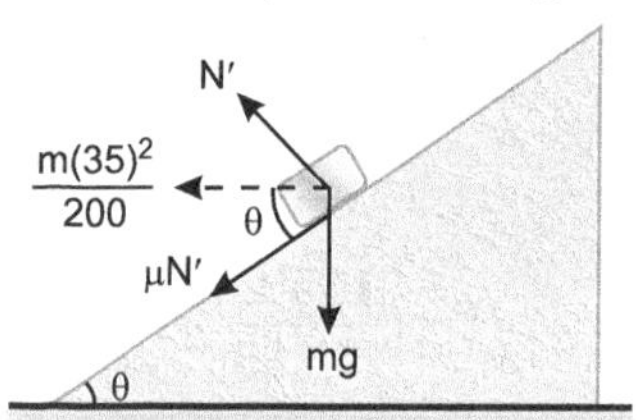

solving (3), (4) and (5), we get

$$\mu = \mathbf{0.18}$$

63. Force diagram on right hand side and on left hand side at all instants is same.

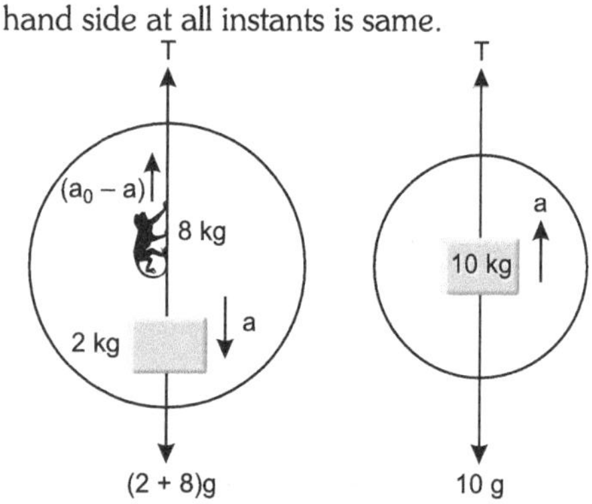

Hence, centre of mass of both sides will move with same acceleration.

$$\therefore \qquad \frac{8(a_0 - a) - 2a}{10} = \frac{10a}{10}$$

Solving this equation we get

$$a = \frac{2}{5} a_0$$

Here a_0 = acceleration of monkey w.r. to string $= 2 \text{ m/s}^2$.

$\therefore$ Acceleration of monkey with respect to ground

$$= (a_0 - a) = 0.6a_0 = 1.2 \text{ m/s}^2$$

$$\therefore \qquad t = \sqrt{\frac{2s}{1.2}} = \sqrt{\frac{2 \times 2}{1.2}} = 1.82 \text{ s}$$

64. (a) From constraint relations we can show that,

$$|a_A| = |2a_B|$$

So, let acceleration of B is a downwards, then acceleration of A will be $2a$ upwards.
Writing equations of motion,
for block A :

$$T - m_A g = m_A(2a)$$

or $\qquad T - 100 = 20a \qquad \ldots(1)$

for block B :

$$m_B g - 2T = m_B(a)$$

$\therefore \qquad 50 - 2T = 5a \qquad \ldots(2)$

Solving Eqs. (1) and (2), we get

$$a = -\frac{10}{3} \text{ m/s}^2$$

$\therefore$ Acceleration of A is $\frac{20}{3}$ m/s^2 downwards and that of B is $\frac{10}{3}$ m/s^2 upwards.

(b) From constraint relations we can show that,

$$|a_B| = |3a_A|$$

So, let a_A is a upwards, then a_B will be $3a$ downwards.
Writing equations of motion,
for block A :

$$3T - m_A g = m_A(a)$$

or $\qquad 3T - 100 = 10a \qquad \ldots(3)$

for block B :

$$m_B g - T = m_B(3a)$$

or $\qquad 50 - T = 15a \qquad \ldots(4)$

Solving Eqs. (3) and (4), we get

$$a = \frac{10}{11} \text{ m/s}^2$$

Therefore acceleration of A is $\frac{10}{11}$ m/s^2 upwards and that of B is $\frac{30}{11}$ m/s^2 downwards.

65. $\qquad 2a_2 = 3a_1$

Considering block I

$$F - 3T - f = 20a_1 \qquad \ldots(1)$$

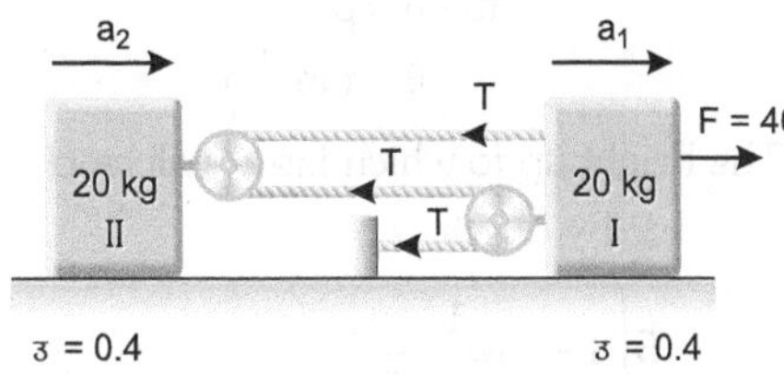

Consider block II

$$2T - f = 20a_2 = 30a_1 \qquad \ldots(2)$$

Solving Eqs. (1) and (2)

$$F - 3\left(\frac{f}{2} + 15a_1\right) - f = 20a_1$$

$$F = 65a_1 + \frac{5}{2} f \qquad \ldots(3)$$

When motion starts

$$t = \frac{\mu mg}{40} = 8 \text{ N} \quad \text{and} \quad a_1 > 0$$

Solving we get,

$$F = 20 \text{ N} \Rightarrow t = \frac{1}{2} \text{ sec}$$

Motion of blocks will begin at $t = \frac{1}{2}$ sec

At $\qquad t = 1$ sec, $\quad F = 40$ N

From Eqs. (3), we get

$$40 = 65a_1 + 20$$

$$a_1 = \frac{20}{65} = \frac{4}{13} \text{ m/s}^2$$

$$a_2 = \frac{3}{2} a_1 = \frac{6}{13} \text{ m/s}^2$$

66. First when the bowl is not rotating

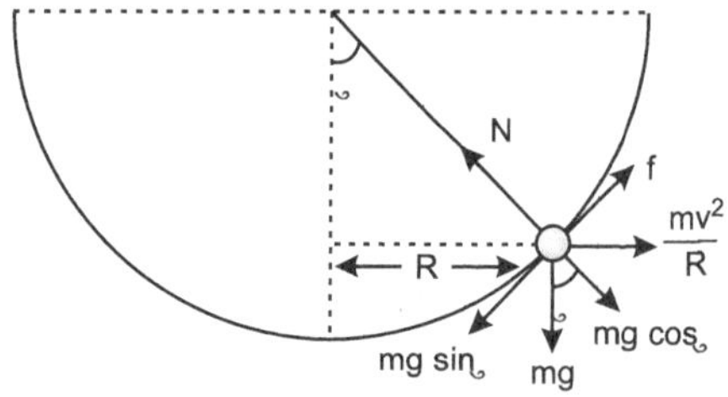

$$f = mg \sin\theta$$
$$N = mg \cos\theta$$
$$f = \mu N$$
$$\mu mg \cos\theta = mg \sin\theta$$
$$\tan\theta = \mu$$
$$\theta = \tan^{-1}\mu$$

The height up to which insect will reach

$$= R(1 - \cos\theta)$$
$$= 5\left(1 - \frac{1}{\sqrt{1+\mu^2}}\right)$$
$$= 5\left(1 - \frac{1}{\sqrt{1.25}}\right) \text{cm} = 0.53 \text{ cm}$$

Now, the bowl start rotating with angular velocity ω.

$$N - mg\cos\theta = \frac{mv^2}{R'}\sin\theta$$
$$f + \frac{mv^2}{R'}\cos\theta = mg\sin\theta$$
$$N = mg\cos\theta + \frac{mv^2}{R'}\sin\theta$$
$$\mu N + \frac{mv^2}{R'}\cos\theta = mg\sin\theta$$
$$\mu\left(mg\cos\theta + \frac{mv^2}{R'}\sin\theta\right) + \frac{mv^2}{R'}\cos\theta = mg\sin\theta$$

Now, when the angular displacement of the insect will be 90° then it will just come out from the bowl put $\theta = 90°$ and $R' = R$ in above equation

$$\mu\frac{mv^2}{R} = mg$$
$$v^2 = \frac{Rg}{\mu}$$
$$R^2\omega^2 = \frac{Rg}{\mu}$$
$$\omega^2 = \frac{g}{\mu R}$$
$$\omega = \sqrt{\frac{g}{\mu R}}$$
$$= \sqrt{\frac{10}{0.5 \times 5 \times 10^{-2}}}$$
$$= \frac{10^2}{5} = 20 \text{ rad/s}$$

67. At the highest point of the bridge the equation of motion of the car is

$$mg - N = m\frac{v^2}{\rho}$$

where N is the normal force acting on the car (and the negative of the required answer), $v = 20 \text{ m/s}^{-1}$ and ρ is the radius of curvature of the bridge there. The most difficult part of the problem is to find this radius of curvature.

If we could find a motion with this trajectory for which the normal acceleration is well known, the radius of curvature could be easily calculated. For a parabolic trajectory the flight of a projectile offers the required analogue. Let the projectile have an initial velocity of v_0 making an angle α with the horizontal.

The range ($d = 100$ m) and height ($h = 5$ m) of the projectile can be expressed using the initial data,

$$d = \frac{2v_0^2 \sin\alpha\cos\alpha}{g}$$

and
$$h = \frac{v_0^2\sin^2\alpha}{2g}$$

The quotient $\frac{h}{d}$ gives $\tan\alpha = \frac{4h}{d}$ (so $\alpha \approx 11.3°$), and the horizontal component of the initial velocity is

$$v_x = v_0\cos\alpha = d\sqrt{\frac{g}{8h}} = 50 \text{ ms}^{-1}$$

Now, the radius of curvature at the highest point can be calculated as

$$\rho = \frac{v_x^2}{g} = 250 \text{ m}$$

So, the normal force at the highest point is

$$N = m\left(g - \frac{v^2}{\rho}\right) = 8.40 \text{ kN}$$

68. In the critical case; the sphere will just touch the horizontal surface as shown in figure

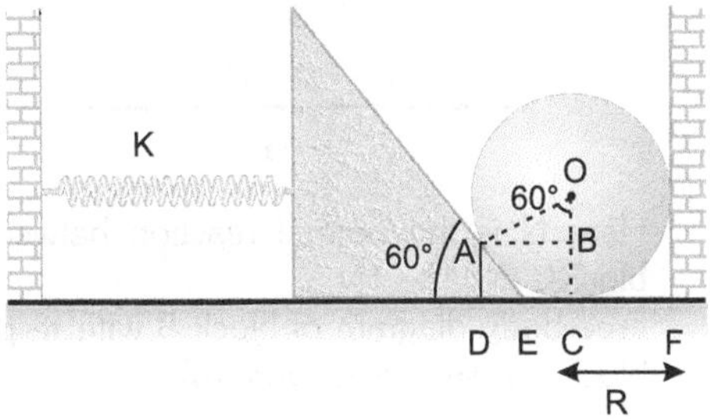

In the figure

$$BC = OC - OB$$

$$= R - R\cos 60° = R - \frac{R}{2}$$

$$BC = \frac{R}{2}$$

$$AB = R\sin 60° = \frac{\sqrt{3}}{2}R$$

$$DE = AD\cot 60° = \frac{AD}{\sqrt{3}} = \frac{BC}{\sqrt{3}} = \frac{R}{2\sqrt{3}}$$

$$\therefore\ EC = DC - DE = AB - DE$$

$$= \frac{\sqrt{3}R}{2} - \frac{R}{2\sqrt{3}} = \frac{R}{\sqrt{3}}$$

$$\therefore\ FE = FC + EC = R + \frac{R}{\sqrt{3}} = R\left(1 + \frac{1}{\sqrt{3}}\right)$$

$\therefore$ Compression in the spring

$$x = FE = R\left(1 + \frac{1}{\sqrt{3}}\right)$$

Now drawing free body diagram of wedge and sphere in equilibrium we get (only those forces are shown which are important for calculation)

N = normal reaction between wedge and sphere

Equilibrium of wedge gives

$$kx = N\sin 60° \quad \ldots(1)$$

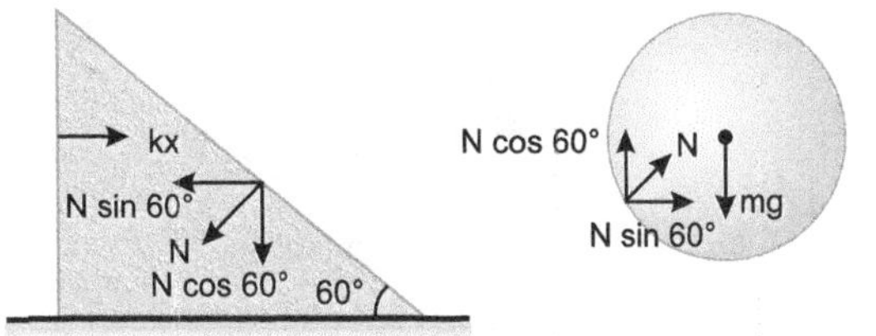

Equilibrium of sphere gives

$$mg = N\cos 60° \quad \ldots(2)$$

Note: Normal reaction between ground and sphere will be zero in critical case.

Dividing (1) by (2) we get

$$\tan 60° = \frac{kx}{mg} \quad \ldots(3)$$

Substituting the values of k and x we get

or $$\sqrt{3} = \eta\left(\frac{mg}{R}\right)\frac{R\left(1 + \frac{1}{\sqrt{3}}\right)}{mg}$$

or $$\eta = \frac{\sqrt{3}}{1 + \frac{1}{\sqrt{3}}}$$

or $$\eta = 1.098$$

So, minimum value of η is 1.098

(ii) Considering equilibrium of (wedge + sphere)

$$N' = kx$$

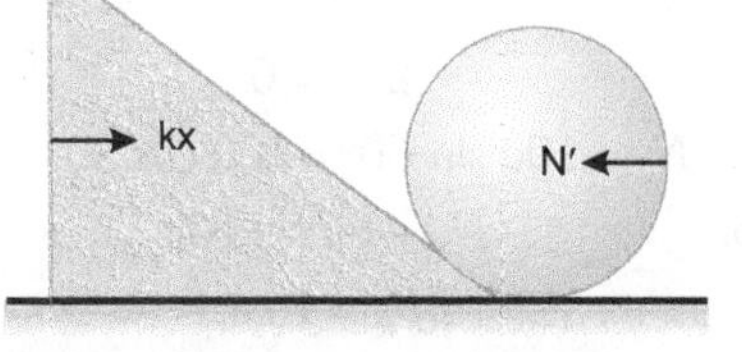

where N' = normal reaction between the sphere and the right wall

$$= mg\tan 60° \quad \text{(from equation 3)}$$

or $$\boldsymbol{N' = \sqrt{3}mg}$$

69. Let acceleration of block A down the slope is a (absolute) and acceleration of block B relative to block A is a_r.

Free body diagram of A with respect to ground is as shown in figure (a).

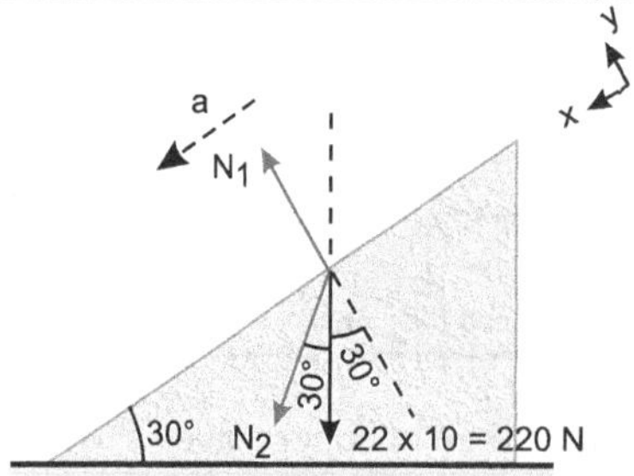

(a)

Note that block A is shown by a dot just for better understanding of force diagram.
Here N_1 = normal reaction between block A and wedge.
and N_2 = normal reaction between blocks A and B.

$$\Sigma F_x = 22a$$

$\therefore$ $220 \sin 30^\circ + N_2 \sin 60^\circ = 22a$

or $\quad 110 + 0.87\, N_2 = 22\, a \quad \ldots(1)$

Free body diagram of B with respect to A is shown in figure (b).

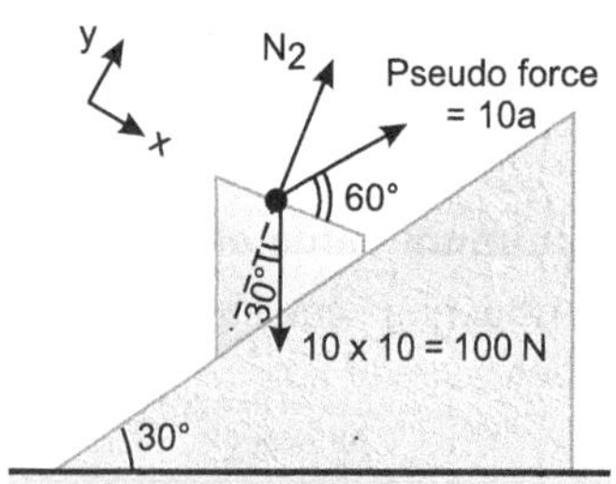

(b)

$$\Sigma F_y = 0$$

$\therefore$ $N_2 + 10a \sin 60^\circ = 100 \cos 30^\circ$

or $\quad N_2 + 8.7a = 87 \quad (2)$

$$\Sigma F_x = 10\, a_r$$

$\therefore$ $10a \cos 60^\circ + 100 \sin 30^\circ = 10\, a_r$

or $\quad 5a + 50 = 10a_r$

or $\quad a - 2a_r = -10 \quad \ldots(3)$

Solving equations (1), (2) and (3), we get

$\boldsymbol{a = 6.28\ \mathbf{m/s^2}}$ (down the wedge)

and $\boldsymbol{a_r = 8.14\ \mathbf{m/s^2}}$

70. Let a be the absolute acceleration of block A down the plane and a_r the relative acceleration of block B with respect to block A.

From constraint equation we can show that

$$a = a_r \quad \ldots(1)$$

Free body diagram of A with respect to ground is as shown in figure (a) (only those forces, which are parallel to plane have been shown).

$$125 + N - T = 25a \quad \ldots(2)$$

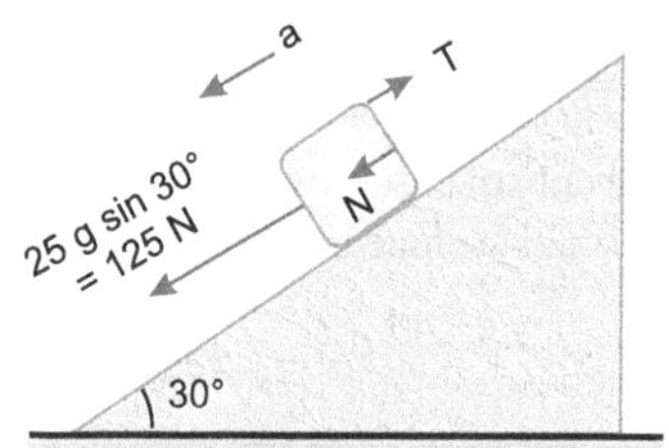

(a)

Here N is the normal reaction between the blocks.
Free body diagram of block B with respect to block A is shown in figure (b)

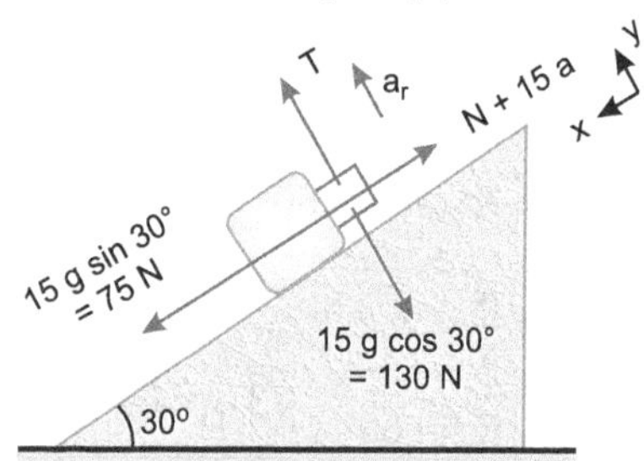

Pseudo force = 15a (up the plane)
(b)

$$\Sigma F_x = 0$$

$\therefore \quad 75 = N + 15a \quad \ldots(3)$

$$\Sigma F_y = 15a_r$$

$\therefore \quad T - 130 = 15a_r \quad \ldots(4)$

Solving equations (1), (2), (3) and (4), we get

$$a = a_r = 1.272\ \text{m/s}^2$$

and $\quad T = 149\ \text{N}$

Absolute acceleration of block B would be the resultant of a and a_r as shown in figure (c).

$\therefore \quad a_B = \sqrt{a^2 + a_r^2}$

$$= \sqrt{2}\,(1.272)\ \text{m/s}^2 = 1.8\ \text{m/s}^2$$

and $\alpha = 45^\circ$ or a_B will make an angle of 15° with horizontal.

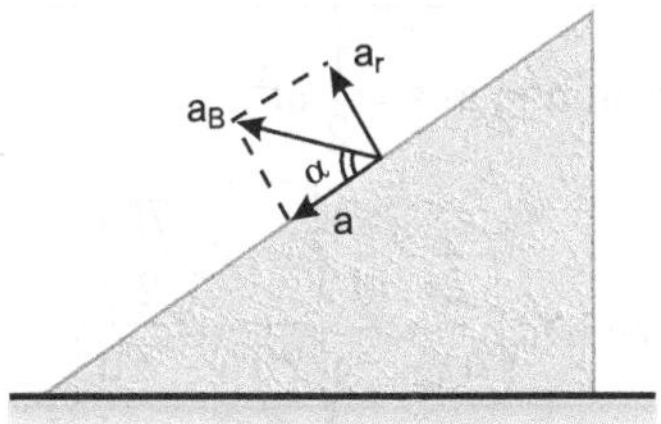

(c)

Hence

Acceleration of $A = a = 1.272$ m/s^2

(down the plane)

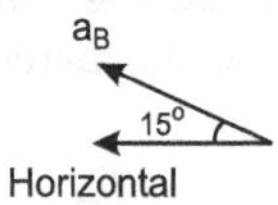

(d)

Acceleration of $B = a_B = 1.8$ m/s^2

(at 15° with horizontal)

and Tension in the string, $T = 149$ N

71. Let acceleration of A is a_1 (down the plane) and acceleration of B is a_2 (vertically downwards) Free body diagram of A relative to ground is shown in figure (a).

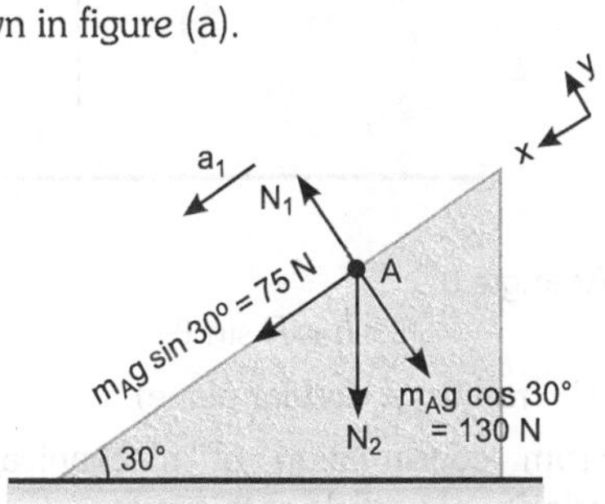

(a)

N_1 = normal reaction between block A and wedge

N_2 = normal reaction between A and B

$$\Sigma F_x = 15a_1$$

$\therefore \quad 75 + N_2 \sin 30° = 15a_1$

or $\quad 75 + 0.5N_2 = 15a_1 \quad \ldots(1)$

Note that block A has been shown as a dot (•) just for better understanding of force diagram. Free body diagram of B relative to ground is shown in figure (b).

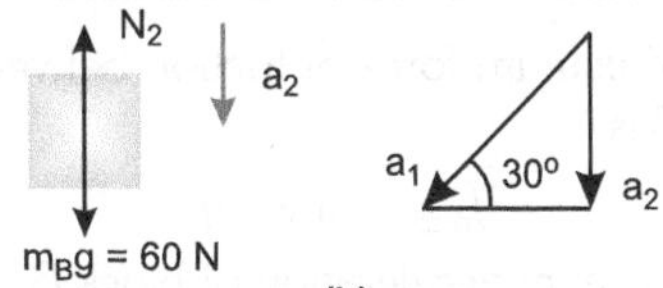

(b)

$$60 - N_2 = 6a_2 \quad \ldots(2)$$

Also $\quad a_2 = a_1 \sin 30°$

or $\quad a_2 = 0.5a_1 \quad \ldots(3)$

solving equations (1), (2) and (3), we get

$$a_1 = 6.36 \text{ m/s}^2$$

The vertical component of acceleration of B relative to A is zero. There is only horizontal component of relative acceleration between the two which is equal to $a_1 \cos 30°$ or 5.5 m/s^2 Hence, **acceleration of A is 6.36 m/s^2 (down the plane) and acceleration of B relative to A is 5.5 m/s^2 in horizontal direction (towards right)**

72. (a) Since the cylinder is not sliding down, total frictional force = net downward force

or frictional force F acting on one side

$$= \frac{mg \sin 30°}{2} = \frac{(10)(9.8)(1/2)}{2}$$

$$\mathbf{F = 24.5 \text{ N}}$$

(b) By symmetry, normal reactions on both sides will be equal. Hence

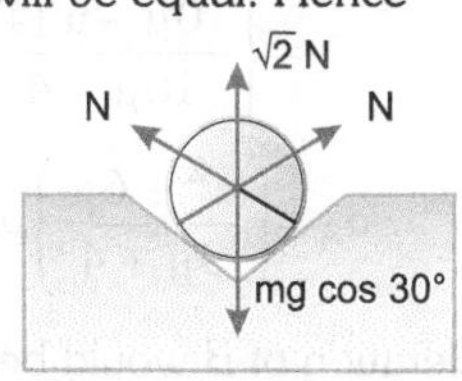

$$\sqrt{2}\, N = mg \cos 30°$$

or $$N = \frac{(10)(9.8)(\sqrt{3}/2)}{\sqrt{2}}$$

$$N = 60 \text{ N}$$

Therefore, total maximum frictional force that can be obtained is

$$f_{max} = 2\mu N = (2)(0.5)(60) = 60 \text{ N}$$

Net downward force is $mg \sin 30°$ or 49 N

Hence, $\quad P = 60 + 49$

$$\mathbf{P = 109 \text{ N}}$$

73. (a) Maximum force of friction between A and B is

$$f_{max} = \mu (m) g$$

$\therefore$ maximum acceleration of block B can be

$$a_{max} = \frac{f_{max}}{4m} = \frac{\mu mg}{4m} = \frac{\mu g}{4}$$

Further $a_{max} = \dfrac{Mg}{m + 4m + M}$

So $$\frac{\mu g}{4} = \frac{Mg}{5m + M}$$

or $$M = \left(\frac{5\mu}{4 - \mu}\right) m$$

So, minimum value of M is $\left(\dfrac{5\mu}{4-\mu}\right) m$

(b) When $M = 2\left\{\dfrac{5\mu}{4-\mu}\right\} m$

or $$M = \left(\frac{10\mu m}{4 - \mu}\right)$$

acceleration of A would be :

$$a_A = \frac{Mg - f_{max}}{M + m} = \frac{\left(\dfrac{10\mu mg}{4-\mu}\right) - \mu mg}{\left(\dfrac{10\mu m}{4-\mu}\right) + m}$$

$$= \left\{\frac{10\mu - \mu (4 - \mu)}{10\mu + 4 - \mu}\right\} g$$

$\Rightarrow$ $$a_A = \left(\frac{\mu^2 + 6\mu}{9\mu + 4}\right) g$$

and acceleration of B would be :

$$a_B = \frac{\mu g}{4}$$

$\therefore$ $$a_{AB} = a_A - a_B = \frac{20\mu - 5\mu^2}{4 (9\mu + 4)} g$$

Now A topples from B when half of its length is outside B *i.e.*, in the position shown below

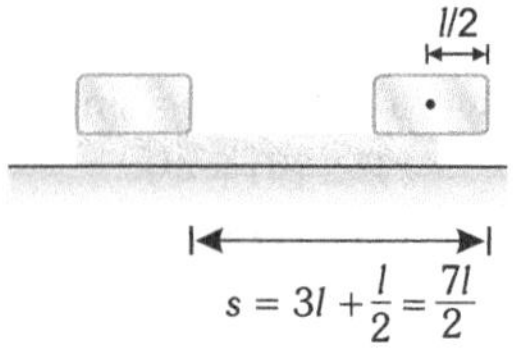

Substituting in $s = \frac{1}{2} at^2$

we get $$t = \sqrt{\frac{2s}{a_{AB}}} = \sqrt{\frac{2\left(\dfrac{7l}{2}\right)}{\dfrac{20\mu - 5\mu^2}{4(9\mu + 4)} g}}$$

$$\boldsymbol{t = \sqrt{\frac{28\,(9\mu + 4)\,l}{(20\mu - 5\mu^2)\,g}}}$$

74. During the motion of the particle over the wedge, let F_{max} be the maximum horizontal thrust exerted by the particle on the wedge. Then

$$\mu_{min} = \frac{F_{max}}{N_2} \quad \text{...(1)}$$

Here N_2 is the normal force between wedge and horizontal plane at that instant. F_{max} and N_2 can be calculated as follows :

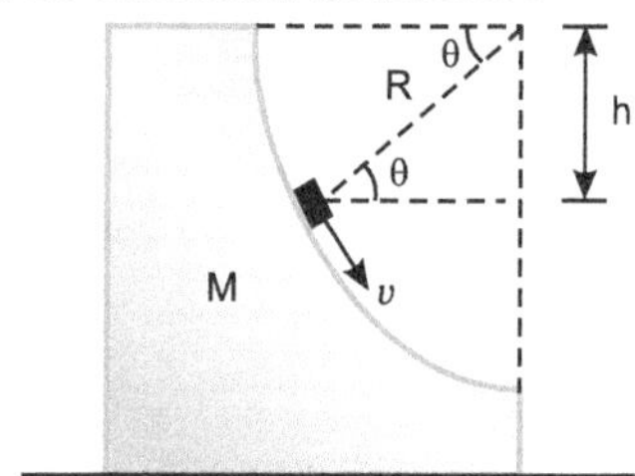

At angle θ

$$h = R \sin \theta$$

(R = radius of circular plane)

From conservation of mechanical energy, velocity of particle at this instant

$$v^2 = 2gh = 2gR \sin \theta$$

Let N_1 be the normal reaction between particle and wedge at this instant.

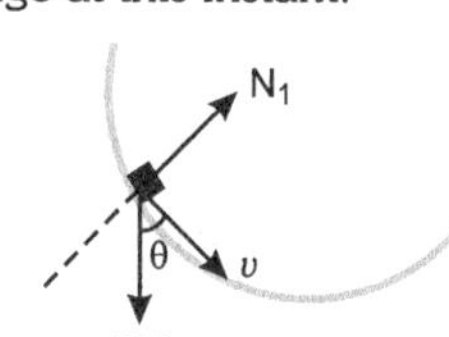

Then

$$N_1 - mg \sin \theta = \frac{mv^2}{R}$$

or $N_1 - mg \sin\theta = \frac{m(2gR \sin\theta)}{R}$

or $N_1 = 3mg \sin\theta$ ...(1)

Horizontal force exerted by the particle on the wedge at this instant is

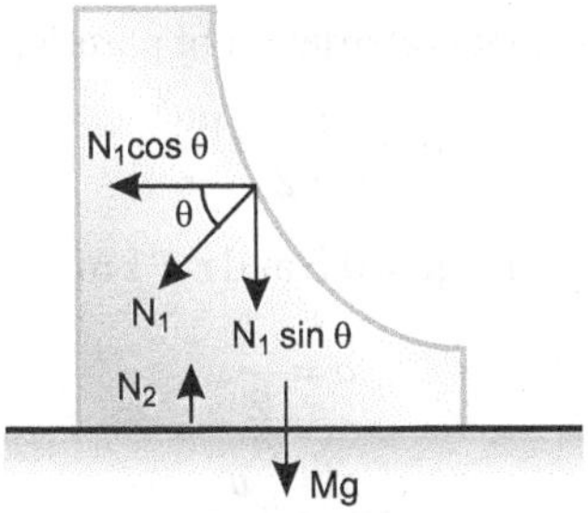

$$F = N_1 \cos\theta = (3mg \sin\theta)(\cos\theta)$$

or $F = \frac{3}{2} mg\,(2 \sin\theta \cos\theta)$

or $F = \frac{3}{2} mg \sin 2\theta$ ...(2)

and $N_2 = Mg + N_1 \sin\theta$ ...(3)

under limiting conditions :

$$F = \mu N_2$$

or $\mu = \frac{F}{N_2} = \frac{\frac{3}{2} mg \sin 2\theta}{Mg + 3mg \sin^2\theta}$

(from equations 1, 2 and 3)

Substituting

$m = 2$Kg and $M = 4$Kg

$\mu = \frac{3 \sin 2\theta}{4 + 6 \sin^2\theta}$ ...(4)

For μ to be maximum $\frac{d\mu}{d\theta} = 0$

or $\frac{(4 + 6\sin^2\theta)(6\cos 2\theta) - (3\sin 2\theta)(12 \sin\theta \cos\theta)}{(4 + 6\sin^2\theta)^2} = 0$

or $12\cos 2\theta + 9\cos 2\theta - 9\cos^2 2\theta - 9 + 9\cos^2 2\theta = 0$

or $21 \cos 2\theta = 9$

or $\cos 2\theta = \frac{9}{21}$

or $2\theta = 64.6^\circ$

or $\theta = 32.3^\circ$

$\therefore \mu_{max} = \frac{3 \sin(64.6^\circ)}{4 + 6\sin^2(32.3^\circ)} = \mathbf{0.474}$

(from equation 4)

75. Before the first collision occurs retardation of A, $a_A = \mu_1 g$ and retardation of B, $a_B = \mu_2 g$

Since $\mu_2 > \mu_1$, therefore, $a_B > a_A$ or B will retard more rapidly. Relative initial velocity between the two is zero, but relative acceleration of A with respect to B is $a_B - a_A$ or $(\mu_2 - \mu_1) g$. Till first collision relative displacement of A is $L - l$. So, if T_1 is the time of first collision, then

$$(L - l) = \frac{1}{2}(\mu_2 - \mu_1) g T_1^2$$

or $T_1 = \sqrt{\frac{2(L-l)}{(\mu_2 - \mu_1) g}}$ ...(1)

Velocities of A and B just before collision are

$v_1 = v_0 - \mu_1 g T_1$ and $v_2 = v_0 - \mu_2 g T_1$

Since $\mu_2 > \mu_1$,

Hence $v_2 < v_1$

Since the collision is elastic and both have equal masses, they exchange their velocities. By repeating the same procedure, we can show that the period of next collision T_2 is equal to T_1 and so on. Hence the time between subsequent collision T is a constant and is given by

$$\boldsymbol{T = \sqrt{\frac{2(L-l)}{(\mu_2 - \mu_1) g}}}$$

76. Consider the point mass moving in the x-y plane around an ellipse with semi-major axes a and b according to the equations.

$$x = a \cos\omega t \quad \text{and} \quad y = b \sin\omega t$$

At $t = 0$, the mass is moving at the end of the major axis with

velocity $v = b\omega$

and acceleration $A = a\omega^2$

On the other end the acceleration is,

$$A = \frac{v^2}{R}$$

So, the radius of curvature is $R = \frac{b^2}{a}$.

Similarly we find the radius of curvature at the end of the minor axes to be $\frac{a^2}{b}$.

Using $F = \frac{mv^2}{R}$, with the given data we obtain,

$\frac{b^2}{a} = 1.25$ m and $\frac{a^2}{b} = 10$ m and hence,

$$\mathbf{2a = 10\ m} \quad \text{and} \quad \mathbf{2b = 5\ m}$$

77. For the particle, tangential retardation

$$a = \mu g = 5 \text{ m/s}^2$$

For disc angular acceleration

$$\alpha = \frac{\mu\left(\frac{m}{2}\right)gr}{\frac{1}{2}mr^2}$$

or $$\alpha = \frac{\mu g}{r} = 5 \text{ rad/s}^2$$

Let after time t, relative motion between particle and disc is stopped, then

$$v = r\omega$$

or $$(u - at) = r(\alpha t)$$

$\therefore$ $$t = \frac{u}{a + r\alpha}$$

Substituting the values we get,

$$t = 1 \text{ sec}$$

From $t = 0$ to $t = 1$ sec particle has rotated an angle,

$$\theta_1 = \frac{s}{r} = \frac{\left(ut - \frac{1}{2}at^2\right)}{r}$$

Substituting the values we get

$$\theta_1 = 7.5 \text{ rad}$$

From $t = 1$ sec to $t = 2$ sec

$$\theta_2 = (\omega)t$$
$$= (5 \times 1) \times 1 = 5 \text{ rad}$$

$\therefore$ $$\theta_{Total} = \theta_1 + \theta_2 = \mathbf{12.5\ rad}$$

78. If after time 't' velocity of particle is v and N is the normal reaction between the particle and the wall. Then,

$$N = \frac{mv^2}{2r}$$

Force of friction,

$$f = \mu N = \frac{\mu mv^2}{2r}$$

$\therefore$ Tangential retardation of particle,

$$a = \frac{f}{m/2} = \frac{\mu v^2}{r}$$

Substituting $\mu = 0.5$ and $r = 1$ m

$$a = \frac{v^2}{2}$$

Now, $$-\frac{dv}{dt} = \frac{v^2}{2}$$

$\therefore$ $$-2\int_{10\,\text{m/s}}^{v} v^{-2}dv = \int_0^t dt$$

$\therefore$ $$2\left(\frac{1}{v} - \frac{1}{10}\right) = t$$

or $$v = \frac{10}{1 + 5t} \qquad \ldots(1)$$

For the disc, angular acceleration,

$$\alpha = \frac{f.r}{\frac{1}{2}mr^2} = \frac{2f}{mr}$$
$$= 2\left(\frac{\mu mv^2}{2r}\right)\left(\frac{1}{mr}\right)$$

Substituting the values we get

$$\alpha = \frac{v^2}{2}$$

Substituting values of v from equation (1), we have

$$\alpha = \frac{50}{(1 + 5t)^2}$$

or $$\frac{d\omega}{dt} = \frac{50}{(1 + 5t)^2}$$

$\therefore$ $$\int_0^\omega d\omega = \int_0^t \frac{50dt}{(1 + 5t)^2}$$

$\therefore$ $$\omega = 10\left[\frac{5t}{1 + 5t}\right] \qquad \ldots(2)$$

Relative motion will stop when,

$$v = r\omega$$

From (1) and (2) with $r = 1$ m

$$\frac{5t}{1+5t} = \frac{1}{1+5t}$$

or $\quad t = \frac{1}{5}$ **sec**

79. $T - m_B a - \mu m_B g \cos\theta - m_B g \sin\theta = 0$

$$T = m_B (a + \mu g \cos\theta + g \sin\theta)$$

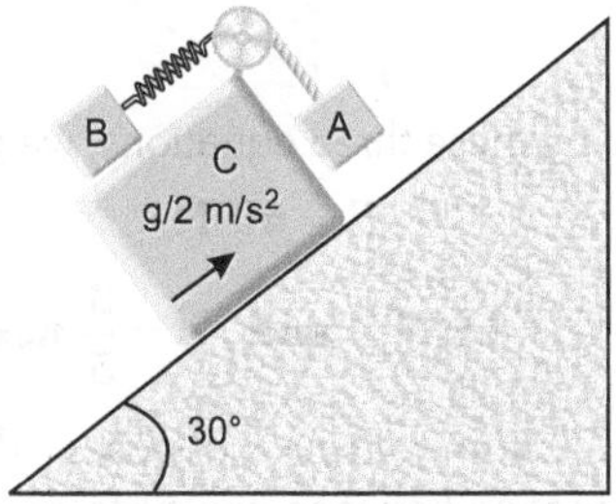

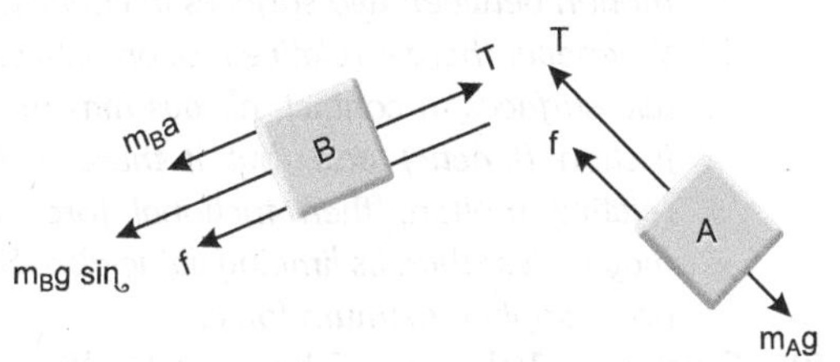

$$T = m_B \left(\frac{g}{2} + \mu g \cos\theta + g \sin\theta\right)$$

$$= m_B g \left(\frac{1}{2} + 0.2 \times \frac{\sqrt{3}}{2} + \frac{1}{2}\right)$$

$$= m_B g \left(1 + \frac{\sqrt{3}}{10}\right) = m_B g \left(\frac{10+\sqrt{3}}{10}\right) \quad \ldots(1)$$

$T = m_A g \cos\theta - \mu N \qquad$ where $N = m_A a$

$$T = m_A g \cos\theta - \mu m_A a$$

$$= m_A (g \cos\theta - \mu a)$$

$$= m_A \left(\frac{g\sqrt{3}}{2} - 0.2 \times \frac{g}{2}\right) \quad \ldots(2)$$

$$m_B g \left(\frac{10+\sqrt{3}}{10}\right) = m_A \frac{g}{2} (\sqrt{3} - 0.2)$$

$$m_B = 0.65 m_A = \mathbf{1.30\ kg}$$

$$T = kx = m_A \frac{g}{2} (\sqrt{3} - 0.2)$$

$$\Rightarrow \quad x = 8.5 \times 10^{-3} \text{ m}$$

Energy stored in the spring

$$= \frac{1}{2} kx^2 = \frac{1}{2} \times 1800 \times (8.5 \times 10^{-3})^2$$

$$= \mathbf{0.065\ J}$$

80. Acceleration of A down the plane,

$$a_A = g \sin 45° - \mu_A g \cos 45°$$

$$= (10)\left(\frac{1}{\sqrt{2}}\right) - (0.2)(10)\left(\frac{1}{\sqrt{2}}\right)$$

$$= 4\sqrt{2} \text{ m/s}^2$$

Similarly acceleration of B down the plane,

$$a_B = g \sin 45° - \mu_B g \cos 45°$$

$$= 10\left(\frac{1}{\sqrt{2}}\right) - (0.3)(10)\left(\frac{1}{\sqrt{2}}\right)$$

$$= 3.5\sqrt{2} \text{ m/s}^2$$

The front face of A and B will come in a line when,

$$s_A = s_B + \sqrt{2}$$

or $\quad \frac{1}{2} a_A t^2 = \frac{1}{2} a_B t^2 + \sqrt{2}$

or $\quad \frac{1}{2} \times 4\sqrt{2} \times t^2 = \frac{1}{2} \times 3.5\sqrt{2} \times t^2 + \sqrt{2}$

Solving this equation we get

$$t = \mathbf{2\ second}$$

Further,

$$s_A = \frac{1}{2} a_A t^2 = \frac{1}{2} \times 4\sqrt{2} \times (2)^2 = 8\sqrt{2} \text{ m}$$

Hence, both the blocks will come in a line after A has travelled a distance $8\sqrt{2}$ m down the plane.

81. Given $m_1 = 20$ kg, $\quad m_2 = 5$ kg, $\quad M = 50$ kg, $\mu = 0.3$ and $g = 10 \text{ m/s}^2$

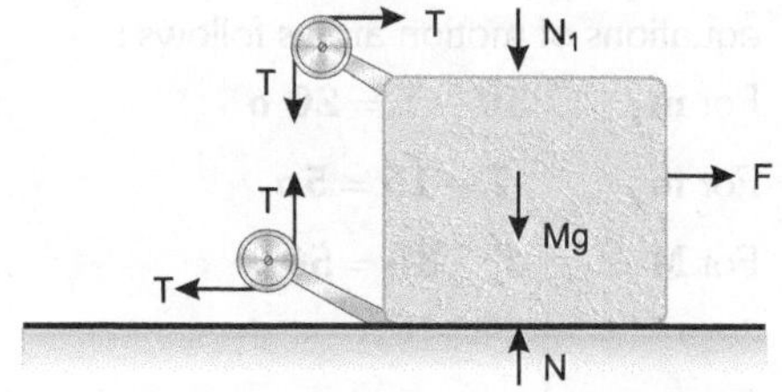

(a) Free body diagram of mass M is given below

(b) The maximum value of f_1 is

$$(f_1)_{max} = (0.3)(20)(10) = 60 \text{ N}$$

The maximum value of f_2 is

$$(f_2)_{max} = (0.3)(5)(10) = 15 \text{ N}$$

Forces on m_1 and m_2 in horizontal direction are as follows

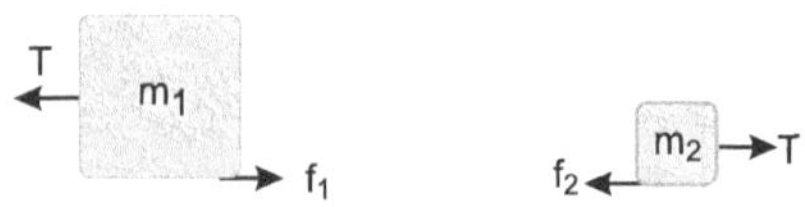

Now there are only two possibilities :

(1) either both m_1 and m_2 will remain stationary (w.r.t. ground) or

(2) both m_1 and m_2 will move (w.r.t. ground)

First case is possible when

$$T \le (f_1)_{max} \quad \text{or} \quad T \le 60 \text{ N}$$

and $\quad T \le (f_2)_{max} \quad$ or $\quad T \le 15 \text{ N}$

These conditions will be satisfied when $T \le 15$ N

say $T = 14$ N then $f_1 = f_2 = T = 14$ N

Therefore, the condition $f_1 = 2f_2$ will not be satisfied. Thus m_1 and m_2 both can't remain stationary.

In the second case, when m_1 and m_2 both move

$$f_2 = (f_2)_{max} = 15 \text{ N}$$

$$\therefore \quad f_1 = 2f_2 = 30 \text{ N}$$

Now since $f_1 < (f_1)_{max}$, there is no relative motion between m_1 and M *i.e.*, all the masses move with same acceleration, say 'a'.

$$\therefore \quad \boldsymbol{f_2 = 15 \text{ N}}$$

and $\quad \boldsymbol{f_1 = 30 \text{ N}}$

Free body diagrams showing the forces which are responsible for motion of the masses and equations of motion are as follows :

For $\mathbf{m_1}$ $\quad \boldsymbol{30 - T = 20a} \quad$ ...(1)

For $\mathbf{m_2}$ $\quad \boldsymbol{T - 15 = 5a} \quad$ ...(2)

For **M** $\quad \boldsymbol{F - 30 = 50a} \quad$ ...(3)

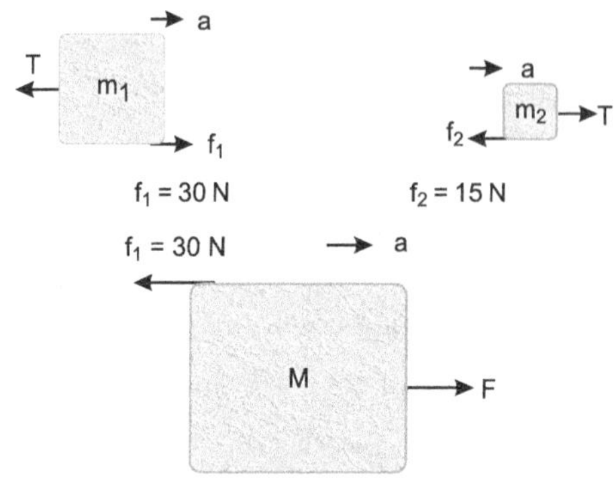

Solving these three equations, we get

$$\boldsymbol{F = 60 \text{ N}}$$

$$\boldsymbol{T = 18 \text{ N} \quad \text{and} \quad a = \frac{3}{5} \text{ m/s}^2}$$

(1) *Friction always opposes the relative motion between two surfaces in contact.*

(2) *Whenever there is relative motion between two surfaces in contact, always maximum friction (kinetic) acts, but if there is no relative motion, then frictional force (f) may be less than its limiting value also. So, don't apply maximum force.*

82. Given $m_1 = 10$ kg, $m_2 = 5$ kg, $\omega = 10$ rad/s

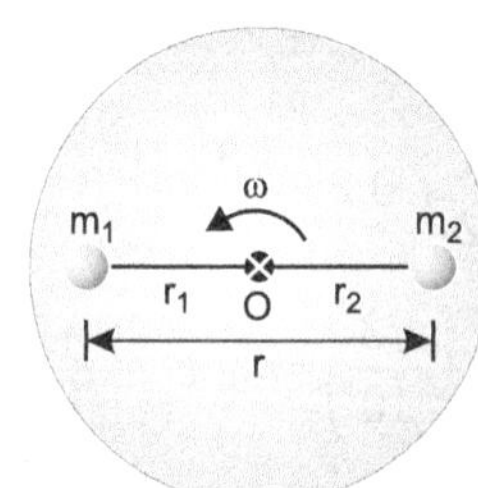

$$r = 0.3 \text{ m}, r_1 = 0.124 \text{ m}$$

$$\therefore \quad r_2 = r - r_1 = 0.176 \text{ m} \quad ...(1)$$

Masses m_1 and m_2 are at rest with respect to rotating table. Let f be the friction between mass m_1 and table.

Free body diagrams of m_1 and m_2 with respect to table (non intertial frame of reference are shown in figures).

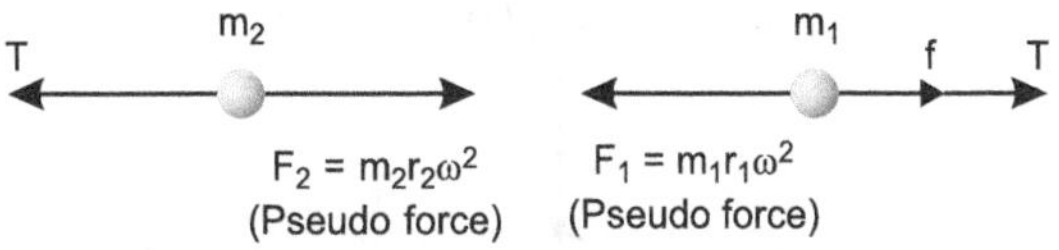

Equilibrium of m_2 gives

$$T = m_2 r_2 \omega^2 \quad ...(1)$$

Since $\quad m_2 r_2 \omega^2 < m_1 r_1 \omega^2 \quad (m_2 r_2 < m_1 r_1)$

Therefore, $\quad m_1 r_1 \omega^2 > T$

and friction on m_1 will be inwards (towards centre)

Equilibrium of m_1 gives—

$$f + T = m_1 r_1 \omega^2 \quad ...(2)$$

From (1) and (2), we get

$$f = m_1 r_1 \omega^2 - m_2 r_2 \omega^2 \quad ...(3)$$

$$= (m_1 r_1 - m_2 r_2)\,\omega^2$$

$$= (10 \times 0.124 - 5 \times 0.176)\,(10)^2 \text{ newton}$$

$\boldsymbol{f = 36\ \mathrm{N}}$

Therefore, frictional force on m_1 is 36N (inwards).

(ii) From equation (3)

$$f = (m_1 r_1 - m_2 r_2)\,\omega^2$$

Masses will start slipping when this force is greater than f_{max} or

$$(m_1 r_1 - m_2 r_2)\,\omega^2 > f_{max} > \mu\, m_1 g$$

∴ Minimum value of ω is

$$\omega_{min} = \sqrt{\frac{\mu m_1 g}{m_1 r_1 - m_2 r_2}}$$

$$= \sqrt{\frac{0.5 \times 10 \times 9.8}{10 \times 0.124 - 5 \times 0.176}}$$

$\boldsymbol{\omega_{min} = 11.67\ \text{rad/s}}$

(iii) From equation (3), frictional force $f = 0$ when $m_1 r_1 = m_2 r_2$

or $\quad \dfrac{r_1}{r_2} = \dfrac{m_2}{m_1} = \dfrac{5}{10} = \dfrac{1}{2}$

and $\quad r = r_1 + r_2 = 0.3$ m

∴ $\quad r_1 = 0.1$ m and $r_2 = 0.2$ m

***i.e.*, mass m_2 should be placed at 0.2 m and m_1 at 0.1 from the centre *O*.**

83. (i) *CP* = *CO* = Radius of circle (R)

∴ $\quad \angle CPO = \angle POC = 60°$

∴ $\angle OCP$ is also 60°

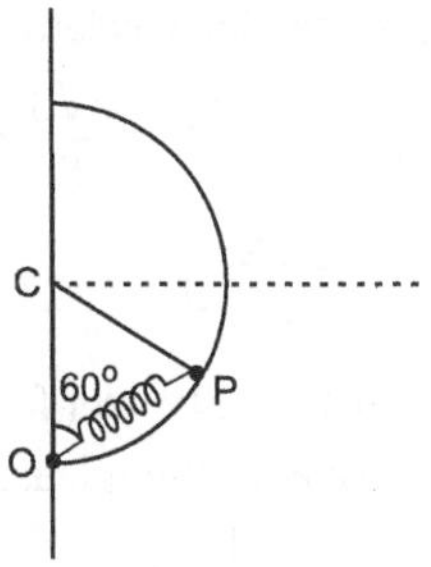

(a)

Therefore, $\Delta\, OCP$ is an equilateral triangle.
Hence $OP = R$
Natural length of spring is $3R/4$.
∴ Extension in the spring,

$$x = R - \frac{3R}{4} = \frac{R}{4}$$

⇒ Spring force,

$$F = kx = \left(\frac{mg}{R}\right)\left(\frac{R}{4}\right) = \frac{mg}{4}$$

The free body diagram of the ring is shown in figure (b).

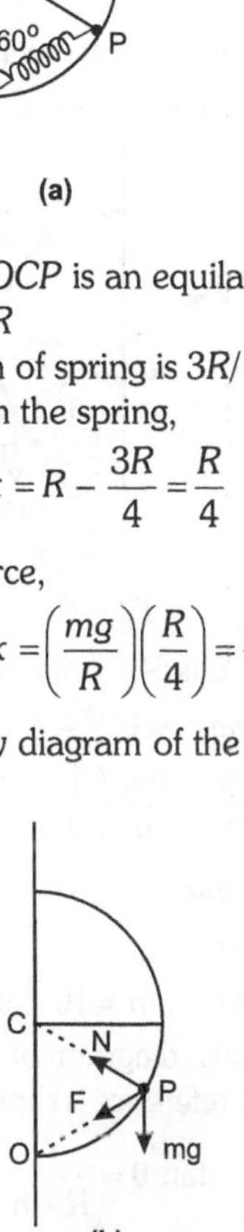

(b)

Here $\quad F = kx = \dfrac{mg}{4}$

and $\quad N$ = Normal reaction.

(ii) **Tangential acceleration, a_T :** The ring will move towards the x-axis just after the release. So net force along x-axis

$$F_x = F \sin 60° + mg \sin 60°$$

$$= \left(\frac{mg}{4}\right)\frac{\sqrt{3}}{2} + mg\left(\frac{\sqrt{3}}{2}\right)$$

$$F_x = \frac{5\sqrt{3}}{8} mg$$

Therefore, tangential acceleration of the ring.

$$a_T = a_x = \frac{F_x}{m} = \frac{5\sqrt{3}}{8} g$$

$$\boldsymbol{a_T = \frac{5\sqrt{3}}{8} g}$$

Normal reaction N : Net force along y-axis on the ring just after the release will be zero.

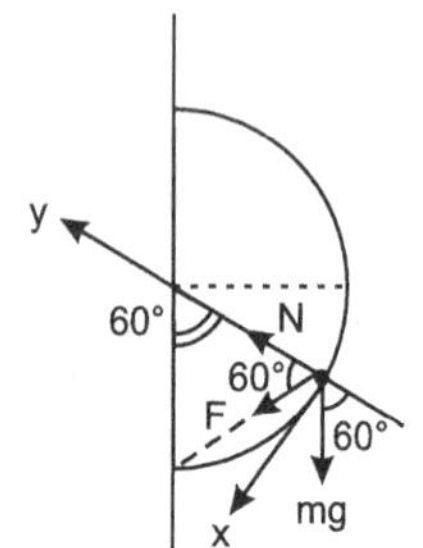

(c)

$$F_y = 0$$

$$\therefore \quad N + F \cos 60° = mg \cos 60°$$

$$\therefore \quad N = mg \cos 60° - F \cos 60°$$

$$= \frac{mg}{2} - \frac{mg}{4}\left(\frac{1}{2}\right) = \frac{mg}{2} - \frac{mg}{8}$$

$$\boldsymbol{N = \frac{3mg}{8}}$$

84. Given $R = 0.1$ m, $m = 10^{-2}$ kg

(i) Free body diagram of particle is ground frame of reference is shown in figure. Hence

$$\tan\theta = \frac{r}{R-h}$$

$$N \cos\theta = \text{mg} \quad \dots(1)$$

and $$N \sin\theta = mr\,\omega^2 \quad \dots(2)$$

Dividing (2) by (1), we obtain

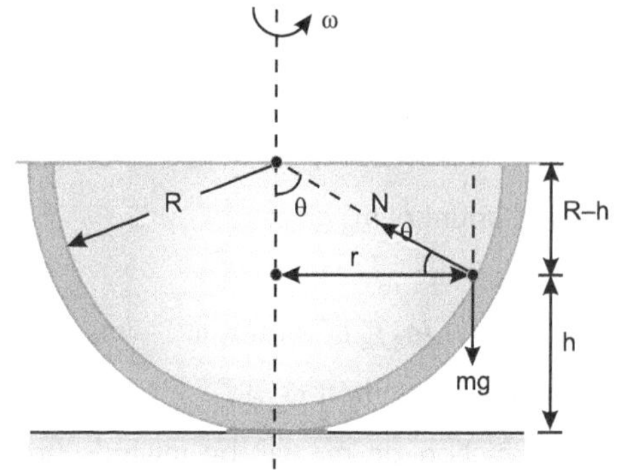

$$\tan\theta = \frac{r\omega^2}{g}$$

or $$\frac{r}{R-h} = \frac{r\omega^2}{g}$$

or $$\boldsymbol{\omega^2 = \frac{g}{R-h}} \quad \dots(3)$$

This is the desired relation between ω and h. From equation (3) we have

$$h = R - \frac{g}{\omega^2}$$

For non-zero value of h

$$R > \frac{g}{\omega^2} \quad \text{or} \quad \omega > \sqrt{g/R}$$

Therefore, minimum value of ω should be

$$\omega_{min} = \sqrt{g/R} = \sqrt{\frac{9.8}{0.1}} \text{ rad/s}$$

or $$\omega_{min} = \mathbf{9.89 \text{ rad/s}}$$

(ii) Equation (3) can be written as

$$h = R - g/\omega^2$$

If R and ω are known precisely, then

$$\Delta h = -\frac{\Delta g}{\omega^2} \quad \text{or} \quad \Delta g = \omega^2 \Delta h$$

(Neglecting the negative sign)

$$(\Delta g)_{min} = (\omega_{min})^2 \cdot \Delta h = (9.89)^2 \times 10^{-4} \text{ m/s}^2$$

$$(\Delta g)_{min} = \mathbf{9.78 \times 10^{-3} \text{ m/s}^2}$$

CHAPTER 3 WORK, POWER AND ENERGY

85. (a) Compression of the spring in position C

$$x = CB = R\theta = 0.3\,\theta \qquad (R = 0.3\text{ m})$$

height difference between C and D is :

$$h = R\,(1 + \cos\theta) = 0.3\,(1 + \cos\theta)$$

From conservation of mechanical energy

$$\frac{1}{2}kx^2 = mgh$$

or $\frac{1}{2}(40)\,(0.3\,\theta)^2 = (0.2)\,(10)\,(0.3)\,(1 + \cos\theta)$

or $\theta^2 = \frac{1}{3}(1 + \cos\theta)$

or $\mathbf{3\theta^2 = 1 + \cos\theta}$

(b) For the above angle, velocity of collar is zero at point D. Height difference between A and D is

$$h = R = 0.3\text{ m}$$

$\therefore$ velocity of collar at point A will be :

$$v = \sqrt{2gh} = \sqrt{2\,(10)\,(0.3)}$$

$$\mathbf{v = 2.45\ m/s}$$

86. Kinetic plus potential energy is the total mechanical energy.

$\therefore$ Rate of dissipation of (kinetic energy + potential energy)

$= -$ [rate of work done by friction]

$= -\,[-f \cdot v] = fv$

Here f = frictional force $= \mu mg\cos\alpha$

and v = velocity of body at time $t = at$

where $a = g\sin\alpha - \mu g\cos\alpha$

$\therefore \quad v = (g\sin\alpha - \mu g\cos\alpha)\,t$

$\therefore$ The desired rate

$= \mu mg\cos\alpha\,(g\sin\alpha - \mu g\cos\alpha)\,t$

$= \mathbf{\mu mg^2\cos\alpha\,(\sin\alpha - \mu\cos\alpha)\,t}$

87. (a) We can see that $\angle ACB = 90^\circ$

From constraint relations we can understand that speeds of both the particles will be same.

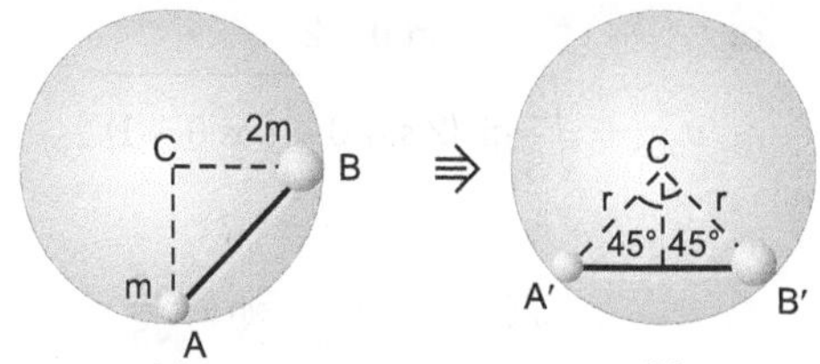

(a) (b)

From conservation of mechanical energy. (Refer figures a and b)

Decrease in potential energy of mass $2m$ = increase in potential energy of mass m + increase in kinetic energies of both.

Hence

$$2mg(h_B) = mg(h_A) + \frac{1}{2}(3m)(v^2)$$

or $\quad 2g(h_B) = g(h_A) + \frac{3}{2}v^2 \qquad \ldots (1)$

Here $\quad h_B = r\cos 45^\circ = 0.707r$

$\quad h_A = r(1 - \cos 45^\circ) = 0.293r$

Substituting the values we have

$$(2)\,(0.707)\,gr = 0.293gr + 1.5v^2$$

or $\quad \mathbf{v = 0.864\sqrt{gr}}$

(b) At any angle θ

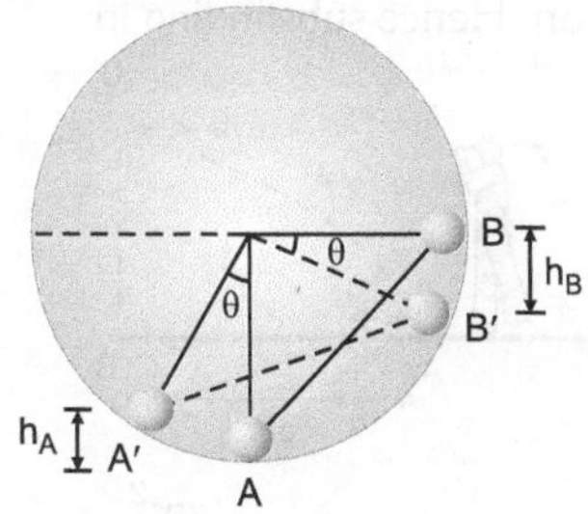

$h_B = r\sin\theta$

and $h_A = r(1 - \cos\theta)$

Hence equation (1) can be written as—

$$2gr \sin\theta = gr(1 - \cos\theta) + \frac{3}{2}v^2$$

or $v^2 = \frac{2}{3}gr(2\sin\theta + \cos\theta - 1) = k$ (say)...(2)

For v to be maximum

$$\frac{dk}{d\theta} = 0 \text{ or } 2\cos\theta - \sin\theta = 0$$

or $\tan\theta = 2$

$$\therefore\ v_{max} = \sqrt{\frac{2}{3}gr\,(2\sin\theta + \cos\theta - 1)}$$

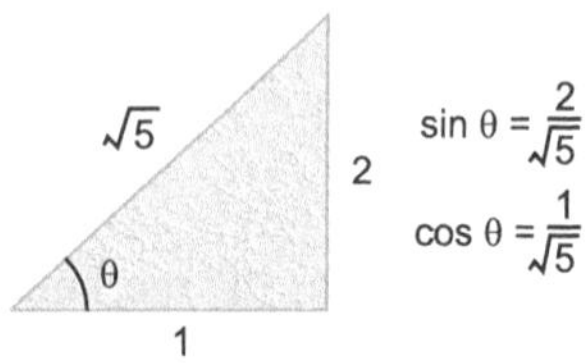

Substituting the values of θ we get

$$\boldsymbol{v_{max} = 0.908\sqrt{gr}}$$

(c) At $\theta = \theta_{max}$, velocity of both the particles will become zero for a moment. Hence substituting $v^2 = 0$ in equation (2) we get

$$2\sin\theta + \cos\theta - 1 = 0$$

Solving this equation we get

$$\boldsymbol{\theta_{max} = 126.9°}$$

88. (i) Let v be the velocity at the highest point.

Then $v^2 = u^2 - 2gh$

or $v^2 = u^2 - 4gR$ ... (1)

After point C path of the ball becomes projectile with initial velocity in horizontal direction. Hence substituting in

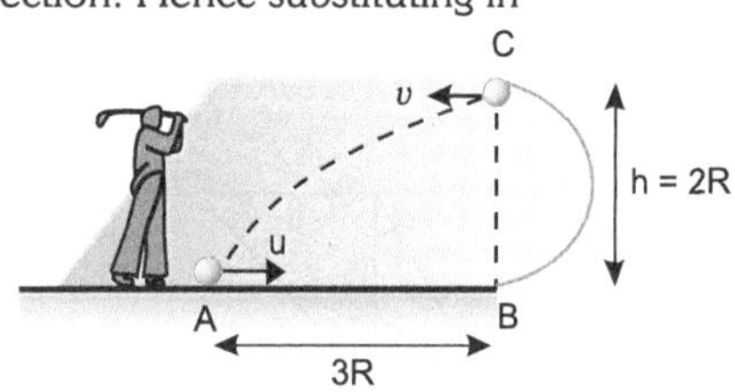

$$y = x\tan\theta - \frac{gx^2}{2v^2\cos^2\theta} \quad ... (2)$$

We get

$$-2R = 3R\tan 0° - \frac{g(3R)^2}{2(u^2 - 4gR)\cos^2 0^0}$$

or $2R = \frac{9gR^2}{2(u^2 - 4gR)}$

or $4u^2 - 16gR = 9gR$

or $u^2 = \frac{25}{4}gR$

or $\boldsymbol{u = \frac{5}{2}\sqrt{gR}}$ $(u > \sqrt{5gR})$

(ii) Minimum value of v to maintain contact at C is $\sqrt{gR}$. Hence substituting $v = \sqrt{gR}$ in equation (2) we get

$$-2R = x\tan 0° - \frac{gx^2}{2(gR)\cos^2 0^0}$$

or $4gR^2 = g\,x^2$ or $x = 2R$

Hence minimum value of x is

$$\boldsymbol{x_{min} = 2R}$$

89. From constraint relations we can show that :

$$3v_A = 2v_B \text{ or } v_B = 1.5\,v_A$$

or $v_A = \frac{2}{3}v_B$ and $s_A = \frac{2}{3}s_B$

Now as the block B moves 1 m vertically downwards, block A will move $\frac{2}{3}$ m along the plane or $\frac{2}{3}\sin 30°$ vertically upwards.

From conservation of mechanical energy.

Decrease in potential energy of block B = increase in potential energy of block A + increase in kinetic energies of both the blocks.

Hence

$$(8)(g)(1) = (10)(g)\left(\frac{2}{3}\sin 30°\right) + \frac{1}{2}(8)(v_B{}^2) + \frac{1}{2}(10)(v_A)^2$$

or $$8 \times 10 = (10)(10)\left(\frac{2}{3} \times \frac{1}{2}\right) + \frac{1}{2}(8)(1.5v_A)^2 + \frac{1}{2}(10)v_A{}^2$$

or $\quad v_A = 1.825$ **m/s**

and $\quad v_B = 2.74$ **m/s**

90. From conservation of mechanical energy, Decrease in gravitational potential energy of rope = increase in kinetic energy of rope.

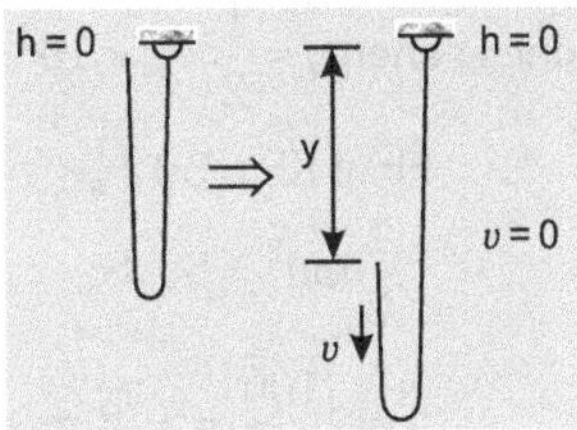

or $\quad U_i - U_t = \text{K.E.} \quad (U = \text{potential energy})$

$$\therefore \quad -(\lambda L)(g)\left(\frac{L}{4}\right) - \left[-(\lambda y)(g)\left(\frac{y}{2}\right) - (\lambda)(L-y)(g)\left(y + \frac{L-y}{4}\right)\right] = \frac{1}{2}\lambda\left(\frac{L-y}{2}\right)v^2.$$

Solving this we get

$$v = \sqrt{2gy\left(\frac{L-y/2}{L-y}\right)}$$

91. This problem can not be solved by the use of the fundamental equation of dynamics since the force F acting on the disc in the region $x_1 < x < x_2$ is not specified. Only we know about this force is that it is perpendicular to Y-axis.
From conservation of energy we get

$$v_2^2 = v_1^2 - 2gh \qquad \ldots(1)$$

or $$v_{2x}^2 + v_{2y}^2 = v_{1x}^2 + v_{1y}^2 - 2gh \qquad \ldots(2)$$

Since the force of field is perpendicular to y-axis it does not affect the v_y projection of the velocity. Hence $v_{2y} = v_{1y}$. Therefore equation (2) can be rewritten as

$$v_{2x}^2 = v_{1x}^2 - 2gh$$

or $$v_2 \cos\alpha_2 = \sqrt{v_1^2\cos^2\alpha_1 - 2gh} \qquad \ldots(3)$$

Solving equation (1) and (3) we get

$$\alpha_2 = \cos^{-1}\sqrt{\frac{v_1^2\cos^2\alpha_1 - 2gh}{v_1^2 - 2gh}}$$

The disc can not overcome the hill if radical of equation (3) is negative or

$$v_1 \cos\alpha_1 < \sqrt{2gh}$$

92. Let v_0 be the speed of the object at point A. Between A and B path of the object is a parabola, where

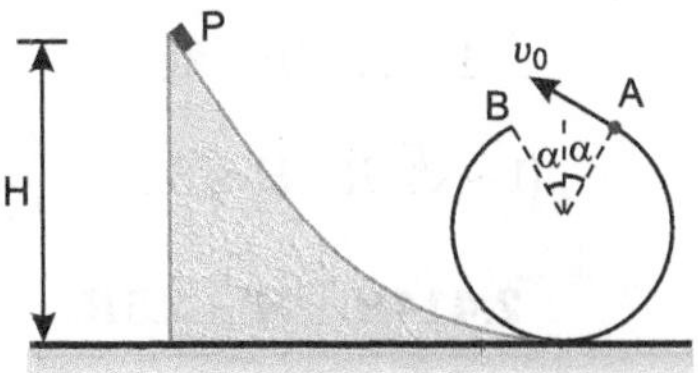

$$AB = \text{Range}$$

or $$2R\sin\alpha = \frac{2v_0^2\sin\alpha\cos\alpha}{g}$$

So $$v_0^2 = \frac{gR}{\cos\alpha} \qquad \ldots(1)$$

Applying conservation of mechanical energy at P and A we get

$$mgH = mgR(1+\cos\alpha) + \frac{1}{2}mv_0^2$$

or $$\frac{H}{R} = 1 + \cos\alpha + \frac{v_0^2}{2gR}$$

but $$\frac{v_0^2}{gR} = \frac{1}{\cos\alpha} \qquad \text{(from equation 1)}$$

$$\therefore \quad \frac{H}{R} = 1 + \cos\alpha + \frac{1}{2\cos\alpha} = k \qquad \text{(say)}$$

or $\quad 2\cos^2\alpha - 2(k-1)\cos\alpha + 1 = 0$

or $$\cos^2\alpha - (k-1)\cos\alpha + \frac{1}{2} = 0$$

or $$\cos\alpha = \frac{1}{2}\left(k - 1 \pm \sqrt{(k-1)^2 - 2}\right)$$

now $\quad (k-1)^2 - 2 \geq 0$ or $k - 1 \geq \sqrt{2}$

or $$k \geq (1+\sqrt{2}) \qquad \ldots(2)$$

On the other hand,

$$\cos\alpha \leq 1$$

i.e., $\frac{1}{2}[(k-1)+\sqrt{(k-1)^2-2}] \leq 1$

or $(k-1)+\sqrt{(k-1)^2-2} \leq 2$

or $\sqrt{(k-1)^2-2} \leq 2-(k-1)$

or $(k-1)^2-2 \leq [2-(k-1)]^2$

or $4k \leq 10$

or $k \leq 2.5$ ...(3)

Hence from (2) and (3) we have

$$1+\sqrt{2} \leq k \leq 2.5$$

or $(1+\sqrt{2})R \leq H \leq 2.5\,R \quad \left(\text{as } k = \frac{H}{R}\right)$

or $\mathbf{2.414R \leq H \leq 2.5R}$

For the limiting values of cosines we have

$$k = 1+\sqrt{2} \quad \text{or} \quad k-1 = \sqrt{2}$$

or $\cos\alpha_1 = \frac{\sqrt{2}}{2}$ i.e., $\alpha_1 = 45°$

and $k = 2.5$ or $k-1 = 1.5$

$$\cos\alpha_2 = \frac{(1.5 \pm 0.5)}{2}$$

$$\cos\alpha_2 = 0.5,\ \cos\alpha_3 = 1$$

or $\alpha_2 = 60°$ and $\alpha_3 = 0°$

Hence $\mathbf{45° \leq \alpha \leq 60°}$

93. (a) When the spring undergoes maximum compression, the relative velocity between the block and the car becomes zero. This means that the entire system moves with the same velocity v. As all forces are internal forces, momentum is conserved

$$(4m+2m)\,v = mv_0$$

or $\boldsymbol{v = \frac{v_0}{6}}$

(b) The kinetic energy of the pair of blocks just after the collision

$$= \frac{1}{2}(2m)\left(\frac{v_0}{2}\right)^2 = \frac{1}{4}mv_0^2$$

Mechanical energy of the system is conserved during compression of the spring and subsequent motion of the system. If x be the maximum compression in the spring, then

$$\frac{1}{2}kx^2 + \frac{1}{2}(6m)\left(\frac{v_0}{6}\right)^2 = \frac{1}{4}mv_0^2$$

Calculating x, we get,

$$\boldsymbol{x = \sqrt{\frac{m}{3k}}\,v_0}$$

94. Initial kinetic energy $= \frac{1}{2} \times 2 \times 10^2 = 100$ J

Final KE = 50 J

$\therefore \quad \frac{1}{2}mv^2 = 50$ J

$\therefore \quad v = \sqrt{\frac{100}{2}} = 5\sqrt{2}$ m/s

When the spring is being compressed, the equation of motion of spring mass system can be written as

$$x = A\sin\omega t \qquad \text{where } \omega = \sqrt{\frac{k}{m}}$$

$\therefore \quad \frac{dx}{dt} = A\omega\cos\omega t$

$\Rightarrow \quad v = v_0\cos\omega t$

$\Rightarrow \quad 5\sqrt{2} = 10\cos\omega t$

$\Rightarrow \quad \omega t = \frac{\pi}{4}$

$\Rightarrow \quad t = \frac{\pi}{4\omega} = \frac{\pi}{4}\sqrt{\frac{m}{k}}$

$$= \frac{\pi}{4}\sqrt{\frac{2}{80}} = \mathbf{0.124\ sec}$$

95. (a) Let at angular deflection θ and let velocity be v, from work energy theorem

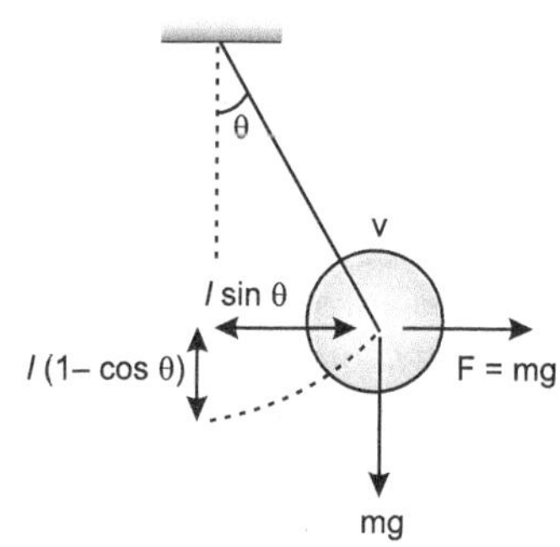

change in kinetic energy

= work done by all forces

$$\frac{1}{2}mv^2 = -mgl\,(1-\cos\theta) + Fl\sin\theta$$

$$= mgl\,[-1+\cos\theta+\sin\theta]$$

Maximum angular deflection

$$v = 0 \Rightarrow \theta = 90°$$

(b) Tension at angular deflection θ

$$T - mg\,(\cos\theta + \sin\theta) = \frac{mv^2}{l}$$

$$\Rightarrow T = mg\cos\theta + mg\sin\theta + 2mg\,(-1+\cos\theta+\sin\theta)$$

$$= 3mg\cos\theta + 3mg\sin\theta - 2mg$$

$$= mg\left[3\sqrt{2}\sin\left(\theta+\frac{\pi}{4}\right) - 2\right]$$

T is maximum at $\theta = \frac{\pi}{4}$

$$\boldsymbol{T_{max} = mg\,[3\sqrt{2} - 2]}$$

96. When the bobs collide at the topmost point, the necessary condition that both of them fall vertically down after collision, are both should have the same velocity in the horizontal direction and the vertical component of velocity for '*Q*' should be zero.

So, if *Q* is projected with velocity 'v_0' at an angle 'θ' with the horizontal, then we have

$$v_0\cos\theta = \sqrt{gl} \quad \ldots(1)$$

and also that it hits '*P*' at the highest point of its trajectory.

So, $$\frac{v_o^2\sin^2\theta}{2g} = 2l$$

or $$v_o^2\sin^2\theta = 4gl$$

or $$v_o\sin\theta = 2\sqrt{gl} \quad \ldots(2)$$

Dividing (2) by (1), we get

$$\tan\theta = 2 \Rightarrow \theta = \tan^{-1}(2) \quad \ldots(3)$$

and on squaring and adding (1) and (2) we get,

$$v_o^2 = 5gl \quad \text{or} \quad v_o = \sqrt{5gl}$$

The distance from '*P*' is projected is

$$d = \frac{\text{Range}}{2} = \frac{v_o^2\sin\theta\cos\theta}{g}$$

$$= \frac{v_o^2\tan\theta}{g\sec^2\theta} = \frac{2v_o^2}{g\,(1+4)} = 2l$$

97. (a) $$h = \left(R + \frac{d}{2}\right)(1-\cos\theta)$$

velocity of ball at angle θ is

$$v = 2gh = 2\left(R+\frac{d}{2}\right)(1-\cos\theta)\,g \quad \ldots(1)$$

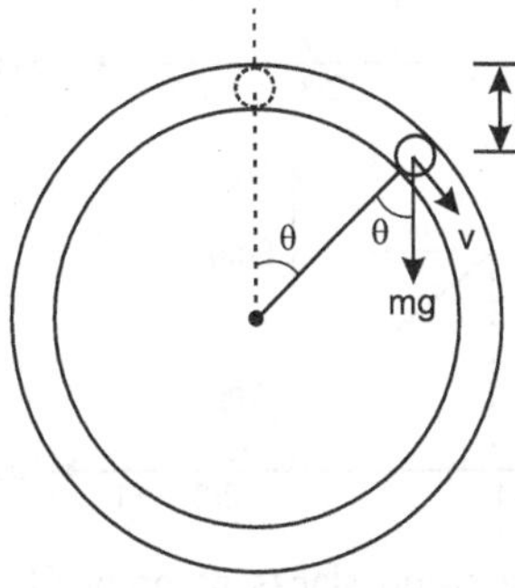

Let *N* be the total normal reaction (away from centre) at angle θ. Then

$$mg\cos\theta - N = \frac{mv^2}{\left(R+\frac{d}{2}\right)}$$

substituting value of v^2 from Eq. (1) we get

$$mg\cos\theta - N = 2mg\,(1-\cos\theta)$$

$$\therefore \quad \boldsymbol{N = mg\,(3\cos\theta - 2)}$$

(b) The ball will lose contact with the inner sphere when

$$N = 0$$

or $$3\cos\theta - 2 = 0$$

or $$\theta = \cos^{-1}\left(\frac{2}{3}\right)$$

After this it makes contact with outer sphere and normal reaction starts acting towards the centre. Thus for $\theta \le \cos^{-1}\left(\frac{2}{3}\right)$:

$$N_B = 0$$

and $$N_A = mg\,(3\cos\theta - 2)$$

and for $$\theta \ge \cos^{-1}\left(\frac{2}{3}\right)$$

$$N_A = 0$$

and $$N_B = mg\,(2 - 3\cos\theta)$$

The corresponding graphs are as follows:

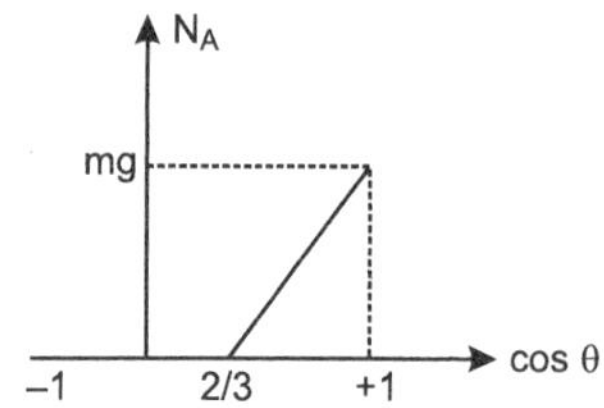

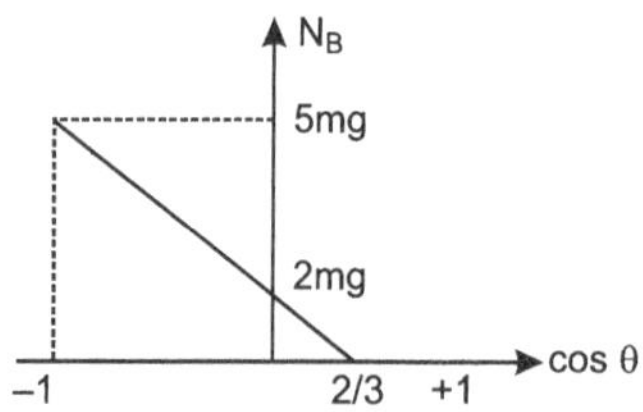

98. Let the string slacks at point Q as shown in figure. From P to Q path is circular and beyond Q, path is parabolic. At point C, velocity of particle becomes horizontal, therefore,

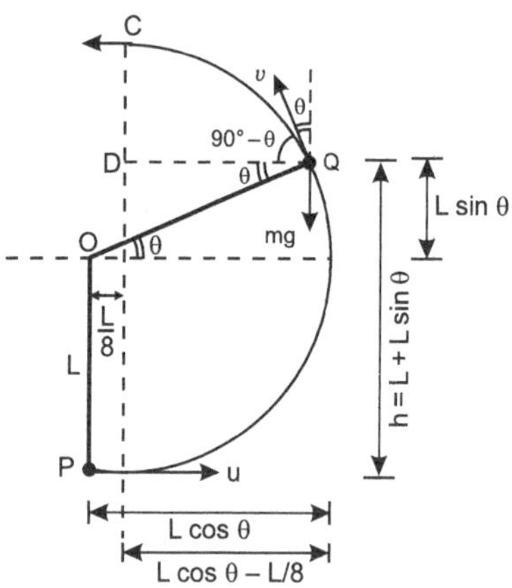

QD = half of the range of the projectile.
Now we have following equations :
(i) $T_Q = 0$

Therefore, $mg \sin\theta = \dfrac{mv^2}{L}$ ...(1)

(ii) $v^2 = u^2 - 2gh$

$= u^2 - 2gL(1 + \sin\theta)$ ...(2)

(iii) $QD = \dfrac{1}{2}$ (Range)

or $\left(L\cos\theta - \dfrac{L}{8}\right) = \dfrac{v^2 \sin 2(90° - \theta)}{2g}$

or $L\left(\cos\theta - \dfrac{1}{8}\right) = \dfrac{v^2 \sin 2\theta}{g}$ ...(3)

Equation (3) can be written as

$$\left(\cos\theta - \frac{1}{8}\right) = \left(\frac{v^2}{gL}\right)\sin\theta\cos\theta$$

from equation (1), substituting value of

$$\left(\frac{v^2}{gL}\right) = \sin\theta$$

we get

$$\left(\cos\theta - \frac{1}{8}\right) = \sin^2\theta.\cos\theta = (1 - \cos^2\theta)\cos\theta$$

or $\cos\theta - \dfrac{1}{8} = \cos\theta - \cos^3\theta$

$\therefore$ $\cos^3\theta = \dfrac{1}{8}$ or $\cos\theta = \dfrac{1}{2}$ or $\theta = 60°$

$\therefore$ From equation (1),

$v^2 = gL\sin\theta = gL\sin 60°$ or $v^2 = \dfrac{\sqrt{3}}{2}gL$

Substituting this value of v^2 in equation (2)

$$u^2 = v^2 + 2gL(1 + \sin\theta)$$

$$= \frac{\sqrt{3}}{2}gL + 2gL\left(1 + \frac{\sqrt{3}}{2}\right)$$

$$= \frac{3\sqrt{3}}{2}gL + 2gL = gL\left(2 + \frac{3\sqrt{3}}{2}\right)$$

$$\boldsymbol{u = \sqrt{gL\left(2 + \frac{3\sqrt{3}}{2}\right)} \approx 2.14\sqrt{gL}}$$

CHAPTER 4

Centre Of Mass, Conservation Of Momentum, Collision, Impulse

99. Let x be the displacement of the platform towards right. Then—

displacement of platform $= x$ (towards right)

displacement of man with respect to ground

$= x_0 - x$ (towards left)

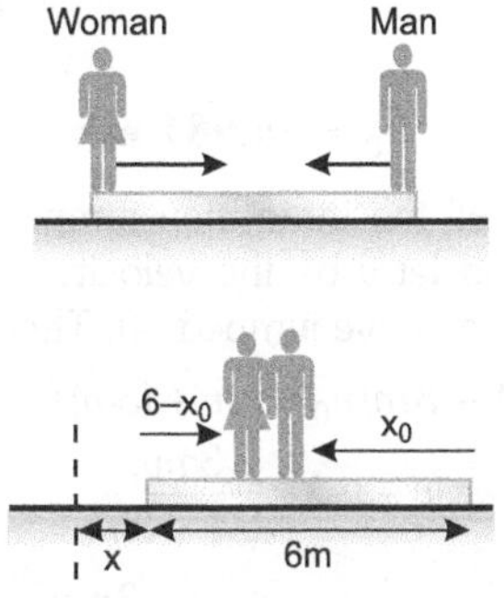

and displacement of woman with respect to ground $= x + 6 - x_0$ (towards right)

Net force on the system in horizontal direction is zero. Hence COM will remain stationary, or

(mass of platform) (displacement of platform) + (mass of woman) (displacement of woman) = (mass of man) (displacement of man)

or $20(x) + 50(x + 6 - x_0) = 60(x_0 - x)$

or $$\boldsymbol{x = \frac{11x_0 - 30}{13}}$$

100. Let x be the displacement of straw when the first insect reaches the opposite end. Hence displacement of insect would be $\left(\frac{3}{2}a - x\right)$.

For center of mass to remain stationary we have—

$$2m(x) = \frac{m}{2}\left(\frac{3}{2}a - x\right) \quad \text{or} \quad x = 0.3a$$

Therefore, the situation is as follows

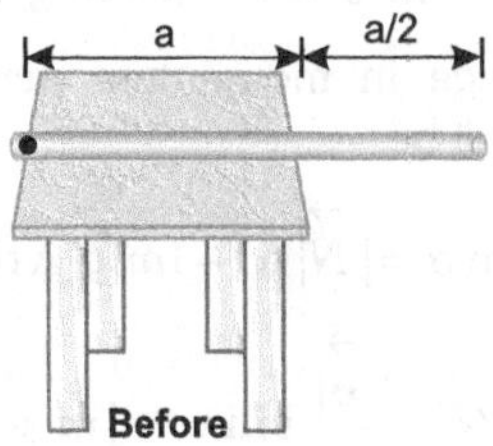

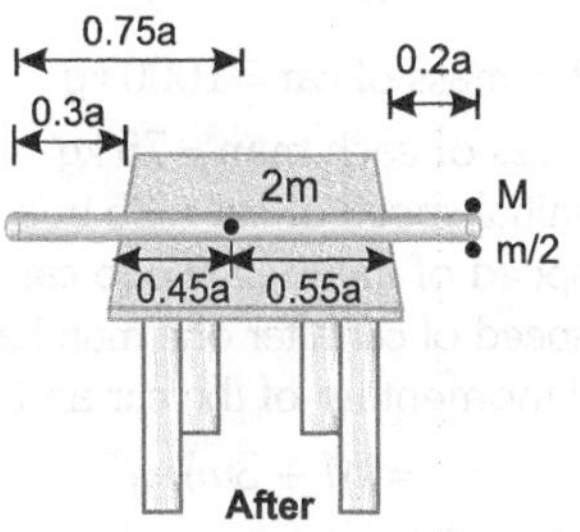

Let M be the mass of second insect.

For the straw not to topple the centre of mass (of straw + two insects) should lie inside the table or

$$2m(0.55a) = \left(M + \frac{m}{2}\right)(0.2\,a)$$

or $$\boldsymbol{M = 5\ m}$$

101. Here the system 'cannon + shell' is non-closed.

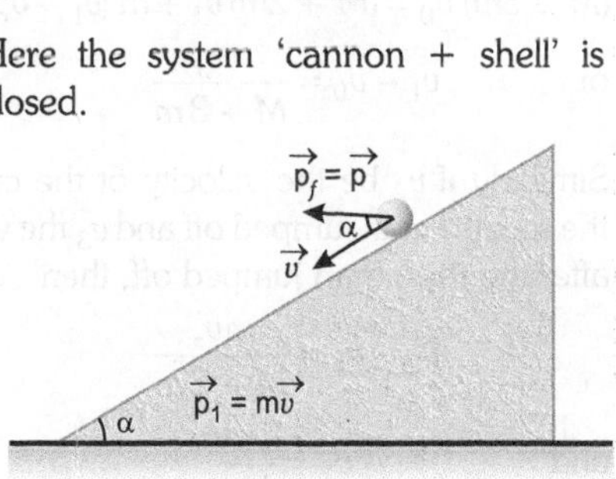

During the time interval Δt, the change in momentum of this system is

$$\Delta \vec{p} = \vec{p}_f - \vec{p}_i = \vec{p} - m\vec{v}$$

This change is caused by two external forces the reaction force $\vec{N}$ (which is perpendicular to the inclined plane) and gravity $m\vec{g}$. Therefore, we can write

$$\vec{p} - m\vec{v} = \vec{N}\cdot\Delta t + m\vec{g}\,\Delta t$$

or change in momentum perpendicular to plane will be

$$|\vec{p}|\sin\alpha = |\vec{N}|\Delta t - |m\vec{g}\,\Delta t|\cos\alpha$$

or $$|\vec{N}| = \frac{|\vec{p}|}{\Delta t}\sin\alpha + |m\vec{g}|\cos\alpha$$

102. (i) M = mass of car = 1000 kg

m = mass of each man = 75 kg

v_0 = initial speed of car = 25 m/s

v_r = speed of men relative to car = 5 m/s

v_1 = speed of car after one man has jumped off

Initial momentum of the car and the men

$$= (M + 3m)\,v_0 \qquad \ldots(1)$$

Velocity of man relative to ground will be $(v_1 - v_r)$

$\therefore$ Momentum of car and two men + the momentum of the man who jumped off

$$= (M + 2m)\,v_1 + m\,(v_1 - v_r) \qquad \ldots(2)$$

Equating (1) and (2) from conservation of linear momentum we have

$$(M + 3m)\,v_0 = (M + 2m)\,v_1 + m\,(v_1 - v_r)$$

or $$v_1 - v_0 = \frac{mv_r}{M + 3m} \qquad \ldots(3)$$

Similarly if v_2 be the velocity of the car after the second man jumped off and v_3 the velocity after the third man jumped off, then

$$v_2 - v_1 = \frac{mv_r}{M + 2m} \qquad \ldots(4)$$

and $$v_3 - v_2 = \frac{mv_r}{M + m} \qquad \ldots(5)$$

Adding (3), (4) and (5) We get

$$v_3 - v_0 = mv_r\left[\frac{1}{M + 3m} + \frac{1}{M + 2m} + \frac{1}{M + m}\right]$$

or $$v_3 = v_0 + mv_r\left[\frac{1}{M + 3m} + \frac{1}{M + 2m} + \frac{1}{M + m}\right]$$

substituting the values, we get

$$v_3 = 25 + (75)(5)\left[\frac{1}{1000 + 225} + \frac{1}{1000 + 150} + \frac{1}{1000 + 75}\right]$$

or $$\boldsymbol{v_3 \approx 25.981 \text{ m/s}}$$

(ii) If all the three men jumped off together and let v be the velocity of car after all three have jumped off. Then

$$(M + 3m)\,v_0 = Mv + (3m)\,(v - v_r)$$

or $$(v - v_0) = \frac{3mv_r}{(M + 3m)}$$

or $$v = v_0 + \frac{3mv_r}{(M + 3m)}$$

substituting the values

$$v = 25 + \frac{(3)(75)(5)}{1000 + 225}$$

or $$\boldsymbol{v \approx 25.918 \text{ m/s}}$$

Note: It is to be noted that although. $v_3 \approx v$ but actually $v_3 > v$ because

$$v_3 = v_0 + mv_r\left[\frac{1}{M + 3m} + \frac{1}{M + 2m} + \frac{1}{M + m}\right]$$

while $$v = v_0 + mv_r\left[\frac{1}{M + 3m} + \frac{1}{M + 3m} + \frac{1}{M + 3m}\right] = v_0 + \frac{3mv_r}{M + 3m}$$

i.e. $\qquad v_3 > v$

But if m is small in comparison with M, $v_3 \approx v$.

103. At maximum angle (of string with vertical) horizontal velocity (v) of ring and particle will be equal.

From conservation of linear momentum,

$$(m + 3m)\, v = mv_0$$

$$\therefore \qquad v = \frac{v_0}{4}$$

Now increase in potential energy of particle = decrease in kinetic energy of system

or $\quad 3mgl\,(1 - \cos\theta_{max}) = \frac{1}{2} mv_0^2 - \frac{1}{2}(4m)\left(\frac{v_0}{4}\right)^2$

Solving this equation we get,

$$\theta_{max} = \cos^{-1}\left(1 - \frac{v_0^2}{8gl}\right)$$

104. From conservation of momentum,

$$3v = \sqrt{(1 \times 3)^2 + (2 \times 2)^2} = 5 \text{ kg-m/s}$$

$$\therefore \qquad v = \left(\frac{5}{3}\right) \text{m/s}$$

$$\Delta \text{KE} = \frac{1}{2} \times 1 \times 9 + \frac{1}{2} \times 2 \times 4 - \frac{1}{2} \times 3 \times \left(\frac{5}{3}\right)^2$$

$$= 4.33 \text{ J}$$

105. (a) Impulse received by m :

$$\vec{J} = m(\vec{v}_f - \vec{v}_i)$$

$$= m(-2\hat{i} + \hat{j} - 3\hat{i} - 2\hat{j}) = m(-5\hat{i} - \hat{j})$$

and impulse received by M :

$$= -\vec{J} = m(5\hat{i} + \hat{j})$$

(b) $Mv = m(5\hat{i} + \hat{j})$

or $\quad v = \frac{m}{M}(5\hat{i} + \hat{j}) = \frac{1}{13}(5\hat{i} + \hat{j})$

(c) e = (Relative velocity of separation/ Relative velocity of approach) in the direction of $-\vec{J}$.

$$= 11/17$$

106. Let v_1 = velocity of solid sphere towards right after collision and

v_2 = velocity of hemisphere towards right.

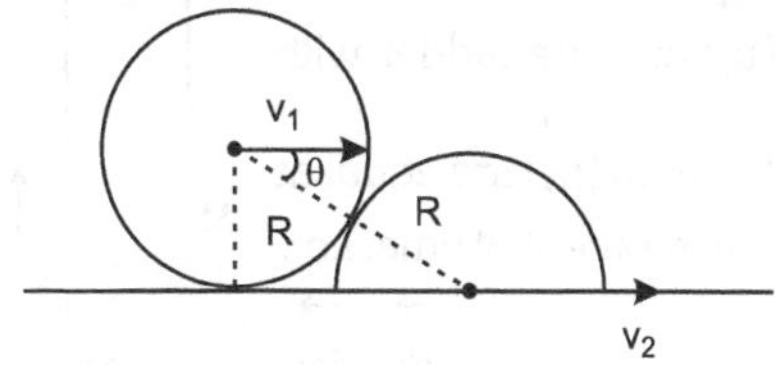

Applying COM along horizontal

$$\Rightarrow \qquad 2Mv_0 = 2Mv_1 + Mv_2$$

$$\Rightarrow \qquad 2v_0 = 2v_1 + v_2$$

$$\Rightarrow \qquad v_2 = 2(v_0 - v_1) \qquad \ldots(1)$$

Applying e along common normal

$$\therefore \qquad e = \frac{(v_2 \cos\theta - v_1 \cos\theta)}{v_0 \cos\theta}$$

$$\therefore \qquad ev_0 = v_2 - v_1 \qquad \ldots(2)$$

From equations (1) and (2)

$$v_1 = \frac{(2 - e)\, v_0}{3} \quad \text{and} \quad v_2 = (1 + e)\frac{2v_0}{3}$$

107. On impact all the balls have the same component velocity parallel to the short side of the table. In rebounding from the short sides the value of this component velocity does not alter, and therefore, since the tables are of equal width, all three balls will reach the

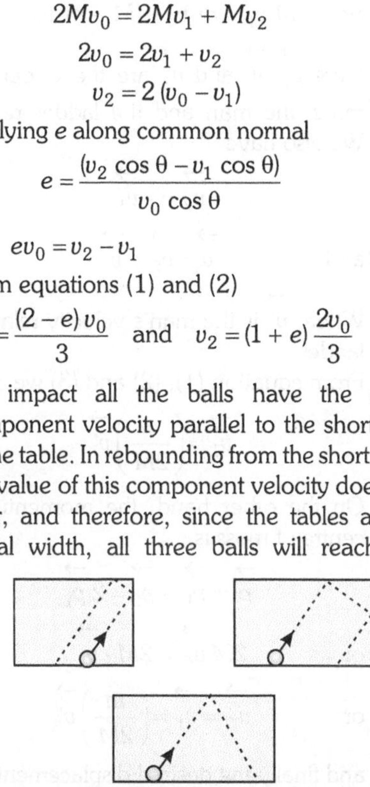

opposite side of the table at the same moment (regardless of whether they have first struck the short sides or not). In rebounding from the long side opposite, the component velocity under consideration changes sign, but remains constant in size; all the balls therefore return to the side from which they started at the same moment.

108. All the bodies of the system are initially at rest. The rope tension is the same both as on the left and the right hand side, at every instant and consequently the momenta of the counter-balancing mass $(\vec{p}_1)$ and the ladder with the man $(\vec{p}_2)$ are equal at any moment of time. i.e.,

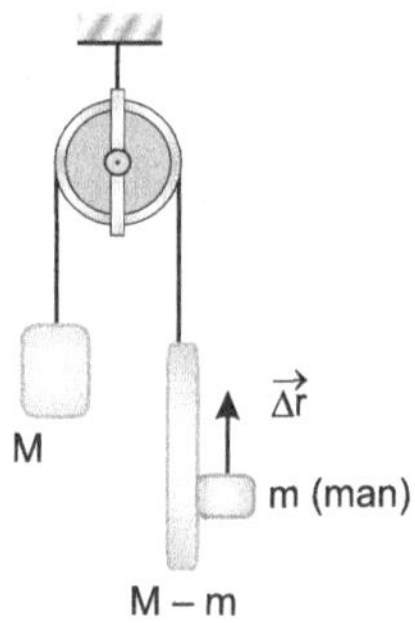

$$\vec{p}_1 = \vec{p}_2$$

or $\quad M\vec{v}_1 = m\vec{v} + (M-m)\vec{v}_2 \qquad \ldots(1)$

Here $\vec{v}_1$, $\vec{v}$ and $\vec{v}_2$ are the velocities of the mass, the man and the ladder respectively. We also have

$$\vec{v}_2 = -\vec{v}_1 \qquad \ldots(2)$$

and $\quad \vec{v} = \vec{v}_2 + \vec{v}' \qquad \ldots(3)$

Where $\vec{v}'$ is the man's velocity relative to the ladder.

From equation (1), (2) and (3) we obtain.

$$\vec{v}_1 = \left(\frac{m}{2M}\right)\vec{v}' \qquad \ldots(4)$$

On the other hand, the momentum of the centre of mass is

$$\vec{p} = \vec{p}_1 + \vec{p}_2 = 2\vec{p}_1$$

or $\quad 2M\vec{v}_C = 2M \cdot \vec{v}_1$

or $\quad \vec{v}_C = \vec{v}_1 = \left(\frac{m}{2M}\right)\vec{v}'$

and finally the desired displacement is

$$\Delta\vec{r}_C = \int \vec{v}_c \cdot dt = \frac{m}{2M}\int \vec{v}' \cdot dt = \frac{m}{2M}\Delta\vec{r}$$

or displacement of centre of mass of system is $\Delta\vec{r}_C = \dfrac{m}{2M}\Delta\vec{r}$

109. (i) Both the particles collide at their highest point.

Hence $\quad \dfrac{T_A}{2} = \dfrac{T_B}{2}$ or $T_A = T_B$

(T = time of flight)

or $\quad \dfrac{2u\sin 45^\circ}{g} = \dfrac{2v\sin 60^\circ}{g}$

or $\quad \dfrac{v}{u} = \dfrac{\sin 45^\circ}{\sin 60^\circ} = \sqrt{\dfrac{2}{3}}$

or $\quad \dfrac{\boldsymbol{v^2}}{\boldsymbol{u^2}} = \dfrac{\mathbf{2}}{\mathbf{3}}$

(ii) $\dfrac{R_A}{2} + \dfrac{R_B}{2} = a \qquad$ (R = Range)

or $\quad R_A + R_B = 2a$

or $\quad \dfrac{u^2 \sin 90^\circ}{g} + \dfrac{v^2 \sin 120^\circ}{g} = 2a$

or $\quad u^2 + \dfrac{\sqrt{3}}{2}v^2 = 2ag$

or $\quad u^2 + \dfrac{\sqrt{3}}{2}\left(\dfrac{2}{3}u^2\right) = 2ag$

or $\quad u^2\left(1 + \dfrac{1}{\sqrt{3}}\right) = 2ag$

simplifying this we get

$$\boldsymbol{u^2 = ag(3 - \sqrt{3})}$$

(iii) At the highest point they collide as follows : after collision, the horizontal component of

velocity of 1 becomes zero as it is given that it falls vertically, and let the horizontal component of velocity of 2 becomes v_0 (towards right) as

then from conservation of linear momentum :

$$mv_0 = mu\cos 45^\circ - mv\cos 60^\circ$$

or $\quad v_0 = \dfrac{u}{\sqrt{2}} - \dfrac{v}{2} = \left(\dfrac{u}{\sqrt{2}} - \dfrac{\sqrt{2}}{2\sqrt{3}}u\right) \qquad \ldots(1)$

$$\left(v = \sqrt{\frac{2}{3}}u\right)$$

also relative velocity of separation = e (relative velocity of approach)

or $\quad v_0 = e\,(u\cos 45^\circ + v\cos 60^\circ)$

or $\quad v_0 = e\left(\dfrac{u}{\sqrt{2}} + \dfrac{\sqrt{2}}{2\sqrt{3}}\cdot u\right) \quad \ldots(2)$

equating (1) and (2), we get

$$\frac{1}{\sqrt{2}} - \frac{\sqrt{2}}{2\sqrt{3}} = e\left(\frac{1}{\sqrt{2}} + \frac{\sqrt{2}}{2\sqrt{3}}\right)$$

or $\quad (2\sqrt{3} - 2) = e\,(2\sqrt{3} + 2)$

simplifying this we get

$$\boldsymbol{e = \frac{\sqrt{3} - 1}{\sqrt{3} + 1}}$$

110. When the ball strikes the wall, its vertical component v_y remains unchanged while its horizontal component $u \cos \theta$ becomes $eu \cos \theta$ in opposite direction. Now since v_y remains unchanged during collision, its time of flight T will also remain unchanged.

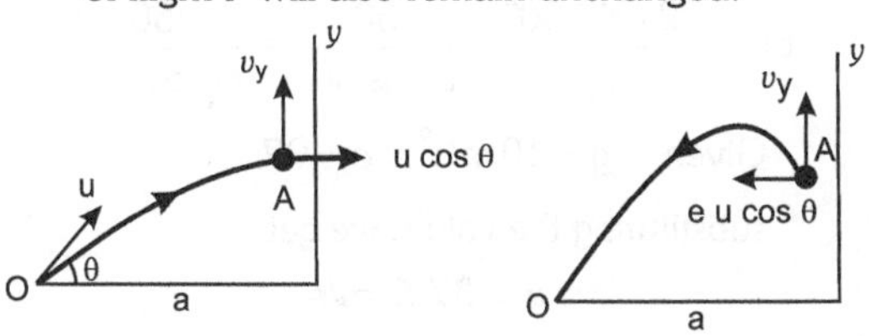

Hence $\quad T = t_{OA} + t_{AO}$

or $\quad \dfrac{2u \sin \theta}{g} = \dfrac{a}{u \cos \theta} + \dfrac{a}{eu \cos \theta}$

multiplying the equation by $u \cos \theta$ get we

$$\frac{2u^2 \sin \theta \cos \theta}{g} = a\left(1 + \frac{1}{e}\right)$$

or $\quad \dfrac{(4ag)(\sin 2\theta)}{g} = a\left(1 + \dfrac{1}{e}\right)$

$$u = 2\sqrt{ag} \quad \text{(given)}$$

or $\quad 4 \sin 2\theta = 1 + \dfrac{1}{e}$

or $\quad \boldsymbol{e = \dfrac{1}{4 \sin 2\theta - 1}}$

Now $e \le 1$. Hence $4 \sin 2\theta - 1 \ge 1$

or $\quad \sin 2\theta \ge \dfrac{1}{2}$

or $\quad 2\theta \ge 30°$ or $\theta \ge 15°$

111. Let v_0 be the velocity with which it strikes the plane.

$$a_x = g \sin \alpha, \quad a_y = -g \cos \alpha$$

Component of velocity along the plane is $v_0 \sin \alpha$ and component perpendicular to plane is $v_0 \cos \alpha$. Collision is elastic hence $v_0 \sin \alpha$ remains unchanged while $v_0 \cos \alpha$ reverses its direction.

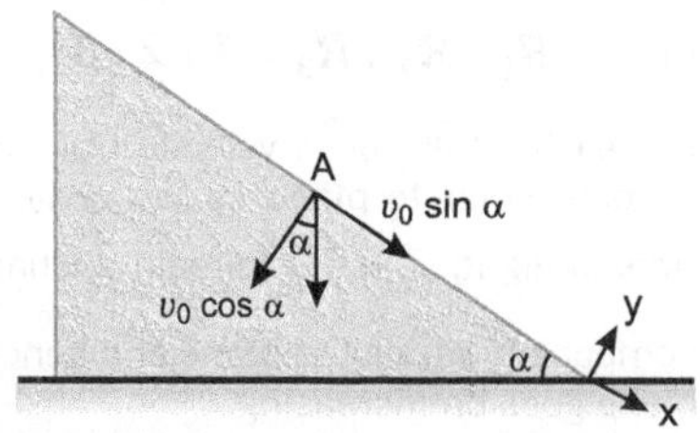

When it strikes at B, its component perpendicular to plane *i.e.*, $v_0 \cos \alpha$ and hence the time of flight T for each collision remains unchanged while component parallel to plane becomes

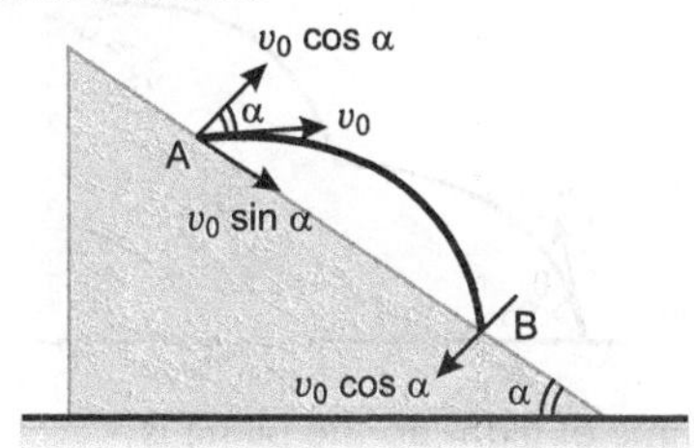

$$v_0 \sin \alpha + (g \sin \alpha)\,T \qquad \{v_x = u_x + a_x t\}$$

here $\quad T = \dfrac{2v_0 \cos \alpha}{g \cos \alpha} = \dfrac{2v_0}{g}$

and $\quad R_1 = AB = s_x = u_x T + \dfrac{1}{2} a_x T^2$

$$= (v_0 \sin \alpha)\frac{2v_0}{g} + \frac{1}{2}(g \sin \alpha)\left(\frac{2v_0}{g}\right)^2$$

$$R_1 = (4\, v_0^2 \sin \alpha)/g$$

next time

$$R_2 = s_x = \{v_0 \sin \alpha + g \sin \alpha \cdot T\}T + \frac{1}{2}(g \sin \alpha)T^2$$

substituting the values—

$$R_2 = \left\{v_0 \sin\alpha + g \sin\alpha \cdot \frac{2v_0}{g}\right\}\left\{\frac{2v_0}{g}\right\} + \frac{1}{2}(g\sin\alpha)\left(\frac{2v_0}{g}\right)^2$$

$$= (8v_0^2 \sin\alpha)/g$$

Similarly we can show that

$$R_3 = (12v_0^2 \sin\alpha)/g$$

or $\boldsymbol{R_1 : R_2 : R_3 = 1 : 2 : 3}$

Note: In each collision, velocity of component perpendicular to plane *i.e.*, $v_0 \cos\alpha$ hence the time of flight, $T = \frac{2v_0}{g}$ remain unchanged while component parallel to plane and hence range on plane goes on increasing.

112. Let v be the velocity of projection.

$$a_x = -g\sin\alpha, \quad a_y = -g\cos\alpha$$

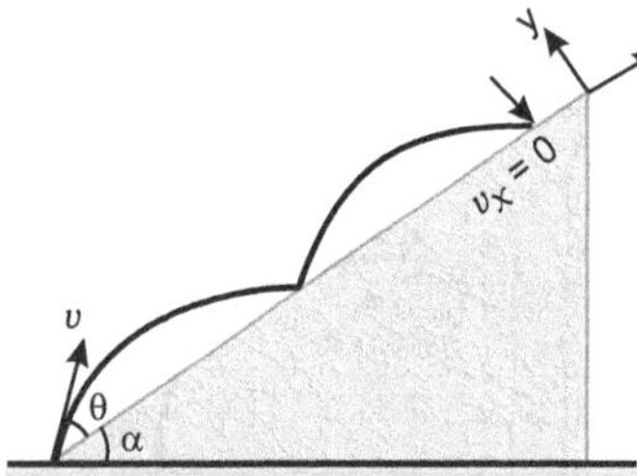

Time of flight up the plane is

$$T = \frac{2v\sin\theta}{g\cos\alpha}$$

Here $v\sin\theta$ is component of velocity perpendicular to the plane which will become $ev\sin\theta$ after first impact, $e^2 v\sin\theta$ after second impact and so on. Hence time of flight will also be T, eT, e^2T etc.

Component of velocity along the plane becomes zero ($v_x = 0$) at the time of second collision or after time $T + eT$ or $(1+e)T$. Hence applying

$$v_x = u_x + a_x t$$

$$0 = v\cos\theta - (g\sin\alpha)(1+e)T$$

or $$0 = v\cos\theta - (g\sin\alpha)(1+e)\frac{2v\sin\theta}{g\cos\alpha}$$

or $$1 + e = \frac{\cos\theta}{2\tan\alpha\sin\theta}$$

or $$1 + e = \frac{1}{2}\cot\alpha\cot\theta$$

or $$\boldsymbol{e = \frac{1}{2}\cot\alpha\cot\theta - 1}$$

113. (a) By symmetry we can say that they will collide at $x = 50$ m.

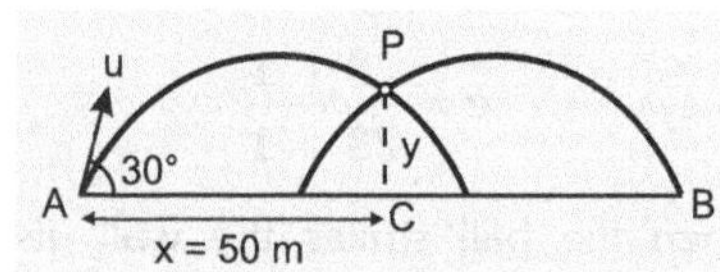

Vertical component of velocity and hence time of flight does not change in collision.

Hence $T_A = t_{AC} + t_{CA}$

or $$\frac{2u\sin 30^\circ}{g} = \frac{50}{u\cos 30^\circ} + \frac{50}{eu\cos 30^\circ}$$

Given $g = 10\,\text{m/s}^2$, $e = 0.7$

substituting the values we get

$$\boldsymbol{u = 37.5 \text{ m/s}}$$

(b) $$y = x\tan 30^\circ - \frac{gx^2}{2u^2\cos^2 30^\circ}$$

$$x = 50\text{m}, u = 37.5\text{ m/s}$$

substituting the values we get $y = 17$ m

Hence coordinates of point P w.r.t. A are **(50m, 17m)**.

114.

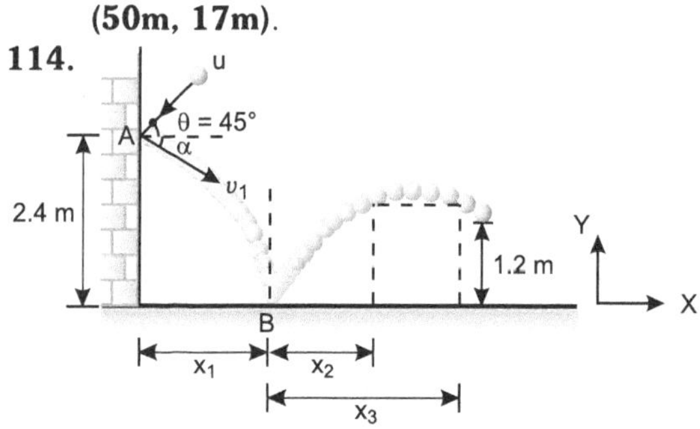

Component of velocity parallel to vertical wall $30\sin 45°$ or $15\sqrt{2}\,\frac{\text{m}}{\text{s}}$ after striking at A remains unchanged, while component perpendicular to wall 30 cos 45° or $15\sqrt{2}\,\frac{m}{s}$ becomes $e_1\, 15\sqrt{2}$ m/s or $7.5\sqrt{2}$ m/s ($e_1 = 0.5$). Hence after striking at A the horizontal component of velocity is $7.5\sqrt{2}$ m/s and vertical component is $15\sqrt{2}$ m/s (vertically downwards). Let t be the time of journey between A and B. Then

$$2.4 = (15\sqrt{2})t + 5t^2 \qquad (h = ut + \frac{1}{2}gt^2)$$

solving this we get—

$$t = 0.11 \text{ s} = t_{AB}$$

$\therefore \quad x_1 = (7.5\sqrt{2})(0.11)$

or $\quad x_1 = 1.17$ m

x and y components of velocity after colliding at B would be

$$v_x = (7.5)(\sqrt{2}) \text{ m/s} = 10.6 \text{ m/s}$$

and $v_y = e_2(15\sqrt{2} + gt)$

$$= 0.3\,(15\sqrt{2} + 10 \times 0.11) = 6.7 \text{ m/s}$$

Now ball will be at height 1.2 m when

$$1.2 = v_y t - \frac{1}{2} g t^2 \quad \text{or} \quad 1.2 = 6.7t - 5t^2$$

or $\quad t = 0.213$ s and 1.127 s

Hence $x_2 = v_x\,(0.213)$

$$= (10.6)(0.213) = 2.26 \text{ m}$$

and $\quad x_3 = (10.6)(1.127) = 11.95$ m

Hence x will have two values $(x_1 + x_2)$ and $(x_1 + x_3)$

or $\quad x = 1.17 + 2.26 = 3.43$ m

and $\quad x = 1.17 + 11.95 = 13.12$ m

or $\quad$ **x = 3.43 m and 13.12 m**

115. Applying conservation of linear momentum in x-direction we get

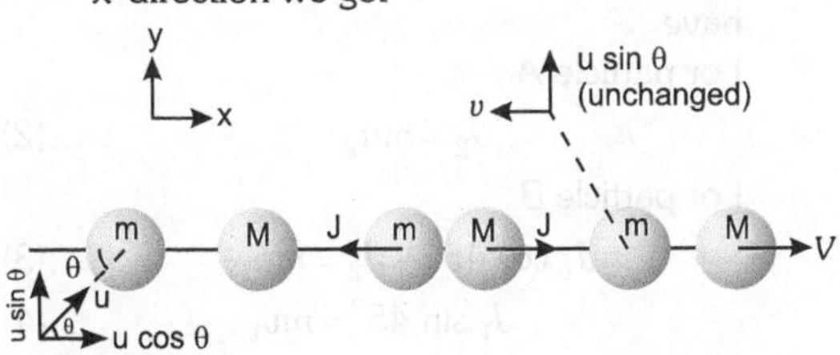

Before impact **At impact** **After impact**

$$mu \cos\theta = MV - mv \qquad \ldots(1)$$

Applying law of restitution, we get

$$e = \frac{\text{relative velocity of separation}}{\text{relative velocity of approach}}$$

$$eu \cos\theta = V + v \qquad \ldots(2)$$

solving (1) and (2) we get

$$\boldsymbol{V = \frac{(1 + e)\, mu \cos\theta}{M + m}}$$

From (1) and (2) we also find

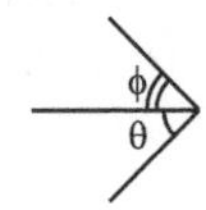

$$v = \frac{(eM - m)u \cos\theta}{(M + m)}$$

Hence $\quad \tan\phi = \frac{u \sin\theta}{v} = \frac{(M + m)\tan\theta}{(eM - m)}$

But the paths of A and B before and after impact are at right angles, therefore,

$$\cot\phi = \tan\theta$$

Hence $\quad \frac{(eM - m)}{(M + m)\tan\theta} = \tan\theta$

or $\quad \tan\theta = \sqrt{\frac{eM - m}{M + m}}$

or $\quad \boldsymbol{\theta = \tan^{-1}\sqrt{\frac{eM - m}{M + m}}}$

116. Let v_1 be the velocity of gun just after the shell is fired. The horizontal velocity of shell with respect to ground is $\left(\frac{u}{\sqrt{2}} - v_1\right)$. Hence from conservation of linear momentum in horizontal direction. We have

$$Mv_1 = m\left(\frac{u}{\sqrt{2}} - v_1\right)$$

or $\quad v_1 = \frac{mu}{\sqrt{2}(M + m)} \qquad \ldots(1)$

During collision vertical component of velocity (v_y) of shell or the time of flight of the shell remains unchanged. Hence

$$T = \frac{2v_y}{g} = \frac{2\frac{u}{\sqrt{2}}}{g} = \frac{\sqrt{2}u}{g} \qquad \ldots(2)$$

Horizontal component of velocity of shell before collision will be $\left(\frac{u}{\sqrt{2}} - v_1\right)$ and after collision $v + e\left\{\frac{u}{\sqrt{2}} - v_1 + v\right\}$ which can be obtained by applying—
relative velocity of separation $= e$ (relative velocity of approach)

Hence

$$T = \frac{L}{\left(\frac{u}{\sqrt{2}} - v_1\right)} + \frac{L + v_1 T}{v + e\left\{\frac{u}{\sqrt{2}} - v_1 + v\right\}}$$

Substituting the values we get

$$\frac{\sqrt{2}u}{g} = \frac{L}{\frac{u}{\sqrt{2}} - \frac{mu}{\sqrt{2}(M+m)}} + \frac{L + \frac{mu}{\sqrt{2}(M+m)}\frac{\sqrt{2}u}{g}}{v + e\left\{\frac{u}{\sqrt{2}} - \frac{mu}{\sqrt{2}(M+m)} + v\right\}}$$

Solving this equation we get

$$\boldsymbol{e \approx 0.5}$$

117. (a) In case of a head on elastic collision between two particles of masses m_1 and m_2 moving with velocities v_1 and v_2 we apply following two equations

$$v'_1 = \left(\frac{m_1 - m_2}{m_1 + m_2}\right)v_1 + \left(\frac{2m_2}{m_1 + m_2}\right)v_2 \quad \ldots(1)$$

and $$v'_2 = \left(\frac{m_2 - m_1}{m_1 + m_2}\right)v_2 + \left(\frac{2m_1}{m_1 + m_2}\right)v_1 \quad \ldots(2)$$

Here v'_1 and v'_2 are the velocities of particles after collision. Equations (1) and (2) are derived from law of conservation of linear momentum and law of conservation of kinetic energies. Here

$m_1 = m, m_2 = 2m, \quad v_2 = 0$ and $v_1 = v$

Hence substituting these values we get

$$v'_2 = \left(\frac{2m}{m + 2m}\right)v = \frac{2}{3}v$$

$\therefore$ Kinetic energy of block B after collision is

$$\text{K.E.} = \frac{1}{2}(2m)v_2'^{\,2} = \frac{4}{9}mv^2$$

Let x be the compression in the spring then

$$\frac{1}{2}kx^2 = \frac{4}{9}mv^2 \quad \text{or} \quad x = \left(\sqrt{\frac{8m}{k}}\right)\cdot\frac{v}{3}$$

The condition for toppling the block A is that

$$x \geq d$$

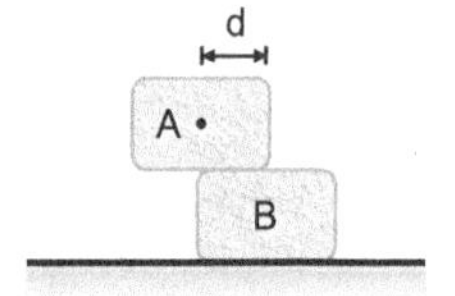

so $$\sqrt{\frac{8m}{k}}\cdot\frac{v_0}{3} = d$$

or $$\boldsymbol{v_0 = 3d\sqrt{\frac{k}{8m}}}$$

(b) When $v = \frac{v_0}{2}$, then amplitude

$$A = x = \sqrt{\frac{8m}{k}}\cdot\frac{v_0}{6}$$

Substituting the value of v_0 we get

$$\boldsymbol{A = \frac{d}{2}}$$

118. The external impulse applied to C causes both strings to jerk exerting internal impulses J_1 and J_2.

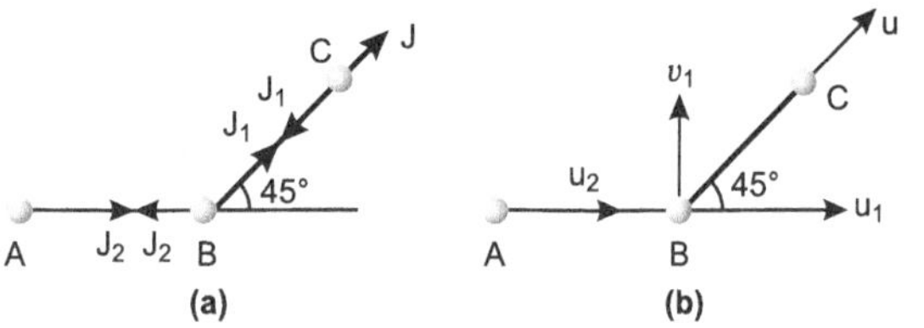

Figure (a) shows the impulses on different particles and figure (b) shows the initial velocity components of each particle.

From constraint relation obviously.

$$u_2 = u_1 \quad \ldots(1)$$

Using impulse = change in momentum we have

For particle A

$$J_2 = mu_2 \quad \ldots(2)$$

For particle B

$$J_1 \cos 45^\circ - J_2 = mu_1 \quad \ldots(3)$$

$$J_1 \sin 45^\circ = mv_1 \quad \ldots(4)$$

For particle C

$$J - J_1 = mu \qquad \ldots(5)$$

Also the velocities of B and C along BC are equal *i.e.*,

$$v_1 \cos 45^\circ + u_1 \cos 45^\circ = u \qquad \ldots(6)$$

Solving these equations we get

$$u_1 = u_2 = \frac{\sqrt{2}J}{7m}, \quad u = \frac{3J}{7m} \quad \text{and} \quad v_1 = \frac{2\sqrt{2}J}{7m}$$

Hence, **the initial speed of A** $= \mathbf{\dfrac{\sqrt{2}J}{7m}}$

the initial speed of C $= \mathbf{\dfrac{3J}{7m}}$

and **initial speed of**

$$\mathbf{B} = \sqrt{u_1^2 + v_1^2} = \frac{\sqrt{10}J}{7m}$$

119. In the figure it is clear that—

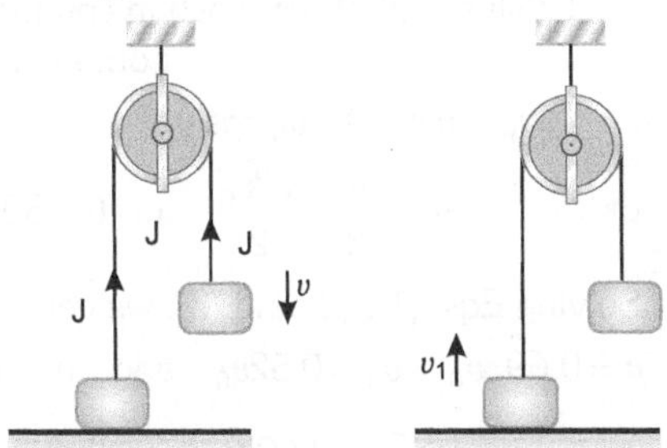

Just when the string jerks **Just after string jerks**

$$v = \sqrt{2gh} \qquad \ldots(1)$$

Using impulse = change in momentum

For mass $2m$

$$J = 2mv_1 \qquad \ldots(2)$$

For mass m

$$J = mv - mv_1 \qquad \ldots(3)$$

Solving equations (1), (2) and (3) we get

$$\mathbf{v_1 = \frac{\sqrt{2gh}}{3}}$$

120. Just after collision velocities of both the balls will be $\sqrt{2gh}$ in opposite directions. Relative acceleration between the two balls is zero and relative velocity of approach is $2\sqrt{2gh}$. Hence they will collide after a time

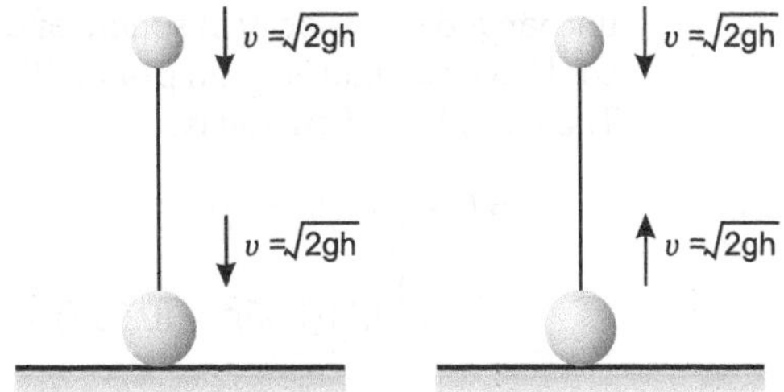

At the time of collision **Just after collision**

$$t = \frac{l}{2\sqrt{2gh}} \qquad \ldots(1)$$

At the time of collision of two balls relative velocity of approach $= 2\sqrt{2gh}$

Hence relative velocity of separation will also be $2\sqrt{2gh}$ (collision is elastic). Hence the string becomes tight after the same time

$$t = \frac{l}{2\sqrt{2gh}}$$

Hence the total time will be **2*t* or** $\mathbf{\dfrac{l}{\sqrt{2gh}}}$

121. (a) In the figure $\beta = \sin^{-1}\left(\dfrac{0.5}{1.2}\right) = 24.62^\circ$

An impulse $\vec{J}$ acts on the sphere in the direction of $A'O$ due to which velocity component of sphere parallel to OA' becomes zero.

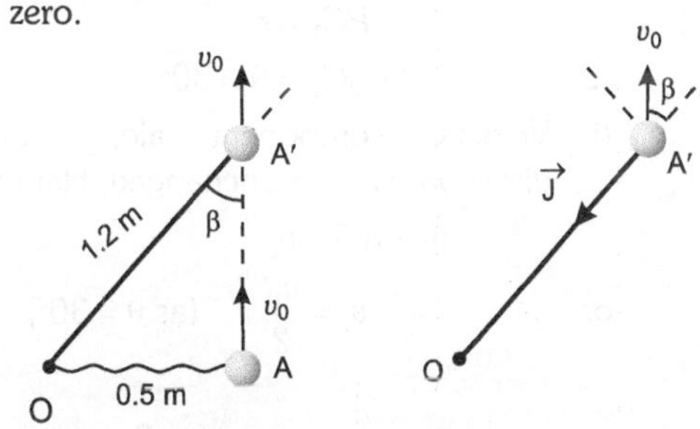

Applying impulse

= change in linear momentum

$$J = mv_0 \cos\beta$$

or $$v_0 = \frac{J}{m\cos\beta} = \frac{3}{2\cos 24.62^\circ}$$

$$\mathbf{v_0 = 1.65\ m/s}$$

(b) Velocity component of sphere perpendicular to OA' will remain unchanged i.e. velocity of sphere after the cord becomes taut is $v_0 \sin \beta$ or 0.687 m/s. Therefore loss of energy is–

$$\Delta E = \frac{1}{2} m(v_i^2 - v_f^2)$$

$$= \frac{1}{2}(2)[(1.65)^2 - (0.687)^2]$$

or $\quad \Delta \boldsymbol{E} = \mathbf{2.25}$ J

122. Ball A is free to move in a plane (after collision). So its velocity can be resolved in two mutually perpendicular directions. Let us resolve it along common tangent and common normal directions. Let v_t and v_n be the corresponding components in these two directions. Ball B is attached to a vertical string. So, just after collision its velocity will be horizontal. Let it be v.

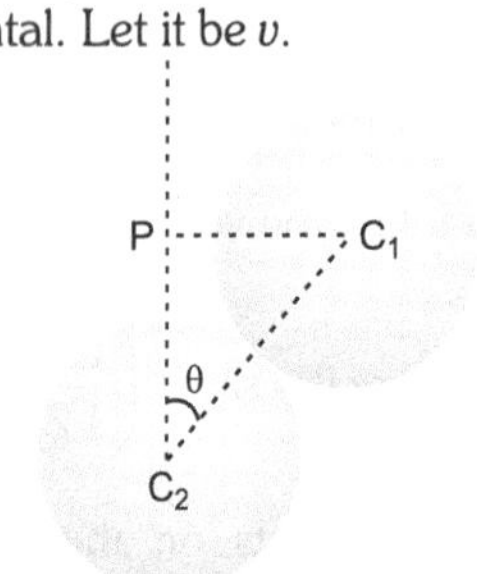

$$C_1C_2 = 2r \quad (r = \text{radius of ball})$$

$$PC_1 = r$$

$$\therefore \quad \angle PC_2C_1 = \theta = 30°$$

(i) Velocity component along tangent direction remains unchanged. Hence,

$$v_0 \sin \theta = v_t$$

or $$v_t = \frac{v_0}{2} \quad (\text{as } \theta = 30°) \quad \ldots(1)$$

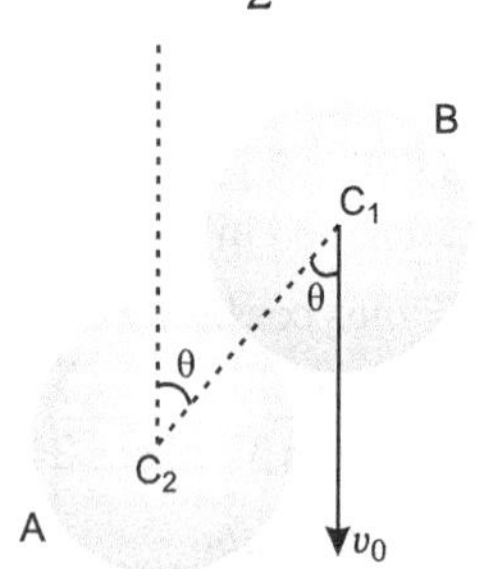

Just before collision

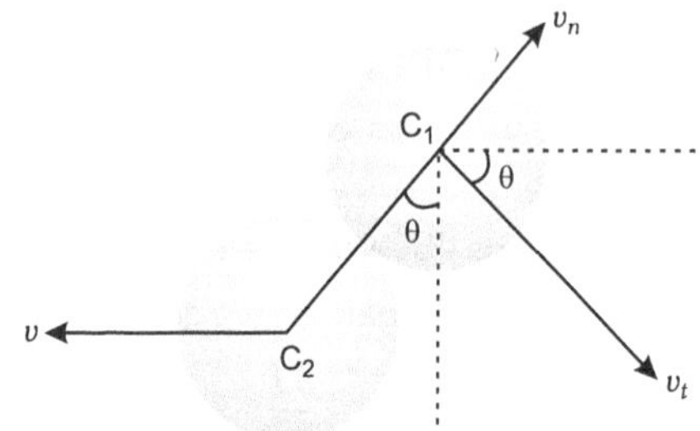

Just after collision

(ii) Linear momentum along horizontal will remain unchanged. Hence,

$$mv - mv_t \cos 30° - mv_n \sin 30° = 0$$

or $$v - \frac{\sqrt{3}v_t}{2} - \frac{v_n}{2} = 0 \quad \ldots(2)$$

(iii) $e = 1$. Hence,

relative speed of separation

= relative speed of approach in common normal direction.

$$\therefore \quad v_n + v \sin \theta = v_0 \cos \theta$$

or $$v_n + \frac{v}{2} = \frac{\sqrt{3}v_0}{2} \quad (\text{as } \theta = 30°) \ldots(3)$$

Solving Eqs. (1), (2) and (3), we get

$v = 0.693v_0$, $v_n = 0.52v_0$ and $v_t = 0.5v_0$

$$\therefore \quad \vec{v_B}' = 0.693v_0 \quad (\text{horizontally})$$

or $$\vec{v_A}' = \sqrt{v_n^2 + v_t^2} = 0.721v_0$$

$$\alpha = \tan^{-1}\left(\frac{v_n}{v_t}\right) = 46.1°$$

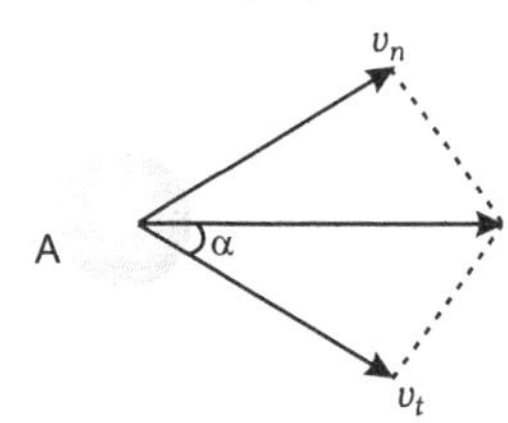

Hence, $\vec{v_A}'$ makes an angle of $(46.1° - 30°)$ or $16.1°$ with horizontal as shown

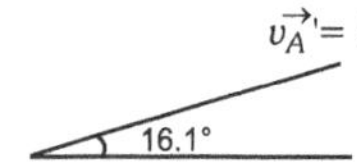

123. Let v_r be the velocity of block m relative to the bigger block M at highest point as shown and v the velocity of block M. Then absolute velocity of block m will be $v_r - v$. Applying conservation of linear momentum in horizontal direction, we have

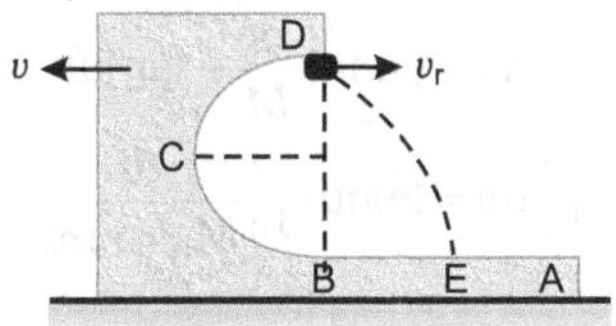

$$mv_0 = Mv - m(v_r - v)$$

or $(1)(20) = (2)v - (1)(v_r - v)$

or $10v - v_r = 10$

...(1)

From conservation of mechanical energy we get

$$\frac{1}{2}mv_0^2 = mg(2R) + \frac{1}{2}Mv^2 + \frac{1}{2}m(v_r - v)^2$$

or $\frac{1}{2}(1)(20)^2 = (1)(10)(2) + \frac{1}{2}(2)v^2 + \frac{1}{2}(1)(v_r - v)^2$

or $(v_r - v)^2 + 2v^2 = 360$ ...(2)

Solving (1) and (2) we get

$$v_r = 7.5 \text{ m/s}$$

Now from $h = \frac{1}{2}gt^2$ we get

$$t_{DE} = t = \sqrt{\frac{2h}{g}} \qquad (h = 2R)$$

$$t = \sqrt{\frac{2 \times 2}{10}} = 0.632 \, s$$

Hence $BE = (v_r)t = 7.5 \times 0.632$ m

or $\mathbf{BE = 4.74\ m}$

124. (a) Let v be the velocity of the car when it has cleared x-distance as shown in figure and let m be the mass of it (including the load) at this instant. Then thrust force on it (due to change in mass) will be qv in backward direction. Hence

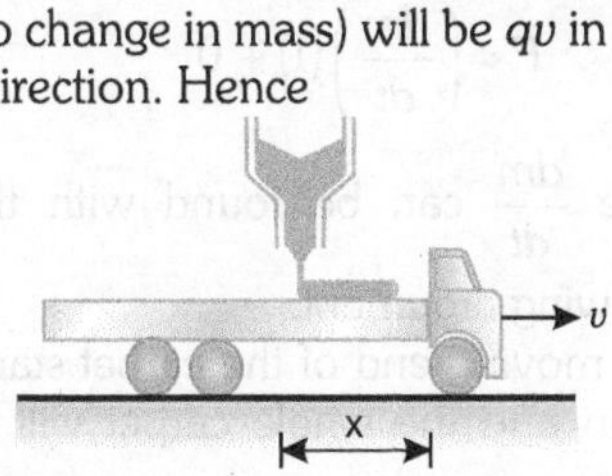

$$F_t = -qv \quad \text{or} \quad ma = -qv$$

or $mv\left(\frac{dv}{dx}\right) = -qv$

From conservation of linear momentum

$$mv = m_0 v_0$$

Hence $m_0 v_0 \left(\frac{dv}{dx}\right) = -qv$

or $m_0 v_0 \frac{dv}{v} = -q\,dx$

or $m_0 v_0 \int_{v_0}^{v_f} \frac{dv}{v} = -q\int_0^L dx$

Solving this we get

$$\boldsymbol{v_f = v_0 e^{\frac{-qL}{m_0 v_0}}}$$

(b) From conservation of linear momentum mass of the car and its load at that instant will be

$$\frac{m_0 v_0}{v_f} \quad \text{or} \quad \boldsymbol{m_0\, e^{\frac{qL}{m_0 v_0}}}$$

125. Initial momentum of the particle is

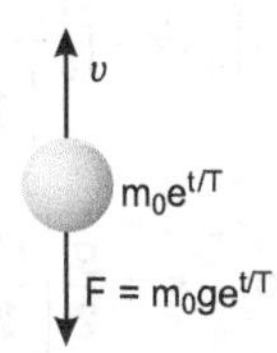

$$p_i = m_0\, gT$$

Mass of the particle at time t is $m = m_0 e^{t/T}$

Therefore force acting on the particle at time t is

$$F = -mg = -m_0 e^{t/T} g$$

Impulse = change in momentum.

or $F \,.\, dt = dP$

Let t_0 be the time when the particle is at highest point. Then

$$\int_0^{t_0} F \,.\, dt = \int_{initial}^{final} dp$$

or $\int_0^{to} (-m_0 g e^{t/T})\, dt = p_{final} - p_{initial}$

or $-m_0 gT\,(e^{t_0/T} - 1) = 0 - m_0 gT$

or $e^{t_0/T} = 2$

or $\boldsymbol{t_0 = T \ln(2)}$

Mass of the particle at highest point is $\boldsymbol{m_0 e^{t_0/T}}$ or $\boldsymbol{2m_0}$ as $\boldsymbol{e^{t_0/T} = 2}$.

126. Let v be the velocity of the rain drop at time t and m the mass of drop. Then

$$m = (m_0 + \mu t)$$

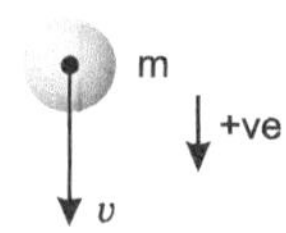

$$F_{ext} = \frac{d\,(mv)}{dt}$$

Here external force is mg.

Hence $$(mg) = \frac{d}{dt}(mv)$$

or $$(m_0 + \mu t)g = \frac{d}{dt}(mv)$$

or $$(m_0 + \mu t)g\,dt = d\,(mv)$$

or $$\int (m_0 + \mu t)g\,dt = \int d\,(mv)$$

or $$\left(m_0 + \frac{\mu t}{2}\right)\cdot gt = mv + C_1$$

at $t = 0$, $v = 0$ $\therefore$ $C_1 = 0$

or $$\left(m_0 + \frac{\mu t}{2}\right)gt = mv = (m_0 + \mu t)\frac{ds}{dt}$$

$$\therefore \int_0^s ds = \int_0^t \frac{\left(1 + \frac{\mu t}{2m_0}\right)}{\left(1 + \frac{\mu t}{m_0}\right)} gt\,dt$$

or $$\boldsymbol{s = \frac{g}{2}\left[\frac{t^2}{2} + \frac{m_0 t}{\mu} - \frac{m_0^2}{\mu^2}\ln\left(1 + \frac{\mu t}{m_0}\right)\right]}$$

Note: The above integration can be done by taking the substitution $x = 1 + \frac{\mu}{m_0}t$.

127. Let n be the frequency with which bullets are fired and m the mass of each bullet. The loss of mass per unit time is

$$\left(-\frac{dM}{dt}\right) = Mn$$

$$u_{relative} = u \qquad \text{(given)}$$

Therefore, thrust force on the cannon is

$$F_t = u_{rel.}\left(-\frac{dM}{dt}\right)$$

or $$F_t = mnu \qquad \text{(backwards)}$$

and frictional force on it is

$$f = \mu Mg \qquad \text{(forwards)}$$

Here M is the instantaneous mass of cannon. At time t

$$M = (M_0 - mnt)$$

$$\therefore \quad f = \mu g\,(M_0 - mnt)$$

Therefore, net force on cannon at time t is

$$F_{net} = F_t - f \qquad \text{(backwards)}$$

or $$M\left(\frac{dv}{dt}\right) = (mnu) - (\mu Mg)$$

or $$dv = (mnu)\frac{dt}{M} - (\mu g)dt$$

or $$\int_0^v dv = (mnu)\int_0^t \frac{dt}{M_0 - mnt} - (\mu g)\int_0^t dt$$

or $$v = (mnu)\left(-\frac{1}{mn}\right)[\ln(M_0 - mnt)]_0^t - \mu gt$$

or $$v = -u\ln\left(\frac{M_0 - mnt}{M_0}\right) - \mu gt$$

or $$v = -u\ln\left(\frac{M}{M_0}\right) - \mu gt \qquad (M_0 - mnt = M)$$

or $$\boldsymbol{v = u\ln\left(\frac{M_0}{M}\right) - \mu gt}$$

128. (a) The co-ordinates of centre of mass (point C) of the moving part is $\frac{3x}{4}$.

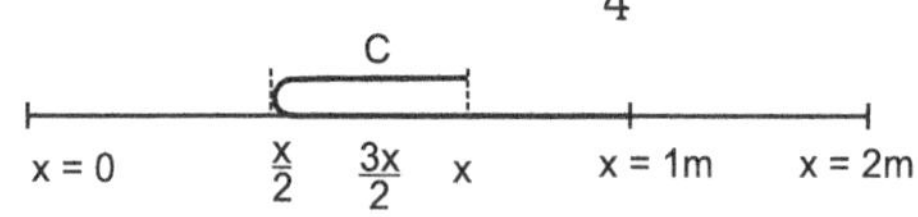

$$\therefore \quad X_c = \frac{3x}{4}$$

$$\therefore \quad v_c = \frac{d}{dt}(X_c) = \frac{3}{4}\frac{dx}{dt} = \frac{3}{4}\text{ m/s}$$

(as $\frac{dx}{dt} = 1$ m / s)

(b) Linear momentum of moving point is,

$$P = mv$$

where $v = 1$ m/s and m is increasing uniformly with time.

The net force on the moving part is thus,

$$F = \frac{dP}{dt} = \frac{dm}{dt}v + \frac{dv}{dt}m$$

or $$F = \left(\frac{dm}{dt}\right)(1) + 0$$

Here $\frac{dm}{dt}$ can be found with the help of following argument.

The moving end of the carpet starts from the origin and the whole carpet will be moving when it reaches $x = 2$ m. This will happen after 2 seconds. Hence,

$$\frac{dm}{dt} = \frac{1}{2} \text{ kg/s}$$

$\therefore F_{min}$ (neglecting all dissipative forces) $= \frac{1}{2}$ N

129. By COM

$$(m+1)u = 1 \times 10 \Rightarrow u = 5 \text{ ms}^{-1}$$

So, the combined body will move up the plane and after a certain time it will come back to its initial position with velocity 5 ms^{-1} and time $t' = 2$ sec.
Now, M is jerked into motion, thread becomes taut, a sharp impulse is exerted by thread to M and combined body.
Let impulse is I

So, $$Mv_0 = I \quad \ldots(1)$$

$$(m+1)v_0 = (m+1)u - I \quad \ldots(2)$$

$$\Rightarrow v_0 = 2 \text{ ms}^{-1}$$

Now, M will move with retardation. Let retardation is 'b'.

So, $T - (m+1)g \cdot \sin 30° = (m+1)b$

$$Mg - T = Mb$$

$$\Rightarrow b = 4 \text{ ms}^2$$

$\therefore$ Maximum height ascended by M,

$$0^2 = 2^2 - 2{\cdot}4 \text{ s}$$

$$\therefore s = \frac{4}{8} = \frac{1}{2} \text{ m}$$

Time taken to reach that height

$$0 = 2 - 4t$$

$$t = 0.5 \text{ sec}$$

$\therefore$ total time $= 2 + 0.5 = 2.5$ sec

130. Consider the reference axes as shown in figure.

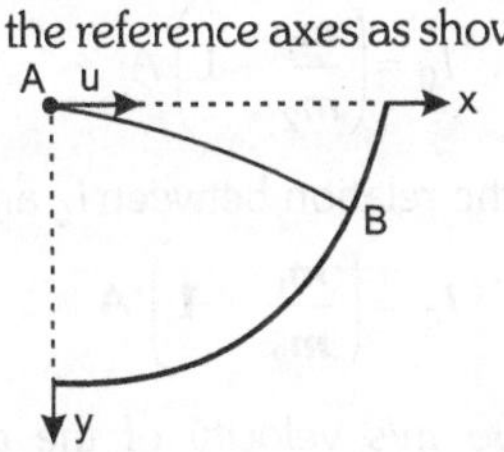

The co-ordinates of the particle as function of time is given as :

$$x = ut = 10t$$

and $$y = \frac{1}{2}gt^2 = 5t^2$$

Eliminating t, we get the equation of trajectory of the particle as

$$y = 5\left(\frac{x}{10}\right)^2 = \frac{x^2}{20} \quad \ldots(1)$$

The equation of the frame can be written as

$$x^2 + y^2 = 21; \quad x \geq 0; \quad y \geq 0 \quad \ldots(2)$$

Solving Eqs. (1) and (2), we get the co-ordinate of the point where the particle strikes the frame.

$$y^2 + 20y - 21 = 0$$

or $$y^2 + 21y - y - 21 = 0$$

or $$y(y+21) - (y+21) = 0$$

or $$(y-1)(y+21) = 0$$

Hence, $y = 1$ $\quad [y \neq -21, \text{ as } y \geq 0]$

and $x = \sqrt{20}$

The particle will strike the frame at B whose co-ordinate is given as $x = \sqrt{20}$, $y = 1$ m.

The y-component of the velocity of the particle just before it strikes the frame at B is

$$u_y^2 = 2 \times g \times y = 2 \times 10 \times 1$$

or $$u_y = \sqrt{20} = 2\sqrt{5} \text{ m/s}$$

x-component is $u_x = u = 10$ m/s.
Resultant velocity is

$$\sqrt{10^2 + (2\sqrt{5})^2} = \sqrt{120} \text{ m/s}$$

The resultant makes an angle θ with the horizontal, where

$$\tan\theta = \frac{2\sqrt{5}}{10} = \frac{1}{\sqrt{5}} = 0.446 \quad \text{or} \quad \theta \approx 24°$$

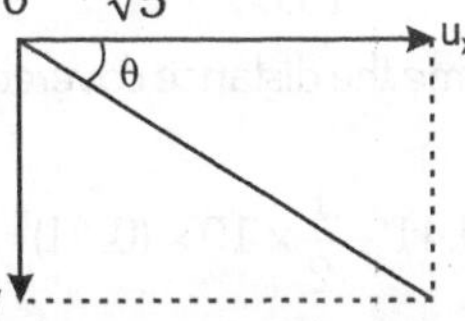

The line AB, which is normal to the surface at B makes an angle α with the horizontal where

$$\tan\alpha = \frac{1}{\sqrt{20}} = 0.223$$

or $$\alpha \approx 13°$$

The resultant velocity vector makes an angle of $\theta - \alpha = 11°$ with the normal (AB).
Since, the particle rebounds elastically, it will return back with the same speed making an angle of 11° on the other side of line AB.

After collision the velocity vector of the particle will make an angle of $(2\alpha - \theta) = 2°$ with the horizontal, as shown in figure.

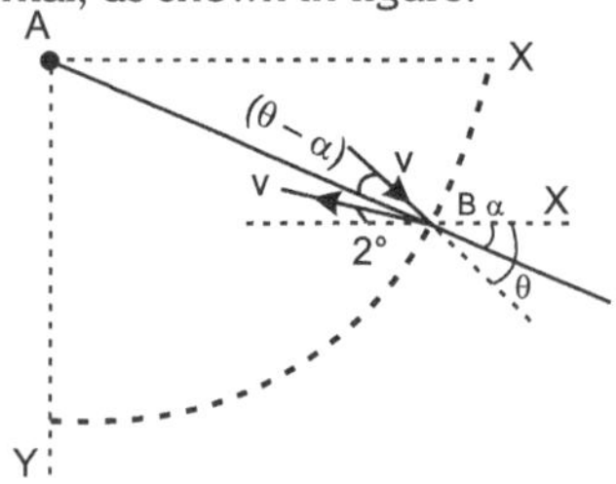

Horizontal component of velocity after collision is

$$v_x = \sqrt{120}\cos 2° = 10.84 \text{ m/s}$$

Vertical component is

$$v_y = -\sqrt{120}\sin 2° = -0.38 \text{ m/s}$$

Negative sign indicates that velocity is in negative Y direction (upwards).

Time, the particle will take to cover a horizontal distance of 4.48 m (see figure) is

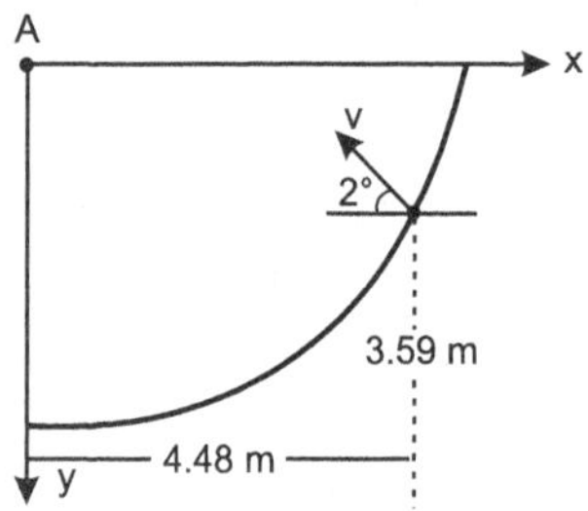

$$t = \frac{4.48}{10.84} = 0.41 \text{ sec}$$

In this time the distance covered along Y-axis is

$$y = -0.38 \times 0.41 + \frac{1}{2} \times 10 \times (0.41)^2 = 0.68 \text{ m}$$

which is less than 3.59 m.

Hence, the particle will not collide with the frame again.

131. Angular speed of particle about centre of the the circle,

$$\omega = \frac{v_2}{R}, \quad \theta = \omega t = \frac{v_2}{R} t$$

$$\vec{v}_p = (-v_2 \sin\theta\,\hat{i} + v_2\cos\theta t\hat{j})$$

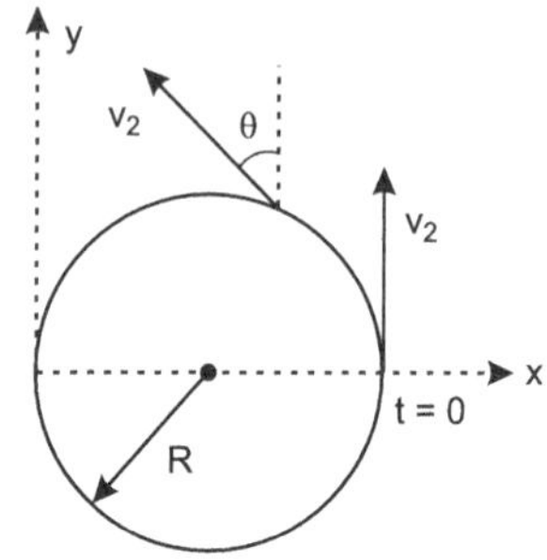

or $\quad \vec{v}_p = \left(-v_2 \sin\frac{v_2}{R} t\,\hat{i} + v_2 \cos\frac{v_2}{R} t\,\hat{j}\right)$

and $\vec{v}_m = v_1\hat{j}$

$\therefore$ linear momentum of particle w.r.t. man as a function of time is

$$\vec{L}_{pm} = m(\vec{v}_p - \vec{v}_m)$$

$$= m\left[\left(-v_2 \sin\frac{v_2}{R} t\right)\hat{i} + \left(v_2\cos\frac{v_2}{R} t - v_1\right)\hat{j}\right]$$

132. (i) $\quad x_1 = v_0 t - A(1 - \cos\omega t)$

$$x_{cm} = \frac{m_1 x_1 + m_2 x_2}{m_2 + m_2} = v_0 t$$

$$\therefore \quad x_2 = v_0 t + \frac{m_1}{m_2} A(1 - \cos\omega t)$$

(ii) $\quad a_1 = \frac{d^2 x_1}{dt^2} = -\omega^2 A\cos\omega t$

The separation $x_2 - x_1$ between the two blocks will be equal to l_0 when $a_1 = 0$ or $\cos\omega t = 0$

$$x_2 - x_1 = \frac{m_1}{m_2} A(1 - \cos\omega t) + A(1 - \cos\omega t)$$

or $\quad l_0 = \left(\frac{m_1}{m_2} + 1\right) A \qquad (\cos\omega t = 0)$

Thus the relation between l_0 and A is,

$$\boldsymbol{l_0 = \left(\frac{m_1}{m_2} + 1\right) A}$$

133. (i) 100 m/s velocity of the cannon ball is relative to ground.

[Unless and until it is mentioned in the question, the velocity is always relative to ground]

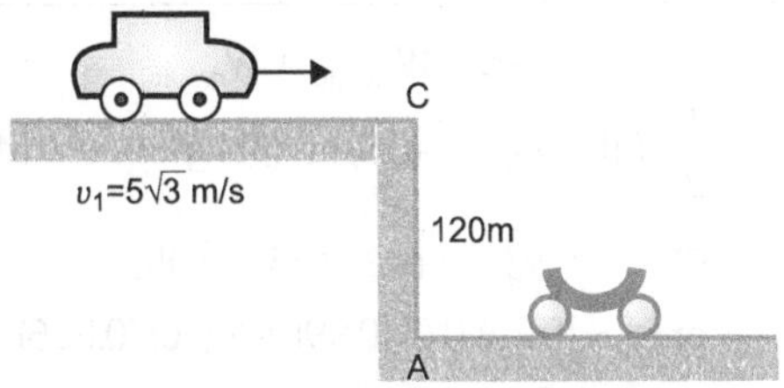

Horizontal component of its velocity,

$$u_x = u \cos 30^\circ$$

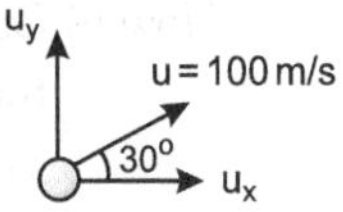

or $$u_x = (100)\frac{\sqrt{3}}{2} \text{ m/s}$$

$$= 50\sqrt{3} \text{ m/s}$$

and vertical component of its velocity,

$$u_y = u \sin 30^\circ$$

or $$u_y = 100\frac{1}{2} \text{ m/s} = 50 \text{ m/s}$$

Vertical displacement of the ball when it strikes the carriage is –120 m or

$$s_y = u_y t + \frac{1}{2} a_y t^2$$

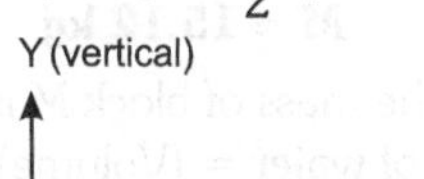

$a_x = 0$ and
$a_y = -g = -10 \text{ m/s}^2$

$$\Rightarrow \quad -120 = (50t) + \left(\frac{1}{2}\right)(-10)\,t^2$$

$$\Rightarrow \quad t^2 - 10t - 24 = 0$$

$$\Rightarrow \quad t = 12 \text{ s or } -2\text{s}$$

Ignoring the negative time, we have

$$\boldsymbol{t_0 = 12 \text{ s}}$$

(ii) When it strikes the carriage, its horizontal component of velocity is still $50\sqrt{3}$ m/s. It sticks to the carriage. Let v_2 be the velocity of (carriage + cannon ball) system after collision. Then applying conservation of linear momentum in horizontal direction

(mass of ball) (horizontal component of its velocity before collision)

= (mass of ball + carriage) (v_2)

$$\therefore \quad (1 \text{ kg})(50\sqrt{3} \text{ m/s}) = (10 \text{ kg})\,(v_2)$$

$$\therefore \quad v_2 = 5\sqrt{3} \text{ m/s}$$

The second cannon ball is fired when the first cannon ball strikes the carriage i.e. after 12 second. In these 12 seconds the car will move forward a distance of $12v_1$ or $60\sqrt{3}$ m. This ball will strike the carriage only when the carriage also covers the same distance of $60\sqrt{3}$ m in next 12 seconds. This is possible only when resistive forces are zero because velocity of car (v_1) = velocity of carriage after first collision (v_2) = $5\sqrt{3}$ m/s.

Hence at the time of second collision :

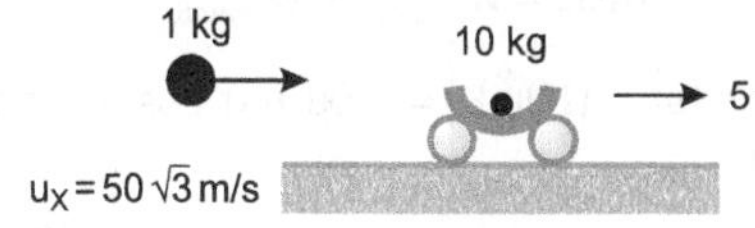

Before collision

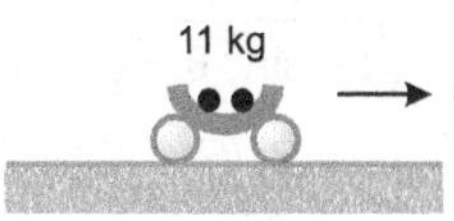

After collision

Horizontal component of velocity of cannon ball $= 50\sqrt{3}$ m/s and horizontal velocity of carriage + first cannon ball = $5\sqrt{3}$ m/s.

Let v be the desired velocity of carriage after second collision.

Then conservation of linear momentum in horizontal direction gives

$$11v = (1)(50\sqrt{3}) + (10)(5\sqrt{3}) = 100\sqrt{3}$$

$$\therefore \quad v = \frac{100\sqrt{3}}{11} \text{ m/s} \quad \text{or } \boldsymbol{v} \approx = \mathbf{15.75 \text{ m/s}}$$

In this particular problem, values are so adjusted that even if we take the velocity of cannon ball with respect to car, we get the same results of both the parts, although the method will be wrong.

134. Let v_1 = velocity of block 2 kg just before collision (up the plane)

v_2 = velocity of block 2 kg just after collision. (down the plane)

and v_3 = velocity of block M just after collision. (up the plane)

Applying work energy theorem (change in kinetic energy = work done by all the forces) at different stages as shown in figures (a), (b) and (c):

Figure (a) :

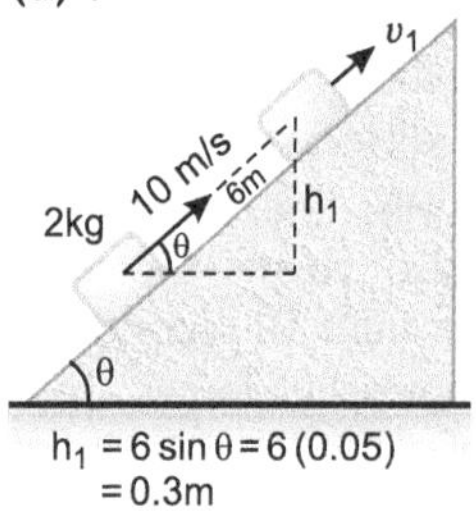

(a)

$$\Delta KE = W_{\text{friction}} + W_{\text{gravity}}$$

$$\left[\frac{1}{2} m \{v_1^2 - (10)^2\}\right] = -6\mu\, mg \cos\theta - mgh_1$$

$(m = 2 \text{ kg})$

or $v_1^2 - 100 = -2\,[6\mu g \cos\theta + gh_1]$

$\cos\theta = \sqrt{1 - \sin^2\theta} = \sqrt{1 - (0.05)^2} \approx 0.99$

$\therefore v_1^2 = 100 - 2\,[(6)(0.25)(10)0.99 + (10)(0.3)]$

or $\qquad v_1 \approx 8 \text{ m/s}$

Figure (b) :

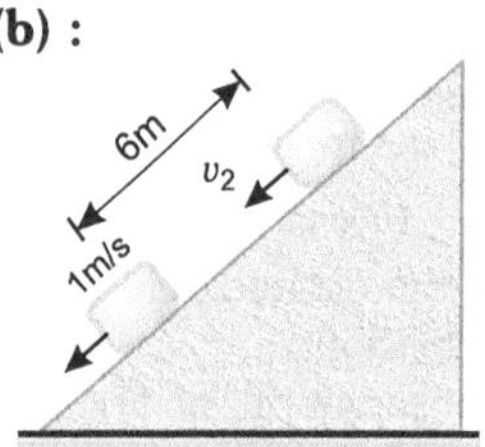

(b)

$$\Delta KE = W_{\text{friction}} + W_{\text{gravity}}$$

$$\frac{1}{2} m\,[(1)^2 - (v_2^2)] = -6\,\mu\, mg \cos\theta + mgh_1$$

$\therefore \qquad 1 - v_2^2 = 2\,[-6\,\mu\, g \cos\theta + gh_1]$

$= 2\,[-(6)\,(0.25)(10)(0.99) + (10)(0.3)] = -23.7$

$\therefore \qquad v_2^2 = 24.7$ or $v_2 \approx 5 \text{ m/s}$

Figure (c) :

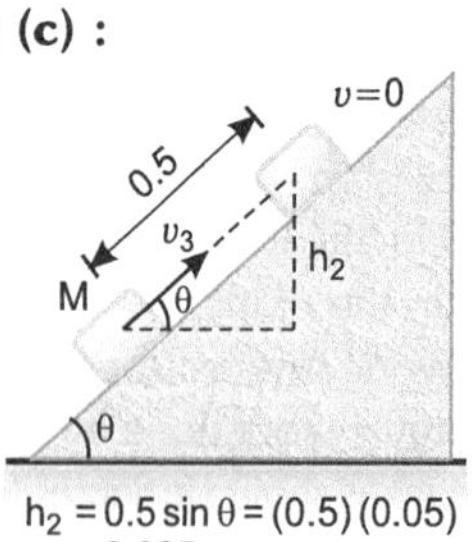

(c)

$$\Delta KE = W_{\text{friction}} + W_{\text{gravity}}$$

$$\frac{1}{2} M\,[0 - v_3^2] = -(0.5)\,(\mu)\,(M)\,g \cos\theta - Mgh_2$$

or $\qquad -v_3^2 = -\mu\, g \cos\theta - 2gh_2$

or $v_3^2 = (0.25)\,(10)\,(0.99) + 2\,(10)\,(0.025)$

or $\qquad v_3^2 = 2.975 \quad \therefore \quad v_3 \approx 1.72 \text{ m/s}$

Now (i) Coefficient of restitution

$$= \frac{\text{relative velocity of separation}}{\text{relative velocity of approach}}$$

$$= \frac{v_2 + v_3}{v_1} = \frac{5 + 1.72}{8} = \frac{6.72}{8}$$

or $\qquad$ $\boldsymbol{e \approx 0.84}$

(ii) Applying conservation of linear momentum before and after collision.

$$2v_1 = Mv_3 - 2v_2$$

$$M = \frac{2\,(v_1 + v_2)}{v_3} = \frac{2\,(8 + 5)}{1.72} = \frac{26}{1.72}$$

$$\boldsymbol{M \approx 15.12 \text{ kg}}$$

Hence, the mass of block M is 15.12 kg

135. **(i)** Mass of water = (Volume) (density)

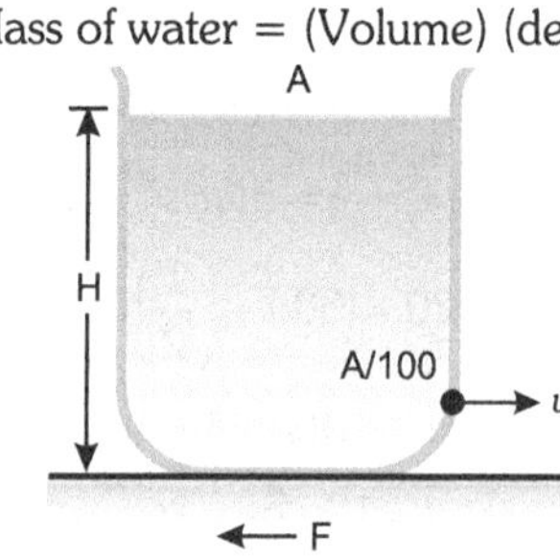

$\therefore \qquad m_0 = (AH)\,\rho$

$\therefore \qquad H = \dfrac{m_0}{A\rho} \qquad \ldots(1)$

Velocity of efflux, $v = \sqrt{2gH} = \sqrt{\dfrac{2m_0\, g}{A\rho}}$

Thrust force on the container due to draining out of liquid from the bottom is given by,
F = (density of liquid) (area of hole) (velocity of efflux)^2

$$\therefore \; F = \rho(A/100)v^2 = \rho(A/100)\left(\frac{2m_0\, g}{A\rho}\right)$$

$$F = \frac{m_0 g}{50}$$

$\therefore$ Acceleration of the container,

$$a = F/m_0 = g/50$$

$$\boldsymbol{a = g/50}$$

(ii) Velocity of efflux when 75% liquid has been drained out *i.e.* height of liquid,

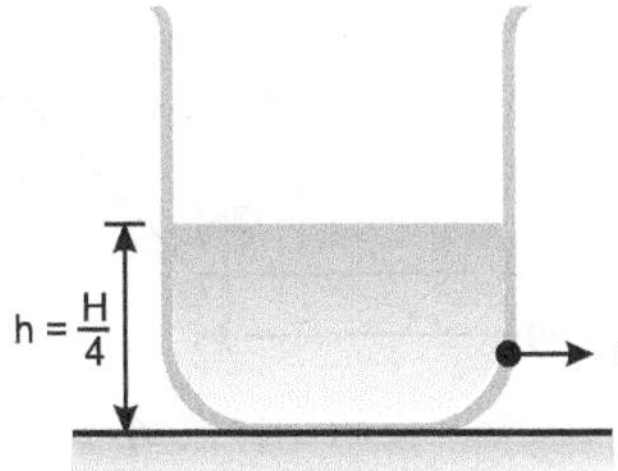

$$h = \frac{H}{4} = \frac{m_0}{4A\rho}$$

$$\therefore \quad \upsilon = \sqrt{2gh} = \sqrt{2g\left(\frac{m_0}{4A\rho}\right)}$$

or $$\boldsymbol{\upsilon = \sqrt{\frac{m_0 g}{2A\rho}}}$$

136. Since all the surfaces are smooth, no external force is acting on the system in horizontal direction. Therefore, the centre of mass of the system in horizontal direction will remain stationary.

$C_1C_2 = 5R$ (in both cases)

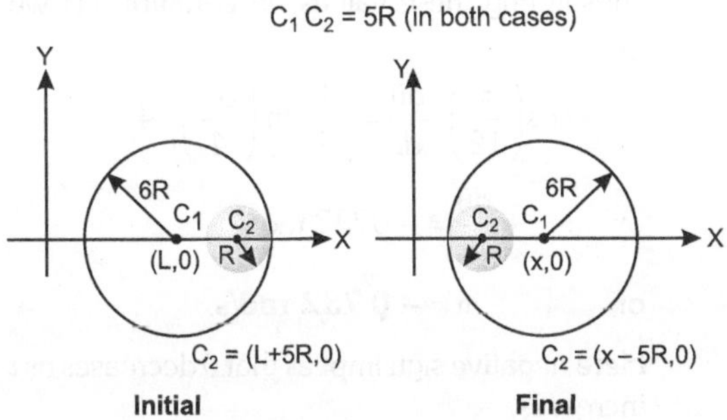

Initial x-coordinate of COM will be given by

$$x_i = \frac{m_1x_1 + m_2x_2}{m_1 + m_2}$$

$$= \frac{(4M)(L) + M(L + 5R)}{4M + M} = (L + R) \quad \ldots(1)$$

Let $(x, 0)$ be the coordinates of the centre of large sphere in final position. Then x-coordinate of COM find will be

$$x_f = \frac{(4M)(x) + M(x - 5R)}{4M + M} = (x - R) \quad \ldots(2)$$

Equating (1) and (2), we have

$$x = L + 2R$$

Therefore, coordinates of large sphere, when the smaller sphere reaches the other extreme position, are $\boldsymbol{(L + 2R, 0)}$

137. Given $m_0 = 10^{-2}$ kg : $A = 10^{-4}$ m^2,

$$\upsilon_0 = 10^3 \text{ m/s}$$

and $\rho_{dust} = \rho = 10^{-3}$ kg/m^3

$m = m_0$ + mass of dust collected so far

$= m_0 \, Ax\rho_{dust}$

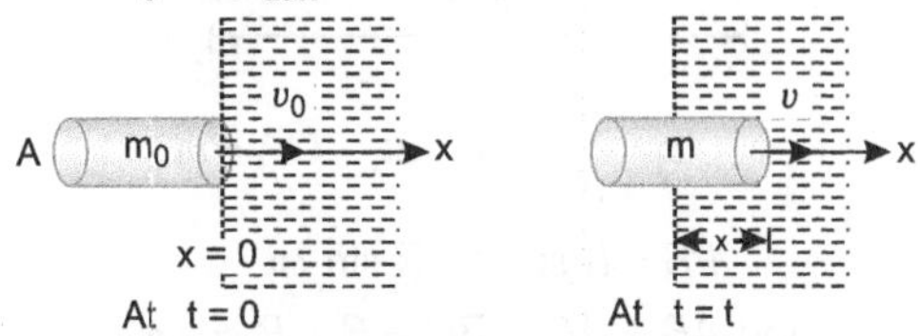

or $$m = m_0 + Ax\rho$$

The linear momentum at $t = 0$ is

$$p_0 = m_0\upsilon_0$$

and momentum at $t = t$ is

$$p_t = m\upsilon = (m_0 + Ax\rho)\upsilon$$

From law of conservation of linear momentum.

$$P_0 = P_t$$

$$\therefore \quad m_0\upsilon_0 = (m_0 + Ax\rho)\,\upsilon$$

or $$m_0\,\upsilon_0 = (m_0 + Ax\rho)\frac{dx}{dt}$$

or $$(m_0 + Ax\rho)\,dx = m_0\upsilon_0\,dt$$

or $$\int_0^x (m_0 + Ax\rho)\,dx = m_0\upsilon_0 \int_0^{150} dt$$

or $$\left(m_0x + A\rho\frac{x^2}{2}\right)_0^x = [m_0\,\upsilon_0\,t]_0^{150}$$

Hence $$m_0x + A\rho\frac{x^2}{2} = 150 m_0\,\upsilon_0$$

Solving this quadratic equation and substituting the values of m_0, A, ρ and υ_0, we get positive value of x as 10^5 m. Therefore

$$\boldsymbol{x = 10^5 \text{ m}}$$

CHAPTER 5

ROTATION

138. (i) Angular velocity of the ring is

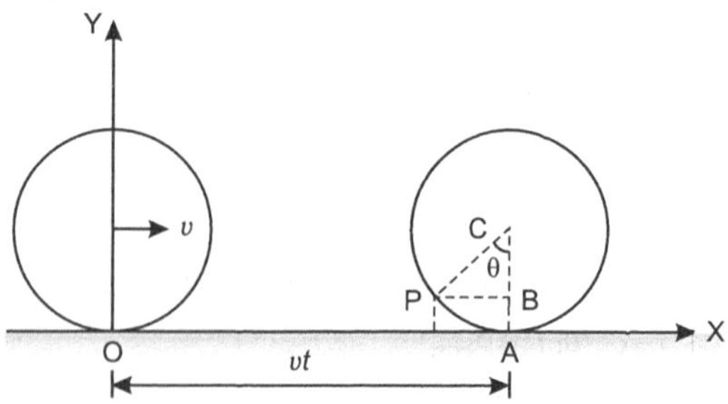

$$\omega = \frac{v}{R}, \quad \theta = \omega t, \quad OA = vt$$

$$PB = R \sin\theta = R \sin\omega t$$

and $AB = AC - BC = R - R\cos\omega t$

$$= R(1 - \cos\omega t)$$

Hence position vector of point P at any time t would be

$$\vec{r}_P = x\,\hat{i} + y\,\hat{j} = (OA - PB)\,\hat{i} + AB\,\hat{j}$$

$$= (vt - R\sin\omega t)\,\hat{i} + R(1 - \cos\omega t)\,\hat{j} \,(v = R\omega)$$

$$\boldsymbol{\vec{r}_P = R\,(\omega t - \sin\omega t)\,\hat{i} + R\,(1 - \cos\,\omega t)\,\hat{j}}$$

(ii) $\boldsymbol{\vec{v}_p = \dfrac{d\vec{r}_p}{dt} = R\omega\,(1 - \cos\omega t)\,\hat{i}}$

$$\boldsymbol{+ R\omega \sin\omega t\,\hat{j}}$$

(iii) $\boldsymbol{\vec{a}_p = \dfrac{d\vec{v}_p}{dt} = R\omega^2\,(\sin\,\omega t\,\hat{i} + \cos\,\omega t\,\hat{j})}$

139. $\angle AOB = 135° = \dfrac{3\pi}{4}$ rad

$$\angle OBA = \pi - \left(\theta + \frac{3\pi}{4}\right) = \left(\frac{\pi}{4} - \theta\right) \text{rad}$$

By sine theorem

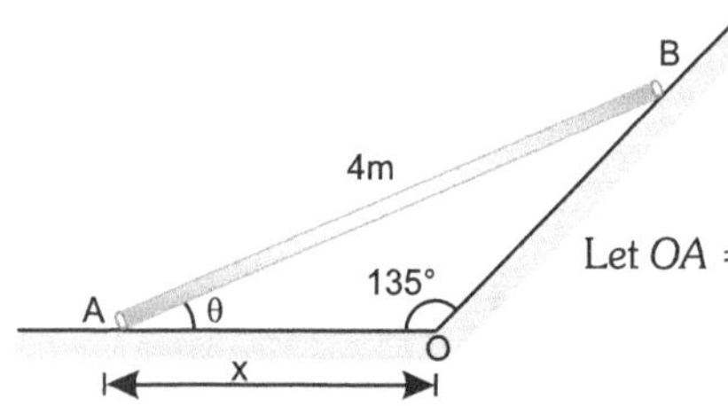

Let $OA = x$

$$\frac{x}{\sin\left(\frac{\pi}{4} - \theta\right)} = \frac{4}{\sin\left(\frac{3\pi}{4}\right)}$$

or $\quad \sin\left(\dfrac{\pi}{4} - \theta\right) = \dfrac{1}{4}\left[\sin\dfrac{3\pi}{4}\right]x$

Differentiating with respect to time we get

$$-\left\{\cos\left(\frac{\pi}{4} - \theta\right)\right\}\frac{d\theta}{dt} = \left\{\frac{1}{4}\sin\frac{3\pi}{4}\right\}\frac{dx}{dt} \quad \ldots(1)$$

The velocity of A is towards left in which direction, x is increasing. Hence we can take

$$\frac{dx}{dt} = 4 \text{ m/s} \quad \text{and} \quad \theta = \frac{\pi}{6}$$

substituting these values in equation (1) we get

$$-\cos\left(\frac{\pi}{12}\right)\cdot\frac{d\theta}{dt} = \left\{\frac{1}{4}\sin\left(\frac{3\pi}{4}\right)\right\}4$$

or $\quad \dfrac{d\theta}{dt} = -\,0.732$ rad/s

or $\quad \omega = \boldsymbol{-\,0.732}$ **rad/s**

Here negative sign implies that θ decreases as t increases.

Again differentiating equation (1) with respect to time, we get

$$\left\{-\cos\left(\frac{\pi}{4} - \theta\right)\frac{d^2\theta}{dt^2}\right\} - \left\{\sin\left(\frac{\pi}{4} - \theta\right)\right\}\left(\frac{d\theta}{dt}\right)^2$$

$$= \left\{\frac{1}{4}\sin\left(\frac{3\pi}{4}\right)\right\}\frac{d^2x}{dt^2}$$

substituting the values

$$\theta = \frac{\pi}{6}, \quad \frac{d\theta}{dt} = -\,0.732$$

and $$\frac{d^2x}{dt^2} = -\,5 \text{ m/s}^2$$

in the above equation. We get

$$\frac{d^2\theta}{dt^2} = \mathbf{0.771\ rad/s^2}$$

140. Since there is no slipping, static friction will act. Normal reaction at bottom will be zero. Cylinder will rotate about point of contact M as shown in figure. Drawing free body diagram of cylinder

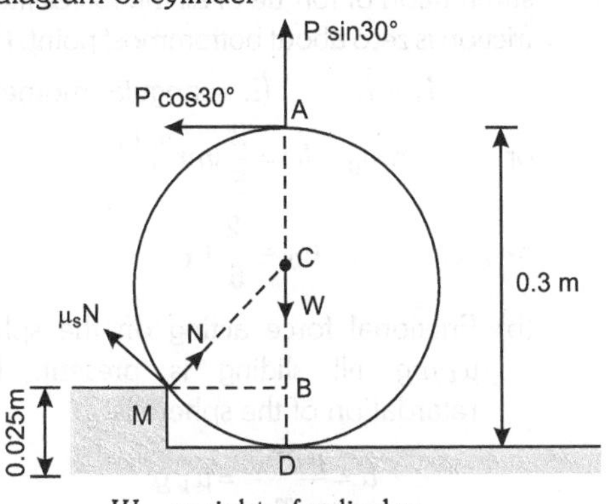

W = weight of cylinder

$= 25\,g = 250$ N

$AB = AD - BD = (0.3 - 0.025)$

$= 0.275$ m

$$MB = \sqrt{(MC)^2 - (CB)^2}$$

$$= \sqrt{(0.15)^2 - (0.15 - 0.025)^2}$$

$= 0.08$ m

Cylinder will rotate about M

when anticlockwise torque > clockwise torque

or $(P \cos 30°)(0.275) + (P \sin 30°)(0.08) > (W)(0.08)$

or $\quad 0.28P > (250)(0.08)$

or $\quad P > 71.4$ N

Therefore, minimum value of P is **71.4 N**

141.

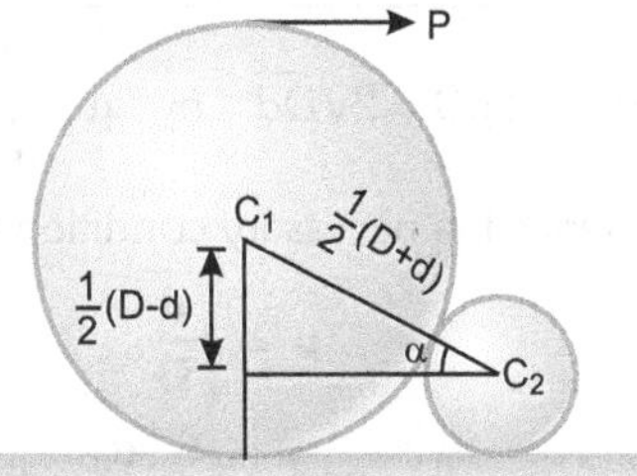

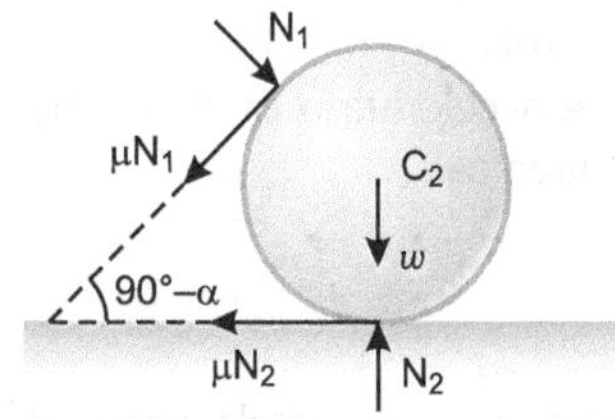

From the above two figures it is clear that

$$\sin\alpha = \left(\frac{D-d}{D+d}\right)$$

and $$\cos\alpha = \sqrt{1 - \sin^2\alpha} = \frac{2\sqrt{dD}}{D+d}$$

Let N_1 be the normal reaction between the two cylinders and N_2 the normal reaction between smaller cylinder and the horizontal plane.

The larger cylinder can be pulled over the smaller one provided the latter neither rolls nor slides.

There will be no rolling provided the sum of the moments of the forces about C_2 is zero.

$\therefore \quad \mu N_1 d = \mu N_2 d$

or $\quad N_1 = N_2 \quad \ldots(1)$

There will be no sliding provided the sum of the resolved parts of the forces in horizontal direction is zero.

$\therefore \quad \mu N_2 + \mu N_1 \cos(90° - \alpha) = N_1 \cos\alpha$

or $\quad \mu N_1 + \mu N_1 \sin\alpha = N_1 \cos\alpha \quad (N_2 = N_1)$

or $\quad \mu(1 + \sin\alpha) = \cos\alpha$

or $$\mu\left(1 + \frac{D-d}{D+d}\right) = \frac{2\sqrt{Dd}}{D+d}$$

or $\quad 2\mu D = 2\sqrt{Dd}$ or $\mu = \sqrt{\dfrac{d}{D}}$

Hence the necessary condition is that

$$\mu \geq \sqrt{\frac{\mathbf{d}}{\mathbf{D}}}$$

142. (i) $\omega = \dfrac{v_0}{r}$ and $\alpha = \dfrac{a_0}{r}$ (in case of pure rolling)

The acceleration of A can be written as :

$$\vec{a}_A = \vec{a}_o + \vec{a}_{A/O}$$

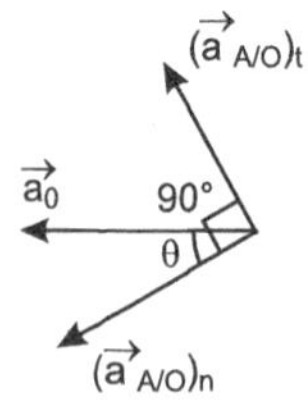

(a)

Here $\vec{a}_{A/O}$ = acceleration of A with respect to O.

or $\quad \vec{a}_A = \vec{a}_O + (\vec{a}_{A/O})_n + (\vec{a}_{A/O})_t$

Here $\quad |\vec{a}_o| = a_o$

$$|(\vec{a}_{A/O})_n| = r_0\,\omega^2 = r_0\left(\frac{v_0}{r}\right)^2$$

and $\quad |(\vec{a}_{A/O})_t| = r_0\,\alpha = r_0\left(\dfrac{a_o}{r}\right)$

Here $(\vec{a}_{A/O})_n$ will be along AO and $(\vec{a}_{A/O})_t$ will be perpendicular to OA

Hence $\vec{a}_o$, $(\vec{a}_{A/o})_n$ and $(\vec{a}_{A/o})_t$ are in the directions shown in figure (a).

$$\therefore |\vec{a}_A| = \sqrt{\left(a_0\cos\theta + \frac{r_0 v_0^2}{r^2}\right)^2 + \left(a_0\sin\theta + \frac{r_0 a_0}{r}\right)^2}$$

(ii) $\vec{a}_C = \vec{a}_o + \vec{a}_{C/O}$

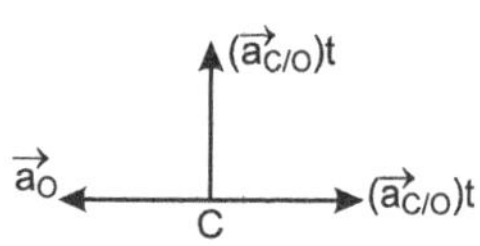

(b)

Here $(\vec{a}_{C/O})_n = r\omega^2$ in the direction of CO and

$(\vec{a}_{C/O})_t = r\alpha = a_o$ opposite to $\vec{a}_o$

Therefore $\vec{a}_0$ and $(\vec{a}_{C/O})_t$ are equal and opposite as shown in figure (b).

Therefore $\quad |\vec{a}_C| = r\omega^2 = r\dfrac{v_0^2}{r^2} = \dfrac{v_0^2}{r}$

143. (a) The sphere has no linear velocity when it starts rolling *i.e.*, it has only angular velocity say ω when pure rolling starts such that

$$\omega r = v_1 \quad \text{or} \quad \omega = \frac{v_1}{r}$$

Angular momentum about the bottommost point will remain conserved because summation of torque of all the forces including friction is zero about bottommost point. Hence

$$L_i = L_f \qquad (L = \text{angular mometum})$$

or $\quad mrv_0 = I\omega = \dfrac{2}{5}(mr^2)\dfrac{v_1}{r}$

$\Rightarrow \quad \mathbf{v_0 = \dfrac{2}{5} v_1}$

(b) Frictional force acting on the sphere is $\mu_k mg$ till sliding is present. Hence retardation of the sphere is

$$a = \frac{\mu_k mg}{m} = \mu_k g$$

Now linear velocity of the sphere has reduced to zero from initial value v_0. Hence

$$0 = v_0 - at_1$$

or $\quad t_1 = \dfrac{v_0}{a} = \dfrac{2v_1}{5\mu_k g}$

(c) From $\quad v^2 = u^2 - 2as$

We have $\quad 0 = v_0^2 - 2(\mu_k g)s$

or $\quad s = \dfrac{v_0^2}{2\mu_k g} = \dfrac{2v_1^2}{25\mu_k g}$

144. Writing the equations:

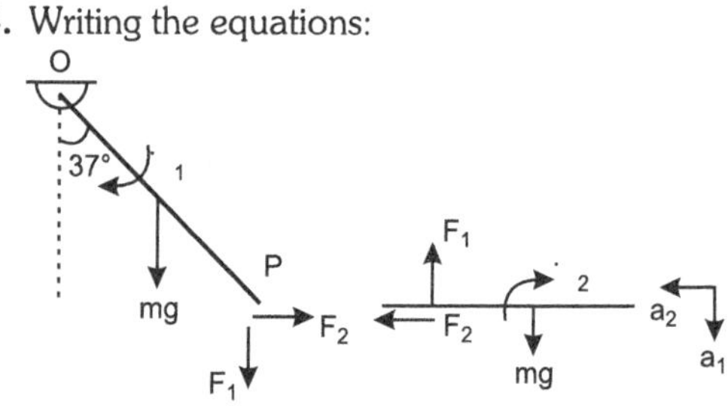

For rod 1:

$$mg\frac{l}{2}\sin 37° + F_1 l \sin 37° - F_2 l \cos 37° = \frac{ml^2}{3}\alpha_1 \quad ...(1)$$

For rod 2:

$$mg - F_1 = ma_1 \quad ...(2)$$

$$F_2 = ma_2 \quad ...(3)$$

$$F_1\frac{l}{2} = \frac{ml^2}{12}\alpha_2 \quad ...(4)$$

At point P, acceleration of both rods should be same.

$\therefore$ $$l\alpha_1 \cos 37° = a_2 \quad ...(5)$$

and $$l\alpha_1 \sin 37° = a_1 - \frac{l}{2}\alpha_2 \quad ...(6)$$

We have six unknown, F_1, F_2, α_1, α_2, a_1 and a_2.

Solving these six equations we get,

$$\alpha_1 = 0.423\frac{g}{l}$$

145. $$\vec{v}_A = 2v_0\hat{i}$$

$$\vec{v}_B = v_0\hat{i} - v_0\hat{j}$$

$\therefore$ $$\vec{v}_{AB} = \vec{v}_A - \vec{v}_B = \vec{v}_0\,\hat{i} + \vec{v}_0\,\hat{j}$$

$\therefore$ $$|\vec{v}_{AB}| = \sqrt{2}v_0$$

$$\tan\theta = 1 \Rightarrow \theta = 45°$$

with the positive x-axis.

146. $f = \mu mg$

t_1 = time when linear velocity is zero, if possible.

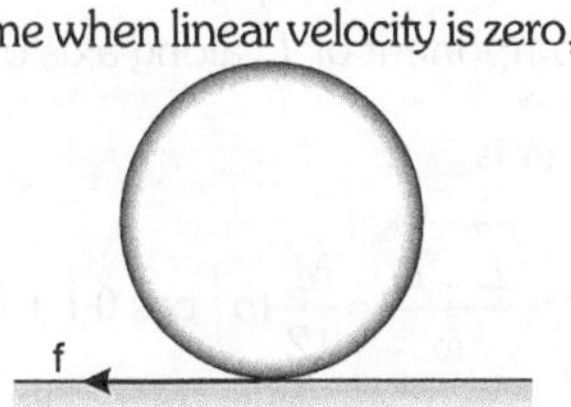

$$v_0 - \mu g t_1 = 0$$

$\therefore$ $$t_1 = \frac{v_0}{\mu g}$$

t_2 = time when angular velocity becomes zero

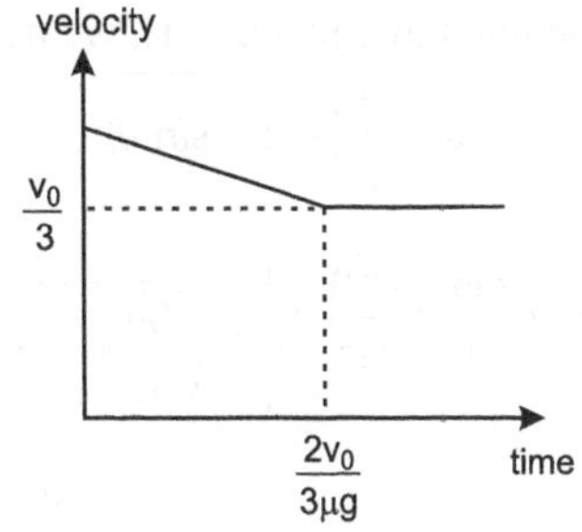

$$\alpha = \frac{\mu mgR}{\frac{1}{2}mR^2} = \frac{2\mu g}{R}$$

$\therefore$ $$\frac{v_0}{R} - \frac{2\mu g}{R}t_2 = 0$$

$\Rightarrow$ $$t_2 = \frac{v_0}{2\mu g}$$

$t_1 > t_2$ so it would not return.

For rolling, $v_t = \omega R$

$$v_0 - \mu g t = \left(-\frac{v_0}{R} + \frac{2\mu g}{R}t\right)R$$

$\therefore$ $$t = \frac{2v_0}{3\mu g}$$

$\therefore$ $$v_t = v_0 - \mu g t = v_0 - \mu g\frac{2v_0}{3\mu g} = \frac{v_0}{3}$$

147. Instantaneous axis of rotation passes through P.

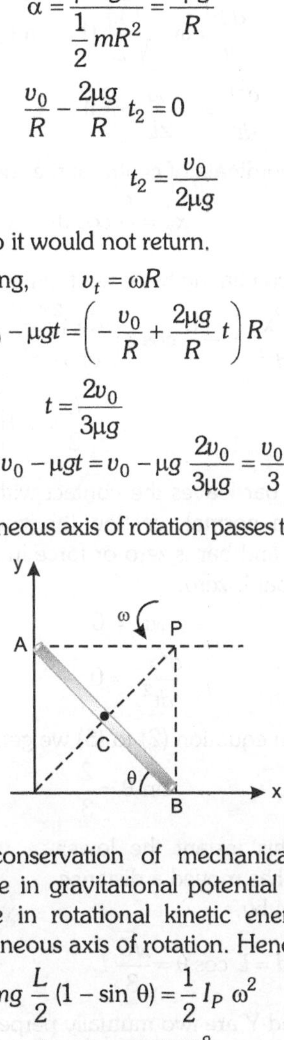

From conservation of mechanical energy, decrease in gravitational potential energy = increase in rotational kinetic energy about instantaneous axis of rotation. Hence,

$$mg\frac{L}{2}(1 - \sin\theta) = \frac{1}{2}I_P\,\omega^2 \quad ...(1)$$

Here, $$I_P = I_C + m(PC)^2$$

$$= \frac{mL^2}{12} + m\left(\frac{L}{2}\right)^2 = \frac{mL^2}{3}$$

Substituting in equation (1), we get,

$$\omega = \sqrt{\frac{3g}{L}(1 - \sin\theta)}$$

Further $\frac{d\omega}{dt} = \frac{d^2\theta}{dt^2} = \left\{\frac{1}{2\sqrt{\frac{3g}{L}(1-\sin\theta)}}\right\}$

$$\left\{\frac{-3g\cos\theta}{L}\right\} \cdot \frac{d\theta}{dt}$$

But $\frac{d\theta}{dt} = \omega = \sqrt{\frac{3g}{L}(1-\sin\theta)}$ ...(2)

$\therefore \quad \frac{d^2\theta}{dt^2} = -\frac{3g}{2L}\cos\theta$ ...(3)

x-coordinate of centre of the rod,

$$x_c = \frac{L}{2}\cos\theta$$

Differentiating twice w.r.t. time, we get

$$\frac{d^2x_c}{dt^2} = -\frac{L}{2}\cos\theta\left(\frac{d\theta}{dt}\right)^2$$

$$-\frac{L}{2}\sin\theta\left(\frac{d^2\theta}{dt^2}\right) \quad ...(4)$$

The bar leaves the contact with vertical wall when normal reaction (horizontal) between wall and bar is zero or force in x-direction of the bar is zero.

$\therefore \quad a_x = 0$

or $\quad \frac{d^2x_c}{dt^2} = 0$ ...(5)

From equation (2) to (5) we get,

$$\sin\theta = \frac{2}{3}$$

At this instant the lower end has moved a distance given by,

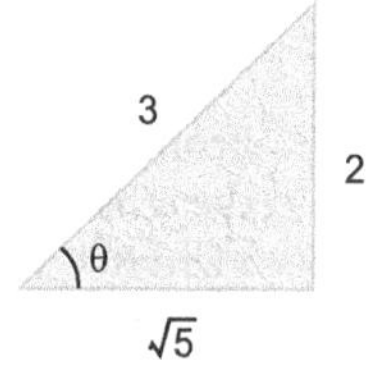

$$d = L\cos\theta = \frac{\sqrt{5}}{3}L$$

148. X and Y are two mutually perpendicular axes lying on the plane of lamina and passing through centre of mass C and z be the axis perpendicular to the lamina and passing through C.

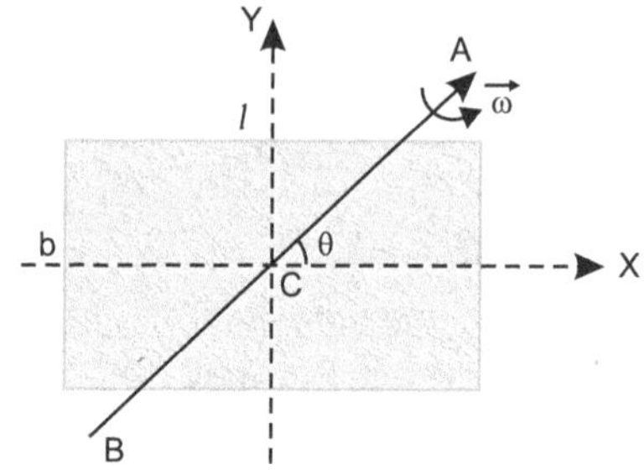

$$I_x = \frac{Mb^2}{12}, \quad I_y = \frac{Ml^2}{12},$$

Hence $\quad I_z = I_x + I_y = \frac{M}{12}(l^2 + b^2)$

$$L_x = I_x\omega_x = \frac{Mb^2}{12}\omega\cos\theta$$

$$L_y = I_y\omega_y = \frac{Ml^2}{12}\omega\sin\theta$$

and $\quad L_z = I_z\omega_z = 0 \quad (\omega_z = 0)$

$\therefore$ Total angular momentum of the lamina is

$$\vec{L} = L_x\,\hat{i} + L_y\,\hat{j} + L_z\,\hat{k}$$

$$\boldsymbol{\vec{L} = \frac{M\omega}{12}(b^2\cos\theta\,\hat{i} + l^2\sin\theta\,\hat{j})}$$

This is the angular momentum of the lamina about C.

Note that $\vec{L}$ is not along $\vec{\omega}$

Component of $\vec{L}$ along axis of rotation :

The component of $\vec{L}$ along axis of rotation or along $\vec{\omega}$ is

$$L_{ACB} = \frac{\vec{L}\cdot\vec{\omega}}{\omega} = \frac{M}{12}(b^2\cos\theta\,\hat{i} + l^2\sin\theta\,\hat{j})$$

$$(\omega\cos\theta\,\hat{i} + \omega\sin\theta\,\hat{j})$$

$$= \frac{M\omega}{12}(b^2\cos^2\theta + l^2\sin^2\theta)$$

$$L_{ACB} = (I_x\cos^2\theta + I_y\sin^2\theta)\,\omega = I_{ACB}\,\omega$$

$$(I_{ACB} = I_x\cos^2\theta + I_y\sin^2\theta)$$

149. Drawing free body diagram of rod immediately after the string breaks.

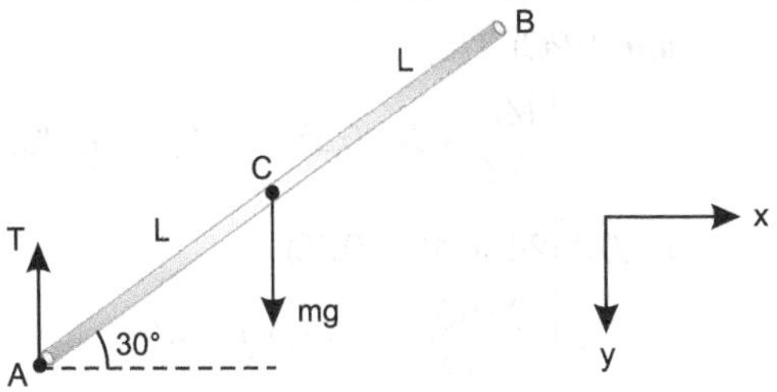

Let a_x and a_y be the linear accelerations of COM and α the angular acceleration of the rod about COM as shown below.

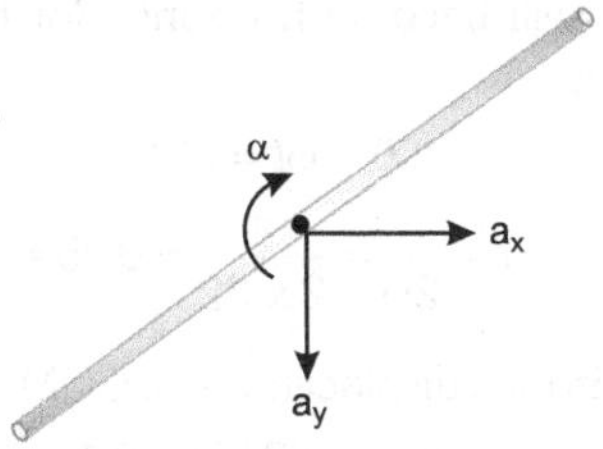

Then $\quad \Sigma F_x = 0$

$\therefore \quad a_x = 0 \quad \ldots(1)$

$$a_y = \frac{\Sigma F_y}{m} = \frac{mg - T}{m} \quad \ldots(2)$$

$$\alpha = \frac{\tau}{I} = \frac{TL \cos 30^\circ}{\frac{m\,(2L)^2}{12}} = \frac{3\sqrt{3}T}{2mL} \quad \ldots(3)$$

Now just after the string breaks, acceleration of point A in vertical direction should be zero.

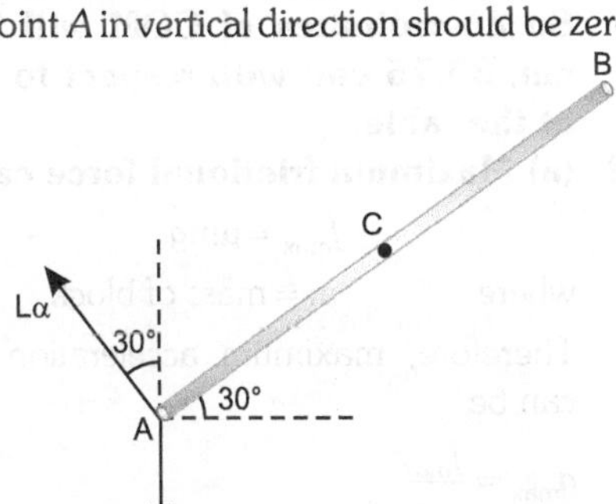

i.e., $\quad a_y = L\alpha \cos 30^\circ \quad \ldots(4)$

Solving equations (1), (2), (3) and (4) we get

$$\boldsymbol{T = \frac{2mg}{11}} \quad \textbf{and} \quad \boldsymbol{\alpha = \frac{3\sqrt{3}}{11}\frac{g}{L}}$$

150. (i) Applying the conservation of mechanical energy between the vertical position (1) and the horizontal position (2)

$$(KE + PE)_1 = (KE + PE)_2$$

$$0 + mg\frac{L}{2} = \frac{1}{2} I_A \omega^2$$

or $\quad mg\dfrac{L}{2} = \dfrac{1}{2}\left(\dfrac{mL^2}{3}\right)\omega^2 \quad$ or $\quad \omega = \sqrt{\dfrac{3g}{L}}$

Considering the free body diagram of the bar at this instant.

Velocity of COM of bar at this instant is

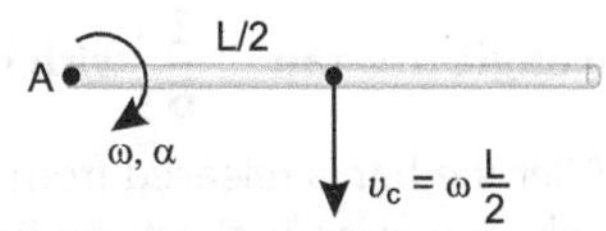

$$v_c = \omega\frac{L}{2} = \frac{\sqrt{3gL}}{2}$$

Let α be the angular acceleration of bar about A at this instant. Then

$$\alpha = \frac{\tau_A}{I_A} = \frac{mg\frac{L}{2}}{\frac{mL^2}{3}} = \frac{3}{2}\frac{g}{L}$$

Now N_x = centripetal force of COM

or $\quad N_x = \dfrac{mv_c^2}{L/2} = \dfrac{3mg}{2}$

and $\quad mg - N_y = ma_y \quad \ldots(1)$

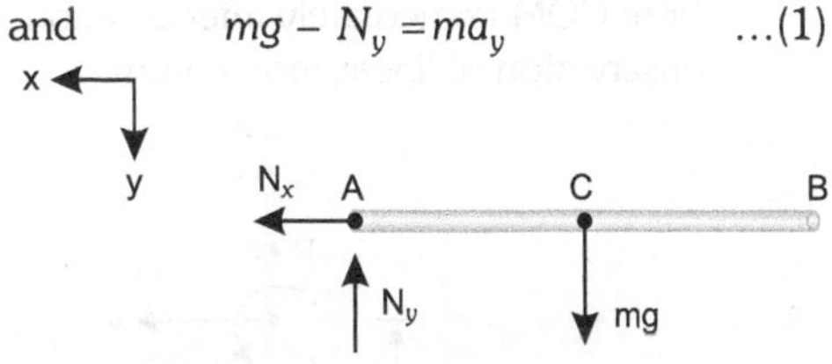

a_y = acceleration of COM in y-direction $= \dfrac{L}{2}\alpha$

or $\quad N_y = mg - m\left(\dfrac{L}{2}\alpha\right)$

$$= mg - \frac{3mg}{4} = \frac{mg}{4} \quad \ldots(2)$$

Therefore, resultant reaction at support A is

$$N = \sqrt{N_x^2 + N_y^2}$$

$$N = \sqrt{\left(\frac{3mg}{2}\right)^2 + \left(\frac{mg}{4}\right)^2} = \frac{\sqrt{37}}{4} mg$$

$$\tan \alpha = \frac{N_y}{N_x}$$

$$= \frac{mg/4}{3mg/2} = \frac{1}{6}$$

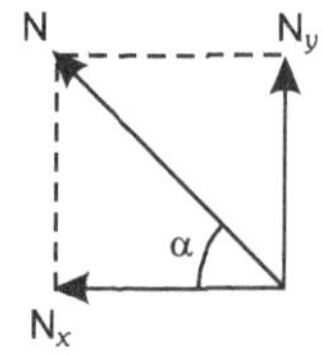

or $\alpha = \tan^{-1}\left(\frac{1}{6}\right)$

Therefore, reaction at pivot is $\frac{\sqrt{37}}{4} mg$ at an angle $\alpha = \tan^{-1}\left(\frac{1}{6}\right)$ with horizontal.

(ii) After the bar is released from the support only one force 'mg' acts on its COM. The angular acceleration thus vanishes and the angular velocity becomes constant. Hence

$$\omega(t) = \sqrt{3\,g/L} \quad \text{and} \quad v_c(t) = \frac{\sqrt{3\,gL}}{2} + gt$$

151. The centre of mass C of the combined system is located at a distance d from the lower end B of the rod such that

$$BC = d = \frac{0.6 \times 30 + 0.3 \times 45}{0.6 + 0.3} = 35 \text{ cm}$$

Let v be the linear velocity (in horizontal direction) of COM and ω the angular velocity about COM immediately after collision. From conservation of linear momentum

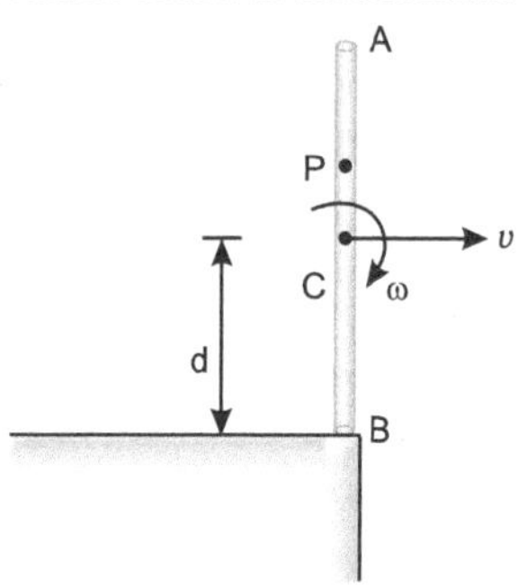

$$mv_0 = (M + m)\,v$$

or $$v = \frac{mv_0}{M+m} = \frac{(0.3)(24)}{0.6 + 0.3} = 8 \text{ m/s} \quad \ldots(1)$$

Similarly applying conservation of angular momentum about COM we have

$$(mv_0)(PC) = I\omega$$

$$= \left[\frac{Ml^2}{12} + M\,(0.35 - 0.30)^2 + m(PC)^2\right]\omega$$

or $$(0.3)(24)(0.45 - 0.35) = \left[\frac{(0.6)(0.6)^2}{12} + (0.6)\,(0.05)^2 + 0.3\,(0.45 - 0.35)^2\right]\omega$$

i.e., $\omega = 32$ rad/s $\quad \ldots(2)$

Rod will become horizontal for the first time when

$$\theta = \omega t = \pi/2$$

or $$t = \frac{\pi}{2\omega} = \frac{\pi}{2 \times 32} \approx 0.05 \text{ s}$$

Horizontal displacement of COM in this time is

$$x = v\,t = (8)(0.05) = 0.4 \text{ m}$$

or $x = 40$ cm

Vertical displacement of COM is

$$y = \frac{1}{2} gt^2 = \frac{1}{2}(10)(0.05)^2 = 0.0125 \text{ m}$$

or $y = 1.25$ cm

Initially COM was at a height of 35 cm from the table. So after $t = 0.05$ s it will be at a height of $(35 - 1.25) = 33.75$ cm from the table.

So co-ordinates of COM will be (40.0 cm, 33.75 cm) with respect to the edge of the table.

152. (a) Maximum frictional force can be

$$f_{max} = \mu mg$$

where m = mass of block

Therefore, maximum acceleration of block can be

$$a_{max} = \frac{f_{max}}{m}$$

$$= \frac{\mu mg}{m} = \mu g$$

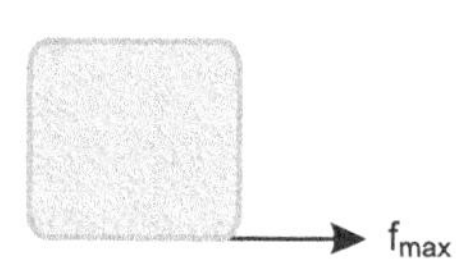

$$= (0.5)\,(9.8)$$

$$= \mathbf{4.9\ m/s^2}$$

If acceleration of car, exceeds this value slipping will start between the block and the car.

(b) In the critical case when the block is about to topple about A, normal reaction will pass through A.

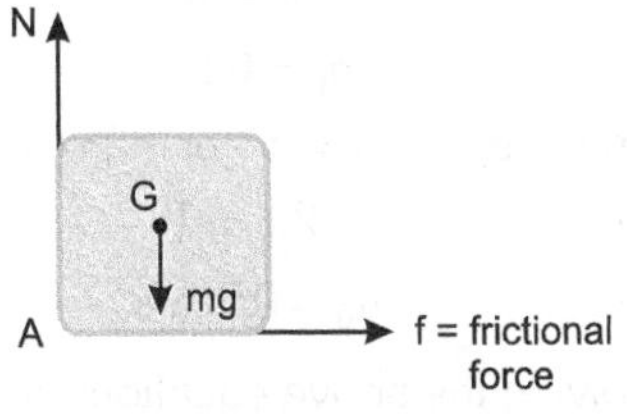

$$f = ma \qquad ...(1)$$

$$N = mg \qquad ...(2)$$

The block topples when torque of f about $G >$ torque of N about G.

Taking the limiting value

$$\tau_f = \tau_N$$

or $$(f)(h) = (N)(b)$$

or $$(ma)(h) = (mg)(b)$$

or $$a = \left(\frac{b}{h}\right)g = \left(\frac{0.6}{0.9}\right)(9.8)$$

$$= \mathbf{6.53\ m/s^2}$$

So, if the acceleration of car exceeds this value, the block will topple about A.

(c) The maximum acceleration/retardation at which the block is not disturbed, is the smaller of the two values, obtained above *i.e.*, 4.9 m/s^2

Hence the maximum retardation can be 4.9 m/s^2

$$u = 72\frac{\text{km}}{\text{hr}} = 20\text{ m/s}$$

So applying

$$v^2 = u^2 - 2as \qquad (v = 0)$$

$$s = \frac{u^2}{2a} \qquad (a = 4.9\text{ m/s}^2)$$

$$= \frac{(20)^2}{(2)(4.9)}$$

or $$\mathbf{s = 40.82\ m}$$

153. Drawing free body diagrams of all the three (block, plank and mass)

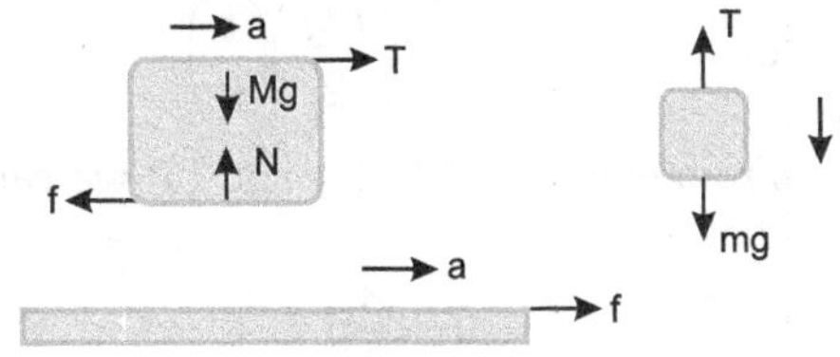

Since we are taking the limiting case. Therefore,

f = maximum friction between the block and the plank.

or $$f = \mu N = \mu Mg = \mu\ (4)\ g$$

$$f = 4\ \mu g \qquad ...(1)$$

Now writing equations of motion for all three

$$mg - T = ma \qquad (m = 1\text{ kg})$$

or $$g - T = a \qquad ...(2)$$

$$T - f = Ma \qquad (M = 4\text{ kg})$$

or $$T - 4\mu g = 4a \qquad ...(3)$$

$$f = Ma \text{ or } f = 4a \text{ or } 4\mu g = 4a$$

or $$a = \mu g \qquad ...(4)$$

Solving equations (2), (3) and (4) for μ, we get

$$\mu = \frac{1}{9}$$

Therefore, minimum value of μ should be $\frac{1}{9}$ or

$$\mu \geq \frac{1}{9}$$

(b) At the time of toppling, normal reaction N will pass through the front edge as shown below

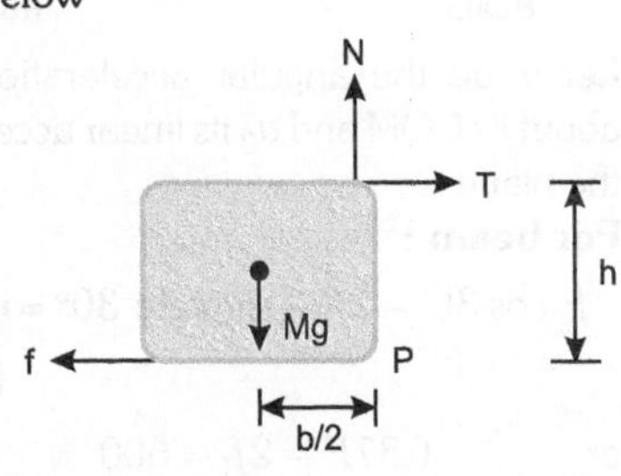

So, torque of T about $P \leq$ torque of Mg about P

or $$T \cdot h \le (Mg)\left(\frac{b}{2}\right)$$

or $$T \le \left(\frac{Mg}{2}\right)\left(\frac{b}{h}\right)$$

From equations (2), (3) and (4) we can find that

$$T = \frac{8}{9}g$$

Therefore, $$\frac{8}{9}g \le \left(\frac{Mg}{2}\right)\left(\frac{b}{h}\right)$$

or $$\frac{b}{h} \ge \frac{16}{9M} \qquad (M = 4\text{ kg})$$

or $$\left(\frac{b}{h}\right) \ge \frac{4}{9}$$

Therefore, minimum value of $\frac{b}{h}$ so that the block does not topple is $\frac{4}{9}$.

154. By symmetry frictional force will be same for both the rollers. Let f_1 and f_2 be the frictional forces between beam and rollers and between rolleres and plane. Drawing free body diagram of beam and any one of the roller.

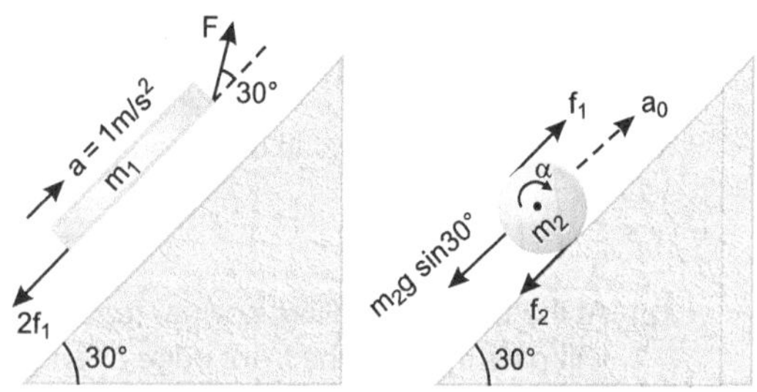

Beam Roller

Let α be the angular acceleration of roller about its COM and a_0 its linear acceleration up the plane.

For beam :

$$F\cos 30° - 2f_1 - m_1 g\sin 30° = m_1 a \qquad (a = 1\text{ m/s}^2)$$

or $$0.87F - 2f_1 = 600 \qquad \ldots(1)$$

For roller :

$$f_1 - f_2 - m_2 g\sin 30° = m_2 a_0$$

or $$f_1 - f_2 - 100 = 20a_0 \qquad \ldots(2)$$

$$\alpha = \frac{\tau}{I} = \frac{(f_1 + f_2)R}{\frac{1}{2}m_2R^2} = \frac{(f_1 + f_2)(0.1)}{\frac{1}{2}(20)(0.1)^2}$$

or $$\alpha = f_1 + f_2 \qquad \ldots(3)$$

As there is no slipping at contacts. Therefore,

$$a_0 = R\alpha$$

or $$a_0 = 0.1\,\alpha \qquad \ldots(4)$$

and $a_0 + R\alpha = a$ or $2a_0 = a$

or $$2a_0 = 1$$

or $$a_0 = 0.5\text{ m/s}^2 \qquad \ldots(5)$$

Solving the above equations we get

$$\boldsymbol{F = 821.84\text{ N}}$$

155. Let a = the linear acceleration of the system down the plane.

α = angular acceleration of each body $= \frac{a}{R}$

F = force in the frame

f_1 = frictional force between sphere and plane

f_2 = frictional force between cylinder and plane

$W = mg = 200$ N = weight of each body

I_1 = moment of inertia of sphere about centre

$$= \frac{2}{5}mR^2 = 8R^2 \qquad (m = 20\text{ kg})$$

I_2 = moment of inertia of cylinder about centre

$$= \frac{1}{2}mR^2 = 10R^2 \qquad (m = 20\text{ kg})$$

Free body diagrams of sphere and cylinder are shown below

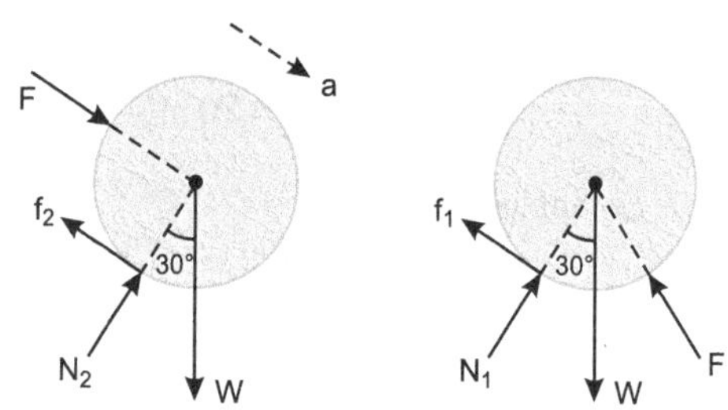

Cylinder Sphere

For sphere :

$$\tau_1 = I_1\alpha$$

or $f_1R = (8R^2)(\alpha) = (8R^2)\left(\frac{a}{R}\right) = 8aR$

or $$f_1 = 8a \quad ...(1)$$

$$W \sin 30° - F - f_1 = ma$$

or $$100 - F - f_1 = 20a \quad ...(2)$$

For cylinder :

$$\tau_2 = I_2\alpha \quad \text{or} \quad f_2R = (10R^2)\left(\frac{a}{R}\right)$$

or $$f_2 = 10a \quad (3)$$

$$W \sin 30° + F - f_2 = ma$$

or $$100 + F - f_2 = 20a \quad ...(4)$$

Solving equations (1), (2), (3) and (4) we get

$$\mathbf{F = 3.45\ N}$$

156. Acceleration of truck $a_0 = 1\ m/s^2$ (towards right)

Let a be the linear acceleration of centre of mass of cylinder (towards left) with respect to truck and α be its angular acceleration (anticlockwise) about its centre of mass.

Drawing free body diagram of cylinder with respect to truck.

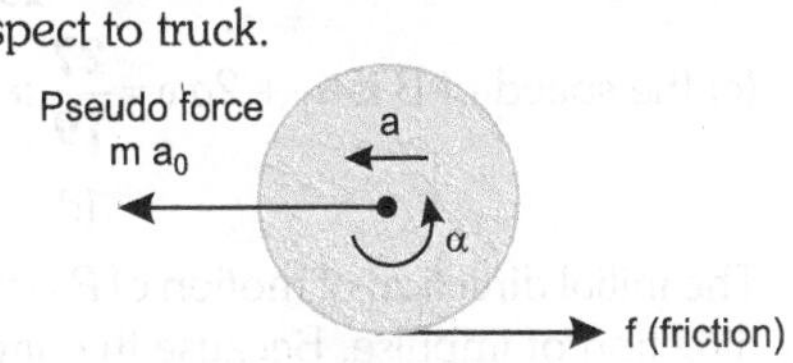

$$ma_0 - f = ma$$

$$\therefore \quad a = a_0 - \frac{f}{m} \quad ...(1)$$

$$\alpha = \frac{\tau}{I} = \frac{fR}{\frac{1}{2}mR^2} = \frac{2f}{mR} \quad ...(2)$$

Since cylinder rolls without slipping on the truck, we have

$$a = R\alpha \quad ...(3)$$

Solving equations (1), (2) and (3) we get

$$a = \frac{2}{3}a_0 = \frac{2}{3}(1) = \frac{2}{3}\ m/s^2$$

with above acceleration, cylinder covers a distance $s = 4$ m on the truck in time t given by

or $$s = \frac{1}{2}at^2 = \frac{1}{2}\left(\frac{2}{3}\right)t^2$$

or $$4 = t^2/3 \quad (s = 4\ m)$$

or $$t = \sqrt{12} = 2\sqrt{3}\ s$$

The linear velocity of the cylinder relative to truck is

$$v_r = at = \left(\frac{2}{3}\right)2\sqrt{3} = \frac{4}{\sqrt{3}}\ m/s \quad \text{(leftwards)}$$

In the same time, truck has acquired a forward velocity.

$$v_T = a_0t = (1)\,2\sqrt{3} = 2\sqrt{3}\ m/s \quad \text{(rightwards)}$$

Hence at the moment cylinder leaves the truck, it is moving with a linear velocity

$$v = v_r - v_T = \frac{4}{\sqrt{3}} - 2\sqrt{3} = -\frac{2}{\sqrt{3}}\ m/s$$

or $$v = \frac{2}{\sqrt{3}}\ m/s \quad \text{(rightwards)}$$

and its angular velocity is

$$\omega = \frac{v_r}{R} = \frac{4}{\sqrt{3} \times 0.5} = \frac{8}{\sqrt{3}}\ rad/s \quad \text{(anticlockwise)}$$

Once the cylinder leaves the truck, there is no external torque on it about centre of mass axis of rotation. Hence its angular velocity and linear horizontal velocity do not change. Therefore, when it touches the ground, it does not perform pure rolling.

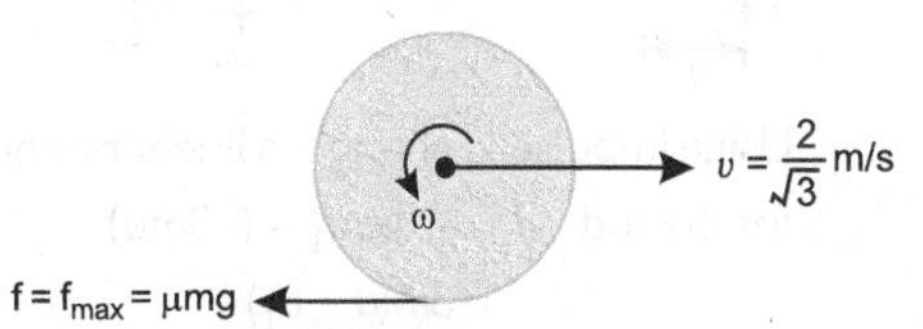

$$\omega = \frac{8}{\sqrt{3}}\ rad/s$$

Friction acts backward.

Linear retardation

$$a' = \frac{f}{m} = \mu g = (0.40\,(10) = 4\ m/s^2$$

and angular retardation

$$\alpha' = \frac{\tau}{I} = \frac{\mu\, mg\, R}{\frac{1}{2}mR^2} = \frac{2\mu g}{R} = \frac{2 \times 4}{0.5} = 16\ rad/s^2$$

Linear velocity becomes zero in time t_1 given by:

$$t_1 = \frac{v}{a'} = \frac{\frac{2}{\sqrt{3}}}{4} = \frac{1}{2\sqrt{3}} \text{ s}$$

and angular velocity becomes zero in time t_2 given by :

$$t_2 = \frac{\omega}{\alpha'} = \frac{\frac{8}{\sqrt{3}}}{16} = \frac{1}{2\sqrt{3}} \text{ s}$$

Since $t_1 = t_2$

i.e., linear velocity and angular velocity become zero simultaneously. Hence the cylinder stops in time $t = \frac{1}{2\sqrt{3}}$ s.

Distance (d) moved by cylinder during this time is :

$$d = \frac{v^2}{2a'} = \frac{\left(\frac{2}{\sqrt{3}}\right)^2}{2\,(4)} = \mathbf{\frac{1}{6}\ m}$$

157. Immediately after impact, let the speed of rod be u_1, the speed of the particle be u_2 (both downwards) and the angular velocity of the rod about G, the centre of mass is ω.

$\vec{J}$ = impulse = $\vec{F}.\Delta t$

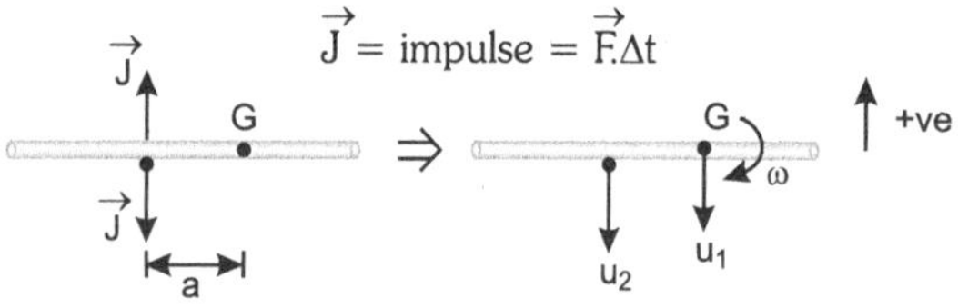

Using impulse = change in linear momentum

for the rod $J = -3mu_1 - (-3mu)$

$$= 3m(u - u_1) \qquad \ldots(1)$$

for the particle

$$-J = -mu_2 - (mu)$$

or $J = m\,(u + u_2) \qquad \ldots(2)$

Using, angular impulse = change in angular momentum of rod

$$J \cdot a = I_G\,\omega = \frac{1}{12}(3m)(4a)^2 \cdot \omega$$

$$= 4ma^2\omega \qquad \ldots(3)$$

In order to use the law of restitution. We need the speed of point C, which is $u_1 - a\omega$ (downwards)

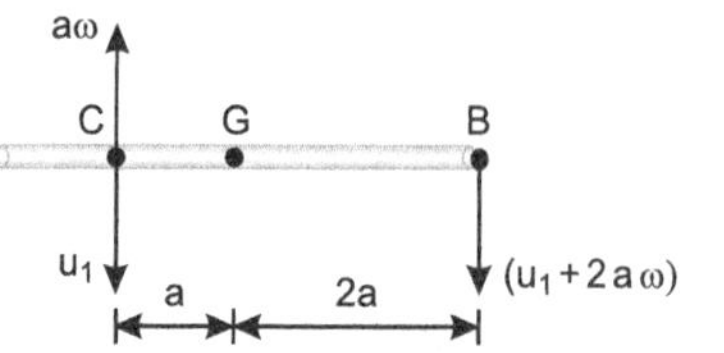

The law of restitution now gives

Relative velocity of separation at point of impact = e (relative velocity of approach)

or $u_2 - (u_1 - a\omega) = e\,(u + u)$

But $e = 1$. Hence

$$2u = u_2 - u_1 + a\omega \qquad \ldots(4)$$

Solving equations (1), (2), (3) and (4) we get

$$u_1 = \frac{3}{19}u,\ u_2 = \frac{29}{19}u, \text{ and } \omega = \frac{12}{19}\frac{u}{a}$$

Therefore,

(a) the speed of the particle is $\mathbf{\frac{29}{19}u}$ (downwards)

(b) the angular speed of the rod is $\mathbf{\frac{12}{19}\frac{u}{a}}$

(c) the speed of B is $u_1 + 2a\omega = \mathbf{\frac{27}{19}u}$ (downwards)

158. The initial direction of motion of P is not in the direction of impulse. Because the direction of impulse is not known. We use two unknown components J_1 and J_2, and an unknown initial speed u for point P.

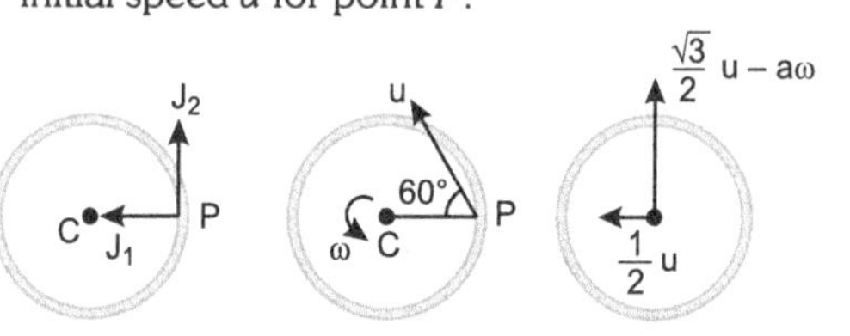

For the motion of the centre of mass :

$$J_1 = m\,(u \cos 60°) = \frac{mu}{2} \qquad \ldots(1)$$

$$J_2 = m\,(u \sin 60° - a\omega)$$

$$= m\left(\frac{\sqrt{3}u}{2} - a\omega\right) \qquad \ldots(2)$$

For the initial rotation

$$J_2 a = I_C \omega = ma^2\omega \qquad \ldots(3)$$

Also given

$$J_1^2 + J_2^2 = (\sqrt{7}\, mv)^2 \qquad \ldots(4)$$

Solving equation (1), (2), (3) and (4) we get

$$u = 4v$$

Therefore, the speed of P is $4v$.

159. Let ω be the angular velocity of cube just after collision about an axis passing through D. From conservation of angular momentum about an axis passing through D

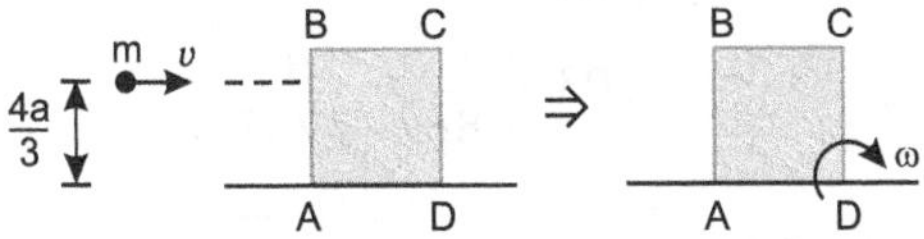

$$mv\left(\frac{4a}{3}\right) = I_D \cdot \omega$$

Here $\qquad I_D = I_{com} + mr^2$

where r = perpendicular distance of axis of rotation passing through D from centre of mass (com) of the cube $= \sqrt{2}\, a$

or $$\frac{4mva}{3} = \left[M\left(\frac{4a^2+4a^2}{12}\right) + M(2a^2)\right]\omega$$

or $$\frac{4mva}{3} = \frac{8}{3}Ma^2 \cdot \omega$$

or $$\omega = \frac{1}{2}\frac{mv}{Ma} \qquad \ldots(1)$$

The cube will topple if its COM is just able to reach in a vertical height h_2 as shown in figure

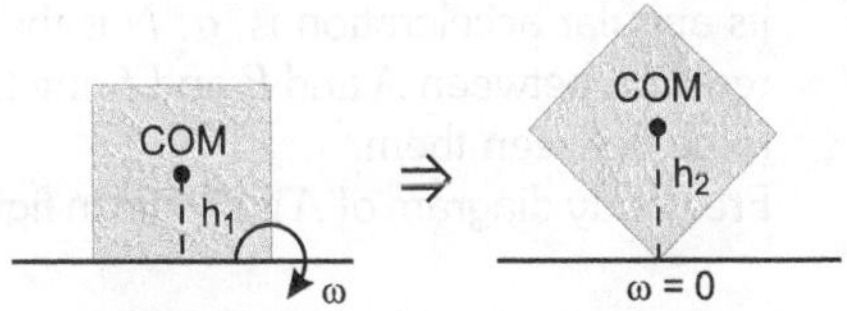

Fig. (a) **Fig. (b)**

(b).

$$h_1 = a \quad \text{and} \quad h_2 = \sqrt{2}\, a$$

Hence, applying conservation of mechanical energy

$$\frac{1}{2} I_D \omega^2 = Mg(h_2 - h_1)$$

or $$\frac{1}{2}\left(\frac{8}{3}Ma^2\right)\cdot\left\{\frac{1}{2}\frac{mv}{Ma}\right\}^2 = Mg(\sqrt{2}-1)\,a$$

$$\left(I_D = \frac{8}{3}Ma^2\right)$$

or $$\frac{1}{3}\left(\frac{m}{M}\right)^2 v^2 = ag(\sqrt{2}-1)$$

or $$\boldsymbol{v = \frac{M}{m}[3ag(\sqrt{2}-1)]^{\frac{1}{2}}}$$

160. $OC = \dfrac{2r}{\pi}, \quad OP = r$

Hence $\quad CP = \sqrt{(OC)^2 + (OP)^2}$

or $\qquad CP = 1.185r$

If there is no slipping, then instantaneous axis of rotation passes through P.

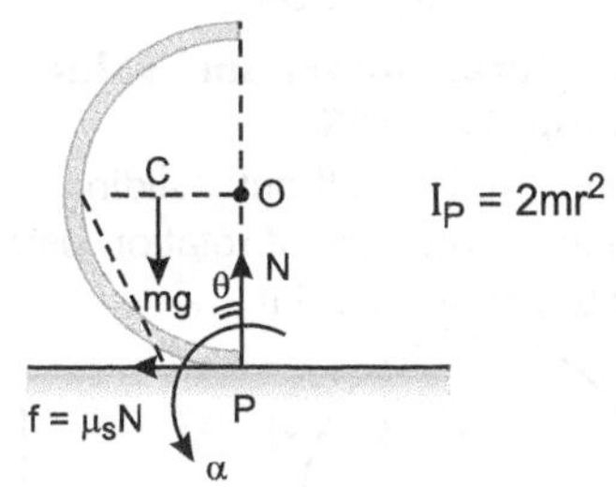

Angular acceleration of ring about P is :

$$\alpha = \frac{\tau_P}{I_P} = \frac{(mg)(CO)}{(2mr^2)}$$

$$= \frac{(mg)\left(\frac{2r}{\pi}\right)}{(2mr^2)} = \frac{g}{\pi r} \qquad \ldots(1)$$

Acceleration of centre of mass

$$a = (CP)(\alpha)$$

$$= (1.185\,r)\left(\frac{g}{\pi r}\right) = \left(\frac{1.185}{\pi}\right) g \qquad \ldots(2)$$

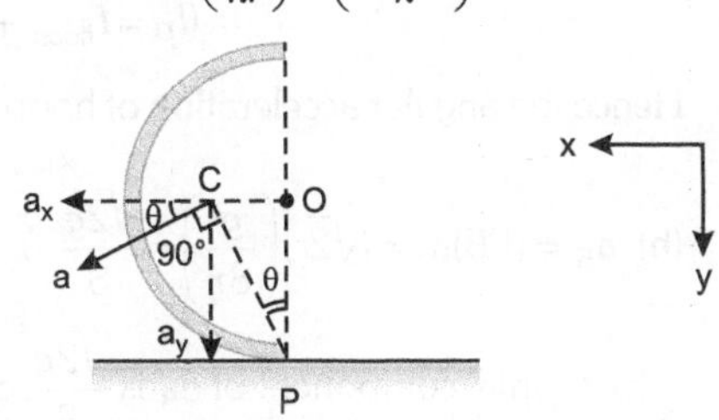

From Newton's law

$$\Sigma Fy = ma_y$$

or $\quad mg - N = m\, a_y$

or $\quad mg - N = m\, a \sin\theta$

or $\quad mg - N = m\left(\dfrac{1.185}{\pi}\right)g\left(\dfrac{2r}{\pi}\right)\left(\dfrac{1}{1.185\,r}\right)$

$$= \frac{2}{\pi^2} mg = 0.2\, mg$$

or $\quad N = mg - 0.2\, mg$

or $\quad N = 0.8\, mg \qquad \ldots(3)$

$$\Sigma F_x = ma_x$$

or $\quad \mu_s N = (m)\,(a\cos\theta)$

or $\quad \mu_s\,(0.8\, mg) = (m)\left(\dfrac{1.185\, g}{\pi}\right)\left(\dfrac{r}{1.185\, r}\right)$

or $\quad \mu_s = 0.398$

Therefore, minimum value of static friction is 0.398.

161. Hoop rolls without sliding. Therefore, instantaneous axis of rotation passes through the bottommost point.

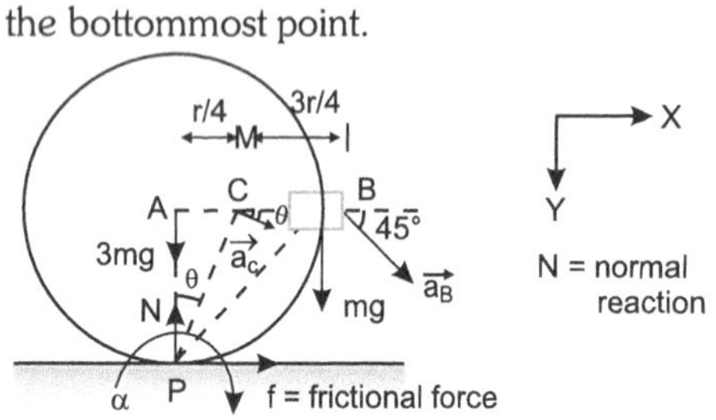

C is the centre of mass of the hoop and the clamp

where $\quad AC = \dfrac{r}{4}$ and $CB = \dfrac{3r}{4}$

(a) $\alpha = \dfrac{\tau_P}{I_P} = \dfrac{mgr}{6mr^2 + 2mr^2}$ or $\alpha = \dfrac{\boldsymbol{g}}{\boldsymbol{8r}}$

$$(I_P = I_{hoop} + I_{block})$$

Hence the angular acceleration of hoop is $\dfrac{g}{8r}$

(b) $a_B = (PB)\alpha = (\sqrt{2}r)\left(\dfrac{g}{8r}\right) = \dfrac{\sqrt{2}g}{8}$

$\therefore$ Horizontal component of $\vec{a}_B$ is $\dfrac{\sqrt{2}g}{8}\cos 45°$

or $\dfrac{\boldsymbol{g}}{\boldsymbol{8}} \rightarrow$ and vertical component of $\vec{a}_B$ is $\dfrac{\sqrt{2}g}{8}\sin 45°$ or $\dfrac{\boldsymbol{g}}{\boldsymbol{8}}\downarrow$

(c) $\tan\theta = \dfrac{AC}{AP} = \dfrac{r/4}{r} = \dfrac{1}{4}$

or $\quad \theta = 14°$

and $\quad PC = \sqrt{r^2 + \dfrac{r^2}{16}} = 1.03\, r$

Acceleration of centre of mass C is

$$a_C = (PC)\alpha$$

$$= (1.03r)\left(\frac{g}{8r}\right) = \frac{g}{7.767}$$

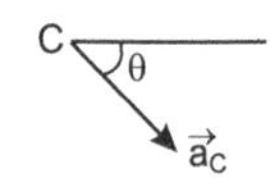

Horizontal component of $\vec{a}_C$ is

$$a_x = a_C\cos\theta = \frac{g}{(7.767)}\cos 14° = 0.125\, g \rightarrow$$

$\therefore \qquad \boldsymbol{f = (4m)\, a_x = 0.5\, mg} \rightarrow$

Similarly vertical component of $\vec{a}_C$ is

$$a_y = a_C\sin\theta = \frac{g}{7.767}\sin 14° = 0.03g\downarrow$$

Now $\quad 4\, mg - N = (4m)a_y$

Hence $\quad N = 4mg - 4ma_y$

$$N = 4m\,(g - a_y)$$

$$N = 4m\,(g - 0.03g)$$

$$\boldsymbol{N = 3.88\, mg}\uparrow$$

162. Let acceleration of A is a (down the plane), linear acceleration of B relative to A is a_r and its angular acceleration is α. N is the normal reaction between A and B and f the frictional force between them.

Free body diagram of A is shown in figure (a).

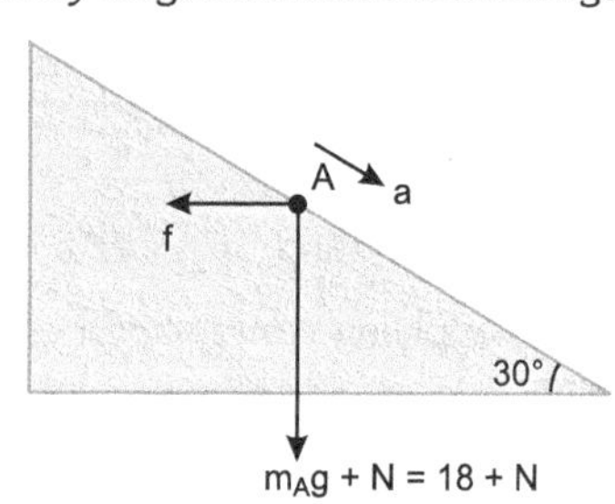

(a)

$$(18 + N) \sin 30° - f \cos 30° = m_A a$$

or $\quad 0.5\,N - 0.87 f + 9 = 1.8a \quad \ldots(1)$

Free body diagram of cylinder relative to A is shown in figure (b).

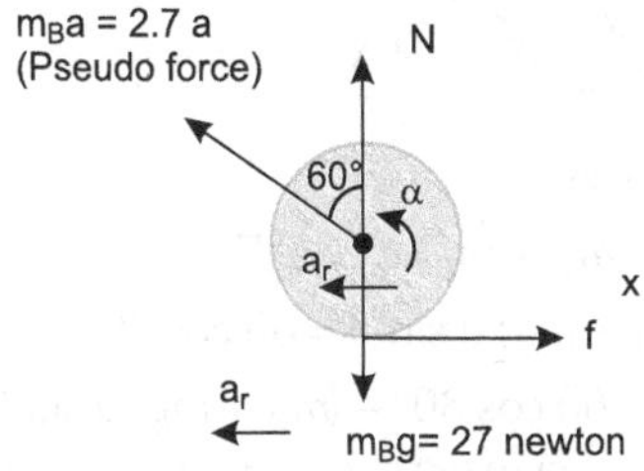

(b)

$$\Sigma F_x = m_B a_r$$

$\therefore \quad 2.7a \sin 60° - f = m_B a_r$

or $\quad 2.34\,a - f = 2.7 a_r \quad \ldots(2)$

$$\Sigma F_y = 0$$

$\therefore \quad N + 2.7a \cos 60° = 27$

or $\quad N + 1.35a = 27 \quad \ldots(3)$

$$\alpha = \frac{\tau}{I} = \frac{f \cdot r}{\frac{1}{2} m_B r^2} = \frac{2f}{m_B r}$$

substituting the values of m_B and r, we get

$$\alpha = 9.75 f \quad \ldots(4)$$

for no sliding $\quad a_r = r\alpha$

or $\quad a_r = 0.076\,\alpha \quad \ldots(5)$

Solving equations (1), (2), (3), (4) and (5), we get

(a) Acceleration of the wedge

$$\boldsymbol{a = 7.1 \text{ m/s}^2}$$

(b) Angular acceleration of the cylinder

$$\boldsymbol{\alpha = 53.9 \text{ rad/s}^2}$$

163. The forces during impact are shown in the figure :

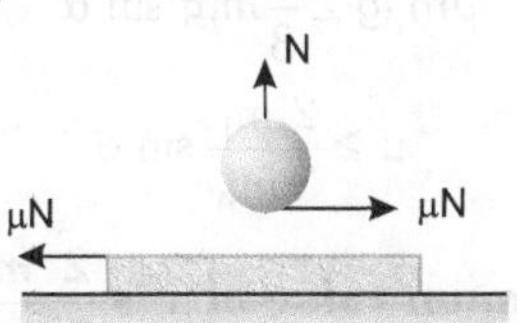

Let the horizontal velocities of the ball and the plank be v_1 and v_2 in opposite directions as shown in figure :

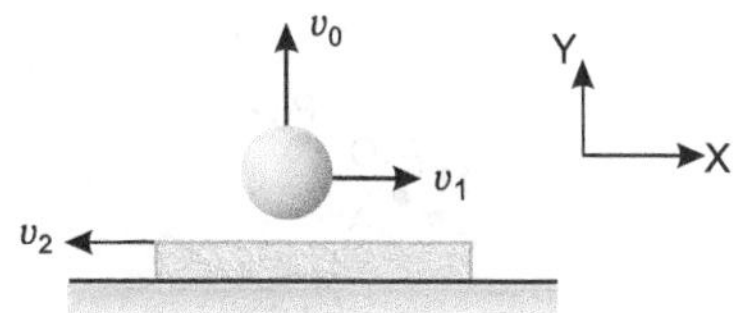

From conservation of linear momentum in horizontal direction

$$m v_1 = M v_2 \quad \ldots(1)$$

Linear impulse on the ball in vertical direction = change in linear momentum in vertical direction. ($J = N dt$)

Hence $\quad J = 2m v_0 \quad \ldots(2)$

Linear impulse on the ball in horizontal direction = change in linear momentum in horizontal direction.

$\therefore \quad \mu J = m v_1 \quad \ldots(3)$

Angular impulse on the ball about COM = change in angular momentum about COM.

$\therefore \quad \mu J \cdot r = I\omega_0 = \frac{2}{5} m r^2 \cdot \omega_0 \quad \ldots(4)$

Solving equations (1), (2), (3) and (4), we get

$$v_1 = \frac{2}{5} r\omega_0 \text{ and } v_2 = \frac{m}{M}\left(\frac{2}{5} r\omega_0\right)$$

Now actual path of the ball is a projectile whose time of flight will be

$$T = \frac{2v_y}{g} = \frac{2v_0}{g}$$

Relative velocity of ball with respect to plank in horizontal direction is

$$v_r = v_1 + v_2 = \left(\frac{M+m}{M}\right) v_1 = \frac{2}{5}\left(\frac{M+m}{M}\right) r\omega_0$$

Therefore, the desired distance is

$$s = v_r T$$

or $$\boldsymbol{s = \frac{4}{5}\left(\frac{M+m}{M}\right)\frac{v_0 r\omega_0}{g}}$$

164. $r = \frac{1}{2}$ m $= 0.5$ m. P and M are the centres of mass of the two parts A and B.

$$MO = PO = \frac{2r}{\pi} = \frac{2\left(\frac{1}{2}\right)}{\pi} = 0.32 \text{ m}$$

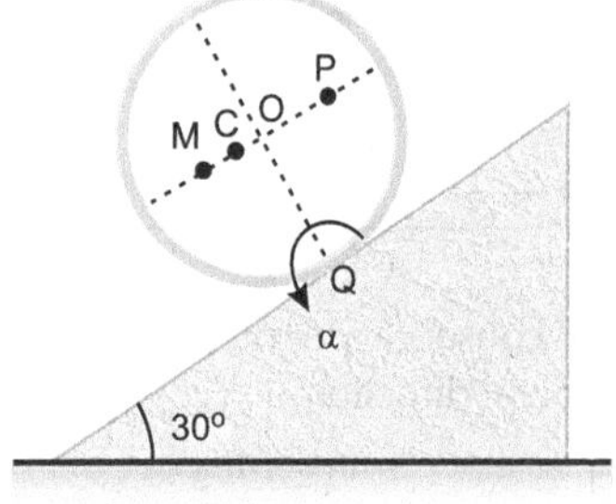

Let C be the centre of mass of the whole ring.

$$MP = 2\,(0.32) = 0.64 \text{ m}$$

Then $MC = \left(\dfrac{1}{3}\right)(0.64) = 0.213 \text{ m}$

$\therefore \quad CO = 0.32 - 0.213 = 0.107 \text{ m}$

$\therefore QC = \sqrt{(OQ)^2 + (CO)^2} = \sqrt{(0.5)^2 + (0.107)^2}$

$QC = 0.511 \text{ m}$

Force diagrams is shown as below :

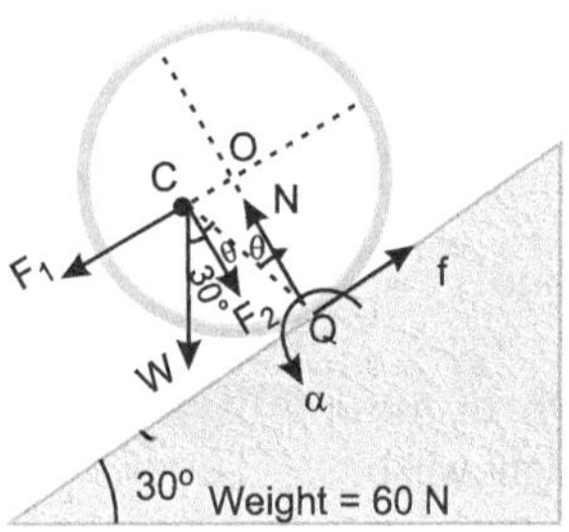

$$\cos\theta = \frac{QO}{CQ} = \frac{0.5}{0.511}$$

$\therefore \quad \theta = 11.9°$

W = 60 N

$F_1 = 60 \sin 30°$

$F_2 = 60 \cos 30°$

(a) The ring will rotate about point Q.

angular acceleration, $\quad \alpha = \dfrac{\tau_Q}{I_Q}$

$$\text{or } \alpha = \frac{(60 \sin 30°)(QO) + (60 \cos 30°)(CO)}{(2m_A + 2m_B)\,r^2}$$

$$= \frac{(60)\,(0.5)\,(0.5) + (60)\,(0.86)\,(0.107)}{(2 \times 4 + 2 \times 2)\left(\dfrac{1}{2}\right)^2}$$

or $\alpha = 6.84$ **rad/s²**

(b) Acceleration of centre of mass, C is :
$a = (QC)\,(\alpha)$ in the direction shown in figure *i.e.*, perpendicular to QC

$$= (0.511)\,(6.84) \text{ m/s}^2 = 3.5 \text{ m/s}^2$$

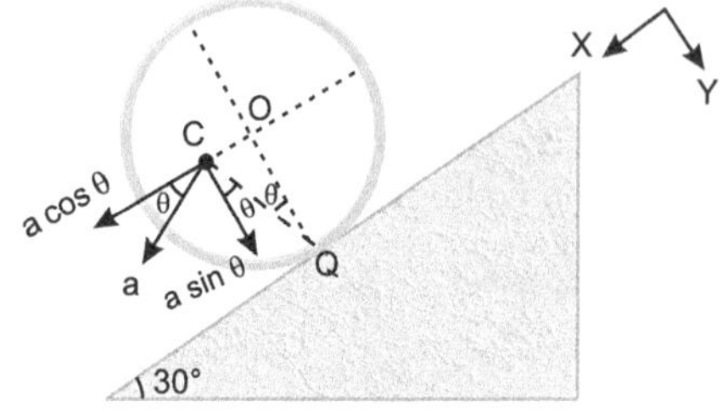

Now $\quad (m_A + m_B)\,a_y = \Sigma F_y$

$$(m_A + m_B)\,a \sin\theta = 60 \cos 30° - N$$

or $\quad N = 60 \cos 30° - (m_A + m_B)\,a \sin\theta$

$$= (60)\,(0.86) - (2 + 4)\,(3.5) \sin 11.9°$$

Hence normal reaction is

$$\boldsymbol{N = 47.27 \text{ N}}$$

(c) $(m_A + m_B)\,a_x = \Sigma F_x$

$\Rightarrow \quad (m_A + m_B) a \cos\theta = 60 \sin 30° - f$

or $\quad f = 60 \sin 30° - (m_A + m_B) a \cos\theta$

$$= (60)(0.5) - (2 + 4)(3.5) \cos 11.9°$$

$\therefore$ frictional force

$$\boldsymbol{f = 9.45 \text{ N}}$$

165. (a) Net torque about point of contact of the left spool should be zero. Therefore,

$$T\,(2R + R) = (m_1 g \sin\alpha)\,(2R)$$

or $\quad \boldsymbol{T = \dfrac{2}{3} m_1 g \sin\alpha}$

(b) Considering equilibrium of block A

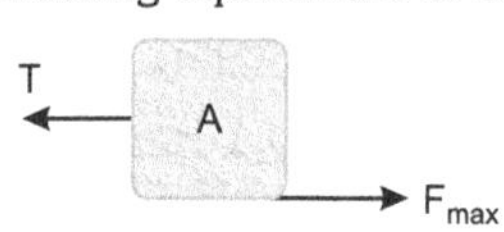

Maximum value of friction $\geq T$

or $\quad f_{max} \geq T$

or $\quad \mu m_A g \geq \dfrac{2}{3} m_1 g \sin\alpha$

or $\quad \mu \geq \dfrac{2}{3}\dfrac{m_1}{m_A} \sin\alpha$

$\therefore$ **minimum value of μ is** $\boldsymbol{\dfrac{2}{3}\dfrac{m_1}{m_A} \sin\alpha}$

(c) Equilibrium of block A gives

$$f = T = \frac{2}{3} m_1 g \sin\alpha$$

Now considering equilibrium of block B

$$kx = f = \frac{2}{3} m_1 g \sin \alpha$$

Hence $$\boldsymbol{x = \frac{2\, m_1 g \sin \alpha}{3k}}$$

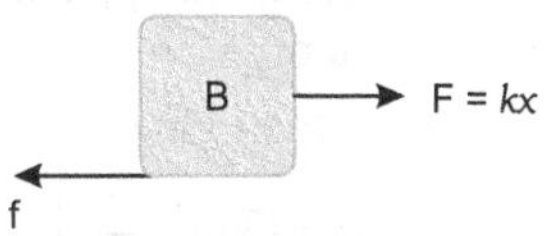

(d) Net torque about point of contact of the right spool should be zero. Hence

$$T'\,(2R - R) = (m_2 g \sin \alpha)(2R)$$

or $$T' = 2m_2 g \sin \alpha$$

But $$T' = kx \text{ or } T$$

Hence $\frac{2}{3} m_1 g \sin \alpha = 2\, m_2 g \sin \alpha$

or $$\boldsymbol{\frac{m_1}{m_2} = 3}$$

166. Angular momentum about end A

$$I\omega_0 = P_0 \frac{3L}{4} \Rightarrow \omega_0 = \frac{9P_0}{4mL}$$

with this angular velocity rod will strike nail and move back.

Time taken to reach N,

$$t_1 = \frac{\pi}{2\omega_0} = \frac{2\pi mL}{9P_0} \quad \ldots(1)$$

Let ω' is the new angular velocity after the collision with nail.

$$t_2 = \frac{\pi}{2\omega'} = \frac{8\pi mL}{27P_0} - \frac{2\pi mL}{9P_0} = \frac{2\pi mL}{27P_0} \quad \ldots(2)$$

$$\Rightarrow \quad \omega' = \frac{27P_0}{4mL}$$

Let impulse of nail be P' then

$$P' \frac{L}{4} = I\,(\omega' + \omega_0)$$

(Again apply conservation of angular momentum)

$$\Rightarrow \quad P' = 12P_0$$

167. Let the angular acceleration of the smaller ball be α_1, that of the larger one α_2, their common horizontal acceleration a_1 and the acceleration of the cart a_2. As the balls are rolling without slipping, we have

$$R\alpha_2 = a_2 - a_1$$

and $$R\alpha_2 = r\alpha_1$$

and because $R = 2r$

$$\alpha_1 = 2\alpha_2 = \frac{a_2 - a_1}{r}$$

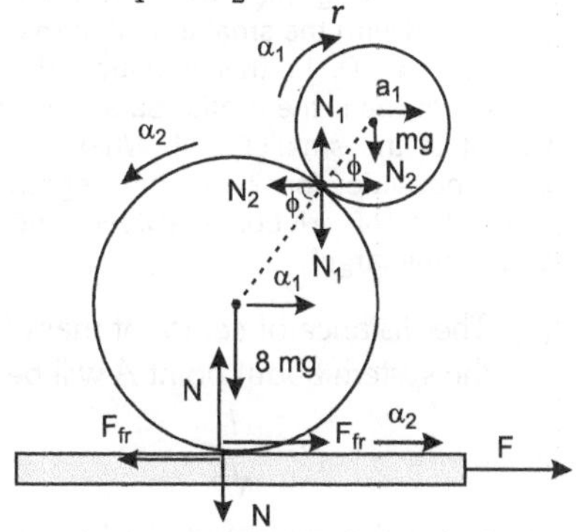

The moment of inertia of the smaller ball is $\frac{2}{5}mr^2$, while that of the larger one with the same density is $\frac{2}{5} \times 8m \times (2r)^2 = \frac{64}{5} mr^2$.

Using the notation of the figure, we can write the following dynamical equations of motion:

$$F - F_{fr} = Ma_2$$

$$8mg + N_1 - N = 0,$$

$$F_{fr} - N_2 = 8ma_1$$

$$mg - N_1 = 0, \quad N_2 = ma_1$$

$$N_1 r \cos \phi - N_2 r \sin \phi = \frac{2}{5} mr^2 \alpha_1$$

$$2rF_{fr} + 2rN_2 r \sin \phi - 2rN_1 r \cos \phi = \frac{64}{5} mr^2 \alpha_2$$

From these equations we can express the force F as

$$F = \left(9m + \frac{7}{2}M\right) \frac{\cos \phi}{1 + \sin \phi}\, g \approx 79\,\text{N}$$

The acceleration of the balls relative to the cart is

$$\Delta a = a_2 - a_1 = \frac{5}{2} \frac{\cos \phi}{(1 + \sin \phi)}\, g$$

At the time t when the balls fall from the cart, the distance they have moved relative to the cart is $\frac{L}{2}$. As their initial velocities are zero,

$$t = \sqrt{\frac{L}{\Delta a}} = 0.55 \text{ s}$$

Note: It is interesting that this stunt can also be performed with the smaller ball in the horizontal position, $\phi = 0$. In this situation the frictional force between the balls balances the entire weight of the smaller ball. What is more, it is even possible for ϕ to be negative, if the coefficient of friction between the balls is sufficiently large!

168. (a) The distance of centre of mass (COM) of the system about point A will be:

$$r = \frac{l}{\sqrt{3}}$$

Therefore the magnitude of horizontal force exerted by the hinge on the body is

$$F = \text{centripetal force}$$

or $$F = (3m) r\omega^2$$

or $$F = (3m)\left(\frac{l}{\sqrt{3}}\right)\omega^2$$

or $$\boldsymbol{F = \sqrt{3}\, ml\, \omega^2}$$

(b) Angular acceleration of system about point A is

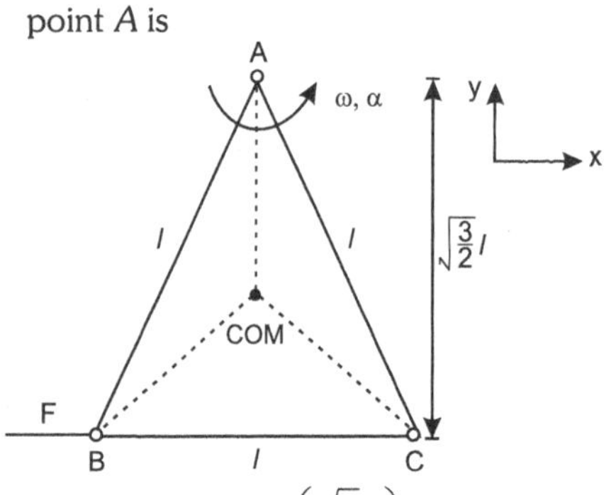

$$\alpha = \frac{\tau_A}{I_A} = \frac{(F)\left(\frac{\sqrt{3}}{2}l\right)}{2ml^2} = \frac{\sqrt{3}F}{4ml}$$

Now acceleration of COM along x-axis is

$$a_x = r\alpha = \left(\frac{l}{\sqrt{3}}\right)\left(\frac{\sqrt{3}F}{4ml}\right)$$

or $$a_x = \frac{F}{4m}$$

Now let F_x be the force applied by the hinge along x-axis. Then:

$$F_x + F = (3m)\, a_x$$

or $$F_x + F = (3m)\left(\frac{F}{4m}\right)$$

or $$F_x + F = \frac{3}{4}F$$

or $$\boldsymbol{F_x = -\frac{F}{4}}$$

Further if F_y be the force applied by the hinge along y-axis. Then:

$$F_y = \text{centripetal force}$$

or $$\boldsymbol{F_y = \sqrt{3}\, ml\, \omega^2}$$

169. Let r be the perpendicular distance of COM of laminar sheet from the line AB and ω the angular velocity of the sheet just after colliding with rubber obstacle for the first time. Obviously the linear velocity of COM before and after collision will be $v_i = (r)\,(1 \text{ rad/s}) = r$ and $v_f = r\omega$. $\vec{v_i}$ and $\vec{v_f}$ will be in opposite directions.

Now Linear impulse on COM = Change in linear momentum of COM

or $$6 = m\,(v_f + v_i) = 30\,(r + r\omega)$$

or $$r\,(1 + \omega) = \frac{1}{5} \quad \ldots(1)$$

Similarly angular impulse about AB = change in angular momentum about AB

Angular impulse = Linear impulse × perpendicular distance of impulse from AB

Hence $6\,(0.5 \text{ m}) = I_{AB}\,(\omega + 1)$

[Initial angular velocity = 1 rad/s]

or $$3 = [I_{COM} + Mr^2]\,(1 + \omega)$$

or $$3 = [1.2 + 30r^2]\,(1 + \omega) \quad \ldots(2)$$

Solving (1) and (2) for r, we get

$$r = 0.4 \text{ m} \quad \text{and} \quad r = 0.1 \text{ m}$$

But at $r = 0.4$ m, ω comes out to be negative (-0.5 rad/s) which is not acceptable. Therefore,

(a) r = distance of COM from $AB = \mathbf{0.1\ m}$

(b) Substituting $r = 0.1$ m in equation (1), we get $\omega = 1$ rad/s *i.e.* the angular velocity with which sheet comes back after the first impact is **1 rad/s**.

(c) Since the sheet returns with same angular velocity of 1 rad/s, the sheet will never come to rest.

170. (a) Let just after collision velocity of COM of rod is v and angular velocity about COM is ω. Then applying following three laws :

(i) External force on the system (rod + mass) in horizontal plane along x-axis is zero.

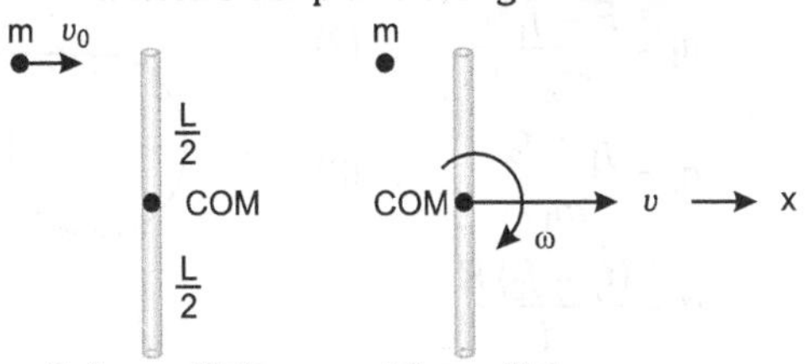

Before collision **After collision**

$\therefore$ Applying conservation of linear momentum in x-direction,

$$mv_0 = Mv \quad \text{... (1)}$$

(ii) Net torque on the system about COM of rod is zero.

$\therefore$ Applying conservation of angular momentum about COM of rod, we get

$$mv_0\left(\frac{L}{2}\right) = I\,\omega \quad \text{or} \quad mv_0\,\frac{L}{2} = \frac{ML^2}{12}\cdot\omega$$

$$\text{or} \quad mv_0 = \frac{ML\omega}{6} \quad \text{...(2)}$$

(iii) Since the collision is elastic, kinetic energy is also conserved.

$$\therefore \quad \frac{1}{2}mv_0^2 = \frac{1}{2}Mv^2 + \frac{1}{2}I\omega^2$$

$$\text{or} \quad mv_0^2 = Mv^2 + \frac{ML^2}{12}\omega^2 \quad \text{...(3)}$$

From equations (1), (2) and (3), we get the following results

$$\frac{\boldsymbol{m}}{\boldsymbol{M}} = \frac{\mathbf{1}}{\mathbf{4}}$$

$$v = \frac{mv_0}{M} \quad \text{and} \quad \omega = \frac{6mv_0}{ML}$$

(b) Point P will be at rest if $x\omega = v$

$$\text{or} \quad x = \frac{v}{\omega} = \frac{mv_0/M}{6mv_0/ML}$$

$$\text{or} \quad x = L/6$$

$$\therefore \quad \boldsymbol{AP} = \frac{\boldsymbol{L}}{\mathbf{2}} + \frac{\boldsymbol{L}}{\mathbf{6}} = \frac{\mathbf{2}}{\mathbf{3}}\boldsymbol{L}$$

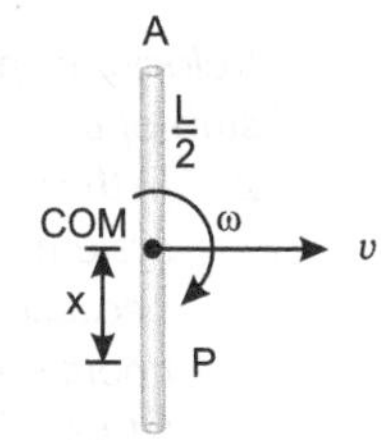

(c) After time $t = \dfrac{\pi L}{3v_0}$

angle rotated by rod,

$$\theta = \omega t = \frac{6mv_0}{ML}\cdot\frac{\pi L}{3v_0} = 2\pi\left(\frac{m}{M}\right)$$

$$\because \quad \frac{m}{M} = \frac{1}{4} \quad \therefore \theta = \frac{\pi}{2}$$

Therefore, situation will be as shown below :

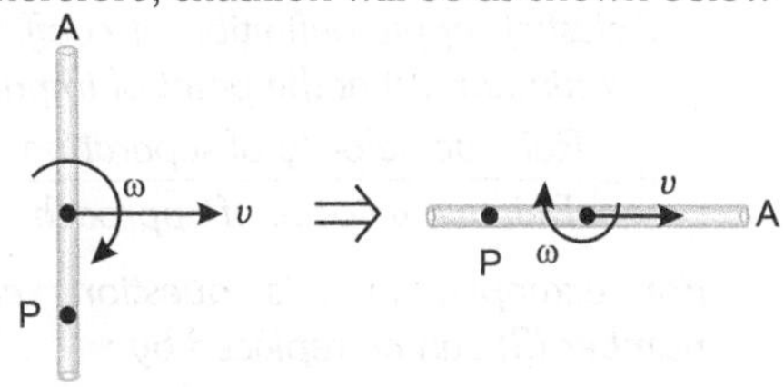

$\therefore$ Resultant velocity of point P will be

$$|\vec{v}_p| = \sqrt{2}v = \sqrt{2}\left(\frac{m}{M}\right)v_0 = \frac{\sqrt{2}}{4}v_0 = \frac{v_0}{2\sqrt{2}}$$

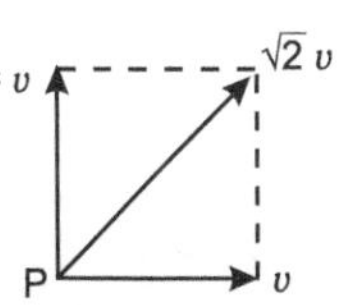

$$\text{or} \quad |\vec{\boldsymbol{v}}_{\boldsymbol{p}}| = \frac{\boldsymbol{v_0}}{\mathbf{2\sqrt{2}}}$$

1. *In a complex type of motion of a rigid body, we need to find two things (a) velocity of centre of mass (b) angular velocity about centre of mass. Because by knowing these two quantities we can describe the motion of any point on the rigid body. For example :*

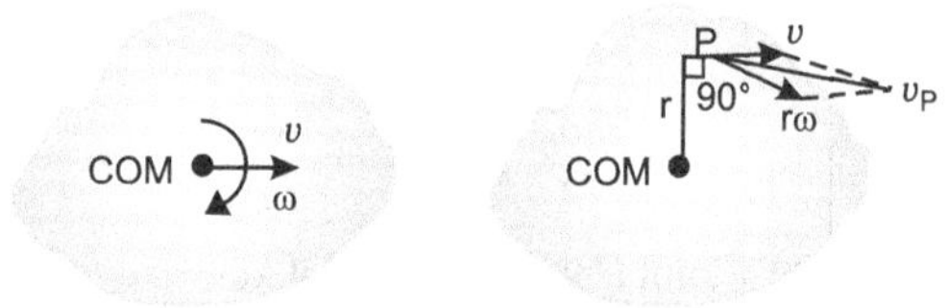

Velocity of point P in the figure is the vector sum of v and $r\omega$

2. *In the problem, angular momentum of the system about any point will be conserved, because torque on the system was zero about any point but we conserved it about COM, because angular velocity 'ω' of rod about COM was required.*
3. *First two equations always hold good (when placed on a 'smooth plane') whether the collision is elastic or not but the third equation i.e., conservation of kinetic energy holds good only when collision is elastic.*
4. *If the collision is inelastic (for even if it is elastic), apply definition of coefficient of restitution (e) at the point of impact*

$$e = \frac{\textit{Relative velocity of separation}}{\textit{Relative velocity of approach}}$$

For example, in this question, equation number (3) can be replaced by :

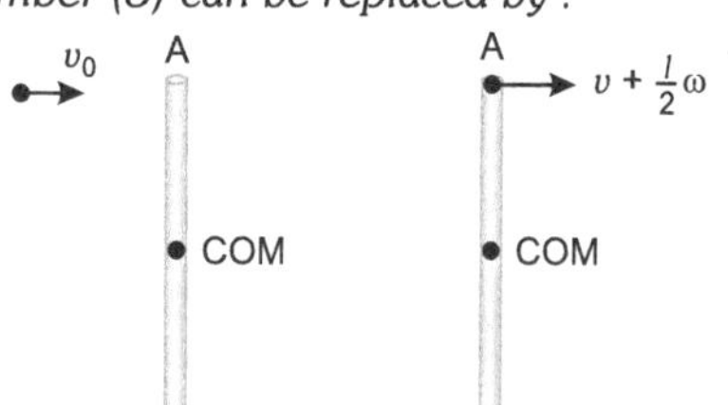

$$v_0 = v + \frac{L}{2}\omega$$

Because collision is elastic, therefore,
$e = 1$ or relative velocity of approach = relative velocity of separation.

171. We can choose any arbitrary directions of frictional forces at different contacts. In the final answer the negative values will show the opposite directions.

Let f_1 = frictional force between plank and cylinder

f_2 = frictional force between cylinder and ground

a_1 = acceleration of plank

a_2 = acceleration of centre of mass of cylinder

and α = angular acceleration of cylinder about its COM.

Directions of f_1 and f_2 are as shown here.

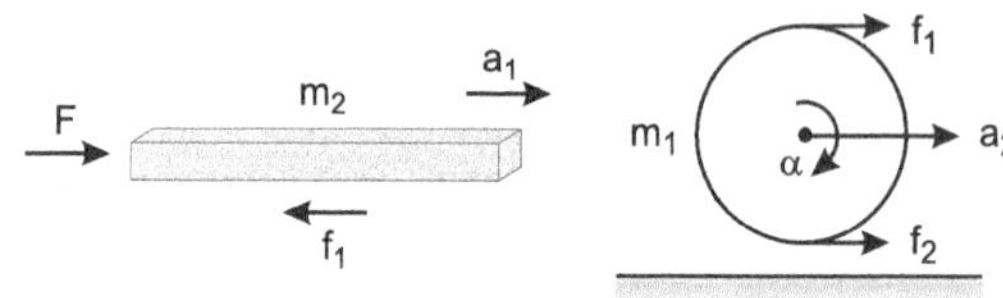

Since there is no slipping anywhere

$$\therefore \qquad a_1 = 2a_2 \qquad \ldots(1)$$

(Acceleration of plank = acceleration of top point of cylinder)

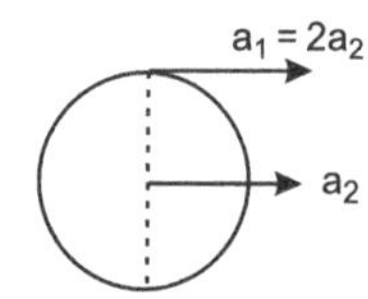

$$a_1 = \frac{F - f_1}{m_2} \qquad \ldots(2)$$

$$a_2 = \frac{f_1 + f_2}{m_1} \qquad \ldots(3)$$

$$\alpha = \frac{(f_1 - f_2)R}{I}$$

(I = moment of inertia of cylinder above COM)

$$\therefore \qquad \alpha = \frac{(f_1 - f_2)R}{\frac{1}{2}m_1R^2}$$

$$\alpha = \frac{2(f_1 - f_2)}{m_1R} \qquad \ldots(4)$$

For no slipping

$$a_2 = R\alpha = \frac{2(f_1 - f_2)}{m_1} \qquad \ldots(5)$$

Solving equations (1), (2), (3) and (5), we get:

(a) $$\boldsymbol{a_1 = \frac{8F}{3m_1 + 8m_2}}$$

and $$\boldsymbol{a_2 = \frac{4F}{3m_1 + 8m_2}}$$

(b) $$\boldsymbol{f_1 = \frac{3m_1F}{3m_1 + 8m_2}},$$

$$\boldsymbol{f_2 = \frac{m_1F}{3m_1 + 8m_2}}$$

Since all are positive, so they are correctly shown in figures.

172.

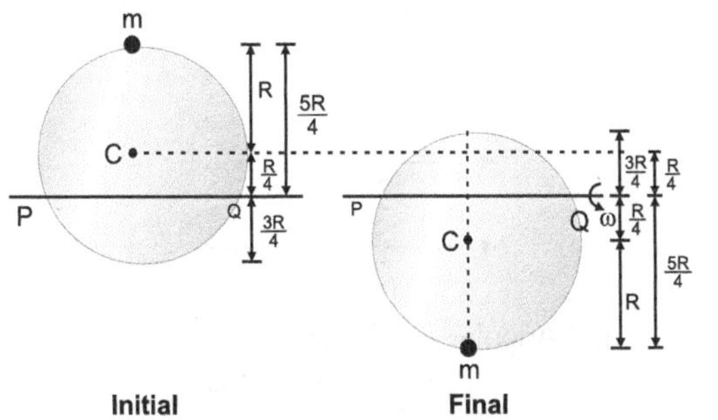

Decrease in potential energy of mass 'm'

$$= mg\left\{2\times\frac{5R}{4}\right\} = \frac{5mgR}{2}$$

Decrease in potential energy of disc

$$= mg\left\{2\times\frac{R}{4}\right\} = \frac{mgR}{2}$$

Therefore, total decrease in potential energy of system $= \frac{5mgR}{2} + \frac{mgR}{2} = 3mgR$

Gain in kinetic energy of system $= \frac{1}{2}I\omega^2$

where I = moment of inertia of system (disc + mass) about axis PQ.

= moment of inertia of disc

+ moment of inertia of mass

$$= \left\{\frac{mR^2}{4} + m\left(\frac{R}{4}\right)^2 + m\left(\frac{5R}{4}\right)^2\right\}$$

$$I = \frac{15mR^2}{8}$$

From conservation of mechanical energy,

Decrease in potential energy = Gain in kinetic energy

$$\therefore \quad 3mgR = \frac{1}{2}\left(\frac{15mR^2}{8}\right)\omega^2$$

$$\Rightarrow \quad \omega = \sqrt{\frac{16g}{5R}}$$

Therefore, linear speed of particle at its lowest point

$$v = \left(\frac{5R}{4}\right)\omega = \frac{5R}{4}\sqrt{\frac{16g}{5R}}$$

or $\quad \boldsymbol{v = \sqrt{5gR}}$

173. Given mass of disc, $m = 2\,\text{kg}$ and radius, $R = 0.1\,\text{m}$

(i) FBD of any one disc is shown in figure :

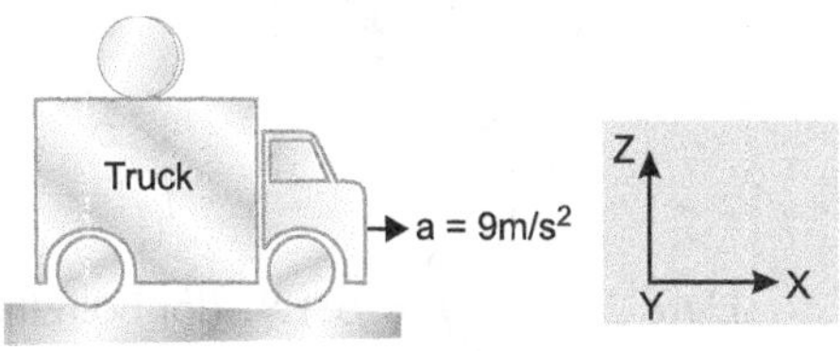

Frictional force on the disc should be in forward direction.

Let a_0 be the linear acceleration of COM of disc and α the angular acceleration about its COM. Then :

$$a_0 = \frac{f}{m} = \frac{f}{2} \qquad \ldots(1)$$

$$\alpha = \frac{\tau}{I} = \frac{f.R}{\frac{1}{2}mR^2}$$

$$= \frac{2f}{mR} = \frac{2f}{2\times 0.1} = 10f \qquad \ldots(2)$$

Since there is no slipping between disc and truck, therefore,

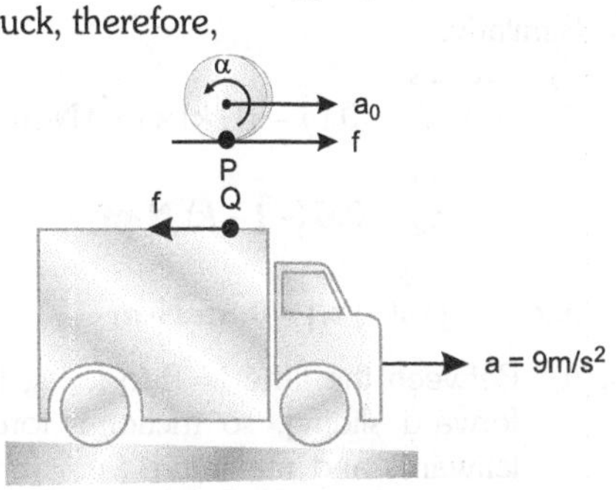

$$a_0 + R\alpha = a$$

or $\quad \left(\frac{f}{2}\right) + (0.1)\,10f) = a$

or $\quad \frac{3}{2}f = a \Rightarrow f = \frac{2a}{3} = \frac{2\times 9.0}{3}\,\text{N}$

$\therefore \quad f = 6\,\text{N}$

Since this force is acting in positive x-direction,

$$\vec{f} = (6\hat{i})\,\text{N}$$

(ii) $\quad \vec{\tau} = \vec{r}\times\vec{f}$

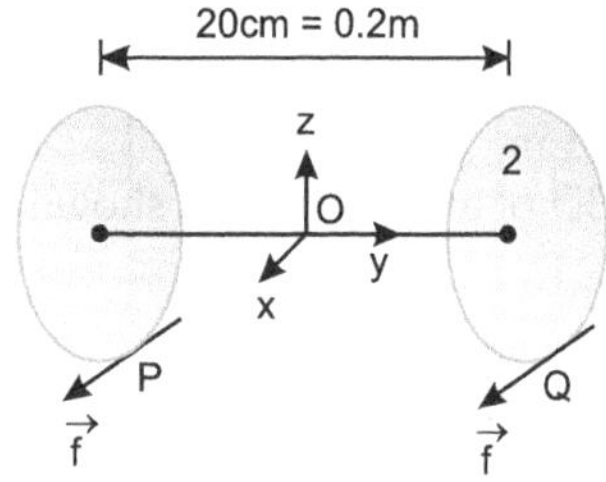

Here $\vec{f} = (6\hat{i})$ N (for both the discs)

$$\vec{r_P} = \vec{r_1} = -0.1\hat{j} - 0.1\hat{k}$$

and $\vec{r_Q} = \vec{r_2} = 0.1\hat{j} - 0.1\hat{k}$

Therefore, frictional torque on disc 1 about point O (centre of mass) :

$$\vec{\tau_1} = \vec{r_1} \times \vec{f} = (-0.1\hat{j} - 0.1\hat{k}) \times (6\hat{i}) \text{ N-m}$$
$$= (0.6\hat{k} - 0.6\hat{j})$$

or $\vec{\tau_1} = \mathbf{0.6\,(\hat{k} - \hat{j})\ N\text{-}m}$

and $|\vec{\tau_1}| = \sqrt{(0.6)^2 + (0.6)^2} = 0.85$ N-m

Similarly,

$$\vec{\tau_2} = \vec{r_2} \times \vec{f} = (0.1\hat{j} - 0.1\hat{k}) \times (6\hat{i}) \text{ N-m}$$

$$\vec{\tau_2} = \mathbf{0.6\,(-\hat{j} - \hat{k})\ N\text{-}m}$$

and $|\vec{\tau_2}| = |\vec{\tau_1}| = \mathbf{0.85\ N\text{-}m}$

174. (i) Between the time $t = 0$ to $t = t_0$, there is forward sliding, so frictional force f is leftwards and maximum *i.e.*, μmg. For time $t > t_0$, frictional force f will become zero, because now pure rolling has started *i.e.*, there is no sliding (no relative motion) between the points of contact.

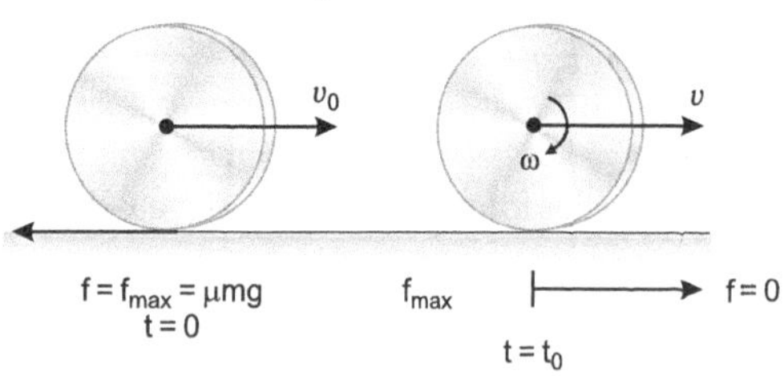

So, for time $t < t_0$

Linear retardation,

$$a = \frac{f}{m} = \mu g \qquad (f = \mu\ mg)$$

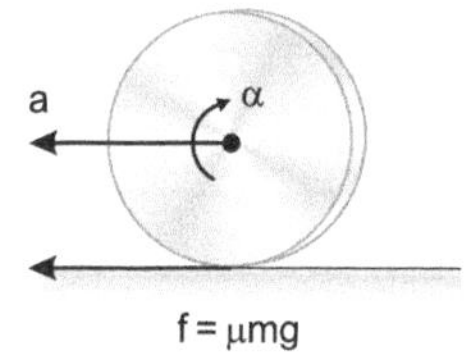

and Angular acceleratio

$$\alpha = \frac{\tau}{I} = \frac{f.R}{\frac{1}{2}mR^2} = \frac{2\mu g}{R}$$

Now let v be the linear velocity and ω, the angular velocity of the disc at time $t = t_0$ then

$$v = v_0 - at_0 = v_0 - \mu g\, t_0 \qquad \ldots(1)$$

and $$\omega = \alpha t_0 = \frac{2\mu g\, t_0}{R} \qquad \ldots(2)$$

For pure rolling to take place

$$v = R\omega$$

i.e., $v_0 - \mu\, gt_0 = 2\mu g t_0$ from (1) and (2)

$$\Rightarrow \qquad t_0 = \frac{v_0}{3\mu g}$$

Substituting the value of $t_0 = \frac{v_0}{3\mu g}$ in equation (1), we have

$$v = v_0 - \mu g\left(\frac{v_0}{3\mu g}\right)$$

$$\boldsymbol{v = \frac{2}{3} v_0} \qquad \ldots(ii)$$

(ii) Work done by friction : For $t \le t_0$, linear velocity of disc at any time t is $v = v_0 - \mu g t$ and angular velocity is $\omega = \alpha t = \frac{2\mu g t}{R}$. From work-energy theorem, work done by friction upto time t = kinetic energy of the disc at time t – kinetic
energy of the disc at time $t = 0$

$$\therefore \quad W = \frac{1}{2}mv^2 + \frac{1}{2}I\omega^2 - \frac{1}{2}mv_0^2$$

$$= \frac{1}{2}m[v_0 - \mu g t]^2$$

$$+\frac{1}{2}\left(\frac{1}{2}mR^2\right)\left(\frac{2\mu gt}{R}\right)^2-\frac{1}{2}mv_0^{\ 2}$$

$$=\frac{1}{2}[mv_0^{\ 2}+m\mu^2g^2t^2-2mv_0\mu gt$$

$$+2m\mu^2g^2t^2-mv_0^{\ 2}]$$

or $\quad \boldsymbol{W=\frac{m\mu gt}{2}[3\mu gt-2v_0]}$

For $t > t_0$, frictional force is zero *i.e.*, work done by friction is zero. Hence the energy will be conserved.

Therefore, total work done by friction over a time t much longer than t_0 is total work done upto time t_0 (because beyond that work done by friction is zero) which is equal to

$$W=\frac{m\mu gt_0}{2}[3\mu gt_0-2v_0]$$

Substituting, $t_0=v_0/3\mu g$, we get

$$W=\frac{mv_0}{6}[v_0-2v_0]$$

$$\boldsymbol{W=-\frac{mv_0^{\ 2}}{6}}$$

175. (a) The cylinder rotates about the point of contact. Hence the mechanical energy of the cylinder will be conserved *i.e.*,

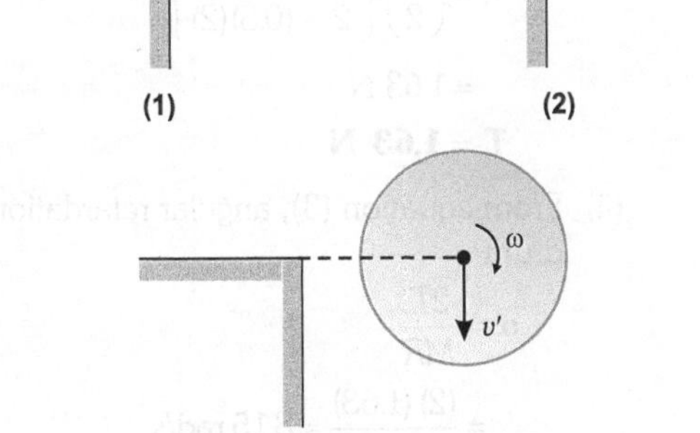

(1) (2)

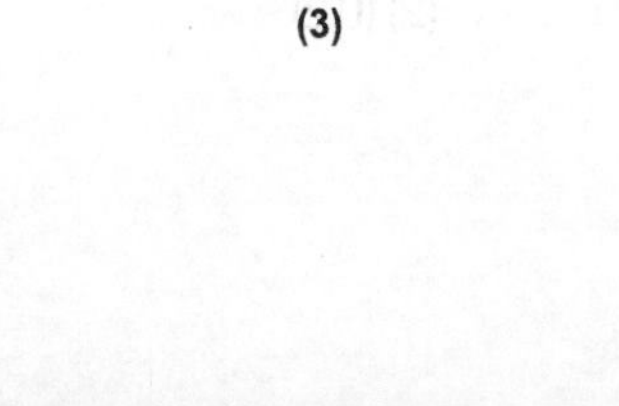

(3)

$\therefore \quad (P.E+K.E)_1=(P.E+K.E)_2$

$$\therefore mgR+0=mgR\cos\theta+\frac{1}{2}I\omega^2+\frac{1}{2}mv^2$$

But $\quad \omega=v/R$

(No slipping at point of contact)

and $\quad I=\frac{1}{2}mR^2$

Therefore, $mgR=mgR\cos\theta$

$$+\frac{1}{2}\left(\frac{1}{2}mR^2\right)(v^2/R^2)+\frac{1}{2}mv^2$$

or $\quad \frac{3}{4}v^2=gR(1-\cos\theta)$

or $\quad \frac{v^2}{R}=\frac{4}{3}g(1-\cos\theta) \quad \ldots(1)$

At the time of leaving contact, normal reaction $N=0$ and $\theta=\theta_c$. Hence

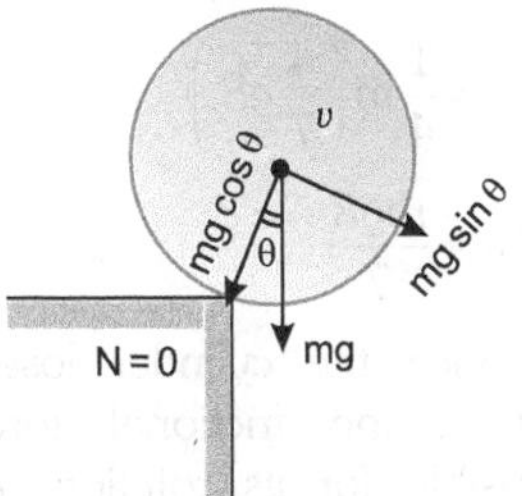

$$mg\cos\theta=\frac{mv^2}{R}$$

or $\quad \frac{v^2}{R}=g\cos\theta \quad \ldots(2)$

From (1) and (2)

$$\frac{4}{3}g(1-\cos\theta_c)=g\cos\theta_c$$

or $\quad \frac{7}{4}\cos\theta_c=1$

or $\quad \cos\theta_c=4/7$

or $\quad \boldsymbol{\theta_c=\cos^{-1}(4/7)}$

(b) $v=\sqrt{\frac{4}{3}gR(1-\cos\theta)}$

(From equation 1)

At the time it leaves the contact

$$\cos\theta=\cos\theta_c=4/7$$

$\therefore \qquad v = \sqrt{\frac{4}{3} gR\,(1 - 4/7)}$

or $\qquad v = \sqrt{\frac{4}{7} gR}$

Therefore, speed of COM of cylinder just before it leaves the contact is $\sqrt{\frac{4}{7} gR}$.

(c) At the moment, when cylinder leaves the contact

$$v = \sqrt{\frac{4}{7} g\,R}$$

Therefore, rotational kinetic energy,

$$K_R = \frac{1}{2} I\omega^2$$

or $\quad K_R = \frac{1}{2}\left(\frac{1}{2}mR^2\right)(v^2/R^2) = \frac{1}{4}mv^2$

$$= \frac{1}{4}m\left(\frac{4}{7}gR\right)$$

or $\quad K_R = \frac{mgR}{7} \qquad \ldots(3)$

Now once the cylinder loses its contact, $N = 0$, *i.e.* the frictional force, which is responsible for its rotation, also vanishes. Hence its rotational kinetic energy now becomes constant, while its translational kinetic energy increases.
Applying conservation of energy at (1) and (3) :
Decrease in gravitational P.E. = Gain in rotational K. E. + translational K. E.
$\therefore$ Translational K.E. (K_T) = Decrease in gravitational P.E. – K_R

or $\quad K_T = (mgR) - \frac{mgR}{7} = \frac{6}{7}mgR \qquad \ldots(4)$

From (3) and (4) we have

$$\frac{K_T}{K_R} = \frac{\frac{6}{7}mgR}{\frac{mgR}{7}} \quad \text{or} \quad \boldsymbol{\frac{K_T}{K_R} = 6}$$

176. **(i)** Given, Mass of block X, $m = 0.5$ kg,
Mass of drum Y, $M = 2$ kg
Radius of drum, $R = 0.2$ m; Angle of inclined plane, $\theta = 30°$

Let 'a' be the linear retardation of block X and α the angular retardation of drum Y. Then

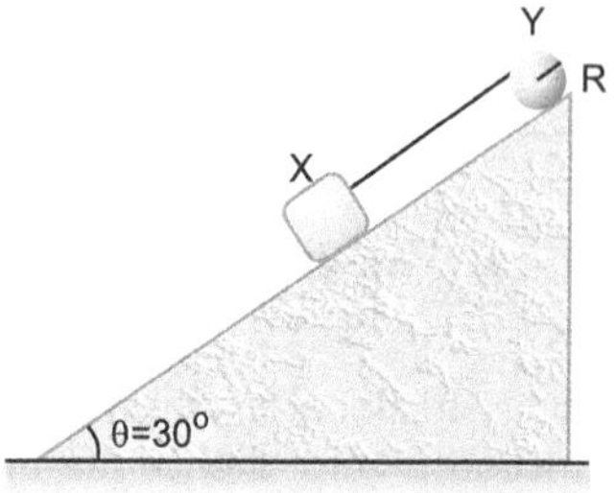

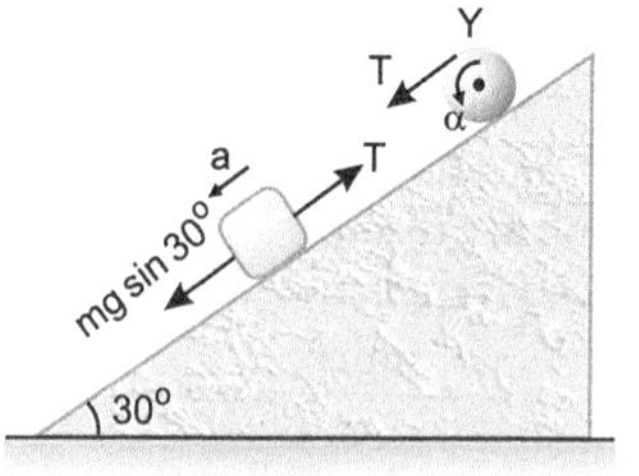

$$a = R\alpha \qquad \ldots(1)$$

$$mg \sin 30° - T = ma$$

$$\Rightarrow \quad mg/2 - T = ma \qquad \ldots(2)$$

$$\alpha = \frac{\tau}{I} = \frac{T \cdot R}{\frac{1}{2}MR^2}$$

or $\quad \alpha = \frac{2T}{MR} \qquad \ldots(3)$

Solving (1), (2) and (3) for T, We get

$$T = \frac{1}{2}\frac{Mmg}{M + 2m}$$

Substituting the values, we get

$$T = \left(\frac{1}{2}\right)\left\{\frac{(2)(0.5)(9.8)}{2 + (0.5)(2)}\right\}$$

$$= 1.63 \text{ N}$$

$$\boldsymbol{T = 1.63 \text{ N}}$$

(ii) From equation (3), angular retardation of drum

$$\alpha = \frac{2T}{MR}$$

$$= \frac{(2)\,(1.63)}{(2)\,(0.2)} = 8.15 \text{ rad/s}^2$$

or linear retardation of block

$$a = R\alpha = (0.2)(8.15)$$
$$= 1.63 \text{ rad/s}^3$$

At the moment when angular velocity of drum is $\omega_0 = 10$ rad/s, the linear velocity of block will be

$$v_0 = \omega_0 R = (10)(0.2) = 2 \text{ m/s}$$

Now the distance (s) travelled by the block unit it comes to rest will be given by

$$s = \frac{v_0^2}{2a}$$

[Using $v_2 = v_0^2 - 2as$ with $v = 0$]

$$= \frac{(2)^2}{2(1.63)} \text{ m}$$

or $\mathbf{s = 1.22\ m}$

177. System is free to rotate but not free to translate. During collision, net torque on the system (rod A + rod B + mass m) about point P is zero.

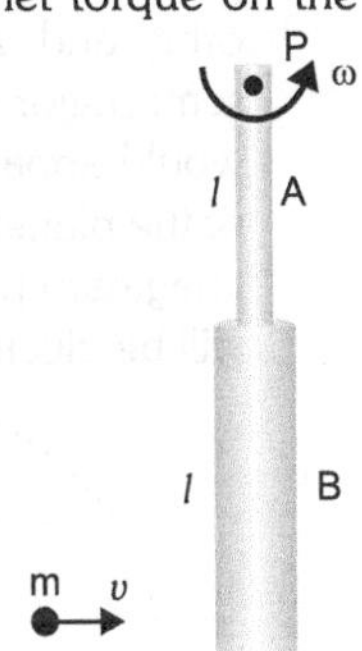

Therefore, angular momentum of system before collision = angular momentum of system just after collision (about P). Let ω be the angular velocity of system just after collision, then

$$L_i = L_f$$

$$\Rightarrow \qquad mv(2l) = I\omega \qquad \ldots(1)$$

Here I = moment of inertia of system about P

$$= m(2l)^2 + m_A (l^2/3) + m_B\left[\frac{l^2}{12} + \left(\frac{l}{2} + l\right)^2\right]$$

Given $l = 0.6$ m, $m = 0.05$ kg, $m_A = 0.01$ kg and $m_B = 0.02$ kg

Substituting the values, we get

$$I = 0.09 \text{ kg-m}^2$$

Therefore, from equation (1)

$$\omega = \frac{2mvl}{I} = \frac{2(0.05)(v)(0.6)}{0.09}$$

$$\omega = 0.67\, v \qquad \ldots(2)$$

Now after collision, as shown in figure mechanical energy will be conserved.

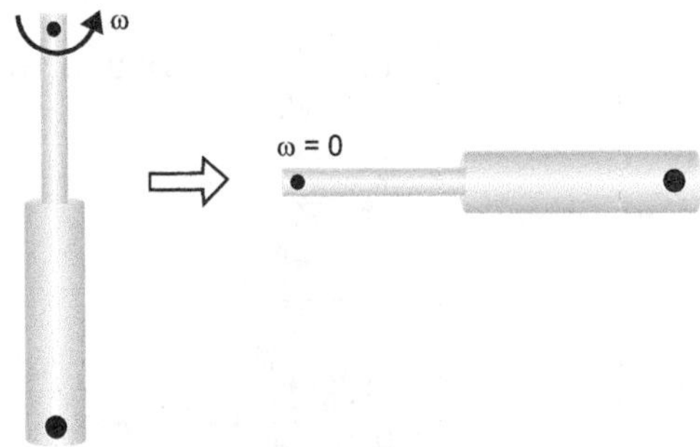

Therefore, decrease in rotational K.E.

= increase in gravitational P.E.

$$\text{or } \frac{1}{2}I\omega^2 = mg\,(2l) + m_A g\left(\frac{l}{2}\right) + m_B g\,(l + l/2)$$

$$\text{or} \qquad \omega^2 = \frac{gl\,(4m + m_A + 3m_B)}{I}$$

$$= \frac{(9.8)(0.6)(4 \times 0.05 + 0.01 + 3 \times 0.02)}{0.09}$$

$$= 17.64 \text{ (rad/s)}^2$$

$$\therefore \qquad \omega = 4.2 \text{ rad/s} \qquad \ldots(3)$$

Equating (2) and (3), we get

$$v = \frac{4.2}{0.67} \text{ m/s} = \mathbf{6.3\ m/s}$$

CHAPTER 6 GRAVITATION

178. $v \propto \frac{1}{\sqrt{r}}$ or $v = \frac{k}{\sqrt{r}}$

Hence $\vec{v}_1 = \frac{k}{\sqrt{r_1}} . \hat{j}$

$$\vec{v}_2 = \frac{k}{\sqrt{r_2}} \quad (-\sin\theta\,\hat{i} + \cos\theta\,\hat{j})$$

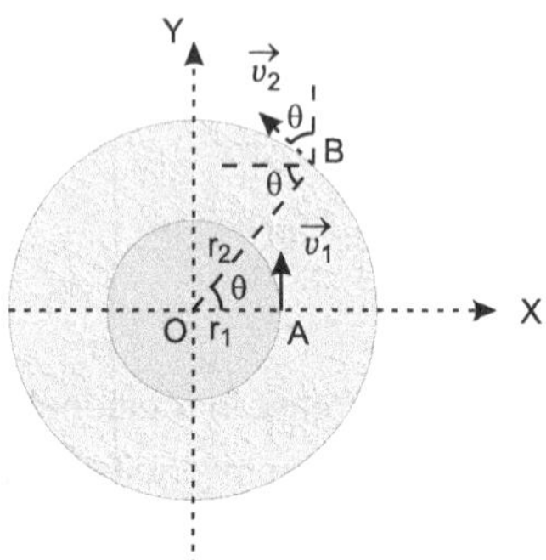

Similarly $\vec{r}_1 = r_1\hat{i}$

and $\vec{r}_2 = r_2\cos\theta\,\hat{i} + r_2\sin\theta\,\hat{j}$

Their angular relative velocity will be zero, if relative linear velocity of B with respect to A is along AB. Or $\vec{v}_2 - \vec{v}_1$ is parallel to $\vec{r}_2 - \vec{r}_1$.

$$\vec{v}_2 - \vec{v}_1 = -\frac{k}{\sqrt{r_2}}\sin\theta\,\hat{i} + \left(\frac{k}{\sqrt{r_2}}\cos\theta - \frac{k}{\sqrt{r_1}}\right)\hat{j}$$

and $\vec{r}_2 - \vec{r}_1 = (r_2\cos\theta - r_1)\,\hat{i} + r_2\sin\theta\,\hat{j}$

These two vectors are parallel if :

$$\frac{-\frac{k}{\sqrt{r_2}}\sin\theta}{r_2\cos\theta - r_1} = \frac{\frac{k}{\sqrt{r_2}}\cos\theta - \frac{k}{\sqrt{r_1}}}{r_2\sin\theta}$$

or $-\sqrt{r_2}\sin^2\theta = \sqrt{r_2}\cos^2\theta - \frac{r_2}{\sqrt{r_1}}\cos\theta$

$$-\frac{r_1}{\sqrt{r_2}}\cos\theta + \sqrt{r_1}$$

or $\cos\theta\left(\frac{r_2}{\sqrt{r_1}} + \frac{r_1}{\sqrt{r_2}}\right) = \sqrt{r_1} + \sqrt{r_2}$

or $\cos\theta = \frac{\sqrt{r_1 r_2}\,(\sqrt{r_1} + \sqrt{r_2})}{r_2\sqrt{r_2} + r_1\sqrt{r_1}}$

or $\boldsymbol{\theta = \cos^{-1}\left[\frac{\sqrt{r_1 r_1}\,(\sqrt{r_1} + \sqrt{r_2})}{r_2\sqrt{r_2} + r_1\sqrt{r_1}}\right]}$

179. Consider an imaginary planet moving along a strongly extended flat ellipse the extreme points of which are located on the planet's orbit and at the centre of the sun. The semi-major axis of the orbit of such a planet would apparently be half the semi-major axis of the planet's orbit. So the time period of the imaginary planet T' according to Kepler's law will be given by :

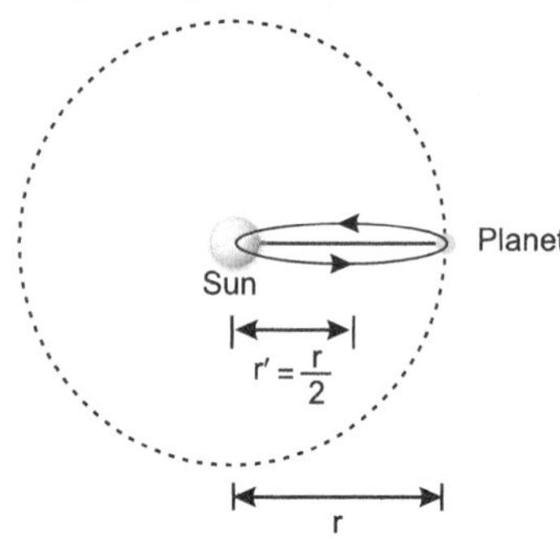

$$\left(\frac{T'}{T}\right) = \left(\frac{r'}{r}\right)^{3/2}$$

or $T' = T\left(\frac{1}{2}\right)^{3/2}$ (as $r' = r/2$)

$\therefore$ time taken by the planet to fall onto the sun is

$$t = \frac{T'}{2} = \frac{T}{2}\left(\frac{1}{2}\right)^{3/2} \Rightarrow \boldsymbol{t = \frac{\sqrt{2}}{8}T}$$

180. Let v be the velocity of the particle at point B. Applying conservation of mechanical energy at point A and B, we have

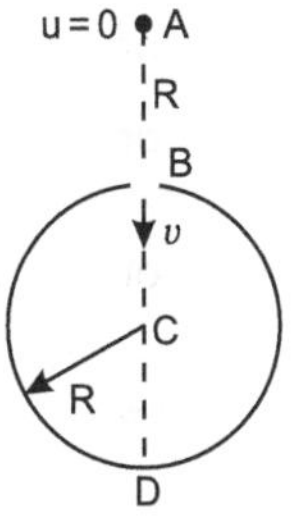

$$-\frac{GMm}{2R} = -\frac{GMm}{R} + \frac{1}{2}mv^2$$

or $$v = \sqrt{\frac{GM}{R}}$$

Inside the shell, gravitational field is zero *i.e.* force on mass m will be zero. Hence the particle will move with constant velocity v. Therefore, the desired time is

$$\boldsymbol{t = \frac{2R}{v} = 2\sqrt{\frac{R^3}{GM}}}$$

181. $v_o = \sqrt{\frac{GM}{r}}$

= orbital speed of satellite ...(1)

where M = Mass of earth

Absolute velocity of particle would be :

$$v_P = v + v_O = \sqrt{\frac{5}{4}}\, v_O = \sqrt{1.25}\, v_O \quad ...(2)$$

Since v_P lies between orbital velocity and escape velocity, path of the particle would be an ellipse with r being the minimum distance. Let r' be the maximum distance and v'_P its velocity at that moment.

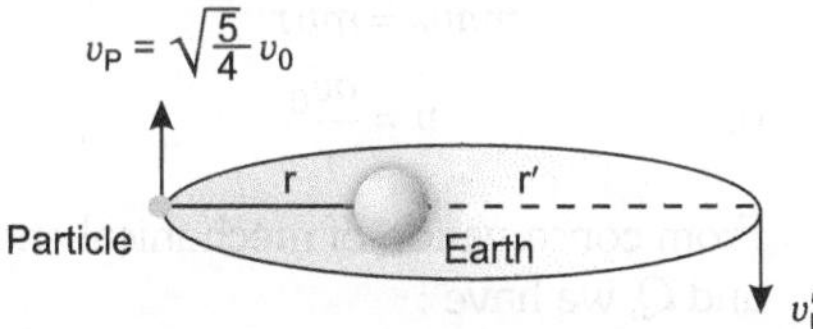

Then from conservation of angular momentum and conservation of mechanical energy, we get :

$$mv_P r = mv'_P r' \quad ...(3)$$

and $$\frac{1}{2}mv_P^2 - \frac{GMm}{r} = \frac{1}{2}mv'^2_P - \frac{GMm}{r'} \quad ...(4)$$

Solving the above equations, (1), (2), (3) and (4) we get

$$r' = \frac{5r}{3} \text{ and } r$$

Hence the maximum and minimum distances are $\frac{5r}{3}$ and r respectively.

182. r_1 = **6930** km = $\mathbf{6.93 \times 10^6}$ m

and r_2 = 7000 km = 7.0×10^6 m

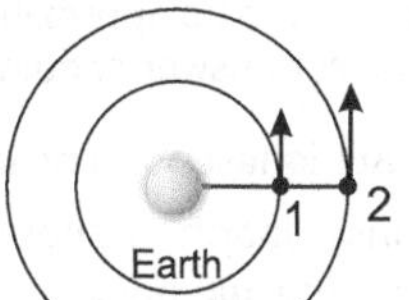

From $$\frac{GMm}{r^2} = mr\omega^2$$

we have $$\omega = \sqrt{\frac{GM}{r^3}}$$

$$\therefore \quad \omega_1 = \sqrt{\frac{6.67 \times 10^{-11} \times 6 \times 10^{24}}{(6.93 \times 10^6)^3}}$$

$$= 1.09 \times 10^{-3} \text{ rad/s} \quad ...(1)$$

$$\omega_2 = \sqrt{\frac{6.67 \times 10^{-11} \times 6 \times 10^{24}}{(7.0 \times 10^6)^3}}$$

$$= 1.08 \times 10^{-3} \text{ rad/s} \quad ...(2)$$

Now two cases are possible.

Case 1 : When both the satellites are revolving in opposite sense. Then they will be at closest distance after a time, say t_1 when :

$$\omega_1 t_1 + \omega_2 t_1 = 2\pi$$

or $$t_1 = \frac{2\pi}{\omega_1 + \omega_2}$$

or $$t_1 = \frac{2\pi}{(1.09 + 1.08) \times 10^{-3}} \text{ s}$$

$$= \frac{2\pi}{2.17 \times 10^{-3} \times 3600} \text{ hrs}$$

$$= \mathbf{0.8 \text{ hrs}}$$

Case 2 : When both the satellites are revolving in the same sense. Then they will be at closest distance after a time t_2 when :

$$\omega_1 t_2 = \omega_2 t_2 + 2\pi \qquad (\omega_1 > \omega_2)$$

or $$t_2 = \frac{2\pi}{\omega_1 - \omega_2}$$

$$= \frac{2\pi}{(0.01 \times 10^{-3}) \times 3600 \times 24} \text{ days}$$

$$= 7.27 \text{ days}$$

Note: If we substitute absolute values of ω_1 and ω_2, answer comes out to be approximately 4.5 days. Hence the correct answer should be **4.5 days.**

183. Let velocity at highest point be v and $R + h = r$

Applying conservation of angular momentum between P and Q, we have

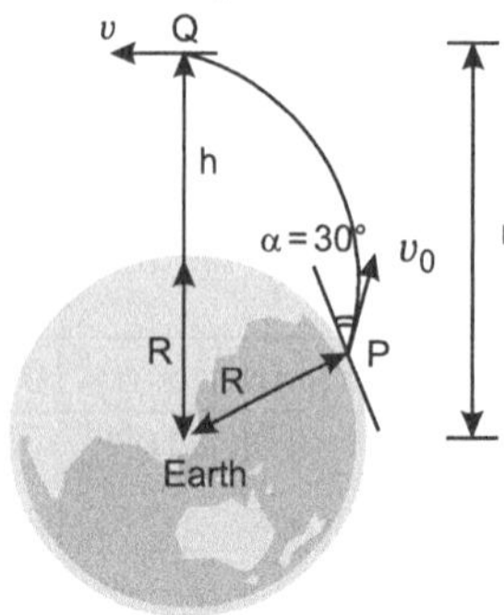

$$m v r = m v_0 R \cos 30°$$

or $$v = \frac{\sqrt{3} v_0 R}{2r} \quad \ldots(1)$$

Applying conservation of mechanical energy between P and Q, we have :

$$\frac{1}{2} m v_0^2 - \frac{GMm}{R} = \frac{1}{2} m v^2 - \frac{GMm}{r}$$

substituting the value of v from equation (1), we get

$$\frac{1}{2} m v_0^2 - \frac{GMm}{R} = \frac{1}{2} m \left\{ \frac{3 v_0^2 R^2}{4r^2} \right\} - \frac{GMm}{r}$$

or $$v_0^2 - \frac{2GM}{R} = \frac{3 v_0^2 R^2}{4r^2} - \frac{2GM}{r}$$

or $$\frac{1.5\,GM}{R} - \frac{2\,GM}{R}$$

$$= \frac{3}{4} \left(\frac{1.5\,GM}{R} \right) \frac{R^2}{r^2} - \frac{2\,GM}{r}$$

$$\left(v_0 = \sqrt{\frac{1.5\,GM}{R}} \text{ is given} \right)$$

or $$-\frac{1}{2R} = \frac{9}{8R} \cdot \frac{R^2}{r^2} - \frac{2}{r}$$

or $$-4r^2 = 9R^2 - 16Rr$$

or $$4r^2 - 16Rr + 9R^2 = 0$$

or $$r = \frac{16R \pm \sqrt{256R^2 - 144R^2}}{8}$$

$\Rightarrow$ $$= \frac{16R \pm 10.58R}{8}$$

$$= 3.323\,R \text{ and } 0.677R$$

but $r \nless R$

Hence $r = 3.323\,R$

or $\boldsymbol{h = r - R = 2.323\,R}$

184. Orbital speed of satellite is

$$v_0 = \sqrt{\frac{GM}{a}} \quad \ldots(1)$$

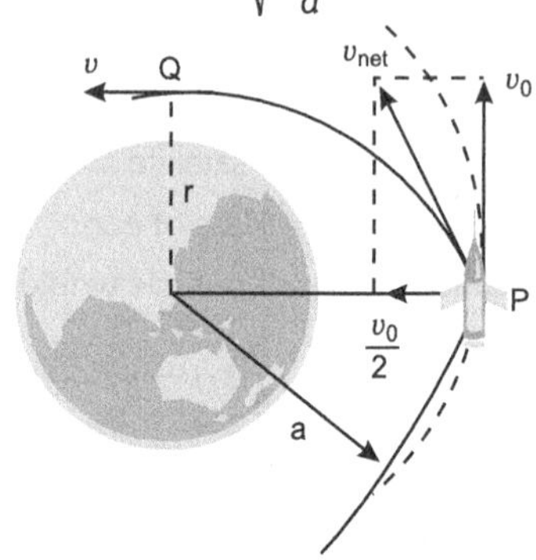

From conservation of angular momentum at P and Q, we have

$$m a v_0 = m v r$$

or $$v = \frac{a v_0}{r} \quad \ldots(2)$$

From conservation of mechanical energy at P and Q, we have :

$$\frac{1}{2} m \left(v_0^2 + \frac{v_0^2}{4} \right) - \frac{GMm}{a} = \frac{1}{2} m v^2 - \frac{GMm}{r}$$

or $$\frac{5}{8} v_0^2 - \frac{GM}{a} = \frac{v^2}{2} - \frac{GM}{r}$$

Substituting values of v and v_0 from (1) and (2), we get

$$\frac{5}{8} \frac{GM}{a} - \frac{GM}{a} = \frac{a^2}{r^2} \cdot \left(\frac{GM}{2a} \right) - \frac{GM}{r}$$

or $$-\frac{3}{8a} = \frac{a}{2r^2} - \frac{1}{r}$$

or $\quad -3r^2 = 4a^2 - 8ar$

or $\quad 3r^2 - 8ar + 4a^2 = 0$

or $$r = \frac{8a \pm \sqrt{64a^2 - 48a^2}}{6}$$

or $$r = \frac{8a \pm 4a}{6}$$

or $\quad r = 2a$ and $\frac{2a}{3}$

Hence the maximum and minimum distances are $2a$ and $\frac{2a}{3}$ respectively.

185. In a double star system both stars revolve about their common centre of mass (COM) with same angular velocity ω but different linear speeds,

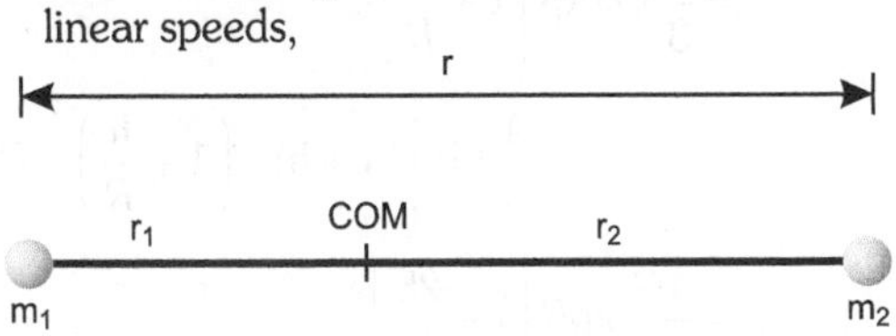

For COM :

$$m_1 r_1 = m_2 r_2 \quad ...(1)$$

$$r_1 + r_2 = r \quad ...(2)$$

Hence $$r_1 = \frac{m_2 r}{m_1 + m_2}$$

and $$r_2 = \frac{m_1 r}{m_1 + m_2}$$

The centripetal force for motion of each star is provided by gravitational force. *i.e.*,

$$m_1 r_1 \omega^2 = m_2 r_2 \omega^2 = \frac{G m_1 m_2}{r^2} \quad ...(3)$$

Solving these equations for the second star *i.e.*,

$$r = \frac{(m_1 + m_2) r_1}{m_2} \quad \text{and} \quad r^2 = \frac{G m_2}{r_1 \omega^2}$$

Eliminating r we get

$$(m_1 + m_2)^2 = \frac{G}{r_1^3 \omega^2} m_2^3$$

Let us take $m_2 = x m_1$, then we get

$$(1 + x)^2 = \frac{m_1 G}{r_1^3 \omega^2} x^3 \quad ...(4)$$

$$\frac{m_1 G}{r_1^3 \omega^2} = \frac{(10^{30})(6.67 \times 10^{-11})(20 \times 3600)^2}{(2.455 \times 10^9)^3 (2\pi)^2}$$

$$= 0.59$$

Substituting this in equation (4) we get

$$(1 + x)^2 = 0.59 x^3 \quad \text{or} \quad x \approx 3$$

Hence mass of second star is

$$m_2 = 3 m_1 = \mathbf{3 \times 10^{30}\ kg}$$

and distance between the two stars is

$$r = \frac{(m_1 + m_2) r_1}{m_2} = \frac{(4 \times 10^{30})(2.455 \times 10^6)}{3 \times 10^{30}} \text{ km}$$

or $\quad \mathbf{r = 3.273 \times 10^6\ km.}$

186. Both the stars rotate about their centre of mass (COM)

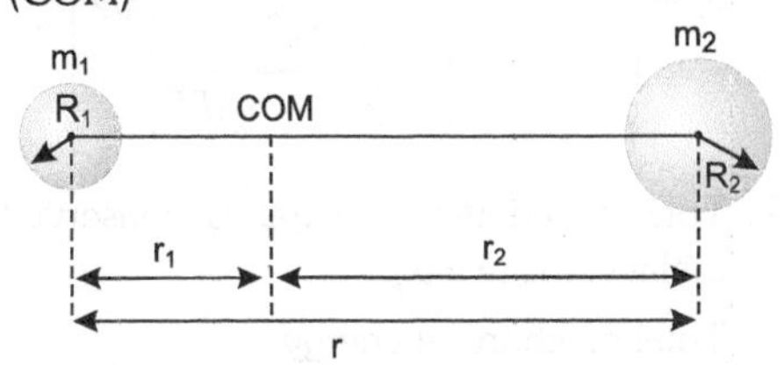

For the position of COM

$$\frac{r_1}{m_2} = \frac{r_2}{m_1} = \frac{r_1 + r_2}{m_1 + m_2} = \frac{r}{m_1 + m_2} \quad (r = r_1 + r_2)$$

Also $$m_1 r_1 \omega^2 = \frac{G m_1 m_2}{r^2}$$

or $$\omega^2 = \frac{G m_2}{r_1 r^2} \quad \left(\omega = \frac{2\pi}{T}\right)$$

But $$r_1 = \frac{m_2 r}{m_1 + m_2}$$

$\therefore$ $$\omega^2 = \frac{G(m_1 + m_2)}{r^3}$$

or $$r = \left\{\frac{G(m_1 + m_2)}{\omega^2}\right\}^{\frac{1}{3}} \quad ...(1)$$

Applying conservation of mechanical energy we have

$$-\frac{G m_1 m_2}{r} = -\frac{G m_1 m_2}{(R_1 + R_2)} + \frac{1}{2} \mu v_r^2 \quad ...(2)$$

Here μ = reduced mass $= \frac{m_1 m_2}{m_1 + m_2}$

and v_r = relative velocity between the two stars. From equation (2) we find that

$$v_r^2 = \frac{2Gm_1m_2}{\mu}\left(\frac{1}{R_1+R_2}-\frac{1}{r}\right)$$

$$= \frac{2Gm_1m_2}{\frac{m_1m_2}{m_1+m_2}}\left(\frac{1}{R_1+R_2}-\frac{1}{r}\right)$$

$$= 2G(m_1+m_2)\left(\frac{1}{R_1+R_2}-\frac{1}{r}\right)$$

Substituting the value of r from equation (1) we get

$$v_r = \sqrt{2G(m_1+m_2)\left[\frac{1}{R_1+R_2} - \left\{\frac{4\pi^2}{G(m_1+m_2)T^2}\right\}^{1/3}\right]}$$

187. Total mechanical energy is conserved. Let initial velocity $= v_o$.

Total mechanical energy

$$E_1 = \frac{1}{2}mv_0^2 - \frac{3GMm}{a/\sqrt{3}} \quad \ldots(1)$$

At final position kinetic energy = 0

Body is at a distance of $\sqrt{(3a)^2+\left(\frac{a}{\sqrt{3}}\right)^2}$ from each star.

Potential energy $= -\dfrac{GMm}{a\sqrt{\frac{28}{3}}}$

Total mechanical energy

$$E_2 = -\frac{3GMm}{a}\sqrt{\frac{3}{28}} \quad \ldots(2)$$

From equation (1) and (2), we get,

$$\frac{1}{2}mv_0^2 = \frac{GMm}{a}\left(3\sqrt{3}-\sqrt{\frac{3}{28}}\right)$$

$$\Rightarrow \quad v_o = \sqrt{\frac{2GM}{a}\left(3\sqrt{3}-\sqrt{\frac{3}{28}}\right)}$$

188. $g_A = G\left(\frac{4/3\,\pi R^3\rho}{R^2}\right) = \frac{4}{3}G\pi R\rho$

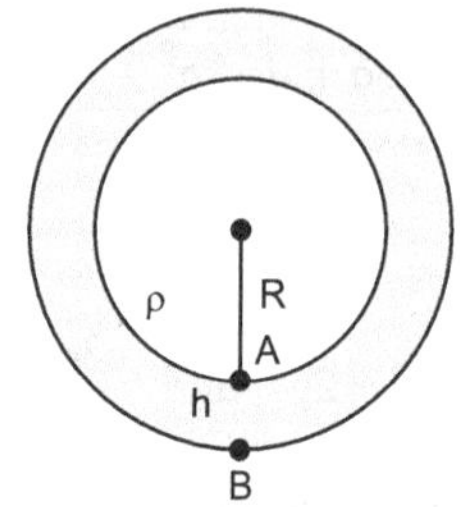

$$g_B = \frac{G\frac{4}{3}\pi R^3\rho}{(R+h)^2} + \frac{4G\pi}{3}\left[\frac{(R+h)^3-R^3}{(R+h)^2}\right]\sigma$$

$$= \frac{4}{3}G\pi R\rho\left[1+\frac{h}{R}\right]^{-2} + \frac{4}{3}G\pi\left[(R+h)-\left(1+\frac{h}{R}\right)^{-2}R\right]\sigma$$

$$= \frac{4G}{3}\pi\rho R\left[1-\frac{2h}{R}\right] + \frac{4G}{3}\pi\left[(R+h)-\left(1-\frac{2h}{R}\right)R\right]\sigma$$

$$g_A - g_B = \frac{4}{3}G\pi R\rho - \frac{4G}{3}\pi\rho R\left[1-\frac{2h}{R}\right] + \frac{4G}{3}\pi\left[(R+h)-\left(1-\frac{2h}{R}\right)R\right]\sigma$$

$$= \frac{4G\pi h}{3}[2\rho-3\sigma]$$

189. According to Kepler's first law the orbit of the rocket is an ellipse with one of its foci at the

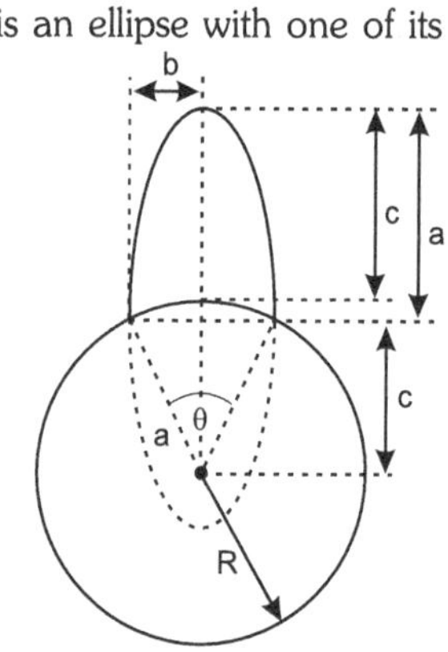

centre of the planet. The launch and return velocities are parallel to each other (though in opposite directions) if the launch and return points are at the ends of the minor axis of the ellipse. But, for an ellipse, the distance from a focus to either end of the minor axis is equal to the length a of its major semi-axis; consequently $a = R$ (see figure).

From Kepler's third law, satellites in orbits having different eccentricities, but the same lengths of major axis, have equal periods and so in our case the period for a full orbit would be the given T_0. The rocket, however, covers only one-half of the ellipse. The time required for this is not half of the full period, but proportional to the fractional area swept by the radius vector joining the rocket to the focus (Kepler's second law). The area of the whole ellipse is

$$A_0 = \pi ab = \pi a^2 \sin\frac{\theta}{2}$$

The swept area for the half orbit is

$$A_1 = \frac{\pi ab}{2} + \frac{1}{2} \times 2bc$$

$$= \frac{1}{2} a^2 \pi \sin\frac{\theta}{2} + a^2 \sin\frac{\theta}{2} \cos\frac{\theta}{2}$$

So, the flight time is

$$T_1 = \frac{A_1}{A_0} T_0 = T_0 \left(\frac{1}{2} + \frac{1}{\pi} \cos\frac{\theta}{2} \right)$$

The maximum distance above the surface of the planet is

$$2a - a - (a - c) = c = R \cos\frac{\theta}{2} \le R$$

If the angle between the launch and arrival points is allowed to approach zero ($\theta \to 0$), the calculated flight time approaches a maximum value of $T_0 \left(\frac{1}{2} + \frac{1}{\pi} \right)$, and the maximum height achieved approaches the radius of the planet ($c \to R$). But, in fact, if the take-off and landing sites are the same ($\theta = 0$), the rocket can reach any arbitrary height, large or small. This implies that the period and maximum height are not continuous functions of θ at the point $\theta = 0$.

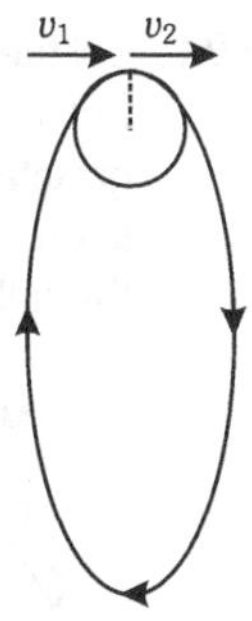

If the launch speed is sufficiently great (equal to or large than the first cosmic speed, $v = \sqrt{Rg}$) and the initial velocity is tangential to the surface of the planet, then the orbit shown in figure is possible. Again the return velocity is parallel to the launch one, but this time in the same direction. The maximum height achieved can be anything, but the period must be at least T_0. These are the orbits corresponding to the special case $\theta = 0$.

190. Speed of the particle at A, v_A

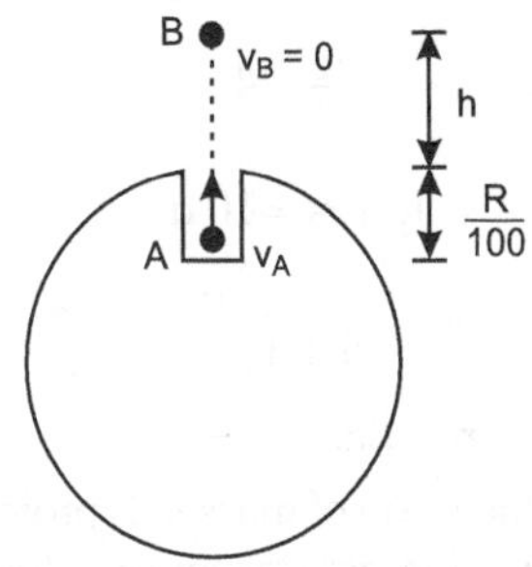

= escape velocity on the surface of earth

$$= \sqrt{\frac{2GM}{R}}$$

At highest point B, $v_B = 0$

Applying conservation of mechanical energy, decrease in kinetic energy = increase in gravitational potential energy

or $$\frac{1}{2} m v_A^2 = U_B - U_A = m(V_B - V_A)$$

or $$\frac{v_A^2}{2} = V_B - V_A$$

$$\therefore \quad \frac{GM}{R} = -\frac{GM}{R+h} - \left[-\frac{GM}{R^3} \left\{ 1.5R^2 - 0.5\left(R - \frac{R}{100} \right)^2 \right\} \right]$$

or $$\frac{1}{R} = -\frac{1}{R+h} + \frac{3}{2R} - \left(\frac{1}{2}\right)\left(\frac{99}{100}\right)^2 \cdot \frac{1}{2}$$

Solving this equation we get

$$h = 99.5R$$

191. Let there are two stars 1 and 2 as shown below : Let P is a point between C_1 and C_2 where gravitational field strength is zero. Or at P field strength due to star 1 is equal and opposite to the field strength due to star 2.

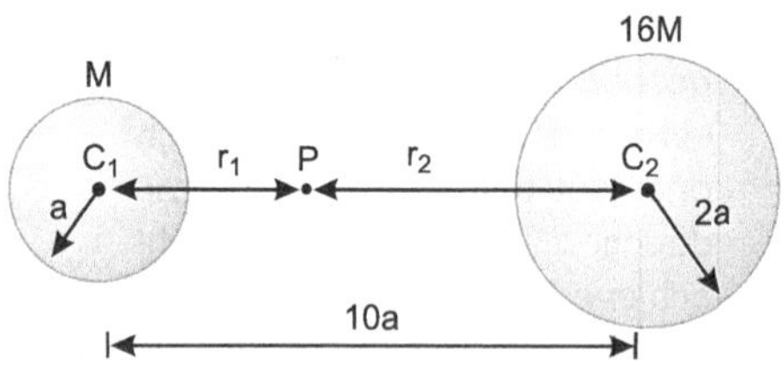

Hence $$\frac{GM}{r_1^2} = \frac{G(16\,M)}{r_2^2}$$

or $$\frac{r_2}{r_1} = 4$$

Also $$r_1 + r_2 = 10\,a$$

$$\therefore \quad r_2 = \left(\frac{4}{4+1}\right)(10\,a) = 8\,a$$

and $r_1 = 2a$.

Now the body of mass m is projected from the surface of larger star towards the smaller one. Between C_2 and P it is attracted towards 2 and between C_1 and P it will be attracted towards 1. Therefore, the body should be projected to just cross P because beyond that the particle is attracted towards the smaller star itself.

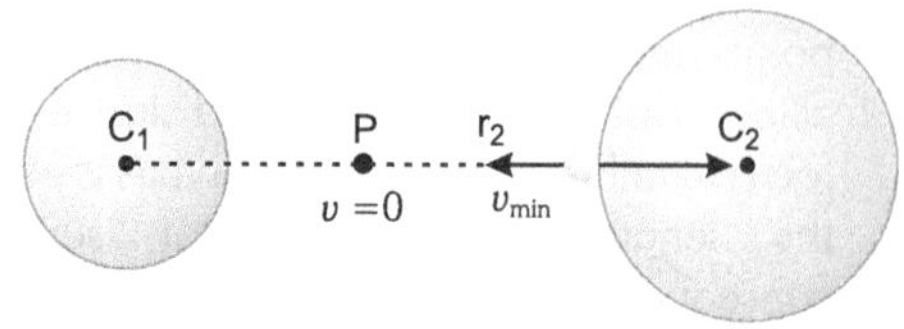

From conservation of mechanical energy,

$\frac{1}{2} m v_{min}^2$ = Potential energy of the body at P − Potential energy at the surface of larger star.

$$\frac{1}{2} m v_{min}^2 = \left[-\frac{GMm}{r_1} - \frac{16\,GMm}{r_2}\right] - \left[-\frac{GMm}{10a-2a} - \frac{16\,GMm}{2a}\right]$$

$$= \left[-\frac{GMm}{2a} - \frac{16\,GMm}{8a}\right] - \left[-\frac{GMm}{8a} - \frac{8\,GMm}{a}\right]$$

or $$\frac{1}{2} m v_{min}^2 = \left(\frac{45}{8}\right)\frac{GMm}{a}$$

$$\therefore \quad \boldsymbol{v_{min} = \frac{3\sqrt{5}}{2}\left(\sqrt{\frac{GM}{a}}\right)}$$

CHAPTER 7 SIMPLE HARMONIC MOTION

192. The pendulum makes $(24 \times 60 \times 60 + 60)$ half oscillations in 24 hour.
The time for half an oscillation is therefore,

$$\frac{T'}{2} = \frac{24 \times 60 \times 60}{60\,(24 \times 60 + 1)}$$

or $$\frac{T'}{2} = \frac{1440}{1441} = \pi\sqrt{\frac{l}{g}} \quad \ldots(1)$$

Let kl be the extra length required for an accurate 1 second beat. Then

$$1 = \pi\sqrt{(l + kl)/g} = \pi\sqrt{l/g}\,(\sqrt{k+1}) \quad \ldots(2)$$

Dividing equation (2) by (1), we get

$$\sqrt{k+1} = \left(\frac{1441}{1440}\right)$$

or $$k = 0.0014$$

∴ percentage increase in length will be 100 k or **0.14%.**

193. (a) $T = \frac{T_1}{2} + \frac{T_2}{2} = \pi\sqrt{\frac{m}{k}} + \pi\sqrt{\frac{m}{2k}}$

$$= \pi\sqrt{\frac{m}{k}}\left(1 + \frac{1}{\sqrt{2}}\right) = 5.36\sqrt{\frac{m}{k}}$$

∴ $$\boldsymbol{T = 5.36\sqrt{\frac{m}{k}}}$$

(b) From conservation of energy

$$\frac{1}{2}kA^2 = \frac{1}{2}(2k)(x_{max}^2)$$

⇒ $$x_{max} = \frac{A}{\sqrt{2}}$$

i.e., the most negative value of x is $\boldsymbol{-\frac{A}{\sqrt{2}}}$

194. The period of oscillations of the pendulum in the direction perpendicular to the rails is

$$T_1 = 2\pi\sqrt{\frac{l}{g}} \qquad (l = \text{length of thread})$$

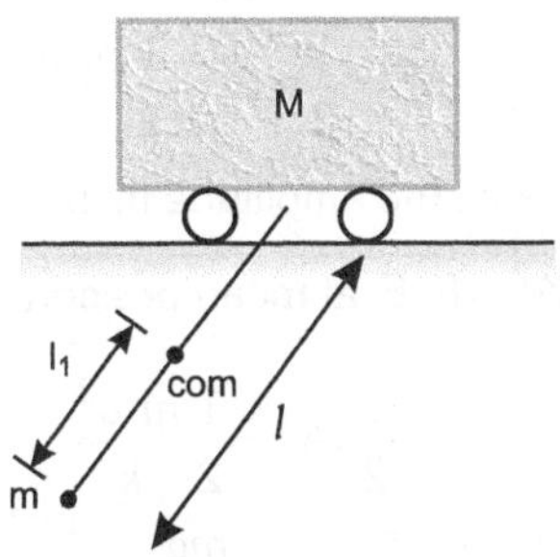

Time period (T_2) in the direction parallel to the rails :
The period of oscillations in the plane parallel to the rails for small oscillations can be found from the condition that the centre of mass (com) of the system remains stationary. The distance of com from m can be found by

$$ml_1 = M\,(l - l_1)$$

∴ $$l_1 = \frac{Ml}{M + m}$$

Thus the ball performs oscillations about COM with time period

$$T_2 = 2\pi\sqrt{\frac{l_1}{g}} = 2\pi\sqrt{\frac{Ml}{(M+m)\,g}}$$

∴ $$\boldsymbol{\frac{T_2}{T_1} = \sqrt{\frac{M}{M+m}}}$$

195. Relative to floor of the truck, block is acted upon by a pseudo force.

$$F = ma \qquad \text{(backwards)}$$

In equilibrium position

$$kx_0 = F \quad \text{or} \quad kx_0 = ma$$

or $$x_0 = \frac{ma}{k}$$

This gives the new position of equilibrium. However, block acquires a velocity, till it comes to its new equilibrium position, which can be obtained by

$$\int_0^v v\,.\,dv = \int_0^{x_0}\left(a - \frac{kx}{m}\right)dx$$

or $$\frac{v^2}{2} = ax_0 - \frac{kx_0^2}{2m} = \frac{1}{2}\frac{ma^2}{k} \qquad \left(x_0 = \frac{ma}{k}\right)$$

The kinetic energy of block at equilibrium position is therefore,

$$\text{KE} = \frac{1}{2}mv^2 = \frac{1}{2}\frac{m^2a^2}{k}$$

Let A be the amplitude then

$$\frac{1}{2}kA^2 = \text{K. E. at mean position}$$

or $$\frac{1}{2}kA^2 = \frac{1}{2}\frac{m^2a^2}{k}$$

or $$A = \frac{ma}{k}$$

substituting the value we have

$$A = \frac{(2)(4)}{40} = \mathbf{0.2\ m}$$

Period of oscillation will remain unchanged. Only the mean position is changed. Therefore,

$$T = 2\pi\sqrt{\frac{m}{k}} = 2\pi\sqrt{\frac{2}{40}}$$

or $$\boldsymbol{T = 1.4\ \mathrm{s}}$$

196. (a) Let F be the thrust and W the weight of the sphere. In equilibrium let x_0 be the compression of the spring, then

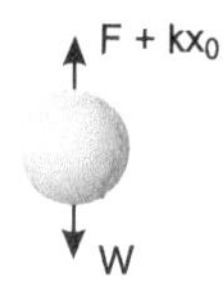

$$F + kx_0 = W$$

or $$k\,x_0 = W - F \qquad \dots(1)$$

If the sphere is further compressed by x, then total energy of the system will be

$$E = -(W - F)\cdot x + \frac{1}{2}k\,(x + x_0)^2 + \frac{1}{2}mv^2 + \frac{1}{4}\rho V v^2$$

Since friction is absent, total energy remains constant, hence

$$\frac{dE}{dt} = 0$$

or $$0 = -(W - F)\cdot\frac{dx}{dt} + k\,(x + x_0)\frac{dx}{dt} + mv\left(\frac{dv}{dt}\right) + \frac{1}{2}\rho V v\left(\frac{dv}{dt}\right) \qquad \dots(2)$$

From (1) and (2), with substitutions $\frac{dx}{dt} = v$ and $\frac{dv}{dt} = a$, we get

$$a = -\frac{k}{\frac{1}{2}\rho V + m}\cdot x$$

$\because$ $$a \propto -x$$

oscillations are simple harmonic, time period of which will be

$$T = 2\pi\sqrt{\left|\frac{x}{a}\right|} = 2\pi\sqrt{\frac{m + \frac{1}{2}\rho V}{k}}$$

$$= 2\pi\sqrt{\frac{0.5 + \frac{1}{2}\times 10^3 \times \frac{4}{3}\times\pi\times(0.08)^3}{500}}$$

$$\boldsymbol{T = 0.352\ \mathrm{s}}$$

(b) When it is accelerated upwards with an acceleration 'a'.

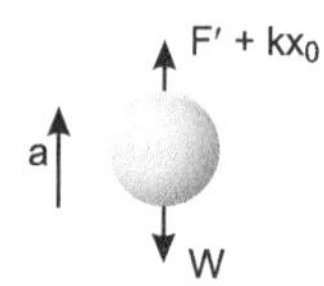

$$F' = \frac{F(g + a)}{g}$$

Now $$F' + k\,x_0 - W = \left(\frac{W}{g}\right)a$$

or $$k\,x_0 = \frac{W}{g}\cdot a + W - F\left(1 + \frac{a}{g}\right)$$

or $$k\,x_0 = (W - F) + \frac{a}{g}(W - F)$$

$$k\,x_0 = (W - F)\left(1 + \frac{a}{g}\right) \qquad \dots(3)$$

When displaced downwards, total energy would be

$$E = -(W - F)\frac{(g + a)}{g}\cdot x + \frac{1}{2}K(x + x_0)^2 + \frac{1}{2}mv^2 + \frac{1}{4}\rho V v^2$$

Substituting $\frac{dE}{dt} = 0$

$$\text{or } 0 = -(W - F)\left(1 + \frac{a}{g}\right)\frac{dx}{dt} + k(x + x_0)\frac{dx}{dt}$$

$$+ mv \cdot \frac{dv}{dt} + \frac{1}{2}\rho v V \cdot \frac{dv}{dt} \quad \ldots(4)$$

From (4) and (3) we get the same result as was obtained in part (a) *i.e.*,

$$\boldsymbol{T = 0.352 \text{ s}}$$

197. Displacement of B down the plane

$$= \frac{\text{displacement of } A}{2}$$

In equilibrium, let the block A is displaced downwards by an amount x_0. Then equilibrium of A gives

$$T = mg \quad \ldots(1)$$

and equilibrium of B gives

$$2T + mg \sin\theta = \frac{k x_0}{2} \quad \ldots(2)$$

Now let A is further depressed by x and its velocity at some moment is v, then velocity of B would be $v/2$.

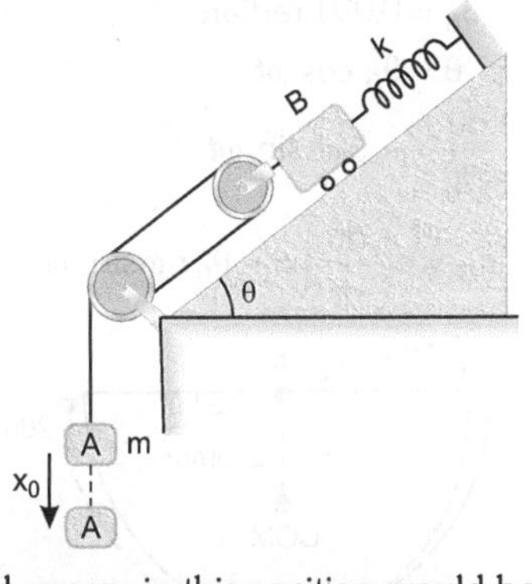

Total energy in this position would be

$$E = -m_A g x - m_B g \frac{x}{2} \sin\theta + \frac{1}{2} m_A v_A^2$$

$$+ \frac{1}{2} m_B v_B^2 + \frac{1}{2} k\left(\frac{x}{2} + \frac{x_0}{2}\right)^2$$

$$\text{or } E = -mgx - \frac{mgx \sin\theta}{2} + \frac{1}{2} mv^2$$

$$+ \frac{1}{2}\left(\frac{v}{2}\right)^2 + \frac{1}{8} k(x + x_0)^2$$

Since friction is absent E = constant, hence $\frac{dE}{dt} = 0$

$$\text{or } 0 = -mg\left(1 + \frac{\sin\theta}{2}\right)\frac{dx}{dt} + mv \cdot \frac{dv}{dt}$$

$$+ \frac{1}{4} mv \frac{dv}{dt} + \frac{1}{4} k(x + x_0)\frac{dx}{dt} \quad \ldots(3)$$

With the substitutions

$$\frac{dx}{dt} = v \quad \text{and} \quad \frac{dv}{dt} = a$$

From equations (1), (2) and (3), we get

$$5m(a) = -kx \quad \text{or} \quad a = -\left(\frac{k}{5m}\right) \cdot x$$

Since $\quad a \propto -x$

oscillations are simple harmonic, angular frequency of which is given by

$$\omega^2 = \frac{k}{5m} \quad \text{or} \quad \omega = \sqrt{\frac{\boldsymbol{k}}{\boldsymbol{5m}}}$$

198. Let x_0 be the elongation in the spring in equilibrium position. Then

$$kx_0 = mg \sin\theta \quad \ldots(1)$$

Let the spring is further stretched by x.
Then total energy of the system is

$$E = -mg\, x \sin\theta + \frac{1}{2} k(x + x_0)^2 + K_T + K_R$$

K_T = translational kinetic energy and
K_R = rotational kinetic energy of cylinder
$K_R = \frac{K_T}{2}$ in case of a cylinder in pure rolling

$$\text{Here } \quad K_R = \frac{K_T}{2} = \frac{\frac{1}{2}mv^2}{2} = \frac{1}{4}mv^2$$

$$\text{or} \quad K_R + K_T = \frac{3}{4}mv^2$$

$$\text{Hence } E = -mgx \sin\theta + \frac{1}{2}k(x + x_0)^2 + \frac{3}{4}mv^2$$

In case of pure rolling total energy is constant.

Hence $\frac{dE}{dt} = 0$

or $0 = -mg\sin\theta\left(\frac{dx}{dt}\right) + k(x + x_0)\frac{dx}{dt}$

$$+\frac{3}{2}mv\cdot\left(\frac{dv}{dt}\right) \quad \ldots(2)$$

With substitutions

$$\frac{dx}{dt} = v \text{ and } \frac{dv}{dt} = a$$

from equations (1) and (2), we get

$$a = -\left\{\frac{k}{(3/2)m}\right\}\cdot x$$

Since $a \propto -x$

oscillations are simple harmonic, time period of which is given by

$$T = 2\pi\sqrt{\left|\frac{x}{a}\right|} = 2\pi\sqrt{\frac{3m}{2k}}$$

$$= 2\pi\sqrt{\frac{3\times 7}{2\times 800}}$$

$$\boldsymbol{T = 0.72\,s}$$

(b) Maximum velocity will be at mean position. From conservation of energy, we have

$$\frac{1}{2}kx^2 = \frac{3}{4}mv_{max}^2$$

or $v_{max} = \left\{\sqrt{\frac{2k}{3m}}\right\}x$

or $v_{max} = \left\{\sqrt{\frac{2\times 800}{3\times 7}}\right\}(10)\frac{mm}{s} = 87.3\frac{mm}{s}$

$\therefore$ $\boldsymbol{v_{max} = 87.3\ mm/s}$

199. $r = 220$ mm $= 0.22$ m

The position of COM of the whole frame from point A can be determined as :

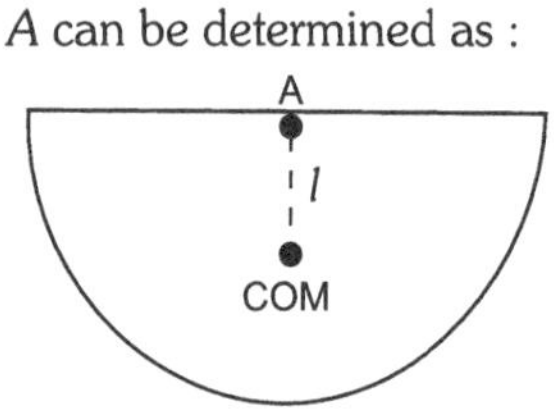

$$l = \frac{(\pi r)\left(\frac{2r}{\pi}\right)}{(\pi r + 2r)} = 0.39r$$

The moment of inertia about A will be :

$$I = (\lambda)(2r)\frac{(2r)^2}{12} + \lambda(\pi r)(r^2)$$

where λ = mass per unit length of wire

$\therefore$ $I = 3.81r^3\lambda$

This is a physical pendulum, time period of which will be given by

$$T = 2\pi\sqrt{\frac{I}{mgl}}$$

or $\omega = \frac{2\pi}{T} = \sqrt{\frac{mgl}{I}}$

$$= \sqrt{\frac{\lambda(2r + \pi r)gl}{I}} = \sqrt{\frac{5.14\lambda rgl}{I}}$$

$$= \sqrt{\frac{(5.14)(\lambda)(r)(9.8)(0.39r)}{(3.81r^3\lambda)}}$$

$$= \sqrt{\frac{5.156}{r}} = \sqrt{\frac{5.156}{0.22}}$$

or $\omega = 4.84$ rad/s

$$\theta_0 \approx \frac{20}{220} \text{ radian}$$

or $\theta_0 = 0.091$ radian

$\therefore$ $\theta = \theta_0\cos\omega t$

or $\frac{d\theta}{dt} = -\theta_0\omega\sin\omega t$

or $v_B = r\left(\frac{d\theta}{dt}\right) = -\theta_0 r\omega\sin\omega t$

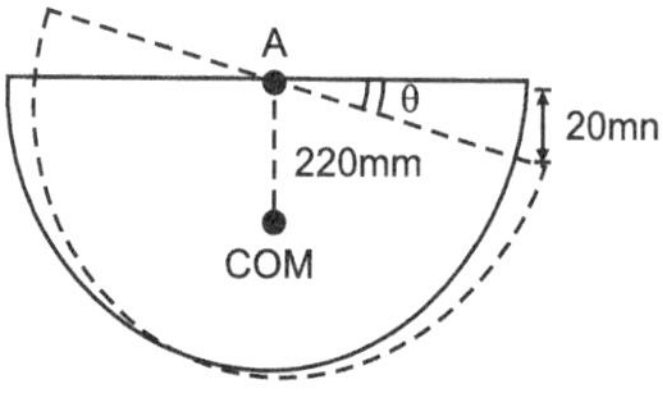

So magnitude of velocity of B after 8 seconds will be

$$|\vec{v}_B| = (0.091)(220)(4.84)\sin(4.84)(8)$$

$$= \mathbf{82.6\ mm/s}$$

200. When displaced slightly by an angle θ, the semicylinder will oscillate about the point of contact with the surface. The torque of weight will provide the necessary restoring torque.

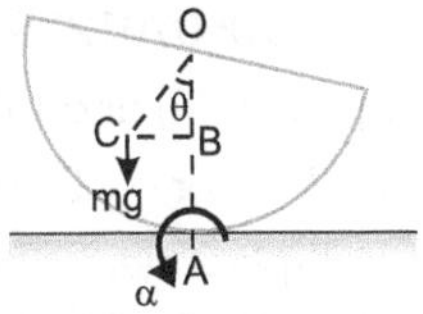

$$OC = \frac{2R}{\pi},\ BC = \frac{2R}{\pi}\sin\theta,\ OB = \frac{2R}{\pi}\cos\theta$$

and $AB = R - \frac{2R}{\pi}\cos\theta$

Now $\tau = -mg\,(BC)$

or $I_A\alpha = -mg\left(\frac{2R}{\pi}\sin\theta\right)$ ...(1)

$$I_O = I_C + m(OC)^2$$

$\therefore$ $I_C = I_O - m(OC)^2$

$\therefore$ $I_A = I_C + m(AC)^2$

Hence $I_A = I_O - m(OC)^2 + m(AC)^2$

$$= I_O - m(OC)^2 + m(BC^2 + AB^2)$$

Substituting the values

$$I_O = mR^2,\quad (OC)^2 = \left(\frac{2R}{\pi}\right)^2,$$

$$(BC)^2 = \left(\frac{2R}{\pi}\sin\theta\right)^2$$

and $AB^2 = \left(R - \frac{2R}{\pi}\cos\theta\right)^2,$

we get $I_A = 2mR^2\left(1 - \frac{2}{\pi}\cos\theta\right)$

So, equation (1) can be written as

$$2mR^2\left(1 - \frac{2}{\pi}\cos\theta\right).\alpha = -mg\left(\frac{2R}{\pi}\sin\theta\right)$$

For small oscillations, θ is small *i.e.*, $\sin\theta \approx \theta$ and $\cos\theta \approx 1$

Hence $2mR^2\left(1 - \frac{2}{\pi}\right)\alpha = -\frac{2mgR}{\pi}.\theta$

Since α is proportional to $-\theta$, oscillations will be simple harmonic in nature, the time period of these oscillations will be given by

$$T = 2\pi\sqrt{\left|\frac{\theta}{\alpha}\right|} \quad \text{or} \quad \boldsymbol{T = 2\pi\sqrt{\frac{R(\pi-2)}{g}}}$$

201. Let '2*a*' be the sides of the square. If it is displaced by θ, then

$$2a = 0.4 \text{ m}$$

$\therefore$ $a = 0.2$ m

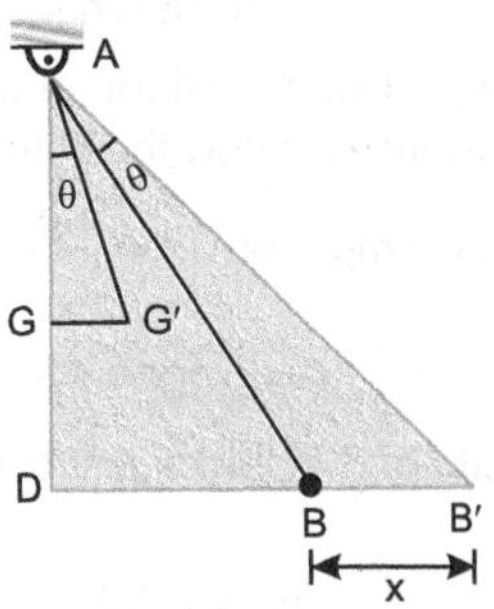

Compression/elongation in each spring :

$$x = (AB)\,\theta = \sqrt{5}\,a\,\theta$$

Net restoring torque about *A* will be :

$$\tau = -\{mg\,(GG') + 2kx\,(AD)\}$$

$$\tau = -\{mg\,(AG.\theta) + 2kx\,(2a)\}$$

$$\tau = -\{mg\,(a\theta) + 2\sqrt{5}\,ka\theta\,(2a)\}$$

Since $\tau \propto -\theta$

$\therefore$ oscillations are simple harmonic in nature

Hence $I_A.\alpha = -\{mga + 4\sqrt{5}\,ka^2\}\theta$

Here $I_A = I_G + m(GA)^2$

$$= m\left(\frac{4a^2 + 4a^2}{12}\right) + ma^2$$

$$= \frac{5}{3}ma^2$$

$$\left(\frac{5}{3}ma^2\right)\alpha = -\{mga + 4\sqrt{5}\,ka^2\}\theta$$

$$\therefore\ f = \frac{1}{2\pi}\sqrt{\left|\frac{\alpha}{\theta}\right|} = \frac{1}{2\pi}\sqrt{\frac{mga + 4\sqrt{5}ka^2}{\frac{5}{3}ma^2}}$$

Substituting the values, we have :

$$f = \frac{1}{2\pi}\sqrt{\frac{(20)(9.8)(0.2) + (4\sqrt{5})(1.4\times10^3)(0.2)^2}{\frac{5}{3}(20)\,(0.2)^2}}$$

or $\boldsymbol{f = 3.2\ \text{Hz}}$

202. $k_{eq} = 4k = 4 \times 800\text{ N/m} = 3200\text{ N/m}$

$$\omega = \sqrt{\frac{k_{eq}}{M+m}} = \sqrt{\frac{3200}{28+3}}$$

or $\quad = 10.16\text{ rad/s} \quad \ldots(1)$

The equilibrium position of the system will be at a distance x below the initial position where

$$k_{eq}.x = mg \quad \text{or} \quad x = \frac{mg}{k_{eq}} = \frac{(3)(9.8)}{3200}$$

or $\quad x = 0.0092\text{ m} \quad \ldots(2)$

Velocity of 3 kg block just before collision will be :

$$v_O = \sqrt{2gh}$$

$\therefore$ From conservation of linear momentum velocity of both the blocks just after collision would be :

$$v = \frac{mv_0}{(M+m)} = \frac{(3)\sqrt{2gh}}{(28+3)}$$

$$= \frac{3\sqrt{2 \times 9.8 \times 2}}{31} = 0.606\text{ m/s} \quad \ldots(3)$$

Now at time $t = 0$, displacement of blocks from mean position is $x = 0.0092$ m and velocity is

$$v = -0.606\text{ m/s}.$$

Hence let the equation of displacement from mean position is

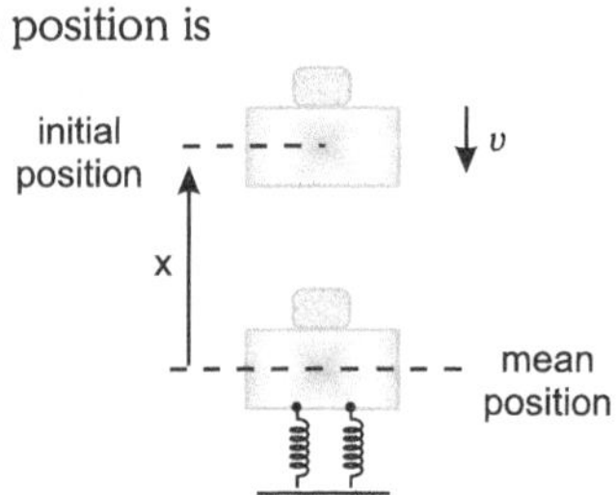

$$x = a\sin(\omega t + \phi)$$

then $\quad v = a\omega\cos(\omega t + \phi)$

$\Rightarrow \quad 0.0092 = a\sin\phi \quad \ldots(4)$

and $\quad -0.606 = 10.16\, a\cos\phi$

$(\omega = 10.16\text{ rad/s})$

or $\quad -0.06 = a\cos\phi \quad \ldots(5)$

Squaring and adding (4) and (5), we get

$$a^2 = 0.00368464$$

or $\quad a = 0.060701235$ m

and from equation (4) and (5), we get

$$\phi = 171.28° \quad \text{or} \quad 2.99\text{ rad}$$

Therefore, displacement from its mean position at any time t would be

$$x = 0.06\sin(10.16t + 2.99)$$

$\therefore$ Displacement at any time from the initial position would be :

$$s = 0.0092 - x$$

$$= \{0.0092 - 0.06\sin(10.16t + 2.99)\}\text{m}$$

or $\mathbf{s = \{9.2 - 60\sin(10.16t + 2.99)\}}$ **mm**

Note: For displacement s from initial position downward direction is positive.

203. Let m be the mass of each disc. Since the upper disc rolls without slipping, P will be the instantaneous axis of rotation.

$$C_1P = R,$$

$$C_2A = L\sin\theta,$$

$$PA = L\cos\theta - R$$

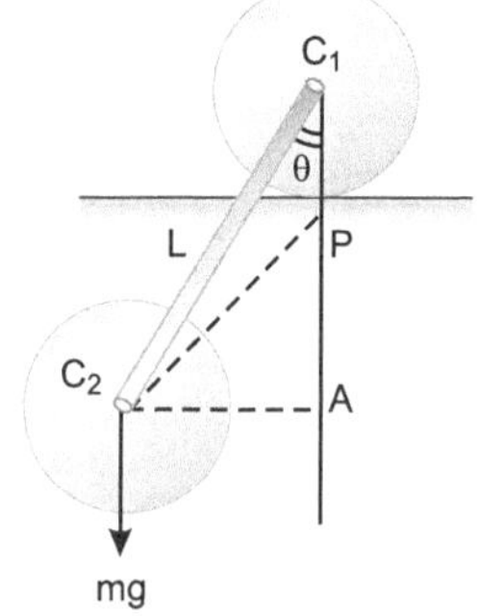

Restoring torque is :

$$\tau = -mg(C_2A)$$

or $\quad I_P.\alpha = -mg(L\sin\theta) \quad \ldots(1)$

I_P = moment of inertia of upper disc + moment of inertia of lower disc

$$\therefore I_P = \left(mR^2 + \frac{1}{2}mR^2\right) + m(C_2P)^2 \frac{1}{2}mR^2 +$$

$$= 2mR^2 + m\{(C_2A)^2 + (PA)^2\}$$

$$= 2mR^2 + m\{(L\sin\theta)^2 + (L\cos\theta - R)^2\}$$

$$I_P = mR^2 + mL^2 - 2mRL\cos\theta$$

So equation (1) can be written as

$$\{mR^2 + mL^2 - 2mRL\cos\theta\}\alpha = -mgL\sin\theta$$

For small oscillations θ is small, so

$$\sin\theta \approx \theta \quad \text{and} \quad \cos\theta \approx 1$$

Hence $(R^2 + L^2 - 2RL).\alpha = -gL.\theta$

Since α is proportional to −θ, oscillations are simple harmonic, time period of which is :

$$T = 2\pi\sqrt{\left|\frac{\theta}{\alpha}\right|} = 2\pi\sqrt{\frac{R_2 + L^2 - 2RL}{gL}}$$

Hence the time period of small oscillations is

$$\boldsymbol{T = 2\pi\sqrt{\frac{R^2 + L^2 - 2RL}{gL}}}$$

204. Let θ_0 be the amplitude of oscillation. Then

$$h = 4R(1 - \cos\theta_0) = 4R\left\{2\sin^2\frac{\theta_0}{2}\right\}$$

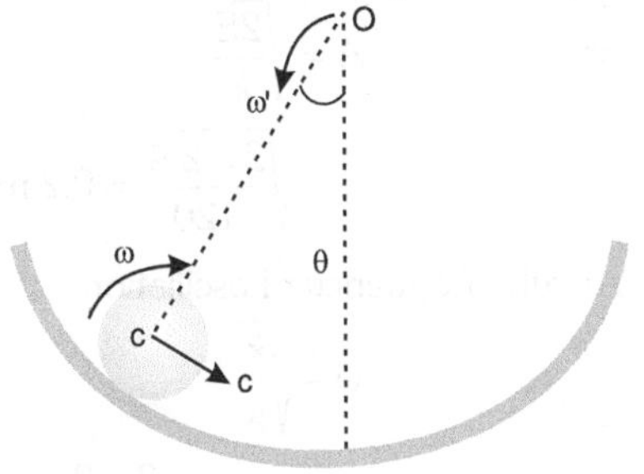

or $$h = 2R\theta_0^2 \quad \ldots(1)$$

$\sin\frac{\theta_0}{2} \approx \frac{\theta_0}{2}$ for small oscillations.

Since there is no slipping, mechanical energy will remain conserved in position 1 and 2.

$$v = 4R\omega'$$

$$\omega = \frac{v}{R} = 4\omega'$$

Here ω′ is the anguler velocity of COM of sphere about C.

∴ From conservation of mechanical energy

$$mgh = \frac{1}{2}mv^2 + \frac{1}{2}I\omega^2$$

or $$mg(2R\theta_0^2) = \frac{1}{2}m(4R\omega')^2 + \frac{1}{2}\left(\frac{2}{5}mR^2\right)(4\omega')^2$$

or $$g\theta_0^2 = \frac{28}{5}R\omega'^2$$

or $$\omega' = \sqrt{\frac{5g}{28R}}\cdot\theta_0 \quad \ldots(2)$$

Now ω′ = maximum angular velocity of centre of mass of sphere about point C

$$= \left(\frac{2\pi}{T}\right)\theta_0 \quad \ldots(3)$$

From (2) and (3), we have

$$\frac{2\pi}{T} = \sqrt{\frac{5g}{28R}} \quad \text{or} \quad \boldsymbol{T = 2\pi\sqrt{\frac{28R}{5g}}}$$

205. Let v and ω be the linear speed and angular speed of centre of mass of cylinder at small displacement θ. Then total energy of the system in this position would be

E = potential energy of link + (rotational + kinetic energy of two cylinders) + kinetic energy of link

or $$E = mgr_0(1 - \cos\theta) + 2\left\{\frac{3}{4}Mv^2\right\} + \frac{1}{2}m(r - r_0)^2\omega^2$$

or $$E = mgr_0(1 - \cos\theta) + \frac{3}{2}M\omega^2r^2 + \frac{1}{2}m(r - r_0)^2\omega^2 \quad (v = r\omega)$$

In case of pure rolling E is constant, hence $\frac{dE}{dt} = 0$

or $$0 = mgr_0\sin\theta\cdot\left(\frac{d\theta}{dt}\right) + 3Mr^2\omega\cdot\frac{d\omega}{dt} + m\omega(r - r_0)^2\cdot\frac{d\omega}{dt}$$

θ is small, so we can substitute sin θ = θ

Further $\frac{d\theta}{dt} = \omega$ and $\frac{d\omega}{dt} = \alpha$

This gives

$$\alpha = -\frac{mgr_0}{3Mr^2 + m(r - r_0)^2}\cdot\theta$$

$$\therefore \quad \boldsymbol{T = 2\pi\sqrt{\left|\frac{\theta}{\alpha}\right|}}$$

$$\boldsymbol{= 2\pi\sqrt{\frac{3Mr^2 + m(r - r_0)^2}{mg\,r_0}}}$$

206. Let at any instant centre of cylinder is displaced by x (towards left).
Then spring attached at C is compressed by x and spring attached at A elongates by $2x$. Let v be the velocity of centre of cylinder and ω its angular velocity.

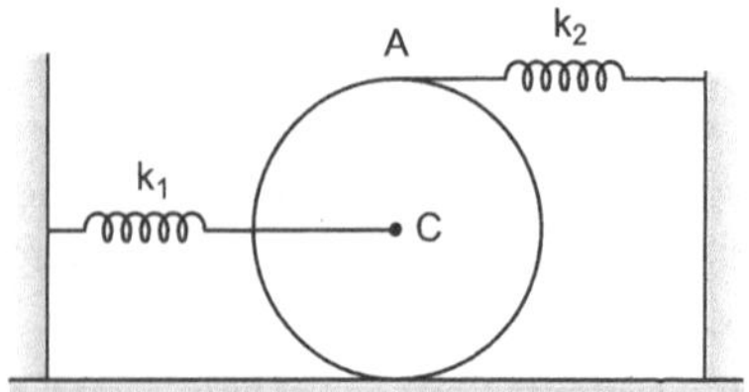

Total mechanical energy in displaced position is

$$E = \frac{1}{2}mv^2 + \frac{1}{2}I_C\omega^2 + \frac{1}{2}k_1x^2 + \frac{1}{2}k_2(2x)^2$$

But $\qquad \omega = \dfrac{v}{R}$

and $\qquad I_C = \dfrac{1}{2}mR^2$

Hence $E = \dfrac{3}{4}mv^2 + \dfrac{1}{2}k_1x^2 + 2k_2x^2$

In case of pure rolling energy is conserved

$$\therefore \quad \frac{dE}{dt} = 0$$

or $\dfrac{3}{2}mv\left(\dfrac{dv}{dt}\right) + k_1x\left(\dfrac{dx}{dt}\right) + 4k_2x\left(\dfrac{dx}{dt}\right) = 0$

$\dfrac{dx}{dt} = v$ and $\dfrac{dv}{dt} = a$ (acceleration), with these substitutions we get,

$$\frac{3}{2}ma = -(k_1 + 4k_2)x$$

Since $\qquad a \propto -x$

oscillations are simple harmonic in nature. Time period of which is given by

$$T = 2\pi\sqrt{\left|\frac{x}{a}\right|} = 2\pi\sqrt{\left|\frac{3m}{2(k_1 + 4k_2)}\right|}$$

Substituting the values we get

$$T = 2\pi\sqrt{\frac{3 \times 1}{2(10 + 4 \times 20)}}$$

or $\qquad \boldsymbol{T = 0.81\ s}$

207. $V_{CM} = \dfrac{2 \times 2 + 3 \times 0}{2 + 3} = 0.8$ m/s

Energy of oscillation,

$$E = \frac{1}{2} \times 2 \times (2)^2 - \frac{1}{2} \times (2 + 3) \times V_{CM}^2$$

or $\quad E = 4 - \dfrac{1}{2} \times 5 \times (0.8)^2 = 2.4$ J

Let A be the maximum extension in the spring, then

$$E = \frac{1}{2}kA^2$$

or $\qquad A = \sqrt{\dfrac{2E}{k}}$

$$= \sqrt{\frac{2 \times 2.4}{120}} = 0.2\ \text{m}$$

Angular frequency of oscillation,

$$\omega = \sqrt{\frac{k}{\mu}}$$

where μ = reduced mass $= \dfrac{3 \times 2}{3 + 2} = 1.2$ kg

$$\therefore \quad \omega = \sqrt{\frac{120}{1.2}} = 10\ \text{rad/s}$$

Amplitude of 2 kg $= \left(\dfrac{3}{3 + 2}\right)A = 0.12$ m

After time t, centre of mass will move a distance $d = 0.8t$.
From centre of mass frame, at $t = 0$, block 2 kg is in its mean position and travelling towards positive x-axis with amplitude 0.12 m and angular frequency 10 rad/s. Hence, displacement of 2 kg after t seconds
= displacement of centre of mass
+ displacement of 2 kg w.r. to centre of mass
$= 0.8t + 1.2 \sin 10t$

208. Let F be the extra tension (net restoring force) in the string when mass m is displaced by 'x' from its mean position. Then,

$$F = k_2x_2 \quad \text{and} \quad 2F = k_1x_1$$

where x_1 and x_2 are further (in displaced position) extensions in the springs.

$$\therefore \quad x_1 = \frac{2F}{k_1} \quad \text{and} \quad x_2 = \frac{F}{k_2} \qquad \ldots(1)$$

Further, $x = 2x_1 + x_2$

$$= \frac{4F}{k_1} + \frac{F}{k_2} = F\left(\frac{4k_2 + k_1}{k_1k_2}\right)$$

or

$$F = \left(\frac{k_1k_2}{4k_2 + k_1}\right)x$$

As $F \propto x$, motion is simple harmonic in nature.

$$k_{eff} = \frac{k_1k_2}{4k_2 + k_1}$$

$$\therefore \quad \mathbf{T = 2\pi\sqrt{\frac{m\,(4k_2 + k_1)}{k_1k_2}}}$$

209. As net force on the system in horizontal direction is zero.

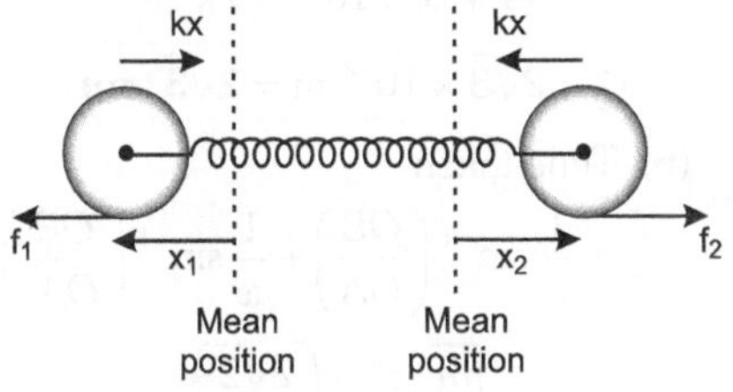

$$m_1x_1 = m_2x_2 \qquad \ldots(1)$$

$$x = x_1 + x_2 \qquad \ldots(2)$$

For cylinder of mass m_1:

$$kx - f_1 = m_1a_1 \qquad \ldots(3)$$

$$\alpha_1 = \frac{f_1r}{\frac{1}{2}m_1r^2} \qquad \ldots(4)$$

For pure rolling to take place,

$$a_1 = r\alpha_1 \qquad \ldots(5)$$

From equation (4) and (5)

$$f_1 = \frac{1}{2}m_1a_1$$

$$\therefore \quad kx = \frac{3}{2}m_1a_1$$

or

$$k(x_1 + x_2) = \frac{3}{2}m_1a_1$$

or

$$k\left(x_1 + \frac{m_1x_1}{m_2}\right) = \frac{3}{2}m_1a_1$$

or

$$a_1 = \frac{2k(m_1 + m_2)}{3m_1m_2}x_1$$

As $a_1 \propto x_1$

motion is simple harmonic in nature.

$$f = \frac{1}{2\pi}\sqrt{\left|\frac{a_1}{x_1}\right|}$$

$$= \mathbf{\frac{1}{2\pi}\sqrt{\frac{2k(m_1 + m_2)}{3m_1m_2}}}$$

210. When 'm' is further displaced from its mean position by 'x'. Then point A is also displaced by x. Displacement of point B in this case will be $\frac{b}{a}\cdot x$. Let F be the extra tension (net restoring force) in this case. Then extra forces on the rod (in the form of extra tension and extra spring force) in displaced position are shown below:

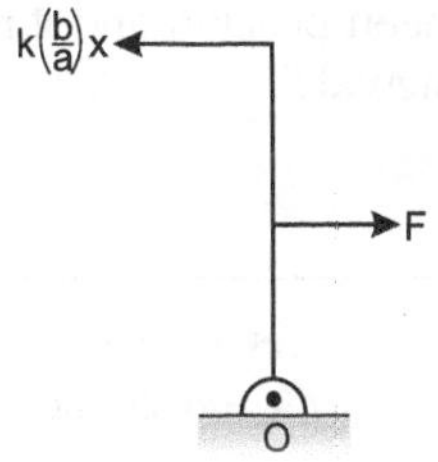

As the rod is massless net extra torque about O should be zero. Hence,

$$F \cdot a = \left\{k\left(\frac{b}{a}\right)\cdot x\right\}\{b\}$$

or

$$F = k\left(\frac{b^2}{a^2}\right)x$$

As $F \propto x$

motion of block is simple harmonic in nature with

$$k_{eff} = k\left(\frac{b^2}{a^2}\right)$$

$$\therefore \quad T = 2\pi\sqrt{\frac{m}{k_{eff}}} = \mathbf{\frac{2\pi a}{b}\sqrt{\frac{m}{k}}}$$

211. When the block is displaced a distance x from its mean position, let F be the extra tension (net restoring force) in the string attached to the block. Then extra tension in the string connecting the spring will be $\frac{F}{4}\cdot$ Let x_0 be the extra extension in the spring in this case, then

$$\frac{F}{4} = kx_0 \qquad \ldots(1)$$

When spring extends further by x_0, the block comes down a distance, $\frac{x_0}{4}$

or $$x = \frac{x_0}{4} = \frac{F}{16k}$$

$\therefore$ $$F = (16k)\,x$$

As $F \propto x$, motion of the block is simple harmonic with

$$k_{eff} = 16k$$

$\therefore$ $$f = \frac{1}{2\pi}\sqrt{\frac{16k}{m}} = \mathbf{\frac{2}{\pi}\sqrt{\frac{k}{m}}}$$

212. Let the particle is at P at an instant t starting from mean position and it returns to P again after interval t'.

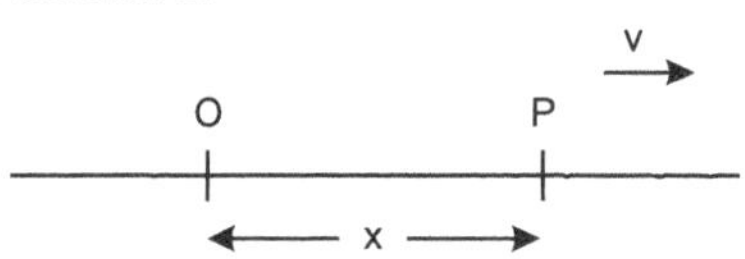

Then $$x = a\sin\omega t \qquad \ldots(1)$$

Also $$x = a\sin\omega(t + t') \qquad \ldots(2)$$

From equation (1)

$$\frac{dx}{dt} = a\omega\cos\omega t = v$$

$\Rightarrow$ $$\cos\omega t = \frac{v}{a\omega} \qquad \ldots(3)$$

from equation (2)

$$x = a\,[\sin\omega t'\cos\omega t + \cos\omega t' \,.\, \sin\omega t]$$

$$= a\left[\sin\omega t'\,\frac{v}{a\omega} + \frac{x}{a}\cos\omega t'\right]$$

Solving this equation for t' we get,

$$t' = \frac{1}{\omega}\tan^{-1}\left(\frac{2vx\omega}{\omega^2x^2 - v^2}\right)$$

213. (a) From conservation of mechanical energy, velocity just before collision:

$$\frac{1}{2}kx_0^2 = \frac{1}{2}kx^2 + \frac{1}{2}mv^2$$

$$v = \sqrt{\frac{k(x_0^2 - x^2)}{m}} = \frac{\sqrt{64\,(16-8)\times 10^{-4}}}{1}$$

$$= 16\sqrt{2}\times 10^{-2}\text{ m/s}$$

$$= 16\sqrt{2}\text{ cm/s}$$

Velocity after collision

$$v' = eV = \frac{1}{\sqrt{2}}\times 16\sqrt{2} = 16\text{ cm/s}$$

COE: $$\frac{1}{2}mv'^2 + \frac{1}{2}kx^2 = \frac{1}{2}kx_0'^2$$

$$\frac{1}{2}mv'^2 = \frac{1}{2}k(x_0'^2 - x^2)$$

$$1\times 256\times 10^{-4} = 64\,(x_0'^2 - x^2)$$

$$4\times 10^{-4} = x_0'^2 - x^2$$

$$(4+8)\times 10^{-4} = x_0'^2$$

$$x_0' = 2\sqrt{3}\times 10^{-2}\text{ m} = \mathbf{2\sqrt{3}\text{ cm}}$$

(b) Time taken

$$= \frac{T}{4} + \frac{1}{\omega}\sin^{-1}\left(\frac{OB}{OA}\right) + \frac{1}{\omega}\sin^{-1}\left(\frac{OB}{OA'}\right) + \frac{T}{4}$$

$$= \pi\sqrt{\frac{m}{k}} + \sqrt{\frac{m}{k}}\sin^{-1}\left(\frac{2\sqrt{2}}{4}\right) + \sqrt{\frac{m}{k}}\sin^{-1}\left(\frac{2\sqrt{2}}{2\sqrt{3}}\right)$$

$$= \pi\sqrt{\frac{m}{k}} + \sqrt{\frac{m}{k}}\frac{\pi}{4} + \sqrt{\frac{m}{k}}\sin^{-1}\left(\sqrt{\frac{2}{3}}\right)$$

$$= \mathbf{\frac{5\pi}{4}\sqrt{\frac{m}{k}} + \sqrt{\frac{m}{k}}\sin^{-1}\left(\sqrt{\frac{2}{3}}\right)}$$

214. (a) Point C should lie below A.

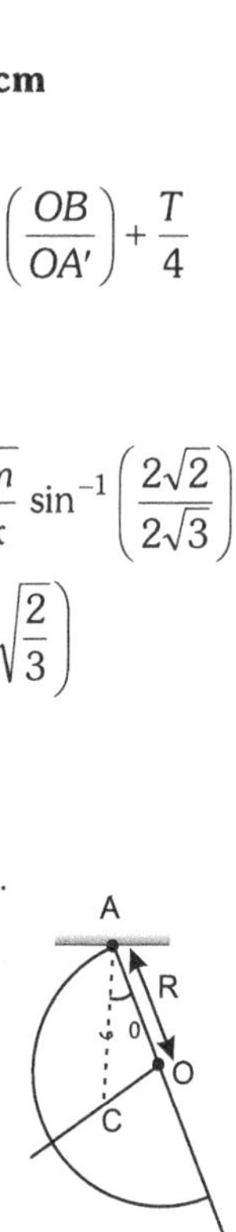

$$OC = \frac{4R}{3\pi}$$

$\Rightarrow$ $$\tan\theta_0 = \frac{OC}{OA} = \frac{4}{3\pi}$$

$$\theta_0 = \tan^{-1}\left(\frac{4}{3\pi}\right)$$

(b) The frequency of oscillation for a compound pendulum is

$$f = \frac{1}{2}\sqrt{\frac{mgd}{I}}$$

where d = distance of the CM from the point of suspension.

I = moment of inertia about the point of suspension.

$$d = \sqrt{R^2 + \left(\frac{4R}{3\pi}\right)^2} = \frac{R}{3\pi}\sqrt{9\pi^2 + 16}$$

$$I = I_C + M(AC)^2$$

$$= [I_0 - M(OC)^2] + M(AC)^2$$

$$= I_0 + M(AO)^2$$

$$= \left(\frac{1}{2}\right)MR^2 + MR^2 = \left(\frac{3}{2}\right)MR^2$$

$$\therefore \quad f = \frac{1}{2\pi}\sqrt{\frac{Mg\left(\frac{R}{3\pi}\right)\sqrt{9\pi^2+16}}{\left(\frac{3}{2}\right)MR^2}}$$

$$= \sqrt{\frac{g\sqrt{9\pi^2+16}}{18\pi^3 R}}$$

215. Let the piston be displaced x downward from equilibrium position and x_1 and x_2 be the deformation of the springs k_1 and k_2 respectively. Using constraint relation

$$x_1 + x_2 = 2x \quad \text{...(1)}$$

Increase in tension is uniform. Hence,

$$k_1x_1 = k_2x_2 \quad \text{...(2)}$$

$$\therefore \quad x_1 = \frac{2k_2x}{k_1+k_2} \text{ and } x_2 = \frac{2k_1x}{k_1+k_2} \quad \text{...(3)}$$

For ideal gas in the cylinder

$$PV^\gamma = \text{constant}$$

$$\frac{dP}{P} + \gamma\frac{dV}{V} = 0$$

$$|dP| = -\frac{\gamma P}{V}\cdot dV$$

Net restoring force

$$= 2\text{ (increase in tension)} + |dP|\, A$$

$$= 2k_1x_1 + |dP|\, A$$

$$\therefore \quad F = \frac{4k_1k_2}{k_1+k_2}x + \frac{\gamma P_0}{V_0}\cdot dV\,.\,A$$

$$F = \frac{4k_1k_2x}{k_1+k_2} + \frac{\gamma P_0A^2x}{V_0}$$

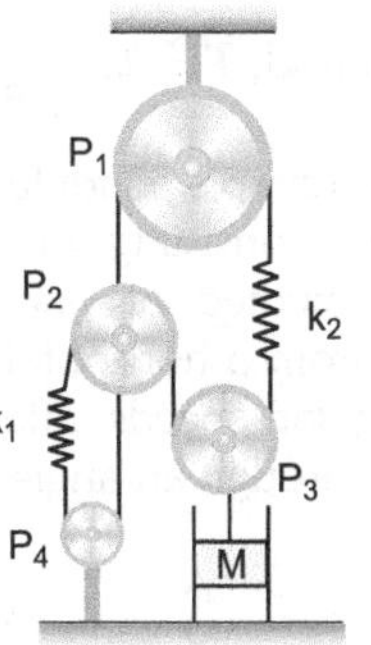

Hence acceleration of the piston

$$a = -\frac{1}{M}\left(\frac{4k_1k_2}{k_1+k_2} + \frac{\gamma P_0A^2}{V_0}\right)x$$

$$\omega^2 = \frac{1}{M}\left(\frac{4k_1k_2}{k_1+k_2} + \frac{\gamma P_0A^2}{V_0}\right)$$

$$\boldsymbol{T = 2\pi\sqrt{\frac{M}{\left(\frac{4k_1k_2}{k_1+k_2} + \frac{\gamma P_0A^2}{V_0}\right)}}}$$

216. Suppose that the mass M undergoes a small displacement x from its equilibrium position and springs of force constant k and $2k$ undergo corresponding extensions x_1 and x_2.

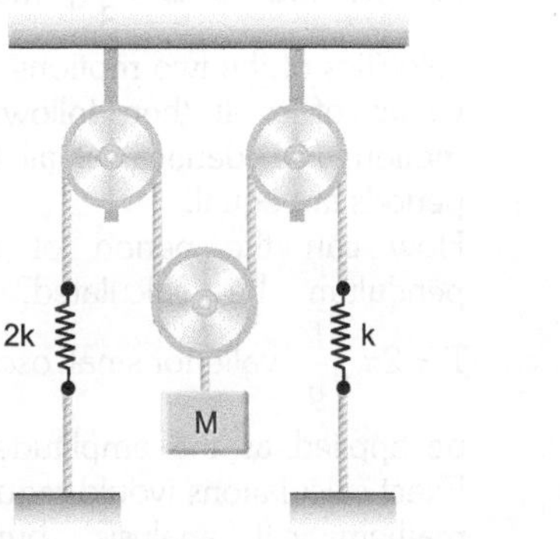

Then, $\quad x_1 + x_2 = 2x$

$$kx_1 = 2kx_2$$

$$\therefore \quad x_1 = \frac{4x}{3}, \quad x_2 = \frac{2x}{3}$$

Net restoring force in displaced position

$$F = -2 \text{ (extra tension)}$$

$$= -2\,(kx_1) = -\frac{8kx}{3}$$

$\therefore$ Time period, $\boldsymbol{T = 2\pi\sqrt{\frac{3M}{8k}}}$

217. Consider a simple pendulum of length L and a pendulum consisting of a uniform rod of length l pivoted at one end. If both are released from a horizontal position, what are their angular speeds after they have each travelled through an angle α?

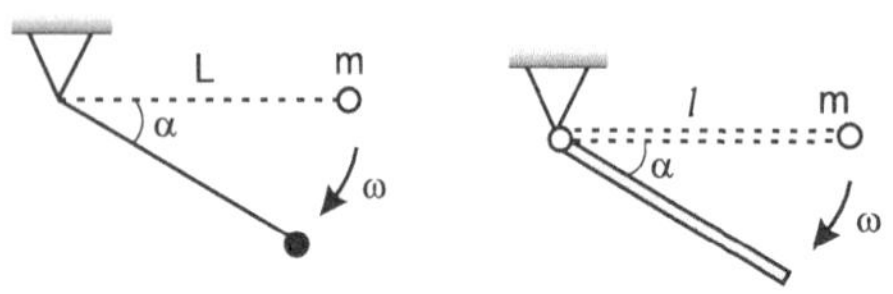

The principle of conservation of energy yields

$$\frac{1}{2}mL^2\omega^2 = mgL \sin\alpha,$$

i.e., $$\omega = \sqrt{\frac{2g}{L}\sin\alpha}$$

for the simple pendulum, and

$$\frac{1}{2}\frac{ml^2}{3}\omega^2 = mg\frac{l}{2}\sin\alpha,$$

i.e., $$\omega = \sqrt{\frac{3g}{l}\sin\alpha}$$

for the rod. If $L = \frac{2}{3}l$, then the angular velocities of the two motions are equal for all values of α. It then follows that the two motions are identical at all times and their periods are equal.

How can the period of this equivalent pendulum be calculated? The formula $T = 2\pi\sqrt{\frac{L}{g}}$, valid for small oscillations, cannot be applied as the amplitude here is large. Exact calculations would require complicated mathematical analysis, but this is not necessary if, instead of calculating the period T, we only wish to determine its dependence on L.

The period of swing of the simple pendulum may depend on its length L, the mass of its bob m, the gravitational acceleration g and the maximum angle of deviation α_{max}. If the dimensions of the quantities involved are taken into consideration, this functional dependence can only be of the form

$$T(L, m, g, \alpha_{max}) = f(\alpha_{max})\sqrt{\frac{L}{g}}$$

To justify this assertion, we note the following points. The dimension of mass is the 'kilogram' and since the 'kilogram' does not occur in the dimensions of any of the other quantities, the period (which has dimension 'seconds') cannot depend upon the mass of the bob. On the other hand, 'seconds' occur only in g, and therefore the required dimension of 'seconds' in T can only be obtained if T is inversely proportional to the square root of g. Finally, in order to settle the 'metre' dimension, the period has to be proportional to the square root of L. The form of the function $f(\alpha_{max})$ cannot be determined via dimensional analysis, since the angle is dimensionless. The only available information is that for small angles $f(\alpha_{max}) \approx 2\pi$.

From the above reasoning, it can be concluded that (with the same initial displacements) the period of a simple pendulum of length $\frac{2}{3}l$ is $\sqrt{\frac{2}{3}}$ times that of a simple pendulum of length l. Thus, the period of a pivoted rod of length l is approximately 82 per cent of that of a simple pendulum of the same length. This conclusion is valid not only for horizontal release, but for any common initial starting position.

218. Let AB be an elastic string of natural length $(a + b)$. Let the particle of mass m be attached to the string at point O such that $OA = a$, $OB = b$ and $a > b$.

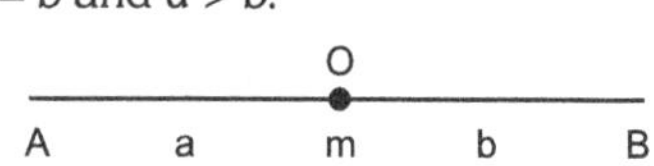

When the particle is held at B, the portion AO of the string is stretched while portion OB is slack and so when the particle is released it moves towards O.

Let the particle be at any point P (at a distance x from O between O and B) at any time t.

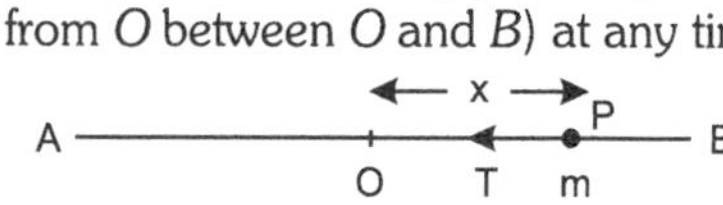

Tension in the string AP is

$$T = Y\frac{x}{a}, \text{ action towards } O.$$

Tension in PB is zero as it is slack.
Equation of motion is

$$m\frac{d^2x}{dt^2} = -\frac{Y}{a}x \quad \text{or} \quad \frac{d^2x}{dt^2} = -\frac{Y}{am}x \qquad \dots(1)$$

This represents a simple harmonic motion with centre at O and amplitude OB.
Let the particle take t_1 time in reaching O from B. Then $t_1 = \frac{1}{4} \times$ time period of SHM represented by (1) $= \frac{1}{4}\cdot\frac{2\pi}{\sqrt{Y/am}} = \frac{\pi}{2}\sqrt{\frac{am}{Y}}$

The particle will have certain kinetic energy at O, due to which it will move towards A, stretching the portion OB. The segment AO becomes slack. Let the particle be at Q (at a distance x_1 from O) at any time t in the segment AO.

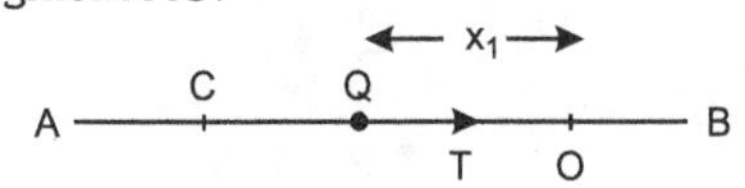

Tension in the string $OB = \frac{Yx_1}{b}$ towards O.
Tension in $AO = 0$.

Hence, $$m\frac{d^2x_1}{dt^2} = -\frac{Yx_1}{b}$$

or $$\frac{d^2x_1}{dt^2} = -\frac{Yx_1}{bm} \qquad \dots(3)$$

The Eq. (3) represents SHM with centre at O and time period

$$\frac{2\pi}{\sqrt{Y/bm}} = 2\pi\sqrt{\frac{bm}{Y}}$$

Let the amplitude of this simple harmonic motion be OC.
Time taken to reach C from O is

$$\frac{1}{4}\cdot 2\pi\sqrt{\frac{bm}{Y}} = \frac{\pi}{2}\sqrt{\frac{bm}{Y}}$$

The required periodic time for complete oscillation between B and C is

$$2\times\left(\frac{\pi}{2}\sqrt{\frac{am}{Y}} + \frac{\pi}{2}\sqrt{\frac{bm}{Y}}\right) = \pi(\sqrt{a} + \sqrt{b})\sqrt{\frac{m}{Y}}$$

219. This is an interesting problem because the maximum displacement of the particle on either side of the mean position is not same through the time taken to cover these distances is same.
Consider Eq. (1) of solution 218

$$\frac{d^2x}{dt^2} = -\frac{Y}{am}x$$

Multiplying both sides by $2\left(\frac{dx}{dt}\right)$ and then integrating, we have

$$\left(\frac{dx}{dt}\right)^2 = -\frac{Y}{am}x^2 + k,$$

where k is a constant.
At point B, $x = b$ and $\frac{dx}{dt} = 0$.

$$\therefore \quad 0 = -\frac{Y}{am}b^2 + k \quad \text{or} \quad k = \frac{Yb^2}{am}$$

Hence, $$\left(\frac{dx}{dt}\right)^2 = \frac{Y}{am}(b^2 - x^2)$$

At the mean position, *i.e.*, at O, $x = 0$

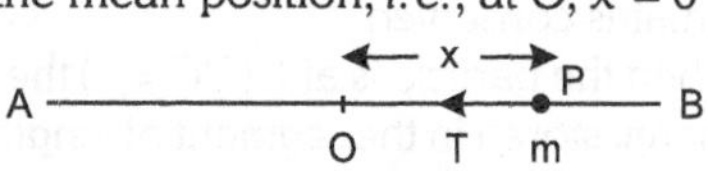

or $$\frac{dx}{dt} = v = \sqrt{\frac{Y}{am}}\,b \qquad \dots(1)$$

This is the velocity of the particle at O. Now from Eq. (3) of solution 218, we have

$$\frac{d^2x_1}{dt^2} = -\frac{Yx_1}{bm}.$$

Multiplying both sides by $2\frac{dx_1}{dt}$ and integrating

$$\left(\frac{dx_1}{dt}\right)^2 = -\frac{Y}{bm}x_1^2 + D,$$

where D is a constant.
But at O, $x_1 = 0$

and $$\left(\frac{dx_1}{dt}\right)^2 = v^2 = \frac{Y}{am}b^2 \qquad \text{[from (1)]}$$

$$\therefore \quad \frac{Y}{am}b^2 = 0 + D \quad \text{or} \quad D = \frac{Y}{am}b^2$$

Hence, $$\left(\frac{dx_1}{dt}\right)^2 = -\frac{Y}{bm}x_1^2 + \frac{Y}{am}b^2$$

$$= \frac{Y}{bm}\left[\frac{b^3}{a} - x_1^2\right]$$

If the extreme postion of the particle is point C such that $OC = c$,

then at $x_1 = c, \frac{dx_1}{dt} = 0$

or $$0 = \frac{Y}{bm}\left(\frac{b^3}{a} - c^2\right) \quad \text{or} \quad c = b\sqrt{\frac{b}{a}}$$

The particle thus oscillates through a distance

$$OC + OB = b\sqrt{\frac{b}{a}} + b = \frac{b(\sqrt{b} + \sqrt{a})}{\sqrt{a}}$$

Alternatively

This problem can be sovled easily by energy considerations. When the string segment of length a is stretched by a distance x, force developed in it will be $F = \frac{Yx}{a}$.

Work done in stretching it to a distance b is

$$W = \frac{Y}{a}\int_a^b x\,dx = \frac{Y}{a}\frac{b^2}{2}$$

This is the potential energy stored in the string when the particle is held at B. This energy remains conserved.

When the particle is at $C(OC = c)$ the potential energy stored in the segment of length $b(OB)$ is

$$\frac{Y}{2}\frac{c^2}{b}$$

Since, the particle's velocity is zero at both B and C, we have from energy conservation

$$\frac{Y}{2}\frac{c^2}{b} = \frac{Y}{a}\frac{b^2}{2} \quad \text{or} \quad c = b\sqrt{\frac{b}{a}}$$

The particle thus oscillates through a distance

$$b + b\sqrt{\frac{b}{a}} = \frac{b(\sqrt{b} + \sqrt{a})}{\sqrt{a}}$$

220. Half of the volume of sphere is submerged. For equilibrium of sphere,

weight = upthrust

$$\therefore \quad V\rho_s g = \frac{V}{2}(\rho)(g)$$

$$\therefore \quad \rho_s = \frac{\rho}{2}$$

When slightly pushed downwards by x, weight will remain as it is while upthrust will increase. The increased upthrust will become the net restoring force (upwards).

$F = -$ (extra upthrust)

$= -$ (extra volume immersed) $(\rho_L)(g)$

or $ma = -(\pi R^2) \times \rho g$ $\quad$ (a = acceleration)

$$\therefore \quad \frac{4}{3}\pi R^3\left(\frac{\rho}{2}\right)a = -(\pi R^2 \rho g)\,x$$

$$\therefore \quad a = -\left(\frac{3g}{2R}\right)x$$

As $a \propto -x$, motion is simple harmonic.

Frequency of oscillation,

$$f = \frac{1}{2\pi}\sqrt{\left|\frac{a}{x}\right|} = \mathbf{\frac{1}{2\pi}\sqrt{\frac{3g}{2R}}}$$

221. Let S be the area of cross-section of the rod. In the displaced position, as shown in figure, weight (W) and upthrust (F_B) both pass through its centre of gravity G.

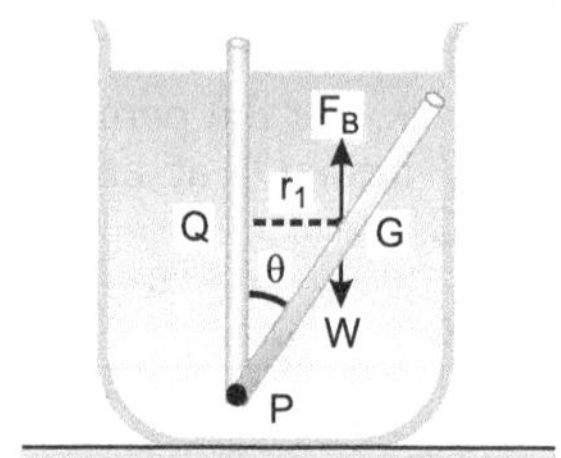

Here W = (volume) × (density of rod) × g

$W = (SL)(d_1)g$

F_B = (volume) × (density of liquid) × g

$= (SL)(d_2)g$

Given that $d_1 < d_2$. Therefore, $W < F_B$

Therefore, net force acting at G will be :

$F = F_B - W = (SLg)(d_2 - d_1)$ upwards.

Restoring torque of this force about point P is :

$$\tau = F \times r_1 = (SLg)(d_2 - d_1)\,(QG)$$

or $$\tau = -(SLg)(d_2 - d_1)\left(\frac{L}{2}\sin\theta\right)$$

Here negative sign shows the restoring nature of torque

or $$\tau = -\left\{\frac{SL^2 g(d_2 - d_1)}{2}\right\}\theta \qquad \ldots(1)$$

As $\sin\theta \approx \theta$ for small values of θ

From equation (1), we see that

$$\tau \propto -\theta$$

Hence motion of the rod will be simple harmonic.

Rewriting equation (1) as

$$I\frac{d^2\theta}{dt^2} = -\left\{\frac{SL^2g(d_2 - d_1)}{2}\right\}\theta \qquad \ldots(2)$$

Here I = moment of inertia of rod about an axis passing through P.

$$I = \frac{ML^2}{3} = \frac{(SLd_1)L^2}{3}$$

Substituting this value of I in equation (1), we have

$$\frac{d^2\theta}{dt^2} = -\left\{\frac{3}{2}\frac{g(d_2 - d_1)}{d_1L}\right\}\theta$$

Comparing this equation with standard differential equation of SHM *i.e.*,

$$\frac{d^2\theta}{dt^2} = -\omega^2\theta$$

The angular frequency of oscillation is

$$\omega = \sqrt{\frac{\mathbf{3\,g(d_2 - d_1)}}{\mathbf{2d_1L}}}$$

222. (i) Given : Mass of each block A and B, $m = 0.1$ kg

Radius of circle, $R = 0.06$ m

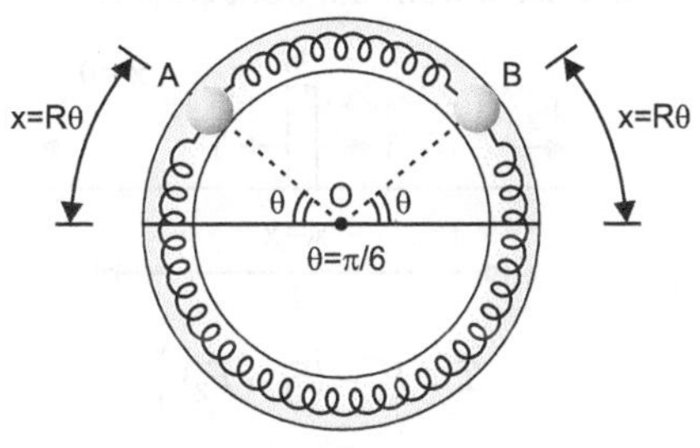

Natural length of spring,

$$l_0 = 0.06\,\pi = \pi R \qquad \text{(Half circle)}$$

and spring constant, $k = 0.1$ N/m

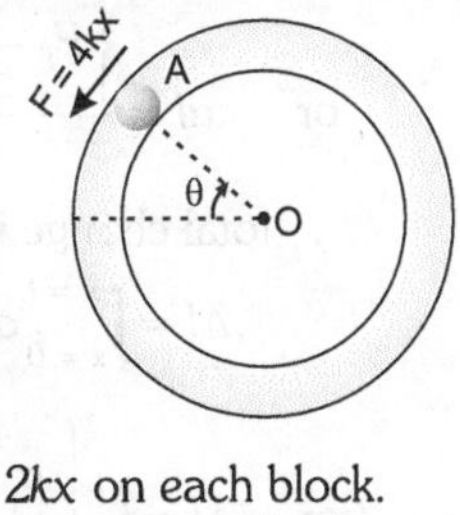

In the stretched position elongation in each spring ($x = R\theta$)

Let us draw *FBD* of A

Spring in lower side is stretched by $2x$ and on upper side compressed by $2x$. Therefore, each spring will exert a force $2kx$ on each block. Hence a restoring force, $F = 4kx$ will act on A in the direction shown in figure.

Restoring torque of this force about origin,

$$\tau = -F \cdot R = -(4kx)R = -(4kR\theta)R$$

or $\quad \tau = -4kR^2 \cdot \theta \qquad \ldots(1)$

Since $\tau \propto -\theta$, each ball executes angular SHM about origin O.

Equation (1) can be rewritten as

$$I\alpha = -4kR^2\theta$$

or $\quad (mR^2)\alpha = -4kR^2\theta \Rightarrow \alpha = -\left(\frac{4k}{m}\right)\theta$

$\therefore$ Frequency of oscillation,

$$f = \frac{1}{2\pi}\sqrt{\left|\frac{\text{acceleration}}{\text{displacement}}\right|}$$

$$= \frac{1}{2\pi}\sqrt{\left|\frac{\alpha}{\theta}\right|} = \frac{1}{2\pi}\sqrt{\left|\frac{4k}{m}\right|}$$

Substituting the values, we have

$$f = \frac{1}{2\pi}\sqrt{\frac{4 \times 0.1}{0.1}} = \frac{\mathbf{1}}{\pi}\ \mathbf{Hz}$$

(ii) In stretched position, potential energy of the system is

$$\text{P.E.} = 2\left(\frac{1}{2}k\right)\{2x\}^2 = 4kx^2$$

and in mean position, both the blocks have kinetic energy only. Hence

$$\text{K.E.} = 2\left\{\frac{1}{2}mv^2\right\} = mv^2$$

From energy conservation:

$$\text{P.E} = \text{K.E.}$$

$\therefore \quad 4kx^2 = mv^2$

$\therefore \quad v = 2x\sqrt{\frac{k}{m}} = 2R\theta\sqrt{\frac{k}{m}}$

Substituting the values

$$v = 2(0.06)(\pi/6)\sqrt{\frac{0.1}{0.1}}$$

or $\quad \mathbf{v = 0.0628\ m/s}$

(iii) Total energy of the system, E = P.E. in stretched position

or $\quad$ = K.E. in mean position

$$E = mv^2 = (0.1)(0.0628)^2\ \text{J}$$

or $\quad \mathbf{E = 3.9 \times 10^{-4}\ J}$

CHAPTER 8 SOLIDS AND FLUIDS

223. Consider an element of length dl and mass dm of the ring.

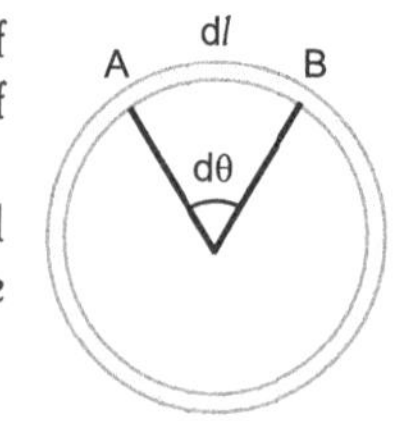

Let S be the cross-sectional area of wire and T the tension.

$$dl = R\,.\,d\theta$$

$$F = 2T\sin\left(\frac{d\theta}{2}\right) \approx T(d\theta) \qquad \left(\sin\frac{d\theta}{2} \approx \frac{d\theta}{2}\right)$$

This F provides the necessary centripetal force.

Hence, $\quad T\,.\,d\theta = (dm)R\omega^2$

$$= (R.\,d\theta.\,S.\,\rho)R\omega^2$$

or $\quad \dfrac{T}{S} = \rho\omega^2 R^2 \qquad \ldots(\text{i})$

Let ΔR be the increase in radius. Then

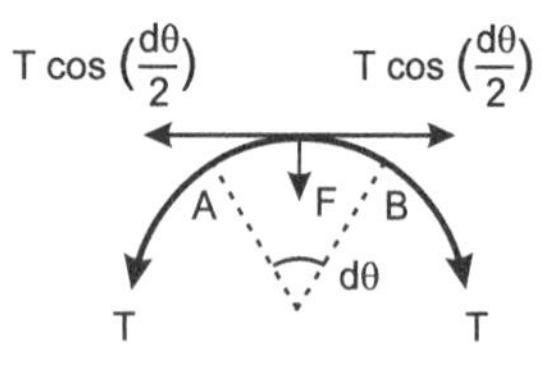

$$\text{longitudinal strain} = \frac{\Delta L}{L} = \frac{\Delta(2\pi R)}{2\pi R} = \frac{\Delta R}{R}$$

Now $\quad Y = \dfrac{\text{stress}}{\text{strain}} = \dfrac{T/S}{\Delta R/R}$

$$Y = \frac{\rho\omega^2 R^3}{\Delta R}$$

Hence $\quad \boldsymbol{\Delta R = \dfrac{\rho\omega^2 R^3}{Y}}$

224. Bulk modulus is given by

$$B = -\frac{dP}{(dV/V)} \quad \text{or} \quad \frac{dV}{V} = -\frac{dP}{B}$$

or $\quad \dfrac{P\,.\,dV}{V} = -\dfrac{P\,dP}{B}$

$\dfrac{P\,dV}{V}$ is the work done per unit volume. The negative sign implies that a decrese in pressure gives rise to increase in volume and *vice-versa.*

Hence volume density of elastic potential energy

$$u = \int_0^{h\rho g}\frac{P\,.\,dP}{B} = \frac{1}{2}\frac{(h\rho g)^2}{B}$$

Substituting the values, we obtain

$$u = \frac{1}{2}\frac{(10^3)^2\,(10^3)^2\,(9.8)^2}{2\times 10^9}$$

$\Rightarrow \quad \boldsymbol{u = 2.4 \times 10^4\ \text{J/m}^3}$

225. Tension in the plank varies linearly from F_1 to F_2. Therefore, tension in the plank at a distance x from the front edge is :

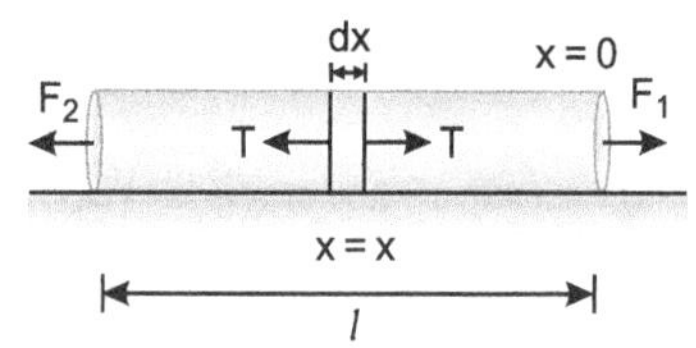

$$T = F_1 - (F_1 - F_2)\frac{x}{l}$$

change in length in element dx is :

$$dl = \frac{T.\,dx}{SY} \qquad \left(\Delta l = \frac{Fl}{AY}\right)$$

or $\quad dl = \dfrac{\left\{F_1 - (F_1 - F_2)\dfrac{x}{l}\right\}}{SY}.\,dx$

$\therefore$ total change in length will be

$$\Delta l = \int_{x=0}^{x=l} dl$$

or $\quad \Delta l = \displaystyle\int_0^l \left\{\frac{F_1 - (F_1 - F_2)\dfrac{x}{l}}{SY}\right\}dx$

or $$\Delta l = \frac{(F_1 + F_2)\,l}{2SY}$$

226. (a) Let Δl be the displacement of the joint towards right.

Strain on first rod $= \dfrac{l_1\alpha_1 t - \Delta l}{l_1}$

or force exerted by the first rod on the joint is

$$F_1 = Y_1 A\left(\frac{l_1\alpha_1 t - \Delta l}{l_1}\right) \quad \ldots(1)$$

Strain on second rod $= \dfrac{l_2\alpha_2 t + \Delta l}{l_2}$

or force exerted by second rod on the joint is

$$F_2 = Y_2 A\left(\frac{l_2\alpha_2 t + \Delta l}{l_2}\right) \quad \ldots(2)$$

In equilibrium $F_1 = F_2$

Solving this, we get

$$\Delta l = \frac{l_1 l_2 t(Y_1\alpha_1 - Y_2\alpha_2)}{(Y_1 l_2 + Y_2 l_1)} \quad \ldots(3)$$

Solving (1), (2) and (3), we get

$$\boldsymbol{F_1 = F_2 = \frac{At\,(l_1\alpha_1 + l_2\alpha_2)}{\left(\frac{l_1}{Y_1} + \frac{l_2}{Y_2}\right)}}$$

(b) $\boldsymbol{l'_1 = l_1 + \Delta l}$

$$\boldsymbol{= l_1 + \frac{l_1 l_2 t\,(Y_1\alpha_1 - Y_2\alpha_2)}{(Y_1 l_2 + Y_2 l_1)}}$$

and $\boldsymbol{l'_2 = l_2 - \Delta l}$

$$\boldsymbol{= l_2 - \frac{l_1 l_2 t(Y_1\alpha_1 - Y_2\alpha_2)}{Y_1 l_2 + Y_2 l_1}}$$

227. (a) The forces acting on the sphere are shown in the figure.

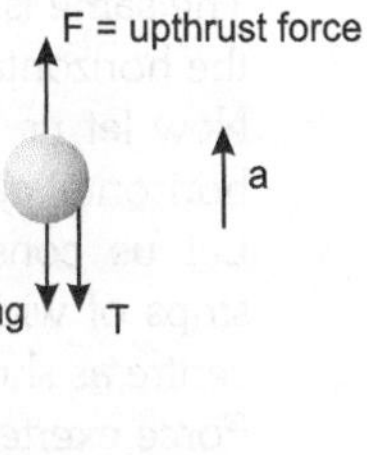

$$F = V\rho_w g_{eff}$$
$$= V\rho_w (g + a)$$
$$= \frac{m}{\rho_s}.\rho_w (g + a)$$

Here ρ_w = density of water = 1000 kg/m^2

and ρ_s = density of sphere = 500 kg/m^3

Equation of motion of sphere is

$$F - mg - T = ma$$

or $$T = F - m(g + a)$$
$$= m\frac{\rho_w}{\rho_s}(g + a) - m(g + a)$$
$$= (g + a)\,m\left(\frac{\rho_w}{\rho_s} - 1\right)$$

Substituting the values, we get

$$\boldsymbol{T = (10 + 2)\,(2)\left(\frac{1000}{500} - 1\right) = 24\ N}$$

(b) When thread snaps,

$$T = 0$$

$\therefore$ absolute acceleration of sphere upwards,

$$a_s = \frac{F - mg}{m}$$

or $$a_s = \frac{m(g + a)\frac{\rho_w}{\rho_s} - mg}{m}$$

or $$a_s = (g + a)\frac{\rho_w}{\rho_s} - g$$
$$= (10 + 2)\left(\frac{1000}{500}\right) - 10 = 14\ \text{m/s}^2$$

$\therefore$ Upward acceleration of sphere relative to tank

$$a_r = a_s - a = 14 - 2$$
$$\boldsymbol{a_r = 12\ m/s^2}$$

228. (a) Level will be maximum when

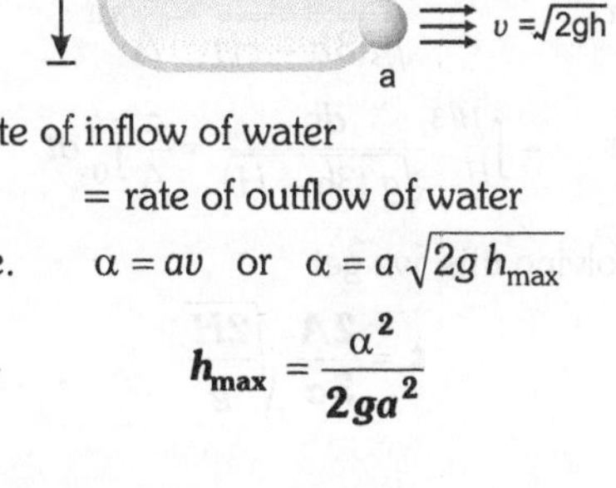

rate of inflow of water

= rate of outflow of water

i.e. $\alpha = av$ or $\alpha = a\sqrt{2g\,h_{max}}$

$\Rightarrow$ $$\boldsymbol{h_{max} = \frac{\alpha^2}{2ga^2}}$$

(b) Let at time t, the level of water be h. Then

$$A\left(\frac{dh}{dt}\right)=\alpha - a\sqrt{2gh}$$

or $$\int_0^h \frac{dh}{\alpha - a\sqrt{2gh}}=\int_0^t \frac{dt}{A}$$

Solving this, we get

$$\boldsymbol{t=\frac{A}{ag}\left[\frac{\alpha}{a}\ln\left\{\frac{\alpha - a\sqrt{2gh}}{\alpha}\right\}-\sqrt{2gh}\right]}$$

229. Let h_f be the final level in the two tanks, then

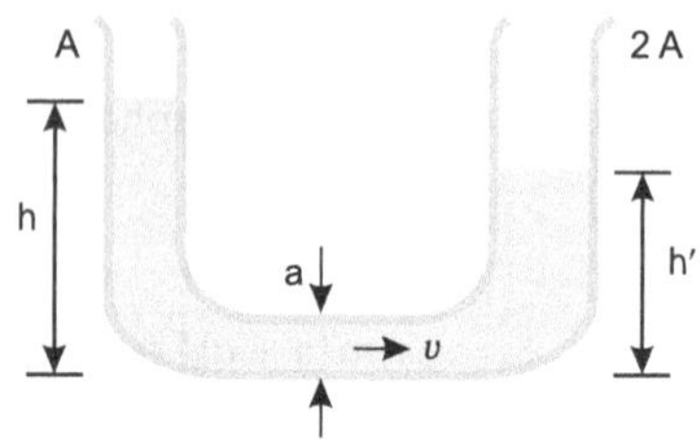

$$(A+2A)h_f = H\cdot A$$

or $$h_f=\frac{H}{3} \qquad ...(1)$$

Let at time t, level of left tank is h and that of the right tank is h'.

Then $$Ah+2Ah'=AH$$

$\therefore$ $$h'=\frac{H-h}{2}$$

$\therefore$ $$\Delta h = h-h' = h-\left(\frac{H-h}{2}\right)=\left(\frac{3h-H}{2}\right)$$

$\therefore$ $$v=\sqrt{2g(\Delta h)}$$

$$A\left(-\frac{dh}{dt}\right)=av=a\sqrt{2g\left(\frac{3h-H}{2}\right)}$$

or $$\frac{-dh}{\sqrt{g(3h-H)}}=\frac{a}{A}dt$$

or $$-\int_H^{H/3}\frac{dh}{\sqrt{g\,(3h-H)}}=\frac{a}{A}\int_0^t dt$$

Solving this, we get

$$\boldsymbol{t=\frac{2A}{3a}\sqrt{\frac{2H}{g}}}$$

230. $\boldsymbol{\frac{\rho_i}{\rho_w}=0.9}$ **i.e. 90% volume of ice is immersed in water.**

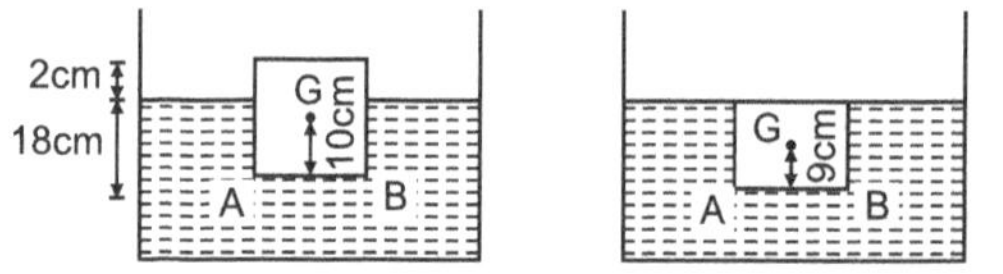

Level of water remains the same when ice melts. Hence Earlier the centre of mass of ice was at a height of 10 cm from the line AB and when it melts, it will be at 9 cm. Hence

$$\Delta h = -1 \text{ cm} = -10^{-2} \text{ m}$$

or $$\Delta U = mg\,\Delta h$$

$$= -(l^3)(\rho_i)\,g(10^{-2}) \quad (m = \text{mass of ice})$$

$$= -(0.2)^3(900)(10)(10^{-2})$$

or $$\boldsymbol{\Delta U = -0.72 \text{ J}}$$

231. Pressure at the bottom of the vertical cylinder is

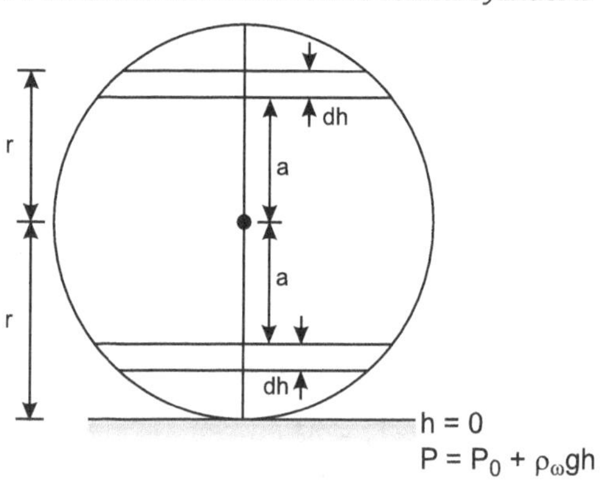

$$P = P_0 + \rho_w\, gh \qquad ...(1)$$

The same is the pressure on the lower part of the horizontal cylinder.

Now let us consider the cross-section of the horizontal cylinder.

Let us consider two symmetrically located strips of width dh at a distance 'a' from the centre as shown in figure.

Force exerted by water on the upper strip is

$$F_1 = \{P-\rho_w g\,(r+a)\}\Delta S$$

and the force on the lower strip is

$$F_2 = \{P-\rho_w g\,(r-a)\}\Delta S$$

Here ΔS is the area of these two strips.

Sum of these two forces,

$$F = F_1 + F_2 = 2\Delta S\,(P - \rho_w g\, r)$$

is independent of a.

Hence the total force on piston will be $(P - \rho_w g\, r)\cdot \pi r^2$ $(\Sigma 2\,\Delta s = \pi r^2)$

In equilibrium, this should be equal to $P_0 \cdot \pi r^2$

i.e. $\quad (P - \rho_w g r)\,\pi r^2 = P_0 \pi r^2$

or $\quad \{P_0 + \rho_w g\,(h - r)\}\,\pi r^2 = P_0 \pi r^2$

$\therefore \quad$ $\boldsymbol{h = r}$

232. $\dfrac{\rho_1}{\rho_2} = \dfrac{1}{2}$ i.e., in equilibrium, block is half immersed in water. Let h' be the increase in level when the block is pressed by an amount h. Then

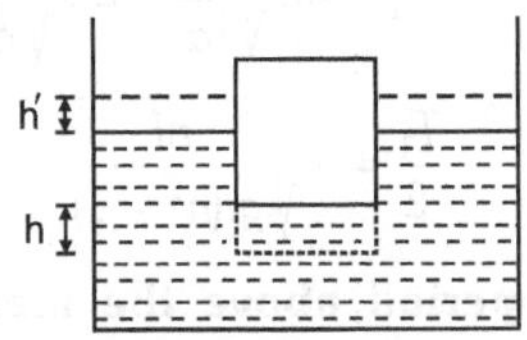

$$(A - l^2)h' = l^2 \cdot h$$

Substituting the values, we get $h' = 0.5h$

We have to immerse further $l/2$. Hence

$$\frac{l}{2} = h + h' = 1.5h \text{ or } h = l/3$$

When the block is depressed by h, extra upthrust

$$F = (h + h')\rho_2 l^2 \cdot g = 1.5\, l^2 \rho_2\, g\, h$$

Work has to be done against this upthrust. Hence

$$W = \int_0^{l/3} F dh = \int_0^{l/3} (1.5\, l^2 \rho_2\, gh)\, dh$$

Substituting the values, we get

$$\boldsymbol{W = 6.75\ J}$$

233. (a) The specific gravity of the cylinder is 0.75, i.e. 75% of its volume or 75% of its length is inside the water or 3 cm of its length is immersed in water as shown.

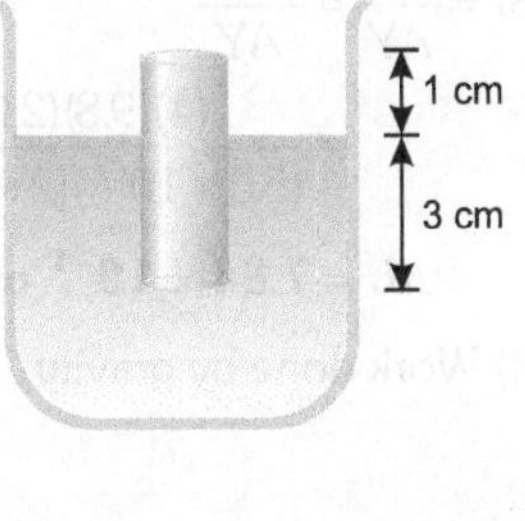

When it is depressed by x, net upward force will be the extra upthrust force i.e.,

$$F = \pi (R)^2 (x) \rho_w g$$

$$= \pi (10^{-2})^2 x (10^3)(10) = \pi x \text{ newton}$$

The same force will have to be applied downwards.

$$\text{Hence} W = \int_{x=0}^{x=10^{-2}\text{ m}} F\,dx = \int_{x=0}^{x=10^{-2}} \pi x \cdot dx$$

$$= \frac{\pi}{2} \times 10^{-4} = \mathbf{1.57 \times 10^{-4}\ J}$$

(b) Reduction in the force = weight of cylinder

$$= (\pi R^2)(h)\rho_c$$

$$= (\pi)(10^{-2})^2(4 \times 10^{-2})(0.75 \times 10^3)(10)$$

$$= \mathbf{9.42 \times 10^{-2}\ N}$$

234. Net force on the block at a height h from the bottom is

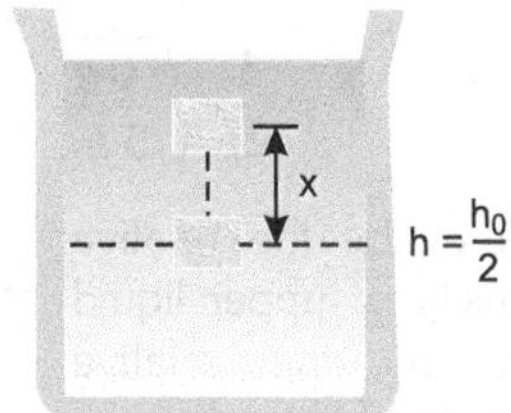

F_{net} = upthrust − weight (upwards)

$$= \left(\frac{m}{5/2\,\rho_0}\right)\rho_0\left(4 - \frac{3h}{h_0}\right) g - mg$$

$$F_{net} = 0 \text{ at } h = \frac{h_0}{2}$$

So, $h = \dfrac{h_0}{2}$ is the equilibrium position of the block.

For $\quad h > \dfrac{h_0}{2}$, weight > upthrust

i.e., net force is downwards and for $h < \dfrac{h_0}{2}$

weight < upthrust

i.e., net force is upwards

For upward displacement x from mean position, net downward force is

$$F = -\left[\left(\frac{m}{5/2\,\rho_0}\right)\rho_0\left\{4 - \frac{3(h+x)}{h_0}\right\}g - mg\right]$$

$$\left(h = \frac{h_0}{2}\right)$$

$$F = -\frac{6mg}{5h_0}x \quad \ldots(1)$$

(because at $h = \frac{h_0}{2}$ upthrust and weight are equal)

Since $F \propto -x$

oscillations are simple harmonic in nature.

Rewriting equation (1)

$$ma = -\frac{6mg}{5h_0}x \quad (a = \text{acceleration of block})$$

or $$a = -\frac{6g}{5h_0}x$$

$\therefore$ $$f = \frac{1}{2\pi}\sqrt{\left|\frac{a}{x}\right|}$$

$$\boldsymbol{f = \frac{1}{2\pi}\sqrt{\frac{6g}{5h_0}}}$$

235. (i) Let ρ_1 be the relative density of upper liquid and ρ_2 the relative density of lower liquid.

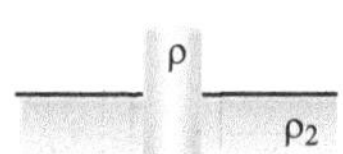

In the first case, $\frac{1}{n}$ part is submerged. Hence

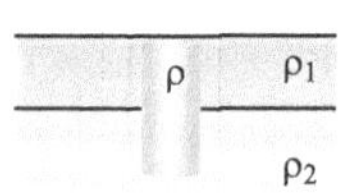

$$\frac{1}{n} = \frac{\rho}{\rho_2} \quad \ldots(1)$$

Given $$\rho = \sqrt{\rho_1\rho_2} \quad \ldots(2)$$

From equations (1) and (2), we get

$$\rho_2 = n\rho \quad \text{and} \quad \rho_1 = \frac{\rho}{n}$$

Now let $\frac{1}{m}$th part is in lower liquid. Then

upthrust = weight

or $$A\left(\frac{1}{m}\right)(n\rho) + A\left(1 - \frac{1}{m}\right)\frac{\rho}{n} = A\cdot\rho$$

or $$\boldsymbol{\frac{1}{m} = \frac{1}{n+1}}$$

(ii) When depressed, the restoring force and hence the time periods will be different for the two portions, above and below the mean position.

$$T = \frac{T_1}{2} + \frac{T_2}{2}$$

Time period below the mean position (T_1) :

when depressed by x restoring force,

$$F = -Ax\left(n\rho - \frac{\rho}{n}\right)g$$

or $$(A\cdot l)\rho a = -Ax\left(\frac{n^2-1}{n}\right)\rho g$$

$\therefore$ $$T_1 = 2\pi\sqrt{\left|\frac{x}{a}\right|} = 2\pi\sqrt{\frac{nl}{g(n^2-1)}}$$

or $$\frac{T_1}{2} = \pi\sqrt{\frac{nl}{g(n^2-1)}} \quad \ldots(3)$$

Time period above the mean position (T_2) :

When the cylinder is x above the mean level, net restoring force downward

$$F = -A\cdot x\rho_2 g$$

or $$(Al)(\rho)a = -Axn\rho g$$

or $$T_2 = 2\pi\sqrt{\left|\frac{x}{a}\right|} = 2\pi\sqrt{\frac{l}{ng}}$$

or $$\frac{T_2}{2} = \pi\sqrt{\frac{l}{ng}} \quad \ldots(4)$$

From equations (3) and (4),

$$T = \frac{T_1}{2} + \frac{T_2}{2}$$

or $$\boldsymbol{T = \pi\left[\sqrt{\frac{l}{ng}} + \sqrt{\frac{nl}{g(n^2-1)}}\right]}$$

236. (a) $$\Delta l = \frac{Fl}{AY} = \frac{(Mg)l}{AY}$$

$$= \frac{(4)(9.8)(2.0)}{(5.0\times10^{-7})(2\times10^{11})}$$

or $$\boldsymbol{\Delta L = 7.84\times10^{-4}\ \text{m}}$$

(b) (i) Work done by gravity

$$W_1 = (Mg)(s)$$

$$W_1 = (4)(9.8)(0.06 \times 10^{-2}) = \mathbf{0.024\ J}$$

(ii) Equivalent force constant of a wire is

$$k = \frac{YA}{l}$$

∴ Work done by the applied force and gravity would be

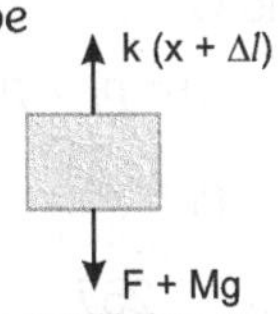

In equilibrium
(F + mg) = k(x + Δl)

$$W_2 = \int_0^{x = 6\times10^{-4}\text{ m}} k\,(x + \Delta l)\,dx$$

$$= k\left[\frac{x^2}{2} + x\cdot\Delta l\right]_0^{x = 6\times10^{-4}\text{ m}}$$

$$= \frac{YA}{l}\left[\frac{x^2}{2} + x\cdot\Delta l\right]_0^{6\times10^{-4}}$$

$$= \frac{(2\times10^{11})(5\times10^{-7})}{2}$$

$$\left[\frac{(6\times10^{-4})^2}{2} + (6\times10^{-4})(7.84\times10^{-4})\right]$$

$$W_2 = 0.033\text{ J}$$

Hence the work done by the applied force would be

$$\mathbf{W_3 = W_2 - W_1 = 0.009\ J}$$

(iii) The work done by the force the wire exerts on the mass will be $-W_2$ or **–0.033 J**

(iv) The change in elastic potential energy = work done by the applied force and force of gravity *i.e.*, **0.033 J.**

237. First of all let us find velocity of efflux at time t.

$$A\left(-\frac{dh}{dt}\right) = a\sqrt{2gh}$$

$$\therefore \quad \int_H^h \frac{dh}{\sqrt{h}} = \int_0^t \frac{-a\sqrt{2g}}{A}\,dt$$

Solving this equation, we get

$$\sqrt{h} = \sqrt{H} - \frac{a}{A}\sqrt{\frac{g}{2}}\,t$$

∴ Velocity of efflux,

$$v = \sqrt{2gh}$$

$$= \sqrt{2g}\left[\sqrt{H} - \frac{a}{A}\sqrt{\frac{g}{2}}\,t\right]$$

Now time to fall a height y is,

$$t_0 = \sqrt{\frac{2y}{g}}$$

Putting, $vt_0 = x$ we get

$$\boldsymbol{t = \frac{A}{a}\sqrt{\frac{2}{g}}\left[\sqrt{H} - \frac{x}{\sqrt{4y}}\right]}$$

238. Using Bernoulli's theorem

$$P_0 + \rho(g + a)\,3 = P_0 + \frac{1}{2}\rho v_x^2$$

$$\Rightarrow \quad v_x = \sqrt{2\,(g + a)\,3}$$

If it falls on the platform in time t then,

$$h = \frac{1}{2}(g + a)\,t^2$$

$$t = \sqrt{\frac{2h}{(g + a)}} = \sqrt{\frac{2\times2}{g + a}}$$

$$\therefore \quad x = v_x t = \sqrt{2\,(g + a)\,3} \times \sqrt{\frac{2\times2}{g + a}}$$

$$= 2\sqrt{6}\text{ m} = \mathbf{4.9\ m}$$

239. Let m is the mass of steel piece and v_1 and v_2 are the volume of the cone and steel piece respectively. From condition of floating

$$(m + 8.8)\times g = (v_1 + v_2)\times g \times 0.8 \quad \ldots(1)$$

$$v_1 = \frac{8.8}{0.5};\quad v_2 = \frac{m}{7.8}$$

$$(m + 8.8)\,g = \left(\frac{8.8}{0.5} + \frac{m}{7.8}\right)\times g \times 0.8$$

$$m\left(1 - \frac{0.8}{7.8}\right) = \frac{8.8\times0.8}{0.5} - 8.8$$

$$m\left(\frac{7}{7.8}\right) = 14.08 - 8.8$$

$$\frac{m\times7}{7.8} = 5.28 \Rightarrow m = \mathbf{5.88\ kg}$$

Tension in the cord

$$T = mg - B$$

$$= m \times g - \frac{m}{7.8} \times 0.5 \times g$$

$$= 5.88\left[1 - \frac{0.5}{7.5}\right] \times 10$$

$$= \mathbf{54.88\ N}$$

240. Suppose that x is the height of the liquid in the tank at a time t.

The rate of volume flow equals,

$$-A\frac{dx}{dt} = \frac{\pi \Delta p a^4}{8\eta l}$$

(by Poiseuille's equation)

$$\text{or} \quad -A\frac{dx}{dt} = \frac{\pi\left[\rho g x + \left(\frac{mg}{A}\right)\right]a^4}{8\eta l}$$

where, Δp = pressure difference between the ends of the capillary tube

$$= \rho g x + \frac{mg}{A}$$

$$\text{or} \quad \frac{dx}{dt} + \lambda x = -\frac{m}{A\rho}\lambda;$$

$$\lambda = \frac{\pi \rho g a^4}{8\eta l A}$$

The solution is

$$x = \left(h + \frac{m}{A\rho}\right)e^{-\lambda t} - \frac{m}{A\rho}$$

241. (a) Flow rate through A

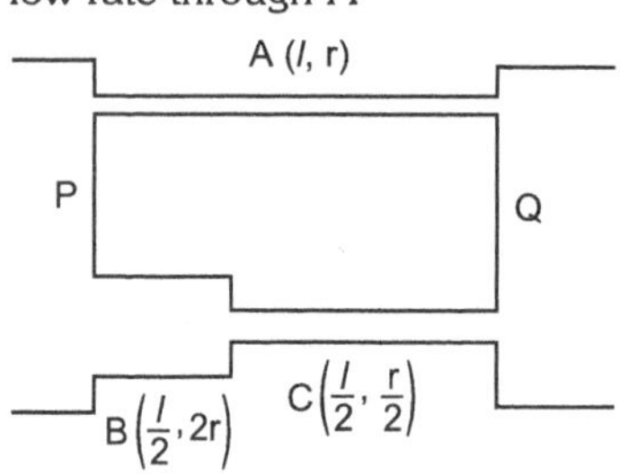

$$v_1 = \frac{\pi p r^4}{8\eta l}$$

B and C will have same flow rate since they are in series, so,

$$v_2 = \frac{\pi p_1 (2r)^4}{8\eta\left(\frac{l}{2}\right)} = \frac{\pi p_2\left(\frac{r}{2}\right)^4}{8\eta\left(\frac{l}{2}\right)}$$

$$\Rightarrow \quad 256 p_1 = p_2$$

where p_1 and p_2 are the pressure difference across B and C

$$\text{but,} \quad p = p_1 + p_2$$

$$\Rightarrow \quad p_1 = \frac{p}{257}$$

(b) $\dfrac{v_1}{v_2} = \dfrac{p}{32p_1} = \dfrac{257}{32}$

242. *FBD* of cylinder

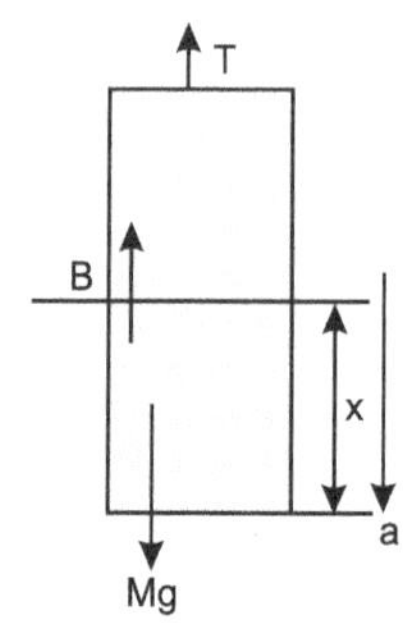

B = buoyant force

$$= \left(\frac{M}{L}\frac{x}{\rho_s}\right)\rho_l g$$

$$Mg - T - \left(\frac{M}{L}\frac{\rho_l}{\rho_s}\right)gx = Ma \quad \ldots(1)$$

For the mass 'm'

$$T - mg = ma \quad \ldots(2)$$

Adding equations (1) and (2)

$$\Rightarrow \left(M - \frac{M}{L}\frac{\rho_l}{\rho_s}x - m\right)g = (m + M)a$$

$$a = \frac{\left(M - \frac{M}{L}\frac{\rho_l}{\rho_s}x - m\right)}{M + m}g$$

$$a = v\frac{dv}{dx}$$

$$v\,dv = a\,dx = \frac{g}{M+m}\left(M - \frac{M}{L}\frac{\rho_l}{\rho_s}x - m\right)dx$$

$$\text{or} \quad \frac{v^2}{2} = \frac{g}{M+m}\left(Mx - \frac{M}{L}\frac{\rho_l}{\rho_s}\frac{x^2}{2} - mx\right) + c$$

at $x = 0$, $v = 0 \Rightarrow c = 0$

$$v = \sqrt{\frac{2g}{m+M}\left(ML - \frac{M\rho_l L}{2\rho_s} - mL\right)}$$

243. Consider the four positions of the cylinder as shown in the figure. Let A_1, A_2, A_3 and A_4 be accelerations of the cylinder in those four situations.

$$F_1 = H\rho g a - \frac{H}{4}\rho g a$$

Here a = area of cross-section of the cylindrical object.

$$\Rightarrow \quad \frac{H}{4}\rho a A_1 = \frac{3}{4}H\rho g a$$

$$\Rightarrow \quad A_1 = 3g \qquad \ldots(1)$$

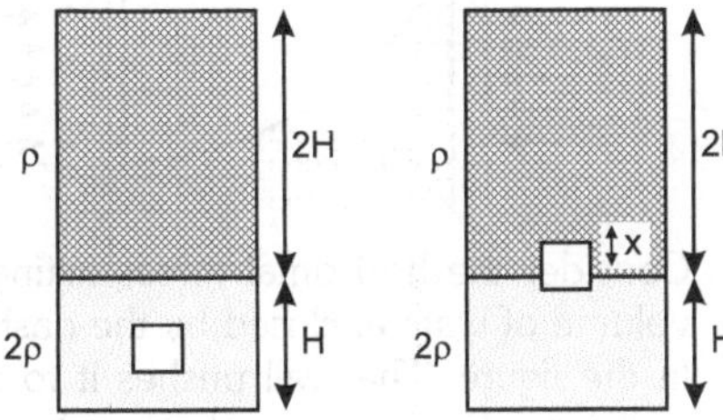

Position-1 Position-2

$$F_2 = \left(\frac{H}{2} - x\right) a\, 2\rho g + x a \rho g - \frac{H}{4}\rho a g$$

$$\Rightarrow \quad \frac{H}{4}\rho a A_2 = \frac{3}{4}H\rho g a - x\rho g a$$

$$\Rightarrow \quad A_2 = 3g - \frac{4}{H}gx \qquad \ldots(2)$$

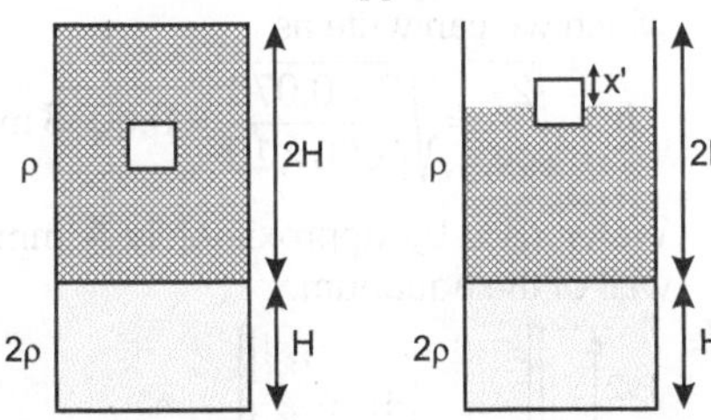

Position-3 Position-4

$$F_3 = \frac{H}{2}a\rho g - \frac{H}{4}\rho a g$$

$$\Rightarrow \quad \frac{H}{4}\rho a A_3 = \frac{H}{4}\rho a g$$

$$\Rightarrow \quad A_3 = g$$

$$F_4 = \left(\frac{H}{2} - x'\right)\rho g a - \frac{H}{4}\rho a g$$

$$\Rightarrow \quad A_4 = g - \frac{4x'}{H}g$$

$$\because \quad \frac{dv}{dt} = \frac{dv}{dx}\frac{dx}{dt} = v\frac{dv}{dx} = A$$

$$\Rightarrow \quad v\,dv = A\,dx$$

$$\Rightarrow \int v\,dv = \int A\,dx = \int_0^{H/2} A_1 dx + \int_0^{H/2} A_2 dx + \int_{H/2}^{2H} A_3 dx + \int_0^{H/2} A_4 dx$$

$$\frac{v_0^2}{2} = 3g\frac{H}{2} + 3g\frac{H}{2} - \frac{4gH}{8} + \frac{3gH}{2} + \frac{gH}{2} - \frac{gH}{2}$$

where v_0 is the velocity of the cylindrical object at the instant, the entire object just comes out of the liquid.

$$v_0^2 = 8gH$$

$$v_f^2 = v_i^2 + 2as$$

$$0 = 8gH - 2gs$$

$$s = 4H$$

244. The cross-sectional edge of the disc of water is a semicircle of radius $r = \frac{1}{2}d$ (see figure). Thus, the curvature of the surface of the water is $2/d$, which corresponds to a pressure of curvature of $\Delta p = \frac{2s}{d}$, where s is the surface tension. (The other component of the curvature is negligible because $D >> d$.)

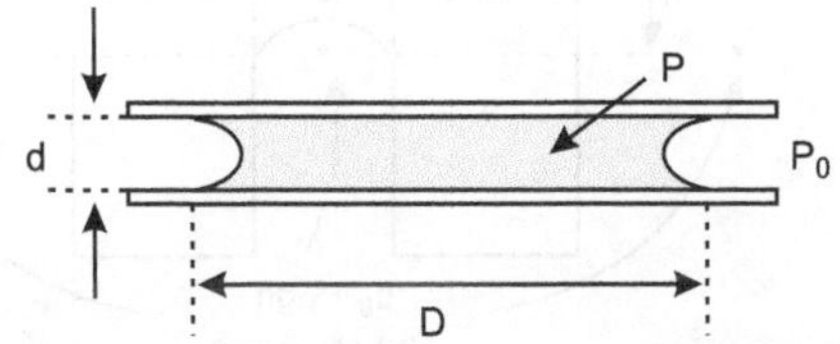

The pressure inside the disc is therefore $p_0 - \frac{2s}{d}$ when the atmospheric pressure is ρ_0. This pressure difference acts over a surface area between the water and each of the glass surfaces of $\frac{\pi D^2}{4}$. This implies that a force,

$$F = \frac{\pi D^2}{4}\frac{2s}{d}$$

'pulls' the glass plates together.

Note: If d is much smaller than D, this force can be quite considerable. It is in fact very difficult (if not impossible) to separate two parallel glass plates by pulling them in a direction perpen- dicular to their common plane when there is water between them. In order to be separated, they have to be slid in a direction parallel to that plane.

245. Because of the surface tension of the liquid, its height between the two objects is not the same as it is outside the objects; in the case of (a) water it is higher, whilst for (b) mercury (which has a negative angle of contact) it is lower.

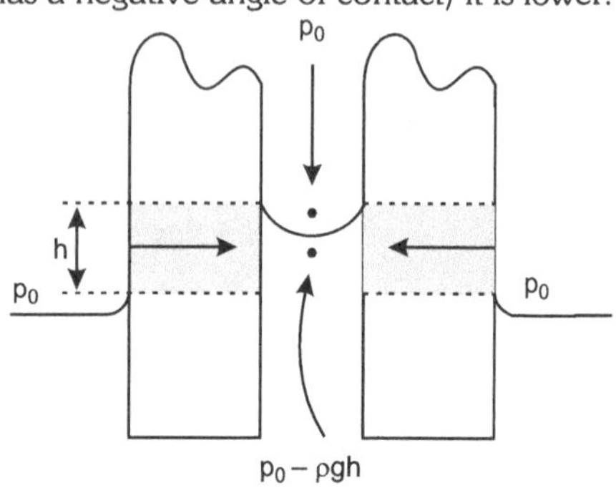

(a) water

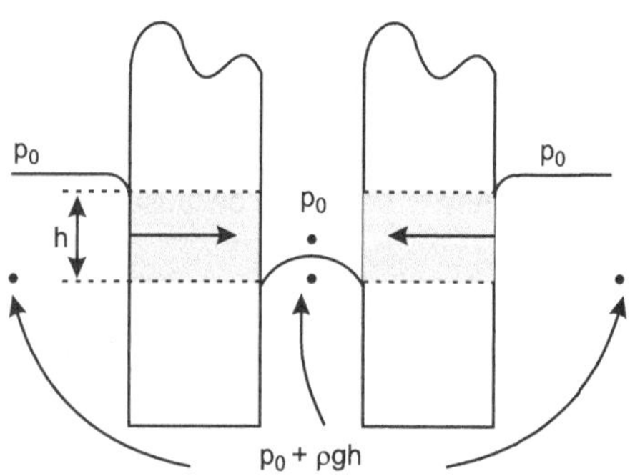

(b) Mercury

Just above the liquid surface between the objects the pressure has to be atmospheric in both cases and correspondingly, just below the surface it has to be less than atmospheric for water and more than atmospheric for mercury. As can be seen from the figure this leads to net inward forces (acting on the shaded areas) in both cases and a tendency for the objects to move towards each other.

246. The pressure of the water changes linearly with the increase in height. At the bottom of the meniscus it is equal to the external atmospheric pressure ρ_0, and at the top to $p_0 - \rho g h$. The average pressure exerted on the wall is $p_{\text{average}} = \rho_0 - \frac{\rho g h}{2}$. The force corresponding to this value, for an aquarium with side walls of length l, is $F_1 = l p_{\text{average}} h$.

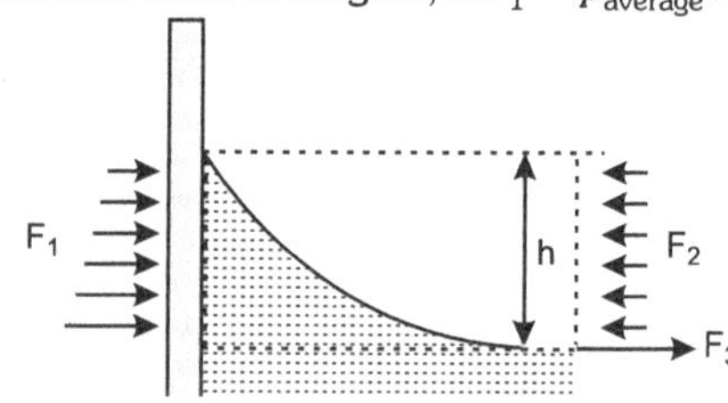

Consider the horizontal forces acting on the volume of wate enclosed by the dashed lines in the figure. The wall pushes it to the right with force F_1, the external air pushes it to the left with force $F_2 = l\rho_0 h$, and the surface tension of the rest of the water pulls it to the right with a force $F_3 = l\gamma$. The resultant of these forces has to be zero, since the volume itself is at rest. This means that

$$\left(p_0 - \frac{1}{2}\rho g h\right) l h - p_0 l h + l\gamma = 0,$$

which we can write as

$$h = \sqrt{\frac{2\gamma}{\rho g}} = \sqrt{\frac{2 \times 0.073}{1000 \times 10}} = 0.0038 \text{ m}$$

Water rises by approximately 4 mm up the wall of the aquarium.

247.

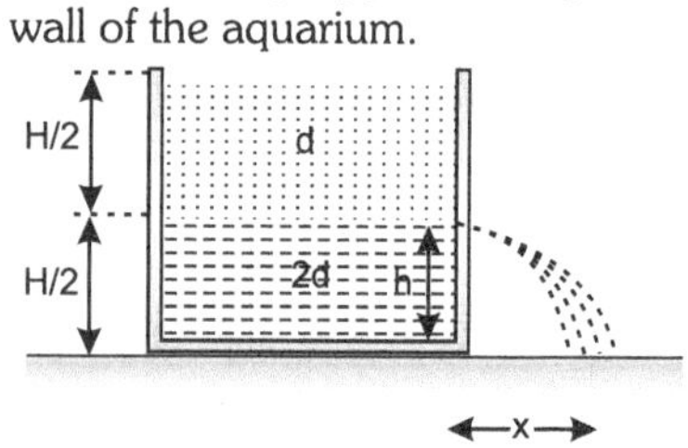

(a) (i) As for floating, $W = Th$

$$V\rho g = V_1 d_1 g + V_2 d_2 g$$

or $L\left(\frac{A}{5}\right)\rho = \left(\frac{3}{4}L\right)\left(\frac{A}{5}\right)d + \left(\frac{1}{4}L\right)\left(\frac{A}{5}\right)2d$

or $\rho = \frac{3}{4}d + \frac{2}{4}d = \boldsymbol{\frac{5}{4}d}$

(ii) Total pressure
$= p_0$ (weight of liquid + weight of solid)/A

Thus, $p = p_0 + \frac{H}{2}dg + \frac{H}{2}2dg + \frac{5}{4}d \times \left(\frac{A}{5} \times L\right) \times g \times \frac{1}{A}$

or $\quad p = p_0 + \frac{3}{2}Hdg + \frac{1}{4}Ldg$

(b) (i) By Bernoulli's theorem for a point just inside and outside the hole

$$p_1 + \frac{1}{2}\rho v_1^2 = p_2 = \frac{1}{2}\rho v_2^2$$

or $p_0 + \frac{H}{2}dg + \left(\frac{H}{2} - h\right)2dg$

$$= p_0 + \frac{1}{2}(2d)v^2$$

or $\quad g(3H - 4h) = 2v^2$

or $\quad v = \sqrt{\left(\frac{g}{2}\right)(3H - 4h)}$

(ii) As at the hole vertical velocity of liquid is zero so time taken by it to reach the ground,

$$t = \sqrt{\left(\frac{2h}{g}\right)}$$

Here, we have $x = vt = \sqrt{\frac{g}{2}(3H - 4h)} \times \sqrt{\frac{2h}{g}}$

$$= \sqrt{h(3H - 4h)} \qquad \ldots(1)$$

(iii) For x to be maximum x^2 must be maximum, thus we have

$$\frac{d}{dh}(x^2) = 0$$

or $\quad \frac{d}{dh}(3Hh - 4h^2) = 0$

or $\quad 3H - 8h = 0,$

or $\quad h = \left(\frac{3}{8}\right)H$

Substituting the value of h in equation (1), we get

$$x_{max} = \sqrt{\frac{3H}{g}\left(3H - \frac{3}{2}H\right)} = \frac{3}{4}H$$

248. Pressure at top of water level in the tank

$$P_A = P_{atm}$$

and pressure at the top of tube

$P_1 = P_A - \frac{1}{2}\rho v^2 - \rho g x; \qquad$ where $v = \sqrt{2gh}$

Thus $\quad P_1 = P_{atm} - \rho g(h + x)$

(a) For siphon to work, P_1 should not be negative. Setting $P_1 = 0$

$$x_{max} = \frac{P_{atm}}{\rho g} - h = \frac{1 \times 10^5}{10^3 \times 9.8} - 5.0$$

$$= 10.2 - 5.0 = \mathbf{5.2\ m}$$

(b) The time of emptying the tank is

$$t = \frac{a_1}{a_2}\sqrt{\frac{2h}{g}}$$

$$= \frac{\pi(0.9)^2}{6.3 \times 10^{-4}}\sqrt{\frac{2 \times 5}{9.8}} \text{ sec}$$

$= \mathbf{68\ min}$ (approx)

249. We know that, shear stress $\sigma = \eta\left(\frac{du}{dy}\right)$

Consider a ring of radius r.

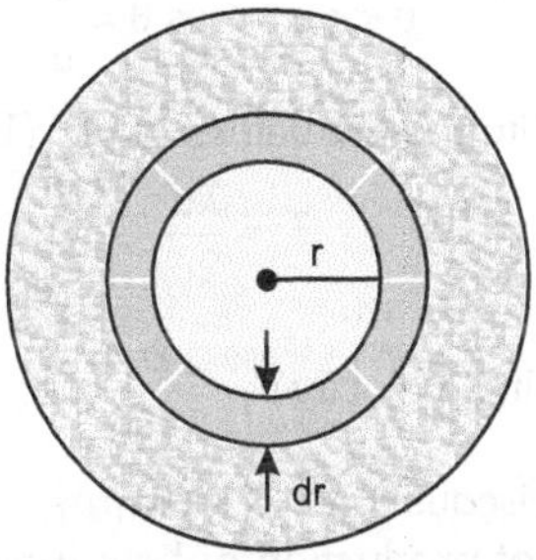

Tangential velocity $= \omega r$.

$\therefore \quad \sigma = \eta\frac{V_{tangential}}{t} = \eta\left(\frac{\omega r}{t}\right)$

Force on ring $= s \times$ Area $= \frac{\eta\omega r}{t}(2\pi r dr)$

$$= \left[\frac{2\pi\eta\,\omega r^2}{t}\right]dr$$

Torque required $\tau =$ force $\times$ radius

$$d\tau = \left[\frac{2\pi\eta\,\omega r^3 dr}{t}\right]$$

$$\therefore \text{Total torque} = \int_0^R d\tau = \frac{2\pi\eta\omega}{t}\left|\frac{r^4}{4}\right|_0^R$$

$$\tau = \left[\frac{2\pi\eta\omega R^4}{4t}\right] = \left[\frac{\pi\eta\omega R^4}{2t}\right]$$

$\therefore$Power required to overcome viscous resistance,

$$P = \tau\omega = \left[\frac{\pi\eta\omega^2 R^4}{2t}\right]$$

250. Free body diagram of the wire is as shown in figure.

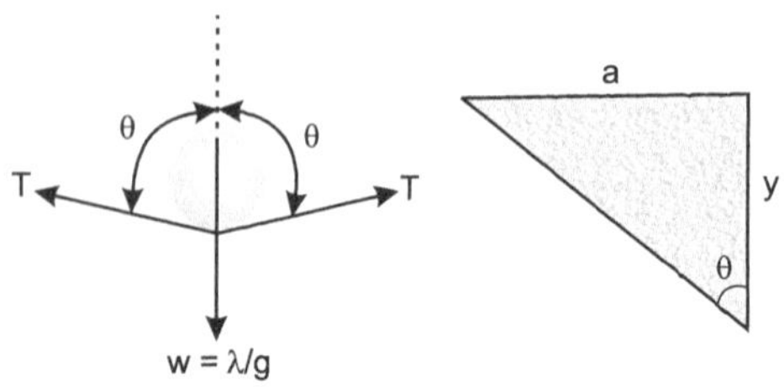

Considering the equilibrium of wire in vertical direction, we have:

$$2Tl \cos\theta = \lambda lg \quad \ldots(1)$$

For, $\quad y << a, \quad \cos\theta \approx \frac{y}{a}$

Substituting the values in Eq. (1) we get

$$T = \frac{\lambda a}{2y}$$

251. Terminal velocity $v_T = \frac{2r^2 g}{9\eta}(\rho_s - \rho_L)$

and viscous force $F = 6\pi\eta r v_T$
Rate of production of heat (power) $= Fv_T$, as viscous force is the only dissipative force. Hence,

$$\frac{dQ}{dt} = Fv_T = (6\pi\eta r v_T)(v_T)$$

$$= 6\pi\eta r v_T^2$$

$$= 6\pi\eta r\left\{\frac{2}{9}\frac{r^2 g}{\eta}(\rho_s - \rho_L)\right\}^2$$

$$= \frac{8\pi g^2}{27\eta}(\rho_s - \rho_L)^2 r^5$$

or $\quad \frac{dQ}{dt} \propto r^5$

252. From equation of continuity (Av = constant)

$$\frac{\pi}{4}(8)^2(0.25) = \frac{\pi}{4}(2)^2(v)$$

Here v is the velocity of water with which water comes out of the syringe (Horizontally).
Solving eq. (1), we get

$$v = 4 \text{ m/s}$$

The path of water after leaving the syringe will be parabola. Substituting proper values in equation of trajectory

$$y = x\tan\theta - \frac{gx^2}{2u^2\cos^2\theta}$$

We have $-1.25 = R\tan 0° - \frac{(10)(R^2)}{(2)(4)^2\cos^2 0°}$

(R = horizontal range)

Solving this equation, we get

$$R = \mathbf{2\ m}$$

253. The bubble will separate from the tube when thrust force due to striking air at B is equal to the force due to excess pressure.

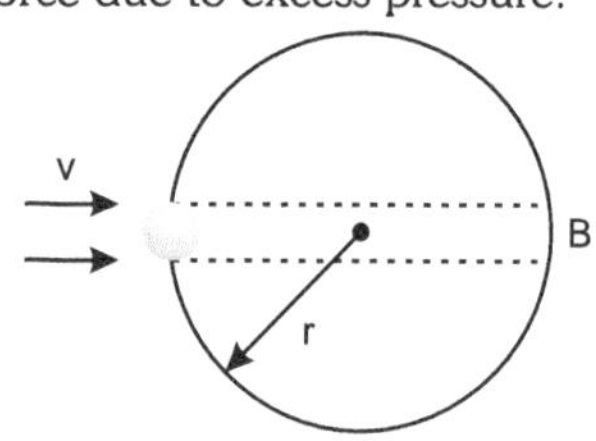

$$\therefore \quad \rho A v^2 = \left(\frac{4T}{r}\right)A$$

(A = area of bubble at B where air strikes)

$$\therefore \quad r = \left(\frac{4T}{\rho v^2}\right)$$

254. When the tube is not there,

$$P + P_0 + \frac{1}{2}\rho v_1^2 + \rho gH = \frac{1}{2}\rho v_2^2 + P_0$$

$$\therefore \quad P + \rho gH = \frac{1}{2}\rho(v_2^2 - v_1^2)$$

$$A_1 v_1 = A_2 v_2$$

or $\quad v_1 = \frac{A_2 v_2}{A_1}$

$$\therefore \; P + \rho g H = \frac{1}{2}\rho\left[v_2^2 - \left(\frac{A_2}{A_1}v_2\right)^2\right]$$

$$= \frac{1}{2}\times\rho\times v_2^2\left[1-\left(\frac{\pi(0.3)^2}{\pi(0.9)^2}\right)^2\right]$$

$$= \frac{1}{2}\times\rho\times(10)^2\left[1-\frac{1}{81}\right]$$

$$= \frac{4\times10^3\rho}{81} = \frac{4\times10^3\times900}{81}$$

$$= \frac{4}{9}\times10^5 \text{ N/m}^2$$

This is also the excess pressure ΔP.

By Poiseuille's equation, the rate of flow of liquid in the capillary tube

$$Q = \frac{\pi(\Delta P)\,a^4}{8\eta l}$$

$$\therefore \quad 8\times10^{-6} = \frac{(\pi a^2)(\Delta P)}{8\eta}\left(\frac{a^2}{l}\right)$$

$$\eta = \frac{(\pi a^2)(\Delta p)\left(\frac{a^2}{l}\right)}{8\times8\times10^{-6}}$$

Substituting the values we have

$$\eta = \frac{(10^{-6})\left(\frac{4}{9}\times10^5\right)(2\times10^{-6})}{8\times8\times10^{-6}}$$

$$= \frac{1}{720}\text{ N-S/m}^2$$

255. (a)

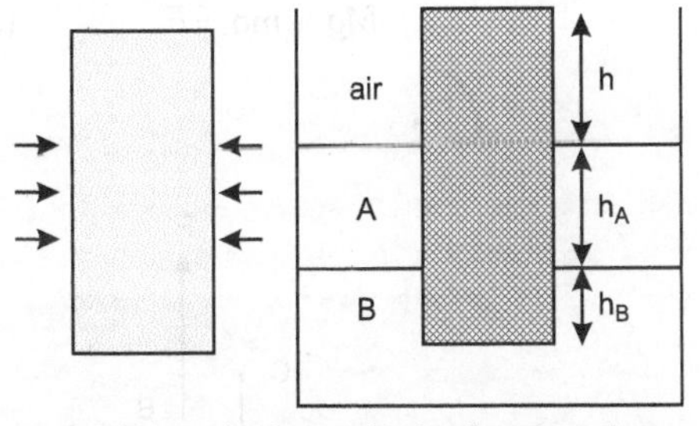

Liquid A is applying the hydrostatic force on cylinder from all the sides. So net force is zero.

(b) In equilibrium:

Weight of cylinder = Net upthrust on the cylinder

Let s be the area of the cross-section of the cylinder, then

weight $= (s)(h + h_A + h_B)\rho_{cylinder}\, g$

and upthrust on the cylinder

= upthrust due to liquid

+ upthrust due to liquid B

$= sh_A\rho_A g + sh_B\rho_B g$

Equating these two

$s(h + h_A + h_B)\rho_{cylinder}\, g = sh_A\rho_A g + sh_B\rho_B g$

or $(h + h_A + h_B)\rho_{cylinder} = h_A\rho_A + h_B\rho_B$

Substituting $h_A = 1.02$ cm,

$h_B = 0.8$ cm,

$\rho_A = 0.7$ g/cm^3

$\rho_B = 1.02$ g/cm^3

and $\rho_{cylinder} = 0.8$ g/cm^3

in the above equation we get:

$$\boldsymbol{h = 0.25 \text{ cm}}$$

(c) Net upward force = extra upthrust

$= sh\rho_B g$

$\therefore$ Net acceleration $a = \dfrac{\text{force}}{\text{mass of cylinder}}$

or $$a = \frac{sh\rho_B g}{s(h + h_A + h_B)\rho_{cylinder}}$$

or $$a = \frac{h\rho_B g}{(h + h_A + h_B)\rho_{cylinder}}$$

substituting the values of h, h_A, h_B, ρ_B and $\rho_{cylinder}$ we get,

$$\boldsymbol{a = \frac{g}{6}} \quad \text{(upwards)}$$

256. Given Length of the wire, $l = 5$ m

Radius of the wire, $r = 2\times10^{-3}$ m

Density of wire, $\rho = 7860$ kg/m^3

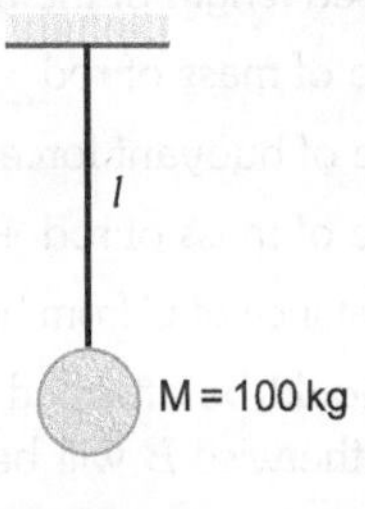

Young's modulus,

$$Y = 2.1 \times 10^{11}\ \text{N/m}^2$$

and Specific heat,

$$s = 420\ \text{J/kg-K}$$

Mass of wire,

$$m = (\text{Density})\ (\text{Volume})$$
$$= (\rho)(\pi r^2 l)$$
$$= (7860)(\pi)(2 \times 10^{-3})^2(5)\ \text{kg}$$
$$= 0.494\ \text{kg}$$

Elastic potential energy stored in the wire,

$$U = \frac{1}{2}\ (\text{Stress})\ (\text{Strain})\ (\text{Volume})$$

$$\left[\because \frac{\text{Energy}}{\text{Volume}} = \frac{1}{2} \times \text{stress} \times \text{strain}\right]$$

or $$U = \frac{1}{2}\left(\frac{Mg}{\pi r^2}\right)\left(\frac{\Delta l}{l}\right)(\pi r^2 l)$$

$$= \frac{1}{2}(Mg).\Delta l \qquad \left(\Delta l = \frac{Fl}{AY}\right)$$

$$= \frac{1}{2}(Mg)\frac{Mgl}{(\pi r^2)Y} = \frac{1}{2}\frac{M^2g^2l}{\pi r^2 Y}$$

Substituting the values, we have

$$U = \frac{1}{2}\frac{(100)^2(10)^2(5)}{(3.14)(2 \times 10^{-3})^2(2.1 \times 10^{11})}\ \text{J}$$

$$= 0.9478\ \text{J}$$

When the bob gets snapped, this energy is utilised in raising the temperature of the wire.

So, $$U = ms\Delta\theta$$

$$\therefore\ \Delta\theta = \frac{U}{ms} = \frac{(0.9478)}{(0.494)(420)}\ {}^\circ\text{C or K}$$

$$\Delta\theta = \mathbf{4.568 \times 10^{-3}\,{}^\circ C}$$

257. Let M = Mass of stick = $\pi R^2 \rho L$

l = Immersed length of the rod

G = Centre of mass of rod

B = Centre of buoyant force (F)

C = Centre of mass of rod + mass (m)

y_{COM} = Distance of C from bottom of the rod

Mass m should be attached to the lower end because otherwise B will be below G and C will be above G and the torque of the couple of two equal and opposite forces F and $(M + m)g$ will be counter clockwise on displacing the rod leftwards. Therefore, the rod can't be in rotational equilibrium. See the figure given below :

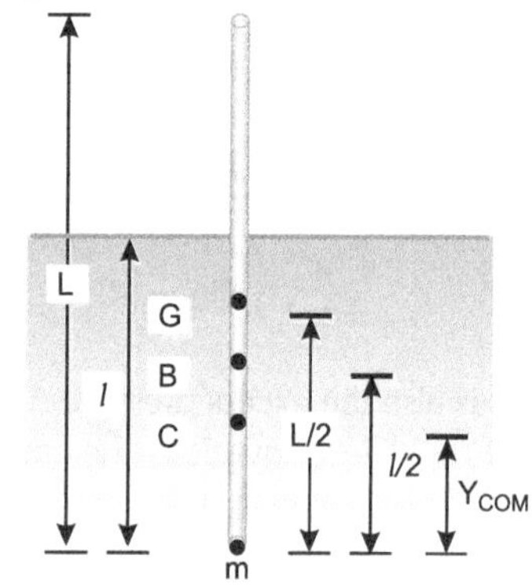

Figure - 1

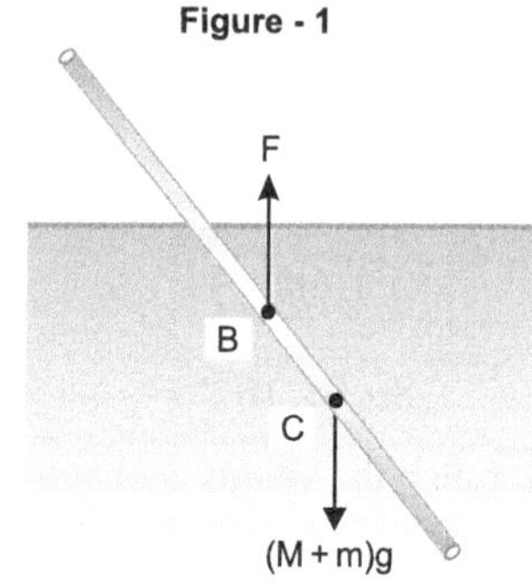

Figure - 2

Now refer figures (1) and (2)

For vertical equilibrium,

$$Mg + mg = F \qquad \text{(upthrust)}$$

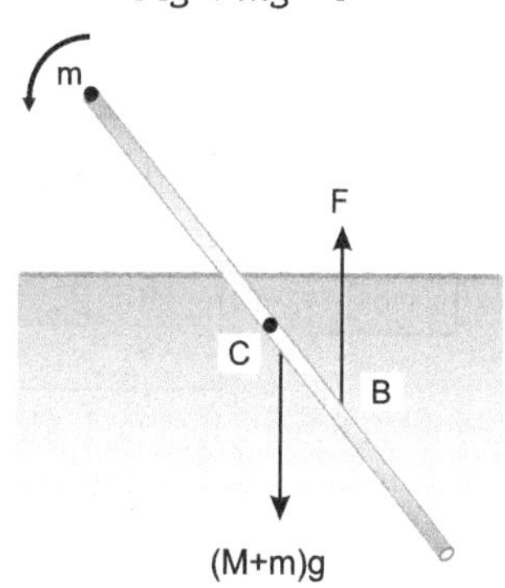

Figure (3)

or $(\pi R^2 L)\rho g + mg = (\pi R^2 l)\sigma g$

$\therefore$ $$l = \left(\frac{\pi R^2 L\rho + m}{\pi R^2 \sigma}\right) \quad \ldots(1)$$

Position of COM (or rod + m) from bottom,

$$y_{COM} = \frac{M \cdot \frac{L}{2}}{M + m} = \frac{(\pi R^2 L\rho)\frac{L}{2}}{(\pi R^2 L\rho) + m} \quad \ldots(2)$$

Centre of buoyancy (B) is at a height of $\frac{l}{2}$ from the bottom.

We can see from figure (2) that for rotational equilibrium of the rod, B should either lie above C or at the same level of B.

Therefore, $\frac{l}{2} \geq y_{COM}$

or $$\frac{\pi R^2 L\rho + m}{2\pi R^2 \sigma} \geq \frac{(\pi R^2 L\rho)\frac{L}{2}}{(\pi R^2 L\rho) + m}$$

or $m + \pi R^2 L\rho \geq \pi R^2 L\sqrt{\rho\sigma}$

or $m \geq \pi R^2 L(\sqrt{\rho\sigma} - \rho)$

$\therefore$ **Minimum value of m is**

$$\pi R^2 L(\sqrt{\rho\sigma} - \rho)$$

258. (i) The change in length due to decrease in temperature,

$\Delta l_1 = L\alpha\Delta\theta = (0.5)(10^{-5})(0 - 100)$

$\Delta l_1 = -0.5 \times 10^{-3}$ m $\quad \ldots(1)$

Negative sign implies that length is decreasing. Now let M be the mass attached to the lower end. Then change in length due to suspension of load is

$$\Delta l_2 = \frac{(Mg)L}{AY} = \frac{(M)(10)(0.5)}{(4 \times 10^{-6})(10^{11})}$$

$\Delta l_2 = (1.25 \times 10^{-5})$ M $\quad \ldots(2)$

Net change in length is zero. Therefore,

$$\Delta l_1 + \Delta l_2 = 0$$

or $(1.25 \times 10^{-5})M = (0.5 \times 10^{-3})$

$\Rightarrow$ $$M = \left(\frac{0.5 \times 10^{-3}}{1.25 \times 10^{-5}}\right) \text{kg}$$

or **M = 40 kg**

(ii) **Energy stored** : At 0°C the natural length of the wire is less than its actual length; but since a mass is attached at its lower end, an elastic potential energy is stored in it. This is given by

$$U = \frac{1}{2}\left(\frac{AY}{L}\right)(\Delta l)^2 \quad \ldots(3)$$

Here $\Delta l = |\Delta l_1| = \Delta l_2 = 0.5 \times 10^{-3}$ m

substituting the values

$$U = \frac{1}{2}\left(\frac{4 \times 10^{-6} \times 10^{11}}{0.5}\right)(0.5 \times 10^{-3})^2$$

U = 0.1 J

→ Comparing the equation

$$Y = \frac{F/A}{\Delta l/L} \quad \text{or} \quad F = \left(\frac{AY}{L}\right).\Delta l$$

with the spring equation $F = k.\Delta x$, we find that equivalent spring constant of a wire is $k = \left(\frac{AY}{L}\right)$.

Therefore, potential energy stored in it should be

$$U = \frac{1}{2}k(\Delta l)^2 = \frac{1}{2}\left(\frac{AY}{L}\right)(\Delta l)^2$$

259. Given : $A_1 = 4 \times 10^{-3}$ m^2,

$A_2 = 8 \times 10^{-3}$ m^2, $h_1 = 2$ m, $h_2 = 5$ m

$v_1 = 1$ m/s and $\rho = 10^3$ kg/m^3

From continuity equation, we have

$$A_1 v_1 = A_2 v_2$$

or $$v_2 = \left(\frac{A_1}{A_2}\right)v_1$$

or $$v_2 = \left(\frac{4 \times 10^{-3}}{8 \times 10^{-3}}\right)(1 \text{ m/s})$$

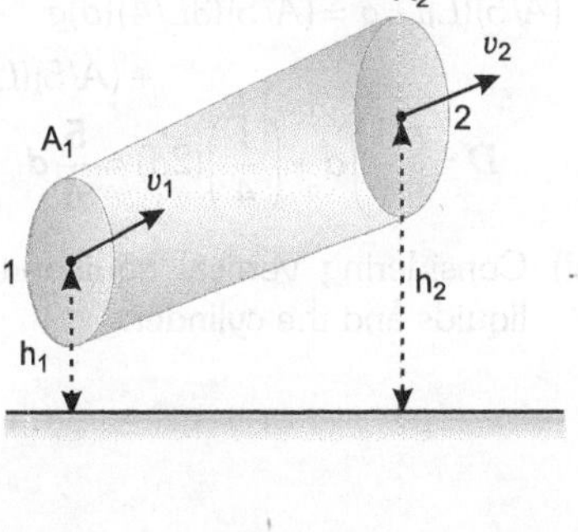

$$v_2 = \frac{1}{2}\ \text{m/s}$$

Applying Bernoulli's equation at section 1 and 2,

$$P_1 + \frac{1}{2}\rho v_1^2 + \rho g h_1 = P_2 + \frac{1}{2}\rho v_2^2 + \rho g h_2$$

or $P_1 - P_2 = \rho g(h_2 - h_1) + \frac{1}{2}\rho\,(v_2^2 - v_1^2)$ ...(1)

(i) Work done for unit volume by the pressure as the fluid flows from P to Q

$$W_1 = P_1 - P_2$$
$$= \rho g(h_2 - h_1) + \frac{1}{2}\rho\,(v_2^2 - v_1^2)$$

[from equation (1)]

$$= \left\{(10^3)(9.8)(5-2) + \frac{1}{2}(10^3)\left(\frac{1}{4} - 1\right)\right\}\ \text{J/m}^3$$

$$= \{29400 - 375\}\ \text{J/m}^3$$

or $\mathbf{W_1 = 29025\ J/m^3}$

(ii) Work done per unit volume by the gravity as the fluid flows from P to Q.

$$W_2 = -\rho g(h_2 - h_1) = -(10^3)(9.8)(5-2)\ \text{J/m}^3$$

or $\mathbf{W_2 = -29400\ J/m^3}$

260. (a) (i) Considering vertical equilibrium of cylinder

Weight of cylinder = Upthrust due to upper liquid + upthrust due to lower liquid

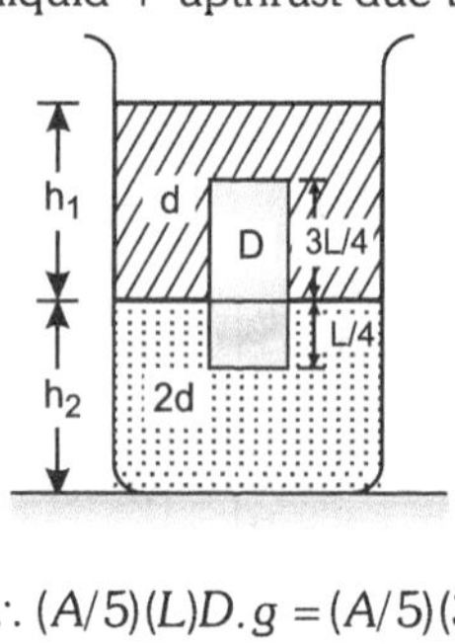

Note that h_1 and $h_2 \neq \frac{H}{2}$

$$\therefore (A/5)(L)D.g = (A/5)(3L/4)(d)g + (A/5)(L/4)(2d)(g)$$

$$\therefore \quad \mathbf{D} = \left(\frac{3}{4}\right)d + \left(\frac{1}{4}\right)(2d) = \mathbf{\frac{5}{4}d}$$

(ii) Considering vertical equilibrium of two liquids and the cylinder.

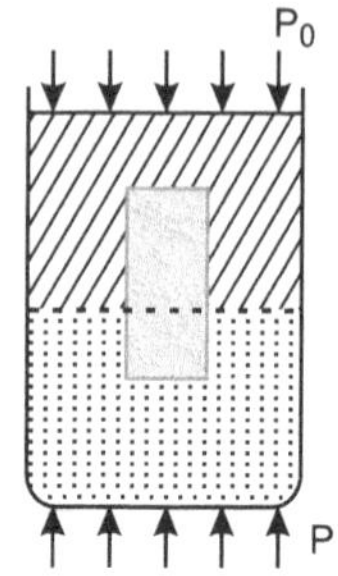

$(P - P_0)A$ = weight of two liquids + weight of cylinder

$$\therefore \quad P = P_0 + \frac{\text{weight of two liquids} + \text{weight of cylinder}}{A} \qquad ...(1)$$

Now Weight of cylinder

$$= (A/5)(L)(D)g = (A/5\,Lg)(5/4\,d)$$
$$= \frac{ALdg}{4}$$

Weight of upper liquid $= \left\{\frac{H}{2}Adg\right\}$

and Weight of lower liquid

$$= \frac{H}{2}A(2d)g = HAgd$$

$\therefore$ Total weight of two liquids $= \frac{3}{2}HAdg$

$\therefore$ From equation (1), pressure at the bottom of the container will be

$$P = P_0 + \frac{(3/2)HAdg + ALdg/4}{A}$$

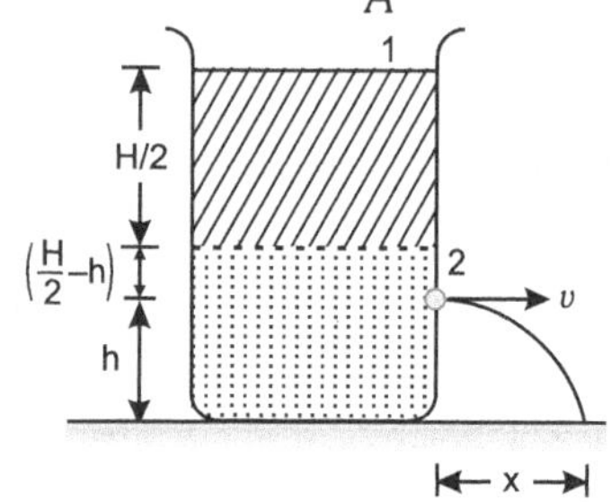

or $\mathbf{P = P_0 + \frac{dg(6H + L)}{4}}$

(b) (i) Applying Bernoulli's theorem at 1 and 2

$$P_0 + dg(H/2) + 2dg(H/2 - h) = P_0 + \frac{1}{2}(2d)v^2$$

Here v is velocity of efflux at 2.
Solving this, we get

$$\mathbf{v = \sqrt{(3H - 4h)g/2}}$$

(iii) Time taken to reach the liquid to the bottom will be

$$t = \sqrt{2h/g}$$

∴ Horizontal distance x travelled by the liquid is

$$x = vt = (\sqrt{(3H - 4h)g/2})\left(\sqrt{\frac{2h}{g}}\right)$$

$$\boldsymbol{x = \sqrt{h(3H - 4h)}}$$

(iii) For x to be maximum— $\frac{dx}{dh} = 0$

or $$\frac{1}{2\sqrt{h(3H - 4h)}}(3H - 8h) = 0$$

⇒ $$h = \frac{3H}{8}$$

Therefore, x will be maximum at $h = 3H/8$

The maximum value of x will be

$$x_m = \sqrt{\left(\frac{3H}{8}\right)\left[3H - 4\left(\frac{3H}{8}\right)\right]}$$

or $$\boldsymbol{x_m = \frac{3}{4}H}$$

261. In elastic collision with the surface, direction of velocity is reversed but its magnitude remains

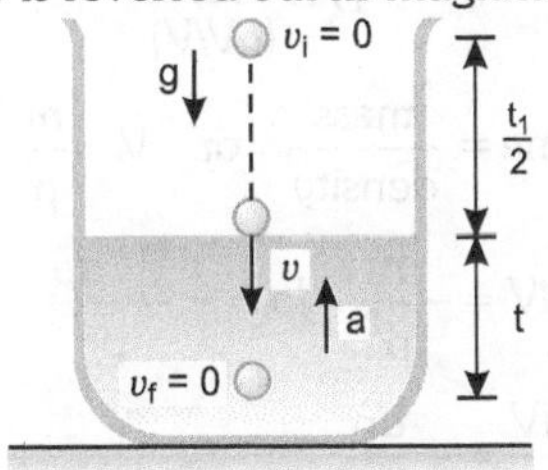

the same. Therefore, time of fall = time of rise.

or $$\text{time of fall} = \frac{t_1}{2}$$

Hence velocity of the ball just before it collides with liquid is

$$v = g\frac{t_1}{2} \quad \ldots(1)$$

Retardation inside the liquid

$$a = \frac{\text{upthrust} - \text{weight}}{\text{mass}}$$

$$= \frac{Vd_L g - Vdg}{Vd} = \left(\frac{d_L - d}{d}\right)g \quad \ldots(2)$$

Here V = Volume of ball

Time taken to come to rest under this retardation will be

$$t = \frac{v}{a} = \frac{g\,t_1}{2a} = \frac{g\,t_1}{2\left(\frac{d_L - d}{d}\right)g} = \frac{d\,t_1}{2(d_L - d)}$$

Same will be the time to come back on the liquid surface. Therefore,

(a) t_2 = time the ball takes to come back to the position from where it was released

$$= t_1 + 2t$$

$$= t_1 + \frac{dt_1}{d_L - d} = t_1\left[1 + \frac{d}{d_L - d}\right]$$

or $$\boldsymbol{t_2 = \frac{t_1 d_L}{d_L - d}}$$

(b) The **motion of the ball is periodic but not simple harmonic** because the acceleration/retardation of the ball is g in air and $\left(\frac{d_L - d}{d}\right)g$ inside the liquid which is not proportional to the displacement, which is necessary and sufficient condition for SHM.

(c) When $d_L = d$, retardation or acceleration inside the liquid becomes zero (upthrust = weight). Therefore,
the ball will continue to move with constant velocity $\boldsymbol{v = gt_1/2}$ inside the liquid.

CHAPTER 9 WAVES

262. $v_s = v_0 = v$

Let u be the speed of sound. Then

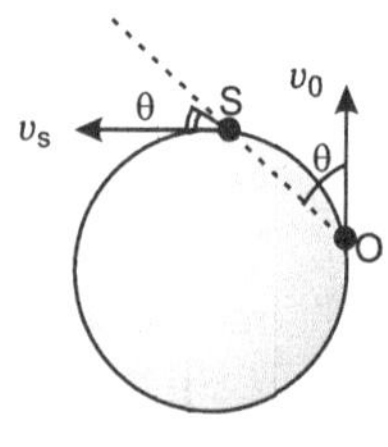

$$f' = f\left(\frac{u + v_0 \cos\theta}{u + v_s \cos\theta}\right)$$

$$= f\left(\frac{u + v \cos\theta}{u + v\cos\theta}\right)$$

or $\qquad$ $\boldsymbol{f' = f}$

263. Given length of pipe $l = 3$ m

Third harmonic implies that

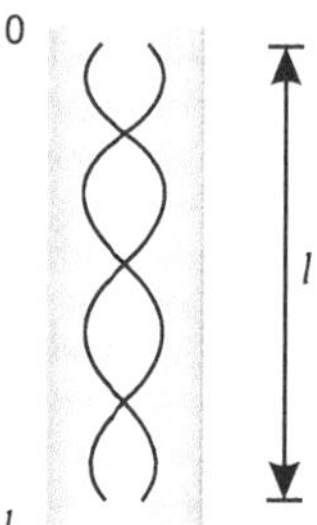

$$3\left(\frac{\lambda}{2}\right) = l$$

or $\lambda = \frac{2l}{3} = \frac{2 \times 3}{3} = 2$ m

The angular frequency is

$$\omega = 2\pi f$$

$$= \frac{2\pi v}{\lambda} = \frac{(2\pi)(332)}{2}$$

or $\qquad \omega = 332\pi$ rad/s

The particle displacement $y(x, t)$ can be written as

$$y(x, t) = A \cos kx \sin \omega t$$

$$k = \frac{2\pi}{\lambda} = \frac{2\pi}{(2l/3)} = \frac{3\pi}{l}$$

and $\qquad \omega = kv = \frac{3\pi v}{l} \qquad \left(v = \frac{\omega}{k}\right)$

$$\therefore \; y(x, t) = A \cos\left(\frac{3\pi x}{l}\right) \cdot \sin\left(\frac{3\pi v}{l}\right)t$$

The longitudinal oscillations of an air column can be viewed as oscillations of particle displacement or pressure wave or density wave. Pressure variation is related to particle displacement as

$$P(x) = -B \cdot \frac{\delta y}{\delta x} \qquad (B = \text{Bulk modulus})$$

$$= \left(\frac{3BA\pi}{l}\right)\sin\left(\frac{3\pi x}{l}\right)\sin\left(\frac{3\pi v}{l}\right)t$$

The amplitude of pressure variation is

$$P_{max} = \frac{3BA\pi}{l}$$

$$v = \sqrt{\frac{B}{\rho}} \quad \text{or} \quad B = \rho v^2$$

$$\therefore \quad P_{max} = \frac{3\rho v^2 A\pi}{l}$$

or $\qquad A = \frac{P_{max}\, l}{3\rho v^2 \pi}$

Here $P_{max} = 1\%$ of $P_0 = 10^3$ N/m^2

Substituting the values

$$A = \frac{(10^3)(3)}{(3)(1.03)(332)^2(\pi)}$$

$$= 0.0028 \text{ m}$$

or $\qquad \boldsymbol{A = 0.28}$ **cm**

According to definition of Bulk modulus (B)

$$B = \frac{-dP}{(dV/V)} \qquad \ldots(1)$$

Volume $= \frac{\text{mass}}{\text{density}}$ or $V = \frac{m}{\rho}$

or $\quad dV = -\frac{m}{\rho^2} . d\rho = -\frac{V . d\rho}{\rho}$

or $\quad \frac{dV}{V} = -\frac{d\rho}{\rho}$

Substituting in equation (1) we get.

$$d\rho = \frac{\rho\,(dp)}{B}$$

or amplitude of density oscillation is

$$d\rho_{max} = \frac{\rho}{B} \cdot P_{max} = \frac{P_{max}}{v^2} \qquad \left(\frac{\rho}{B} = \frac{1}{v^2}\right)$$

$$= \frac{10^3}{(332)^2} = \mathbf{9 \times 10^{-3}\ kg/m^3}$$

264. Sound level (in dB)

$$L = 10 \log_{10}\left(\frac{I}{I_0}\right)$$

where $I_0 = 10^{-12}$ watt/m^2

$L = 60$ dB

Hence $I = (10^6) I_0 = 10^{-6}$ watt /m^2

Intensity, $I = \frac{\text{Power}}{\text{Area}} = \frac{P}{4\pi r^2}$

$\Rightarrow \quad P = I(4\pi r^2)$

$$P = (10^{-6})(4\pi)(500)^2 = 3.14 \text{ watt}$$

$$\therefore \text{Time}, \quad t = \frac{(1.0 \times 10^3)\left(\frac{30}{100}\right)}{3.14}$$

$$\boldsymbol{t = 95.5\ \mathrm{s}}$$

265. (a) (i) In 3 second, wave will travel a distance of 3 cm. The formation of the reflected pulse from a fixed support is similar to the overlap of two inverted pulses travelling in opposite directions as shown in figure :

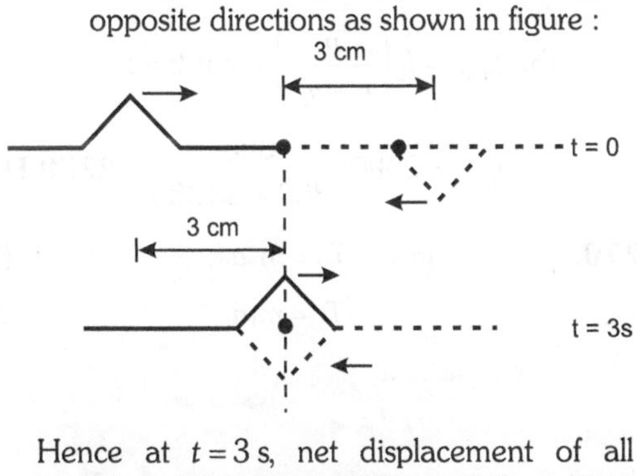

Hence at $t = 3$ s, net displacement of all particles of the string will be zero. *i.e.*, string will be straight as shown below

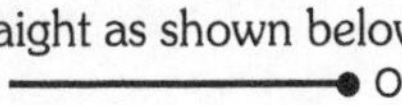

(ii) At $t = 4$ s, wave will travel a distance of 4 cm Hence

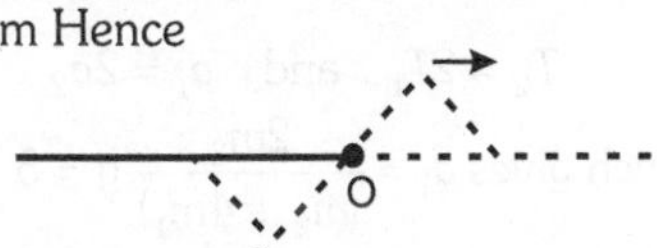

Therefore, the shape of the pulse will be as shown below :

(b) (i) The formation of the reflected pulse from a free support is similar to the overlap of two pulses of same nature travelling in opposite directions as shown below :

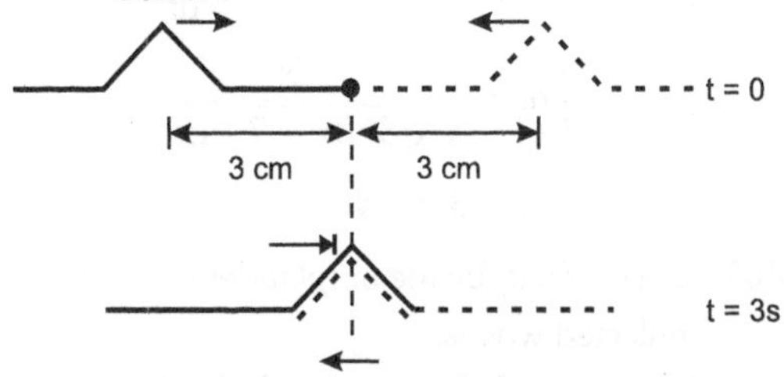

Therefore, net displacements of all points will become twice as shown below :

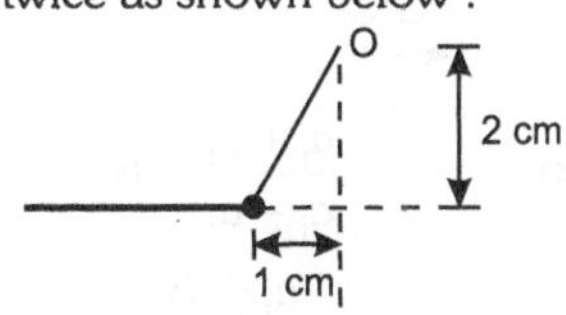

(ii) At $t = 4$ s, two pulses will cross each other completely and hence resultant pulse will be as shown below :

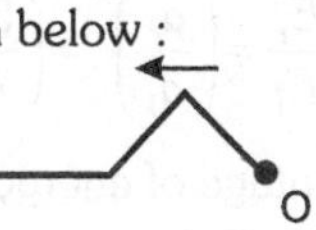

266. At a distance x from the bottom, speed of wave pulse is

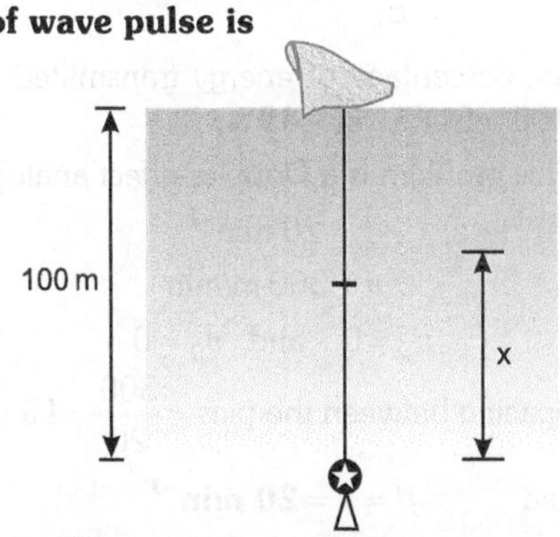

$$v = \sqrt{\frac{T}{\mu}} \quad (\mu = \text{mass per unit length of cable})$$

where T = (weight of diver + weight of string − buoyant forces on both)

$= \{120 \times 10 + (1.1\, x)(10)\}$
$\quad - \{0.08 + \pi(10^{-2})^2 x\} \times 10^3 \times 10$

$= 1200 + 11x - 800 - 3.14x$

$= (400 + 7.86x)$

$\therefore \quad v = \sqrt{\dfrac{(400 + 7.86x)}{1.1}}$

$v = \sqrt{(363.6 + 7.14x)} = \dfrac{dx}{dt}$

$\therefore \quad \int_0^t dt = \int_0^{100} \dfrac{dx}{\sqrt{363.6 + 7.14x}}$

or $\quad \boldsymbol{t = 3.85\ s}$

267. Let a_i and a_r be the amplitudes of incident and reflected waves.

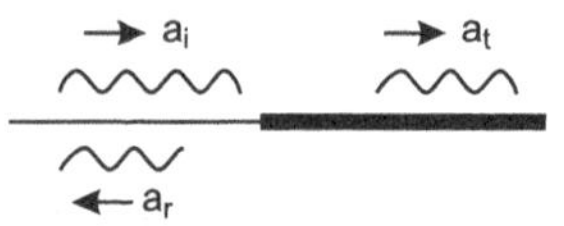

Then $\quad \dfrac{a_i + a_r}{a_i - a_r} = 6 \quad$ (given)

Hence $\quad \dfrac{a_r}{a_i} = \dfrac{5}{7}$

Now $\quad \dfrac{E_r}{E_i} = \left(\dfrac{a_r}{a_i}\right)^2 = \left(\dfrac{5}{7}\right)^2 = 0.51$

or percentage of energy reflected is

$\dfrac{E_r}{E_i} \times 100$ or 51%.

So, percentage of energy transmitted will be $(100 - 51)\%$ or **49%.**

268. This problem is a Doppler-effect analogy.

(a) Here, $\quad f = 20\ \text{min}^{-1}$

$v = 300$ m/min

$v_s = 0 \quad$ and $\quad v_0 = 0$

Spacing between the pies $= \dfrac{300}{20} = \mathbf{15\ m}$

and $\quad f' = f = \mathbf{20\ min^{-1}}$

(b) $v_s = 30$ m/min

spacing between the pies will be

$\dfrac{300 - 30}{20}$ or **13.5 m**

and $\quad f' = f\left(\dfrac{v}{v - v_s}\right) = (20)\left(\dfrac{300}{300 - 30}\right)$

$= \mathbf{22.22\ min^{-1}}$

269. The graph shows the situation shown in figure.

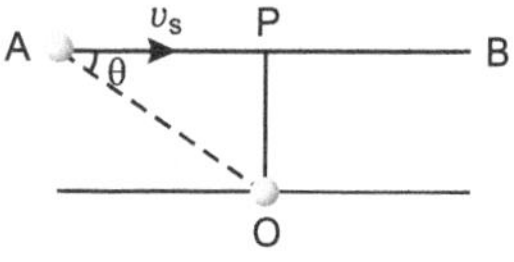

At P, apparent frequency = natural frequency = 2000 Hz

(a) For region AP

$$f' = f_0\left(\frac{v}{v - v_s \cos\theta}\right)$$

and for region PB

$$f' = f_0\left(\frac{v}{v + v_s \cos\theta}\right)$$

$f'_{min} = f_0\left(\dfrac{v}{v + v_s}\right) \quad$ when $\cos\theta = 1$

i.e. $\quad 1800 = 2000\left(\dfrac{300}{300 + v_s}\right)$

Solving this, we get

$\boldsymbol{v_s = 33.33\ m/s}$

(b) $f_{max} = f_0\left(\dfrac{v}{v - v_s}\right)$ at $\cos\theta = 1$

or $\quad f_{max} = 2000\left(\dfrac{300}{300 - 33.33}\right) = \mathbf{2250\ Hz}$

270. $\quad m_2 g - T_2 = m_2 a_2 \quad$ ...(1)

$T_1 = m_1 a_1 \quad$ ...(2)

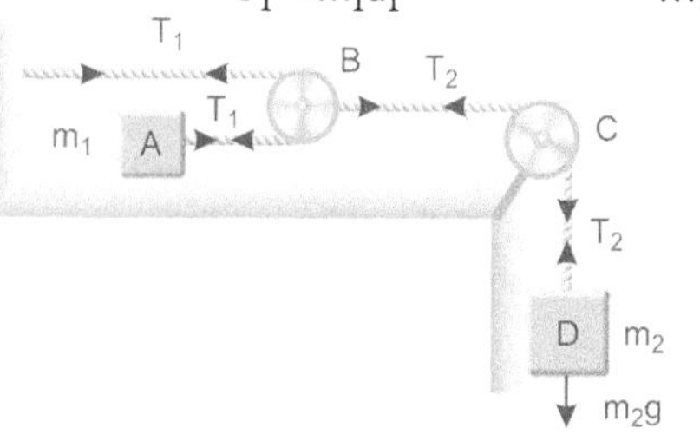

$T_2 = 2T_1 \quad$ and $\quad a_1 = 2a_2$

which gives $a_1 = \dfrac{2m_2}{(m_2 + 4m_1)} g = 5\ \text{m/s}^2$

and $\qquad T_1 = 37.5\text{ N}$

speed of transverse pulse

$$v = \sqrt{\frac{T}{\mu}} = \sqrt{\frac{37.5}{15\times 10^{-3}}} = 50\text{ ms}^{-1}$$

From frame of the pulley B, speed of the pulse

$$= v + v_A - v_B$$

$$v + a_1 t - a_2 t = -\frac{dx}{dt}$$

$$-dx = (50 + 2.5t)\, dt$$

Integrating:

$$-\int_{10}^{0} dx = \int_0^t (50 + 5t)\, dt$$

$$10 = 50t + \frac{5t^2}{2}$$

$$5t^2 + 100t - 20 = 0$$

which gives $t =$ **0.2 sec.**

271. (a) Comparing with the equation of a travelling wave

$$y = a\sin(kx - \omega t)$$

$$k = 15\pi \quad \text{and} \quad \omega = 6000\pi$$

$\therefore$ velocity of the sound $v = \frac{\omega}{k}$

$$= \frac{6000\pi}{15\pi} = 400\text{ ms}^{-1}$$

as $\qquad v = \sqrt{\frac{B}{\rho}}$

Hence, $\rho = \frac{B}{v^2} = \frac{1.6\times 10^5}{(400)^2} = \mathbf{1\ kg/m^3}$

(b) Pressure amplitude $P_0 = BAk$

Hence $\quad A = \frac{P_0}{Bk}$

$$= \frac{24\pi}{1.6\times 10^5 \times 15\pi}$$

$$= 10^{-5}\text{ m} = \mathbf{10\ \mu m}$$

(c) Intensity received by the person

$$I = \frac{W}{4\pi R^2} = \frac{W}{4\pi (10)^2}$$

Also, $\quad I = \frac{p_0^2}{2\rho v}$

$$\frac{W}{4\pi (10)^2} = \frac{(24\pi)^2}{2\times 1\times 400}$$

$$W = \mathbf{288p^3\ watt}$$

272. The wave emitted by the source after t_1 sec from starting position will be heard the receiver after $t = 9$ sec.

$\therefore \qquad V_s = 3t_1$

$$V_R = 2t$$

$\Rightarrow \qquad f = f_0\left(\frac{V - V_R}{V + V_s}\right)$

Equation for solving t_1 is

$$330(9 - t_1) = \frac{1}{2}\times 3\times t_1^2 + \frac{1}{2}\times 2(9)^2$$

$$t_1 = 8.43\text{ sec}$$

$\Rightarrow \qquad f = 1000\left(\frac{330 - 18.00}{330 + 25.29}\right)$

$$= 878.156\text{ Hz}$$

273. Time period of the wave is

$$T = (0.05 - 0.01) = 0.04\text{ s}$$

Let the equation of the wave moving in positive x-direction is

$$y = a\sin(\omega t - kx + \phi) \qquad \ldots(1)$$

Here $\qquad a = 4\text{ mm}$

$$\omega = \frac{2\pi}{T}$$

From the figure it is clear that $y = 0$ at $t = 0$ for a particle at $x = 0$. Substituting in (1) we get.

$$\phi = 0$$

It can also be observed that $y = 3$ mm at $t = 0$ for a particle at $x = 0.09$ m. Substituting in (1) we get

$$3 = 4\sin(-kx)$$

or $\qquad \frac{3}{4} = -\sin kx$

or $\qquad kx = \pi + \sin^{-1}\left(\frac{3}{4}\right)$

$$= 3.989\text{ radian}$$

or $\qquad \left(\frac{2\pi}{\lambda}\right)(0.09) = 3.989$

$(k = \frac{2\pi}{\lambda}$ and $x = 0.09$ m)

or $\qquad \lambda = \mathbf{0.142\ m}$

and wave speed

$$v = \frac{\lambda}{T} = \frac{0.142}{0.04}\ \text{m/s}$$

or $\qquad \boldsymbol{v} = \mathbf{3.55\ m/s}$

274. Amplitude at a distance x is

$$A = a \sin kx$$

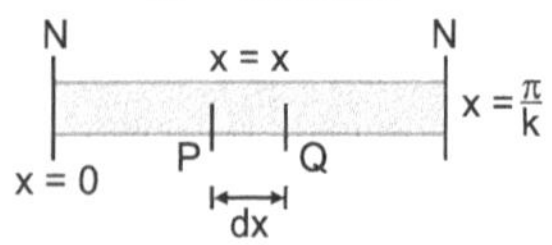

First node can be obtained at $x = 0$,

and the second at $x = \pi / k$

At position x, mass of the element PQ is

$$dm = (\rho S)dx$$

its amplitude is $A = a \sin kx$

Hence mechanical energy stored in this element is

$$dE = \frac{1}{2}(dm)A^2\omega^2$$

(energy of particle in SHM)

or $$dE = \frac{1}{2}(\rho S A^2\omega^2).dx$$

$$= \frac{1}{2}(\rho S a^2\omega^2 \sin^2 kx)\,dx$$

Therefore, total energy stored between two adjacent nodes will be

$$E = \int_{x=0}^{x=\pi/k} dE$$

Solving this, we get

$$\boldsymbol{E = \frac{\pi S\rho\omega^2 a^2}{4k}}$$

275. For first overtone

$$l = \lambda$$

$\therefore$ $$f = \frac{v}{\lambda} = \frac{v}{l}$$

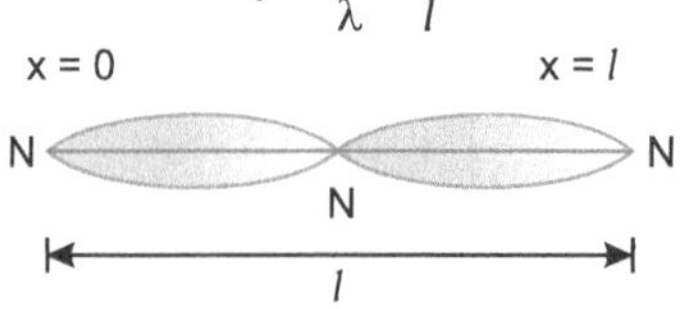

$$k = \frac{2\pi}{\lambda} = \frac{2\pi}{l}$$

Amplitude of oscillations at $x = x$ can be written as

$$A = a_0 \sin kx$$

Let ρ be the density and S be the cross-sectional area of the wire. Then

$$m = \text{mass per unit length} = \rho S$$

Mass of element PQ of the string is

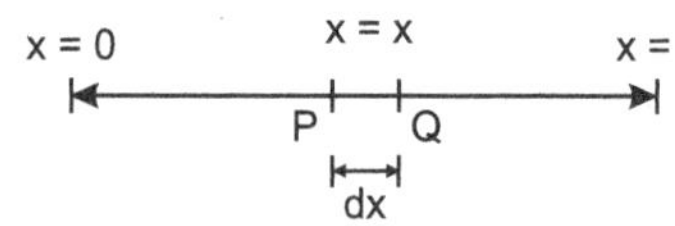

$$dm = (\rho S.dx)$$

energy stored in this element is

$$dE = \frac{1}{2}(dm)A^2.\omega^2 \qquad \text{(energy in SHM)}$$

$$= \frac{1}{2}(\rho S.dx)(a_0^2 \sin^2 kx).\omega^2$$

$\therefore$ Total energy stored in the wire will be

$$E = \int_{x=0}^{x=l} dE$$

$$E = \frac{1}{4}(\rho S\omega^2 a_0^2)\int_0^l 2\sin^2 kx\,dx$$

$$= \frac{1}{4}(\rho S\omega^2 a_0^2)\left[x - \frac{1}{2k}\sin 2kx\right]_0^l$$

$$= \frac{1}{4}(\rho S\omega^2 a_0^2)\left[l - \frac{1}{2k}\sin 2k.l\right] \qquad (2kl = 4\pi)$$

$$E = \frac{1}{4}(\rho S\omega^2 a_0^2 l) \qquad \ldots(1)$$

Here, $v = \sqrt{\frac{T}{m}} = \sqrt{\frac{T}{\rho S}} = f\lambda = \left(\frac{\omega}{2\pi}\right) l$

$$\therefore \qquad \rho S = \frac{4\pi^2 T}{\omega^2 l^2}$$

or $$\rho S\omega^2 l = \frac{4\pi^2 T}{l} \qquad \ldots(2)$$

Substituting the value of $\rho S\omega^2 l$ from equation (2) in equation (1), we get

$$\boldsymbol{E = \frac{a_0^2\pi^2 T}{l}}$$

276. Let the pulse emitted at time t is received at time t_0 $(t < t_0)$. Then

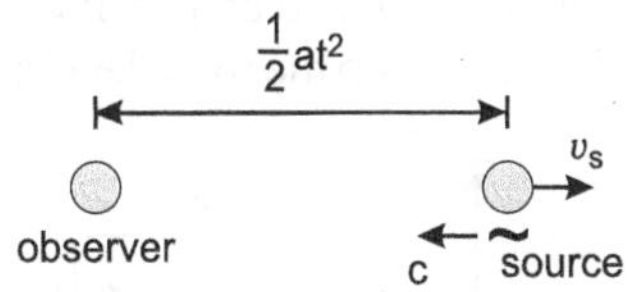

$$t_0 = t + \frac{\frac{1}{2}at^2}{c}$$

or $$at^2 + 2ct - 2ct_0 = 0$$

or $$t = \frac{\sqrt{c^2 + 2act_0} - c}{a}$$

Velocity of source at this time t would be

$$v_s = at = (\sqrt{c^2 + 2act_0}) - c$$

$\therefore$ $$f' = f\left(\frac{c}{c + v_s}\right)$$

$$f' = f\left(\frac{c}{c + \sqrt{c^2 + 2act_0} - c}\right)$$

$$= \frac{f}{\sqrt{1 + \frac{2at_0}{c}}}$$

or $$\boldsymbol{f' = \frac{f}{\sqrt{1 + \frac{2at_0}{c}}}}$$

277. Let t be the time when the sound is emitted which is received at $t = 4$ s. Then

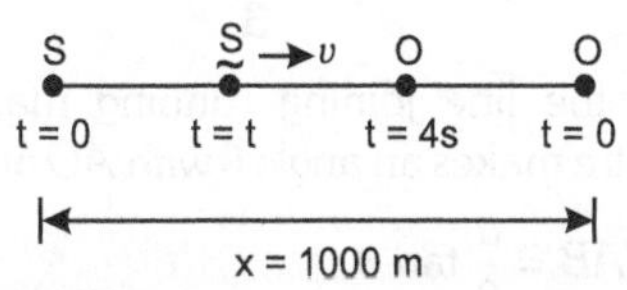

$$\frac{1}{2}at^2 + \frac{1}{2}a(4)^2 + v(4 - t) = 1000$$

or $$\frac{1}{2}(10)t^2 + \frac{1}{2}(10)(16) + 300(4 - t) = 1000$$

or $$5t^2 - 300t + 280 = 0$$

or $$t = 0.95 \text{ s} \quad \text{and} \quad 59 \text{ s}$$

Therefore the pulse emitted at $t = 0.95$ s is received at $t = 4$ s. Now

Speed of sound at $t = 0.95$ s is

$$v_s = a\,t = (10)\,(0.95) \text{ m/s}$$

or $$v_s = 9.5 \text{ m/s}$$

and speed of observer at $t = 4$ s is

$$v_0 = a\,t = (10)\,(4) \text{ m/s}$$

or $$v_0 = 40 \text{ m/s}$$

Since both of them are approaching towards each other

Hence $$f' = f\left(\frac{v + v_0}{v - v_s}\right) = 2000\left(\frac{300 + 40}{300 - 9.5}\right)$$

$$= \mathbf{2340.79\ Hz}$$

278. In this problem two cases are possible :

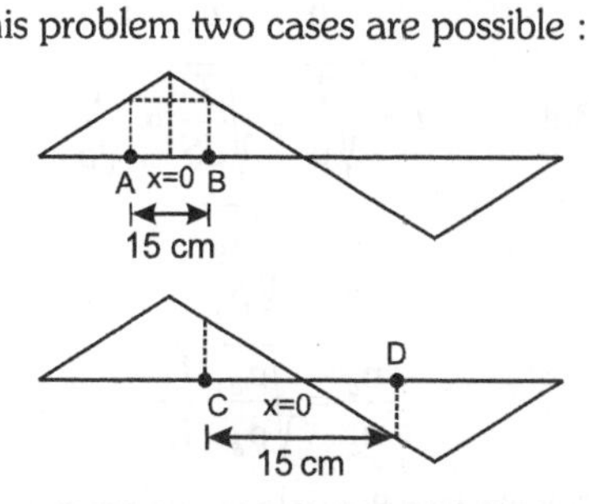

(a) **Case-1** is that A and B have the same displacement amplitude and **case-2** is that C and D have the same amplitude *viz* 3.5 mm. In case 1, if $x = 0$ is taken at antinode then

$$A = a \cos kx$$

In case-2, if $x = 0$ is taken at node, then

$$A = a \sin kx$$

But since nothing is given in the question. Hence from both the cases, result should be same. This is possible only when

$$a \cos kx = a \sin kx$$

or $$kx = \frac{\pi}{4}$$

or $$a = \frac{A}{\cos kx} = \frac{3.5}{\cos \pi/4} = \mathbf{4.95\ mm}$$

(b) $$kx = \frac{\pi}{4}$$

or $$\left(\frac{2\pi}{\lambda}\right)\left(\frac{15}{2}\right) = \frac{\pi}{4}$$

or $$\frac{\lambda}{2} = 30 \text{ cm}$$

$\therefore$ Number of loops $= \dfrac{120 \text{ cm}}{\lambda/2} = \dfrac{120}{30} = 4$

Hence it corresponds to **third overtone**.

279. Let n_a loops are formed in aluminium wire and n_s in steel. Then

$$f_a = f_s$$

or $$n_a\left(\frac{v_a}{2l_a}\right) = n_s\left(\frac{v_s}{2l_s}\right)$$

or $$\frac{n_a}{n_s} = \left(\frac{v_s}{v_a}\right)\left(\frac{l_a}{l_s}\right)$$

But $$v = \sqrt{\frac{T}{\mu}} = \sqrt{\frac{T}{\rho S}} \propto \frac{1}{\sqrt{\rho}}$$

Therefore, $$\frac{v_s}{v_a} = \sqrt{\frac{\rho_a}{\rho_s}}$$

$\therefore$ $$\frac{n_a}{n_s} = \sqrt{\frac{\rho_a}{\rho_s}} \cdot \frac{l_a}{l_s}$$

Substituting the values, we have

$$\frac{n_a}{n_s} = \sqrt{\frac{2.6 \times 10^3}{1.04 \times 10^4}} \cdot \frac{0.6}{0.9} = \frac{1}{3}$$

i.e., at lowest frequency, one loop is formed in aluminium wire and three loops are formed in steel wire as shown below.

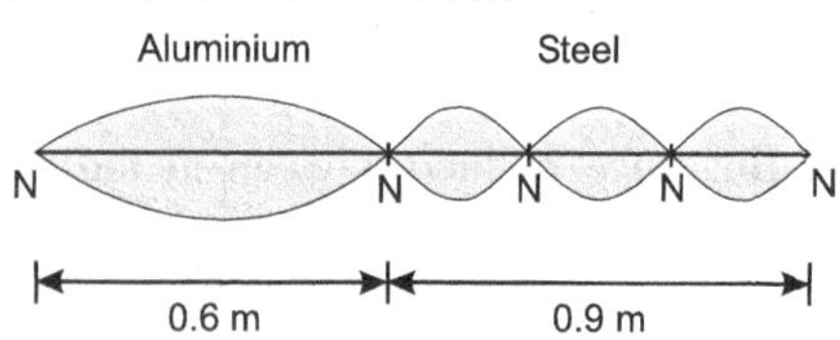

$$f_{min} = n_a \frac{v_a}{2l_a} = n_a \frac{\sqrt{T/\rho_a S}}{2l_a}$$

$$= (1)\frac{\sqrt{(100)/(2.6 \times 10^3 \times 10^{-6})}}{2 \times 0.6}$$

$(T = mg = 100 \text{ N})$

$$= \mathbf{163.4\ Hz}$$

Total number of nodes are **five** as shown in figure.

280. Let y be the length of second liquid poured in A.

Now let the first liquid comes down a level x in arm A and rises up the same level x in arm B.

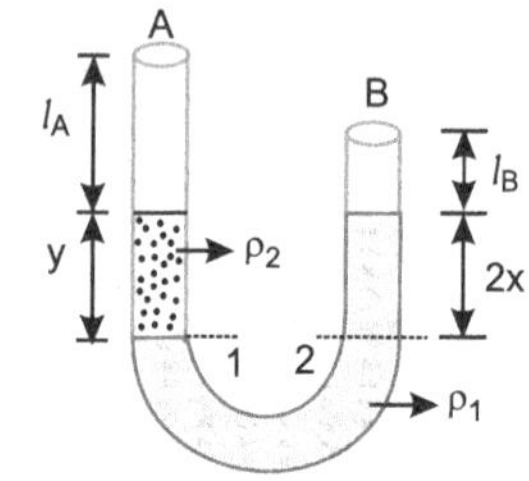

Then Pressure at 1 = Pressure at 2

so, $$\rho_2 g y = \rho_1 g(2x)$$

or $$x = \frac{\rho_2}{2\rho_1} \cdot y = \frac{y}{4} \qquad \left(\frac{\rho_2}{\rho_1} = \frac{1}{2}\right)$$

Length of air column in arm A is

$$l_A = (l_1 - h) - (y - y/4) = l_1 - h - \frac{3y}{4}$$

and length of air column in arm B is

$$l_B = (l_2 - h) + \frac{y}{4}$$

Now first overtone of A is in unison with fundamental tone of B, so

$$3\left\{\frac{v}{4l_A}\right\} = \frac{v}{4l_B} \quad \text{or} \quad l_A = 3l_B$$

or $$l_1 - h - \frac{3y}{4} = 3\left\{(l_2 - h) + \frac{y}{4}\right\}$$

or $$\boldsymbol{y = \frac{2(l_1 - 3l_2 + 2h)}{3}}$$

281. Let the line joining running man, car and centre makes an angle θ with AO at any time t.

$$AB = \frac{a}{2}\tan\theta$$

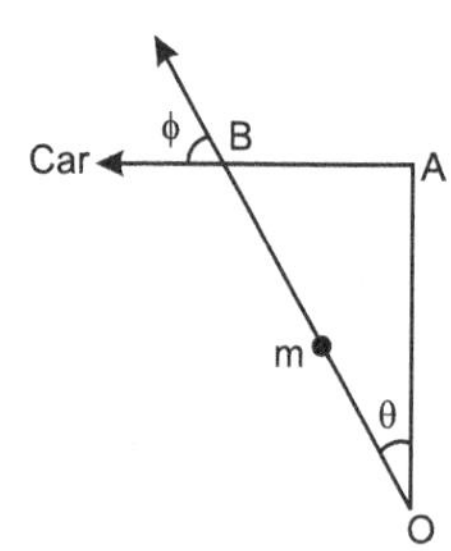

$$v_{car} = \frac{d}{dt}(AB)$$

$$= \frac{a}{2}\frac{d\theta}{dt}\sec^2\theta$$

But $$\frac{d\theta}{dt} = 22 \text{ rad/s}$$

$$v_{car} = 11a\sec^2\theta$$

Now $\quad v' = \left(\dfrac{\upsilon - \upsilon_0 \cos \phi}{\upsilon} \right) v$

where υ is the velocity of sound and ϕ is the angle between the velocity vector of the observer car and the line joining man in car and centre.

$$v' = \left(\frac{\upsilon - \upsilon_0 \sin \theta}{\upsilon} \right) v$$

$$= \frac{\upsilon - 11a \sec^2 \theta \sin \theta}{\upsilon}$$

But $\sec^2 \theta \sin \theta$ is a increasing function in the range $0 \le \theta \le \dfrac{\pi}{2}$

v_1' will be minimum when $\sec^2 \theta \sin \theta$ is maximum which is when $\theta = 45°$

$$v_{min} = \frac{\upsilon - 11a\sqrt{2}}{\upsilon} v$$

$$= \frac{330 - 11 \times 10}{330} \times 300$$

$$v_{min} = 200 \text{ Hz}$$

282. (a) Distance between two nodes is $\lambda/2$ or π/k. The volume of string between two nodes is therefore,

$$V = \frac{\pi}{k} s \qquad \ldots(1)$$

Energy density (energy per unit volume) of each wave will be,

$$u_1 = \frac{1}{2} \rho\omega^2 (8)^2 = 32\rho\omega^2$$

and $\quad u_2 = \dfrac{1}{2} \rho\omega^2 (6)^2 = 18\rho\omega^2$

$\therefore$ Total mechanical energy between two consecutive nodes will be,

$$E = (u_1 + u_2) V$$

$$= 50 \frac{\pi}{k} \rho\omega^2 s$$

(b) $y = y_1 + y_2$

$$= 8 \sin(\omega t - kx) + 6 \sin(\omega t + kx)$$

$$= 2 \sin(\omega t - kx) + \{6 \sin(\omega t - kx) + 6 \sin(\omega t + kx)\}$$

$$= 2 \sin(\omega t - kx) + 12 \cos kx \sin \omega t$$

Thus, the resultant wave will be a sum of standing wave and a travelling wave.

Energy crossing through a node per second = power of travelling wave

$\therefore \quad P = \dfrac{1}{2} \rho\omega^2 (2)^2 s\upsilon$

$$= \frac{1}{2} \rho\omega^2 (4)(s)\left(\frac{\omega}{k}\right)$$

$$= \frac{2\rho\omega^3 s}{k}$$

283. Fundamental frequency,

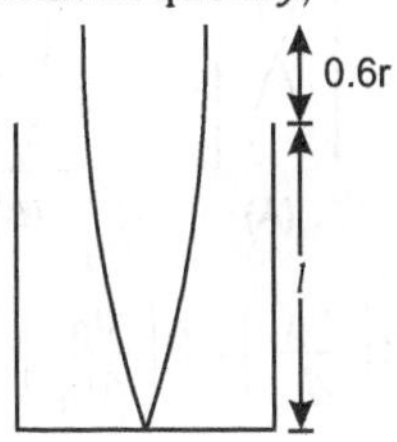

$$f = \frac{\upsilon}{4(l + 0.6r)}$$

$\therefore$ speed of sound $\upsilon = 4f(l + 0.6r)$

or $\quad \upsilon = (4)(480)[(0.16) + (0.6)(0.025)]$

$= \mathbf{336\ m/s}$

284. $l = \dfrac{\lambda}{2}$ or $\lambda = 2l$, $k = \dfrac{2\pi}{\lambda} = \dfrac{\pi}{l}$

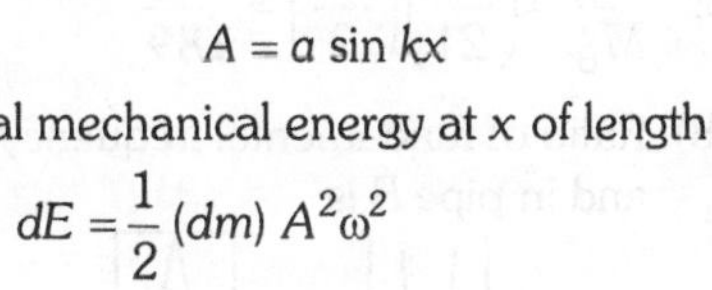

The amplitude at a distance x from $x = 0$ is given by

$$A = a \sin kx$$

Total mechanical energy at x of length dx is

$$dE = \frac{1}{2}(dm) A^2\omega^2$$

$$= \frac{1}{2}(\mu dx)(a \sin kx)^2 (2\pi f)^2$$

or $\quad dE = 2\pi^2 \mu f^2 a^2 \sin^2 kx \, dx \qquad \ldots(1)$

Here $\quad f = \dfrac{\upsilon^2}{\lambda^2} = \dfrac{\left(\dfrac{T}{\mu}\right)}{(4l^2)}$ and $k = \dfrac{\pi}{l}$

Substituting these values in Eq. (1) and integrating it from $x = 0$ to $x = l$, we get total energy of string,

$$\boldsymbol{E = \frac{\pi^2 a^2 T}{4l}}$$

285. (a) Frequency of second harmonic in pipe A = frequency of third harmonic in pipe B

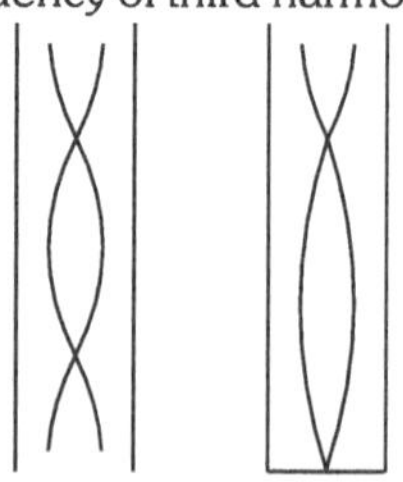

(A) (B)

$$\therefore \quad 2\left(\frac{v_A}{2l_A}\right) = 3\left(\frac{v_B}{4l_B}\right)$$

or $$\frac{v_A}{v_B} = \frac{3}{4} \qquad (\text{as } l_A = l_B)$$

or $$\frac{\sqrt{\frac{\gamma_A R T_A}{M_A}}}{\sqrt{\frac{\gamma_B R T_B}{M_B}}} = \frac{3}{4}$$

or $$\sqrt{\frac{\gamma_A}{\gamma_B}}\sqrt{\frac{M_B}{M_A}} = \frac{3}{4} \qquad (\text{as } T_A = T_B)$$

$$\therefore \quad \frac{M_A}{M_B} = \frac{\gamma_A}{\gamma_B}\left(\frac{16}{9}\right)$$

$$= \left(\frac{5/3}{7/5}\right)\left(\frac{16}{9}\right) \quad \left(\gamma_A = \frac{5}{3} \text{ and } \gamma_B = \frac{7}{5}\right)$$

or $$\frac{M_A}{M_B} = \left(\frac{25}{21}\right)\left(\frac{16}{9}\right) = \mathbf{\frac{400}{189}}$$

(b) Ratio of fundamental frequency in pipe A and in pipe B is:

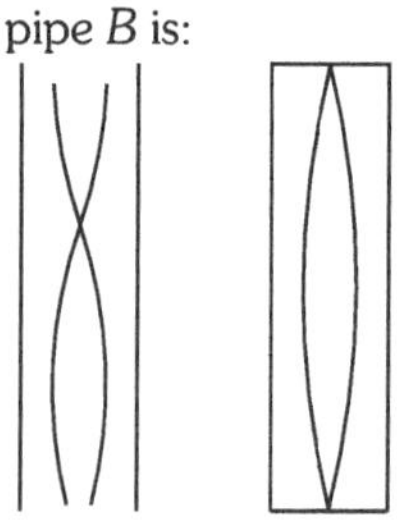

(a) (b)

$$\frac{f_A}{f_B} = \frac{v_A/2l_A}{v_B/2l_B}$$

$$= \frac{v_A}{v_B} \qquad (\text{as } l_A = l_B)$$

$$= \frac{\sqrt{\frac{\gamma_A R T_A}{M_A}}}{\sqrt{\frac{\gamma_B R T_B}{M_B}}}$$

$$= \sqrt{\frac{\gamma_A}{\gamma_B}\cdot\frac{M_B}{M_A}} \qquad (\text{as } T_A = T_B)$$

substituting $\frac{M_B}{M_A} = \frac{189}{400}$ from part (a), we get

$$\frac{f_A}{f_B} = \sqrt{\frac{25}{21}\times\frac{189}{400}} = \mathbf{\frac{3}{4}}$$

286. Velocity of sound in water is

$$v_w = \sqrt{\frac{B}{\rho}} = \sqrt{\frac{2.088\times 10^9}{10^3}} = 1445 \text{ m/s}$$

Frequency of sound in water will be

$$f_0 = \frac{v_w}{\lambda_w} = \frac{1445}{14.45\times 10^{-3}} \text{ Hz}$$

$$f_0 = 10^5 \text{ Hz}$$

(a) Frequency of sound detected by receiver (observer) at rest would be

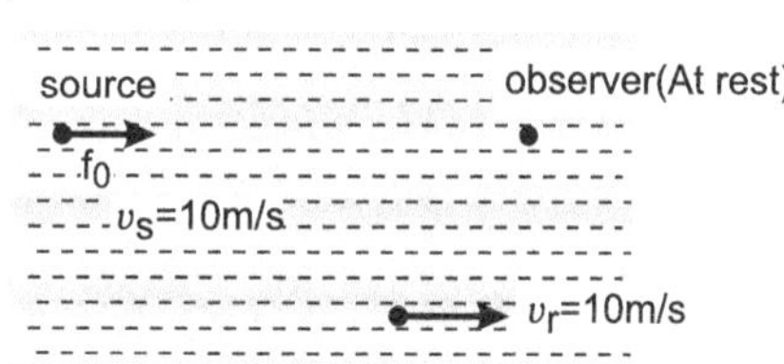

$$f_1 = f_0\left(\frac{v_w + v_r}{v_w + v_r - v_s}\right)$$

$$= (10^5)\left(\frac{1445 + 2}{1445 + 2 - 10}\right) \text{Hz}$$

$$\boldsymbol{f_1 = 1.0069 \times 10^5 \text{ Hz}}$$

(b) Velocity of sound in air is

$$v_a = \sqrt{\frac{\gamma R T}{M}}$$

$$= \sqrt{\frac{(1.4)(8.31)(20+273)}{28.8 \times 10^{-3}}}$$

$$= 344 \text{ m/s}$$

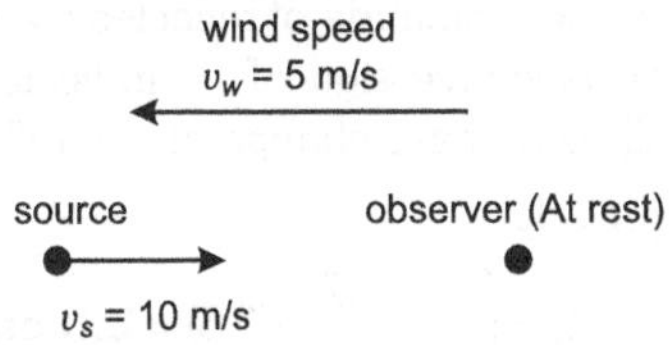

Frequency does not depend on the medium. Therefore, frequency in air is also $f_0 = 10^5$ Hz.

$\therefore$ Frequency of sound detected by receiver (observer) in air would be

$$f_2 = f_0 \left(\frac{v_a - v_w}{v_a - v_w - v_s} \right)$$

$$= 10^5 \left[\frac{344 - 5}{344 - 5 - 10} \right] \text{Hz}$$

$$\boldsymbol{f_2 = 1.0304 \times 10^5 \text{ Hz}}$$

287.

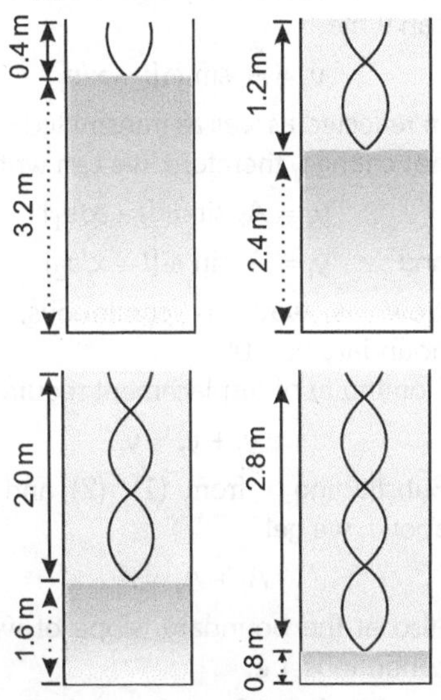

Speed of sound, $v = 340$ m/s

Let l_0 be the length of air column corresponding to the fundamental frequency. Then

$$\frac{v}{4l_0} = 212.5$$

or $$l_0 = \frac{v}{4(212.5)} = \frac{340}{4(212.5)} = 0.4 \text{ m}$$

In closed pipe only odd harmonics are obtained. Now let l_1, l_2, l_3, l_4 etc. be the lengths corresponding to the 3rd harmonic, 5th harmonic, 7th harmonic, etc. Then

$$3\left(\frac{v}{4l_1}\right) = 212.5 \Rightarrow l_1 = 1.2 \text{ m},$$

$$5\left(\frac{v}{4l_2}\right) = 212.5 \Rightarrow l_2 = 2.0 \text{ m},$$

$$\text{and } 7\left(\frac{v}{4l_3}\right) = 212.5 \Rightarrow l_3 = 2.8 \text{ m},$$

$$\text{and } 9\left(\frac{v}{4l_4}\right) = 212.5 \Rightarrow l_4 = 3.6 \text{ m}$$

or heights of water level are $(3.6 - 0.4)$ m, $(3.6 - 1.2)$ m, $(3.6 - 2.0)$ m and $(3.6 - 2.8)$ m.

$\therefore$ **Heights of water level are 3.2 m, 2.4 m, 1.6 m and 0.8 m.**

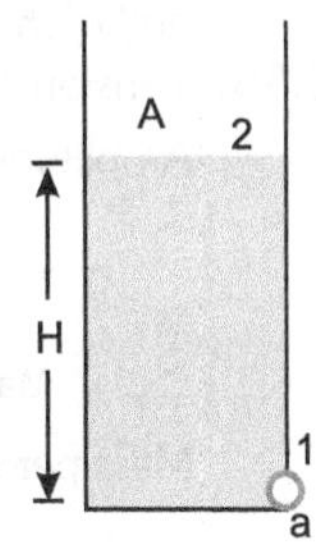

Let A and a be the area of cross-sections of the pipe and hole respectively. Then

$$A = \pi (2 \times 10^{-2})^2 = 1.26 \times 10^{-3} \text{ m}^2$$

and $$a = \pi (10^{-3})^2 = 3.14 \times 10^{-6} \text{ m}^2$$

Velocity of efflux, $v = \sqrt{2gH}$

continuity equation at 1 and 2 gives

$$a\sqrt{2gH} = A\left(\frac{-dH}{dt}\right)$$

$\therefore$ Rate of fall of water level in the pipe,

$$\left(\frac{-dH}{dt}\right) = \frac{a}{A}\sqrt{2gH}$$

Substituting the values, we get

$$\frac{-dH}{dt} = \frac{3.14 \times 10^{-6}}{1.26 \times 10^{-3}} \sqrt{2 \times 10 \times H}$$

or $$\boldsymbol{-\frac{dH}{dt} = (1.11 \times 10^{-2})\sqrt{H}}$$

Between first two resonances, the water level falls from 3.2 m to 2.4 m.

$$\therefore \quad \frac{dH}{\sqrt{H}} = -(1.11 \times 10^{-2})\, dt$$

or $$\int_{3.2}^{2.4} \frac{dH}{\sqrt{H}} = -(1.11 \times 10^{-2}) \int_0^t dt$$

or $$2[\sqrt{2.4} - \sqrt{3.2}] = -(1.11 \times 10^{-2}) \,.\, t$$

or $$t \approx \mathbf{43\ second}$$

Rate of fall of level at a height h is $\left(\frac{-dh}{dt}\right) = \frac{a}{A}\sqrt{2gh} \propto \sqrt{h}$ i.e. rate decreases as the height of water (or any other liquid) decreases in the tank. That is why the time required to empty the first half of the tank is less than the time required to empty the rest half of the tank.

288. Tension $T = 80$ N

Amplitude of incident wave, $A_i = 3.5$ cm

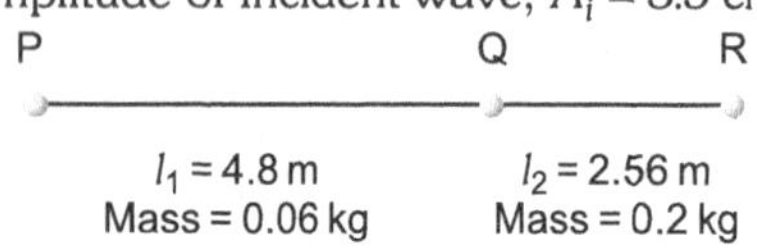

Mass per unit length of wire PQ is

$$m_1 = \frac{0.06}{4.8} = \frac{1}{80} \text{ kg/m}$$

and mass per unit length of wire QR is

$$m_2 = \frac{0.2}{2.56} = \frac{1}{12.8} \text{ kg/m}$$

(a) Speed of wave in wire PQ is

$$v_1 = \sqrt{T/m_1} = \sqrt{\frac{80}{1/80}} = 80 \text{ m/s}$$

and speed of wave in wire QR is

$$v_2 = \sqrt{T/m_2} = \sqrt{\frac{80}{1/12.8}} = 32 \text{ m/s}$$

$\therefore$ Time taken by the wave pulse to reach from P to R is

$$t = \frac{4.8}{v_1} + \frac{2.56}{v_2} = \left(\frac{4.8}{80} + \frac{2.56}{32}\right) \text{s} = \mathbf{0.14\ s}$$

(b) The expressions for reflected and transmitted amplitudes (A_r and A_t) in terms of v_1, v_2 and A_i are as follows :

$$A_r = \frac{v_2 - v_1}{v_2 + v_1} A_i \text{ and } A_t = \frac{2v_2}{v_1 + v_2} A_i$$

Substituting the values, we get

$$A_r = \left(\frac{32 - 80}{32 + 80}\right)(3.5) = \mathbf{-\ 1.5\ cm}$$

i.e., the amplitude of reflected wave will be 1.5 cm. Negative sign of A_r indicates that there will be a phase change of π in reflected wave.

Similarly

$$A_t = \left(\frac{2 \times 32}{32 + 80}\right)(3.5) = \mathbf{2.0\ cm}$$

i.e. the amplitude of transmitted wave will be **2.0 cm.**

The expressions of A_r and A_t are derived as below. In the opinion of author, you can directly use these relations in examination. But if time permits, you derive the relations separately.

Derivation

Suppose the incident wave of amplitude A_i and angular frequency 'ω' is travelling in positive x-direction with velocity v_1 then we can write

$$y_i = A_i \sin \omega [t - x/v_1] \qquad \ldots(1)$$

In reflected as well as transmitted wave, 'ω' will not change, therefore, we can write

$$y_r = A_r \sin \omega [t + x/v_1] \qquad \ldots(2)$$

and $$y_t = A_t \sin \omega [t - x/v_2] \qquad \ldots(3)$$

Now as wave is continuous, so at the boundary ($x = 0$)

Continuity of displacement requires

$$y_i + y_r = y_t \qquad \text{for } x = 0$$

Substituting from (1), (2) and (3) in the above, we get

$$A_i + A_r = A_t \qquad \ldots(4)$$

Also at the boundary, slope of wave will be continuous, *i.e.*

$$\frac{\partial y_i}{\delta x} + \frac{\partial y_r}{\delta x} = \frac{\partial y_t}{\partial x} \qquad \text{for } x = 0$$

which gives

$$A_i - A_r = \left(\frac{v_1}{v_2}\right) A_t \qquad \ldots(5)$$

Solving (4) and (5) for A_r and A_t we get the required equations, *i.e.*

$$A_r = \frac{v_2 - v_1}{v_2 + v_1} A_i$$

and $$A_t = \frac{2v_2}{v_2 + v_1} A_i$$

289. (a) Frequency of second overtone of the closed pipe

$$= 5\left(\frac{v}{4L}\right) = 440 \text{ Hz} \qquad \text{(Given)}$$

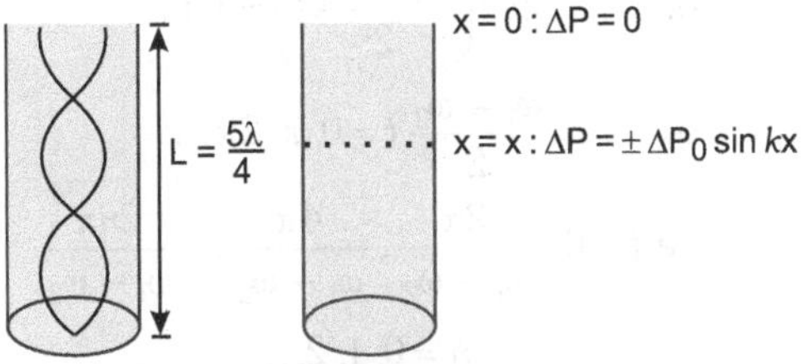

$$\therefore \quad L = \frac{5v}{4 \times 440} \text{ m}$$

Substituting v = speed of sound in air = 330 m/s

$$\boldsymbol{L} = \frac{5 \times 330}{4 \times 440} = \frac{\mathbf{15}}{\mathbf{16}} \textbf{ m}$$

$$\lambda = \frac{4L}{5} = \frac{4\,(15/16)}{5} = \frac{3}{4} \text{ m}$$

(b) Open end is displacement antinode. Therefore, it would be a pressure node

or at $x = 0$; $\Delta P = 0$

Pressure amplitude at $x = x$ can be written as

$$\Delta P = \pm\, \Delta P \sin kx$$

where $$k = \frac{2\pi}{\lambda} = \frac{2\pi}{3/4} = \frac{8\pi}{3} \text{ m}^{-1}$$

Therefore, pressure amplitude at $x = \frac{L}{2} = \frac{15/16}{2}$ m or (15/32) m will be

$$\Delta P = \pm\, \Delta P_0 \sin\left(\frac{8\pi}{3}\right)\left(\frac{15}{32}\right)$$

$$= \pm\, \Delta P_0 \sin\left(\frac{5\pi}{4}\right)$$

$$\boldsymbol{\Delta P} = \pm \frac{\boldsymbol{\Delta P_0}}{\sqrt{\mathbf{2}}}$$

(c) Open end is a pressure node *i.e.* $\Delta P = 0$

Hence $\boldsymbol{P_{max} = P_{min}}$ = **Mean pressure (P_0)**

(d) Closed end is a displacement node or pressure antinode.

Therefore, $\boldsymbol{P_{max} = P_0 + \Delta P_0}$

and $\boldsymbol{P_{min} = P_0 - \Delta P_0}$

290. Let l_1 and l_2 be the lengths of closed and open pipes respectively.

Fundamental frequency of closed organ pipe is given by

$$f_1 = \frac{v}{4l}$$

v = speed of sound in air = 330 m/s

But $f_1 = 110$ Hz is given

Therefore, $\frac{v}{4l_1} = 110$ Hz

$$\therefore \quad l_1 = \frac{v}{4 \times 110} = \frac{330}{4 \times 110} \text{ m} = 0.75 \text{ m}$$

First overtone of closed organ pipe will be

$$f_3 = 3f_1 = 3(110) \text{ Hz} = 330 \text{ Hz}$$

This produces a beat frequency of 2.2 Hz with first overtone of open organ pipe. Therefore, first ovetone frequency of open organ pipe is either

$$(330 + 2.2) \text{ Hz} = 332.2 \text{ Hz}$$

or $$(330 - 2.2) \text{ Hz} = 327.8 \text{ Hz}$$

If it is 332.2 Hz, then

$$2\left(\frac{v}{2l_2}\right) = 332.2 \text{ Hz}$$

or $$l_2 = \frac{v}{332.2} = \frac{330}{332.2} \text{ m} = 0.99 \text{ m}$$

and if it is 327.8 Hz, then

$$2\left(\frac{v}{2l_2}\right) = 327.8 \text{ Hz}$$

or $$l_2 = \frac{v}{327.8} \text{ m} = \frac{330}{327.8} \text{ m} = 1.0067 \text{ m}$$

Therefore, length of the closed organ pipe is l_1 = 0.75 m while length of open pipe is either

l_2 = 0.99 m or 1.0067 m.

291. (a) Speed of longitudinal travelling wave in the rod will be

$$v = \sqrt{\frac{Y}{\rho}} = \sqrt{\frac{2 \times 10^{11}}{8000}} = 5000 \text{ m/s}$$

Amplitude at antinode $= 2A$

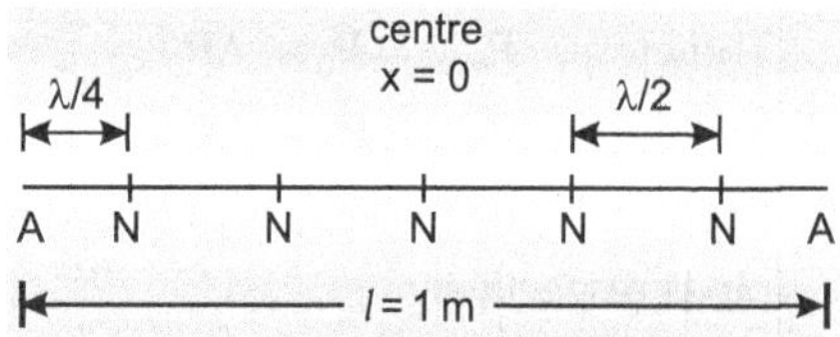

(Here A is the amplitude of constituent waves)

$= 2 \times 10^{-6}$ m

$$\therefore \quad A = 10^{-6} \text{ m} \Rightarrow l = \frac{5\lambda}{2}$$

$$\Rightarrow \quad \lambda = \frac{2l}{5} = \frac{(2)(1.0)}{5} \text{ m} = 0.4 \text{ m}$$

Hence the equation of motion at a distance x from the mid-point will be given by

$$y = A \sin kx \sin \omega t$$

Here $k = 2\pi / \lambda = 2\pi / 0.4 = 5\pi$

$$\omega = 2\pi f = 2\pi \frac{v}{\lambda}$$

$$= 2\pi \left(\frac{5000}{0.4} \right) \text{rad/s} = 25000\,\pi$$

$$\therefore \quad y = (10^{-6}) \sin (5\pi x) \sin (25000\,\pi t)$$

Therefore, y at a distance $x = 2$ cm $= 2 \times 10^{-2}$ m is

$$y = (10^{-6}) \sin (5\pi \times 2 \times 10^{-2}) \sin (25000\,\pi t)$$

or $\boldsymbol{y = 10^{-6} \sin (0.1\,\pi) \sin (25000\,\pi t)}$

(b) The equations of constituent waves are

$$y_1 = A \sin (\omega t - kx)$$

and $\quad y_2 = A \sin (\omega t + kx)$

or $\quad \boldsymbol{y_1 = 10^{-6} \sin (25000\,\pi t - 5\pi x)}$

and $\quad \boldsymbol{y_2 = 10^{-6} \sin (25000\,\pi t + 5\pi x)}$

292. (i) If the detector is at $x = 0$, the two radio waves can be represented as

$$y_1 = A \sin \omega_1 t$$

and $\quad y_2 = A \sin \omega_2 t \qquad (A_1 = A_2 = A)$

By the principle of superposition

$$y = y_1 + y_2 = A \sin \omega_1 t + A \sin \omega_2 t$$

$$y = 2A \cos \left(\frac{\omega_1 - \omega_2}{2} t \right) \sin \left(\frac{\omega_1 + \omega_2}{2} t \right)$$

$$= A_0 \sin \left(\frac{\omega_1 + \omega_2}{2} t \right)$$

Here $A_0 = 2A \cos \left(\frac{\omega_1 - \omega_2}{2} t \right)$

Since $I \propto (A_0)^2 \propto 4A^2 \cos^2 \left(\frac{\omega_1 - \omega_2}{2} t \right)$

So, intensity will be maximum when

$$\cos^2 \left(\frac{\omega_1 - \omega_2}{2} t \right) = \text{maximum} = 1$$

or $$\cos \left(\frac{\omega_1 - \omega_2}{2} t \right) = \pm 1$$

or $$\frac{\omega_1 - \omega_2}{2} t = 0, \pi, 2\pi \ldots$$

$$i.e.\ t = 0, \frac{2\pi}{\omega_1 - \omega_2}, \frac{4\pi}{\omega_1 - \omega_2} \ldots \frac{2n\pi}{\omega_1 - \omega_2}$$

$$n = 0, 1, 2, \ldots$$

Therefore, time interval betwen any two successive maxima is $\frac{2\pi}{\omega_1 - \omega_2}$ or $\frac{2\pi}{10^3}$ s

or $\quad \boldsymbol{6.28 \times 10^{-3}}$ **s.**

(ii) The detector can detect if resultant intensity $\geq 2A^2$,

or the resultant amplitude $\geq \sqrt{2}A$

Hence $$2A \cos \left(\frac{\omega_1 - \omega_2}{2} t \right) \geq \sqrt{2}A$$

$$\cos \left(\frac{\omega_1 - \omega_2}{2} t \right) \geq \frac{1}{\sqrt{2}}$$

Therefore, the detector lies idle, when value of $\cos \left(\frac{\omega_1 - \omega_2}{2} t \right)$ is between 0 and $\frac{1}{\sqrt{2}}$

or when $\frac{\omega_1 - \omega_2}{2} t$ is between $\pi/2$ and $\pi/4$ or t lies between

$$\frac{\pi}{\omega_1 - \omega_2} \text{ and } \frac{\pi}{2(\omega_1 - \omega_2)}$$

$$\therefore \quad t = \frac{\pi}{\omega_1 - \omega_2} - \frac{\pi}{2(\omega_1 - \omega_2)}$$

$$= \frac{\pi}{2(\omega_1 - \omega_2)} = \frac{\pi}{2 \times 10^3} \text{ s}$$

$$t = 1.57 \times 10^{-3} \text{ s}$$

Hence the detector lies idle for a time of $\boldsymbol{1.57 \times 10^{-3}}$ s in each cycle.

CHAPTER 10 HEAT AND THERMODYNAMICS

293. Process AB is isochoric (V = constant). Hence

$$\Delta W_{AB} = 0$$

$$\Delta W_{BCD} = P_0V_0 + \frac{\pi}{2}(P_0)\left(\frac{V_0}{2}\right)$$

$$= \left(\frac{\pi}{4} + 1\right)P_0V_0$$

$$\Delta W_{DA} = -\frac{1}{2}\left(\frac{P_0}{2} + P_0\right)(2V_0 - V_0)$$

$$= -\frac{3}{4}P_0V_0$$

$$\Delta U_{AB} = nC_V\Delta T = (2)\left(\frac{3}{2}R\right)(T_B - T_A)$$

$$\left(n = 2,\ C_V = \frac{3}{2}R\right)$$

$$= 3R\left(\frac{P_0V_0}{2R} - \frac{P_0V_0}{4R}\right)$$

$$= \frac{3}{4}P_0V_0 = \Delta Q_{AB} \qquad \left(T = \frac{PV}{nR}\right)$$

$$\Delta U_{BCD} = nC_V\Delta T = (2)\left(\frac{3}{2}R\right)(T_D - T_B)$$

$$= (3R)\left(\frac{2P_0V_0}{2R} - \frac{P_0V_0}{2R}\right) = \frac{3}{2}P_0V_0$$

Hence $\Delta Q_{BCD} = \Delta U_{BCD} + \Delta W_{BCD}$

$$= \left(\frac{\pi}{4} + \frac{5}{2}\right)P_0V_0$$

$$\Delta U_{DA} = nC_V\Delta T$$

$$= (2)\left(\frac{3}{2}R\right)(T_A - T_D)$$

$$= (3R)\left(\frac{P_0V_0}{4R} - \frac{2P_0V_0}{2R}\right)$$

$$= -\frac{9}{4}P_0V_0$$

$\therefore \quad \Delta Q_{DA} = \Delta U_{DA} + \Delta W_{DA}$

$$= -\frac{9}{4}P_0V_0 - \frac{3}{4}P_0V_0$$

$$= -3P_0V_0$$

Net work done is

$$W_{net} = \left(\frac{\pi}{4} + 1 - \frac{3}{4}\right)P_0V_0$$

$$= 1.04\,P_0V_0$$

and heat absorbed is

$$Q_{ab} = \Delta Q_{+ve}$$

$$= \left(\frac{3}{4} + \frac{\pi}{4} + \frac{5}{2}\right)P_0V_0 = 4.03\,P_0V_0$$

Hence efficiency of the cycle is

$$\eta = \frac{W_{net}}{Q_{ab}} \times 100$$

$$= \frac{1.04\,P_0V_0}{4.03\,P_0V_0} \times 100 = \mathbf{25.8\%}$$

294. For isothermal process $\Delta U_1 = 0$

$$\Delta W_1 = nRT_1 \ln\left(\frac{V_2}{V_1}\right) = \Delta Q_1 \qquad \ldots(1)$$

In the second step, gas is heated isochorically.

So, $\quad \Delta W_2 = 0$

and $\quad \Delta Q_2 = \Delta U_2 = nC_V(T_3 - T_2)$

$$= \frac{nR}{\gamma - 1}(T_3 - T_2) \qquad \left(C_V = \frac{R}{\gamma - 1}\right)$$

or $\quad \Delta Q_2 = \dfrac{P_3V_3 - P_2V_2}{\gamma - 1} \qquad \ldots(2)$

$\therefore \quad \Delta Q = \Delta Q_1 + \Delta Q_2$

$$= nRT_1 \ln\left(\frac{V_2}{V_1}\right) + \frac{nR(T_3 - T_2)}{\gamma - 1}$$

Given that $\Delta Q = 80$ kJ, $\quad n = 3,$

$T_1 = T_2 = 273$ K,

$V_2 = 5V_1 = V_3,$

$$P_3 = P_1$$

and $nRT_3 = P_3V_3 = 5P_1V_1 = 5nRT_1$

We get

$$\Delta Q = nRT_1 \ln(5) + \frac{4nRT_1}{\gamma - 1}$$

or $$\gamma = \frac{4nRT_1}{\Delta Q - nRT_1 \ln(5)} + 1$$

Substituting the values we get

$$\gamma = \frac{(4)(3)(8.31)(273)}{80{,}000 - (3)(8.31)(273)(\ln 5)} + 1$$

or $\gamma = \mathbf{1.4}$

295. The initial and final positions of the two gases are shown in figure below :

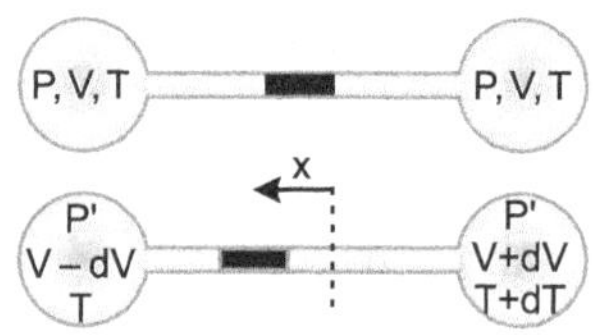

Let the mercury is displaced by x. Then

$$dV = x\,S \qquad \ldots(1)$$

Where $S = 5\ \text{mm}^2 = 5 \times 10^{-2}\ \text{cm}^2$

In equilibrium, pressure on both sides is say P'. Then

$$P'(V - dV) = nRT \qquad \ldots(2)$$

and $$P'(V + dV) = nR(T + dT) \qquad \ldots(3)$$

Dividing (3) by (2), we get

$$\frac{V + dV}{V - dV} = \frac{T + dT}{T}$$

Here $V = 1\ \text{litre} = 10^3\ \text{cm}^3$

$T = 293$ K and $dT = 0.1$ K

Solving, we get

$$dV = 0.17\ \text{cm}^3$$

Hence $x = \dfrac{dV}{S} = \dfrac{0.17}{5 \times 10^{-2}} = \mathbf{3.4\ cm}$

296. $$P = \frac{\alpha T - \beta T^2}{V} \qquad (P = \text{constant})$$

Hence $$V = \frac{\alpha T - \beta T^2}{P}$$

or $$dV = \left(\frac{\alpha - 2\beta T}{P}\right).dT$$

$$W = \int P dV = \int_{T_1}^{T_2} P\left(\frac{\alpha - 2\beta T}{P}\right) dT$$

or $$W = [\alpha T - \beta T^2]_{T_1}^{T_2}$$

$$= \alpha(T_2 - T_1) - \beta(T_2^2 - T_1^2)$$

297. The work done by the gas is

$$W = W_1 + W_2$$

Here W_1 is work done against the force of atmospheric pressure and

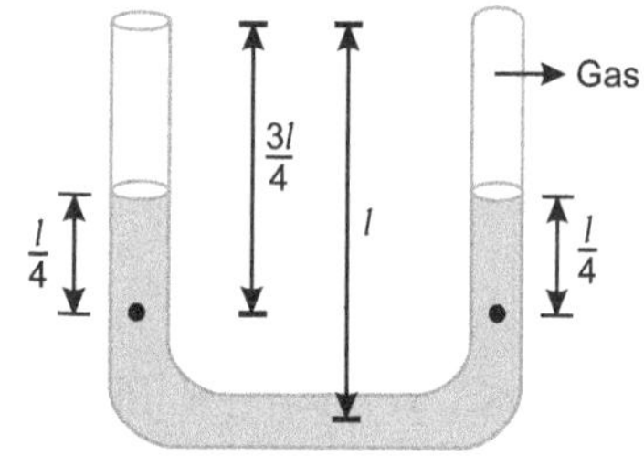

W_2 = work done against the force of gravity. The mercury gas interface is shifted by $(l + l + l) - l/2$ or $\dfrac{5l}{2}$ upon the complete displacement of mercury. Hence

$$W_1 = \frac{5}{2} P_0\, S.l$$

The work done against the force of gravity is equal to the change in the potential energy of mercury as a result of its displacement.

COM of the vertical portion of the mercury of mass $\dfrac{m}{2}$ rises $3l/4$ while that of horizontal portion of mass $\dfrac{m}{2}$ rises by l. Hence

$$W_2 = \left(\frac{m}{2}\right)(g)\left(\frac{3l}{4}\right) + \left(\frac{m}{2}\right)g\,(l) = \frac{7}{8}mgl$$

Here, $m = 2l\rho_{Hg}S$

Hence $$W = W_1 + W_2 = \frac{5}{2}P_0Sl + \frac{7}{8}mgl$$

$$W = \left[\frac{5}{2}P_0S + \frac{7}{8}\rho_{Hg}(2l)gS\right]l$$

Substituting the values

$$W = \left(\frac{5}{2} \times 10^5 \times 10^{-4} + \frac{7}{8} \times 13.6 \times 10^3 \times 2 \times 0.25 \times 9.8 \times 10^{-4}\right) 0.25$$

$\mathbf{W \approx 7.7\ J}$

298. Let A be the area of the pond. Transfer of heat will take place from bottom to top.

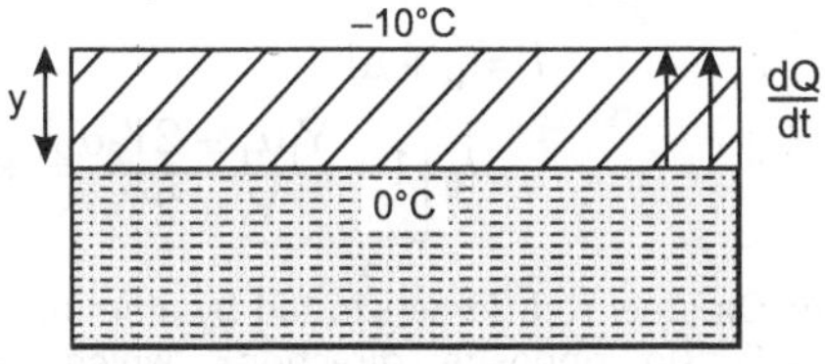

K = thermal conductivity of ice = 2 W/m°C

ρ = density of ice = 900 kg/m^3

L = latent heat = 80 cal/g
$= (80 \times 10^3 \times 4.18)$ J/kg

Now, rate of heat flow $\dfrac{dQ}{dt} = L \cdot \left(\dfrac{dm}{dt}\right)$

or $$\frac{\text{Temperature difference}}{\text{Thermal resistance}} = L\rho A\left(\frac{dy}{dt}\right)$$

or $$\frac{10}{y/KA} = L\rho A\left(\frac{dy}{dt}\right)$$

or $$y \cdot dy = \frac{10K}{L\rho} \cdot dt$$

or $$dt = \frac{L\rho}{10K} y dy$$

or $$\int_0^t dt = \frac{L\rho}{10K}\int_{4\text{ cm}}^{8\text{ cm}} y\, dy = \frac{L\rho}{10.K}\left(\frac{y^2}{2}\right)_{4\times 10^{-2}\text{ m}}^{8\times 10^{-2}\text{ m}}$$

or $$t = \frac{L\rho}{10K}(64-16)\frac{(10^{-4})}{2}$$

Substituting the values, we have

$$t = \frac{80 \times 4.18 \times (10^3)(48)(10^{-4})(900)}{10 \times 2 \times 2}$$

or $\mathbf{t = 36115.2\ s = 10.03\ hr}$

299. Let m_1 and m_2 be the masses of iron rod and aluminium rod respectively. Then

$m_1 = LA\rho_{\text{iron}}$ and $m_2 = lA\rho_{\text{Al}}$ (ρ = density)

The centre of mass of iron rod is at its centre, *i.e.* at a distance $y_1 = \dfrac{L}{2}$ from the top and the centre of mass of aluminium rod is at a distance $y_2 = \left(L + \dfrac{l}{2}\right)$ from the top.

From definition of centre of mass, the centre of mass of iron and aluminium rods from the top is given by :

$$y = \frac{m_1y_1 + m_2y_2}{m_1+m_2} = \frac{(m_1)\left(\frac{L}{2}\right) + m_2\left(L + \frac{l}{2}\right)}{m_1+m_2}$$

$$\therefore \quad \frac{dy}{d\theta} = \frac{\left(\frac{m_1}{2}\right)\frac{dL}{d\theta} + m_2\left(\frac{dL}{d\theta} + \frac{1}{2}\frac{dl}{d\theta}\right)}{m_1+m_2} \quad \ldots(1)$$

Here, $\dfrac{dL}{d\theta} = L\alpha_{\text{iron}}$ and $\dfrac{dl}{d\theta} = l\alpha_{\text{Al}}$

$\therefore$ Equation (1) can be written as

$$\frac{dy}{d\theta} = \frac{\left(\frac{LA\rho_{\text{iron}}}{2}\right)(L\alpha_{\text{iron}}) + (lA\rho_{\text{Al}})(L\alpha_{\text{iron}}) + \left(\frac{lA\rho_{\text{Al}}}{2}\right)(l\alpha_{\text{Al}})}{LA\rho_{\text{iron}} + lA\rho_{\text{Al}}}$$

or $$\frac{dy}{d\theta} = \frac{\frac{L^2}{2}\rho_{\text{iron}}\,\alpha_{\text{iron}} + Ll\,\rho_{\text{Al}}\,\alpha_{\text{iron}} + \frac{l^2\rho_{\text{Al}}\,\alpha_{\text{Al}}}{2}}{L\rho_{\text{iron}} + l\rho_{\text{Al}}}$$

Substituting the values, we get

$$\mathbf{\frac{dy}{d\theta} = 8.06 \times 10^{-6}\ m/^\circ C}$$

300. Let θ_1 and θ_2 be the temperatures of the vessels at any time t. Then the temperature difference will be

$$\theta = \theta_1 - \theta_2$$

or $$\left(-\frac{d\theta}{dt}\right) = \left(-\frac{d\theta_1}{dt}\right) + \left(\frac{d\theta_2}{dt}\right) \quad \ldots(1)$$

with time, θ and θ_1 will decrease while θ_2 will increase

$$\frac{dQ}{dt} = \frac{KA(\theta_1 - \theta_2)}{l} = \frac{KA\theta}{l} \quad \ldots(2)$$

The vessels are adiabatic and volume is constant. Hence heat will transfer from A to B through the rod. Work done by both the gases will be zero. Temperature of A will decrease while that of B will increase.

Hence $\left(-\frac{d\theta_1}{dt}\right) = \frac{dQ/dt}{nC_{V_1}} = \frac{2}{3nR} \cdot \frac{dQ}{dt}$

and $\frac{d\theta_2}{dt} = \frac{dQ/dt}{nC_{V_2}} = \frac{2}{5nR} \cdot \frac{dQ}{dt} \quad \ldots(3)$

From (1), (2) and (3)

$$\left(-\frac{d\theta}{dt}\right) = \frac{16}{15nR} \cdot \frac{dQ}{dt} = \frac{16}{15nR} \cdot \frac{KA}{l}\theta$$

or $\frac{d\theta}{\theta} = -\frac{16KA}{15nRl} \cdot dt$

or $\int_{T_1 - T_2}^{\theta} \frac{d\theta}{\theta} = \frac{-16KA}{15nRl} \int_0^t dt$

or $\theta = (T_1 - T_2)\, e^{-\frac{16KA}{15nRl}t}$

301. At increased temperature, let Δl_1 and Δl_2 be the increase in length of aluminium and steel respectively (if they are free). Then

$$\Delta l_1 = l_0 \alpha_1 \theta$$

and $\Delta l_2 = l_0 \alpha_2 \theta$

Suppose $\Delta l_1 < \Delta l_2$

Therefore, the composite rod will increase in between of Δl_1 and Δl_2. Say it is Δl, where

$$\Delta l_1 < \Delta l < \Delta l_2$$

Due to this, aluminium rod has a length $(\Delta l - \Delta l_1)$ more than its natural length at temperature θ and steel rod (s) will have a length $(\Delta l_2 - \Delta l)$ less than its natural length at temperature θ.

Due to this steel rods will exert a force F_2 on aluminium rod from two sides, which in equilibrium be balanced by internal restoring force F_1. Thus,

$$F_1 = 2F_2$$

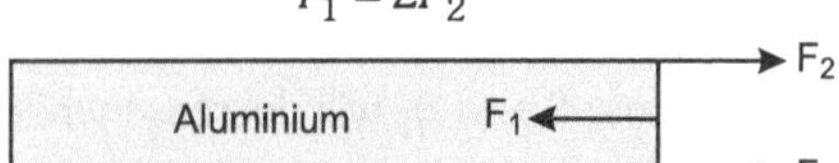

$$\therefore \quad Y_1 A\left(\frac{\Delta l - \Delta l_1}{l_0}\right) = 2Y_2 A\left(\frac{\Delta l_2 - \Delta l}{l_0}\right)$$

Solving this we get,

$$\Delta l = \frac{Y_1 \Delta l_1 + 2Y_2 \Delta l_2}{Y_1 + 2Y_2}$$

$$= \frac{Y_1 l_0 \alpha_1 \theta + 2Y_2 l_0 \alpha_2 \theta}{Y_1 + 2Y_2}$$

$$\therefore \quad l = l_0 + \Delta l$$

$$= l_0\left[1 + \frac{Y_1\alpha_1 + 2Y_2\alpha_2}{Y_1 + 2Y_2} \cdot \theta\right]$$

Note: On steel rod, force exerted by aluminium rod is F_2 (in opposite direction), which is being balanced by its own restoring force.

302. Let H be the rate of heat flow. Then let us take a strip of length dx at a distance x where temperature is T. Then thermal resistance of strip,

$$dR = \frac{dx}{KA} = \frac{dx}{K_0(1 + \alpha T)A}$$

$$(-dT) = H \cdot dR = \frac{H \cdot dx}{K_0 A(1 + \alpha T)}$$

$$\therefore \int_{2T_0}^{T_0} -\frac{K_0 A}{H}(1 + \alpha T)\, dT = \int_0^{L_0} dx$$

Solving this equation we get,

$$H = \frac{K_0 A T_0}{L_0}\left[1 + \frac{3\alpha T_0}{2}\right]$$

303. $\frac{k_1 A(T_1 - T)}{l} - \frac{k_2 A(T - T_2)}{l} = C\frac{dT}{dt}$

$$\Rightarrow \quad [(k_1T_1 + k_2T_2) - (k_1 + k_2)T] = \frac{lC}{A}\frac{dT}{dt}$$

$$\Rightarrow \quad \left(1 - \frac{(k_1 + k_2)T}{(k_1T_1 + k_2T_2)}\right) = \frac{lC}{A(k_1T_1 + k_2T_2)}\frac{dT}{dt}$$

$$\Rightarrow \quad (1 - \lambda T) = \lambda' \frac{dT}{dt},$$

where $\lambda = \frac{k_1 + k_2}{k_1T_1 + k_2T_2}$

and $\lambda' = \frac{lC}{A(k_1T_1 + k_2T_2)}$

$$\Rightarrow \quad \int_{T_0}^{T} \frac{dT}{1 - \lambda T} = \int_0^t \frac{dt}{\lambda'}$$

$$\Rightarrow \quad \ln\left(\frac{1-\lambda T}{1-\lambda T_0}\right)=\frac{-\lambda}{\lambda'}t$$

$$\Rightarrow \quad T=\frac{[1-(1-\lambda T_0)\,e^{\frac{-\lambda}{\lambda'}t}]}{\lambda}$$

By putting values of λ, λ' and t_0 we can see that

$$\frac{\lambda t_0}{\lambda'}=1$$

$$\Rightarrow \quad T_{t=t_0}=\frac{1}{\lambda}-\frac{(1-\lambda T_0)\,e^{-1}}{\lambda}$$

$$\Rightarrow \quad T_{t=t_0}=\frac{1}{\lambda}\left(1-\frac{1}{e}\right)+\frac{T_0}{e}$$

$$\Rightarrow T_{t=t_0}=\frac{T_0}{e}+\left(1-\frac{1}{e}\right)\frac{(k_1T_1+k_2T_2)}{k_1+k_2}$$

304. (a) $PT^2=\text{constant}$

$$(2P_1)\,T_B^2=(P_1)\,T_A^2$$

$$T_A=T_B\sqrt{2}$$

$$T_A=T_C=600\text{ K}$$

During the process $A\to B$

$$PT^2=C$$

$$P^3V^2=k$$

$$P=\frac{k^{1/3}}{V^{2/3}}$$

$$W_{AB}=\int_A^B PdV$$

$$=k^{1/3}\left[\frac{V^{\frac{-2}{3}+1}}{\frac{-2}{3}+1}\right]_{V_A}^{V_B}$$

$$=3\,[k^{1/3}\cdot(V_B^{1/3}-V_A^{1/3})]$$

$$W_{AB}=3\,(P_BV_B-P_AV_A)=3nR\,(T_B-T_A)$$
$$=3\times 3\times R\,(300-600)=-\,2700R$$

(b) Heat evolved in different processes are

$$Q_{AB}=W_{AB}+\Delta U$$
$$=W_{AB}+nC_V\,\Delta T$$
$$=-\,2700R+3\times\frac{5}{2}R\times(-\,300)$$
$$=-\,2700R-15\times 150R$$
$$=-\,4950R$$

$$Q_{BC}=nC_P\,\Delta T=3\times\frac{7}{2}R\,(300R)$$
$$=21\times 150R=3150R$$
$$Q_{CA}=\Delta U+W=0+W_{CA}$$

[As CA is an isothermal process]

$$Q_{CA}=nRT_A\ln\frac{V_A}{V_C}=nRT_A\ln\frac{P_C}{P_A}$$

$$Q_{CA}=3\times R\times 600\ln(2)=1247.4R$$

305. We have, for an adiabatic process

$$TV^{\gamma-1}=\text{constant}$$

Differentiating with respect to T

$$V^{\gamma-1}.1+T\,(\gamma-1)\,V^{\gamma-2}\frac{dV}{dT}=0$$

$$\Rightarrow \quad \frac{dV}{dT}=-\frac{V}{T\,(\gamma-1)}$$

Also, $\left.\frac{dV}{dT}\right|_{T_0,\,V_0}=m$

$$\therefore \quad \frac{V_0}{T_0\,(\gamma-1)}=m$$

or $\gamma-1=\dfrac{V_0}{T_0m}$

and $C_V=\dfrac{R}{\gamma-1}=\dfrac{RT_0m}{V_0}$

and $C_P=C_V+R=\left(1+\dfrac{T_0m}{V_0}\right)R$

306. Under isobaric process,

$$W=P\Delta V=nR\Delta T$$

$$\Delta U=nC_V\,\Delta T=\frac{nR\Delta T}{\gamma-1}$$

$$Q=W+\Delta U,$$

$$\Delta=\frac{n\gamma R\Delta T}{\gamma-1} \qquad \ldots(1)$$

At T_1, distance between nodes = 3 cm

$$\frac{\lambda_1}{2}=3 \Rightarrow \lambda_1=6\text{ cm}$$

$$\Rightarrow \quad v_1=n\lambda_1=300\text{ m/s}$$

at $T_2=\dfrac{\lambda_2}{2}=4 \Rightarrow \lambda_2=8\text{ cm}$

$$\Rightarrow \quad v_2=n\lambda_2=400\text{ m/s}$$

$$V = \sqrt{\frac{\gamma RT}{M}} = \sqrt{\frac{\gamma P}{\rho}}$$

$$v \propto \sqrt{T}$$

i.e., $$\frac{v_1}{v_2} = \sqrt{\frac{T_1}{T_2}}$$

hence $$\frac{T_1}{T_2} = \left(\frac{v_1}{v_2}\right)^2 = \frac{9}{16}$$

Hence, $$T_2 = \left(\frac{16}{9}\right) T_1 = 512 \text{ K}$$

$$\text{Density} = \frac{\text{mass}}{\text{volume}}$$

and $$v = 300 = \sqrt{\frac{\gamma \times 1.5 \times 10^3}{2.8 \times 10^{-2}}}$$

which gives

$$\gamma = \frac{(300)^2 \times 2.8 \times 10^{-2}}{1.5 \times 10^3} = 1.68$$

$$n = \frac{PV}{RT}$$

$$= \frac{1.5 \times 10^3 \times 10^2 \times 10^{-3}}{8.31 \times 288}$$

$$= 6.27 \times 10^{-2}$$

Substituting the values of n, γ and ΔT in equation (1).
Amount of the heat supplied

$$Q = \frac{6.27 \times 1.68 \times 8.31 \times 224 \times 10^{-2}}{(1.68 - 1)}$$

$$= 288.34 \text{ J}$$

307. Escape velocity from the surface of the planet

$$v_p = \sqrt{2g_p R_p}$$

Given $$v_p = \frac{v_e}{\sqrt{6}} = \sqrt{\frac{2g_e R_e}{6}}$$

$$\sqrt{\frac{g_e R_e}{3}} = \sqrt{\frac{2g_p R_e}{36}}$$

$$\Rightarrow \quad g_p = 6g_e$$

Pressure exerted by the atmospheric column of height h on the surface of the planet

$$P = \rho g_p h$$

Therefore, $$\frac{P}{\rho} = g_p h$$

Hence, speed of the sound

$$v = \sqrt{\frac{\gamma P}{\rho}} = \sqrt{\gamma g_p h}$$

$$= \sqrt{6\gamma g_e h} = \sqrt{10 g_e h}$$

308. (a) The cycle is shown in the adjacent diagram:

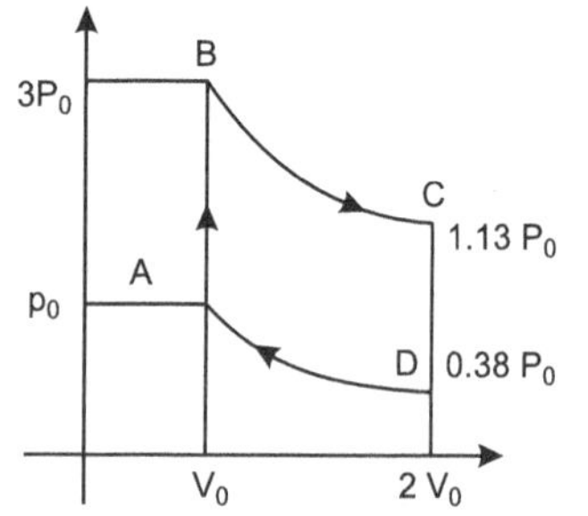

The pressures and temperatures at A, B, C and D are found by using the equation of the process and the equation of state.

$$\frac{P_0}{T_0} = \frac{P_B}{3T_0} \Rightarrow P_B = 3P_0$$

$$P_B V_B^\gamma = P_C V_C^\gamma$$

$$\Rightarrow \quad P_C = P_B \left(\frac{V_B}{V_C}\right)^\gamma = 1.13 P_0$$

Similarly using $P_A V_A^\gamma = P_D V_D^\gamma$ we can find

$$P_D = 0.38 P_0$$

The temperatures are: (as $T \propto PV$)

$$T_A = T_0, \quad T_B = 3T_0,$$

$$T_C = 2.26 T_0, \quad T_D = 0.76 T_0$$

(b) $$Q_{AB} = \frac{5}{2} R(3T_0 - T_0) = 5RT_0$$

$$Q_{BC} = Q_{DA} = 0$$

$$Q_{CD} = -\frac{5}{2} R\,(1.5T_0) = -3.75 RT_0$$

$$Q_{\text{Total}} = 1.25 RT_0$$

Efficiency,

$$\eta = \frac{1.25 RT_0}{5RT_0} \times 100 = 25\%$$

309. 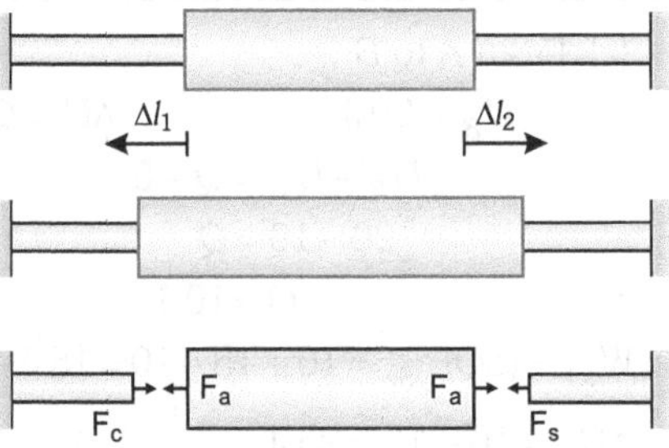

In equilibrium,

$$F_a = F_c$$

$$\left(\frac{\Delta l_a - \Delta l_1 - \Delta l_2}{l_a}\right)(Y_a A_a) = \left(\frac{\Delta l_c + \Delta l_1}{l_c}\right)(Y_c A_c)$$

$$\therefore \left[\frac{(50)(20\times 10^{-6})(60) - \Delta l_1 - \Delta l_2}{50}\right](90)(10)^2$$

$$= \left[\frac{(25)(16\times 10^{-6})(60) + \Delta l_1}{25}\right](100)(5)^2$$

$$\therefore 1.8(6\times 10^{-2} - \Delta l_1 - \Delta l_2)$$

$$= (2.4\times 10^{-2} + \Delta l_1)$$

$$\therefore \quad 2.8\Delta l_1 + 1.8\Delta l_2 = 8.4\times 10^{-2} \qquad \ldots(1)$$

Similarly :

$$F_a = F_s$$

$$\therefore \left[\frac{(50)(20\times 10^{-6})(60) - \Delta l_1 - \Delta l_2}{50}\right](90)(10)^2$$

$$= \left[\frac{(25)(12\times 10^{-6})(60) + \Delta l_2}{25}\right](200)(5)^2$$

$$\therefore 0.9(6\times 10^{-2} - \Delta l_1 - \Delta l_2)$$

$$= (1.8\times 10^{-2} + \Delta l_2)$$

$$0.9\Delta l_1 + 1.9\Delta l_2 = 3.6\times 10^{-2} \qquad \ldots(2)$$

Solving Eqs. (1) and (2), we get

$$\Delta l_1 = 2.56\times 10^{-2} \text{ cm}$$

and $\quad \Delta l_2 = 0.68\times 10^{-2}$ cm

Stress in three portions,

$$\sigma_c = \left[\frac{(25)(16\times 10^{-6})(60) + 2.56\times 10^{-2}}{25}\right] \times (100\times 10^9)\ \text{N/m}^2$$

$$= 19.84\times 10^7\ \text{N/m}^2$$

$$\sigma_a = \left[\frac{(50)(20\times 10^{-6})(60) - 2.56\times 10^{-2} - 0.68\times 10^{-2}}{50}\right](90\times 10^9)$$

$$= 4.97\times 10^7\ \text{N/m}^2 \text{ (compressive)}$$

and $$\sigma_s = \left[\frac{(25)(12\times 10^{-6})(60) + 0.68\times 10^{-2}}{25}\right] \times (200\times 10^9)$$

$$= 19.84\times 10^8\ \text{N/m}^2 \quad \text{(compressive)}$$

310. (a) First law of thermodynamics for the given process from state 1 to state 2

$$Q_{12} - W_{12} = U_2 - U_1$$

Here, $\quad Q_{12} = +10P_0V_0$ J

$W_{12} = 0 \quad$ (Volume remains constant)

$$U_2 - U_1 = nC_V(T_2 - T_1)$$

$$nC_V(T_2 - T_1) = 10P_0V_0$$

For an ideal gas

$$P_0V_0 = nRT_0$$

and $\quad C_P - C_V = R$

$$\therefore \quad C_V = C_P - R = \frac{5R}{2} - R = \frac{3R}{2}$$

$$\therefore \quad n\left(\frac{3R}{2}\right)(T_2 - T_0) = 10nRT_0$$

$$T_2 = \frac{23}{3}T_0$$

As $P \propto T$ for constant volume

$$P_2 = \frac{23}{3}P_0$$

(b) 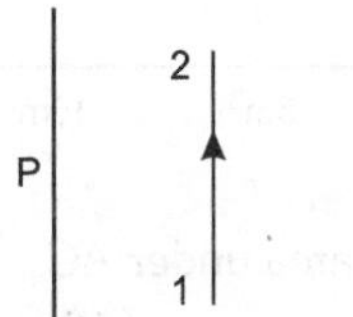

311. Initial pressure of gas

$$P_1 = P_0 + \frac{H}{2}dg,$$

where P_0 is atmospheric pressure

Initial volume of gas $V_1 = \dfrac{V}{2}$

Initial temperature $T_1 = 300$ K

When half of mercury comes out of the cylinder, final pressure of gas

$$P_2 = P_0 + \frac{H}{4} dg$$

and final volume of gas

$$V_2 = \frac{V}{2} + \frac{V}{4} = \frac{3V}{4}$$

we have $$\frac{P_1V_1}{T_1} = \frac{P_2V_2}{T_2}$$

$$\therefore\ T_2 = \frac{P_2V_2}{P_1V_1} T_1$$

$$= \frac{\left(P_0 + \frac{H}{4} dg\right)\cdot\left(\frac{3V}{4}\right) \times 300}{\left(P_0 + \frac{H}{2} dg\right)\cdot \frac{V}{2}}$$

Putting $P_0 = 0.76dg$ (atmospheric pressure) we get

$$\mathbf{T_2 = 337.5\ K}$$

312. (a) W_{AC} is less than W_{ABC} as area under graph is less.

(b) For process A to C

$$Q = 200 \text{ J}$$

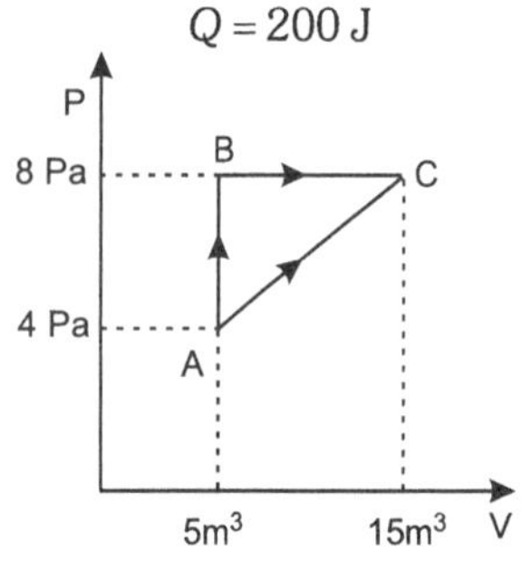

Work done

$$W_{AC} = \text{area under } AC$$

$$= (10 \times 4) + \left(\frac{10 \times 4}{2}\right) = 60 \text{ J}$$

From first law of thermodynamics.

$$\Delta U = Q - W_{AC}$$

$$U_C - U_A = 200 - 60$$

$$\therefore\ U_C = U_A + 140$$

$$= 10 + 140 = 150 \text{ J}$$

(c) From A to B

$$U_B = 20 \text{ J} \qquad \therefore \Delta U = Q - W_{AB}$$

$$U_B - U_A = Q - 0$$

$$20 - 10 = Q$$

$$\therefore\ Q = 10 \text{ J}$$

313. $W_{net} = (\Sigma Q)_{cycle} = 10 + 15 - 10 = 15$ J

$$\Delta Q_{in} = 10 + 15 = 25 \text{ J}$$

$$\eta = \frac{15}{25} \times 100 = 60\%$$

314. (a) Given $\frac{V_B}{V_A} = 2$ and $\frac{V_D}{V_C} = 2$

In the **process *AB*** (isothermal)

$$\Delta U_{AB} = 0$$

Hence $\Delta Q_{AB} = \Delta W_{AB}$

$$= nRT_1 \ln \frac{V_B}{V_A} = RT_1 \ln (2) \quad (n = 1)$$

In **process *BC*** (adiabatic)

$$\Delta Q_{BC} = 0$$

hence $\Delta W_{BC} = -\ \Delta U_{BC}$

$$= -\ nC_V (T_C - T_B) = C_V (T_1 - T_2)$$

In **process *CD*** (isothermal)

$$\Delta U_{CD} = 0$$

hence $\Delta Q_{CD} = \Delta W_{CD}$

$$= nRT_2 \ln\left(\frac{V_D}{V_C}\right) = RT_2 \ln (2)$$

In **process *DE*** (adiabatic)

$$\Delta Q_{DE} = 0$$

hence $\Delta W_{DE} = -\ \Delta U_{DE}$

$$= -\ nC_V (T_E - T_D) = C_V (T_2 - T_3)$$

In adiabatic process, we know that

$$TV^{\gamma - 1} = \text{constant} \quad \text{or} \quad T \propto V^{1-\gamma}$$

Hence in process BC, DE and FA:

$$\frac{T_1}{T_2} = \left(\frac{V_B}{V_C}\right)^{1-\gamma} \qquad \ldots(1)$$

$$\frac{T_2}{T_3} = \left(\frac{V_D}{V_E}\right)^{1-\gamma} \qquad \ldots(2)$$

and $$\frac{T_3}{T_1} = \left(\frac{V_F}{V_A}\right)^{1-\gamma} \qquad \ldots(3)$$

Multiplying equations (1), (2) and (3), we get

$$1 = \left(\frac{V_B}{V_A} \cdot \frac{V_D}{V_C} \cdot \frac{V_F}{V_E}\right)^{1-\gamma}$$

or $$1 = \left(2 \times 2 \times \frac{V_F}{V_E}\right)^{1-\gamma} = \left(\frac{4V_F}{V_E}\right)^{1-\gamma}$$

or $$\frac{4V_F}{V_E} = 1 \Rightarrow \frac{V_F}{V_E} = \frac{1}{4}$$

Now in **process *EF*** (isothermal)

$$\Delta U_{EF} = 0$$

and $$= -2RT_3 \ln(2)$$

In **process *FA*** (adiabatic)

$$\Delta Q_{FA} = 0$$

$$\therefore \quad \Delta W_{FA} = -\Delta U_{FA}$$

$$= -nC_V(T_A - T_F)$$

$$= C_V(T_3 - T_1)$$

Three quantities *viz.* ΔQ, ΔU and ΔW in tabular form in different processes are shown below :

Process	ΔQ	ΔU	ΔW
AB	$RT_1 \ln(2)$	0	$RT_1 \ln(2)$
BC	0	$C_V(T_2 - T_1)$	$C_V(T_1 - T_2)$
CD	$RT_2 \ln(2)$	0	$RT_2 \ln(2)$
DE	0	$C_V(T_3 - T_2)$	$C_V(T_2 - T_3)$
EF	$-2RT_3 \ln(2)$	0	$-2RT_3 \ln(2)$
FA	0	$C_V(T_1 - T_3)$	$C_V(T_3 - T_1)$

(a) From the table

$$\mathbf{W_{net} = R(T_1 + T_2 - 2T_3)\ln(2)}$$

(b) Heat absorbed by the gas

$$\mathbf{Q_{ab} = Q_{+ve} = R(T_1 + T_2)\ln(2)}$$

(c) Efficiency of the cycle

$$\eta = \frac{\text{net work done}}{\text{heat absorbed}}$$

$$= \frac{R(T_1 + T_2 - 2T_3)\ln(2)}{R(T_1 + T_2)\ln(2)}$$

$$= \mathbf{1 - \frac{2T_3}{T_1 + T_2}}$$

315. **Process *A-B*** is an isothermal process *i.e.*

T = constant

Hence $P \propto \frac{1}{V}$ or P-V graph will be a rectangular hyperbola with increasing P and decreasing V.

$\rho \propto \frac{1}{V}$. Hence ρ-V graph is also a rectangular hyperbola with decreasing V and hence increasing ρ.

$$\rho \propto P \qquad \left[\rho = \frac{PM}{RT}\right]$$

Hence ρ-P graph will be a straight line passing through origin, with increasing ρ and P.

Process *B-C* is an isochoric process, because P-T graph is a straight line passing through origin

i.e. V = constant

Hence P-V graph will be a straight line parallel to P-axis with increasing P.

Since V = constant hence ρ will also be constant

Hence ρ-V graph will be a dot.

ρ-P graph will be a straight line parallel to P-axis with increasing P, because

$$\rho = \text{constant}$$

Process *C-D* is inverse of A-B and D-A is inverse of B-C.

Different values of P, V, T and ρ in tabular form are shown below

	P	V	T	ρ
A	P_0	V_0	T_0	ρ_0
B	$2P_0$	$\frac{V_0}{2}$	T_0	$2\rho_0$
C	$4P_0$	$\frac{V_0}{2}$	$2T_0$	$2\rho_0$
D	$2P_0$	V_0	$2T_0$	ρ_0

Here $$V_0 = nR\left(\frac{T_0}{P_0}\right)$$

and $$\rho_0 = \frac{P_0 M}{RT_0}$$

The corresponding graphs are as follows :

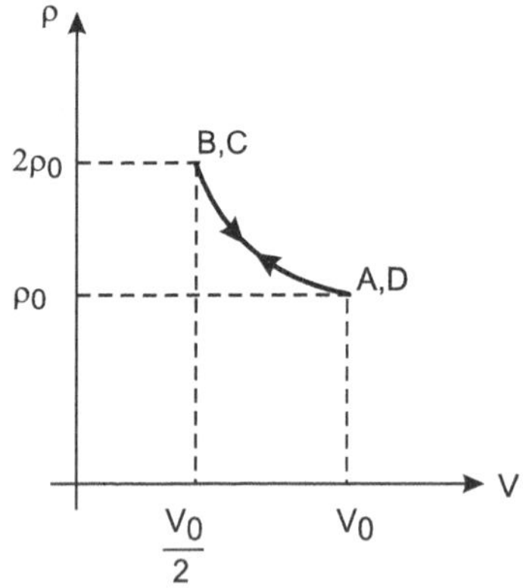

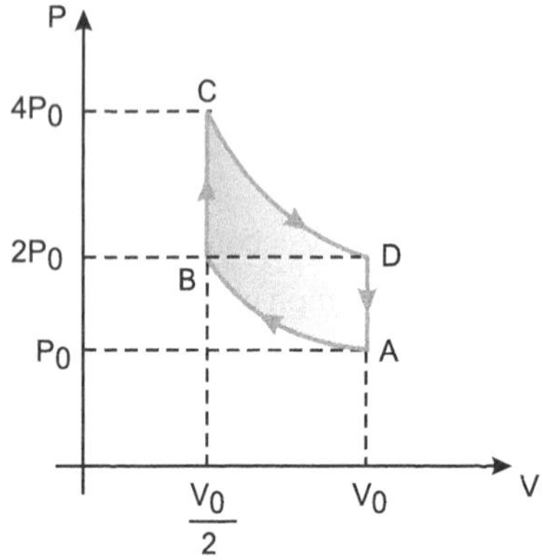

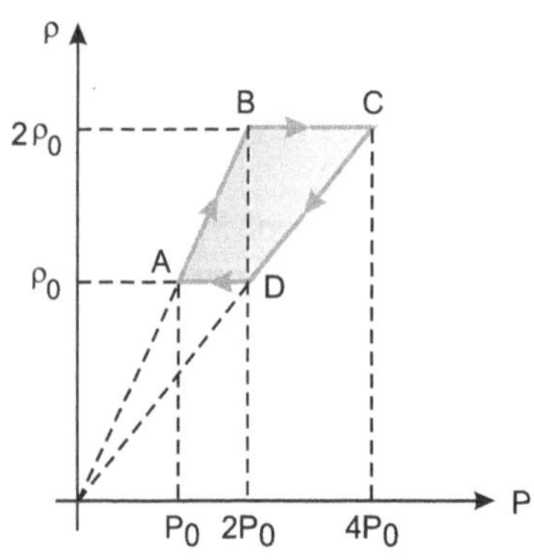

316. **Let $V_C = V_D = V_0$(minimum)**

Then $\quad V_A = 8\sqrt{2}\,V_0 \quad$ (maximum)

$$T \propto PV$$

Hence temperature at A and D will be maximum and at B will be minimum. (In adiabatic compressin temperature is increased. Hence $T_c > T_B$)

So, let $\quad T_B = T_0$

Then $\quad T_A = T_D = 4T_0$

Process *AB* is isobaric ($P =$ constant)

$\therefore \quad V \propto T$

or $\quad V_B = T_B\left(\dfrac{V_A}{T_A}\right)$

$$= (T_0)\left(\frac{8\sqrt{2}V_0}{4T_0}\right) = 2\sqrt{2}V_0$$

Process *BC* is adiabatic, so

$$T_C = T_B\left(\frac{V_B}{V_C}\right)^{\gamma - 1}$$

$$= T_0\left(\frac{2\sqrt{2}V_0}{V_0}\right)^{(5/3)-1} = 2T_0$$

(γ for He is 5/3)

So, finally $T_A = 4T_0$, $T_B = T_0$,

$T_C = 2T_0$ and $T_D = 4T_0$

Now in **process *AB*** (isobaric)

$$\Delta W_{AB} = \Delta Q_{AB} - \Delta U_{AB}$$
$$= nC_p\Delta T - nC_V\Delta T = nR\Delta T$$
$$= nR(T_B - T_A) = -3nRT_0$$

and ΔQ_{AB} is negative (released)

In **process *BC*** (adiabatic)

$$\Delta Q_{BC} = 0$$

$$\Delta W_{BC} = -\Delta U_{BC} = -nC_V\Delta T$$
$$= n\left(\frac{3}{2}R\right)(T_B - T_C) = -\frac{3}{2}nRT_0$$

In **process *CD*** (isochoric)

$$\Delta W_{CD} = 0$$

and $\quad \Delta Q_{CD} = \Delta U_{CD} = nC_V\,(T_D - T_C)$

$$= n\left(\frac{3}{2}R\right)(2T_0) = 3nRT_0$$

In **process *DA*** (isothermal)

$$\Delta U_{DA} = 0$$

Hence $\quad \Delta Q_{DA} = \Delta W_{DA} = nRT_D\,\ln\left(\dfrac{V_A}{V_D}\right)$

$$= nR(4T_0)\ln(8\sqrt{2})$$
$$= 14nRT_0\ln(2)$$

$\therefore$ Total work done is

$$W_{net} = -\ 3nRT_0 - \frac{3}{2}nRT_0 + 14nRT_0 \ln(2)$$

$$W_{net} = nRT_0\left[14\ln(2) - \frac{9}{2}\right] = 5.202\,nRT_0$$

and heat absorbed is :

$$Q_{ab} = Q_{+ve} = 3nRT_0 + 14nRT_0 \ln(2)$$
$$= 12.702\,nRT_0$$

Hence efficiency of the cycle is

$$\eta = \frac{W_{net}}{Q_{ab}} \times 100$$

$$= \frac{5.202\,nRT_0}{12.702\,nRT_0} \times 100 = \mathbf{41\%}$$

317. Suppose initially pressure and volume of each gas is P_0 and V_0 and n_1 and n_2 are the number of moles in A and B respectively. Similarly T_1 and T_2 are the initial temperature of sample A and B then

$$P_0V_0 = n_1RT_1 = n_2RT_2 \quad \ldots(1)$$

Work done in isothermal process is

$$W_1 = n_1RT_1\, ln\left(\frac{2V_0}{V_0}\right) \quad \text{(volume is doubled)}$$

or $$W_1 = n_1RT_1\, ln\,(2) \quad \ldots(2)$$

Let T_2' be the temperature of sample B after expansion. Then

$$W_2 = \frac{n_2RT_2 - n_2RT_2'}{\gamma - 1} \quad \ldots(3)$$

For adiabatic process

$$T\,V^{\gamma-1} = \text{constant}$$

or $$T_2V_0^{\gamma-1} = T_2'\,(2\,V_0)^{\gamma-1}$$

or $$T_2' = T_2\,(2)^{1-\gamma}$$

Substituting this in equation (3) we get

$$W_2 = \frac{n_2RT_2\,(1 - 2^{1-\gamma})}{\gamma - 1} \quad \ldots(4)$$

Given that $W_1 = W_2$

Hence equating (2) and (4) we get

$$n_1RT_1\,\ln(2) = \frac{n_2RT_2\,(1 - 2^{1-\gamma})}{\gamma - 1}$$

But $n_1\,RT_1 = n_2RT_2$ (from equation 1)

Hence $1 - 2^{1-\gamma} = (\gamma - 1)\ln(2)$ **Proved**

318. Given that

$$\frac{P}{\rho^\gamma} = \text{constant}$$

$\therefore$ $\rho = KP^{1/\gamma}$ where K is a constant.

A = area of cross-section of element

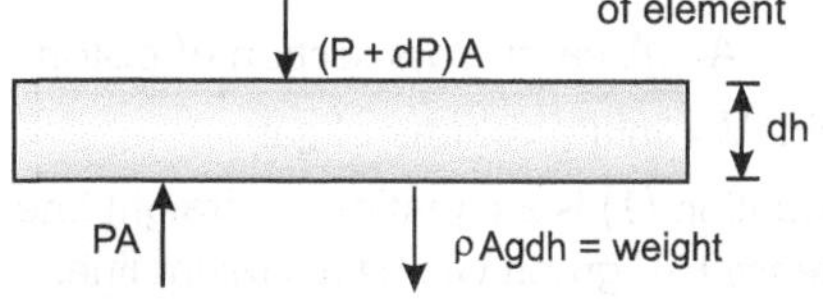

Considering the free body diagram of an element in the atmosphere

Equilibrium of this element gives

$$(P + dP)\,A + \rho Agdh = PA$$

or $$\frac{dP}{dh} = -\ \rho g$$

or $$\frac{dP}{dh} = -\ K\,P^{\frac{1}{\gamma}}g \quad \ldots(1)$$

From gas equations

$$P = \frac{\rho}{M}RT = \frac{KP^{1/\gamma}}{M}RT \quad \ldots(2)$$

Differentiating (2) with respect to h, we get

$$\frac{\gamma - 1}{\gamma}P^{-1/\gamma}\left(\frac{dP}{dh}\right) = \frac{KR}{M}\cdot\frac{dT}{dh} \quad \ldots(3)$$

From (1) and (3), we have

$$\frac{dT}{dh} = -\ \frac{Mg}{R}\left(\frac{\gamma - 1}{\gamma}\right) = \text{constant}$$

Hence $$dT = -\left(\frac{Mg}{R}\right)\left(\frac{\gamma - 1}{\gamma}\right)dh$$

substituting the values, $dh = 10^3$ m, $M = 0.029$ kg, $\gamma = 1.4$ (for diatomic gas), we get

$$dT = -\ \frac{(0.029)(9.8)(1.4 - 1)}{8.31 \times 1.4} \times 10^3$$

$$= \mathbf{-\ 9.8\ K}$$

319. When the piston moves upward by Δx : pressure inside the gas is

$$P = P_0 + \frac{mg}{A} + \frac{k\,\Delta x}{A}$$

or $$P = P_0 + \frac{mg}{A} + \frac{k\,\Delta V}{A^2}$$

$$P = P_0 + \frac{mg}{A} + \frac{k}{A^2}(V - V_i) \quad \ldots(1)$$

Here k = force constant of spring

P_0 = atmospheric pressure,

m = mass of piston

A = Area of cross-section of piston

and V_i = initial volume of gas

Equation (1) is a equation of straight line. Hence P-V graph will be a straight line.

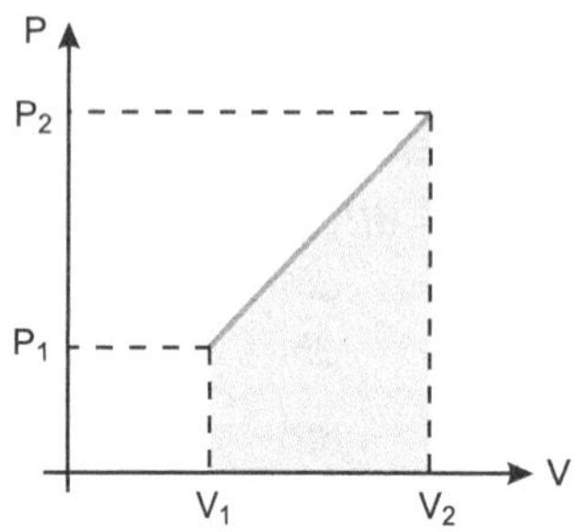

The work done will be the area under P-V graph or the hatched area.

Hence $\boldsymbol{W = \frac{(P_1 + P_2)(V_2 - V_1)}{2}}$

320. Let T_1 and T_2 be the initial and final temperatures then

$$P_0V = nRT_1 \quad \ldots(1)$$

and $$\left(P_0 + \frac{W}{A}\right)(V - Ah) = nRT_2 \quad \ldots(2)$$

Since the gas is thermally insulated, the entire work done on the gas is spent to change its internal energy. Work done on the gas is Wh. Hence

$$Wh = \Delta U = nC_V\Delta T = \frac{3}{2}nR(T_2 - T_1) \quad \ldots(3)$$

$C_V = \frac{3}{2}R$ for a monoatomic gas

Equation (2) – (1) gives :

$$nR(T_2 - T_1) = \frac{W}{A}.V - Wh - P_0Ah$$

Substituting $(T_2 - T_1)$ from equation (3), we get

$$\frac{2}{3}Wh = \frac{WV}{A} - Wh - P_0Ah$$

or $$Ah = \frac{WV}{\left(P_0A + \frac{5}{3}W\right)}$$

Substituting value of Ah in equation (2), we get

$$T_2 = \frac{\left(P_0 + \frac{W}{A}\right)\left[V - \frac{WV}{P_0A + \frac{5}{3}W}\right]}{nR}$$

or $$T_2 = \frac{(P_0A + W)\left(P_0AV + \frac{2}{3}WV\right)}{nAR\left(P_0A + \frac{5}{3}W\right)}$$

or $$\boldsymbol{T_2 = \frac{(P_0A + W)(3P_0AV + 2WV)}{nAR(3P_0A + 5W)}}$$

321. Total heat given to the system is $Q - Q'$. So from first law of thermodynamics,

$Q - Q'$ = total work done by the gas in both the chambers (W) + change in internal energies of both the gases (ΔU) ...(1)

Here W = sum of potential energies stored in the springs

$$W = 2\left[\frac{1}{2}k\left(\frac{l}{2}\right)^2\right] = \frac{kl^2}{4} \quad \ldots(2)$$

Since the temperature of left part remains constant (piston does not conduct heat), internal energy of left part does not change. ΔU of right part can be given as

$$\Delta U = nC_V\Delta T = \frac{3}{2}nR\Delta T \quad \ldots(3)$$

ΔT can be found from the condition of equilibrium at the end of the process.

pressure on right side = pressure on left side

or $$\frac{nR(T + \Delta T)}{A(l + l/2)} = \frac{nRT}{A(l - l/2)} + \frac{2k\left(\frac{l}{2}\right)}{A}$$

simplyfying this we get :

or $$\Delta T = \frac{3kl^2}{2nR} + 2T \quad \ldots(4)$$

From equations (1), (2), (3) and (4), we get

$$Q' = Q - \frac{kl^2}{4} - \frac{3}{2}nR\left(\frac{3kl^2}{2nR} + 2T\right)$$

or $$\boldsymbol{Q' = Q - \frac{5}{2}kl^2 - 3nRT}$$

322. (a) Heat capacity at constant pressure will be

$$C_p = \frac{(dQ)_P}{d\theta} = \frac{dW + dU}{d\theta}$$

$$= \left[\frac{dW}{d\theta} + \frac{dU}{d\theta}\right] = \left[\frac{PdV}{d\theta} + \frac{dU}{d\theta}\right]$$

$$= \left[\frac{(P)(\gamma V . d\theta)}{d\theta} + \frac{dU}{d\theta}\right] \qquad (\gamma = 3\alpha)$$

$$= \left[(P)(V)(3\alpha) + \frac{dU}{d\theta}\right] \qquad (V = \frac{m}{\rho})$$

$$\therefore \quad C_p = \left[\frac{3P\alpha m}{\rho} + \frac{dU}{d\theta}\right] \qquad \ldots(1)$$

Heat capacity at constant volume will be

$$C_v = \frac{(dQ)_V}{d\theta} = \frac{dW + dU}{d\theta}$$

$$= \frac{0 + dU}{d\theta} = \frac{dU}{d\theta} \qquad \ldots(2)$$

From (1) and (2), we get

$$\boldsymbol{C_p - C_v = \frac{3P\alpha m}{\rho}}$$

(b) The above expression, *i.e.* $C_p - C_v = \dfrac{3P\alpha m}{\rho}$ is for m kg of substance.

We have to find this value for 1 mol or 27×10^{-3} kg. Hence substituting the given values, we get

$$C_p - C_v = \frac{(3)(1.01 \times 10^5)(23 \times 10^{-6})(27 \times 10^{-3})}{2700}$$

$$= \mathbf{69.69}\ \boldsymbol{\mu}\mathbf{J/mol\text{-}K}$$

323. Let θ be the rise in temperature, γ_{Hg} and γ_w the coefficients of volume expansion of mercury and water respectively.

Then,

$\therefore$ volume of water outflow

= final volume of water

+ final volume of mercury

– volume of vessel.

Initial volumes of water and mercury are $\dfrac{500}{1}$ cc and $\dfrac{1000}{13.6}$ cc respectively. Then

volume of water outflow

$$= \frac{500}{1}(1 + \gamma_w\theta) + \frac{1000}{13.6}(1 + \gamma_{Hg}\theta) - \left(\frac{500}{1} + \frac{1000}{13.6}\right)$$

or $$\frac{3.52}{\rho_{final}} = 500\gamma_w . \theta + \frac{1000}{13.6}\gamma_{Hg} . \theta$$

or $$\frac{3.52}{1}(1 + \gamma_w\theta) = 500\gamma_w\theta + 73.53\gamma_{Hg}\theta$$

or $$496.48\gamma_w\theta + 73.53\gamma_{Hg}\theta = 3.52 \qquad \ldots(1)$$

Another equation can be formed from calorimetry

$$21200 = 500 \times 1 \times \theta + 1000 \times 0.03 \times \theta$$

or $$\theta = 40°\text{C}$$

Substituting the values of θ and γ_w in equation (1)

we get

$$\gamma_{Hg} = \mathbf{1.84 \times 10^{-4}\ per\ °C.}$$

324. Initial pressure,

$$P = P_{atm} + \frac{(mg)}{A}$$

$$= 1.05 \times 10^5 + \frac{9 \times 10}{0.09}$$

$$= 1.05 \times 10^5 + \frac{90 \times 10^2}{9}$$

$$= 1.05 \times 10^5 + 1 \times 10^3$$

$$= 1.06 \times 10^5\ \text{N/m}^2$$

If n is the number of moles of the gas

$$n = \frac{PV}{RT}$$

$$= \frac{1.06 \times 10^5 \times 0.027}{8.31 \times 300} = 1.15$$

When heat is given to the system piston will move under constant pressure.

$$\Delta Q = nC_P \,\Delta T$$

$$2.5 \times 10^4 = 1.15 \times \frac{5}{2} \times 8.31 \times (\Delta T)$$

$$\Delta T = 1046 \text{ K}$$

Final temperature $= 1046 + 300 = 1346$ K

Final pressure will be same as initial

$$= 1.06 \times 10^5 \text{ N/m}^2$$

$$W = Q - \Delta U = nC_P \,\Delta T - nC_V \,\Delta T = nR \,\Delta T$$

$$= (1.15)(8.31)(1046) \text{ J} = 10^4 \text{ J}$$

325. Initial charge on the capacitor

$$q_0 = CV_0 = 75 \times 10^{-3} \times \frac{640}{3} = 16 \text{ C}$$

The charge on the capacitor decays as

$$q = q_0 e^{-t/RC}$$

At $\quad t = 2.5 \ln(4)$ minutes

$$= 150 \ln(4) \text{ sec}$$

$$q = 16 \times e^{-\ln(4)} = 4 \text{ C}$$

$\because \quad RC = 150$ s

Total heat dissipated in the resistor in the given time

$$= \frac{q_0^2 - q^2}{2C} = 1.6 \text{ kJ}$$

= heat imparted to the gas

(a) Work done by the gas at constant pressure

$$= P\Delta V = nR\Delta T \approx 0.182 \text{ kJ}$$

(b) Increment in the internal energy

$$\Delta U = Q - W = 1.6 - 0.182 = 1.418 \text{ kJ}$$

(c) $\gamma = \dfrac{C_P}{C_V} = \dfrac{nC_P\Delta T}{nC_V\Delta T} = \dfrac{Q}{\Delta U} = 1.13$

326.

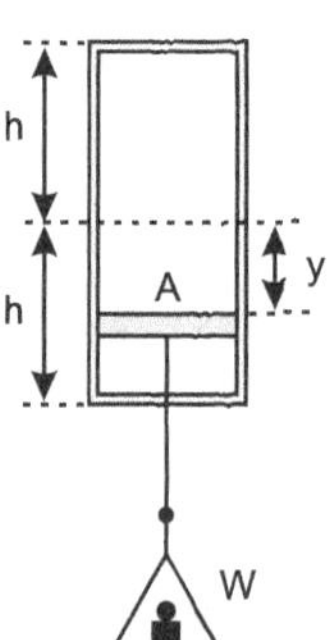

Let A denote the cross-sectional area of the piston and y the vertical displacement between its initial and final equilibrium positions. (See fig.)

The decrease in potential energy of the weight W increases the internal energy of the air inside the cylinder. Conservation of energy between the initial and the final states gives

$$Wy = \frac{5}{2}[p_1 A(h - y) + p_2 A(h + y)] - [2p_0 Ah] \quad \ldots[1(a)]$$

where p_1 is the final pressure in the lower part of the cylinder and p_2 that in the upper part. The internal energy of a gas made up of diatomic molecules has been written in the form $\frac{5}{2}pV$. If W is very large, the decrease in its potential energy (and the corresponding increase in the internal energies of the gases) is very large, and the initial internal energy of the air can be neglected. Thus

$$Wy = \frac{5}{2}[p_1 A(h - y) + p_2 A(h + y)] \quad \ldots[1(b)]$$

When the load is finally at rest,

$$(p_1 - p_2) A = W \quad \ldots(2)$$

The temperatures and the masses of the gases in the two halves are identical, and so their internal energies must be equal:

$$\frac{5}{2} p_1 A(h - y) = \frac{5}{2} p_2 A(h + y) \quad \ldots(3)$$

Equations (1b), (2) and (3) yield $y = \sqrt{\frac{5}{7}}\, h$ for the displacement of the piston, *i.e.*, the gas in the lower part is compressed to $1 - \sqrt{\frac{5}{7}} \approx 15$ per cent of its original volume.

Note: The surprising result is that the volume of air in the lower part does not tend to zero, however large the weight is, even though gases are supposed to be compressible! The large load increases the internal energy, and hence the temperature, of the enclosed gas. This causes considerable increases in not only the absolute pressures, but also in the difference between the upper and lower pressures.

327. (a) $\quad V_1 = 4 \times 10^{-3}$ m^3

$$P_1 = 2 \times 10^5 \text{ Pa}$$

$$V_2 = 2 \times 10^{-3} \text{ m}^3$$

For adiabatic compression.

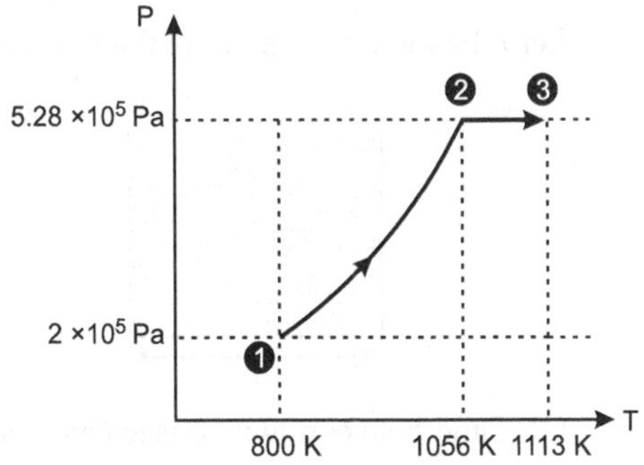

$$P_2 = P_1\left(\frac{V_1}{V_2}\right)^{\gamma} = 2\times 10^5\left(\frac{4\times 10^{-3}}{2\times 10^{-3}}\right)^{1.4}$$

$$= 5.28\times 10^5 \text{ Pa}$$

$$T_1 = \frac{P_1V_1}{nR} = \frac{2\times 10^5\times 4\times 10^{-3}}{\frac{1}{R}\times R}$$

$$= 800 \text{ K}$$

$$T_2 = \frac{P_2V_2}{nR}$$

$$= \frac{5.28\times 10^5\times 2\times 10^{-3}}{\frac{1}{R}\times R}$$

$$= 1056 \text{ K}$$

In the second process heat given is 200 J.

$$\Delta Q = nC_P\Delta T$$

$$200 = \frac{1}{R}\times\frac{7}{2}R\,\Delta T$$

$$\Rightarrow \quad \Delta T = 57 \text{ K}$$

$$T_3 = 1113 \text{ K}$$

(b) $W_{\text{total}} = W_{\text{adibatic}} + W_{\text{isobaric}}$

$$= \frac{1}{R}\,\frac{R\,(1056-800)}{1-1.4} + \frac{1}{R}\cdot R\,(57)$$

$$= -583 \text{ J}$$

328. At constant pressure, $V \propto T$

or $$\frac{V_2}{V_1} = \frac{T_2}{T_1}$$

or $$\frac{Ah_2}{Ah_1} = \frac{T_2}{T_1}$$

$$\therefore \quad h_2 = h_1\left(\frac{T_2}{T_1}\right)$$

$$= (1.0)\left(\frac{400}{300}\right)\text{m} = \mathbf{\frac{4}{3}\ m}$$

As there is no heat loss, process is adiabatic. For adiabatic process,

$$T_fV_f^{\gamma-1} = T_iV_i^{\gamma-1}$$

$$\therefore \quad T_f = T_i\left(\frac{V_i}{V_f}\right)^{\gamma-1} = (400)\left(\frac{h_i}{h_f}\right)^{1.4-1}$$

$$= 400\left(\frac{4}{3}\right)^{0.4} = \mathbf{448.8\ K}$$

329. When the temperature is increased, volume of the cube will increase while density of liquid will decrease. The depth upto which the cube is submerged in the liquid remains the same, hence the upthrust will not change.

$$F = F'$$

$$\therefore \quad V_i\rho_L g = V_i'\rho_L' g \quad (V_i = \text{volume immersed})$$

$$\therefore \quad (Ah_i)(\rho_L)(g) = A(1+2\alpha_s\Delta T)(h_i) \times\left(\frac{\rho_L}{1+\gamma l\,\Delta T}\right)g$$

Solving this equation, we get

$$\gamma_l = 2\alpha_s$$

330. Rate of heat conduction through rod = rate of heat lost from the right end of the rod.

$$\therefore \quad \frac{KA(T_1-T_2)}{L} = eA\sigma\,(T_2^4 - T_s^4) \quad \ldots(1)$$

Given that $T_2 = T_s + \Delta T$

$$\therefore \quad T_2^4 = (T_s+\Delta T)^4 = T_s^4\left(1+\frac{\Delta T}{T_s}\right)^4$$

Using binomial expansion we have,

$$T_2^4 \cong T_s^4\left(1+4\frac{\Delta T}{T_s}\right) \quad \text{as } \Delta T << T_s$$

$$\therefore \quad T_2^4 - T_s^4 = 4(\Delta T)(T_s^3)$$

Substituting in Eq. (1), we have

$$\frac{K(T_1-T_s-\Delta T)}{L} = 4e\sigma T_s^3\,.\,\Delta T$$

or $$\frac{K(T_1-T_s)}{L} = \left(4e\sigma T_s^3 + \frac{K}{L}\right).\,\Delta T$$

$\therefore \quad \Delta T = \dfrac{K(T_1 - T_s)}{(4e\sigma L T_s^3 + K)}$

Comparing with the given relation, proportionality constant $= \dfrac{K}{4e\sigma L T_s^3 + K}$.

331. Decrease in kinetic energy

= increase in internal energy of the gas

$$\therefore \quad \frac{1}{2} m v_0^2 = nC_v \, \Delta T = \left(\frac{m}{M}\right)\left(\frac{3}{2}R\right)\Delta T$$

$$\therefore \quad \Delta T = \frac{M v_0^2}{3R}$$

332. (a) Rate of heat loss per unit area due to radiation

$$I = e\sigma(T^4 - T_0^4)$$

Here $T = 127 + 273 = 400$ K

and $T_0 = 27 + 273 = 300$ K

$$\therefore \quad I = 0.6 \times \frac{17}{3} \times 10^{-8} [(400)^2 - (300)^4]$$

$= \mathbf{595\ watt/m^2}$

(b) Let θ be the temperature of the oil. Then rate of heat flow through conduction

= rate of heat loss due to radiation

$$\therefore \quad \frac{\text{temperature difference}}{\text{thermal resistance}} = (595)A$$

$$\therefore \quad \frac{(\theta - 127)}{\left(\dfrac{l}{kA}\right)} = (595)A$$

Here A = area of disc; k = thermal conductivity and l = thickness (or length) of disc

$$\therefore \quad (\theta - 127)\frac{k}{l} = 595$$

$$\therefore \quad \theta = 595\left(\frac{l}{k}\right) + 127$$

$$= \frac{595 \times 10^{-2}}{0.167} + 127$$

$= \mathbf{162.6°C}$

333. Volume of the box $= 1\ \text{m}^3$

Pressure of the gas $= 100\ \text{N/m}^2$

Let T be the temperature of the gas. Then

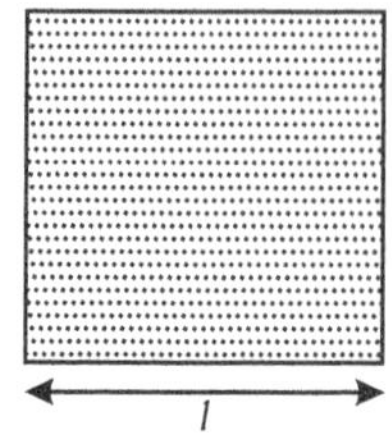

(a) Time between two consecutive collisions with one wall $= \dfrac{1}{500}$ s. This time should be equal to $\dfrac{2l}{v_{rms}}$, where l is the side of the cube.

$$\frac{2l}{v_{rms}} = \frac{1}{500}$$

or $\quad v_{rms} = 1000$ m/s $\quad$ (as $l = 1$ m)

or $\quad \sqrt{\dfrac{3RT}{M}} = 1000$

$$\therefore \quad T = \frac{(1000)^2 M}{3R}$$

$$= \frac{(10)^6 (4 \times 10^{-3})}{3(25/3)} = \mathbf{160\ K}$$

(b) Average kinetic energy per atom $= \dfrac{3}{2}kT$

$$= \frac{3}{2}(1.38 \times 10^{-23})(160)\ \text{J}$$

$= \mathbf{3.312 \times 10^{-21}\ J}$

(c) From $PV = nRT = \dfrac{m}{M}RT$

we get mass of helium gas in the box,

$$m = \frac{PVM}{RT}$$

substituting the values we get,

$$m = \frac{(100)(1)(4)}{(25/3)(160)} = \mathbf{0.3\ g}$$

334. Given : Number of moles, $n = 2$

$C_v = \dfrac{3}{2}R$ and $C_p = \dfrac{5}{2}R$ $\quad$ (Monoatomic)

$$T_A = 27°\text{C} = 300\ \text{K}$$

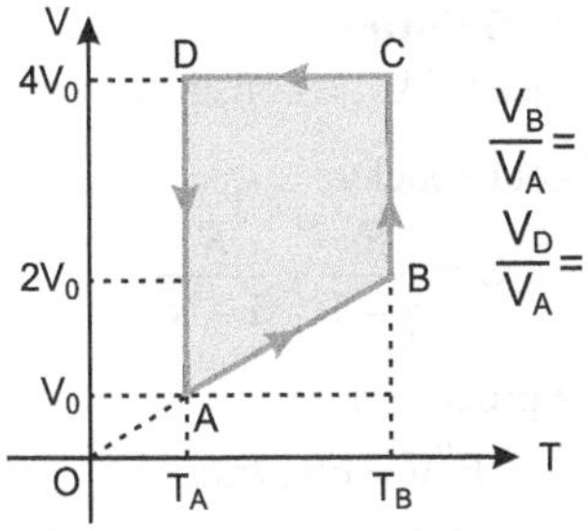

$\frac{V_B}{V_A} = 2$ and $\frac{V_D}{V_A} = 4$

Let $\quad V_A = V_0$

then $\quad V_B = 2V_0$

and $\quad V_D = V_C = 4V_0$

(a) Process $A \to B$

$$V \propto T \Rightarrow \frac{T_B}{T_A} = \frac{V_B}{V_A}$$

$$\therefore T_B = T_A\left(\frac{V_B}{V_A}\right) = (300)(2) = 600 \text{ K}$$

$$\therefore \quad \mathbf{T_B = 600 \text{ K}}$$

(b) Process $A \to B$

$$V \propto T \Rightarrow P = \text{constant}$$

$$\therefore \quad Q_{AB} = nC_P\, dT = nC_P\,(T_B - T_A)$$

$$= (2)\left(\frac{5}{2}R\right)(600 - 300)$$

$$\mathbf{Q_{AB} = 1500\, R \text{ (absorbed)}}$$

Process $B \to C$

$$T = \text{constant} \quad \therefore \quad dU = 0$$

$$\therefore \quad Q_{BC} = W_{BC} = nRT_B \ln\left(\frac{V_C}{V_B}\right)$$

$$= (2)(R)(600) \ln\left(\frac{4V_0}{2V_0}\right)$$

$$= (1200R) \ln(2)$$

$$= (1200\,R)(0.693)$$

or $\mathbf{Q_{BC} \approx 831.6\, R}$ **(absorbed)**

Process $C \to D$

$$V = \text{constant}$$

$$\therefore \quad Q_{CD} = nC_V dT = nC_V\,(T_D - T_C)$$

$$= n\left(\frac{3}{2}R\right)(T_A - T_B)$$

$$(T_D = T_A \text{ and } T_C = T_B)$$

$$= (2)\left(\frac{3}{2}R\right)(300 - 600)$$

$\mathbf{Q_{CD} = -\,900\, R}$ **(released)**

Process $D \to A$

$$T = \text{constant} \Rightarrow dU = 0$$

$$\therefore \quad Q_{DA} = W_{DA} = nRT_D \ln\left(\frac{V_A}{V_D}\right)$$

$$= 2\,(R)\,(300) \ln\left(\frac{V_0}{4V_0}\right)$$

$$= 600\, R \ln\left(\frac{1}{4}\right)$$

$$\mathbf{Q_{DA} \approx -\,831.6\, R \text{ (released)}}$$

(c) In the complete cycle

$$dU = 0$$

Therefore, from conservation of energy

$$W_{net} = Q_{AB} + Q_{BC} + Q_{CD} + Q_{DA}$$

$$W_{net} = 1500\,R + 831.6R - 900R - 831.6R$$

or $\quad \mathbf{W_{net} = W_{total} = 600\, R}$

335. (a) Number of moles,

$$n = 2, T_1 = 300 \text{ K}$$

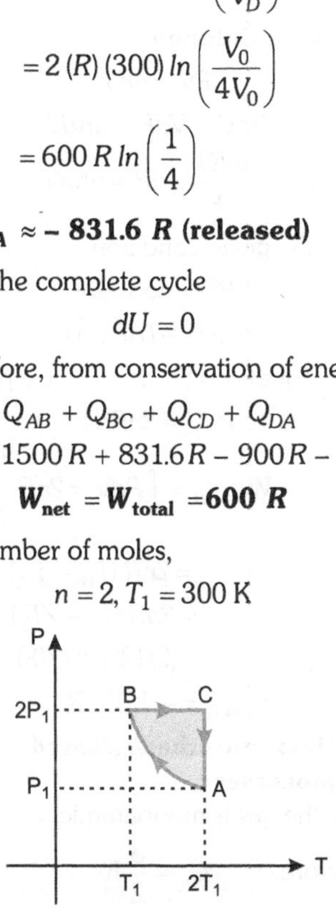

During the process $A \to B$

$$PT = \text{constant}$$

or $\quad P^2V = \text{constant} = K$ (say)

$$\therefore \quad P = \frac{\sqrt{K}}{\sqrt{V}}$$

$$\therefore \quad W_{A\to B} = \int_{V_A}^{V_B} P.dV = \int_{V_A}^{V_B} \frac{\sqrt{K}}{\sqrt{V}}\, dV$$

$$= 2\sqrt{K}\,[\sqrt{V_B} - \sqrt{V_A}]$$

$$= 2\,[\sqrt{KV_B} - \sqrt{KV_A}]$$

$$= 2[\sqrt{(P_B^2V_B)V_B} - \sqrt{(P_A^2V_A)V_A}] \qquad (K = P^2V)$$

$$= 2[P_BV_B - P_AV_A] = 2[nRT_B - nRT_A]$$

$$= 2nR[T_1 - 2T_1] = (2)(2)(R)[300 - 600]$$

$$= -1200R$$

$\therefore$ Work done on the gas in the process AB is 1200 R.

$$\mathbf{W_{A\to B} = -1200\,R}$$

Alternate Solution

$$PV = nRT$$

$$\therefore \quad PdV + VdP = nRdT$$

$$\text{or} \quad PdV + \frac{(nRT)}{P}.dP = nRdT \qquad \ldots(1)$$

From the given condition

$$PT = \text{constant}$$

$$PdT + TdP = 0 \qquad \ldots(2)$$

From equations (1) and (2), we get

$$PdV = 2nRdT$$

$$\therefore \quad W_{A\to B} = \int PdV = 2nR\int_{T_A}^{T_B} dT$$

$$= 2nR(T_B - T_A)$$

$$= 2nR(T_1 - 2T_1)$$

$$= (2)(2)(R)(300 - 600)$$

$$\text{or} \quad W_{A\to B} = -1200\,R$$

(b) **Heat absorbed/released in different processes :**

since the gas is monoatomic,

therefore, $C_v = \frac{3}{2}R$

and $C_P = \frac{5}{2}R$ and $\gamma = 5/3$

Process $A \to B$

$$\Delta U = nC_V\Delta T = (2)\left(\frac{3}{2}R\right)(T_B - T_A)$$

$$= (2)\left(\frac{3}{2}R\right)(300 - 600) = -900\,R$$

$$\therefore \quad Q_{A\to B} = W_{A\to B} + \Delta U$$

$$= (-1200\,R) - (900\,R)$$

$$\mathbf{Q_{A\to B} = -2100\,R \text{ (released)}}$$

Alternate Solution :

In the process PV^x = constant

Molar heat capacity,

$$C = \frac{R}{\gamma - 1} + \frac{R}{1 - x}$$

Here the process is

$$P^2V = \text{constant}$$

$$\text{or} \quad PV^{1/2} = \text{constant}$$

$$i.e. \quad x = \frac{1}{2}$$

$$\therefore \quad C = \frac{R}{(5/3) - 1} + \frac{R}{1 - 1/2}$$

$$C = 3.5\,R$$

$$\therefore \quad Q_{A\to B} = nC\Delta T$$

$$= (2)(3.5\,R)(300 - 600)$$

$$\text{or} \quad Q_{A\to B} = -2100\,R$$

Process $B \to C$: Process is isobaric

$$\therefore \quad Q_{B\to C} = nC_P\Delta T$$

$$= (2)\left(\frac{5}{2}R\right)(T_C - T_B)$$

$$= 2\left(\frac{5}{2}R\right)(2T_1 - T_1)$$

$$= (5R)(600 - 300)$$

$$\mathbf{Q_{B\to C} = 1500\,R} \qquad \text{(absorbed)}$$

Process $C \to A$: Process is isothermal

$$\therefore \quad \Delta U = 0$$

$$\text{and} \quad Q_{C\to A} = W_{C\to A} = nRT_c \ln\left(\frac{P_C}{P_A}\right)$$

$$= nR(2T_1)\ln\left(\frac{2P_1}{P_1}\right)$$

$$= (2)(R)(600)\ln(2)$$

$$\mathbf{Q_{C\to A} = 831.6\,R} \text{ (absorbed)}$$

In first law of thermodynamics, ($dQ = dU + dW$) we come across three terms dQ, dU and dW.

$dU = nC_VdT$ for all the processes whether it is isobaric, isochoric or else,

and $dQ = nCdT$ where $C = \frac{R}{\gamma - 1} + \frac{R}{1 - x}$ in the process PV^x = constant.

In both the terms we require $dT\ (= T_f - T_i)$ only. The third terms dW is obviously $dQ - dU$. Therefore, if in any process change in temperature (dT) and $P - V$ relation is known, then the above method is the simplest one. Note that even if we have V-T or T-P relation, it can be converted into PV relation by the equation $PV = nRT$.

336. (a) The *P-V* diagram for the complete process will be as shown :

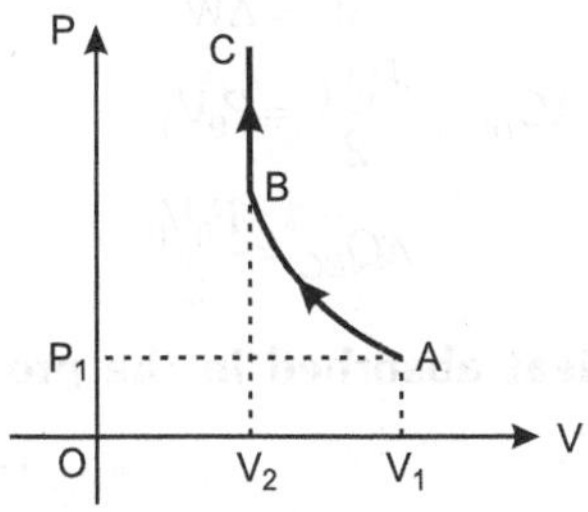

Process $A \to B$ is adiabatic compression, and Process $B \to C$ is isochoric.

(b) **(i) Total work done by the gas**

Process A-B

$$W_{AB} = \frac{P_A V_A - P_B V_B}{\gamma - 1} \quad \left[W_{\text{adiabatic}} = \frac{P_i V_i - P_f V_f}{\gamma - 1}\right]$$

$$= \frac{P_1 V_1 - P_2 V_2}{(5/3) - 1}$$

$\gamma = 5/3$ for monoatomic gas

$$= \frac{P_1 V_1 - P_1 \left(\frac{V_1}{V_2}\right)^{\gamma} V_2}{2/3} \quad \left[\begin{array}{l} P_1 V_1^{\gamma} = P_2 V_2^{\gamma} \\ \therefore P_2 = P_1 \left(\frac{V_1}{V_2}\right)^{\gamma} \end{array}\right]$$

$$= \frac{3}{2} P_1 V_1 \left[1 - \left(\frac{V_1}{V_2}\right)^{\gamma - 1}\right]$$

$$= -\frac{3}{2} P_1 V_1 \left[\left(\frac{V_1}{V_2}\right)^{\frac{5}{3} - 1} - 1\right]$$

$$= -\frac{3}{2} P_1 V_1 \left[\left(\frac{V_1}{V_2}\right)^{2/3} - 1\right]$$

Process B-C

$$W_{BC} = 0 \qquad (V = \text{constant})$$

$$\therefore \mathbf{W_{Total}} = W_{AB} + W_{BC}$$

$$= -\frac{3}{2} P_1 V_1 \left[\left(\frac{V_1}{V_2}\right)^{2/3} - 1\right]$$

(ii) Total change in internal energy

Process A-B

$$Q_{AB} = 0 \qquad \text{(Process is adiabatic)}$$

$$\therefore \Delta U_{AB} = -W_{AB} = \frac{3}{2} P_1 V_1 \left[\left(\frac{V_1}{V_2}\right)^{2/3} - 1\right]$$

Process B-C

$$W_{BC} = 0$$

$$\therefore \Delta U_{BC} = Q_{BC} = Q \qquad \text{(Given)}$$

$$\therefore \Delta \boldsymbol{U}_{\textbf{Total}} = \Delta U_{AB} + \Delta U_{BC}$$

$$= \frac{3}{2} P_1 V_1 \left[\left(\frac{V_1}{V_2}\right)^{2/3} - 1\right] + Q$$

(iii) Final temperature of the gas

$$\Delta U_{\text{Total}} = nC_V \Delta T = 2\left(\frac{R}{\gamma - 1}\right)(T_C - T_A)$$

$$\therefore \frac{3}{2} P_1 V_1 \left[\left(\frac{V_1}{V_2}\right)^{2/3} - 1\right] + Q$$

$$= \frac{2R}{(5/3) - 1}\left(T_C - \frac{P_A V_A}{2R}\right)$$

$$\text{or } \frac{3}{2} P_1 V_1 \left[\left(\frac{V_1}{V_2}\right)^{2/3} - 1\right] + Q$$

$$= 3R\left(T_C - \frac{P_1 V_1}{2R}\right)$$

$$\therefore \boldsymbol{T_C} = \frac{\boldsymbol{Q}}{\boldsymbol{3R}} + \frac{\boldsymbol{P_1 V_1}}{\boldsymbol{2R}} \left(\frac{\boldsymbol{V_1}}{\boldsymbol{V_2}}\right)^{\mathbf{2/3}} = \boldsymbol{T}_{\text{final}}$$

337. (a) *ABCA* is a clockwise cyclic process.

$\therefore$ Work done by the gas

$$W = + \text{Area of triangle } ABC$$

$$= \frac{1}{2} \text{(base) (height)}$$

$$= \frac{1}{2}(2V_0 - V_0)(3P_0 - P_0)$$

$$\boldsymbol{W = P_0V_0}$$

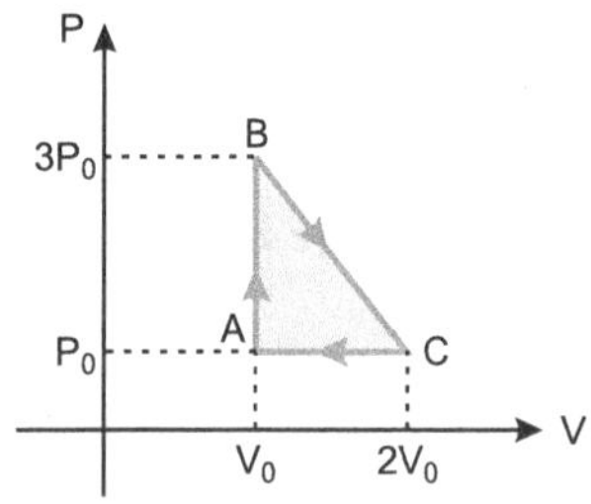

(b) Number of moles, $n = 1$ and gas is monoatomic, therefore,

$$C_V = \left(\frac{3}{2}\right)R \text{ and } C_P = \left(\frac{5R}{2}\right)$$

$$\Rightarrow \quad \frac{C_V}{R} = \frac{3}{2} \text{ and } \frac{C_P}{R} = \frac{5}{2}$$

(i) Heat rejected in path CA : (process is isobaric)

$$\therefore \Delta Q_{CA} = C_P dT = C_P(T_f - T_i)$$

$$= C_P\left(\frac{P_fV_f}{R} - \frac{P_iV_i}{R}\right)$$

$$= \frac{C_P}{R}(P_fV_f - P_iV_i)$$

Substituting the values

$$\Delta Q_{CA} = \frac{5}{2}(P_0V_0 - 2P_0V_0) = -\frac{5}{2}P_0V_0$$

Therefore, heat rejected in the process CA is $\boldsymbol{\frac{5}{2}P_0V_0}$

(ii) Heat absorbed in path AB : (process is isochoric)

$$\therefore \quad \Delta Q_{AB} = C_V\Delta T = C_V(T_f - T_i)$$

$$= C_V\left(\frac{P_fV_f}{R} - \frac{P_iV_i}{R}\right)$$

$$= \frac{C_V}{R}(P_fV_f - P_iV_i)$$

$$= \frac{3}{2}(P_fV_f - P_iV_i)$$

$$= \frac{3}{2}(3P_0V_0 - P_0V_0)$$

$$\Rightarrow \quad \Delta Q_{AB} = 3P_0V_0$$

$\therefore$ **Heat absorbed in the process *AB* is $\boldsymbol{3P_0V_0}$**

(c) Let ΔQ_{BC} be the heat absorbed in the process BC :

Total heat absorbed,

$$\Delta Q = \Delta Q_{CA} + \Delta Q_{AB} + \Delta Q_{BC}$$

$$\Delta Q = \left(-\frac{5}{2}P_0V_0\right) + (3P_0V_0) + \Delta Q_{BC}$$

$$\Delta Q = \Delta Q_{BC} + \frac{P_0V_0}{2}$$

Change in internal energy, $\Delta U = 0$

$$\therefore \quad \Delta Q = \Delta W$$

$$\therefore \quad \Delta Q_{BC} + \frac{P_0V_0}{2} = P_0V_0$$

$$\Rightarrow \quad \Delta Q_{BC} = \frac{P_0V_0}{2}$$

$\therefore$ **Heat absorbed in the process *BC* is $\boldsymbol{\frac{P_0V_0}{2}}$.**

(d) Maximum temperature of the gas will be somewhere between B and C. Line BC is a straight line. Therefore, P-V equation for the process BC can be written as :

$$P = -mV + C \qquad (y = mx + c)$$

Here $m = \frac{2P_0}{V_0}$ and $C = 5P_0$

$$\therefore \quad P = -\left(\frac{2P_0}{V_0}\right)V + 5P_0$$

Multiplying the equation by V,

$$PV = -\left(\frac{2P_0}{V_0}\right).V^2 + 5P_0V$$

($PV = RT$ for $n = 1$)

$$\therefore \quad RT = -\left(\frac{2P_0}{V_0}\right)V^2 + 5P_0V$$

$$\text{or} \quad T = \frac{1}{R}\left[5P_0V - \frac{2P_0}{V_0}.V^2\right] \qquad \ldots(1)$$

For T to be maximum,

$$\frac{dT}{dV} = 0$$

$$\Rightarrow \quad 5P_0 - \frac{4P_0}{V_0}.V = 0 \Rightarrow V = \frac{5V_0}{4}$$

i.e. at $V = \frac{5V_0}{4}$ (on line BC), temperature of the gas is maximum. From equation (1), this maximum temperature will be

$$T_{max} = \frac{1}{R}\left[5P_0\left(\frac{5V_0}{4}\right) - \frac{2P_0}{V_0}\left(\frac{5V_0}{4}\right)^2\right]$$

$$\boldsymbol{T_{max} = \frac{25}{8}\frac{P_0V_0}{R}}$$

338. In the first part of the question ($t \le t_1$)
At $t = 0; T_X = T_0 = 400$ K and at $t = t_1$;

$$T_X = T_1 = 350 \text{ K}$$

Temperature of atmosphere,

$T_A = 300$ K (constant).

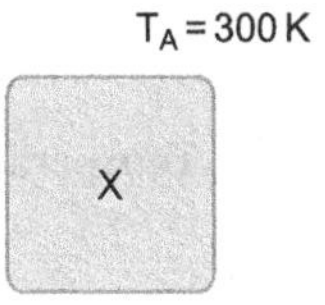

This cools down according to Newton's law of cooling. Therefore, rate of cooling $\propto$ temperature difference.

$$\therefore \quad \left(-\frac{dT}{dt}\right) = k\,(T - T_A)$$

$$\Rightarrow \quad \frac{dT}{T - T_A} = -k.dt$$

$$\Rightarrow \quad \int_{T_0}^{T_1}\frac{dT}{T - T_A} = -k\int_0^{t_1} dt$$

$$\Rightarrow \quad ln\left(\frac{T_1 - T_A}{T_0 - T_A}\right) = -kt_1$$

$$\Rightarrow \quad kt_1 = -ln\left(\frac{350 - 300}{400 - 300}\right)$$

$$\Rightarrow \quad kt_1 = ln\,(2) \qquad ...(1)$$

In the second part, body X cools by radiation (according to Newton's law) as well as by conduction ($t > t_1$).

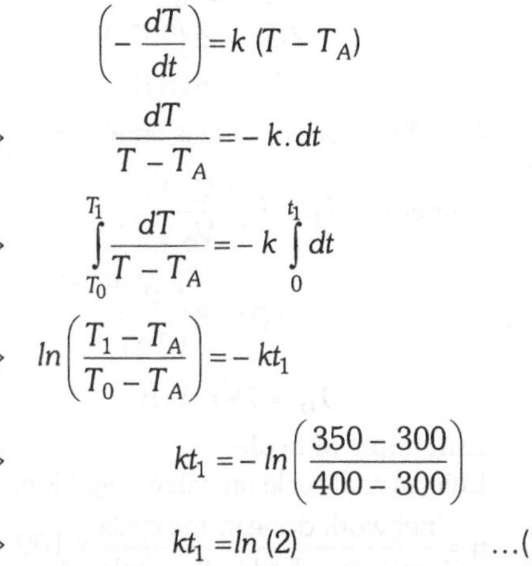

Therefore, Rate of cooling
= (cooling by radiation)
+ (cooling by conduction)

$$\therefore \left(-\frac{dT}{dt}\right) = k\,(T - T_A) + \frac{KA}{CL}(T - T_A) \qquad ...(2)$$

In conduction,

$$\frac{dQ}{dt} = \frac{KA\,(T - T_A)}{L} = C\left(-\frac{dT}{dt}\right)$$

$$\therefore \quad \left(-\frac{dT}{dt}\right) = \frac{KA}{LC}(T - T_A)$$

where C = heat capacity of body X

$$\therefore \quad \left(-\frac{dT}{dt}\right) = \left(k + \frac{KA}{CL}\right)(T - T_A) \qquad ...(3)$$

Let at $t = 3t_1$, temperature of X becomes T_2
Therefore, from equation (3) we have

$$\int_{T_1}^{T_2}\frac{dT}{T - T_A} = -\left(k + \frac{KA}{LC}\right)\int_{t_1}^{3t_1} dt$$

$$\text{or} \quad ln\left(\frac{T_2 - T_A}{T_1 - T_A}\right) = -\left(k + \frac{KA}{LC}\right)(2t_1)$$

$$= -\left(2kt_1 + \frac{2KA}{LC}t_1\right)$$

$$\text{or} \quad ln\left(\frac{T_2 - 300}{350 - 300}\right) = -2\,ln\,(2) - \frac{2KAt_1}{LC}$$

$kt_1 = ln\,(2)$ from equation (1)

$$\therefore \quad \boldsymbol{T_2 = \left(300 + 12.5e^{\frac{-2KAt_1}{CL}}\right) \textbf{kelvin}}$$

339. Number of gram moles of He,

$$n = \frac{m}{M} = \frac{2 \times 10^3}{4} = 500$$

(i) $V_A = 10 \text{ m}^3 \quad P_A = 5 \times 10^4 \text{ N/m}^2$

$$\therefore \quad T_A = \frac{P_AV_A}{nR} = \frac{(10)\,(5 \times 10^4)}{(500)\,(8.31)} \text{ K}$$

or $\boldsymbol{T_A = 120.34}$ **K**

Similarly, $V_B = 10 \text{ m}^3, P_B = 10 \times 10^4 \text{ N/m}^2$

$$\therefore \quad T_B = \frac{(10)\,(10 \times 10^4)}{(500)\,(8.31)} \text{ K}$$

$\therefore \quad T_B =$ **240.68 K**

$V_C = 20 \text{ m}^3, \quad P_C = 10 \times 10^4 \text{ N/m}^2$

$$\therefore \quad T_C = \frac{(20)\,(10 \times 10^4)}{(500)\,(8.31)} \text{ K}$$

$$T_C = \mathbf{481.36\ K}$$

and $V_D = 20\ \text{m}^3, \quad P_D = 5 \times 10^4\ \text{N/m}^2$

$$\therefore \quad T_D = \frac{(20)(5 \times 10^4)}{(500)(8.31)}\ \text{K}$$

$$T_D = \mathbf{240.68\ K}$$

(ii) No, it is not possible to tell afterwards which sample went through the process *ABC* or *ADC*. But during the process if we note down the work done in both the processes, then the process which require more work goes through process *ABC*.

(iii) In the process *ABC*

$$\Delta U = nC_V \Delta T = n\left(\frac{3}{2}R\right)(T_C - T_A)$$

$$= (500)(3/2)(8.31)(481.36 - 120.34)\ \text{J}$$

$$\Delta U = 2.25 \times 10^6\ \text{J}$$

and ΔW = Area under

$BC = (20 - 10)(10) \times 10^4\ \text{J} = 10^6\ \text{J}$

$$\therefore \Delta Q_{ABC} = \Delta U + \Delta W = (2.25 \times 10^6 + 10^6)\ \text{J}$$

$$\Delta Q_{ABC} = \mathbf{3.25 \times 10^6\ J}$$

In the process *ADC*
ΔU will be same (because it depends on initial and final temperatures only).

ΔW = Area under *AD*

$$= (20 - 10)(5 \times 10^4)\ \text{J} = 0.5 \times 10^6\ \text{J}$$

$$\therefore \quad \Delta Q_{ADC} = \Delta U + \Delta W$$

$$= (2.25 \times 10^6 + 0.5 \times 10^6)\ \text{J}$$

$$\Delta Q_{ADC} = \mathbf{2.75 \times 10^6\ J}$$

340. The corresponding *P-V* diagram is as shown :

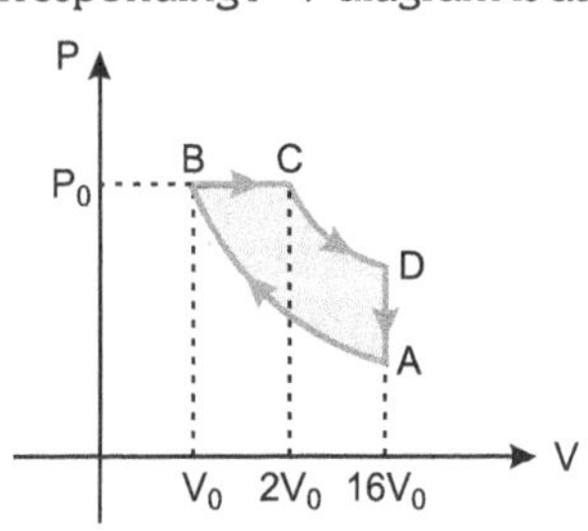

Given : $T_A = 300\ \text{K}$,

$n = 1, \gamma = 1.4, V_A/V_B = 16$

and $V_C/V_B = 2$

Let $V_B = V_0$

and $P_B = P_0$

Then $V_C = 2V_0$

and $V_A = 16\ V_0$

Temperature at B

Process A-B is adiabatic. Hence

$$T_A V_A^{\gamma - 1} = T_B V_B^{\gamma - 1}$$

or

$$T_B = T_A \left(\frac{V_A}{V_B}\right)^{\gamma - 1}$$

$$= (300)(16)^{1.4 - 1}$$

$$T_B = \mathbf{909\ K}$$

Temperature at D

$B \to C$ is an isobaric process (P = constant)

$$\therefore \quad T \propto V$$

$$V_C = 2V_B$$

$$\therefore \quad T_C = 2T_B = (2)(909)\ \text{K}$$

$$T_C = 1818\ \text{K}$$

Now the process *C-D* is adiabatic.

Therefore, $T_D = T_C \left(\frac{V_C}{V_D}\right)^{\gamma - 1}$

$$= (1818)\left(\frac{2}{16}\right)^{0.4 - 1}$$

$$T_D = \mathbf{791.4\ \ K}$$

Efficiency of cycle

Efficiency of cycle (in percentage) is defined as

$$\eta = \frac{\text{net work done in the cycle}}{\text{heat absorbed in the cycle}} \times 100$$

or

$$\eta = \frac{W_{\text{Total}}}{Q_{+\text{ve}}} \times 100$$

$$= \frac{Q_{+\text{ve}} - Q_{-\text{ve}}}{Q_{+\text{ve}}} \times 100$$

$$= \left(1 - \frac{Q_1}{Q_2}\right) \times 100 \qquad \ldots(1)$$

where Q_1 = Negative heat in the cycle (heat released)

and Q_2 = Positive heat in the cycle (heat absorbed)

In the cycle

$$Q_{AB} = Q_{CD} = 0 \qquad \text{(Adiabatic process)}$$

$$Q_{DA} = n\, C_V \Delta T = (1)\left(\frac{5}{2}R\right)(T_A - T_D)$$

$(C_V = \frac{5}{2}R$ for a diatomic gas)

$$= \frac{5}{2} \times 8.31\,(300 - 791.4)\text{ J}$$

$$= -\,10208.8\text{ J}$$

and $Q_{BC} = nC_P\Delta T$

$$= (1)\left(\frac{7}{2}R\right)(T_C - T_B)$$

$(C_P = \frac{7}{2}R$ for a diatomic gas)

$$= \left(\frac{7}{2}\right)(8.31)(1818 - 909)\text{ J}$$

or $\quad Q_{BC} = 26438.3$ J

Therefore, substituting $Q_1 = 10208.8$ J and $Q_2 = 26438.3$ J in equation (1), we get

$$\therefore \quad \eta = \left(1 - \frac{10208.8}{26438.3}\right) \times 100$$

or $\quad \eta = \mathbf{61.4\%}$

341. Density of a liquid varies with temperature as

$$\rho_{t°C} = \left(\frac{\rho_{0°C}}{1 + \gamma t}\right)$$

Here γ is the coefficient of volume expansion of temperature.

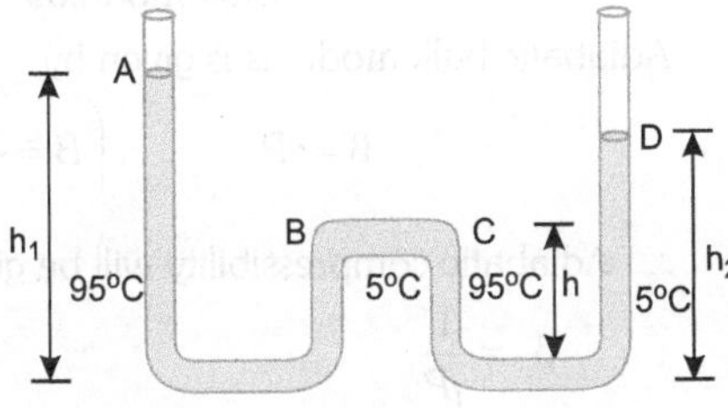

In the figure

$$h_1 = 52.8\text{ cm},$$

$$h_2 = 51\text{ cm} \quad \text{and} \quad h = 49\text{ cm}$$

Now pressure at B = pressure at C.

Therefore, $P_0 + h_1\rho_{95°}g - h\,\rho_{5°}g$

$$= P_0 + h_2\rho_{5°}g - h\rho_{95°}g$$

$$\Rightarrow \quad \rho_{95°}(h_1 + h) = \rho_{5°}(h_2 + h)$$

$$\Rightarrow \quad \frac{\rho_{95°}}{\rho_{5°}} = \frac{h_2 + h}{h_1 + h}$$

$$\Rightarrow \quad \frac{\dfrac{\rho_{0°}}{(1 + 95\gamma)}}{\dfrac{\rho_{0°}}{(1 + 5\gamma)}} = \frac{h_2 + h}{h_1 + h}$$

$$\Rightarrow \quad \frac{1 + 5\gamma}{1 + 95\gamma} = \frac{51 + 49}{52.8 + 49} = \frac{100}{101.8}$$

Solving this equation, we get

$$\gamma = 2 \times 10^{-4}\text{ per °C}$$

$\therefore$ Coefficient of linear expansion of temperature,

$$\alpha = \frac{\gamma}{3} = \mathbf{6.7 \times 10^{-5}\ per\ °C}$$

342. Given $T_1 = 27°\text{C} = 300\text{ K}, V_1 = V, V_2 = 2V$

(i) Final temperature

In adiabatic process

$$TV^{\gamma - 1} = \text{constant}$$

$$\therefore \quad T_1V_1^{\gamma - 1} = T_2V_2^{\gamma - 1}$$

or

$$T_2 = T_1\left(\frac{V_1}{V_2}\right)^{\gamma - 1}$$

$$= 300\left(\frac{V}{2V}\right)^{5/3 - 1}$$

$(\gamma = \frac{5}{3}$ for a monoatomic gas)

$$\mathbf{T_2 \approx 189\ K}$$

(ii) Change in internal energy :

$$\Delta U = nC_V\Delta T \qquad n = 2\text{ (Given)}$$

$$= (2)\left(\frac{3}{2}R\right)(T_2 - T_1)$$

$$C_V = \frac{3}{2}R \text{ for a monoatomic gas}$$

$$= 2\left(\frac{3}{2}\right)(8.31)\,(189 - 300)\text{ J}$$

$$\mathbf{\Delta U = -\ 2767\ J}$$

(iii) Work done

Process is adiabatic, therefore, $\Delta Q = 0$

and from first law of thermodynamics,

$$\Delta Q = \Delta W + \Delta U$$

$$\Delta W = -\Delta U = -(-2767\ \text{J})$$

or $\Delta \boldsymbol{W}$ = **2767 J**

343. (a) Number of moles of gas A are $n_A = 1$ (Given)

Let the number of moles of gas B be $n_B = n$

The internal energy of the mixture = internal energy of gas A + internal energy of gas B.

and since $U = \dfrac{nRT}{\gamma - 1}$, therefore,

$$(n_A + n_B)\frac{R}{\gamma_{\text{mixture}} - 1}T = n_A \frac{R}{\gamma_A - 1}T + n_B \frac{R}{\gamma_B - 1}T \qquad \ldots(1)$$

Since the mixture obeys the law

$PV^{19/13}$ = constant (in adiabatic process)

Therefore,

$$\gamma_{\text{mixture}} = 19/13 \qquad (PV^{\gamma} = \text{constant})$$

Substituting the values in equation (1), we have

$$\frac{(1+n)}{(19/13) - 1} = \frac{1}{(5/3) - 1} + \frac{n}{(7/5) - 1}$$

Solving this, we get $\boldsymbol{n}$ = **2**

Note: For γ_{mixture} we can directly use the formula :

$$\frac{n}{\gamma_{\text{mixture}} - 1} = \frac{n_1}{\gamma_1 - 1} + \frac{n_2}{\gamma_2 - 1}$$

(b) Molecular weight of the mixture will be given by

$$M = \frac{n_A M_A + n_B M_B}{n_A + n_B} = \frac{(1)(4) + 2(32)}{1 + 2} = 22.67$$

Speed of sound in a gas is given by

$$v = \sqrt{\frac{\gamma RT}{M}}$$

Therefore, in the mixture of the gas

$$v = \sqrt{\frac{(19/13)(8.31)(300)}{(22.67 \times 10^{-3})}}\ \text{m/s}$$

or $\boldsymbol{v}$ = **401 m/s**

Note: It is a common mistake that the students generally put $M = 22.67$ in $v = \sqrt{\dfrac{\gamma RT}{M}}$.

Note that if you are writing the value of $R = 8.31$ J/mole-K, substitute

$$M = 22.67 \times 10^{-3}\ \text{kg}.$$

(c) $v \propto \sqrt{T}$

or $\quad v = KT^{1/2} \qquad \ldots(2)$

$$\Rightarrow \quad \frac{dv}{dT} = \frac{1}{2}KT^{-1/2}$$

$$\Rightarrow \quad dv = K\left(\frac{dT}{2\sqrt{T}}\right)$$

$$\Rightarrow \quad \frac{dv}{v} = \frac{K}{v}\cdot\left(\frac{dT}{2\sqrt{T}}\right)$$

$$\frac{K}{v} = \frac{1}{\sqrt{T}} \qquad \text{(from 2)}$$

$$\Rightarrow \quad \frac{dv}{v} = \frac{1}{\sqrt{T}}\left(\frac{dT}{2\sqrt{T}}\right) = \frac{1}{2}\left(\frac{dT}{T}\right)$$

$$\Rightarrow \quad \frac{dv}{v} \times 100 = \frac{1}{2}\left(\frac{dT}{T}\right) \times 100$$

$$= \frac{1}{2}\left(\frac{1}{300}\right) \times 100 = 0.167$$

Therefore, percentage change in speed is 0.167%.

(d) Compressibility $= \dfrac{1}{\text{Bulk modulus}} = \beta$ (say)

Adiabatic bulk modulus is given by

$$\beta = \gamma P \qquad \left(B = -\frac{dP}{dV/V}\right)$$

$\therefore$ Adiabatic compressibility will be given by

$$\beta = \frac{1}{\gamma P}$$

and $\beta' = \dfrac{1}{\gamma P'} = \dfrac{1}{\gamma P(5)^{\gamma}} \qquad [PV^{\gamma} = \text{constant}]$

because $PV^{\gamma} = P'(V/5)^{\gamma} \Rightarrow P' = P(5)^{\gamma}$

$$\therefore \quad \Delta\beta = \beta - \beta' = \frac{1}{\gamma P}\left[1 - \left(\frac{1}{5}\right)^{\gamma}\right]$$

$$= \frac{V}{\gamma (n_A + n_B) RT}\left[1-\left(\frac{1}{5}\right)^{\gamma}\right] \quad \left(P = \frac{nRT}{V}\right)$$

$$= \frac{V}{\left(\frac{19}{13}\right)(1+2)(8.31)(300)}\left[1-\left(\frac{1}{5}\right)^{19/13}\right]$$

$$\left(\gamma = \gamma_{\text{mixture}} = \frac{19}{13}\right)$$

$$\Delta\beta = \mathbf{8.27 \times 10^{-5}\ V}$$

344. (i) In a cyclic process $\Delta U = 0$

Therefore, $\Delta Q = \Delta W$

or $Q_1 + Q_2 + Q_3 + Q_4$

$= W_1 + W_2 + W_3 + W_4$

Hence $W_4 = (Q_1 + Q_2 + Q_3 + Q_4) - (W_1 + W_2 + W_3)$

$= \{(5960 - 5585 - 2980 + 3645) - (2200 - 825 - 1100)\}$

or $W_4 = \mathbf{765\ J}$

(ii) Efficiency,

$$\eta = \frac{\text{Total work done in the cycle}}{\text{Heat absorbed (positive heat) by the gas during the cycle}} \times 100$$

$$= \left(\frac{W_1 + W_2 + W_3 + W_4}{Q_1 + Q_4}\right) \times 100$$

$$= \left\{\frac{2200 - 825 - 1100 + 765}{5960 + 3645}\right\} \times 100$$

$$= \frac{1040}{9605} \times 100$$

$$\eta = \mathbf{10.82\%}$$

Note : From energy conservation :

$$W_{net} = Q_{+ve} - Q_{-ve} \quad \text{(in a cycle)}$$

$$\therefore \quad \eta = \frac{W_{net}}{Q_{+ve}} \times 100 = \frac{(Q_{+ve} - Q_{-ve})}{Q_{+ve}} \times 100$$

$$= \left(1 - \frac{Q_{-ve}}{Q_{+ve}}\right) \times 100$$

In the above question

$Q_{-ve} = |Q_2| + |Q_3|$

$= (5585 + 2980)\ \text{J} = 8565\ \text{J}$

and $Q_{+ve} = Q_1 + Q_4$

$= (5960 + 3645)\ \text{J} = 9605\ \text{J}$

$$\therefore \quad \eta = \left(1 - \frac{8565}{9605}\right) \times 100$$

$$\eta = \mathbf{10.82\%}$$

345. Given temperature of the mixture,

$$T = 27°\text{C} = 300\ \text{K}$$

Let m be the mass of the neon gas in the mixture. Then mass of argon would be $(28 - m)$

Number of gram moles of neon, $n_1 = \frac{m}{20}$

Number of gram moles of Argon,

$$n_2 = \frac{(28 - m)}{40}$$

From Dalton's law of partial pressures, Total pressure of the mixture (P) = Pressure due to Neon (P_1) + Pressure due to Argon (P_2)

or Substituting the values

$$1.0 \times 10^5 = \left(\frac{m}{20} + \frac{20 - m}{40}\right)\frac{(8.314)(300)}{0.02}$$

Solving this equation, we get

$m = 4.074$ g and $28 - m = 23.926$ g

Therefore, in the mixture, 4.074 g Neon is present and the rest *i.e.* 23.926 g argon is present.

346. Given $T_A = 1000$ K,

$$P_B = (2/3)P_A, \quad P_C = \frac{1}{3}P_A$$

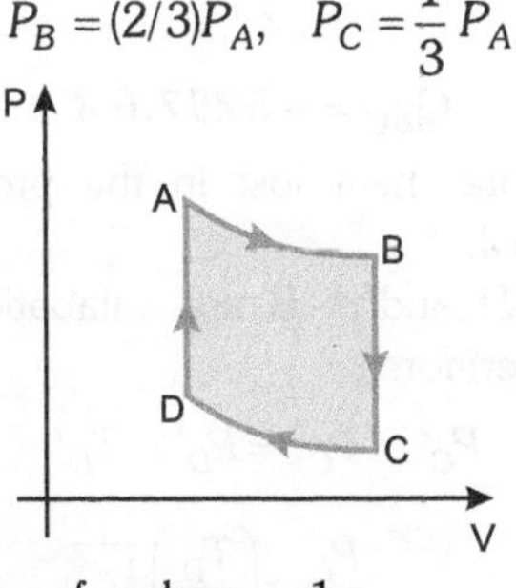

Number of moles, $n = 1$

$$\gamma = \frac{C_P}{C_V} = \frac{5}{3}R \quad \text{(monoatomic)}$$

(i) $A \rightarrow B$ is an adiabatic process, therefore,

$$P_A^{1-\gamma} T_A^{\gamma} = P_B^{1-\gamma} T_B^{\gamma}$$

$$\therefore T_B = T_A\left(\frac{P_A}{P_B}\right)^{\frac{1-\gamma}{\gamma}} = (1000)\left(\frac{3}{2}\right)^{\frac{1-5/3}{5/3}}$$

$$= (1000)\left(\frac{3}{2}\right)^{-2/5} = (1000)\left(\frac{2}{3}\right)^{2/5}$$

$$T_B = (1000)\,(0.85)$$

$$\therefore T_B = 850\text{ K} \qquad \left[\because \text{ Given } \left(\frac{2}{3}\right)^{2/3} = 0.85\right]$$

Now work done in the process $A–B$ will be

$$W_{AB} = \frac{R}{1-\gamma}(T_B - T_A)$$

$$= \frac{8.31}{1-5/3}(850 - 1000)$$

or $\boldsymbol{W_{AB} = 1869.75}$ **J**

(ii) $B–C$ is an isochoric process (V = constant)

$$\therefore \quad \frac{T_B}{T_C} = \frac{P_B}{P_C}$$

$$\therefore \quad T_C = \left(\frac{P_C}{P_B}\right)T_B = \left(\frac{1/3)\,P_A}{(2/3)\,P_A}\right)850\text{ K}$$

$$T_C = 425\text{ K}$$

Therefore,

$$Q_{BC} = nC_V\Delta T = (1)\left(\frac{3}{2}R\right)(T_C - T_B)$$

$$= \left(\frac{3}{2}\right)(8.31)(425 - 850)$$

$$\boldsymbol{Q_{BC} = -5297.6\text{ J}}$$

Therefore, heat lost in the process BC is 5297.6 J.

(iii) $C–D$ and $A–B$ are adiabatic processes. Therefore,

$$P_C^{1-\gamma} T_C^{\gamma} = P_D^{1-\gamma} T_D^{\gamma}$$

$$\Rightarrow \quad \frac{P_C}{P_D} = \left(\frac{T_D}{T_C}\right)^{\frac{\gamma}{1-\gamma}} \qquad \ldots(1)$$

$$P_A^{1-\gamma} T_A^{\gamma} = P_B^{1-\gamma} T_B^{\gamma}$$

$$\Rightarrow \quad \frac{P_A}{P_B} = \left(\frac{T_B}{T_A}\right)^{\frac{\gamma}{1-\gamma}} \qquad \ldots(2)$$

Multiplying (1) and (2), we get

$$\frac{P_C P_A}{P_D P_B} = \left(\frac{T_D T_B}{T_C T_A}\right)^{\frac{\gamma}{1-\gamma}} \qquad \ldots(3)$$

Processes $B–C$ and $D–A$ are isochoric (V = constant)

Therefore,

$$\frac{P_C}{P_B} = \frac{P_C}{T_B} \quad \text{and} \quad \frac{P_A}{P_D} = \frac{T_A}{T_D}$$

Multiplying these two equations, we get

$$\frac{P_C P_A}{P_D P_B} = \frac{T_C T_A}{T_B T_D} \qquad \ldots(4)$$

From (3) and (4), we have

$$\left(\frac{T_D T_B}{T_C T_A}\right)^{\frac{\gamma}{1-\gamma}} = \frac{T_C T_A}{T_B T_D}$$

or $$\left(\frac{T_C T_A}{T_D T_B}\right)^{\frac{1-\gamma}{\gamma}} = \left(\frac{T_C T_A}{T_B T_D}\right)$$

$$\Rightarrow \quad \frac{T_C T_A}{T_D T_B} = 1$$

or $$T_D = \frac{T_C T_A}{T_B} = \frac{(425)\,(1000)}{850}\text{ K}$$

or $\boldsymbol{T_D = 500}$ **K**

347. Given number of moles, $n = 2$

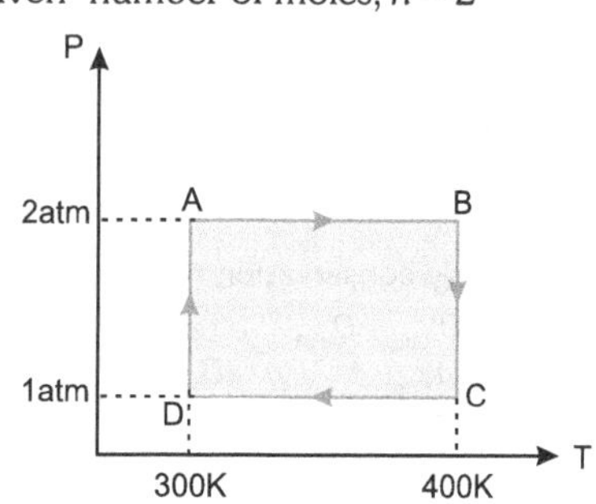

Process AB and CD are isobaric.

Hence $\quad Q_{AB} = -Q_{CD}$

because $\quad (\Delta T)_{AB} = +100\text{ K}$

whereas $\quad (\Delta T)_{CD} = -100\text{ K}$

and $$Q_{isobaric} = nC_P\Delta T$$

or $$Q_{AB} + Q_{CD} = 0$$

Process BC is isothermal ($\Delta U = 0$)

$$\therefore \quad Q_{BC} = W_{BC} = nRT_B \ ln\left(\frac{P_B}{P_C}\right)$$

$$= (2)\,(8.31)\,(400)\ ln\left(\frac{2}{1}\right)$$

or $$Q_{BC} = 4608 \text{ J}$$

Similarly, process DA is also isothermal hence

$$Q_{DA} = W_{DA} = nRT_D \ ln\left(\frac{P_D}{P_A}\right)$$

or $$Q_{DA} = -\ 3456 \text{ J}$$

(a) Net heat in the process

$$Q = Q_{AB} + Q_{BC} + Q_{CD} + Q_{DA}$$
$$= (4608 - 3456) \text{ J}$$

$\mathbf{Q = 1152 \ J}$

(b) From first law of thermodynamics,

$$\Delta U = 0 \quad \text{(in complete cycle)}$$

$$\therefore \quad \Delta Q = \Delta W$$

Hence net work done in the cycle,

$$W = Q = 1152 \text{ J}$$

$\mathbf{W} = \mathbf{1152}$ J

(c) Since $T_i = T_f$, therefore, net change in internal energy,

$\Delta \mathbf{U} = \mathbf{0}$

348. Let at any time temperature of the disc be θ. At this moment rate of heat flow

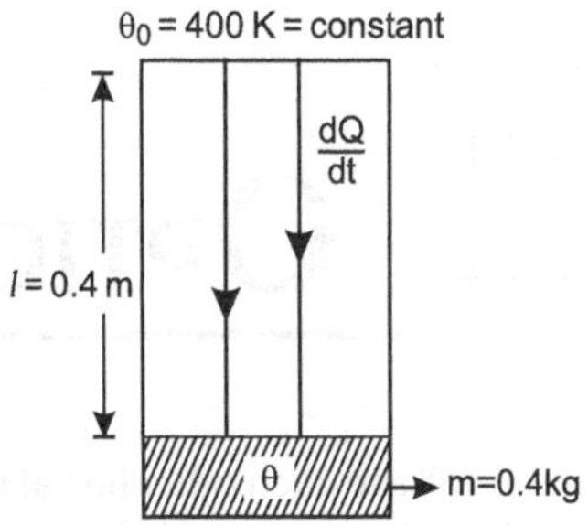

$$\frac{dQ}{dt} = \frac{KA(\Delta\theta)}{l} = \frac{KA}{l}(\theta_0 - \theta) \qquad \ldots(1)$$

This heat is utilised in increasing the temperature of the disc. Hence

$$\frac{dQ}{dt} = ms\,\frac{d\theta}{dt} \qquad \ldots(2)$$

Equating (1) and (2), we have

$$ms.\frac{d\theta}{dt} = \frac{KA}{l}(\theta_0 - \theta)$$

Therefore, $$\frac{d\theta}{\theta_0 - \theta} = \frac{KA}{msl}\,dt$$

or $$\int_{300\text{K}}^{350\text{K}} \frac{d\theta}{\theta_0 - \theta} = \frac{KA}{msl}\int_0^t dt$$

or $$[-\ln(\theta_0 - \theta)]_{300\text{ K}}^{350\text{ K}} = \frac{KA}{msl}\,t$$

$$\therefore \quad t = \frac{msl}{KA}\ln\left(\frac{\theta_0 - 300}{\theta_0 - 350}\right)$$

Substituting the values, we have

$$t = \frac{(0.4)\,(600)\,(0.4)}{(10)\,(0.04)}\ln\left(\frac{400-300}{400-350}\right)$$

$\mathbf{t = 166.32 \ s}$

CHAPTER 11 OPTICS

349. A and B will see each other first at the moment when the extreme ray of light from A strikes B after being reflected from the mirror and conversely, when a ray of light from B strikes A

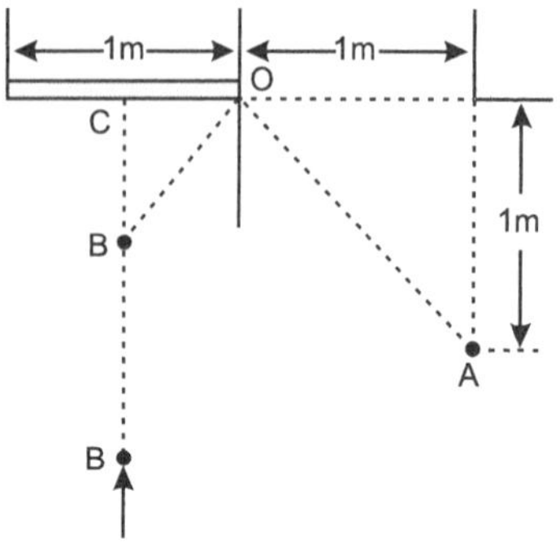

(See figure). Since, the angle of incidence equals the angle of reflection, it is evident that rays BO and AO must strike the mirror at the same angle.

Therefore with the measurements of the mirror given and the given position of A, we must have

$$BC = CO = \frac{1}{2}\text{ m},$$

i.e., A and B will catch sight of each other when B is at a distance of 1/2 m from the mirror.

350. $$\angle PQR = 180 - 2(i - r)$$

$\angle PCR$ (external) $= 180 + 2r$

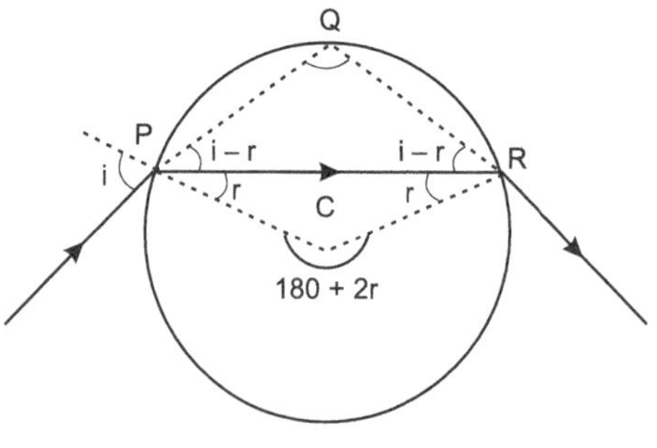

From the property of a circle,

$$\angle PCR = 2(\angle PQR)$$

This gives,

$$2i - r = 90° \qquad \ldots(1)$$

Also $$\mu = \frac{\sin i}{\sin r}$$

$\therefore$ $$\sqrt{3} = \frac{\sin i}{\sin r} \qquad \ldots(2)$$

Solving equations (1) and (2) we get,

$$\angle i = 60°$$

351. Incident ray suffers two reflections and two refractions. In reflection, deviation is $(180 - 2r)$ and in refraction deviation is $(i - r)$. Hence, total deviation,

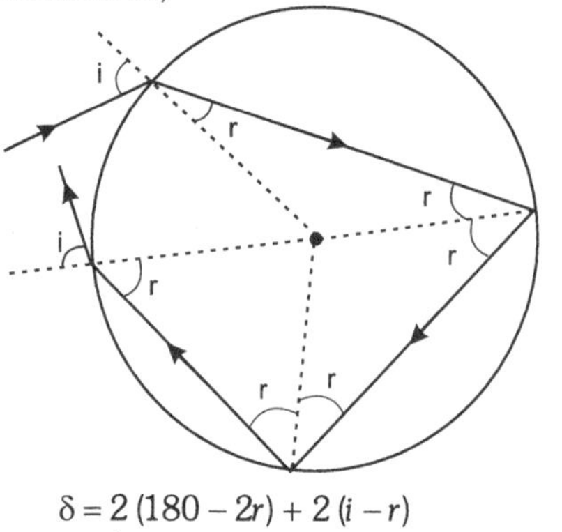

$$\delta = 2(180 - 2r) + 2(i - r)$$
$$= 360 + 2i - 6r$$

Substituting,

$$r = \sin^{-1}\left(\frac{\sin i}{\mu}\right)$$

$$\delta = 360 + 2i - 6\sin^{-1}\left(\frac{\sin i}{\mu}\right) \qquad \ldots(1)$$

For δ to be minimum,

$$\frac{d\delta}{di} = 0$$

Differentiating (1) with respect to i and putting $\frac{d\delta}{di} = 0$ we get the required condition.

352. $CI_1 = CQ - I_1Q = (4 - 3\cot 74°) = 3.14$ cm

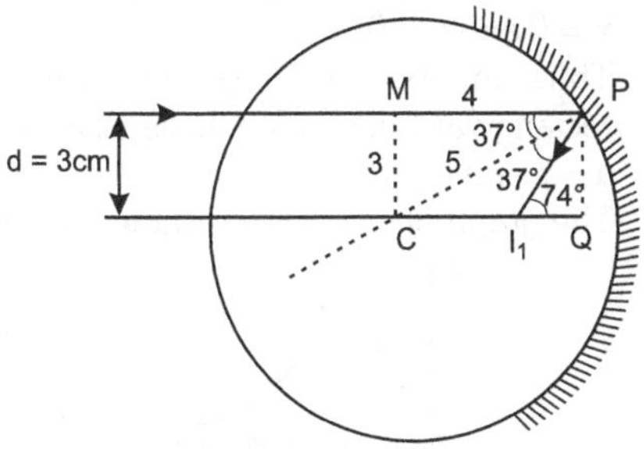

Similarly by making another figure (with $d = 0.5$ cm) we can show that,

$$CI_2 = 2.51 \text{ cm}$$

$$\therefore \quad I_1I_2 = (3.14 - 2.51) \text{ cm}$$

$$= 0.63 \text{ cm}$$

353. Point of intersection of two curves, $y = \frac{x^2}{4}$ and $y = x + 3$ is, (–2, 1)

$$\frac{dy}{dx} = \frac{x}{2}$$

i.e., $$\left(\frac{dy}{dx}\right)_{x=-2} = -1$$

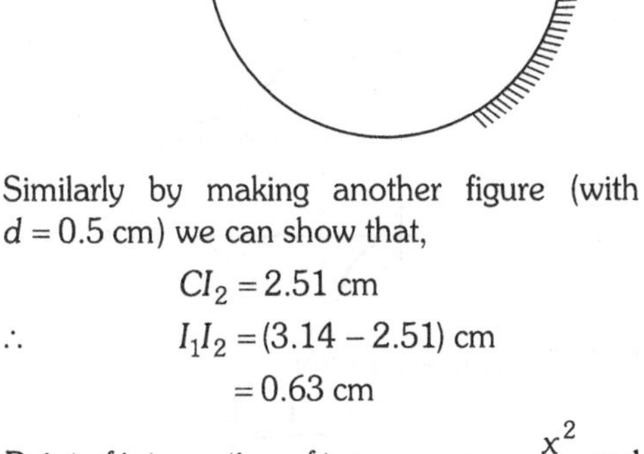

$\tan\beta$ = slope of incident ray $RP = 1$ or $\beta = 45°$ and

$$\tan\alpha = \left|\left(\frac{dy}{dx}\right)_{x=-2}\right| = 1$$

therefore $\alpha = 45°$.

From geometry we can show that

$$\angle i = \angle r = 0°$$

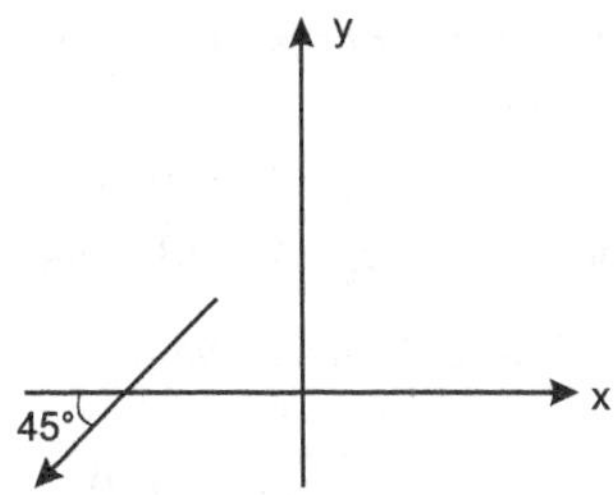

i.e., the ray is incident normally. Hence, the reflected ray will make an angle 45° with x-axis as shown. Hence, the desired unit vector will be,

$$\hat{n} = \frac{1}{\sqrt{2}}(-\hat{i} - \hat{j})$$

354. Let object is at a depth h below the water surface. Then in the first case, for the lens, object will appear at a depth $\frac{h}{\mu_1}$ below the water surface. So applying $\frac{1}{v} - \frac{1}{u} = \frac{1}{f}$ we get,

$$\frac{1}{x} + \frac{\mu_1}{h} = \frac{1}{f} \qquad \ldots(1)$$

In the second case, lens is immersed in water. So, its focal length will change.

$$\frac{1}{f_\omega} = \frac{\left(\frac{\mu_2}{\mu_1} - 1\right)}{(\mu_2 - 1)f}$$

(f_ω = focal length in water)

For the lens, object will appear at the same distance (i.e., h). So applying $\frac{1}{v} - \frac{1}{u} = \frac{1}{f}$.

$$\frac{1}{v} + \frac{1}{h} = \frac{1}{f_\omega} = \frac{\left(\frac{\mu_2}{\mu_1} - 1\right)}{(\mu_2 - 1)f} \qquad \ldots(2)$$

Now refraction will take place at plane surface ($R = \infty$). So applying,

$$\frac{\mu_2}{v} - \frac{\mu_1}{u} = \frac{\mu_2 - \mu_1}{R}$$

we get $$\frac{1}{x'} - \frac{\mu_1}{v} = 0 \qquad \ldots(3)$$

Solving equations (1), (2) and (3) we get the desired result.

355. The velocity of object can be divided into two components.

$$u_x = 2.5 \text{ cm/s}$$

and $$u_y = 2.5\sqrt{3} \text{ cm/s}$$

For image distance, we have,

$$\frac{1}{v} - \frac{1}{50} = \frac{1}{-100}$$

$$\therefore \quad v = 100 \text{ cm}$$

Since, object is between focus and pole image will be virtual and erect. After time dt position of object and image will be as shown in figure.

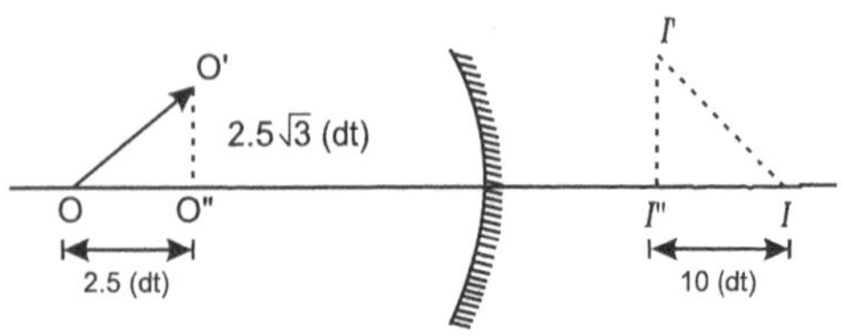

Differentiating $\frac{1}{v} + \frac{1}{u} = \frac{1}{f}$ w.r.to time we get,

$$-v^{-2} \cdot \left(\frac{dv}{dt}\right) - u^{-2}\left(\frac{du}{dt}\right) = 0$$

$\therefore$ Component of v_I along x-direction,

$$v_x = -\left(\frac{v^2}{u^2}\right) \cdot u_x = -10 \text{ cm/s}$$

We can assume $m = 2$ to be constant in the given time interval.
Hence,

$$I'I'' = m\,(O'O'')$$
$$= (2)\,(2.5\sqrt{3})\,dt = 5\sqrt{3}dt$$

$\therefore$ y-component of velocity of image,

$$v_y = \frac{I'I''}{dt} = 5\sqrt{3} \text{ cm/s}$$

$\therefore$ Velocity of image is,

$$\vec{v} = (-10\hat{i} + 5\sqrt{3}\hat{j}) \text{ cm/s}$$

356. Let us take $x = u - f$ and $y = v - f$ substituting in the mirror formula,

$$\frac{1}{v} + \frac{1}{u} = \frac{1}{f}$$

or $$\frac{1}{y+f} + \frac{1}{x+f} = \frac{1}{f}$$

On simplifying we get,

$$xy = f^2$$

This is the equation of a rectangular hyperbola whose asymptotes are the co-ordinate axes, $x = 0$ and $y = 0$.
Thus in the initial system, we have a rectangular hyperbola whose asymptotes are $u = f$ and $v = f$.
The graph of v versus u is shown in figure.

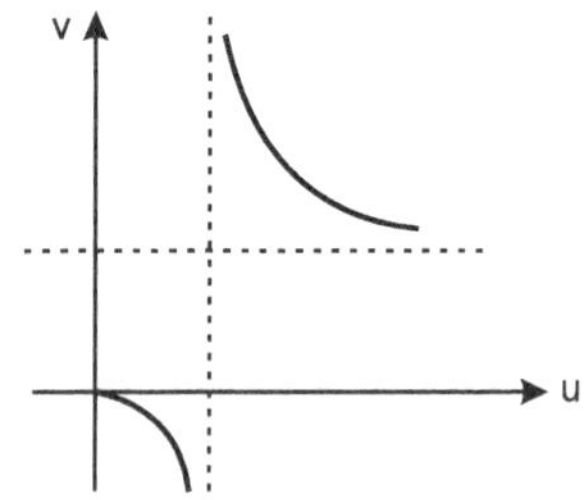

357. For concave mirror,

$$\frac{1}{v} + \frac{1}{u} = \frac{1}{f}$$

or $$\frac{1}{v} - \frac{1}{110} = -\frac{1}{100}$$

or $$v = -1100 \text{ cm}$$

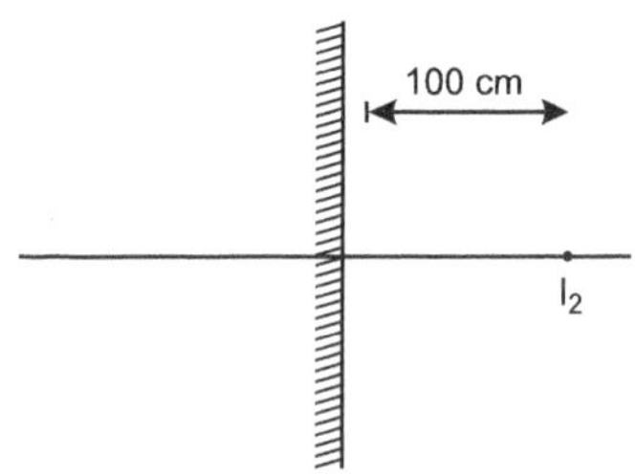

(a)

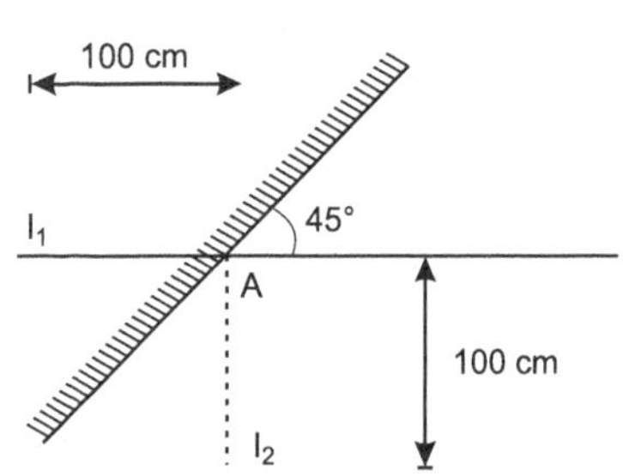

(b)

Thus first image I_1 is formed at 100 cm behind the plane mirror. This acts as virtual object for the plane mirror. If plane mirror would had been in vertical position as shown in figure (a), the image of I_1 had been formed at I_2, 100 cm in front of plane mirror. But since the plane mirror has been rotated 45° in clockwise direction. Hence all the reflected rays will rotate 90° in the same sense *i.e.*, clockwise and the image will be formed 100 cm vertically below A.

358. Shift due to the slab $= \left(1 - \frac{1}{\mu}\right) t = \frac{20}{3}$ cm

(in the direction of ray of light)

∴ For the mirror

$$u = -\left(20 + 10 - \frac{20}{3}\right) = -\frac{70}{3} \text{ cm}$$

Using the mirror formula,

$$\frac{1}{v} + \frac{1}{u} = \frac{1}{f} \text{ we have,}$$

$$\frac{1}{v} - \frac{3}{70} = -\frac{1}{9}$$

$$\therefore \quad v = -14.6 \text{ cm}$$

Now the slab will further produce a shift of 20/3 cm or 6.7 cm in the direction of ray of light. Hence, the final image will be formed at a distance of $(14.6 + 6.7) = 21.3$ cm from the mirror (towards right).

359. Total deviation of the ray is 108° as shown in figure. Hence

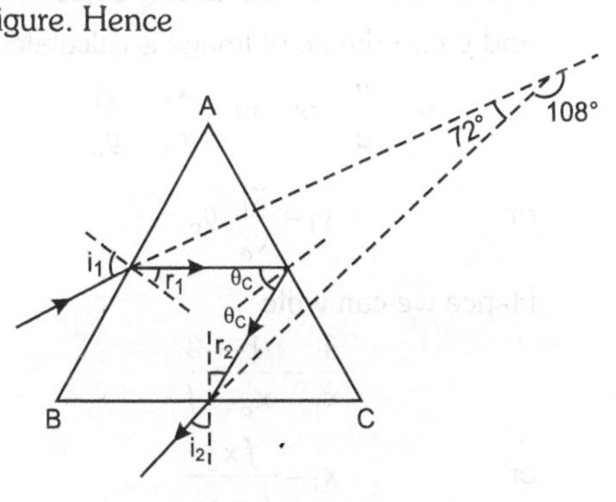

$$\delta_{AB} + \delta_{AC} + \delta_{BC} = 108° \qquad ...(1)$$

$$r_1 + \theta_c = A = 60°$$

and $\quad r_2 + \theta_c = C = 60°$

this implies that $\quad r_1 = r_2 \qquad ...(2)$

i.e. $\quad i_1 = i_2 \qquad ...(3)$

Now $\quad \delta_{AB} = i_1 - r_1$

$$\delta_{AC} = 180° - 2\theta_c$$

and $\quad \delta_{BC} = i_2 - r_2$

Hence $108° = 180° + (i_1 + i_2) - (r_1 + r_2) - 2\theta_c$

or $\quad 108° = 180° + 2i_1 - 2(r_1 + \theta_c)$

$$(i_1 = i_2 \text{ and } r_1 = r_2)$$

$$= 180° + 2i_1 - 2(60°)$$

$$\therefore \quad i_1 = 24° \qquad ...(4)$$

Now $\quad \mu = \frac{\sin i_1}{\sin r_1} = \frac{\sin 24°}{\sin(60° - \theta_c)}$

$$\Rightarrow \mu \sin(60° - \theta_c) = \sin 24° \qquad \left(\mu = \frac{1}{\sin\theta_c}\right)$$

Hence $\quad \frac{\sin(60° - \theta_c)}{\sin\theta_c} = \sin 24°$

or $\quad \frac{\sin(60° - \theta_c)}{\sin\theta_c \sin 60°} = \frac{\sin 24°}{\sin 60°}$

$$\frac{\sin 60° \cos\theta_c - \cos 60° \sin\theta_c}{\sin\theta_c \sin 60°} = \frac{\sin 24°}{\sin 60°}$$

or $\cot\theta_c - \cot 60° = \frac{0.407}{0.866} = 0.469$

or $\quad \cot\theta_c = 1.047$

$$\mu = \frac{1}{\sin\theta_c} = \sqrt{1 + \cot^2\theta_c} = \mathbf{1.45}$$

360. Applying Snell's law at point P

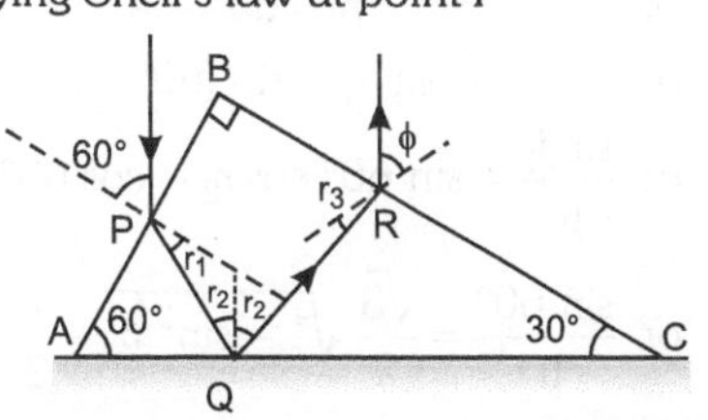

$$\mu = \frac{\sin i}{\sin r_1}$$

or $\quad \sin r_1 = \frac{\sin i}{\mu} = \frac{\sin 60°}{1.5} = \frac{\sqrt{3}/2}{1.5}$

or $\quad r_1 = 35.3°$

But $\quad r_1 + r_2 = 60°$

or $\quad r_2 = 60° - 35.3° = 24.7°$

But $\quad r_2 + r_3 = 30°$

$\therefore \quad r_3 = 30° - 24.7° = 5.3°$

Now applying Snell's law at R, we get

$$\mu = \frac{\sin \phi}{\sin r_3}$$

or $\sin \phi = \mu \sin r_3 = (1.5) \sin (5.3)^\circ$

or $\phi = \mathbf{7.96}^\circ$

361. Let μ be the refractive index of the prism.

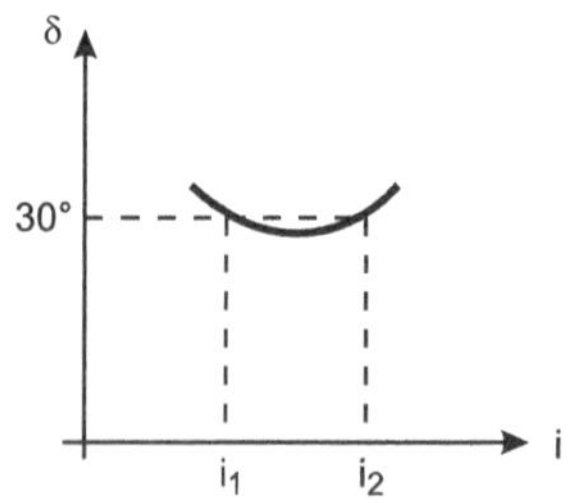

The deviation (δ) versus angle of incidence (i) graph varies as follows :
At two values of i_1 and i_2, deviation is 30°.

$$\delta = (i_1 + i_2) - A$$

Hence $(i_1 + i_2) = \delta + A = 30^\circ + 60^\circ$

or $(i_1 + i_2) = 90^\circ$ ...(1)

Also $(i_1 - i_2) = 30^\circ$ ...(2)

Solving these equations, we get

$$i_1 = 60^\circ \text{ and } i_2 = 30^\circ$$

Now $r_1 + r_2 = A = 60^\circ$

or $r_1 = 60^\circ - r_2$

or $\sin r_1 = \sin (60^\circ - r_2)$

or $\frac{\sin i_1}{\mu} = \sin 60^\circ \cos r_2 - \cos 60^\circ \sin r_2$

or $\frac{\sin 60^\circ}{\mu} = \frac{\sqrt{3}}{2}\sqrt{1 - \sin^2 r_2} - \frac{1}{2}\cdot\frac{\sin i_2}{\mu}$

$(i_1 = 60^\circ, i_2 = 30^\circ)$

or $\frac{\sqrt{3}}{2\mu} = \frac{\sqrt{3}}{2}\sqrt{1 - \left(\frac{\sin i_2}{\mu}\right)^2} - \frac{1}{4\mu}$

or $\frac{\sqrt{3}}{2\mu} = \frac{\sqrt{3}}{4\mu}\sqrt{4\mu^2 - 1} - \frac{1}{4\mu}$

or $2\sqrt{3} = \sqrt{3}\,(\sqrt{4\mu^2 - 1}) - 1$

$$\sqrt{3}\,(\sqrt{4\mu^2 - 1}) = 1 + 2\sqrt{3}$$

or $\sqrt{4\mu^2 - 1} = 2.577$

$\Rightarrow \quad 4\mu^2 - 1 = 6.643$

or $\mu = \mathbf{1.382}$

362. Assuming the optical centre of the lens to be origin and X and Y-axis as shown in figure :

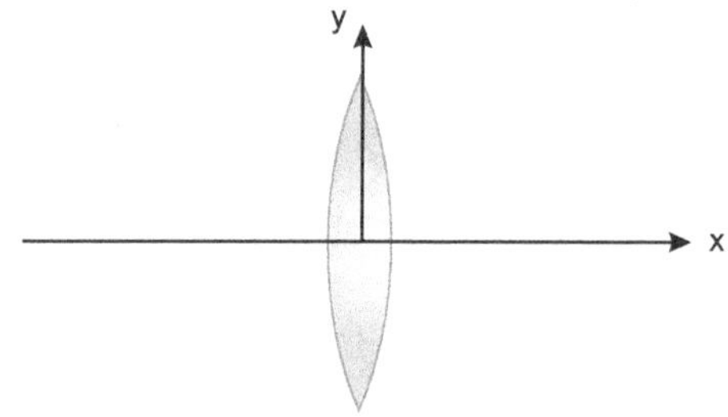

Co-ordinates of points A, B and C (in centimetres) are :

$$A = [(-45 - 8\cos 45^\circ), (15 - 8\sin 45^\circ)] = (-50.66, 9.34)$$

$$C = (-45, 15),$$

and $B = [(-45 + 8\cos 45^\circ), (15 + 8\sin 45^\circ)] = (-39.34, 20.66)$

Now the x-coordinate of image is calculated from the formula :

$$\frac{1}{v} - \frac{1}{u} = \frac{1}{f}$$

Here $v = x_i$ = x-coordinate of image

and $u = x_o$ = x-coordinate of object

and y-coordinate of image is calculated by

$$m = \frac{v}{u} \quad \text{or} \quad m = \frac{x_i}{x_o} = \frac{y_i}{y_o}$$

or $y_i = \frac{x_i}{x_o}\cdot y_o$

Hence we can write

$$\frac{1}{x_i} - \frac{1}{x_o} = \frac{1}{f}$$

or $x_i = \frac{f x_o}{f + x_o}$ ...(1)

and $y_i = \frac{f y_o}{f + x_o}$ ...(2)

substituting the coordinates of A, B and C, we can find corresponding coordinates of their images.

For A

$$x_i = \frac{(20)(-50.66)}{20-50.66} = 33.0 \text{ cm}$$

$$y_i = \frac{(20)(9.34)}{(20)-50.66} = -6.1 \text{ cm}$$

For C

$$x_i = \frac{(20)(-45)}{(20)-(45)} = +36 \text{ cm}$$

$$y_i = \frac{(20)(15)}{(20)-(45)} = -12 \text{ cm}$$

For B

$$x_i = \frac{(20)(-39.34)}{(20)-(39.34)} = 40.7 \text{ cm}$$

and $$y_i = \frac{(20)(20.66)}{(20)-(39.34)}$$

$$= -21.36 \text{ cm}$$

Hence the coordinates of images of *A*, *C* and *B* in centimetres are *A* (33.0, – 6.1), *C* (36, – 12) and *B* (40.7, – 21.36).

(b) Length of image AB is

$$L = \sqrt{(33.0-40.7)^2 + (-6.1+21.36)^2}$$

or $$\boldsymbol{L = 17.1 \text{ cm}}$$

363. Focal length 'f' of the mirror is $\frac{R}{2} = \frac{40}{2}$ cm or 20 cm.

First refraction would take place from the bottom of the glass slab (glass water boundary) and since the ray of light is travelling from a denser medium to a rarer medium, we can use

$$\frac{\mu_2}{v} - \frac{\mu_1}{u} = \frac{\mu_2-\mu_1}{R}$$

or $$\frac{4/3}{v_1} - \frac{3/2}{-45} = \frac{4/3-3/2}{\infty} = 0$$

[$R = \infty$ for a plane surface]

or $$v_1 = -40 \text{ cm}$$

Now reflection will take place from mirror. So, we can use

$$\frac{1}{v} + \frac{1}{u} = \frac{1}{f}$$

or $$\frac{1}{v_2} + \frac{1}{-(40+20)} = -\frac{1}{20}$$

or $$v_2 = -30 \text{ cm}$$

Now again refraction will take place at water-glass interface. So, applying

$$\frac{\mu_2}{v} - \frac{\mu_1}{u} = \frac{\mu_2-\mu_1}{R}$$

or $$\frac{3/2}{v_3} - \frac{4/3}{+10} = 0$$

or $$v_3 = 11.25 \text{ cm}$$

and finally refraction would take place at glass-air interface. So, applying

$$\frac{\mu_2}{v} - \frac{\mu_1}{u} = \frac{\mu_2-\mu_1}{R}$$

or $$\frac{1}{v_4} - \frac{3/2}{-(45-11.25)} = 0$$

or $$v_4 = \mathbf{-22.5 \text{ cm}}$$

So, final image will be formed at 22.5 cm below the upper face of the glass slab or **at the middle of the glass slab.**

364. We will have to first calculate the shift produced by the three slabs. Let the shifts be Δx_1, Δx_2 and Δx_3. Then

$$\Delta x_1 = \left(1 - \frac{\mu_w}{\mu_1}\right)(45) \text{ cm}$$

$$= \left(1 - \frac{4/3}{1.5}\right)(45) = 5 \text{ cm}$$

$$\Delta x_2 = \left(\frac{\mu_w}{\mu_2} - 1\right)(24) = 8 \text{ cm}$$

and $$\Delta x_3 = \left(1 - \frac{\mu_w}{\mu_3}\right)(54) = 6 \text{ cm}$$

Here Δx_1 and Δx_3 are towards the mirror while Δx_2 is away from the mirror. Hence net shift is

$\Delta x = (5 + 6 - 8)$ cm or 3 cm towards the mirror.

So, apparent distance of object from the mirror is

$u = (20 + 45 + 24 + 54 + 10) - 3$ or 150 cm.

Radius of curvature of mirror is also 150 cm. *i.e.* rays will fall normal on the mirror and they will be retraced back to their original path.

So, image will coincide with the object.

365. The first image formed by the first lens will be at a distance f from it as shown in figure (a) :

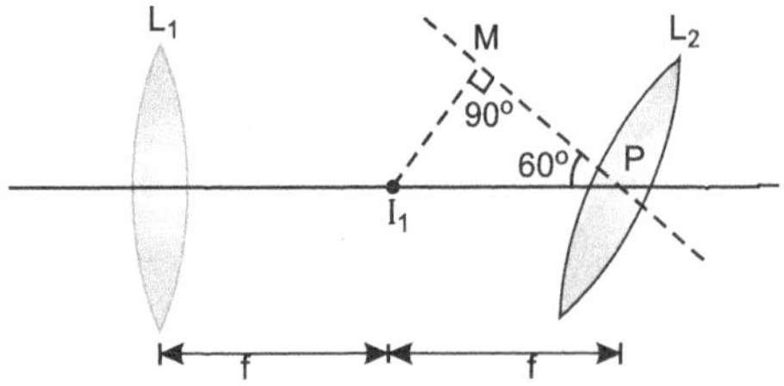

(a)

This image will act as an object for the second lens Here

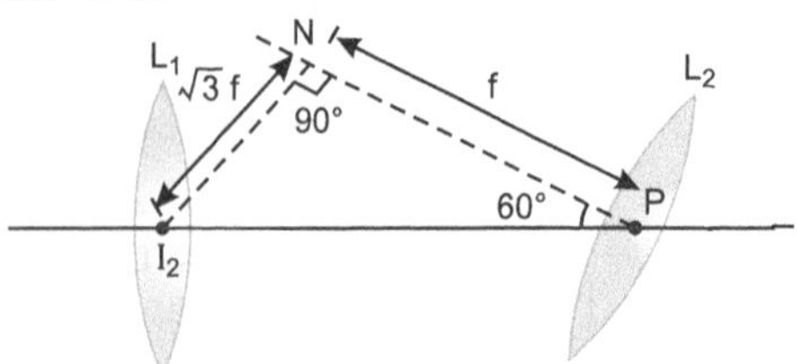

(b)

$$PM = f\cos 60° = \frac{f}{2}$$

and $$MI_1 = f\sin 60° = \frac{\sqrt{3}f}{2}$$

For L_2 $$\frac{1}{v} - \frac{1}{-(PM)} = \frac{1}{f}$$

or $$\frac{1}{v} + \frac{2}{f} = \frac{1}{f}$$

or $$v = -f$$

Height from the optic axis of I_2 will be

$$h = \left(\frac{v}{u}\right)(MI_1) = \left(\frac{-f}{-f/2}\right)\left(\frac{\sqrt{3}f}{2}\right) = \sqrt{3}f$$

The final image is shown in figure (b) :

$$PI_2 = f\sec 60° = 2f$$

i.e. final image is formed at the optical centre of the first lens L_1 or **the desired coordinates are (0, 0).**

366. Let x be the distance between lens and screen at some moment and y the distance between object and lens. Since image is real. Therefore, x is positive and y is negative. So, applying lens formula $\left(\frac{1}{v} - \frac{1}{u} = \frac{1}{f}\right)$ we get :

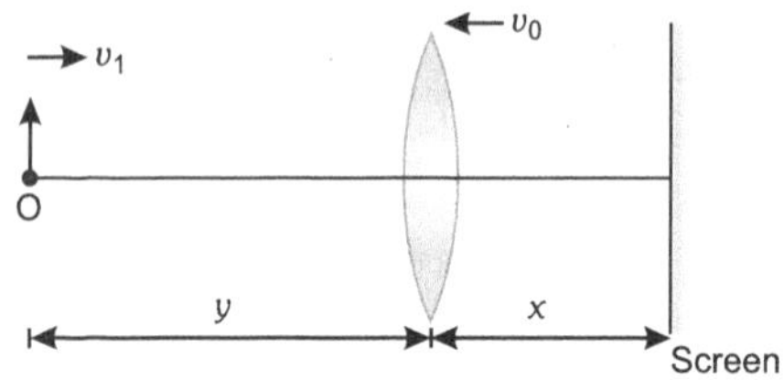

$$\frac{1}{x} - \frac{1}{-y} = \frac{1}{f}$$

or $$\frac{1}{x} + \frac{1}{y} = \frac{1}{f} \qquad \ldots(1)$$

Now lens is moving away from the screen. Therefore, y is increasing i.e. x should decrease to keep the right hand side to be constant. Differentiating (1) with respect to time, we get

$$-\frac{1}{x^2}\cdot\left(\frac{dx}{dt}\right) - \frac{1}{y^2}\cdot\frac{dy}{dt} = 0 \qquad \ldots(2)$$

Here $\frac{dy}{dt}$ = rate with which y is increasing $= v_0$

and $\left(-\frac{dx}{dt}\right)$ = rate with which x is decreasing

or $$\left(-\frac{dx}{dt}\right) = v_1 + v_0 \qquad \ldots(3)$$

Here v_1 = speed of object towards lens or screen.

$$\left(-\frac{dx}{dt}\right) = \left(\frac{x^2}{y^2}\right)\cdot\left(\frac{dy}{dt}\right)$$

from equation (2)

or $$\left(-\frac{dx}{dt}\right) = \frac{1}{n^2}\cdot v_0 \qquad \ldots(4)$$

$\therefore$ $$v_1 + v_0 = \frac{1}{n^2}\cdot v_0$$

from equations (3) and (4)

or $$v_1 = \left(\frac{1}{n^2} - 1\right)v_0$$

or $$v_1 = \left(\frac{1-n^2}{n^2}\right)v_0$$

Therefore, speed of object towards the screen is $\left(\frac{1-n^2}{n^2}\right)v_0$

Note: y can be decreased by moving the object away from the screen with speed $v_1 < v_0$. But in the given condition, this is not possible, because in that case

$$v_0 - v_1 = \left(-\frac{dx}{dt}\right) = \frac{v_0}{n^2}$$

or $$v_1 = v_0\left(1 - \frac{1}{n^2}\right)$$

which comes out to be negative as $n < 1$

367. Let x be the distance between object and mirror at time t. Then :

$$x = d + (v_1 - v_2)\,t \qquad \ldots(1)$$

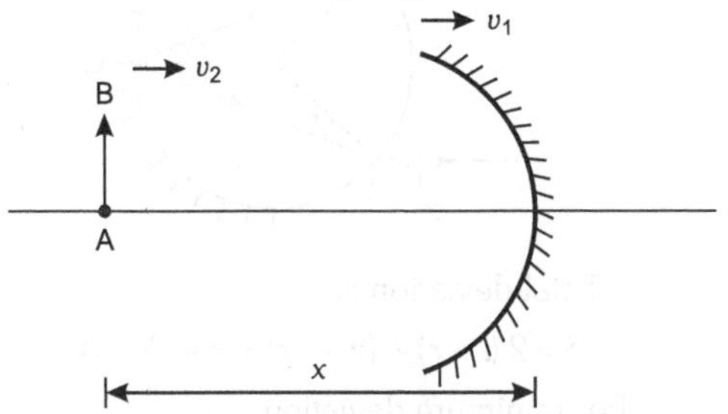

Let y be the distance between image (screen) and mirror at this moment. Object is always beyond focus i.e. image is always real. Hence x and y are both negative. Applying mirror formula $\left(\frac{1}{v} + \frac{1}{u} = \frac{1}{f}\right)$ we get

$$\frac{1}{-x} - \frac{1}{y} = -\frac{2}{R} \qquad \left(\frac{1}{f} = \frac{2}{R}\right)$$

or $$\frac{1}{x} + \frac{1}{y} = \frac{2}{R} \qquad \ldots(2)$$

x is increasing therefore, y should decrease to keep the right hand side of equation (2) to be constant. Differentiating equation (2) with respect to time we get :

$$-\frac{1}{x^2}\cdot\left(\frac{dx}{dt}\right) - \frac{1}{y^2}\left(\frac{dy}{dt}\right) = 0$$

Here $\frac{dx}{dt}$ = rate at which x is increasing

$$= v_1 - v_2$$

and $\left(-\frac{dy}{dx}\right)$ = rate at which y is decreasing

$$= v_3 - v_1 \qquad (v_3 = \text{speed of screen} > v_1)$$

$$\therefore \quad \left(-\frac{dy}{dt}\right) = v_3 - v_1 = \left(\frac{y^2}{x^2}\right)\cdot\frac{dx}{dt}$$

or $$v_3 = v_1 + \left(\frac{y^2}{x^2}\right)\cdot\frac{dx}{dt}$$

Here $\frac{y}{x} = \frac{R}{2x - R}$ from equation (2)

$\therefore$ speed of screen

$$\boldsymbol{v_3 = v_1 + \left(\frac{R}{2x - R}\right)^2 (v_1 - v_2)}$$

where $x = d + (v_1 - v_2)\,t$

Note: v_3 is in the direction of v_1 and v_2

368. Given that $\mu > \mu'$. Hence the ray diagram of the ray will be as shown in figure.

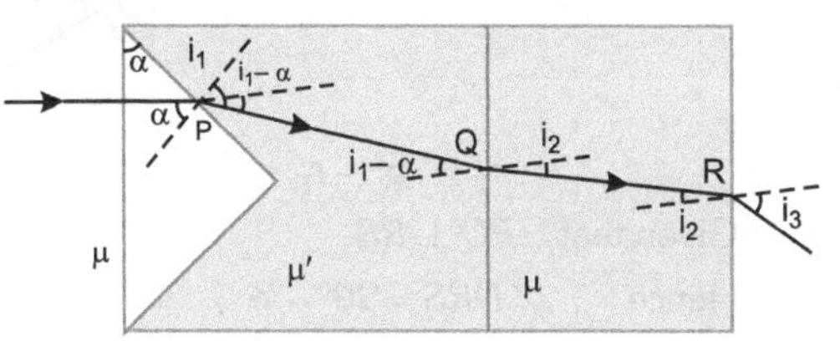

Applying Snell's law at P ($\mu \sin i$ = constant or μi = constant for small angles)

$$\mu' \sin i_1 = \mu \sin \alpha$$

or $$\mu' i_1 = \mu\alpha$$

or $$i_1 = \frac{\mu\alpha}{\mu'} \qquad \ldots(1)$$

Applying Snell's law at Q

$$(i_1 - \alpha)\mu' = \mu i_2$$

or $$i_2 = (i_1 - \alpha)\frac{\mu'}{\mu} \qquad \ldots(2)$$

and finally Snell's law at R gives

$$(i_3)(1) = (i_2)\,\mu$$

or $$i_3 = \mu i_2 \qquad \ldots(3)$$

Now total deviation is

$$\delta = \delta_P - \delta_Q + \delta_R$$
$$= (i_1 - \alpha) - \{(i_1 - \alpha) - i_2\} + (i_3 - i_2)$$
$$= i_3 \text{ (it is shown in figure also)}$$
$$= \mu i_2 \text{ (from equation 3)}$$
$$= \mu\,(i_1 - \alpha)\frac{\mu'}{\mu}$$

or $\mu' (i_1 - \alpha)$ (from equation 2)

$$= \mu' \left(\frac{\mu\alpha}{\mu'} - \alpha \right)$$

or $\delta = (\mu - \mu')\,\alpha$

369. The ray diagram of the beam of light is as shown in figure

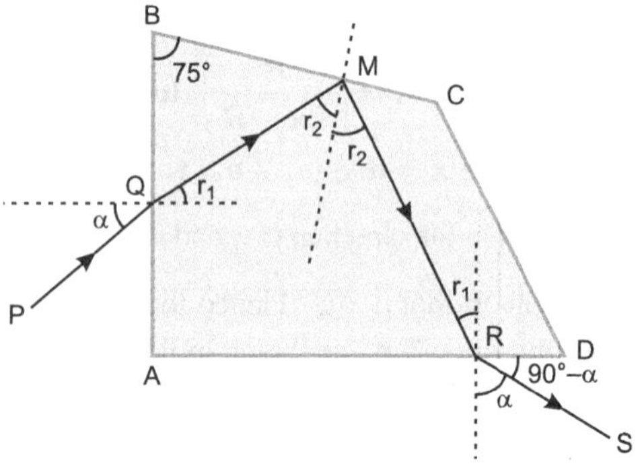

$$r_2 = 75° - r_1$$

Given that $PQ \perp RS$

Hence $\angle DRS = 90° - \alpha$

or the angle of emergent ray with the normal at face AD will be α. Hence the ray MR will make an angle of r_1 with normal.
Total deviation is 90° *i.e.*

$$90° = \delta_Q + \delta_M + \delta_R$$

or $90° = (\alpha - r_1) + [180° - 2(75° - r_1)] - (\alpha - r_1)$

or $r_1 = 30°$

or $r_2 = 75° - 30° = 45°$

Now it is given that the ray is reflected completely at face BC. Hence

$$r_2 \geq \theta_c \quad \text{or} \quad \sin r_2 \geq \sin \theta_c$$

or $\sin 45° \geq \frac{1}{\mu} \Rightarrow \frac{1}{\sqrt{2}} \geq \frac{1}{\mu}$

$\therefore$ $\mu \geq \sqrt{\mathbf{2}}$

Further Snell's law at Q gives

$$\mu = \frac{\sin \alpha}{\sin r_1} \quad \text{or} \quad \sin \alpha = \mu \sin r_1$$

Since $\mu \geq \sqrt{2}$

Hence $\sin \alpha \geq \sqrt{2} \sin r_1$

or $\sin \alpha \geq \sqrt{2} \sin 30°$

or $\sin \alpha \geq \frac{1}{\sqrt{2}}$

or $\alpha \geq \mathbf{45°}$

370. (a) Deviation at $P = i - r$

deviation at $Q = \pi - 2r$

and deviation at $R = i - r$

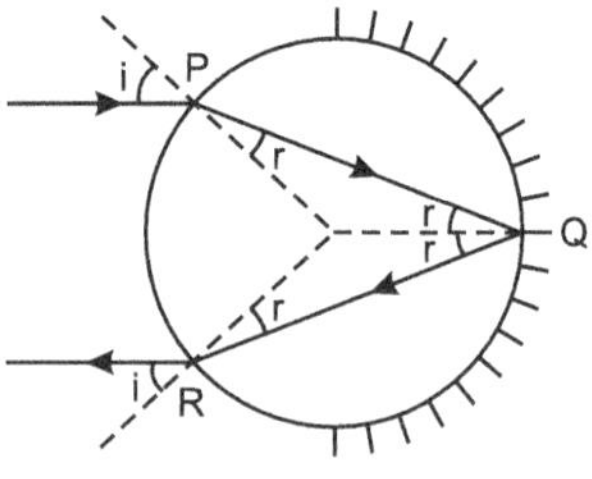

$\therefore$ Total deviation is

$$\delta = 2(i - r) + (\pi - 2r) = \pi + 2i - 4r$$

For minimum deviation

$$\frac{d\delta}{di} = 0 \quad \text{or} \quad 2 - 4\frac{dr}{di} = 0$$

or $\frac{dr}{di} = \frac{1}{2}$ ...(1)

Also $\mu \sin r = \sin i$

or $\mu \cos r \,.\, dr = \cos i \,.\, di$

or $\frac{dr}{di} = \frac{\cos i}{\mu \cos r}$ ...(2)

$\therefore$ $\frac{1}{2} = \frac{\cos i}{\mu \cos r}$

or $\mu^2 \cos^2 r = 4 \cos^2 i$

$\Rightarrow$ $\mu^2 (1 - \sin^2 r) = 4 \cos^2 i$

or $\mu^2 \left(1 - \frac{\sin^2 i}{\mu^2} \right) = 4 \cos^2 i$

or $\mu^2 - \sin^2 i = 4 \cos^2 i$

or $\mu^2 - (1 - \cos^2 i) = 4 \cos^2 i$

or $\left(\frac{7}{4} - 1 \right) = 3 \cos^2 i$

$\Rightarrow$ $\frac{3}{4} = 3 \cos^2 i$

or $\cos i = \dfrac{1}{2} \Rightarrow \boldsymbol{i = 60°}$

(b) $\sin r = \dfrac{\sin i}{\mu} = \dfrac{\sqrt{3}/2}{\sqrt{7}/2} = \sqrt{\dfrac{3}{7}}$

or $r = 41°$

Hence deviation is

$\delta = \pi + 2i - 4r = 180° + 2 \times 60° - 4 \times 41°$

or $\delta = \mathbf{136°}$

371. The ray becomes parallel after refraction from the lens (because the object is placed at first focus of the lens). It strikes the mirror at $Q(x_1, y_1)$

$$\left(\frac{dy}{dx}\right)_{(x_1, y_1)} = \tan\alpha = \frac{1}{16y_1} \quad \ldots(1)$$

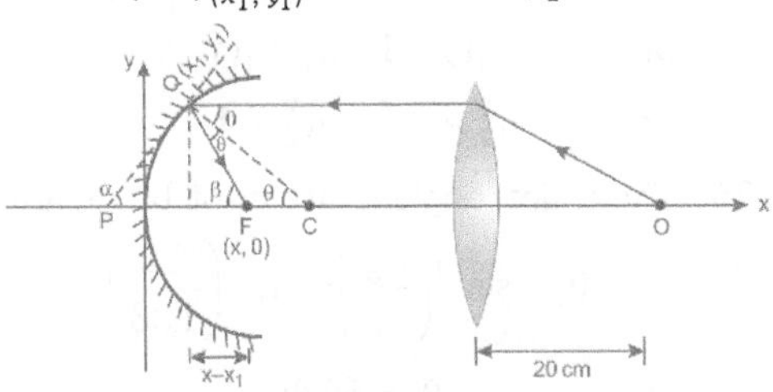

$$\alpha = 90° - \theta \quad \ldots(2)$$

and $\beta = 2\theta = 180° - 2\alpha \quad \ldots(3)$

The reflected ray passes through $F(x, 0)$

where $\tan\beta = \dfrac{y_1}{x - x_1}$

$$\tan 2\theta = \tan(180° - 2\alpha) = \tan\beta$$

or $\tan\beta = -\tan 2\alpha$

or $\tan\beta = \dfrac{-2\tan\alpha}{1 - \tan^2\alpha} \quad \ldots(4)$

Substituting the values

$$\frac{y_1}{x - x_1} = \frac{-2\left(\frac{1}{16y_1}\right)}{1 - \left(\frac{1}{16y_1}\right)^2}$$

or $y_1 - \dfrac{1}{(16)^2 y_1} = -\dfrac{x}{8y_1} + \dfrac{x_1}{8y_1}$

or $y_1^2 - \dfrac{1}{(16)^2} = -\dfrac{x}{8} + \dfrac{x_1}{8}$

But $y_1^2 = \dfrac{x_1}{8}$

Hence $\dfrac{x}{8} = \dfrac{1}{(16)^2}$

or $x = \dfrac{8}{(16)^2} = 0.03125 \text{ m} = 3.125 \text{ cm}$

Note that x is independent of (x_1, y_1). Hence, **all rays are focussed at *F***

or $\boldsymbol{x = 3.125}$ **cm.**

372. The ray diagram is shown in the figure :

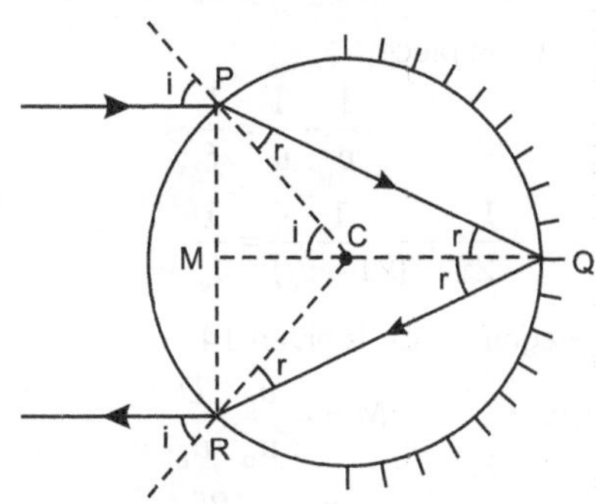

Total deviation,

$$\delta = \delta_P + \delta_Q + \delta_R$$

or $180° = (i - r) + (180 - 2r) + (i - r)$

or $r = \dfrac{i}{2} \quad \ldots(1)$

Now in triangle PMC

$$PC = \text{radius} = 2 \text{ m}$$

and $PM = \dfrac{PR}{2} = 1 \text{ m}$

Hence $\sin i = \dfrac{PM}{PC} = \dfrac{1}{2}$

or $i = 30°$ and $r = 15°$

$\therefore$ Refractive index $\mu = \dfrac{\sin i}{\sin r}$

$$= \frac{\sin 30°}{\sin 15°} = 2\cos 15°$$

$$\mu = \mathbf{1.932}$$

373. Let focal length of objective and eye piece are f_o and f_e respectively

For objective,

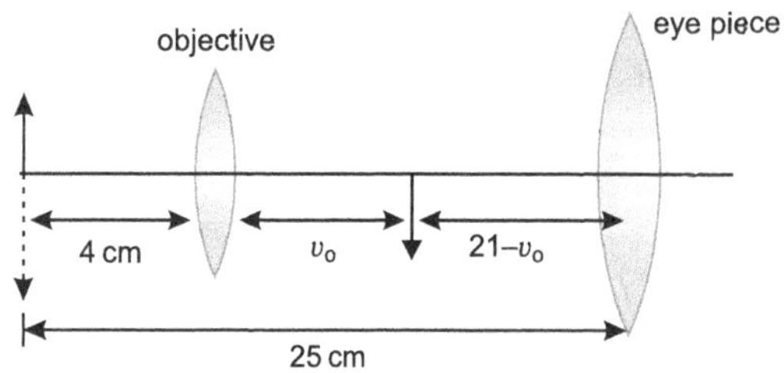

$$\frac{1}{v}-\frac{1}{u}=\frac{1}{f}$$

$$\frac{1}{v_o}+\frac{1}{4}=\frac{1}{f_o} \qquad \ldots(1)$$

For eye piece,

$$\frac{1}{v}-\frac{1}{u}=\frac{1}{f}$$

$$\frac{1}{-25}+\frac{1}{(21-v_o)}=\frac{1}{f_e} \qquad \ldots(2)$$

Magnification is given 14

i.e. $$M=-\frac{v_o}{u_o}\cdot\frac{D}{u_e}$$

or $$14=\frac{v_o}{4}\cdot\frac{25}{(21-v_o)} \qquad \ldots(3)$$

solving equations (1), (2) and (3), we get

$\boldsymbol{f_o} = \mathbf{3.136}$ cm and $\boldsymbol{f_e} = \mathbf{8.75}$ cm

374. Figure displays the given problem.

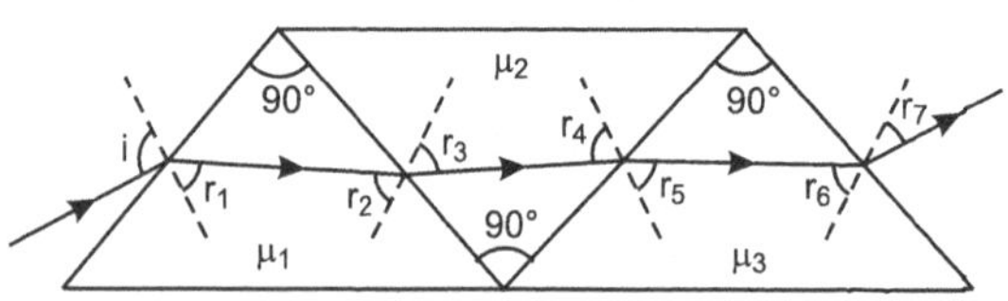

We have

$$r_1+r_2=90° \qquad \ldots(1)$$

$$r_3+r_4=90° \qquad \ldots(2)$$

and $$r_5+r_6=90° \qquad \ldots(3)$$

Since the emergent ray is parallel to incident ray. We have

$$r_7=90°-i \qquad \ldots(4)$$

Writing the Snell's equations of refraction

$$\sin i=\mu_1 \sin r_1$$

$$\mu_1 \sin r_2=\mu_2 \sin r_3$$

$$\mu_2 \sin r_4=\mu_3 \sin r_5$$

and $$\mu_3 \sin r_6=\sin r_7$$

Using equations (1) to (4) in the above equations. we get

$$\sin i=\mu_1 \sin r_1 \qquad \ldots(5)$$

$$\mu_1 \cos r_1=\mu_2 \sin r_3 \qquad \ldots(6)$$

$$\mu_2 \cos r_3=\mu_3 \sin r_5 \qquad \ldots(7)$$

$$\mu_3 \cos r_5=\cos i \qquad \ldots(8)$$

(equation 5)^2 + (equation 7)^2

– (equation 6)^2 – (equation 8)^2

gives

$$\sin^2 i+\mu_2^2\cos^2 r_3-\mu_1^2\cos^2 r_1-\mu_3^2\cos^2 r_5$$

$$=\mu_1^2\sin^2 r_1+\mu_3^2\sin^2 r_5-\mu_2^2\sin^2 r_3-\cos^2 i$$

or $$\mu_2^2+1=\mu_1^2+\mu_3^2$$

or $$\boldsymbol{\mu_1^2+\mu_3^2-\mu_2^2=1}$$

375. The critical angle for water-oil boundary is

$$\theta_1=\sin^{-1}\left(\frac{\mu_o}{\mu_\omega}\right)=\sin^{-1}\left(\frac{1.2}{1.33}\right)$$

or $$\theta_1=64.45° \qquad \ldots(1)$$

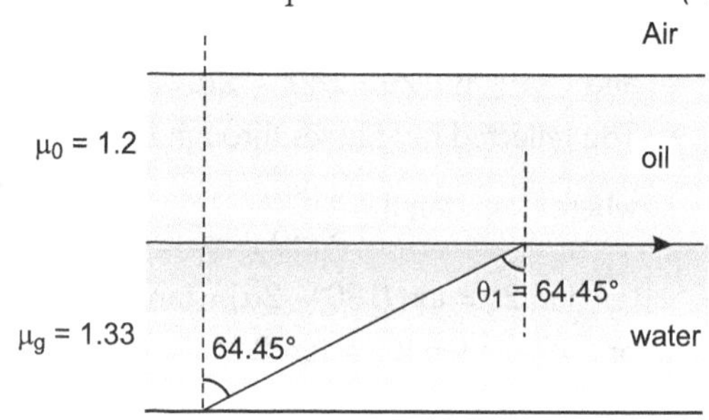

(a)

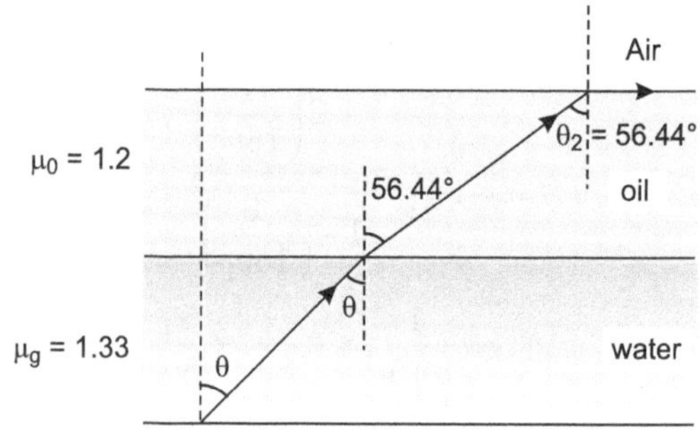

(b)

The critical angle for oil-air boundary is

$$\theta_2 = \sin^{-1}\left(\frac{1}{\mu_o}\right) = \sin^{-1}\left(\frac{1}{1.2}\right)$$

or $\theta_2 = 56.44°$ ...(2)

For the light ray to emerge from the oil-air surface at critical angle θ_2, the incident angle θ at the water-oil boundary should be given by

$$1.33 \sin\theta = 1.2 \sin\theta_2 \quad \text{(Snell's law)}$$

or $1.33 \sin\theta = 1.2 \sin(56.44°)$

or $\theta = 48.83°$

The solid angle subtended by vertex of a cone of angle α is given by

$$\omega = \int_0^{\alpha} \frac{(2\pi r \sin\theta)(r\, d\theta)}{r^2}$$

$$= \int_0^{\alpha} 2\pi \sin\theta \cdot d\theta = 2\pi(1 - \cos\alpha)$$

$\therefore$ percentage of light escaping from the lake surface

$$= \frac{2\pi}{4\pi}(1 - \cos 48.83°) \times 100$$

$$= \mathbf{17\%}$$

(b) Percentage of light escaping from water surface

$$= \frac{2\pi}{4\pi}(1 - \cos 64.45°) \times 100 = 28.5\%$$

$\therefore$ percentage of light totally reflected at water-oil surface

$= (100 - 28.5)\%$ or **71.5%**.

376. $\delta_1 = (n_1 - 1)\theta = 0.044$ radian

and $\delta_1 = \dfrac{y_1}{45.50}$

$\Rightarrow$ $y_1 = 45.5 \times \delta_1 = 2$ cm

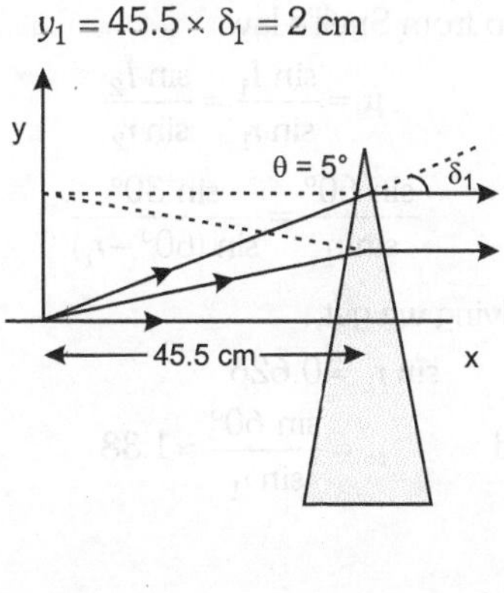

Due to refraction with lower part of the prism

$$\delta_2 = (n_2 - 1)\theta = 0.03 \text{ radian}$$

$$y_2 = 45.5\,\delta_2 = 1.3 \text{ cm}$$

For the lens:

$$\frac{1}{v} - \frac{1}{u} = \frac{1}{f}$$

$$\Rightarrow \quad \frac{1}{v} = \frac{1}{f} + \frac{1}{u}$$

$$= \frac{1}{-30} + \frac{1}{-90} = \frac{-4}{90}$$

$$v = -\frac{90}{4} = -22.5 \text{ cm}$$

$$m = \frac{I}{O} = \frac{v}{u}$$

$$\therefore \quad I_1 = O_1 \frac{v}{u}$$

$$I_1 = 2\left(\frac{-22.5}{-90}\right) = +0.5$$

$$I_2 = 1.3\left(\frac{-22.5}{-90}\right) = +0.325$$

$\therefore$ Co-ordinates of two images of the point source are (67.5 cm, 0.5 cm) and (67.5 cm, 0.325 cm).

377. (a) As the prism is thin and we are only considering paraxial rays, the presence of prism will only cause lateral shifting of the image. Thus the x co-ordinate of image is not affected by prism.

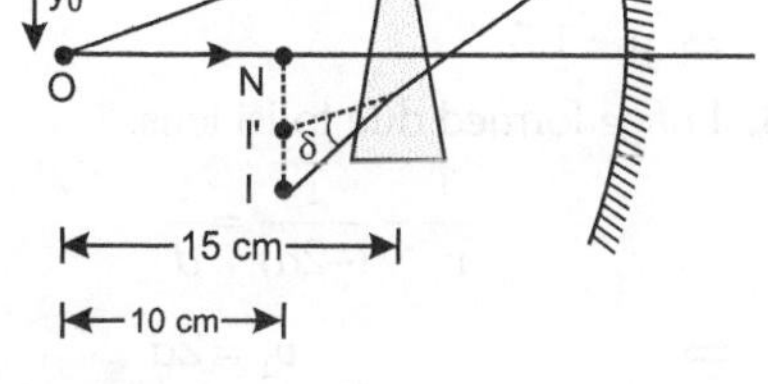

For the mirror

$$\frac{1}{u} + \frac{1}{v} = \frac{2}{R}$$

$$= -\frac{1}{30} + \frac{1}{-(30-10)} = \frac{2}{R}$$

$\Rightarrow \quad R = -24$ cm

(b) The image of the object as 'seen' by the mirror will be laterally shifted by a distance

$$y_0 = \delta (15 \text{ cm})$$
$$= (\mu_1 - 1)\left(\frac{\pi}{180}\right)(15 \text{ cm})$$
$$= (\mu - 1)\frac{\pi}{12} \text{ cm}$$

$(\because$ Prism angle $= 1° = \frac{\pi}{180}$ radian$)$

Mirror will form the image of O' at I' where

$$\frac{NI'}{OO'} = \left|\frac{v}{u}\right| = \frac{2}{3}$$

$$\Rightarrow \quad NI' = \frac{2}{3}(\mu - 1)\frac{\pi}{12}$$

$$NI' = (\mu - 1)\frac{\pi}{18}$$

But since the ray again passes through the prism it again suffers a deviation δ.

$\therefore$ Final image is formed at I where

$$II' = \delta (15 \text{ cm} - 10 \text{ cm})$$
$$= (\mu - 1)\left(\frac{\pi}{180}\right)(5 \text{ cm})$$
$$= (\mu - 1)\frac{\pi}{36} \text{ cm}$$

y co-ordinate of the final image I

$$-NI = -(NI' + II')$$
$$= -(\mu - 1)\frac{\pi}{12} \text{ cm} = -\frac{\pi}{24} \text{ cm} \quad \text{(Given)}$$

$\Rightarrow \mu = 1.5$

378. Image formed due to Ist lens.

$$\frac{1}{v_1} - \frac{1}{(-2d)} = \frac{1}{d}$$

$$\Rightarrow \quad v_1 = 2d$$

For the second lens,

$$u_2 = d$$

$$\therefore \quad \frac{1}{v_2} - \frac{1}{d} = \frac{1}{f_2} = \frac{2}{3d}$$

$$v_2 = \frac{3}{5}d$$

$\therefore$ x co-ordinates of image

$$2d + d + \frac{3}{5}d = 3d + \frac{3d}{5}$$
$$= \frac{18}{5}d = \frac{18}{5} \times 10 = 36 \text{ cm}$$

Magnification due to Ist lens,

$$m_1 = \frac{v_1}{u_1} = \left(\frac{2d}{-2d}\right) = -1$$

Magnification due to IInd lens,

$$m_2 = \frac{v_2}{u_2} = \frac{(3/5)\,d}{d} = \frac{3}{5}$$

'y' co-ordinate of Ist image $= 2\Delta$

y co-ordinate of final image

$$= \frac{3}{5} \times 2\Delta = \frac{6\Delta}{5} = \frac{6}{5} \times 0.005$$
$$= 0.006 \text{ cm}$$

Co-ordinates of final image are (36 cm, 0.006 cm).

379. For a prism, angle of deviation

$$\delta = I_1 + I_2 - A \quad \text{...(1)}$$

and $\quad r_1 + r_2 = A \quad$ ...(2)

Hence, $\delta = 30°$, $I_1 = 60°$, $A = 60°$

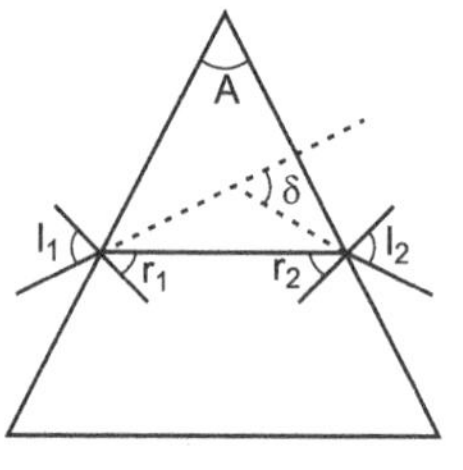

Putting these values in (1) and (2), we get

$$I_2 = 30° \quad \text{and} \quad r_1 + r_2 = 60°$$

Also from Snell's law

$$\mu = \frac{\sin I_1}{\sin r_1} = \frac{\sin I_2}{\sin r_2}$$

$$\Rightarrow \quad \frac{\sin 60°}{\sin r_1} = \frac{\sin 30°}{\sin (60° - r_1)}$$

Solving we get,

$$\sin r_1 = 0.626$$

and $\quad \mu = \frac{\sin 60°}{\sin r_1} \approx 1.38$

Now, if the ray is incident normally on this prism angle of incidence at the second surface = 60° (as shown)

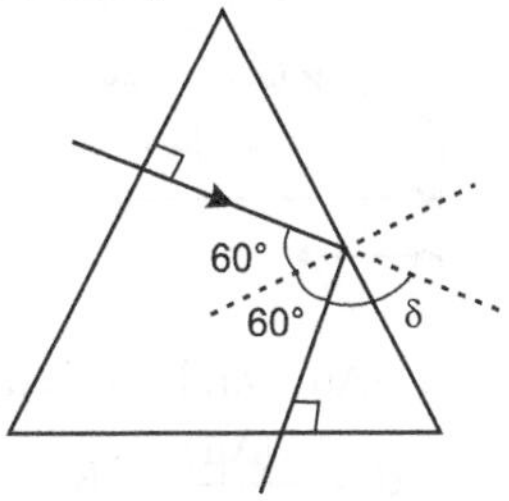

But, $\sin 60° > \frac{1}{\mu}$

∴ Total internal reflection will take place.

At the third surface this ray is again incident normally (as shown). Hence, the ray suffer deviation only at the second surface, which is equal to $180° - 2 \times 60° = 60°$.

380. Consider the light beam as consisting of parallel light rays. They cross the vertical plane face of the quarter-cylinder without changing their direction, and strike the curved surface of the cylinder at various angles of incidence. The normals at the points of incidence of the rays are radii of the cylinder.

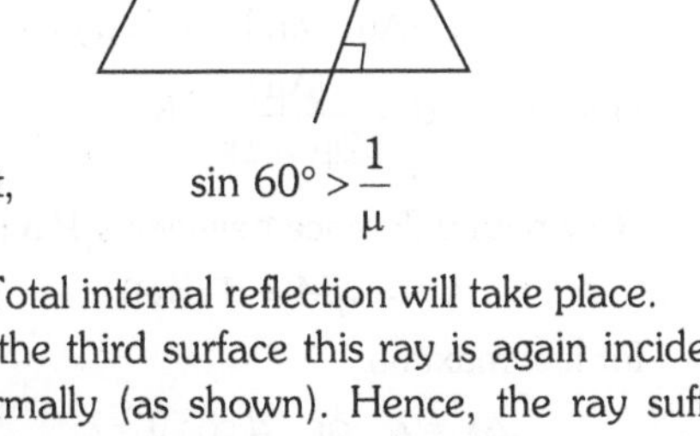

The higher the position of a light ray entering the quarter-cylinder, the larger is its angle of incidence at the cylinder's curved surface. The angle of incidence for the ray shown in the figure is the critical angle for total internal reflection. Therefore only light rays closer to the table than this one can leave the quarter-cylinder (refracted to different extents). The limiting case is determined using the figure:

$$\sin \alpha_h = \frac{1}{n} = \frac{2}{3}$$

and $$\frac{R}{R + x} = \cos \alpha_h$$

which yield $x = 1.71$ cm. This is the closest to the quarter-cylinder that light can reach the table.

381. According to Snell's law, $n \sin \alpha$ is constant along the light ray's trajectory.

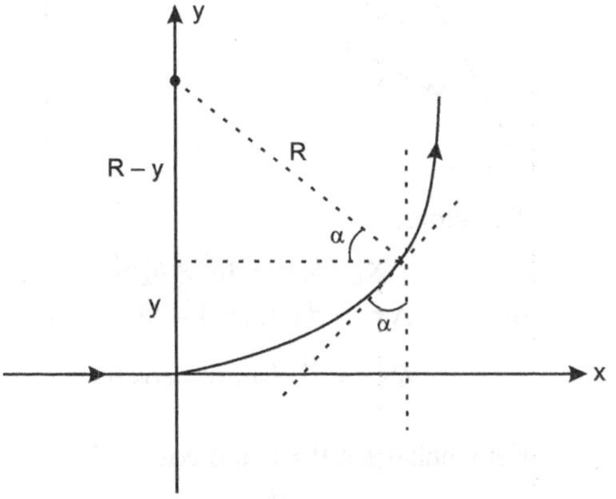

Place the origin of the co-ordinate system at the point where the light ray enters the medium. In this case, the angle of incidence for the first 'plate' starting at $y = 0$ is 90° and the refractive index is n_0, which gives the above constant as $n(y) \sin \alpha = n_0$.

The light travels along a circular arc of radius R and we first examine its relationship to co-ordinate y. From figure it is clear that

$$n_0 = n \sin \alpha = n(y) \frac{R - y}{R}.$$

This gives the space-dependence of the refractive index as

$$n(y) = \frac{R}{R - y} n_0$$

The material with the greatest known refractive index is diamond, but even the refractive index of this material does not reach the value $n_{max} = 2.5$. It is this limit that sets the maximum angular size of the arc the light ray can cover. If the refractive index changes from $n_0 = 1$ to $n_{max} = 2.5$ then the maximum value of y is $\frac{3}{5} R$, corresponding to an arc of angular size 66.4°.

382. For small θ, we can write

$$\angle s_1 s_2 M \approx \frac{\pi}{4} - \theta$$

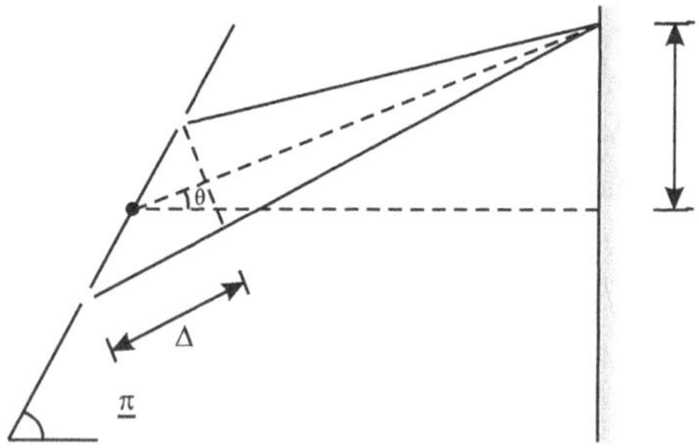

Therefore,

$$\Delta x_1 = s_2 P - s_1 P \approx s_2 M$$

or $$\Delta x_1 = d\cos(\pi/4 - \theta) \qquad (d = s_1 s_2)$$

$$\Delta x_1 = \frac{d}{\sqrt{2}}(\sin\theta + \cos\theta)$$

For small θ, $\sin\theta \approx \theta$ and $\cos\theta \approx 1$

Hence $$\Delta x_1 = \frac{d}{\sqrt{2}}(1+\theta) \qquad \dots(1)$$

Let $$ss_1 - ss_2 = \Delta x_2$$

Then, net path difference at P is

$$\Delta x = \Delta x_1 - \Delta x_2$$

$$\Delta x = \frac{d}{\sqrt{2}}(1+\theta) - \Delta x_2 \qquad \dots(3)$$

For maximum intensity

$$\Delta x = n\lambda \qquad \dots(4)$$

Substituting $\theta \approx \frac{y}{D}$ and putting the proper values in equation (3) and (4), we get fringe width $\omega = y_{n+1} - y_n = (\sqrt{2}\,\lambda) \times \frac{D}{d}$ or

$\omega = (\sqrt{2})(5 \times 10^{-7})(10^3)$ m $= \mathbf{0.7\ mm}$

383. The parallel rays converge at two points s_1 and s_2 from the two parts L_1 and L_2

Distance between s_1 and s_2 is

$$d = f_2 - f_1 = \frac{R/2}{\mu_2 - 1} - \frac{R/2}{\mu_1 - 1}$$

$$= \frac{(\mu_1 - \mu_2)}{2(\mu_1 - 1)(\mu_2 - 1)} . R$$

we can write

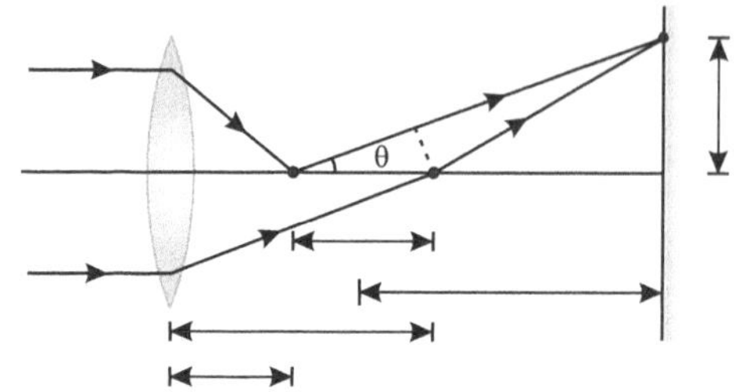

$$\mu_1 - \mu_2 = \Delta\mu \quad \text{and} \quad \mu_1 \approx \mu_2 \approx \mu$$

Hence $$d = \frac{(\Delta\mu)}{2(\mu - 1)^2} . R \qquad \dots(1)$$

Now path difference between s_1P and s_2P is

$$\Delta x \approx s_1 M = d\cos\theta$$

for first maxima

$$\Delta x = \lambda \quad \text{or} \quad d\cos\theta = \lambda$$

or $$\cos\theta = \frac{\lambda}{d}$$

or $$1 - 2\sin^2\left(\frac{\theta}{2}\right) = \frac{\lambda}{d} \qquad \sin\frac{\theta}{2} \approx \frac{\theta}{2}$$

Hence $$1 - 2\left(\frac{\theta}{2}\right)^2 = \frac{\lambda}{d}$$

or $$1 - \frac{\theta^2}{2} = \frac{\lambda}{d} \qquad \left(\theta \approx \frac{y}{D}\right)$$

so $$1 - \frac{y^2}{2D^2} = \frac{\lambda}{d}$$

or $$y = \sqrt{\left(\frac{d-\lambda}{d}\right) 2D^2}$$

or $$y = D\sqrt{2\left(\frac{d-\lambda}{d}\right)}$$

$$= D\sqrt{2\left(1 - \frac{\lambda}{d}\right)}$$

Substituting the value of d from equation (1). we get

$$\boldsymbol{y = D\sqrt{2\left\{1 - \frac{2(\mu - 1)^2\lambda}{\Delta\mu R}\right\}}}$$

384. Light is reflected from two surfaces of the plate P. But they are mutually out of phase because one is reflected by a denser medium while the other by a rarer medium. So, the two sourcs s_1 and s_2 can be assumed at a distance of $2l$ and $2l + 2d$ from the screen as shown in figure.

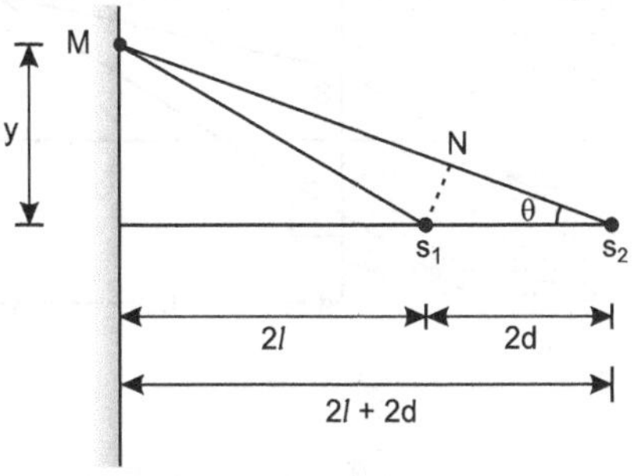

Path difference

$$\Delta x = s_2M - s_1M \approx s_2N = 2d\cos\theta$$

or $$\Delta x = 2d\left\{1 - 2\sin^2\frac{\theta}{2}\right\}$$

$$= 2d\left[1 - \left(\frac{\theta^2}{2}\right)\right] \qquad \left\{\sin\frac{\theta}{2} \approx \frac{\theta}{2}\right\}$$

or $$\Delta x = 2d\left(1 - \frac{(y/2l)^2}{2}\right) \qquad \left\{\theta \approx \frac{y}{2l}\right\}$$

or $$\Delta x = 2d\left(1 - \frac{y^2}{8l^2}\right)$$

For dark fringes

$\Delta x = n\lambda$ (since the rays are out of phase)

or $$2d\left(1 - \frac{y^2}{8l^2}\right) = n\lambda$$

or $$y^2 = \frac{4l^2}{d}(2d - n\lambda)$$

or $$r_i^2 = \frac{4l^2}{d}(2d - i\lambda) \qquad \ldots(1)$$

and $$r_k^2 = \frac{4l^2}{d}(2d - k\lambda) \qquad \ldots(2)$$

Subtracting (1) from (2), we get

$$\lambda = \frac{\boldsymbol{d\,(r_k^2 - r_i^2)}}{\boldsymbol{4l^2\,(i - k)}}$$

385. (a) Path difference at angle θ is

$$\Delta x = d\sin(90° - \theta) = d\cos\theta$$

Corresponding phase difference between 1 and 2 is

$$\phi' = \left(\frac{2\pi}{\lambda}\right).\Delta x$$

or $$\phi' = \left(\frac{2\pi}{\lambda}\right)d\cos\theta \qquad \text{(2 lags)}$$

∴ Net phase difference will be

$$\phi_{net} = \phi' + \phi$$
$$= \left(\frac{2\pi}{\lambda}d\cos\theta\right) + \phi$$

For maximum intensity

$$\phi_{net} = 2n\pi \qquad n = 0, \pm 1, \pm 2, \ldots$$

or $$\left(\frac{2\pi}{\lambda}d\cos\theta\right) + \phi = 2n\pi$$

or $$\theta = \boldsymbol{\cos^{-1}\left(n - \frac{\phi}{2\pi}\right)\frac{\lambda}{d}} \qquad n = 0, \pm 1, \pm 2, \ldots$$

(b) In the direction θ = π, path difference is

$$\Delta x = d$$

Hence corresponding phase difference is

$$\phi' = \left(\frac{2\pi}{\lambda}\right).\Delta x = \left(\frac{2\pi}{\lambda}\right).d \qquad \text{(1 lags)}$$

∴ Net phase difference is

$$\phi_{net} = \phi' - \phi = \left(\frac{2\pi}{\lambda}\right).d - \phi$$

For maximum intensity

$$\phi_{net} = 2n\pi \qquad n = 0, \pm 1, \pm 2, \ldots$$

or $$\left(\frac{2\pi}{\lambda}\right)d - \phi = 2n\pi \qquad \ldots(1)$$

In the direction θ = 0°, net phase difference is

$$\phi_{net} = \left(\frac{2\pi}{\lambda}\right).d + \phi$$

For minimum intensity

$$\phi_{net} = (2n + 1)\pi, \qquad n = 0, \pm 1, \pm 2, \ldots$$

or $$\left(\frac{2\pi}{\lambda}\right).d + \phi = (2n + 1)\pi \qquad \ldots(2)$$

From equations (1) and (2), we get

$$\phi = \frac{\pi}{\boldsymbol{2}} \quad \text{and} \quad \frac{\boldsymbol{d}}{\lambda} = \boldsymbol{n + \frac{1}{4}}$$

where $n = 0, \pm 1, \pm 2, \ldots$

386. (a) Using $\frac{1}{f} = (\mu - 1)\left(\frac{1}{R_1} - \frac{1}{R_2}\right)$

we get $f = 20$ cm.

Since source lies in focal plane of lens so all the emergent rays will be parallel. Focal length of lens = 20 cm.

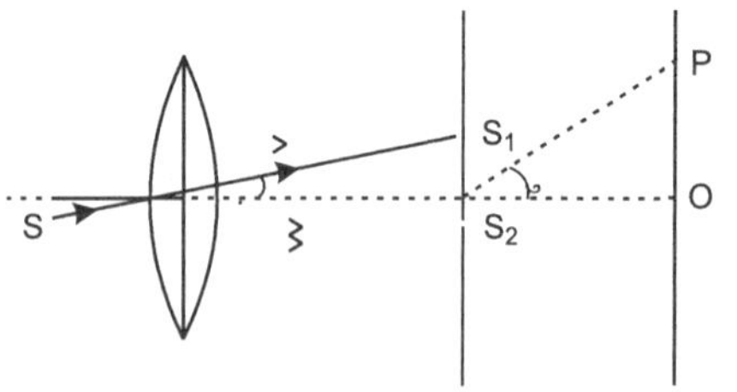

$\therefore$ Inclination of emergent rays from principal axis :

$$\sin\alpha \approx \tan\alpha = \frac{d/2}{20} = \frac{d}{40}$$

Initial path difference $= d\sin\alpha$

Initial phase difference $\phi = \frac{2\pi}{\lambda} d\sin\alpha$

For the central maxima

by $\quad d\sin\alpha = d\sin\theta$

or $\quad \theta = \alpha$

So, central maximum will be shifted by a distance

$$y = D\tan\theta = (1)\frac{d}{40} = \frac{d}{40}$$

Fringe width $= \lambda\frac{D}{d} = \frac{\lambda}{d}$ (as $D = 1$ m)

(b) $\frac{I_0}{I_{max}} = \cos^2\left(\frac{\phi}{2}\right)$

$$= \cos^2\left(\frac{\pi d\sin\alpha}{\lambda}\right) = \cos^2\left(\frac{\pi d^2}{40\lambda}\right)$$

387. (a) Let central maxima lies at point P at a distance y from the point O. Then optical path difference Δx at point P is

$$\Delta x = \frac{yd}{D} + (\mu_g - 1)t - \mu_w d\sin\theta$$

For central maxima $\Delta x = 0$

$$y = \frac{D}{d}[(\mu_w d\sin\theta) - (\mu_g - 1)t]$$

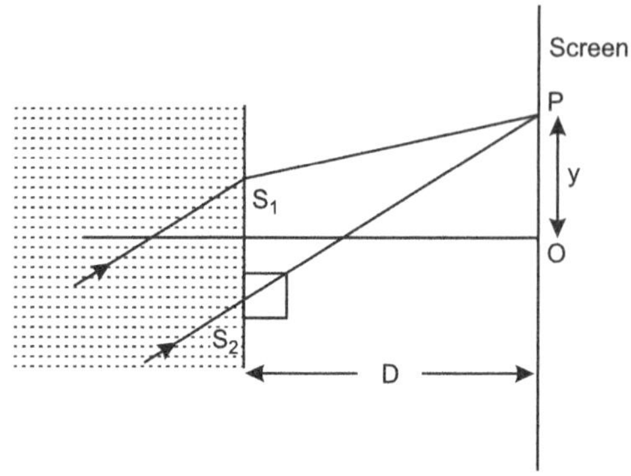

$$= \frac{1}{3\times10^{-4}}\left[\frac{4}{3}\times3\times10^{-4}\times\frac{1}{2} - 0.5\times0.41\times10^{-3}\right]$$

$$= \frac{1}{3\times10^{-4}}[2\times10^{-4} - 2.05\times10^{-4}]$$

$$= -\frac{5\times10^{-6}}{3\times10^{-4}} = -1.66\times10^{-2}\text{ m}$$

$$= -1.66\text{ cm}$$

So central maxima lies at a distance 1.66 cm below the central line.

(b) At point O, optical path difference is

$(\mu_g - 1)t - \mu_w d\sin\theta = -5\times10^{-6}$ m

$$= -10\lambda$$

So intensity at O is maximum or,

$$\frac{I_O}{I_{max}} = 1$$

388. Let P and Q' would have been the position of first and second minima (last two), had the

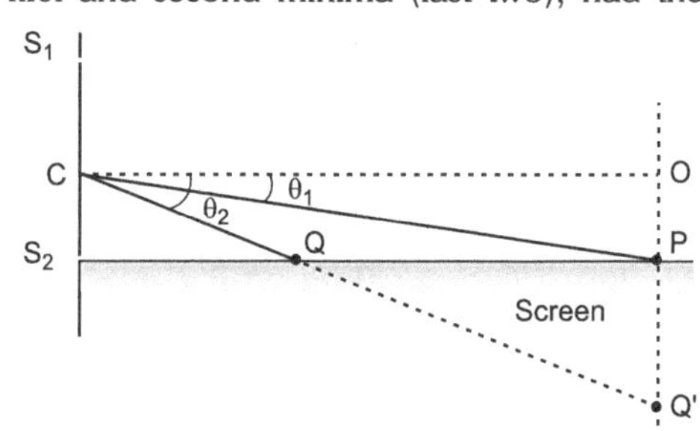

screen be perpendicular to S_2P.

Since, the angular positions of minima do not depend on the position of the screen.

$\therefore$ Second minima would be formed at Q on the screen.

$$OP = \frac{d}{2}$$

if $S_1S_2 = d$

Let $S_2P = X_1$

For minima at P

$$d \sin \theta_1 = \frac{\lambda_1}{2}$$

$$\Rightarrow \quad d (\tan \theta_1) = \frac{\lambda_1}{2}$$

$$\therefore \quad d\left[\frac{d/2}{X_1}\right] = \frac{\lambda_1}{2}$$

$$\therefore \quad S_2P = X_1 = \frac{d^2}{\lambda_1}$$

For minima at Q

Let $S_2Q = X_2$

$$d \sin \theta_2 = \frac{3}{2}\lambda_1$$

$$\sin \theta_2 = \tan \theta_2 = \frac{3\lambda_1}{2d}$$

and $$\tan \theta_2 = \frac{(d/2)}{X_2}$$

By comparison

$$\frac{3\lambda_1}{2d} = \frac{d}{2X_1}$$

$$X_2 = \frac{d^2}{3\lambda_1}$$

$$\therefore \quad PQ = X_2 - X_1$$

$$= \frac{d^2}{\lambda_1}\left[1 - \frac{1}{3}\right] = \frac{2d^2}{3\lambda_1}$$

In the second case

Given that fringe width $= \frac{D\lambda_2}{d}$

$$PQ = 600\left(\frac{D\lambda_2}{d}\right)$$

$$\Rightarrow \quad \frac{2d^2}{3\lambda_1} = 600\frac{D\lambda_2}{d}$$

$$\Rightarrow \quad d^3 = 900\lambda_1\lambda_2 D$$

$$= 900 \times 4000 \times 6000 \times 10^{-20} \times 1$$

$$= 216 \times 10^{-12}$$

$$d = 6 \times 10^{-4} \text{ m} = 0.6 \text{ mm}$$

389. Applying Snell's law on face AB,

$$(1) \sin 45° = (\sqrt{2}) \sin r$$

$$\therefore \quad \sin r = \frac{1}{2}$$

or $$r = 30°$$

i.e., ray becomes parallel to AD inside the block.

Now applying,

$$\frac{\mu_2}{v} - \frac{\mu_1}{u} = \frac{\mu_2 - \mu_1}{R} \text{ on face } CD,$$

$$\frac{1.514}{OE} - \frac{\sqrt{2}}{\infty} = \frac{1.414 - \sqrt{2}}{0.4}$$

Solving this equation, we get

$$OE = 6.06 \text{ m}$$

390. Differentiating the lens formula $\frac{1}{v} - \frac{1}{u} = \frac{1}{f}$ with respect to time, we get

$$-\frac{1}{v^2}\cdot\frac{dv}{dt} + \frac{1}{u^2}\cdot\frac{du}{dt} = 0 \qquad (\text{as } f = \text{constant})$$

$$\therefore \quad \left(\frac{dv}{dt}\right) = \left(\frac{v^2}{u^2}\right)\cdot\frac{du}{dt} \qquad \ldots(1)$$

Further, substituting proper values in lens formula we have,

$$\frac{1}{v} + \frac{1}{0.4} = \frac{1}{0.3} \qquad (u = -0.4 \text{ m}, f = 0.3 \text{ m})$$

or $$v = 1.2 \text{ m}$$

Putting the values in Eq. (1)

Magnitude of rate of change of position of image = 0.09 m/s

Lateral magnification $m = \frac{v}{u}$

$$\therefore \quad \frac{dm}{dt} = \frac{u\cdot\frac{dv}{dt} - v\cdot\frac{du}{dt}}{u^2}$$

$$= \frac{(-0.4)(0.09) - (1.2)(0.01)}{(0.4)^2}$$

$$= -0.3 \text{ per second}$$

$\therefore$ Magnitude of rate of change of lateral magnification = 0.3 per second.

391. For refraction at first surface,

$$\frac{\mu_2}{v_1} - \frac{\mu_1}{-\infty} = \frac{\mu_2 - \mu_1}{+R} \qquad \ldots(1)$$

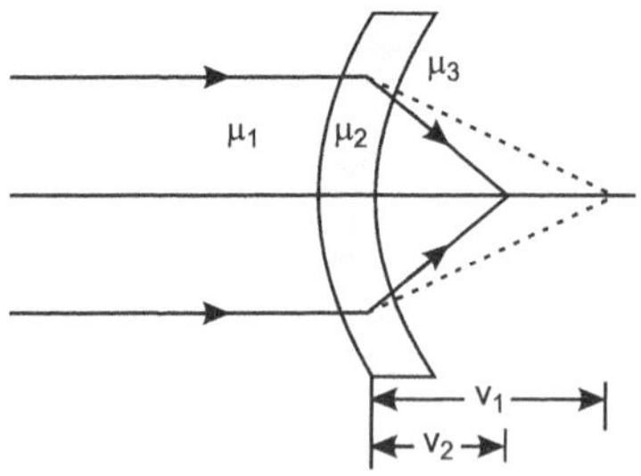

For refraction at 2nd surface,

$$\frac{\mu_3}{v_2} - \frac{\mu_2}{v_1} = \frac{\mu_3 - \mu_2}{+R} \qquad \ldots(2)$$

Adding Eqs. (1) and (2), we get

$$\frac{\mu_3}{v_2} = \frac{\mu_3 - \mu_1}{R} \quad \text{or} \quad v_2 = \frac{\mu_3 R}{\mu_3 - \mu_1}$$

Therefore, focal length of the given lens system is $\frac{\mu_3 R}{\mu_3 - \mu_1}$.

392. (a) $\sin i_1 = \mu \sin r_1$

or $\sin 60° = \sqrt{3} \sin r_1$

$\therefore$ $\sin r_1 = \frac{1}{2}$

or $r_1 = 30°$

Now $r_1 + r_2 = A$

$\therefore$ $r_2 = A - r_1 = 30° - 30° = 0°$

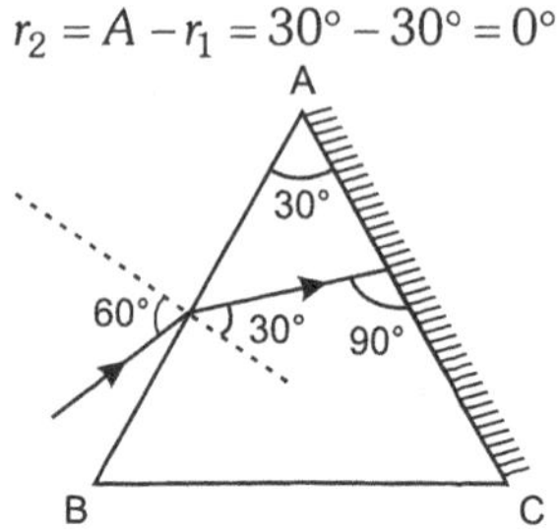

Therefore, ray of light falls normally on the face AC and angle of emergence $i_2 = 0°$.

(b) Multiple reflection occurs between surfaces of film. Intensity will be maximum if interference takes place in the transmitted wave.

For maximum thickness

$$\Delta x = 2\mu t = \lambda \qquad (t = \text{thickness})$$

$$\therefore \quad t = \frac{\lambda}{2\mu} = \frac{6600}{2 \times 2.2} = 1500 \text{ Å}$$

393. Let R be the radius of curvature of both the surfaces of the equiconvex lens. In the first case :

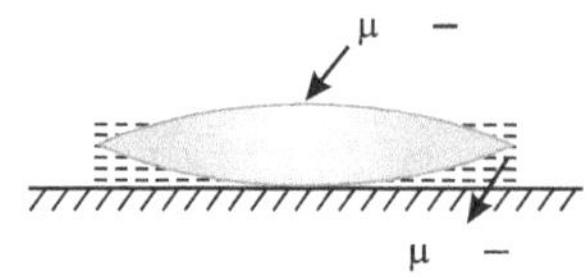

Let f_1 be the focal length of equiconvex lens of refractive index μ_1 and f_2 the focal length of planoconcave lens of refractive index μ_2. The focal length of the combined lens system will be given by :

$$\frac{1}{F} = \frac{1}{f_1} + \frac{1}{f_2}$$

$$= (\mu_1 - 1)\left(\frac{1}{R} - \frac{1}{-R}\right) + (\mu_2 - 1)\left(\frac{1}{-R} - \frac{1}{\infty}\right)$$

$$= \left(\frac{3}{2} - 1\right)\left(\frac{2}{R}\right) + \left(\frac{4}{3} - 1\right)\left(-\frac{1}{R}\right)$$

$$= \frac{1}{R} - \frac{1}{3R} = \frac{2}{3R}$$

or $F = \frac{3R}{2}$

Now image coincides with the object when ray of light retraces its path or it falls normally on the plane mirror. This is possible only when object is at focus of the lens system.

Hence $F = 15$ cm

(Distance of object = 15 cm)

or $\frac{3R}{2} = 15$ cm or $R = 10$ cm

In the second case, let μ be the refractive index of the liquid filled between lens and mirror and let F' be the focal length of new lens system. Then

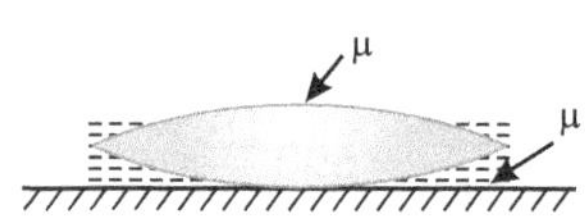

$$\frac{1}{F'} = (\mu_1 - 1)\left(\frac{1}{R} - \frac{1}{-R}\right) + (\mu - 1)\left(\frac{1}{-R} - \frac{1}{\infty}\right)$$

or $\dfrac{1}{F'} = \left(\dfrac{3}{2} - 1\right)\left(\dfrac{2}{R}\right) - \dfrac{(\mu - 1)}{R}$

or $= \dfrac{1}{R} - \dfrac{\mu - 1}{R} = \dfrac{(2-\mu)}{R}$

$\therefore$ $F' = \dfrac{R}{2-\mu} = \dfrac{10}{2-\mu}$ $(R = 10 \text{ cm})$

Now the image coincides with object when it is placed at 25 cm distance.

Hence $F' = 25$ or $\dfrac{10}{2-\mu} = 25$

or $50 - 25\mu = 10$

or $25\mu = 40$

$\therefore$ $\mu = \dfrac{40}{25} = 1.6$ or $\mu = \mathbf{1.6}$

394. Rays coming from object AB first refract from the lens and then reflect from the mirror.

Refraction from the lens :

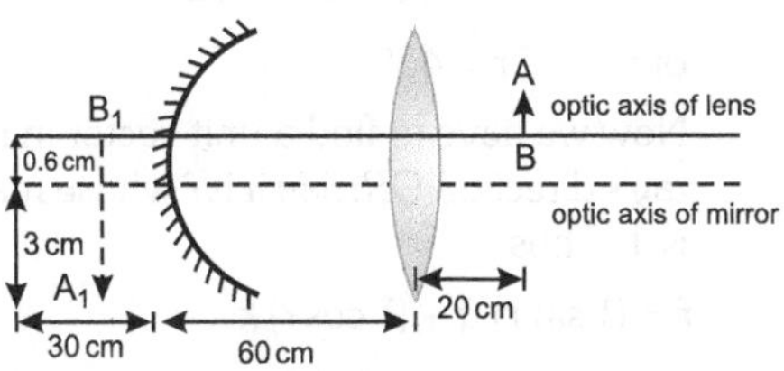

(a)

$u = -20$ cm, $f = +15$ cm

Using lens formula,

$$\frac{1}{v} - \frac{1}{u} = \frac{1}{f}$$

or $\dfrac{1}{v} - \dfrac{1}{-20} = \dfrac{1}{15}$

$\therefore$ $v = +60$ cm

and linear magnification,

$$m_1 = \frac{v}{u} = \frac{+60}{-20} = -3$$

i.e. first image formed by the lens will be at 60 cm from it (or 30 cm from mirror) towards left and 3 times magnified but inverted. Length of first image A_1B_1 would be $1.2 \times 3 = 3.6$ cm (inverted).

Reflection from mirror :

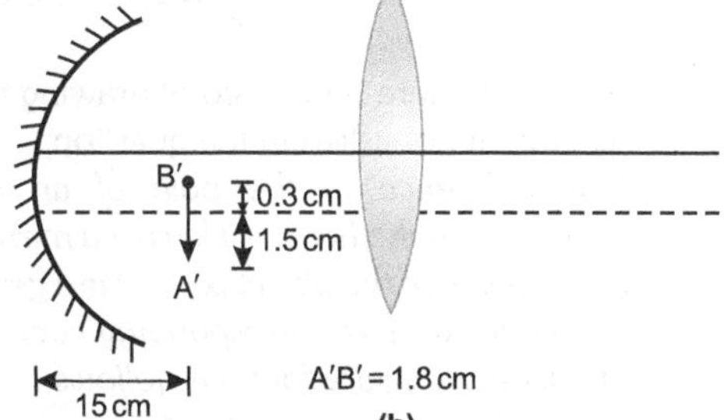

(b)

Image formed by lens (A_1B_1) will behave like a virtual object for mirror at a distance of 30 cm from it as shown in figure (a). Therefore,

$u = +30$ cm, $f = -30$ cm

Using mirror formula, $\dfrac{1}{v} + \dfrac{1}{u} = \dfrac{1}{f}$

or $\dfrac{1}{v} + \dfrac{1}{30} = -\dfrac{1}{30}$

$\therefore$ $v = -15$ cm

And linear magnification,

$$m_2 = -\frac{v}{u} = -\frac{(-15)}{(+30)} = +\frac{1}{2}$$

i.e. final image **$A'B'$ will be located at a distance of 15 cm from the mirror** (towards right).

and since magnification is $+\dfrac{1}{2}$, length of final image would be

$$3.6 \times \frac{1}{2} = 1.8 \text{ cm}$$

Point B_1 is 0.6 cm above the optic axis of mirror, therefore, its image B' would be $(0.6)\left(\dfrac{1}{2}\right) = 0.3$ cm above optic axis. Similarly, point A_1 is 3 cm below the optic axis, therefore, its image A' will be $3 \times \dfrac{1}{2} = 1.5$ cm below the optic axis as shown in figure (b).

Total magnification of the image,

$$\boldsymbol{m} = m_1 \times m_2 = (-3)\left(+\frac{1}{2}\right) = -\frac{\mathbf{3}}{\mathbf{2}}$$

$$\therefore\ A'B' = (m)(AB) = \left(-\frac{3}{2}\right)(1.2) = -1.8 \text{ cm}$$

Note that, there is no need of drawing the ray diagram if not asked in the question.

With reference to the pole of an optical instrument (whether it is a lens or a mirror) the coordinates of the object (x_o, y_o) are generally known to us. The corresponding coordinates of image (x_i, y_i) are found as follows :

x_i is obtained using $\dfrac{1}{v} \pm \dfrac{1}{u} = \dfrac{1}{f}$

Here v is actually x_i and u is x_o, i.e., the above formula can be written as

$$\frac{1}{x_i} \pm \frac{1}{x_o} = \frac{1}{f}$$

Similarly, y_i is obtained from $m = \dfrac{I}{O}$

Here, I is y_i and O is y_o, i.e. the above formula can be written as $m = y_i/y_o$ or $y_i = my_o$.

395. Incident ray

$$\vec{A} = 6\sqrt{3}\,\hat{i} + 8\sqrt{3}\,\hat{j} - 10\hat{k}$$
$$= (6\sqrt{3}\,\hat{i} + 8\sqrt{3}\,\hat{j}) + (-10\hat{k})$$
$$= \vec{QO} + \vec{PQ} \qquad \text{(As shown in figure)}$$

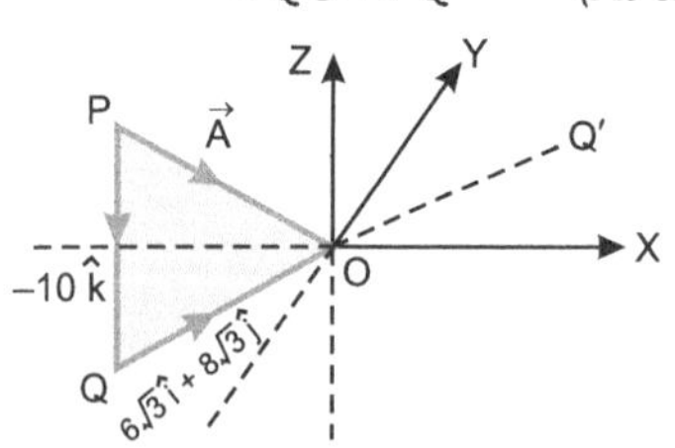

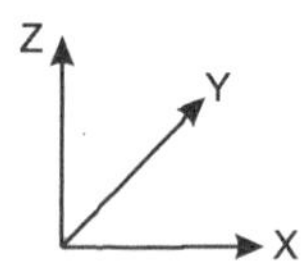

Note that $\vec{QO}$ is lying on $x-y$ plane.

Now QQ' and Z-axis are mutually perpendicular. Hence we can show them in two-dimensional figure as below :

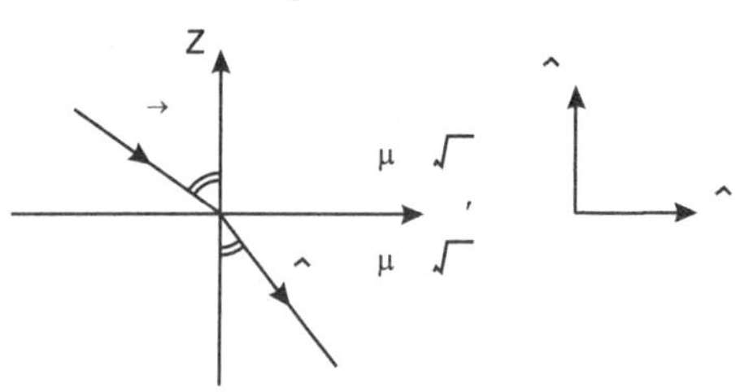

Vector $\vec{A}$ makes an angle i with Z-axis given by

$$i = \cos^{-1}\left\{\frac{10}{\sqrt{(10)^2 + (6\sqrt{3})^2 + (8\sqrt{3})^2}}\right\}$$
$$= \cos^{-1}\left\{\frac{1}{2}\right\}$$

hence $\qquad i = 60°$

Unit vector in the direction of QOQ' will be

$$\hat{q} = \frac{6\sqrt{3}\,\hat{i} + 8\sqrt{3}\,\hat{j}}{\sqrt{(6\sqrt{3})^2 + (8\sqrt{3})^2}} = \frac{1}{5}(3\hat{i} + 4\hat{j})$$

Snell's law gives

$$\frac{\sqrt{3}}{\sqrt{2}} = \frac{\sin i}{\sin r} = \frac{\sin(60°)}{\sin r}$$

$$\therefore \qquad \sin r = \frac{\sqrt{3}/2}{\sqrt{3}/\sqrt{2}} = \frac{1}{\sqrt{2}}$$

or $\qquad r = 45°$

Now we have to find a unit vector in refracted ray's direction OR. Say it is $\hat{r}$ whose magnitude is 1. Thus

$$\hat{r} = (1 \sin r)\,\hat{q} - (1 \cos r)\,\hat{k}$$
$$= \frac{1}{\sqrt{2}}(\hat{q} - \hat{k}) = \frac{1}{\sqrt{2}}\left[\frac{1}{5}(3\hat{i} + 4\hat{j}) - \hat{k}\right]$$
$$\boldsymbol{\hat{r} = \frac{1}{5\sqrt{2}}(3\hat{i} + 4\hat{j} - 5\hat{k})}$$

396. $$n_1 = 1.20 + \frac{10.8 \times 10^4}{\lambda^2}$$

and $$n_2 = 1.45 + \frac{1.80 \times 10^4}{\lambda^2}$$

Here λ is in nm.

(a) The incident ray will not deviate at BC if

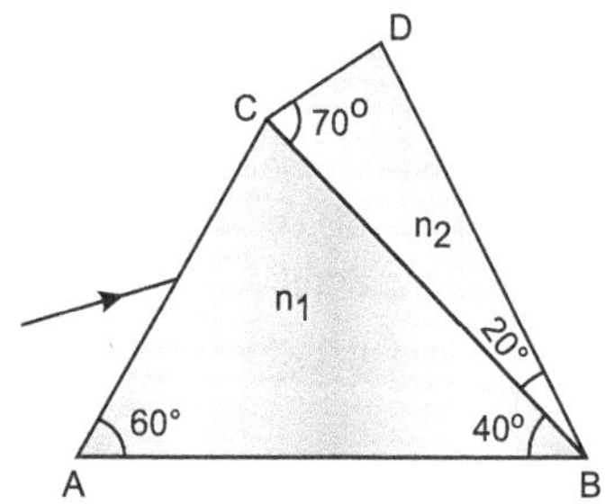

$$n_1 = n_2$$

$$\Rightarrow \quad 1.20 + \frac{10.8 \times 10^4}{\lambda_0^2} = 1.45 + \frac{1.80 \times 10^4}{\lambda_0^2} \quad (\lambda = \lambda_0)$$

$$\Rightarrow \quad \frac{9 \times 10^4}{\lambda_0^2} = 0.25$$

or $$\lambda_0 = \frac{3 \times 10^2}{0.5}$$

or $$\lambda_0 = \mathbf{600\ nm}$$

(b) The given system is a part of an equilateral prism of prism angle 60° as shown in figure.

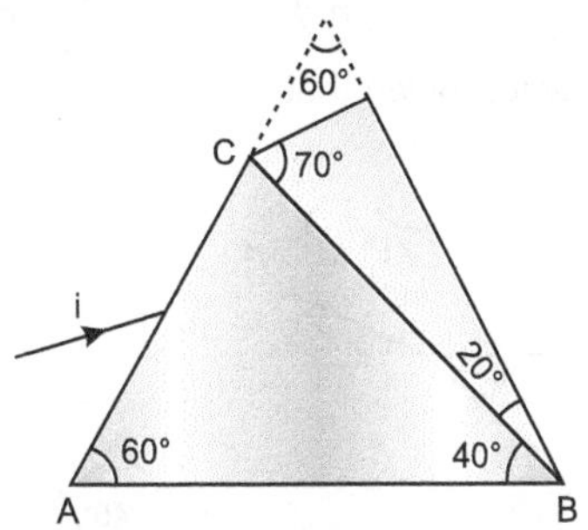

At minimum deviation,

$$r_1 = r_2 = \frac{60^\circ}{2} = 30^\circ = r \quad \text{(say)}$$

$$\therefore \quad n_1 = \frac{\sin i}{\sin r}$$

$$\therefore \quad \sin i = n_1 \cdot \sin 30^\circ$$

$$\sin i = \left\{1.20 + \frac{10.8 \times 10^4}{(600)^2}\right\}\left(\frac{1}{2}\right)$$

$$= \frac{1.5}{2} = \frac{3}{4} \quad (\lambda = \lambda_0 = 600\text{ nm})$$

or $$\boldsymbol{i = \sin^{-1}(3/4)}$$

397. From lens maker's formula,

$$\frac{1}{f} = (\mu - 1)\left(\frac{1}{R_1} - \frac{1}{R_2}\right)$$

we have $$\frac{1}{0.3} = \left(\frac{3}{2} - 1\right)\left(\frac{1}{R} - \frac{1}{-R}\right)$$

(Here $R_1 = R$ and $R_2 = -R$)

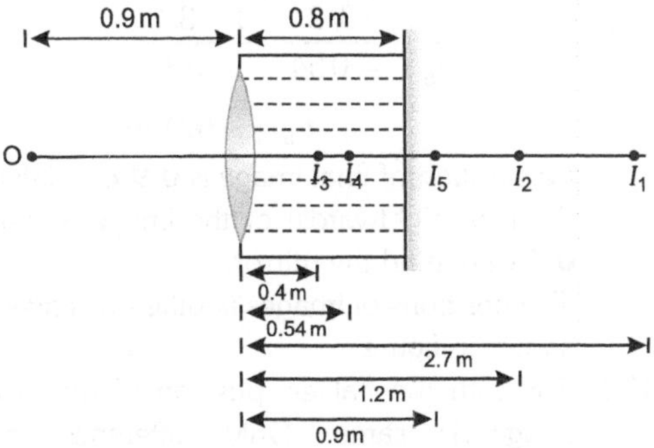

$$\therefore \quad R = 0.3\text{ m}$$

Now applying $\frac{\mu_2}{v} - \frac{\mu_1}{u} = \frac{\mu_2 - \mu_1}{R}$ at air glass surface, we get

$$\frac{3/2}{v_1} - \frac{1}{-(0.9)} = \frac{3/2 - 1}{0.3} \quad \therefore \ v_1 = 2.7\text{ m}$$

i.e., first image I_1 will be formed at 2.7 m from the lens. This will act as the virtual object for glass water surface. Therefore, applying $\frac{\mu_2}{v} - \frac{\mu_1}{u} = \frac{\mu_2 - \mu_1}{R}$ at glass water surface, we have

$$\frac{4/3}{v_2} - \frac{3/2}{2.7} = \frac{4/3 - 3/2}{-0.3}$$

$$\therefore \quad v_2 = 1.2\text{ m}$$

i.e., second image I_2 is formed at 1.2 m from the lens or 0.4 m from the plane mirror. This will act as a virtual object for mirror. Therefore, third real image I_3 will be formed at a distance of 0.4 m in front of the mirror after reflection from it. Now this image will work as a real object for water-glass interface. Hence applying

$$\frac{\mu_2}{v} - \frac{\mu_1}{u} = \frac{\mu_2 - \mu_1}{R}$$

We get $$\frac{3/2}{v_4} - \frac{4/3}{-(0.8 - 0.4)} = \frac{3/2 - 4/3}{0.3}$$

$$\therefore \quad v_4 = -\ 0.54\text{ m}$$

i.e. fourth image is formed to the right of the lens at a distance of 0.54 m from it. Now finally applying the same formula for glass-air surface,

$$\frac{1}{v_5} - \frac{3/2}{-0.54} = \frac{1-3/2}{-0.3}$$

$\therefore$ $\quad v_5 = -\mathbf{0.9\ m}$

i.e. position of final image is **0.9 m** relative to the lens (rightwards) or the image is formed **0.1 m** behind the mirror.

The positions of images at different stages are shown in figure.

398. For both the halves, position of object and image is same. Only difference is of magnification. Magnification for one of the halves is given as 2 (> 1). This can be for the first one, because for this, $|v| > |u|$. Therefore, magnification, $|m| = |v/u| > 1$

So, for the first half

$$|v/u| = 2 \quad \text{or} \quad |v| = 2|u|$$

Let $u = -x$, then $v = +2x$

and $\quad |u| + |v| = 1.8$ m

i.e. $\quad 3x = 1.8$ m

or $\quad x = 0.6$ m

Hence $\quad u = -0.6$ m

and $\quad v = +1.2$ m

Using $\quad \frac{1}{f} = \frac{1}{v} - \frac{1}{u} = \frac{1}{1.2} - \frac{1}{-0.6} = \frac{1}{0.4}$

$\therefore$ $\quad f = \mathbf{0.4\ m}$

For the second half

$$\frac{1}{f} = \frac{1}{1.2-d} - \frac{1}{-(0.6+d)}$$

or $$\frac{1}{0.4} = \frac{1}{1.2-d} + \frac{1}{0.6+d}$$

Solving this, we get

$$d = \mathbf{0.6\ m}$$

Magnification for the second half will be

$$m_2 = \frac{v}{u} = \frac{0.6}{-(1.2)} = -\frac{1}{2}$$

and magnification for the first half is

$$m_1 = \frac{v}{u} = \frac{1.2}{-(0.6)} = -2$$

The ray diagram is as follows :

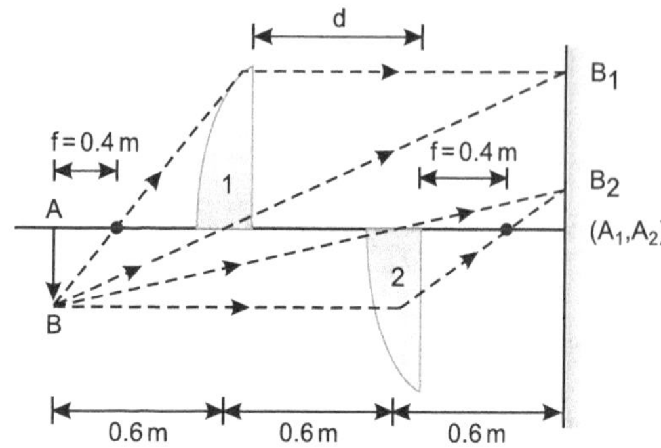

399. (i) Critical angle θ_c at face AC will be given by

$$\theta_c = \sin^{-1}\left(\frac{n_1}{n}\right) \quad \text{or} \quad \sin\theta_c = \frac{n_1}{n}$$

Now it is given that

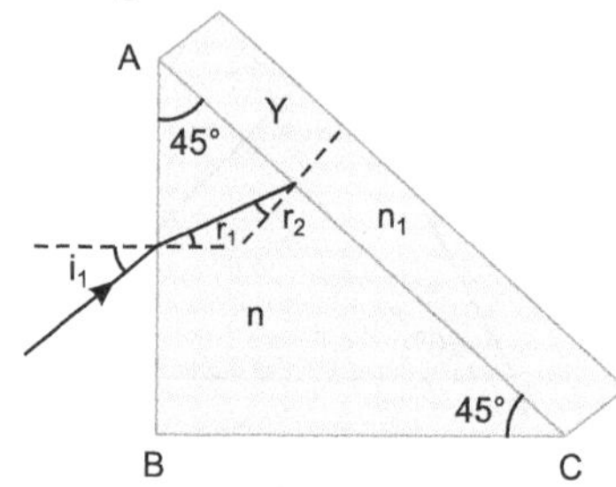

$r_2 = \theta_c$ or $r_1 = A - r_2 = (45° - \theta_c)$

Applying Snell's law at face AB, we have

$$n = \frac{\sin i_1}{\sin r_1} \quad \text{or} \quad \sin i_1 = n \sin r_1$$

$\therefore$ $\quad i_1 = \sin^{-1}\{n \sin r_1\}$

Substituting value of r_1, we get

$i_1 = \sin^{-1}\{n \sin(45° - \theta_c)\}$

$= \sin^{-1}\{n(\sin 45° \cos\theta_c - \cos 45° \sin\theta_c)\}$

$= \sin^{-1}\left\{\frac{n}{\sqrt{2}}\left(\sqrt{1-\sin^2\theta_c} - \sin\theta_c\right)\right\}$

$$\left(\sin\theta_c = \frac{n_1}{n}\right)$$

$= \sin^{-1}\left\{\frac{n}{\sqrt{2}}\left(\sqrt{1-\frac{n_1^2}{n^2}} - \frac{n_1}{n}\right)\right\}$

$$i_1 = \sin^{-1}\left\{\frac{1}{\sqrt{2}}(\sqrt{n^2 - n_1^2} - n_1)\right\}$$

Therefore, required angle of incidence (i_1) at face AB for which the ray strikes at AC at critical angle is

$$\boldsymbol{i_1 = \sin^{-1}\left\{\frac{1}{\sqrt{2}}(\sqrt{n^2 - n_1^2} - n_1)\right\}}$$

(ii) The ray will pass undeviated through face AC when either (a) $n_1 = n$ or

(b) $r_2 = 0°$ *i.e.* ray falls normally on face AC.

Here $n_1 \neq n$ (because $n_1 < n$ is given)

$\therefore$ $r_2 = 0°$

or $r_1 = A - r_2 = 45° - 0° = 45°$

Now applying Snell's law at face AB, we have

$$n = \frac{\sin i_1}{\sin r_1} \quad \text{or} \quad 1.352 = \frac{\sin i_1}{\sin 45°}$$

$$\therefore \quad \sin i_1 = (1.352)\left(\frac{1}{\sqrt{2}}\right) = 0.956$$

$$\therefore \quad i_1 = \sin^{-1}(0.956) = 73°$$

Therefore, required angle of incidence is

$$\boldsymbol{i_1 = 73°}$$

400. (a) Refractive index is a function of y. It varies along y-axis *i.e.* the boundary separating two media is parallel to x-axis or normal at any point will be parallel to y-axis.

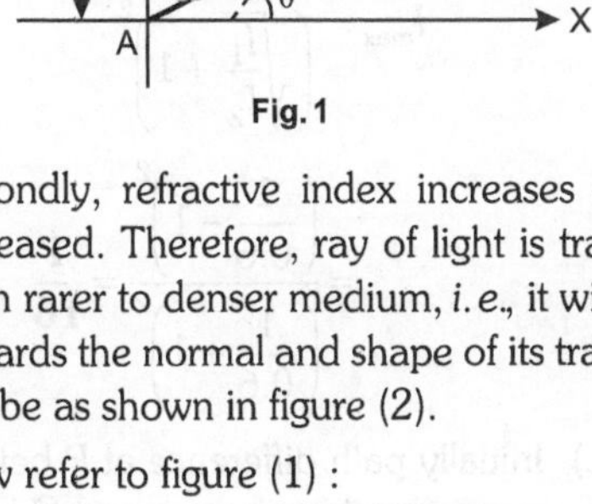

Fig. 1

Secondly, refractive index increases as y is increased. Therefore, ray of light is travelling from rarer to denser medium, *i.e.*, it will bend towards the normal and shape of its trajectory will be as shown in figure (2).

Now refer to figure (1) :

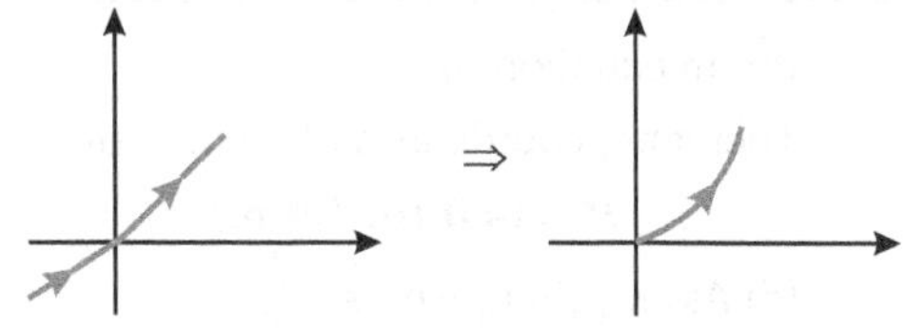

Fig. 2

Let i be the angle of incidence at any point B on its path

$$\theta = 90° - i$$

or $\tan\theta = \tan(90° - i) = \cot i$

or **slope = cot i**

(b) but $\tan\theta = \dfrac{dy}{dx}$

$$\therefore \quad \frac{dy}{dx} = \cot i \qquad \text{...(1)}$$

Applying Snell's law at A and B,

$n_A \sin i_A = n_B \sin i_B$

$n_A = 1$, because $y = 0$

$\sin i_A = 1$

because $i_A = 90°$

(Grazing incidence)

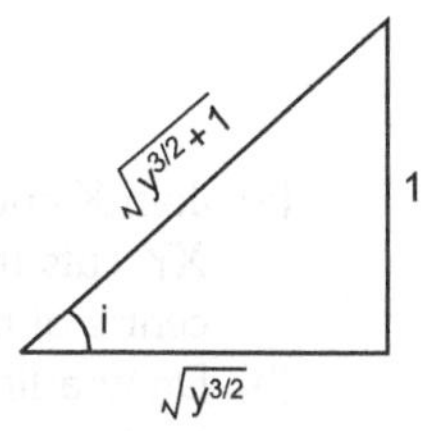

Fig. 3.

$$n_B = \sqrt{ky^{3/2} + 1} = \sqrt{y^{3/2} + 1}$$

because $k = 1.0\,(\text{m})^{-3/2}$

$$\therefore \quad (1)(1) = \left(\sqrt{y^{3/2} + 1}\right)\sin i$$

$$\text{or} \quad \sin i = \frac{1}{\sqrt{y^{3/2} + 1}}$$

$$\therefore \quad \cot i = \sqrt{y^{3/2}} \quad \text{or} \quad y^{3/4} \qquad \text{...(2)}$$

Equating (1) and (2), we get

$$\frac{dy}{dx} = y^{3/4} \quad \text{or} \quad y^{-3/4}dy = dx$$

$$\text{or} \quad \int_0^y y^{-3/4}dy = \int_0^x dx$$

$$\text{or} \quad \boldsymbol{4y^{1/4} = x} \qquad \text{...(3)}$$

$\therefore$ The required equation of trajectory is $4y^{1/4} = x$.

(c) At point P, where the ray emerges from the slab

$y = 1.0$ m

$\therefore$ $x = 4.0$ m

(From equation 3)

Therefore, coordinates of point P are

$$\boldsymbol{P \equiv (4.0\ \text{m}, 1.0\ \text{m})}$$

(d) As $n_A \sin i_A = n_P \sin i_P$

and as $n_A = n_P = 1$

Therefore, $i_p = i_A = 90°$, *i.e.*, **the ray will emerge parallel to the boundary at P** *i.e.* at grazing emergence.

401. Steps (i) In case of a lens :

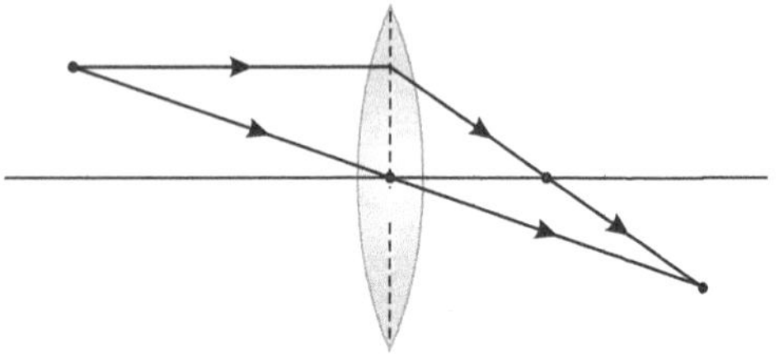

(a) Join X and Y. The point O, where the line XY cuts the optic axis AB, is the optical centre of the lens.

(b) Draw a line parallel to AB from point X. Let it cuts the lens at L. Join L and Y. The point F where the line LY cuts the optic axis AB is the focus of the lens F.

Note: As the image is inverted, lens should be a convex because a concave lens always forms a virtual and erect image.

Steps (ii) In case of a concave mirror :

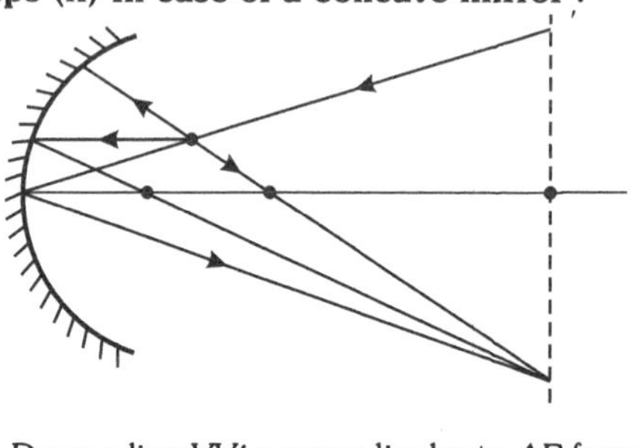

(a) Draw a line YY' perpendicular to AB from point Y. Let it cuts the line AB at point P. Locate a point Y' such that $PY = PY'$.

(b) Extend the line XY'. Let it cuts the line AB at point O. Then O is the pole of the mirror.

(c) Join X and Y. The point C, where the line XY cuts the optic axis AB, is the centre of curvature of the mirror.

(d) The centre point F of OC is the focus of the mirror.

402. Let n_1 bright fringe corresponding to wavelength $\lambda_1 = 500$ nm coincides with n_2 bright fringe corresponding to wavelength $\lambda_2 = 700$ nm.

$$\therefore \quad n_1 \frac{\lambda_1 D}{d} = n_2 \frac{\lambda_2 D}{d}$$

$$\text{or} \quad \frac{n_1}{n_2} = \frac{\lambda_2}{\lambda_1} = \frac{7}{5}$$

This implies that 7th maxima of λ_1 coincides with 5th maxima of λ_2. Similarly 14th maxima of λ_1 will coincide with 10th maxima of λ_2 and so on.

$$\therefore \text{Minimum distance} = \frac{n_1 \lambda_1 D}{d}$$

$$= 7 \times 5 \times 10^{-7} \times 10^3$$

$$= 3.5 \times 10^{-3}\ \text{m}$$

$$= 3.5\ \text{mm}$$

403. (a) Shape of the interference fringes will be circular.

(b) Intensity of light reaching on the screen directly from the source $I_1 = I_0$ (say) and intensity of light reaching on the screen after reflecting from the mirror is

$I_2 = 36\%$ of $I_0 = 0.36 I_0$

$$\therefore \quad \frac{I_1}{I_2} = \frac{I_0}{0.36 I_0} = \frac{1}{0.36}$$

$$\text{or} \quad \sqrt{\frac{I_1}{I_2}} = \frac{1}{0.6}$$

$$\therefore \quad \frac{I_{min}}{I_{max}} = \frac{\left(\sqrt{\frac{I_1}{I_2}} - 1\right)^2}{\left(\sqrt{\frac{I_1}{I_2}} + 1\right)^2}$$

$$= \frac{\left(\frac{1}{0.6} - 1\right)^2}{\left(\frac{1}{0.6} + 1\right)^2} = \boldsymbol{\frac{1}{16}}$$

(c) Initially path difference at P between two waves reaching from S and S' is

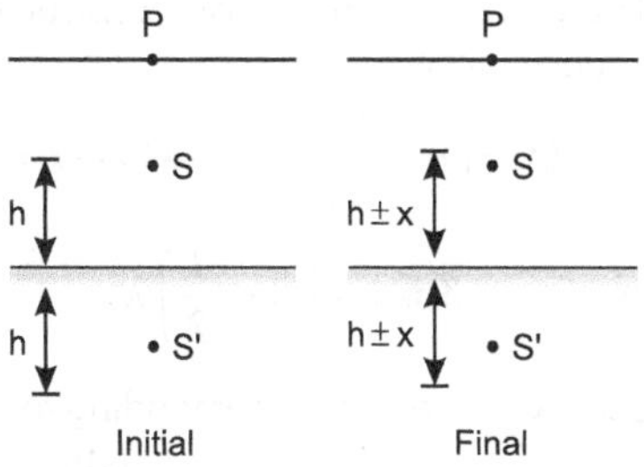

Therefore for maximum intensity at P:

$$2h = \left(n - \frac{1}{2}\right)\lambda \qquad \ldots(1)$$

Now let the source S is displaced by x (away or towards mirror) then new path difference will be $2h + 2x$ or $2h - 2x$. So for maximum intensity at P.

$$2h + 2x = \left[n + 1 - \frac{1}{2}\right]\lambda \qquad \ldots(2)$$

or $$2h - 2x = \left[n - 1 - \frac{1}{2}\right]\lambda \qquad \ldots(3)$$

Solving (1) and (2) or (1) and (3), we get

$$x = \frac{\lambda}{2} = \frac{600}{2} = \mathbf{300\ nm}$$

Note: Here we have taken the condition of maximum intensity at P as:

path difference $\Delta x = \left(n - \frac{1}{2}\right)\lambda$

because the reflected beam suffers a phase difference of π.

404. Given $y_1 = 40$ cm

$D_1 = 2\text{ m} = 200$ cm

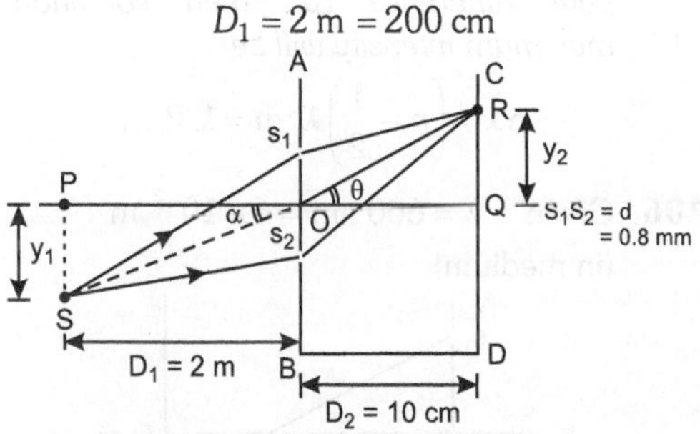

$D_2 = 10$ cm

$$\tan\alpha = \frac{y_1}{D_1} = \frac{40}{200} = \frac{1}{5}$$

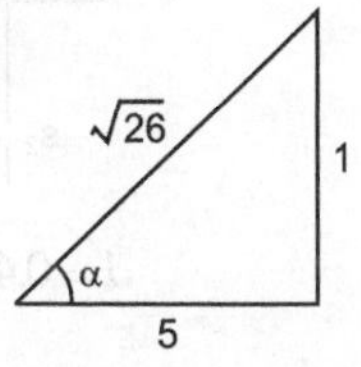

$\therefore \qquad \alpha = \tan^{-1}(1/5)$

$$\sin\alpha = \frac{1}{\sqrt{26}} \approx \frac{1}{5} = \tan\alpha$$

Path difference between Ss_1 and Ss_2 is

$$\Delta x_1 = Ss_1 - Ss_2$$

or $\quad \Delta x_1 = d\sin\alpha = (0.8\text{ mm})\left(\frac{1}{5}\right)$

or $\quad \Delta x_1 = 0.16\text{ mm} \qquad \ldots(1)$

Now let at point R on the screen, central bright fringe is observed (*i.e.* net path difference = 0). Path difference between s_2R and s_1R would be

$$\Delta x_2 = s_2R - s_1R$$

or $\quad \Delta x_2 = d\sin\theta \qquad \ldots(2)$

Central bright fringe will be observed when net path difference is zero.

or $\quad \Delta x_2 - \Delta x_1 = 0$

$$\Delta x_2 = \Delta x_1$$

or $\quad d\sin\theta = 0.16$

or $\quad (0.8)\sin\theta = 0.16$

or $\quad \sin\theta = \frac{0.16}{0.8} = \frac{1}{5}$

$$\tan\theta = \frac{1}{\sqrt{24}} \approx \sin\theta = \frac{1}{5}$$

Hence $\tan\theta = \frac{y_2}{D_2} = \frac{1}{5}$

$\therefore \quad y_2 = \frac{D_2}{5} = \frac{10\text{ cm}}{5} = \mathbf{2\ cm}$

Therefore, central bright fringe is observed at 2 cm above point *Q* on side *CD*.

Alternate solution :

Δx at R will be zero if

$$\Delta x_1 = \Delta x_2$$

or $\quad d\sin\alpha = d\sin\theta$

or $\quad \alpha = \theta$

or $\quad \tan\alpha = \tan\theta$

$$\frac{y_1}{D_1} = \frac{y_2}{D_2}$$

or $\quad y_2 = \frac{D_2}{D_1} \cdot y_1 = \left(\frac{10}{200}\right)(40)$ cm

or $\quad y_2 = 2$ cm

(ii) The central bright fringe will be observed at point Q. If the path difference created by the liquid slab of thickness $t = 10$ cm or 100 mm is equal to Δx_1, so that the net path difference at Q becomes zero.

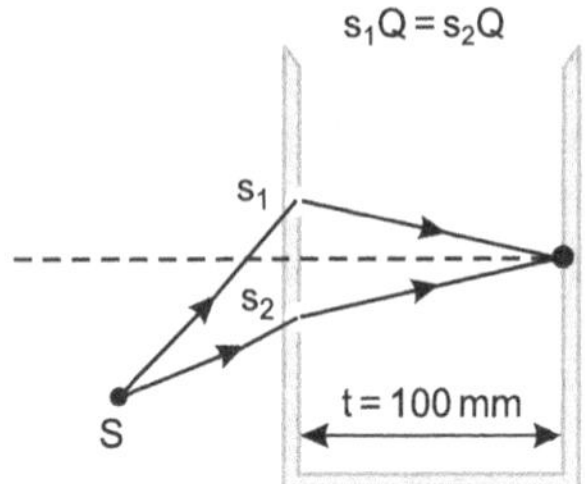

So $\quad (\mu - 1)\, t = \Delta x_1$

or $\quad (\mu - 1)(100) = 0.16$

or $\quad \mu - 1 = 0.0016$

or $\quad \mu = \mathbf{1.0016}$

405. Incident ray AB is partly reflected as ray 1 from the upper surface and partly reflected as ray 2 from the lower surface of the layer of thickness t and refractive index $\mu_1 = 1.8$ as shown in figure. Path difference between the two rays would be

$$\Delta x = 2\mu_1 t = 2\,(1.8)\, t = 3.6\, t$$

Ray 1 is reflected from a denser medium, therefore, it undergoes a phase change of π, whereas the ray 2 gets reflected from a rarer medium, therefore, there is no change in phase of ray 2.

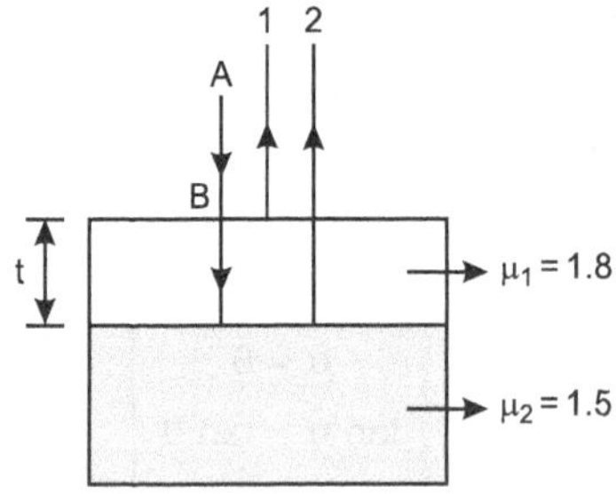

Hence phase difference between rays 1 and 2 would be $\Delta\phi = \pi$.

Therefore, condition of constructive interference will be :

$$\Delta x = \left(n - \frac{1}{2}\right)\lambda, \qquad \text{where } n = 1, 2, 3, \ldots$$

or $\quad \mathbf{3.6\, t} = \left(\boldsymbol{n} - \frac{\mathbf{1}}{\mathbf{2}}\right)\lambda$

Least value of t is corresponding to $n = 1$ or

$$t_{min} = \frac{\lambda}{2 \times 3.6} \quad \text{or} \quad t_{min} = \frac{648}{7.2} \text{ nm}$$

or $\quad \boldsymbol{t_{min} = 90}$ **nm**

(i) For a wave (whether it is sound or electromagnetic), a medium is denser or rarer is decided from the speed of wave in that medium. In denser medium, speed of wave is less. For example, water is rarer for sound, while denser for light compared to air because speed of sound in water is more than in air, while speed of light is less.

(ii) In transmission/refraction, no phase change takes place. In reflection, there is a change of phase of π when it is reflected by a denser medium and phase change is zero if it is reflected by a rarer medium.

(iii) If two waves in phase interfere having a path difference of Δx, then condition of maximum intensity would be $\Delta x = n\lambda : n = 0, 1, 2, \ldots$

But if two waves, which are already out of phase (a phase difference of π), interfere with path difference Δx, then condition of maximum intensity will be

$$\Delta x = \left(n - \frac{1}{2}\right)\lambda : n = 1, 2, \ldots$$

406. Given $\quad \lambda = 600 \text{ nm} = 6 \times 10^{-7}$ m (in medium)

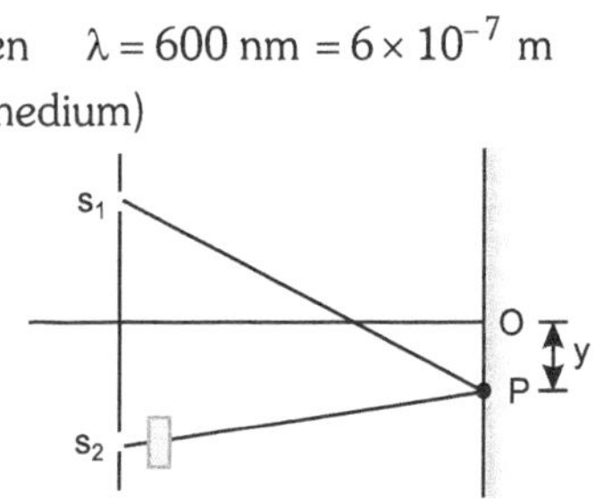

$$d = 0.45 \text{ mm} = 0.45 \times 10^{-3} \text{ m}$$

$$D = 1.5 \text{ m}$$

Thickness of glass sheet,

$$t = 10.4\ \mu m = 10.4 \times 10^{-6} \text{ m}$$

Refractive index of medium, $\mu_m = 4/3$ and refractive index of glass sheet,

$$\mu_g = 1.5$$

(a) Let central maximum is obtained at a distance y below point O. Then

$$\Delta x_1 = s_1P - s_2P = \frac{yd}{D}$$

Path difference due to glass sheet

$$\Delta x_2 = \left(\frac{\mu_g}{\mu_m} - 1\right)t$$

Net path difference will be zero when

$$\Delta x_1 = \Delta x_2 \quad \text{or} \quad \frac{yd}{D} = \left(\frac{\mu_g}{\mu_m} - 1\right)t$$

$$\therefore \quad y = \left(\frac{\mu_g}{\mu_m} - 1\right)t\frac{D}{d}$$

Substituting the values, we have

$$y = \left(\frac{1.5}{4/3} - 1\right)\frac{(10.4 \times 10^{-6})(1.5)}{0.45 \times 10^{-3}}$$

$$y = 4.33 \times 10^{-3} \text{ m}$$

or we can say $\mathbf{y = -\ 4.33\ mm.}$

(b) At O,

$$\Delta x_1 = 0 \quad \text{and} \quad \Delta x_2 = \left(\frac{\mu_g}{\mu_m} - 1\right)t$$

$\therefore$ Net path difference,

$$\Delta x = \Delta x_2$$

Corresponding phase difference, $\Delta\phi$ or simply $\phi = \frac{2\pi}{\lambda} \cdot \Delta x$

Substituting the values, we have

$$\phi = \frac{2\pi}{6 \times 10^{-7}}\left(\frac{1.5}{4/3} - 1\right)(10.4 \times 10^{-6})$$

$$= \left(\frac{13}{3}\right)\pi$$

Now $\quad (\phi) = I_{max} \cos^2\left(\frac{\phi}{2}\right)$

So $\quad I = I_{max} \cos^2\left(\frac{13\pi}{6}\right)$

$$\boldsymbol{I = \frac{3}{4} I_{max}}$$

(c) At O : Path difference is

$$\Delta x = \Delta x_2 = \left(\frac{\mu_g}{\mu_m} - 1\right)t$$

For maximum intensity at O

$$\Delta x = n\lambda \qquad \text{(Here } n = 1, 2, 3, \ldots)$$

$$\therefore \lambda = \frac{\Delta x}{1}, \frac{\Delta x}{2}, \frac{\Delta x}{3}, \ldots \text{ and so on}$$

$$\Delta x = \left(\frac{1.5}{4/3} - 1\right)(10.4 \times 10^{-6} \text{ m})$$

$$= \left(\frac{1.5}{4/3} - 1\right)(10.4 \times 10^{3} \text{ nm})$$

$$\Delta x = 1300 \text{ nm}$$

$\therefore$ Maximum intensity will be corresponding to

$$\lambda = 1300 \text{ nm}, \frac{1300}{2} \text{ nm}, \frac{13000}{3} \text{ nm}, \frac{1300}{4} \text{ nm}, \ldots$$

$$= 1300 \text{ nm}, 650 \text{ nm}, 433.33 \text{ nm}, 325 \text{ nm}, \ldots$$

The wavelengths in the range 400 to 700 nm are 650 nm and 433.33 nm

407. Given $\lambda = 0.5$ mm, $d = 1.0$ mm, $D = 1$ m

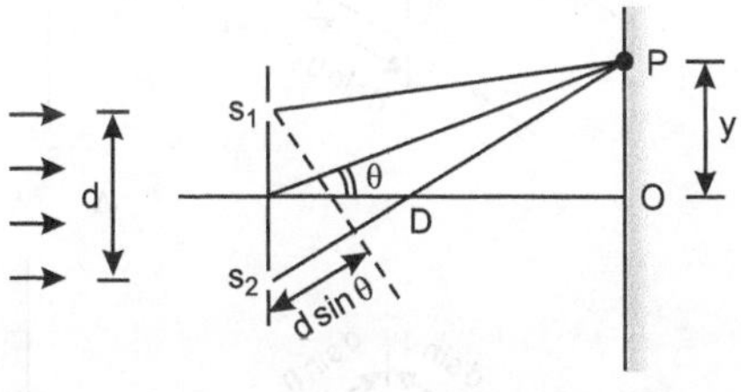

(a) When the incident beam falls normally : Path difference between the two rays s_2P and s_1P is

$$\Delta x = s_2P - s_1P \approx d \sin\theta$$

For minimum intensity,

$$d \sin\theta = (2n - 1)\lambda/2, \qquad n = 1, 2, 3, \ldots$$

$$\text{or} \quad \sin\theta = \frac{(2n-1)\lambda}{2d}$$

$$= \frac{(2n-1)\,0.5}{2\times 1.0} = \frac{(2n-1)}{4}$$

As $\sin\theta \le 1$, therefore, $\frac{2n-1}{4} \le 1$ or

$n \le 2.5$

So, n can be either 1 or 2.

When $\quad n = 1, \sin\theta_1 = 1/4$

or $\quad \tan\theta_1 = \frac{1}{\sqrt{15}}$

$\quad n = 2, \sin\theta_2 = 3/4$

or $\quad \tan\theta_2 = \frac{3}{\sqrt{7}}$

$\therefore \quad y = D\tan\theta = \tan\theta \qquad (D = 1\text{ m})$

So, the position of minima will be :

$$y_1 = \tan\theta_1 = \frac{1}{\sqrt{15}}\text{ m} = \mathbf{0.26\ m}$$

$$y_2 = \tan\theta_2 = \frac{3}{\sqrt{7}}\text{ m} = \mathbf{1.13\ m}$$

and as minima can be on either side of centre O. **Therefore, there will be four minimas at positions $\pm$ 0.26 m and $\pm$ 1.13 m on the screen.**

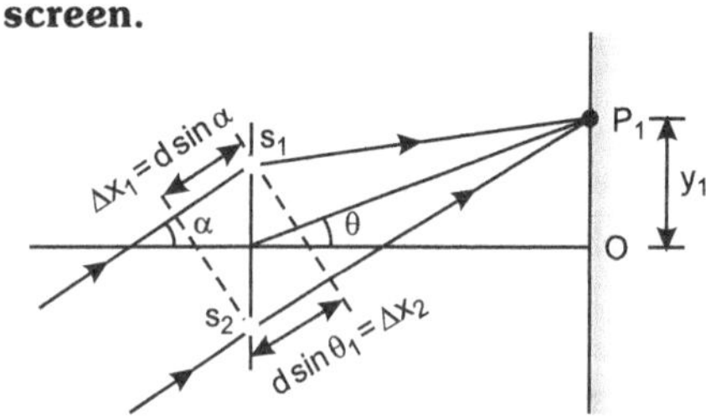

Δ Δ Δ

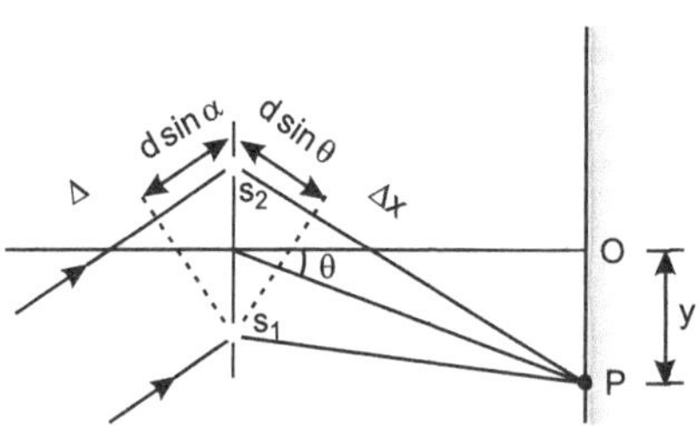

Δ Δ Δ

(b) When $\alpha = 30°$ path difference between the rays before reaching s_1 and s_2 is

$\Delta x_1 = d\sin\alpha = (1.0)\sin 30° = 0.5\text{ mm} = \lambda$

So, there is already a path difference of λ between the rays.

Position of central maximum : Central maximum is defined as a point where net path difference is zero.

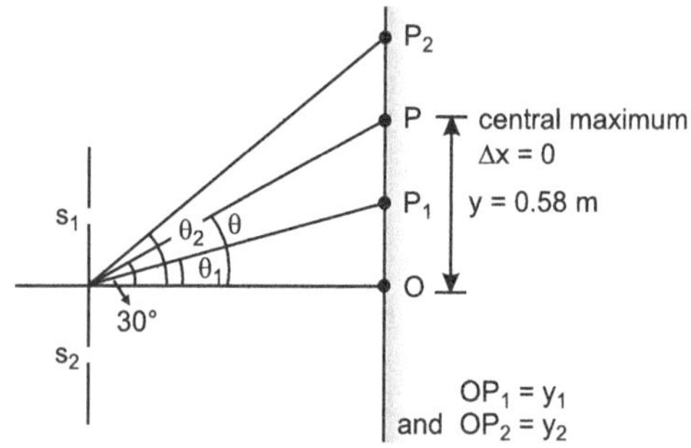

So $\quad \Delta x_1 = \Delta x_2$

or $\quad d\sin\alpha = d\sin\theta$

or $\quad \theta = \alpha = 30°$

or $\quad \tan\theta = \frac{1}{\sqrt{3}} = \frac{y_0}{D}$

$\therefore \quad y_0 = \frac{1}{\sqrt{3}}\text{ m} \qquad (D = 1\text{ m})$

or $\quad y_0 = 0.58\text{ m}$

At point P, $\Delta x_1 = \Delta x_2$

Above point P, $\Delta x_2 > \Delta x_1$

and below point P, $\Delta x_1 > \Delta x_2$

Now let P_1 and P_2 be the two minimas on either side of central maxima.

Then for P_2

$$\Delta x_2 - \Delta x_1 = \lambda/2$$

or $\quad \Delta x_2 = \Delta x_1 + \lambda/2 = \lambda + \lambda/2 = 3\lambda/2$

or $\quad d\sin\theta_2 = 3\lambda/2$

or $\quad \sin\theta_2 = \frac{3\lambda}{2d} = \frac{(3)(0.5)}{(2)(1.0)} = \frac{3}{4}$

$\therefore \quad \tan\theta_2 = \frac{3}{\sqrt{7}} = \frac{y_2}{D}$

or $\quad y_2 = \frac{3}{\sqrt{7}}\text{ m} = \mathbf{1.13\ m}$

Similarly for P_1

$$\Delta x_1 - \Delta x_2 = \lambda/2$$

or $\quad \Delta x_2 = \Delta x_1 - \lambda/2 = \lambda - \lambda/2 = \lambda/2$

or $\quad d\sin\theta_1 = \lambda/2$

or $\sin \theta_1 = \dfrac{\lambda}{2d} = \dfrac{0.5}{(2)(1.0)} = \dfrac{1}{4}$

$\therefore \quad \tan \theta_1 = \dfrac{1}{\sqrt{15}} = \dfrac{y_1}{D}$

or $y_1 = \dfrac{1}{\sqrt{15}}$ m $= \mathbf{0.26\ m}$

Therefore, y-coordinates of the first minima on either side of the central maximum are $y_1 = 0.26$ m and $y_2 = 1.13$ m

In this problem $\sin\theta \approx \tan\theta \approx \theta$ *is not valid as* θ *is large.*

408. $\mu_1 = 1.4, \mu_2 = 1.7$ and let t be the thickness of each glass plate.

Path difference at O, due to insertion of glass plate, will be

$\Delta x = (\mu_2 - \mu_1) t = (1.7 - 1.4) t = 0.3 t$ ...(1)

Now since 5th maxima (earlier) lies below O and 6th minima lies above O.

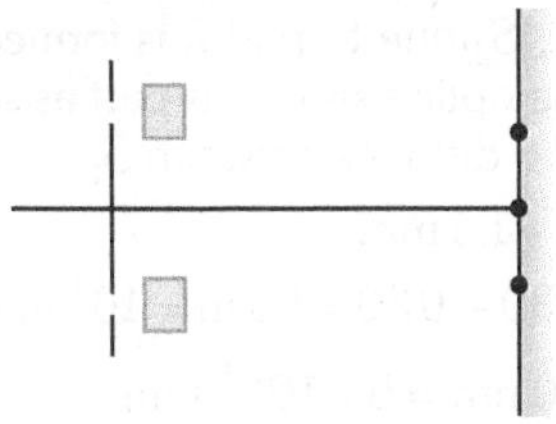

This path difference should lie between 5λ and $5\lambda + \lambda/2$

So, let $\Delta x = 5\lambda + \Delta$ ...(2)

where $\Delta < \lambda/2$

Due to the path difference Δx, the phase difference at O will be

$$\phi = \frac{2\pi}{\lambda}\Delta x = \frac{2\pi}{\lambda}(5\lambda + \Delta)$$

$$= \left(10\pi + \frac{2\pi}{\lambda}.\Delta\right) \quad ...(3)$$

Intensity at O is given, $\dfrac{3}{4} I_{max}$ and since

$$I(\phi) = I_{max} \cos^2\left(\frac{\phi}{2}\right)$$

$$\therefore \quad \frac{3}{4} I_{max} = I_{max} \cos^2\left(\frac{\phi}{2}\right)$$

or $\dfrac{3}{4} = \cos^2\left(\dfrac{\phi}{2}\right)$ ...(4)

From equations (3) and (4), we find that

$$\Delta = \lambda/6$$

i.e. $\Delta x = 5\lambda + \lambda/6$

$= (31/6)\lambda = 0.3 t \quad (\Delta x = 0.3 t)$

$$\therefore \quad t = \frac{31\lambda}{6(0.3)} = \frac{(31)(5400 \times 10^{-10})}{1.8} \text{ m}$$

or $\mathbf{t = 9.3 \times 10^{-6}}$ **m = 9.3 μm**

409. (i) Path difference due to the glass slab,

$$\Delta x = (\mu - 1) t = (1.5 - 1) t = 0.5 t$$

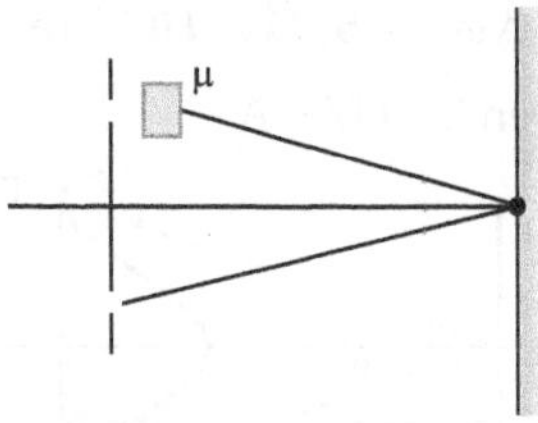

Due to this slab, 5 red fringes have been shifted upwards. Therefore,

$$\Delta x = 5\lambda_{red}$$

or $0.5 t = (5)(7 \times 10^{-7}$ m)

$\therefore$ t = thickness of glass slab

$= \mathbf{7 \times 10^{-6}}$ **m**

(ii) Let μ' be the refractive index for green light then

$$\Delta x' = (\mu' - 1) t$$

Now the shifting is of 6 fringes of red light. Therefore,

$$\Delta x' = 6\lambda_{red} \Rightarrow (\mu' - 1) t = 6\lambda_{red}$$

$$\therefore \quad (\mu' - 1) = \frac{(6)(7 \times 10^{-7})}{7 \times 10^{-6}} = 0.6$$

$\therefore$ $\mathbf{\mu' = 1.6}$

(iii) In part (i), shifting of 5 bright fringes was equal to 10^{-3} m. Which implies that

$5\omega_{red} = 10^{-3}$ m [Here ω = Fringe width]

$$\therefore \quad \omega_{red} = \frac{10^{-3}}{5} \text{ m} = 0.2 \times 10^{-3} \text{ m}$$

Now since

$$\omega = \frac{\lambda D}{d} \quad \text{or} \quad \omega \propto \lambda$$

$$\therefore \quad \frac{\omega_{green}}{\omega_{red}} = \frac{\lambda_{green}}{\lambda_{red}}$$

$$\therefore \quad \omega_{green} = \omega_{red} \frac{\lambda_{green}}{\lambda_{red}}$$

$$= (0.2 \times 10^{-3}) \left(\frac{5 \times 10^{-7}}{7 \times 10^{-7}} \right)$$

$$\omega_{green} = 0.143 \times 10^{-3} \text{ m}$$

$$\therefore \quad \Delta \omega = \omega_{green} - \omega_{red}$$

$$= (0.143 - 0.2) \times 10^{-3} \text{ m}$$

$$\Delta \omega = \mathbf{-5.71 \times 10^{-5} \text{ m}}$$

410. (i) Given $\lambda = 6000$ Å

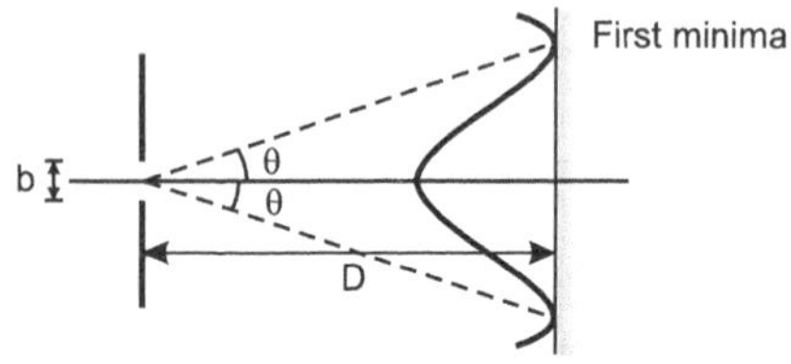

Let b be the width of slit and D the distance between screen and slit.
First minima is obtained at

$$b \sin \theta = \lambda$$

or $\quad b\theta = \lambda \quad (\sin \theta \approx \theta)$

or $\quad \theta = \dfrac{\lambda}{b}$

Angular width of first maxima $= 2\theta = \dfrac{2\lambda}{b} \propto \lambda$

Angular width will decrease by 30% when λ is also decreased by 30%.
Therefore, new wavelength,

$$\lambda' = \left\{ (6000) - \left(\frac{30}{100} \right) 6000 \right\} \text{ Å}$$

$$\lambda' = \mathbf{4200 \text{ Å}}$$

(ii) When the apparatus is immersed in a liquid of refractive index μ, the wavelength is decreased μ times. Therefore,

$$4200 \text{ Å} = \frac{6000 \text{ Å}}{\mu}$$

$$\therefore \quad \mu = \frac{6000}{4200} \quad \text{or} \quad \mu = \mathbf{1.429}$$

411. (i) For the lens, $u = -0.15$ m; $f = +0.10$ m

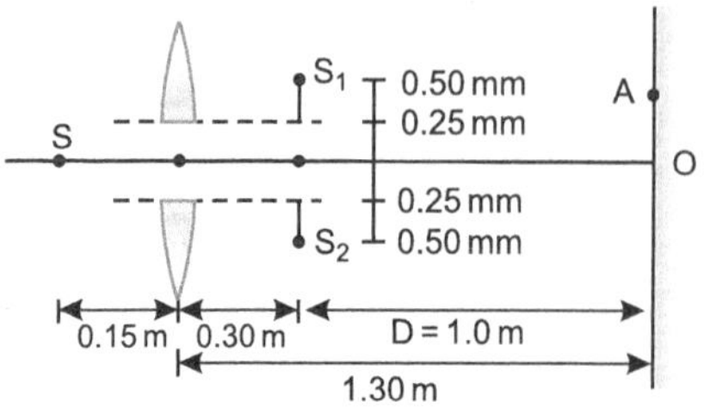

Therefore, using $\dfrac{1}{v} - \dfrac{1}{u} = \dfrac{1}{f}$ we have

$$\frac{1}{v} = \frac{1}{u} + \frac{1}{f} = \frac{1}{(-0.15)} + \frac{1}{(0.10)}$$

or $\quad v = 0.3$ m

Linear magnification, $m = \dfrac{v}{u} = \dfrac{0.3}{-0.15} = -2$

Hence two images S_1 and S_2 of S will be formed at 0.3 m from the lens as shown in figure. Image S_1 due to part 1 will be formed at 0.5 mm above its optic axis ($m = -2$). Similarly, S_2 due to part 2 is formed 0.5 mm below the optic axis of this part as shown.
Hence d = distance between S_1 and $S_2 = 1.5$ mm

$$D = 1.30 - 0.30 = 1.0 \text{ m} = 10^3 \text{ mm}$$

$$\lambda = 500 \text{ nm} = 5 \times 10^{-4} \text{ mm}$$

Therefore, fringe width,

$$\omega = \frac{\lambda D}{d} = \frac{(5 \times 10^{-4})(10^3)}{(1.5)} \text{ mm}$$

$$= \frac{1}{3} \text{ mm}$$

Now as the point A is at the third maxima

$$OA = 3\omega = 3\,(1/3) \text{ mm}$$

or $\quad \mathbf{OA = 1 \text{ mm}}$

Note: The language of the question is slightly confusing. The third intensity maximum may be understood as second order maximum (zero order, first order and the second order). In that case $OA = 2\omega = 2\,(1/3) = 0.67$ mm.

(ii) If the gap between L_1 and L_2 is reduced, d will decrease. Hence the fringe width ω will increase or **the distance *OA* will increase.**

CHAPTER 12 CURRENT ELECTRICITY

412. Since wires *MACN* and *MBDN* are connected in parallel between points *M* and *N*, the fall of potential is one and the same for the whole length of these wires. The fall of potential is distributed uniformly along each of these wires. Therefore if length *MA* equals length *MB*, the potentials at *A* and *B* will be the same and current will not pass along wire *AB*. The same thing can be said of sections *MC* and *MD*. If they are of the same length as one another, current will not pass along *CD*. But wires *AB* and *CD* will be at different potentials and if points *E* and *F* upon these wires are connected, current will pass along *EF* and consequently also along *AE*, *BE*, *FC* and *FD*. This will happen wherever points *E* and *F* are chosen along wires *AB* and *CD*, since the potential at *E* will always be higher than that at *F*.

413. If we connect a voltmeter between points *A* and *B*, we find which of the two points has the higher potential. Suppose that we find that the potential at *A* is higher than that at *B*. Then we bring the magnetic needle, mounted on a vertical pivot, up under the corresponding wire, *e.g.*, the upper one. The deflection of the magnetic needle's North pole tells us the direction of flow of current through the wire. For example, if the needle's North pole is deflected towards us from the plane of the paper, the current in this wire is flowing through *A* from right to left. Hence it follows that the source of current in our example is to the right of *A*.

414. The galvanometer should be connected to the arm to which is normally connected the unknown resistance R_x, and a switch should replace the galvanometer on the bridge's diagonal. Resistances r_1 and r_2 should be selected to that the galvanometer shows the same deflection whether the switch is open or closed. This will mean that there is no current across the bridge's diagonal and consequently the ratio obtains :

$$\frac{R_G}{R} = \frac{r_1}{r_2},$$

therefore $R_G = R\frac{r_1}{r_2}$

415. As you know, a voltmeter connected directly to the source of current shows not the e.m.f. but the potential at the source's terminals

$$V = E - Ir,$$

Where r is the internal resistance of the source. Since r is unknown, it is not possible to determine the e.m.f. from the readings of the voltmeter and ammeter with the sliding contact of the rheostat in only one position. But if the sliding contact is moved and the current and potential difference are measured for this new position then we shall obtain the self-evident equation

$$E = V_1 + I_1 r = V_2 + I_2 r.$$

Hence $r = \dfrac{V_1 - V_2}{I_2 - I_1}$

and the unknown e.m.f.

$$E = V_1 + I_1 r = V_1 + I_1 \frac{V_1 - V_2}{I_2 - I_1}$$

416. It is possible if $\dfrac{E_2}{E_1} < \dfrac{r_2}{(r_1 + R)}$. If one cell of e.m.f. E_1 and internal resistance r_1 be connected to an external resistance R, then the current passing through the circuit

$$I_1 = \frac{E_1}{r_1 + R}$$

If a second cell of e.m.f. E_2 and internal resistance r_2 is then added in series, the current

$$I_2 = \frac{E_1 + E_2}{r_1 + r_2 + R}$$

Clearly

$$I_2 < I_1 \text{ if } \frac{E_1 + E_2}{r_1 + r_2 + R} < \frac{E_1}{r_1 + R},$$

hence $$\frac{E_2}{E_1} < \frac{r_2}{r_1 + R}$$

In other words for this inequality to obtain, the internal resistance of the second cell must be sufficiently great.

417. There are two circuits in parallel connecting points A and B, each consisting of half one side of the hexagon, a rhombus and then half another side of the hexagon, connected in series. The resistance of each rhombus is R, therefore the resistance of one circuit is $2R$. Therefore the resistance of the whole frame is R. The fact that the vertices of the rhombi are linked at O plays no part in this, since the fact that the two circuits under consideration are identical means that the vertices of both rhombi are at the same potential. Therefore we can calculate the resistances of the two circuits without taking into consideration their link at O.

418. When a positive charge is introduced into a capacitor, a negative charge is induced on the inside faces of the capacitor plates and it remains there, while a positive charge appears on the outside faces. The positive charge on the earthed plate escapes into the earth. The charge on the other plate also escapes into the earth, passing through the galvanometer. So the galvanometer will show a deflection. But when the positive charge begins to escape from the capacitor, the negative charge will begin to flow into the earth from the plates. The charge flowing from one of the plates will pass through the galvanometer, which will then show a deflection in the opposite direction.

419. The simplified circuit can be redrawn as follows

$$\therefore \quad i = \frac{E}{\left(\frac{63}{144}\right) r} = \frac{10}{\left(\frac{63}{144}\right)}$$

$$= \mathbf{22.85\ A} \quad \text{or} \quad \mathbf{\frac{1440}{63}}\ \text{A}$$

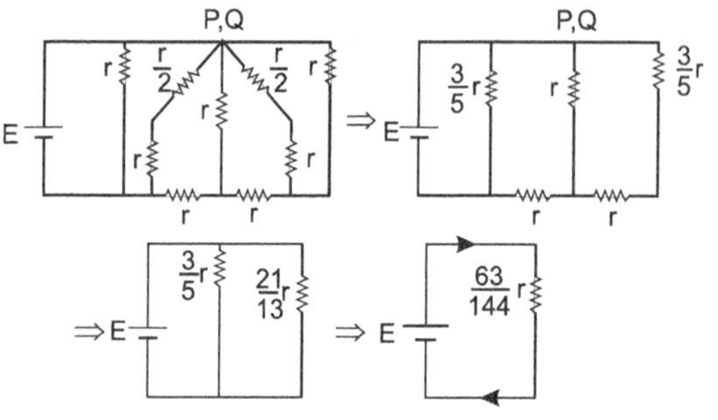

(a)

as $r = 1\Omega$

Now current distribution in different branches would be as shown in figure (b).

Here $$\frac{(i_2 + i_3)}{i_i} = \frac{3/5\, r}{21/13\, r} = \frac{39}{105}$$

or $$(i_2 + i_3) = \left(\frac{39}{105 + 39}\right)\left(\frac{1440}{63}\right)$$

$$= \frac{390}{63}\ \text{A}$$

and $$i_1 = \left(\frac{105}{105 + 39}\right)\left(\frac{1440}{63}\right)$$

$$= \frac{1050}{63}\ \text{A}$$

also $$\frac{i_2}{i_3} = \frac{(3/5 + 1)\, r}{r} = \frac{8}{5}$$

$$\therefore \quad i_2 = \left(\frac{8}{8+5}\right)\left(\frac{390}{63}\right) = \frac{240}{63}\ \text{A}$$

and $$i_3 = \left(\frac{5}{8+5}\right)\left(\frac{390}{63}\right) = \frac{150}{63}\ \text{A}$$

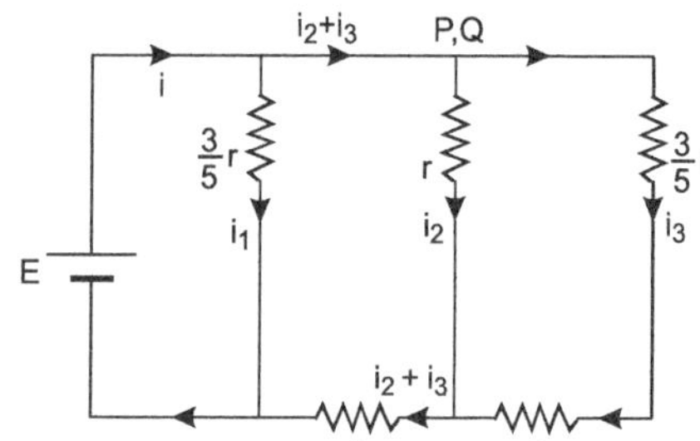

(b)

Finally currents in all the branches are follows as shown in figure :

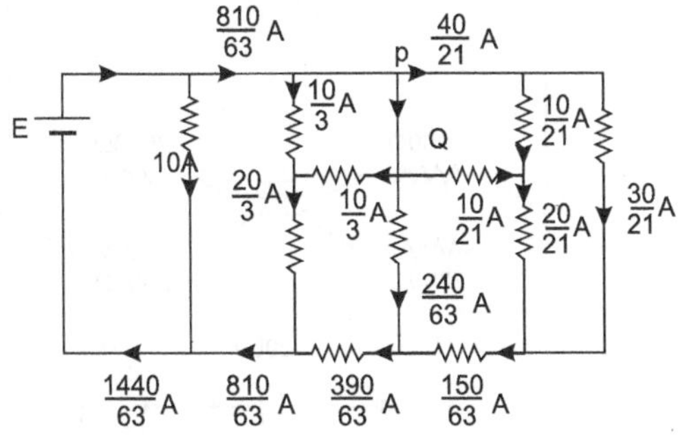

(c)

Hence the current in branch *PQ* is

$$i_{PQ} = \frac{10}{3} + \frac{10}{21} + \frac{240}{63}$$

$$= \frac{480}{63} \text{ A} = \mathbf{7.62\ A}$$

420. (a) Heat produced per second in the resistance is $H = i^2r = (0.3)^2(500) = 45$ J

As temperature remains constant, whole heat supplied to the gas is used in doing work against gravitation and atmospheric pressure. If v is the speed of piston, then work done per unit time is

$$W = (P_0.S + mg)\,v$$

Hence $\quad 45 = (P_0.S + mg)\,v$

or $\quad v = \dfrac{45}{P_0S + mg}$

$$= \frac{45}{(10^5)(10^{-3}) + (10)(10)}$$

$$= \mathbf{0.225\ m/s}$$

(b) In this case, whole heat is used to raise the temperature of the gas.

Hence $\quad i^2rt = nC_V d\theta$

or $\quad d\theta = \dfrac{(i^2rt)}{n\left(\dfrac{3}{2}R\right)}$

but $\quad i^2r = 45$ J/sec

$$d\theta = \frac{(45)(30)}{(2)\left(\dfrac{3}{2}\right)(8.31)} = \mathbf{54.15°\ C}$$

421. (a) EMF of the cell $= 1.55$ V(reading of potentiometer)

Now $\quad V = E - ir$

So $\quad 1.40 = 1.55 - \left(\dfrac{1.55}{280 + r}\right)\cdot r$

Solving this equation we get $\mathbf{r = 30\ \Omega}$

(b) $i = \dfrac{E}{R + r} = \dfrac{1.55}{5 + 30}$ A $= \mathbf{0.044\ A}$

422. Let $R_{BP} = x$ then $R_{PC} = (a - x)$

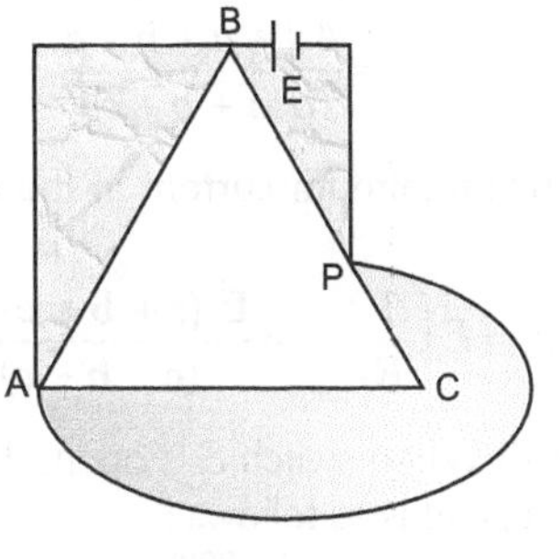

Also $\quad R_{AB} = c, \quad R_{AC} = b$

and $\quad R_{AP} = d \quad$ (given)

Hence the total resistance R of the circuit will be given by

$$\frac{1}{R} = \frac{1}{R_{AB} + R_{BP}} + \frac{1}{R_{AC} + R_{PC}} + \frac{1}{R_{AP}}$$

or $\quad \dfrac{1}{R} = \dfrac{1}{c + x} + \dfrac{1}{b + (a - x)} + \dfrac{1}{d} = y \quad$ (say)

current in the circuit will be minimum when $\dfrac{1}{R}$

or y is minimum or $\dfrac{dy}{dx} = 0$

or $\quad \dfrac{-1}{(c + x)^2} + \dfrac{1}{(b + a - x)^2} = 0$

or $\quad \left(\dfrac{c + x}{b + a - x}\right)^2 = 1$

or $\quad \dfrac{c + x}{b + a - x} = 1$

or $\quad 2x = (a + b) - c$

or $\quad x = \dfrac{(a + b) - c}{2}$

So, minimum value of $\frac{1}{R}$ is

$$\left(\frac{1}{R}\right)_{min} = \frac{1}{c + \left(\frac{a+b-c}{2}\right)} + \frac{1}{b + a - \left(\frac{a+b-c}{2}\right)} + \frac{1}{d}$$

or
$$\left(\frac{1}{R}\right)_{min} = \frac{2}{a+b+c} + \frac{2}{a+b+c} + \frac{1}{d}$$

$$= \frac{4d + a + b + c}{d(a+b+c)}$$

Hence minimum current in the circuit would be

$$i_{min} = E\left(\frac{1}{R}\right)_{min} = \boldsymbol{\frac{E\,(a+b+c+4d)}{(a+b+c)\,d}}$$

423. (a) (i) When switch S is open, the equivalent circuit is as follows :

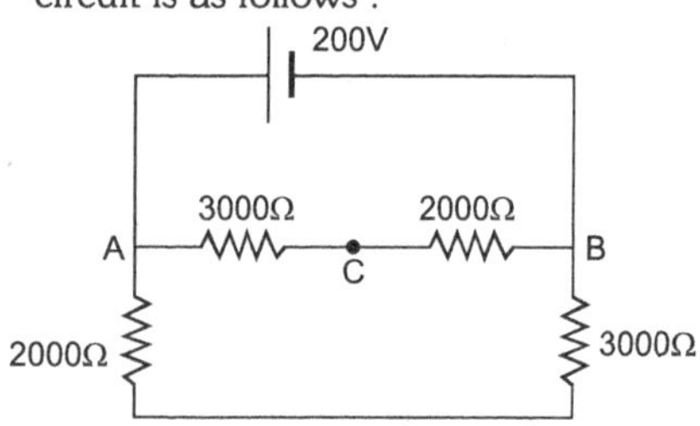

Potential difference across AB is 200 V

and $\frac{V_{AC}}{V_{CB}} = \frac{V_1}{V_2} = \frac{3000}{2000} = \frac{3}{2}$

Hence $\boldsymbol{V_1} = \left(\frac{3}{3+2}\right)(200) = \mathbf{120\ V}$

and $\boldsymbol{V_2} = \left(\frac{2}{3+2}\right)(200) = \mathbf{80\ V}$

(ii) When switch S is closed, equivalent circuit is as follows :

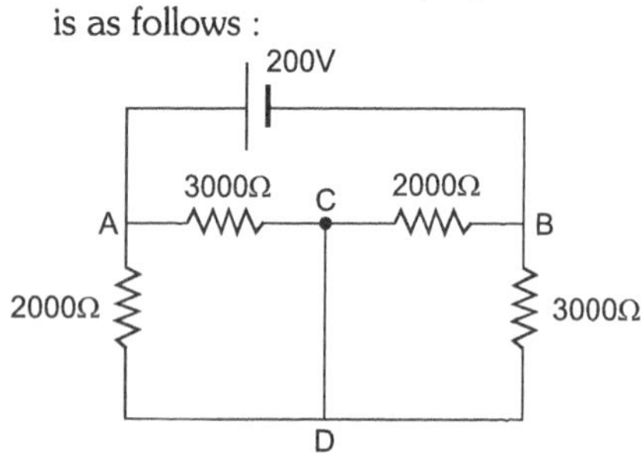

This circuit can be simplified as :

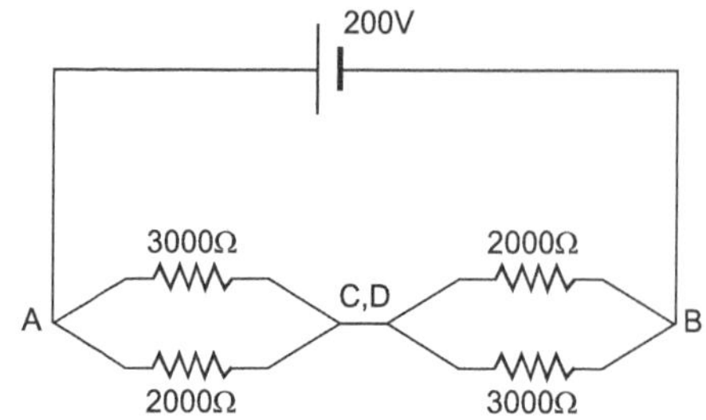

OR

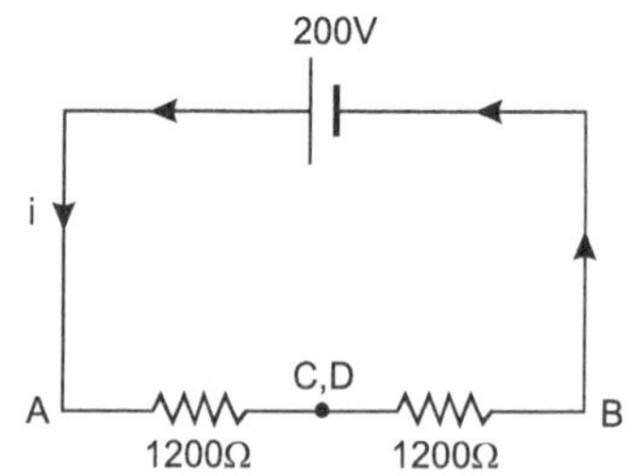

$\therefore \quad V_{AC} = V_{CB}$

or $\quad \boldsymbol{V_1 = V_2 = 100\ V}$

and $\quad i = \frac{200}{1200 + 1200}$

$= \frac{1}{12}$ A

(b) Current $i = \frac{1}{12}$ A will be distributed as shown in figure

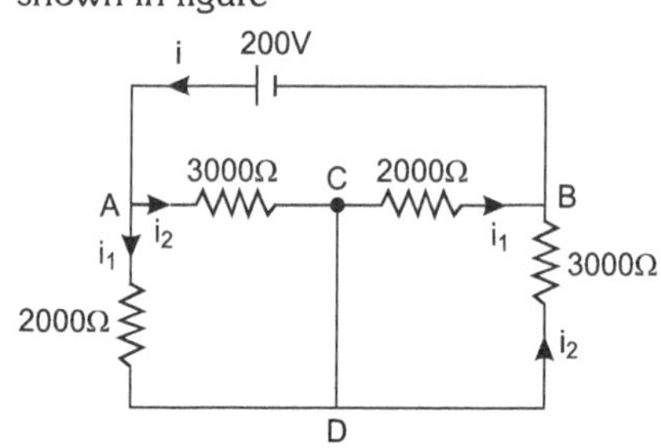

$$i_1 = \left(\frac{3}{3+2}\right)\left(\frac{1}{12}\right) = \frac{3}{60}\text{ A}$$

and
$$i_2 = \left(\frac{2}{3+2}\right)\left(\frac{1}{12}\right) = \frac{2}{60}\text{ A}$$

Hence current in CD will be :

$$i_1 - i_2 = \frac{\mathbf{1}}{\mathbf{60}}\ \textbf{A from D to C.}$$

424. (a) Current time graph in the first case will be a straight line as shown in figures.

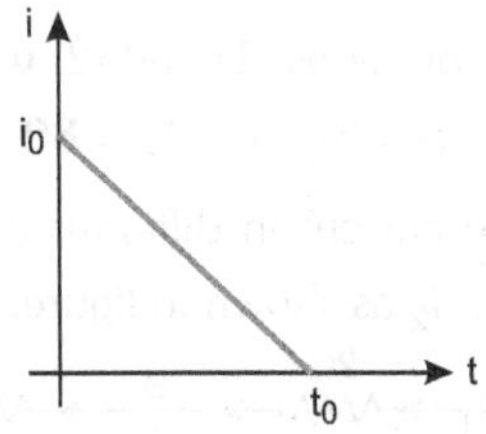

Now since $(dq) = i\,(dt)$

hence area under $i - t$ graph gives the charge flown through the resistance.

Hence $$q = \frac{1}{2}(i_0)\,t_0$$

or $$i_0 = \frac{2q}{t_0}$$

or $$\frac{i_0}{t_0} = \frac{2q}{t_0^2} \qquad \ldots(1)$$

Now from the graph, $i - t$ equation can be written as

$$i = i_0 - \left(\frac{i_0}{t_0}\right)t$$

or $$i = \left(\frac{2q}{t_0}\right) - \left(\frac{2q}{t_0^2}\right)t$$

Since $dH = i^2 R dt$

Hence $$H = \int_0^{t_0} dH$$

or $$H = \int_0^{t_0}\left\{\frac{2q}{t_0} - \frac{2q}{t_0^2}\cdot t\right\}^2 R\,dt$$

Solving this we get

$$\boldsymbol{H = \frac{4}{3}\frac{q^2 R}{t_0}}$$

(b) In the second case current decreases exponentially from a peak value i_0 to zero.

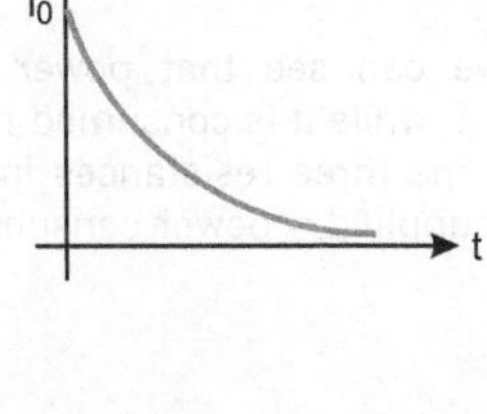

Comparing this with radioactive disintegration.

Half life is $t_{1/2} = t_0$ (given)

Hence, disintegration constant

$$\lambda = \frac{\ln(2)}{t_{1/2}} = \frac{\ln(2)}{t_0} \qquad \ldots(2)$$

So $$i = i_0\, e^{-\lambda t}$$

Now $$q = \int_0^{\infty} i\,dt$$

or $$q = \int_0^{\infty} i_0\, e^{-\lambda t}\,dt$$

or $$q = \left(\frac{i_0}{-\lambda}\right)[e^{-\lambda t}]_0^{\infty}$$

or $$i_0 = q\lambda = \frac{q\ln(2)}{t_0} \qquad \ldots(3)$$

Therefore, $$H = \int_0^{\infty} i^2 R\cdot dt$$

$$= \int_0^{\infty}\left\{\frac{q\ln(2)}{t_0}\right\}^2 e^{-2\lambda t}R\cdot dt$$

$$H = \frac{q^2\{\ln(2)\}^2}{t_0^2}\cdot R\left(\frac{1}{2\lambda}\right)$$

$$= \frac{q^2\{\ln(2)\}^2}{t_0^2}\cdot R\left(\frac{1}{2}\right)\cdot\frac{t_0}{\ln(2)}$$

$$= \boldsymbol{\left\{\frac{1}{2}\ln(2)\right\}\frac{q^2 R}{t_0}}$$

425. Let at time t, temperature of conductor be T.

Then $$C\cdot\left(\frac{dT}{dt}\right) = \frac{V^2}{R} - q$$

or $$C\cdot\left(\frac{dT}{dt}\right) = \frac{V^2}{R} - \alpha\,(T - T_0)$$

or $$\frac{dT}{\dfrac{V^2}{R} - \alpha\,(T - T_0)} = \frac{dt}{C}$$

or $$\int_{T_0}^{T}\frac{dT}{\dfrac{V^2}{R} - \alpha\,(T - T_0)} = \int_0^t \frac{dt}{C}$$

solving this equation we get,

$$\boldsymbol{T = T_0 + \left(1 - e^{\frac{-\alpha t}{C}}\right)\frac{V^2}{\alpha R}}$$

426. Let ρ be the resistance per unit length of the wire. If the fault exists at a distance x from the end A and r is the resistance of the leak.

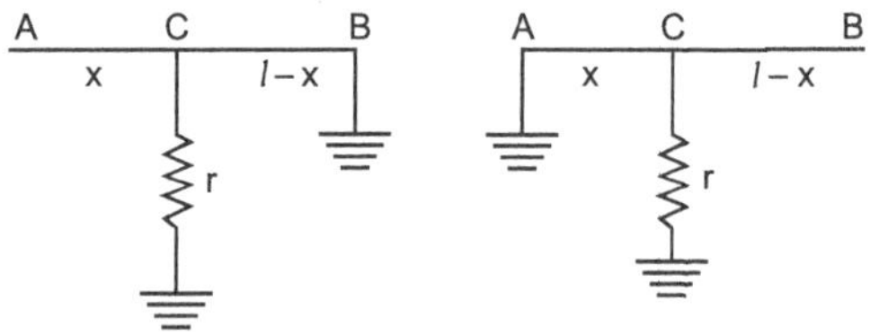

(a) **When *B* is earthed** : The resistance between A and the earth through the cable is

$$x\rho + \frac{1}{1/r + 1/(l - x)\rho} = l_1\rho \qquad \text{(given)}$$

or $$\frac{1}{\rho\,(l_1 - x)} = \frac{1}{r} + \frac{1}{(l - x)\,\rho} \qquad \ldots(1)$$

(b) **When *A* is earthed** : The resistance between B and the earth through the cable is

$$(l - x)\,\rho + \frac{1}{1/r + 1/\rho x} = l_2\rho$$

or $$\frac{1}{\rho\,(l_2 - l + x)} = \frac{1}{r} + \frac{1}{x\rho} \qquad \ldots(2)$$

Eliminating r from equations (1) and (2) we get

$$\frac{1}{\rho\,(l_1 - x)} - \frac{1}{(l - x)\rho} = \frac{1}{\rho\,(l_2 - l + x)} - \frac{1}{\rho x}$$

or $$\frac{\boldsymbol{x}}{\boldsymbol{l - x}} = \sqrt{\frac{\boldsymbol{l_1\,(l - l_2)}}{\boldsymbol{l_2\,(l - l_1)}}}$$ **Hence proved.**

427. The condition required for melting of the wire should be

$$i^2R = \alpha S\,(T_{\text{melt}} - T_0)$$

Now $$i_1^2R_1 = \alpha S_1\,(T_{\text{melt}} - T_0)$$

or $$\frac{i_1^2\,(\rho l)}{d_1^2} = \alpha\,(4d_1)l\,(T_{\text{melt}} - T_0)$$

(ρ = specific resistance of the wire)

or $$i_1^2 = \frac{4\alpha d_1^3}{\rho}\,(T_{\text{melt}} - T_0) \qquad \ldots(1)$$

and similarly

$$i_2^2 = \frac{4\alpha d_2^3}{\rho}\,(T_{\text{melt}} - T_0) \qquad \ldots(2)$$

From equations (1) and (2) we get,

$$i_2 = 8i_1 \quad \text{or} \quad \boldsymbol{i_2 = 80\text{ A}}$$

428. Let the current in different branches be i_1, i_2 and $i_1 - i_2$ as shown in figure.

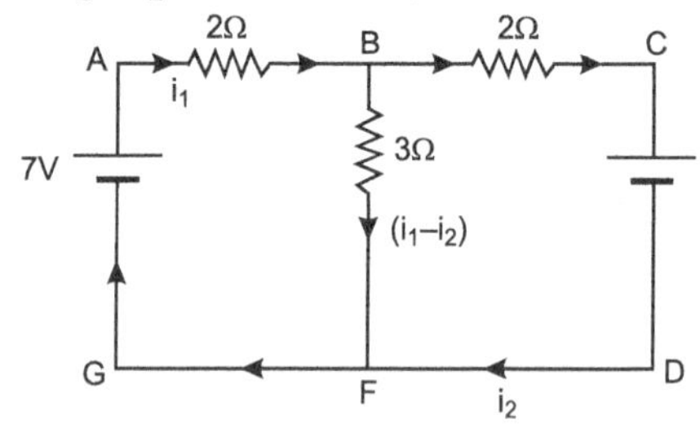

Applying Kirchhoff's second law in loop *ABFGA* we get

$$7 = 2i_1 + 3\,(i_1 - i_2)$$

or $$7 = 5i_1 - 3i_2 \qquad \ldots(1)$$

Similarly applying Kirchhoff's second law in loop ACDGA, we get

$$7 - 1 = 2i_1 + 2i_2$$

or $$3 = i_1 + i_2 \qquad \ldots(2)$$

solving these two equations we get

$$i_1 = 2\text{ A} \quad \text{and} \quad i_2 = 1\text{ A}$$

now power supplied/consumed by a battery, is given by

$$P = \pm\,Ei$$

Plus sign is used when power in supplied (i.e. when current flows from negative terminal to positive terminal inside the battery) and negative sign is used when power is consumed (i.e. when current flows from positive terminal to negative terminal inside the battery)

Here power will be supplied by the battery of emf $E_1 = 7$ V and which is equal to $+\,7i_1$, or **+ 14 watt** and will be consumed by the battery of emf $E_2 = 1$ V which is equal to $-1i_2$ or **−1 watt.**

Note: Here we can see that power is supplied by battery E_1 while it is consumed by the battery E_2 and all the three resistances in the circuit and power supplied = power consumed

429. We know that net emf of a battery is equal to the potential difference across the terminals of the battery when no current is drawn from the battery *i.e.*

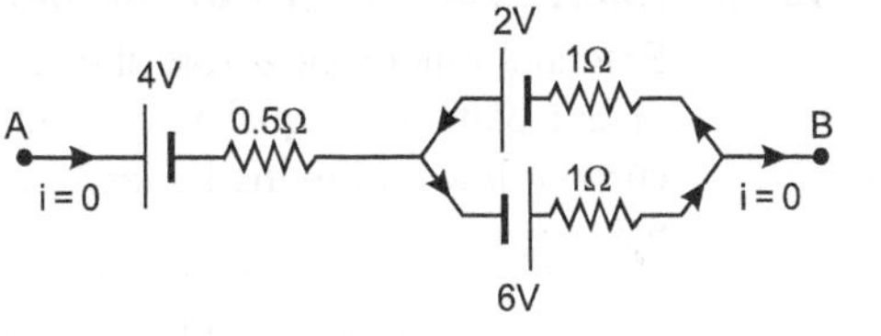

$$E = V \qquad \text{when } i = 0$$

In this case also let the current drawn from the battery is zero *i.e.*,

But internal current $i = \dfrac{2+6}{1+1} = 4$ A will flow in the circuit.

Now $\quad E = V_A - V_B$

where $\quad V_A - 4 - 2 + (1)(4) = V_B$

or $\quad V_A - V_B = 2\text{V}$

Hence equivalent emf is 2V.

430. (a) Current flowing through resistance 5 Ω is 11 ampere

power dissipated $= i^2 R$

$= (121)5 = 605$ watt

(b) $V_B + 8\text{V} + 3\text{V} + 12\text{V} - 12\text{V} - 5\text{V} = V_C$

$V_B + 11\text{V} - 5\text{V} = V_C$

$6\text{V} = V_C - V_B$

(c) Both batteries are being charged.

431. After connecting A and O to the terminals of a battery we find, there is symmetry about OA. Therefore the current distribution will the same as shown in the figure.

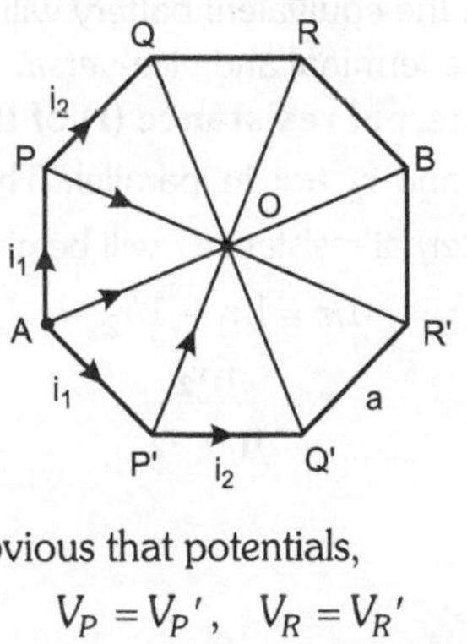

It is obvious that potentials,

$$V_P = V_P', \quad V_R = V_R'$$

$$V_Q = V_Q'$$

$\therefore$ On superimposing P and P' and Q and Q' etc. we get the simplified arrangement of resistors as shown in figure,

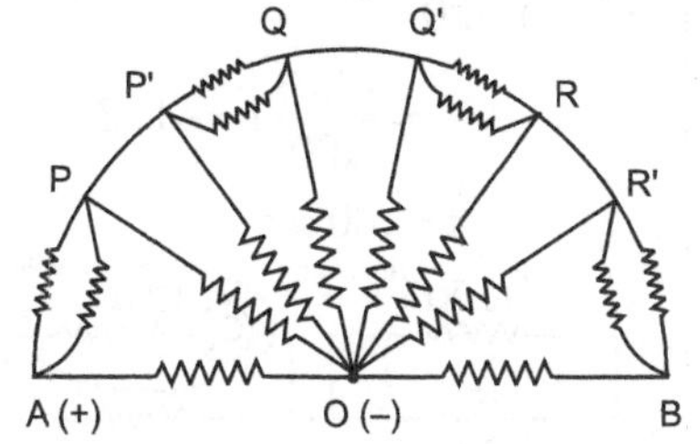

Hence $\quad R_{eq} = \dfrac{47}{105} r_0$

432. (a) Let i_1 and i_2 be the currents in two loops respectively

$\therefore \; 10 - 10i_1 - R(i_1 - i_2) + 5 = 0 \qquad$ (for loop 1)

$(10 + R)\, i_2 - Ri_1 = -5 \qquad$ (for loop 2)

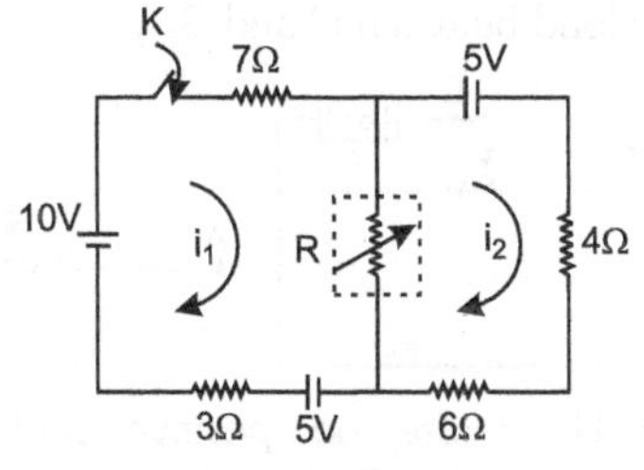

solving $\quad i_1 = \dfrac{15 + 2R}{10 + 2R}$,

$$i_2 = \frac{5 + 2R}{10 + 2R}$$

Power dissipated in R,

$$P = (i_1 - i_2)^2 R = \frac{25}{(5+R)^2} \times R$$

$\Rightarrow$ For maximum power dissipation $\dfrac{dP}{dR} = 0$

$\Rightarrow \quad R = 5\,\Omega$

(b) $\quad R = R_0 - \left(\dfrac{dR}{d\theta}\right)\Delta\theta$

$$5 = 15 - \frac{1}{2}\Delta\theta$$

$\Rightarrow \quad \Delta\theta = 20°\text{C}$

$\Rightarrow$ temperature at that instant = 30° C

(c) According to Newton's law:

$$\frac{d\theta}{dt} = -k\,(\theta - 20°)$$

$$\int_{50}^{30} \frac{d\theta}{\theta - 20} = -kt$$

$$= \frac{-\ln 3}{100} t = -\ln 3$$

$\therefore$ $t = 100$ sec

433.

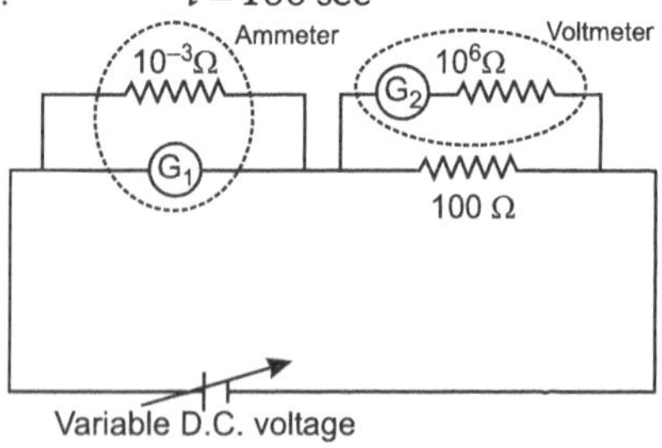

434. The rheostate is as shown in figure. Battery should be connected between *A* and *B* and the load between *C* and *B*.

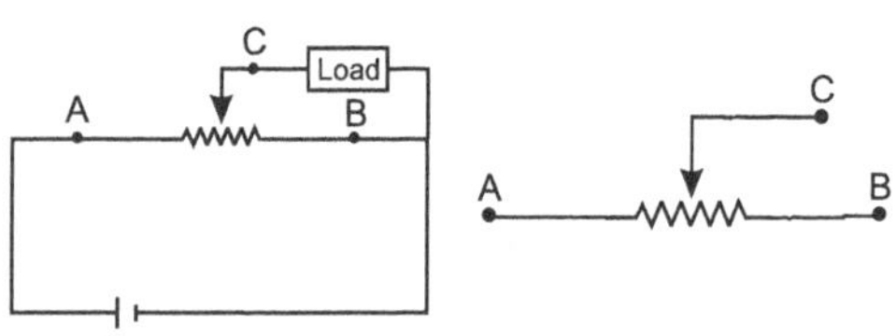

435. (a) There are no positive and negative terminals on the galvanometer because only zero deflection is needed.

(b)

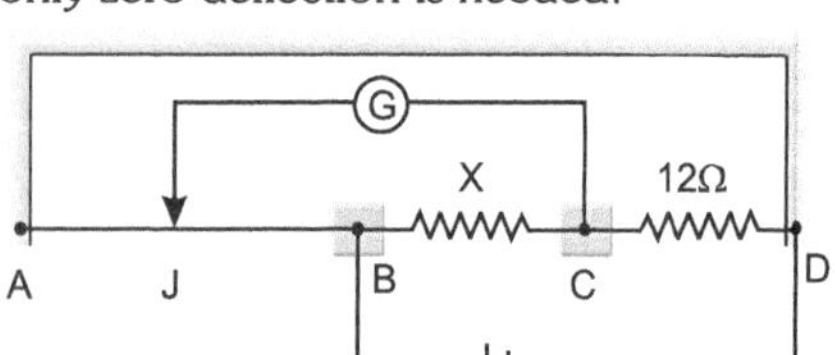

(c) $AJ = 60$ cm

$\therefore$ $BJ = 40$ cm

If no deflection is taking place. Then the Wheatstone bridge is be said to be balanced. Hence

$$\frac{X}{12} = \frac{R_{BJ}}{R_{AJ}}$$

or $$\frac{X}{12} = \frac{40}{60} = \frac{2}{3}$$

or $$\mathbf{X = 8\,\Omega}$$

436. (i) Equivalent emf (*V*) of the battery : Potential difference across the terminals of the battery is equal to its emf when current drawn from the battery is zero. In the given circuit :

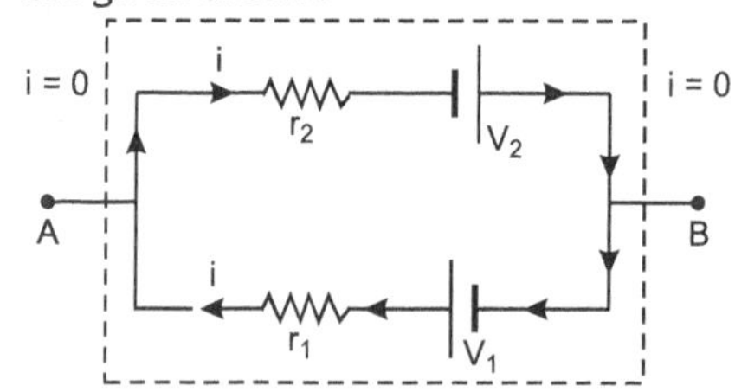

Current in the internal circuit

$$i = \frac{\text{Net emf}}{\text{Total resistance}} = \frac{V_1 + V_2}{r_1 + r_2}$$

Therefore, potential difference between *A* and *B* would be

$$V_A - V_B = V_1 - ir_1$$

$$\therefore \quad V_A - V_B = V_1 - \left(\frac{V_1 + V_2}{r_1 + r_2}\right) r_1$$

$$= \frac{V_1 r_2 - V_2 r_1}{r_1 + r_2}$$

So, the equivalent emf of the battery is

$$\boldsymbol{V = \frac{V_1 r_2 - V_2 r_1}{r_1 + r_2}}$$

Note that if $V_1 r_2 - V_2 r_1$ then $V = 0$

if $V_1 r_2 > V_2 r_1$ then $V_A - V_B =$ Positive i.e. A side of the equivalent battery will become the positive terminal and vice-versa.

(ii) **Internal resistance (*r*) of the battery:** r_1 and r_2 are in parallel. Therefore, the internal resistance *r* will be given by

$$1/r = 1/r_1 + 1/r_2$$

or $$\boldsymbol{r = \frac{r_1 r_2}{r_1 + r_2}}$$

437. The given circuit actually forms-a balanced Wheatstone bridge (including the voltmeter) as shown in figure

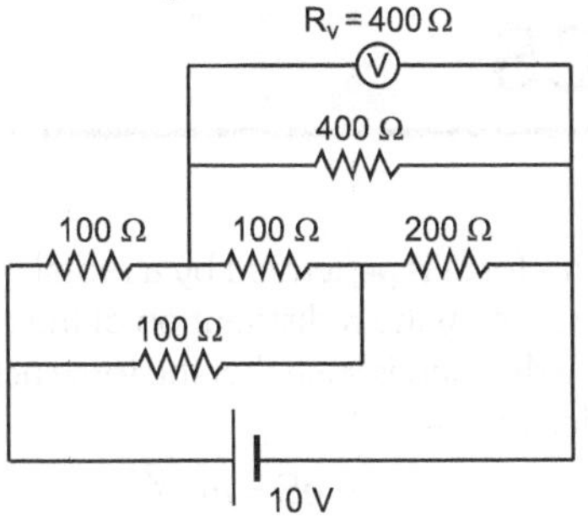

(a)

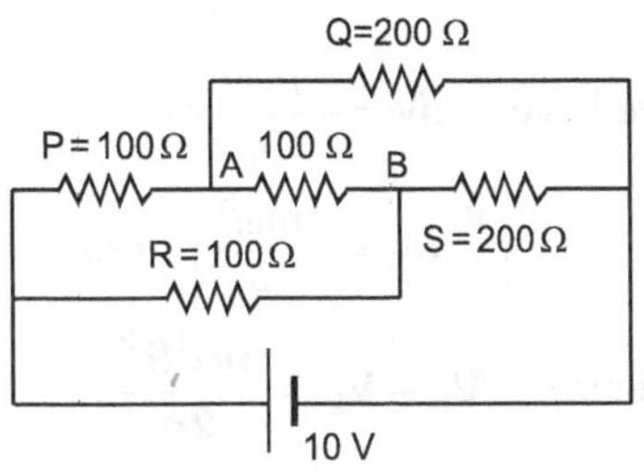

(b)

Here we see that $\frac{P}{Q} = \frac{R}{S}$

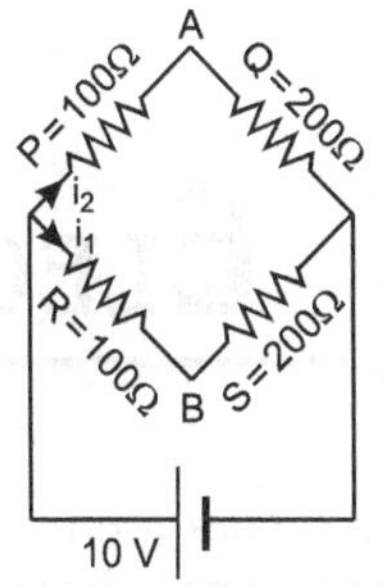

(c)

(Bridge is balanced)

Therefore, resistance between A and B can be ignored and equivalent simple circuit can be drawn as shown in figure (c).

The voltmeter will read the potential difference across resistance Q.

$$\text{Currents } i_1 = i_2 = \frac{10}{100 + 200}$$

$$= \frac{1}{30}\text{ A}$$

$\therefore$ Potential difference across voltmeter

$$= Qi_1 = (200)\left(\frac{1}{30}\right)\text{V} = \frac{20}{3}\text{ V}$$

Therefore, reading of voltmeter will be $\mathbf{\frac{20}{3}}$ **V**

CHAPTER 13 ELECTROSTATICS

438. The work goes on increasing the accumulator's energy. Since the plates of the capacitor are connected all the time to the accumulator terminals, the potential difference applied to them remains constant and consequently their charge must decrease. For the capacitor's charge

$$Q = CV = \left(\frac{\varepsilon_0 SV}{d}\right)$$

where C is the capacitor's capacity, V is the potential difference, S is the area of the capacitor plates and d is the distance between them.

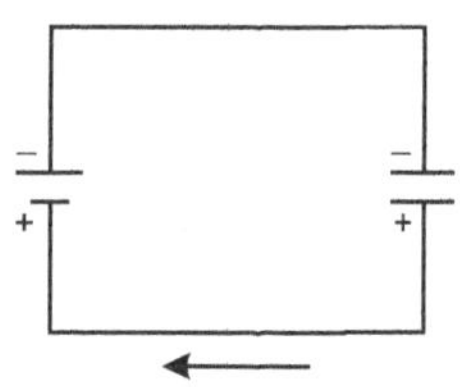

If V remains constant and d increases, Q must decrease. A partial discharge of the capacitor takes place and current flows through the circuit in the direction shown by the arrow in figure, as a result of which the accumulator is charged.

When the capacitor plates are moved apart, the energy W of the electric field in the capacitor decreases. In fact $W = \frac{1}{2}CV^2$ and if V is constant and C decreases, then W also decreases. The energy which is released in the capacitor also goes on charging the accumulator. Thus all the work expended on moving the plates apart goes on increasing the accumulator's energy and so too does a part of the energy stored in the capacitor.

439. For an electron that is inside the disc at a distance r from the axis to move along a circle, there should be a force putting it to the axis. According to Newton's second law

$$F = m\omega^2 r$$

This force is generated by a radial electric field caused by the redistribution of the electrons in the disc and is such that the force acting on the electron is—

$$F = eE = mr\omega^2$$

Substituting $\quad E = -\dfrac{dV}{dr}$

we have $\quad dV = -\dfrac{m\omega^2}{e} r dr$

or $\quad \int_{V_1}^{V_2} dV = -\dfrac{m\omega^2}{e}\int_0^R r\,dr$

Hence $\quad \mathbf{V_1 - V_2 = \dfrac{m\omega^2 R^2}{2e}}$

i.e. $V_1 > V_2$ or potential at centre is more than the potential at edge.

440. (i) Let $\vec{v}_1$ and $\vec{v}_2$ be the velocities of the first and second balls after the given time interval say Δt. The angle between $\vec{v}_1$ and the initial velocity $\vec{v}$ is 60°. Hence

$$|\Delta \vec{p}_1| = |q_1 \vec{E}\, \Delta t| = 2m_1 v \sin 60° \qquad \ldots(1)$$

Here we use the condition that $v_1 = 2v$ which implies that the change in momentum $\Delta \vec{p}_1$ of the first ball occurs in a direction perpendicular to the initial velocity $\vec{v}$. Because after time Δt component of momentum in the direction of initial velocity $\vec{v}$ is $2m_1 v \cos 60°$ or $m_1 v$ which is equal to the initial momentum. So, we can conclude that the electric field is in a direction perpendicular to the initial velocity of both.

For the second ball also, momentum in initial direction should not change. Hence

$$m_2 v_2 \cos 30° = m_2 v$$

or $\quad \boldsymbol{v_2} = v \sec 30° = \dfrac{\mathbf{2}}{\sqrt{\mathbf{3}}}\boldsymbol{v}$

(ii) For the second ball

$$|\Delta \vec{p_2}| = |q_2 \vec{E}\, \Delta t| = \frac{2}{\sqrt{3}} m_2 v \sin 30° \quad \ldots(2)$$

Dividing equation (1) by equation (2) we get

$$\frac{q_1}{q_2} = \frac{\sqrt{3}m_1}{m_2} \frac{\sin 60°}{\sin 30°}$$

or $\quad \dfrac{\boldsymbol{q_2}}{\boldsymbol{m_2}} = \dfrac{q_1}{m_1} \cdot \dfrac{\sin 30°}{\sin 60°} \cdot \dfrac{1}{\sqrt{3}}$

$$= \frac{1}{3} \cdot \frac{q_1}{m_1} = \frac{\boldsymbol{\alpha_1}}{\mathbf{3}}$$

441. In the figure shown

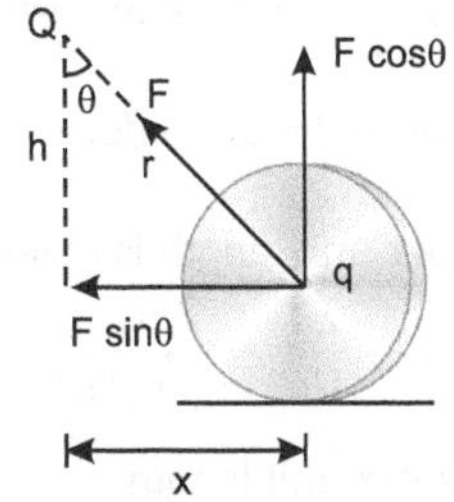

$$F = \frac{1}{4\pi\varepsilon_0} \cdot \frac{Qq}{r^2} \qquad (r^2 = x^2 + h^2)$$

$$F = \frac{1}{4\pi\varepsilon_0} \cdot \frac{Qq}{(x^2 + h^2)}$$

Restoring torque about point of contact is

$$\tau = -(F \sin \theta) R$$

or $\quad \tau = -\left\{\dfrac{1}{4\pi\varepsilon_0} \cdot \dfrac{Qq}{x^2 + h^2}\right\}\left(\dfrac{x}{\sqrt{x^2 + h^2}}\right)R$

Since $\quad x << h$

$\therefore \quad x^2 + h^2 \approx h^2$

$\therefore \quad \tau = -\left(\dfrac{QqR}{4\pi\varepsilon_0 h^3}\right)x \qquad (x = h\theta)$

or $\quad \tau = -\left(\dfrac{QqR}{4\pi\varepsilon_0 h^2}\right)\theta$

Since $\quad \tau \propto -\theta$

oscillations are simple harmonic in nature.

Now $\quad \tau = I\alpha \qquad \left(I = \dfrac{3}{2}MR^2\right)$

or $\quad \dfrac{3}{2}MR^2 \cdot \alpha = -\left(\dfrac{QqR}{4\pi\varepsilon_0 h^2}\right)\theta$

$\therefore \quad T = 2\pi\sqrt{\left|\dfrac{\theta}{\alpha}\right|}$

or $\quad \boldsymbol{T = 2\pi h\sqrt{\dfrac{6\pi\varepsilon_0 MR}{Qq}}}$

442. Equivalent capacitance in the displaced position x is :

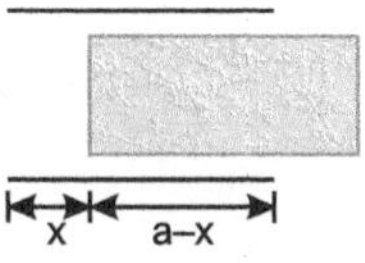

$$C = \frac{\varepsilon_0 a x}{d} + \frac{4\varepsilon_0 (a - x)\, a}{d}$$

$$C = \frac{\varepsilon_0 a}{d}[x + 4a - 4x]$$

or $\quad C = \dfrac{\varepsilon_0 a}{d}[4a - 3x]$

Force on dielectric due to electric field will be

$$F_e = \left|\frac{dU}{dx}\right| = \frac{1}{2}V^2\left|\frac{dC}{dx}\right|$$

$$= \frac{V^2}{2}\left(\frac{3\varepsilon_0 a}{d}\right) = \frac{3}{2}\frac{\varepsilon_0 a V^2}{d} \qquad \text{(inwards)}$$

Net force on the slab will be

$$F = F_0 - kx - \frac{3}{2}\frac{\varepsilon_0 a V^2}{d}$$

In equilibrium $F = 0$

Hence $\quad F_0 - kx - \dfrac{3}{2}\dfrac{\varepsilon_0 a V^2}{d} = 0$

or $\quad \boldsymbol{x = \dfrac{1}{k}\left[F_0 - \dfrac{3}{2}\dfrac{\varepsilon_0 a V^2}{d}\right]}$

This is also the amplitude of oscillations. Now if the slab is further displaced by x_0 only spring force will increase, other two forces are constant. Hence net restoring force will be the extra spring force kx_0 or the time period will be same viz; $\quad 2\pi\sqrt{\dfrac{M}{k}}$.

So, $\quad \boldsymbol{T = 2\pi\sqrt{\dfrac{M}{k}}}$

443. Let l be the natural length of the spring in equilibrium

electrostatic force = spring force

or $$\frac{q^2}{(4\pi\varepsilon_0)(4l^2)} = kl \qquad \ldots(1)$$

here q = charge on each ball and k = force constant of spring

when further displaced by x, net restorting force will be

$$F = k(x+l) - \frac{1}{4\pi\varepsilon_0}\frac{q^2}{(2l+x)^2}$$

$$= kx + kl - \frac{q^2}{4\pi\varepsilon_0}\left[\frac{1}{4l^2\left(1+\frac{x}{2l}\right)^2}\right]$$

$$= kx + kl - \frac{q^2}{4\pi\varepsilon_0(4l^2)}\left(1+\frac{x}{2l}\right)^{-2}$$

$$= kx + kl - \frac{q^2}{4\pi\varepsilon_0(4l^2)}\left(1-\frac{x}{l}\right) \qquad (x << l)$$

$$= kx + kl - \frac{q^2}{4\pi\varepsilon_0(4l^2)} + \frac{q^2}{4\pi\varepsilon_0(4l^2)}\frac{x}{l}$$

substituting $kl = \dfrac{q^2}{4\pi\varepsilon_0(4l^2)}$ from equation (1)

$F = 2kx$

Since it is restoring in nature, we can write

$F = -2kx$

Thus it is the case as if the stretched spring has double the force constant. Hence

$$\frac{\boldsymbol{f'}}{\boldsymbol{f}} = \frac{\sqrt{2k}}{\sqrt{k}} = \sqrt{2} \qquad \left(f = \frac{1}{2\pi}\sqrt{\frac{k}{m}}\right)$$

444. Electrostatic force is in vertical direction. Hence, two balls and COM will always lie on the same vertical line. Net acceleration of COM is still g and initial velocity is v in horizontal direction.

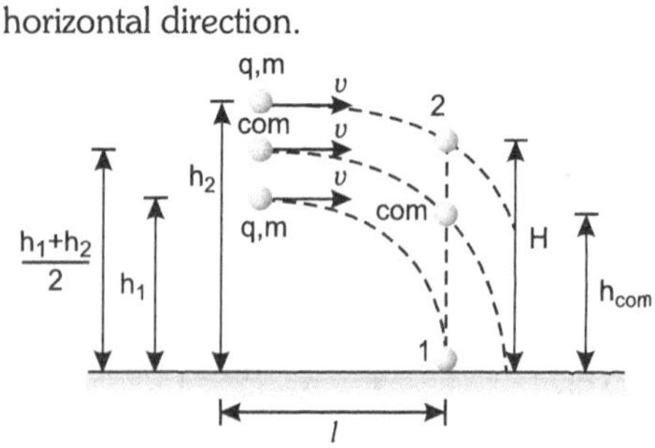

$$t = \frac{l}{v}$$

In this time COM will fall a height

$$h = \frac{1}{2}gt^2 = \frac{1}{2}g\left(\frac{l}{v}\right)^2 \qquad \ldots(1)$$

Let H be the height of the second ball at the moment the first ball hits the ground

Then $$h_{COM} = \frac{H}{2}$$

or $$\left(\frac{h_1+h_2}{2}\right) - h = \frac{H}{2}$$

or $$H = (h_1+h_2) - 2h$$

or $$\boldsymbol{H = h_1 + h_2 - g\left(\frac{l}{v}\right)^2}$$

as $2h = g\left(\dfrac{l}{v}\right)^2$ from equation (1)

445. Let V be the potential difference between the two.

Capacity per unit length is given by

$$C = \frac{\lambda}{V} = \frac{2\pi\varepsilon_0}{ln\,(D_2/D_1)}$$

λ = charge per unit length

or $$\lambda = \frac{2\pi\varepsilon_0 V}{ln\,(D_2/D_1)}$$

Now electric field strength at the core is

$$E = \frac{\lambda}{2\pi\varepsilon_0(D_1/2)} = \frac{\lambda}{\pi\varepsilon_0 D_1}$$

$$= \frac{2V}{D_1 ln\,(D_2/D_1)} \qquad \left(E = \frac{\lambda}{2\pi\varepsilon_0 r}\right)$$

For E to be minimum

$$\frac{dE}{dD_1} = 0$$

or $$-2V\frac{\{(ln D_2 - ln D_1) - 1\}}{[D_1(ln D_2 - ln D_1)]^2} = 0$$

or $$\ln D_2 - \ln D_1 = 1$$

or $$\ln\left(\frac{D_2}{D_1}\right) = 1$$

or $$\frac{D_2}{D_1} = e \quad \text{or} \quad \boldsymbol{D_1 = \frac{D_2}{e}}$$

446. Electric field at a distance r at an angle θ_1 due to dipole $\vec{p}_1$ is

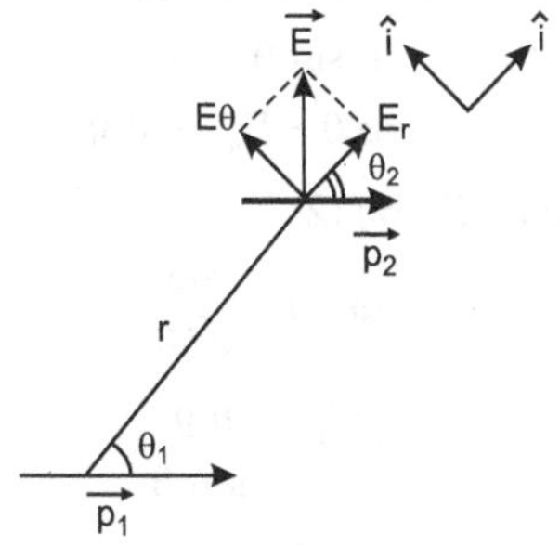

$$\vec{E} = E_r\,\hat{i} + E_\theta\hat{j}$$

Here $\quad E_r = \dfrac{2k\,p_1\cos\theta_1}{r^3}$

where $\quad k = \dfrac{1}{4\pi\varepsilon_0}$

and $\quad E_\theta = \dfrac{kp_1\sin\theta_1}{r^3}$

Hence

$$\vec{E} = \frac{2k\,p_1\cos\theta_1}{r^3}\hat{i} + \frac{k\,p_1\sin\theta_1}{r^3}\hat{j} \qquad \ldots(1)$$

$\vec{p}_2$ in vector form can be written as

$$\vec{p}_2 = p_2\cos\theta_2\,\hat{i} - p_2\sin\theta_2\hat{j} \qquad \ldots(2)$$

Now $\quad U = -\vec{p}_2\cdot\vec{E}$

or $\quad U = -\dfrac{2k\,p_1\,p_2\cos\theta_1\cos\theta_2}{r^3} + \dfrac{kp_1p_2\sin\theta_1\sin\theta_2}{r^3}$

Substituting $k = \dfrac{1}{4\pi\varepsilon_0}$ we get

$$\boldsymbol{U = \frac{p_1p_2}{4\pi\varepsilon_0 r^3} \times (\sin\theta_1\,\sin\theta_2 - 2\cos\theta_1\,\cos\theta_2)}$$

447. If a charge q is given to the outermost sphere C, some of the charge will go to the outer surface of sphere A. Some will go to the inner surface of C and the rest will go to the outer surface of C. The charge q_1 on sphere A will induce a charge $-q_1$ on the inner surface of sphere B. Charge q_2 on the inner surface of sphere C will induce a charge $-q_2$ on the outer surface of sphere B.

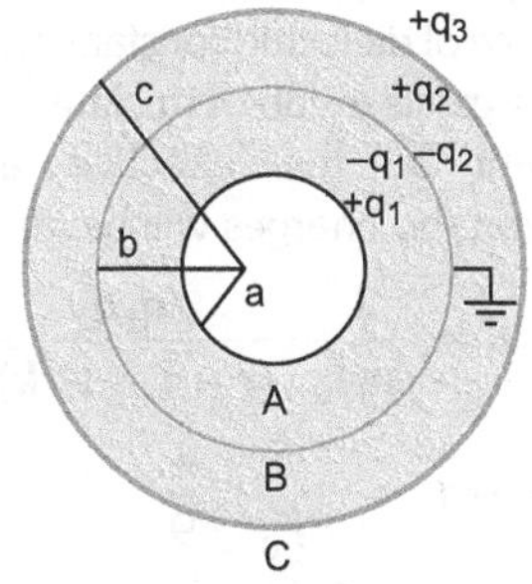

In this way three capacitors are formed. One between the outer surface of A and inner surface of B. Second between the outer surface of B and inner surface of C and third by the outer surface of C.

Their capacities are

$$C_1 = 4\pi\varepsilon_0\frac{ab}{b-a};$$

$$C_2 = 4\pi\varepsilon_0\frac{bc}{c-b} \quad \text{and} \quad C_3 = 4\pi\varepsilon_0 c$$

Hence the combined capacity is

$$C = C_1 + C_2 + C_3$$

$$\mathbf{C} = 4\pi\varepsilon_0\left[\frac{ab}{b-a} + \frac{bc}{c-b} + c\right]$$

$$= \boldsymbol{4\pi\varepsilon_0\left[\frac{ab}{b-a} + \frac{c^2}{c-b}\right]}$$

448. (a) The repulsive force between the charges in air is

$$F_0 = \frac{1}{4\pi\varepsilon_0}\cdot\frac{q_1q_2}{r^2}$$

If the space between the charges is completely filled with a dielectric of dielectric constant k, the force becomes

$$F = \frac{1}{4\pi\varepsilon_0}\cdot\frac{q_1q_2}{kr^2}$$

Now let us suppose that force between the two charges remains the same i.e. F when they are placed at a distance r' in air. Then

$$\frac{1}{4\pi\varepsilon_0}\frac{q_1 q_2}{kr^2}=\frac{1}{4\pi\varepsilon_0}\frac{q_1q_2}{r'^2}$$

or $$r' = r\sqrt{k}$$

This gives equivalent air separation due to presence of dielectric constant k and thickness r. If there exists a slab of thickness t and dielectric constant k, the effective air separation between the charges will be : $(r-t)+t\sqrt{k}$

$$\therefore \quad \mathbf{F}=\frac{\mathbf{1}}{\mathbf{4\pi\varepsilon_0}}\cdot\frac{\mathbf{q_1q_2}}{\mathbf{(r-t+t\sqrt{k})^2}}$$

(b) Given $\frac{F}{F_0}=\frac{4}{9}$ where $t=\frac{r}{2}$

Hence $$\frac{4}{9}=\frac{\frac{1}{4\pi\varepsilon_0}\cdot\frac{q_1q_2}{\left(r-\frac{r}{2}+\frac{r}{2}\sqrt{k}\right)^2}}{\frac{1}{4\pi\varepsilon_0}\cdot\frac{q_1q_2}{r^2}}$$

Solving this we get $\boldsymbol{k}=\mathbf{4}$

449. $r=3$ cm, $l=100$ cm

$$F=\frac{1}{4\pi\varepsilon_0}\cdot\frac{q^2}{r^2}$$

$$\cos\theta=\frac{r/2\sec 30°}{l}$$

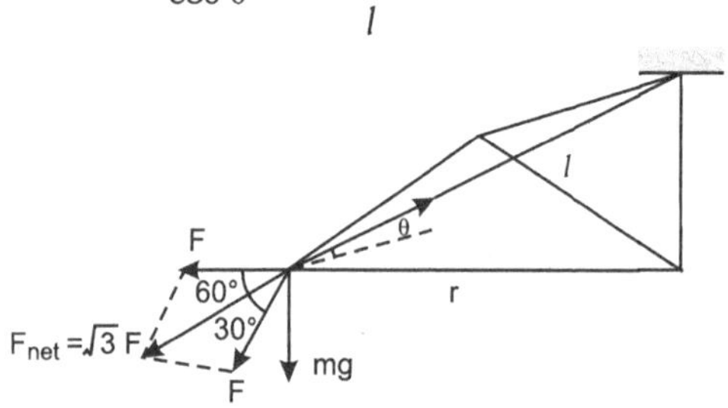

Here θ is the angle which each string makes with horizontal.

$$=\frac{\left(\frac{3}{2}\right)\left(\frac{2}{\sqrt{3}}\right)}{100}$$

or $$\theta=89°$$

Each particle is in equilibrium under three forces :

(1) F_{net} which is resultant of two forces of magnitude F acting at 60°

(2) weight (mg) and

(3) tension (T)

Equilibrium of particle gives

$$T\sin\theta=mg \quad \ldots(1)$$

and $$T\cos\theta=F_{net}=\sqrt{3}F \quad \ldots(2)$$

Dividing (1) by (2)

$$\tan\theta=\frac{mg}{\sqrt{3}F}$$

or $$\sqrt{3}F=\frac{mg}{\tan\theta}$$

or $$\sqrt{3}\left\{\frac{1}{4\pi\varepsilon_0}\frac{q^2}{r^2}\right\}=\frac{mg}{\tan\theta}$$

or $$(\sqrt{3})(9\times10^9)\frac{q^2}{(3\times10^{-2})^2}=\frac{(10^{-3})(10)}{\tan 89°}$$

or $$\boldsymbol{q}=\mathbf{3.174\times10^{-9}\ C}$$

450. (a) For conservative force field,

$$dU=-dW \quad (U=\text{potential energy})$$

or $U_f-U_i=-W_{i\to f}$

or $W_{i\to f}=U_i-U_f=q(V_i-V_f)$

Hence $(5\times10^{-5})=(2\times10^{-6})\,[2(a)(0.1)^2-0]$

or $\boldsymbol{a}=\mathbf{1.25\times10^3\ V/m^2}$

(b) For any plane parallel to X-Y plane, $Z=$ constant.

Hence $V=ax^2+ay^2+$ constant

For equipotential surface,

$V=$ constant say V_0.

Hence $V_0-\text{constant}=ax^2+ay^2=k$

$(k=\text{constant})$

Which is an equation of a circle.

(c) Substituting $a=1.25\times10^3$ V/m^2, $V=6250$ V and $Z=\sqrt{2}$ m in given equation we get

$$6250=(1.25\times10^3)(x^2+y^2+2\times2)$$

or $$x^2+y^2=1$$

This is an equation of circle of radius **1 m.**

451. Total charge q_0 on the ring,

$$q_0=\int_{-\pi/2}^{+\pi/2}dq$$

or $12.0 \times 10^{-6} = \int_{-\pi/2}^{+\pi/2} (\lambda_0 \cos\theta)(Rd\theta)$

$$= \lambda_0 R\,[\sin\theta]_{-\pi/2}^{+\pi/2} = 2\lambda_0 R$$

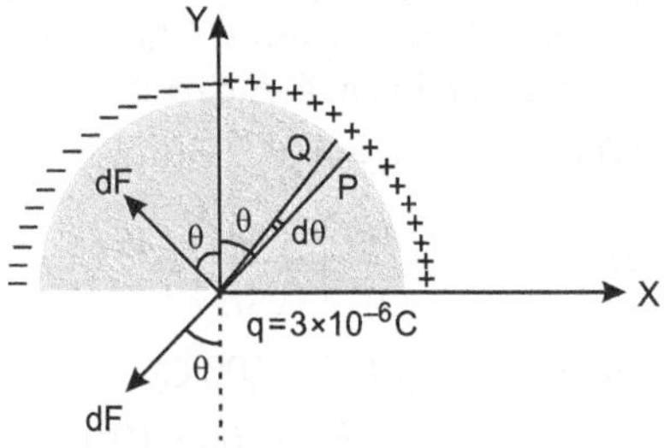

$$\therefore \quad \lambda_0 = \frac{12.0\times 10^{-6}}{2\times 0.6} = 10^{-5}\ \text{C/m}$$

Small charge dq on element PQ,

$$dq = (\lambda_0 \cos\theta)(Rd\theta)$$

$$\therefore F_{net} = \int_0^{\pi/2} 2\,(dF \sin\theta)$$

$$= \int_0^{\pi/2} 2\left[\frac{1}{4\pi\varepsilon_0}\cdot q\,\frac{(dq)}{R^2}\cdot \sin\theta\right]$$

$$= \int_0^{\pi/2} 2\left[\frac{1}{4\pi\varepsilon_0}\,\frac{(q)(\lambda_0\cos\theta)(Rd\theta)}{R^2}\cdot\sin\theta\right]$$

$$= \frac{\lambda_0 q}{(4\pi\varepsilon_0)R}\int_0^{\pi/2}\sin 2\theta\,.\,d\theta$$

$$= \frac{\lambda_0 q}{2\,(4\pi\varepsilon_0)R}\,[-\cos 2\theta]_0^{\pi/2}$$

$$= \frac{\lambda_0 q}{(4\pi\varepsilon_0)R}$$

$$= \frac{(9\times 10^9)(10^{-5})(3\times 10^{-6})}{0.6} = 0.45\ \text{N}$$

Since, this force will be in negative x-axis. So we can write.

$$\vec{F} = (-0.45\ \text{N})\,\hat{i}$$

452. (a) Applying Kirchhoff's second law

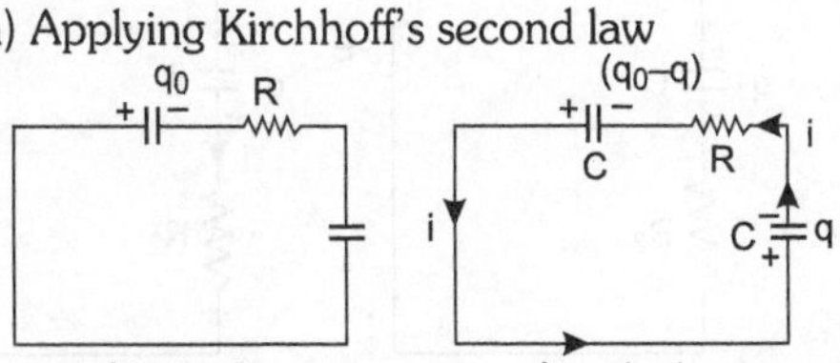

$$\left(\frac{q_0 - q}{C}\right) - iR - \frac{q}{C} = 0 \qquad \left(i = \frac{dq}{dt}\right)$$

or $\quad \dfrac{1}{C}(q_0 - 2q) = R\,\dfrac{dq}{dt}$

or $\quad \dfrac{dq}{q_0 - 2q} = \dfrac{dt}{CR}$

or $\quad \displaystyle\int_0^q \frac{dq}{q_0 - 2q} = \int_0^t \frac{dt}{CR}$

This gives $q = \dfrac{q_0}{2}\left(1 - e^{-\frac{2}{CR}t}\right)$ ... (1)

Substituting $q_0 = 10\ \text{mC} = 10^{-2}\ \text{C}$;

$$C = 2\,\mu\text{F} = 2\times 10^{-6}\ \text{F}$$

and $\quad \Delta t = t = 1\,\mu\text{s} = 10^{-6}\ \text{s}$

$$R = 5\,\Omega$$

the charge flowing through the resistance in 1μ s will be

$$q = \left(\frac{10^{-2}}{2}\right)\left(1 - e^{\frac{-2\times 10^{-6}}{2\times 10^{-6}\times 5}}\right)$$

or $\quad \boldsymbol{q = 0.9}$ **mC**

(b) Current flowing through the resistance at any time t will be

$$i = \frac{dq}{dt} = \frac{q_0}{CR}\,e^{-\frac{2}{CR}t} \text{From equation (1)]}$$

$$\therefore \quad H = \int_0^t i^2 R dt$$

$$= \int_0^t \left(\frac{q_0}{CR}\right)^2 R e^{-\frac{4}{CR}t}\,dt$$

$$= \left(\frac{q_o^2}{C^2R}\right)\left(-\frac{CR}{4}\right)\left[e^{-\frac{4}{CR}\cdot t}\right]_0^t$$

$$= \frac{q_o^2}{4C}\left[1 - e^{-\frac{4}{CR}\cdot t}\right]$$

Substituting the values we have

$$H = \frac{(10^{-2})^2}{4\times 2\times 10^{-6}}\left[1 - e^{-\frac{4\times 10^{-6}}{2\times 10^{-6}\times 5}}\right]$$

or $\quad \boldsymbol{H = 4.12}$ **J**

453. (a) It is basically a problem of discharging in a *C-R* circuit. Under ideal conditions perfect insulater is filled between the electrodes of a capacitor and charge remains constant in both the electrodes. But due to small conductivity of dielectric filled in between, capacitor starts discharging.

Hence charge at any time t will be given by

$$q = q_0 e^{-t/\tau_C}$$

Here τ_C is the time constant of the *C-R* circuit

or $\tau_C = CR$

where $C = 4\pi\varepsilon_0 K\left(\dfrac{ab}{b-a}\right)$

and $R = \int_a^b \rho \dfrac{dr}{4\pi r^2} = \dfrac{\rho}{4\pi}\left(\dfrac{b-a}{ab}\right)$

Hence $\tau_C = \rho\varepsilon_0 K$

or $\boldsymbol{q = q_0 e^{-t/\rho\varepsilon_0 K}}$

(b) Amount of heat generated = electrical potential energy stored in the capacitor

or $H = \dfrac{1}{2}\dfrac{q_0^2}{C}$

where $C = 4\pi\varepsilon_0 K\left(\dfrac{ab}{b-a}\right)$

$\therefore$ $\boldsymbol{H = \left(\dfrac{1}{a} - \dfrac{1}{b}\right)\dfrac{q_0^2}{8\pi\varepsilon_0 K}}$

454. (a) When S_1 and S_2 are both closed simultaneously, then

power dissipated in R_1 :

$$P_{R_1} = \frac{E^2}{R_1} = 0.2 \text{ watt} \qquad \text{(given)}$$

Hence $R_1 = \dfrac{E^2}{0.2} = \dfrac{(100)^2}{(0.2)} = 50{,}000\,\Omega$

or $R_1 = 50\,k\,\Omega$

initial current in R_2 :

$$i_0 = \frac{E}{R_2} = 10^{-2}\text{A} \qquad \text{(given)}$$

Hence $R_2 = \dfrac{100}{10^{-2}} = 10{,}000\,\Omega$

or $R_2 = 10\text{ k}\Omega$

$\therefore$ $\boldsymbol{R_1 = 50\text{ k}\Omega}$

and $\boldsymbol{R_2 = 10\text{ k}\Omega}$

(b) When capacitor gets fully charged and S_2 is suddenly opened. Then it is a case of discharging in *C-R* circuit.

Where $q = q_0 e^{-t/\tau_C}$

or $i = \left(-\dfrac{dq}{dt}\right) = \left(\dfrac{q_0}{\tau_C}\right)e^{-t/\tau_C}$

Here $q_0 = CE = (100\text{ C})$

$\tau_C = C(R_1 + R_2) = (60 \times 10^3)\text{ C}$

$\therefore$ $i = \dfrac{100\text{ C}}{(60\times 10^3)\text{ C}}\, e^{-\frac{t}{(60\times 10^3)\text{ C}}}$

or $i = \left(\dfrac{1}{600}\right)e^{-\frac{t}{(6\times 10^4)\text{ C}}}$

It is given that at $t = 5$ s,

$$i = 0.74 \times 10^{-3}\text{ A}$$

So $(0.74\times 10^{-3}) = \left(\dfrac{1}{600}\right)e^{-\frac{5}{(6\times 10^4)\text{ C}}}$

or $\dfrac{5}{(6\times 10^4)\text{ C}} = 0.812$

or $C = 1.0 \times 10^{-4}$ F

or $\boldsymbol{C = 100\ \mu\text{F}}$

(c) Charge remaining in the capacitor when S_1 is opened and S_2 is closed is

$$q_i = (CE)\,e^{-t/\tau_C} = CE\,e^{\frac{-t}{C(R_1+R_2)}}$$

$$= (10^{-4})(100)\,e^{\frac{-5}{(10^{-4})(60\times 10^3)}}$$

$$= 4.34 \times 10^{-3}\text{ C}$$

Now charge at time t can be calculated as follows :

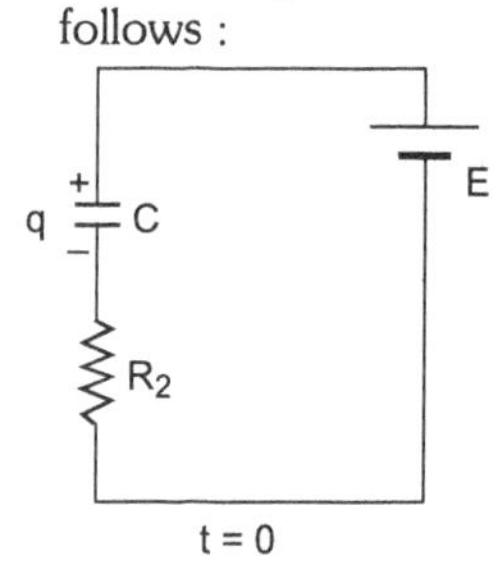

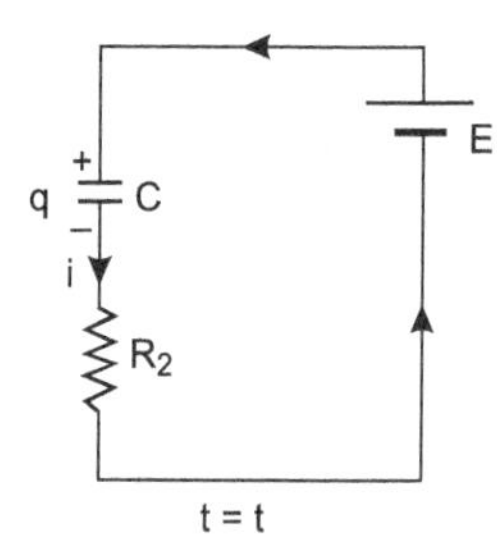

Applying Krichhoff's second law we get

$$\frac{q}{C} + iR_2 - E = 0 \qquad \left(i = \frac{dq}{dt}\right)$$

or $$E - \frac{q}{C} = \left(\frac{dq}{dt}\right) \cdot R_2$$

or $$\frac{dq}{E - \frac{q}{C}} = \frac{dt}{R_2}$$

or $$\int_{q_i}^{q} \frac{dq}{E - \frac{q}{C}} = \int_0^t \frac{dt}{R_2}$$

Solving this we get

$$q = CE - (CE - q_i)\, e^{\frac{-t}{CR_2}}$$

substituting the values we get

$$q = (10^{-4})(100) - (10^{-4} \times 100 - 4.34 \times 10^{-3})\, e^{-\frac{t}{(10^{-4})(10^4)}}$$

$$= 10^{-2} - (10^{-2} - 0.434 \times 10^{-2})\, e^{-t}$$

or $$\mathbf{q = (10 - 5.66\, e^{-t})\ mC}$$

455. Let the charge $-q_3$ is so placed between q_1 and q_2 that the distance between q_1 and $-q_3$ is a and between q_2 and $-q_3$ is b. Net force on $-q_3$ is zero.

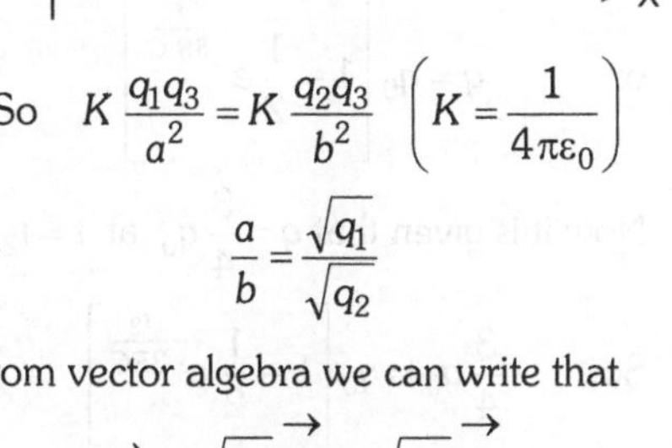

So $$K\frac{q_1 q_3}{a^2} = K\frac{q_2 q_3}{b^2} \qquad \left(K = \frac{1}{4\pi\varepsilon_0}\right)$$

or $$\frac{a}{b} = \frac{\sqrt{q_1}}{\sqrt{q_2}}$$

From vector algebra we can write that

$$\vec{r}_3 = \frac{\sqrt{q_1}\,\vec{r}_2 + \sqrt{q_2}\,\vec{r}_1}{\sqrt{q_1} + \sqrt{q_2}}$$

Secondly net force on q_1 is zero. So

$$\frac{Kq_1q_3}{a^2} = \frac{Kq_1q_2}{(a+b)^2}$$

or $$q_3 = \frac{q_2}{\left(1 + \frac{b}{a}\right)^2}$$

or $$q_3 = \frac{q_2}{\left(1 + \frac{\sqrt{q_2}}{\sqrt{q_1}}\right)^2}$$

$$= \frac{q_1 q_2}{(\sqrt{q_1} + \sqrt{q_2})^2} \qquad \left(\frac{b}{a} = \frac{\sqrt{q_2}}{\sqrt{q_1}}\right)$$

or with sign we can write

$$\mathbf{q_3 = -\frac{q_1 q_2}{(\sqrt{q_1} + \sqrt{q_2})^2}}$$

456. At a point $P(x, y, z)$ on the sphere a unit vector perpendicular to surface can be written as

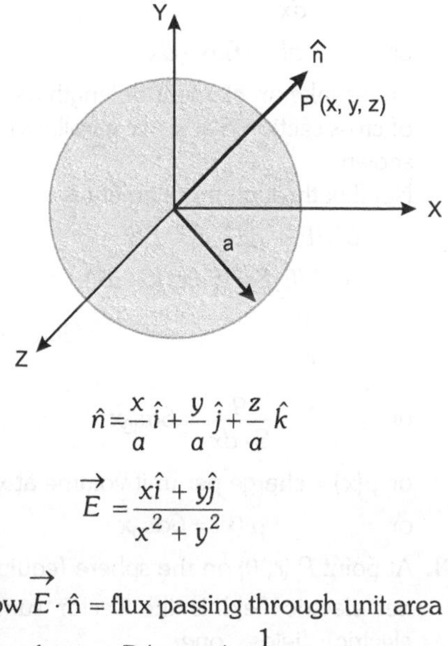

$$\hat{n} = \frac{x}{a}\hat{i} + \frac{y}{a}\hat{j} + \frac{z}{a}\hat{k}$$

$$\vec{E} = \frac{x\hat{i} + y\hat{j}}{x^2 + y^2}$$

Now $\vec{E} \cdot \hat{n}$ = flux passing through unit area on the sphere at $P(x, y, z) = \frac{1}{a}$

This is independent of the co-ordinates of P. Hence total flux passing through the whole sphere.

$$\phi = \left(\frac{1}{a}\right)(\text{Area of sphere})$$

$$= \left(\frac{1}{a}\right)(4\pi a^2) = 4\pi a$$

$$= \frac{q}{\varepsilon_0} \quad \text{(according to Gauss theorem)}$$

So $\quad \boldsymbol{q = 4\pi\varepsilon_0 a}$

457. Since the potential V depends on x-coordinates only. Electric field is in X-direction only

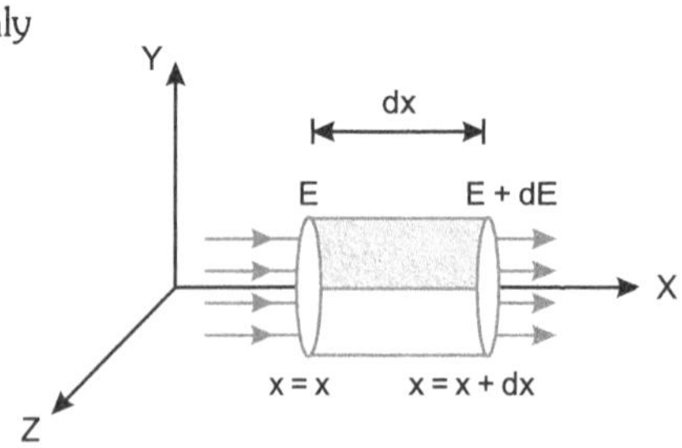

$$E = -\frac{dV}{dx} = 3\alpha x^2$$

and $\quad \dfrac{dE}{dx} = 6\alpha x$

or $\quad dE = 6\alpha x \cdot dx$

Let us take an element of length dx and area of cross section S at $x = x$ parallel to X-axis as shown

Net flux through this element is

$$\phi = (E + dE) \cdot S - ES$$

$$\phi = (dE)S = (6\alpha x)(S \cdot dx)$$

$$= \frac{q}{\varepsilon_0} \quad \text{(Gauss theorem)}$$

or $\quad \dfrac{q}{S \cdot dx} = 6\alpha\varepsilon_0 x$

or $\rho(x)$ = charge per unit volume at $x = x$

or $\quad \rho(x) = 6\alpha\varepsilon_0 x$

458. At point $P(r, \theta)$ on the sphere (equipotential) net electric field is the vector sum of two electric fields one due to dipole and the other is external field E_0. Electric field due to dipole has two components E_r and E_θ. Being an equipotential surface,

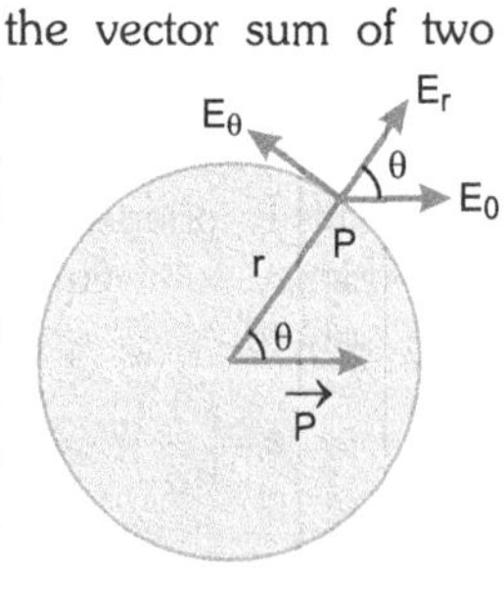

net field at P should be perpendicular to the sphere. So, tangential component of electric field should be zero.

or $\quad E_\theta = E_0 \sin\theta$

or $\quad \left\{\dfrac{1}{4\pi\varepsilon_0} \cdot \dfrac{P \sin\theta}{r^3}\right\} = E_0 \sin\theta$

or $\quad \boldsymbol{r = \left(\dfrac{P}{4\pi\varepsilon_0 E_0}\right)^{1/3}}$

459. In the first case capacitor is discharged through a resistance R. So,

$$q = q_0 e^{-t/CR}$$

Given $\quad q = \dfrac{q_0}{2} \quad$ at $t = t_1$

Hence $\quad \dfrac{q_0}{2} = q_0 e^{-\frac{t_1}{CR}}$

or $\quad t_1 = CR \ln(2) \quad \ldots(1)$

In the second case capacitor is again charged exponentially from $\dfrac{q_0}{2}$ to q_0.

So, equation of instantaneous charge at any time t will be as follows

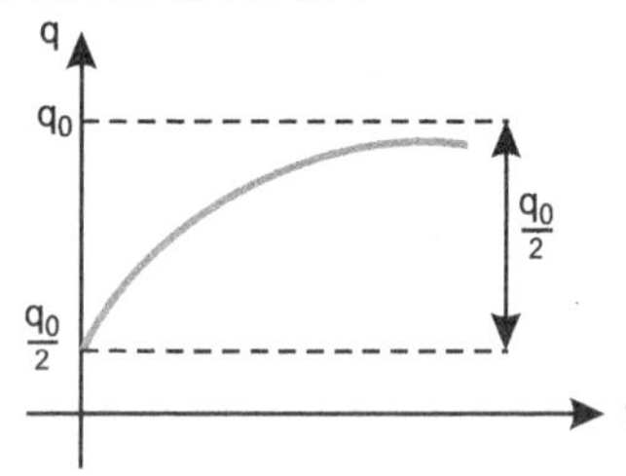

$$q = \frac{q_0}{2} + \frac{q_0}{2}\left[1 - e^{-\frac{t}{(R + 2R)C}}\right]$$

or $\quad q = q_0\left[1 - \dfrac{1}{2} e^{-\frac{t}{3RC}}\right]$

Now it is given that $q = \dfrac{3}{4} q_0$ at $t = t_2$

So, $\quad \dfrac{3}{4} q_0 = q_0\left[1 - \dfrac{1}{2} e^{-\frac{t_2}{3RC}}\right]$

or $\quad \dfrac{3}{4} = 1 - \dfrac{1}{2} e^{-\frac{t_2}{3RC}}$

or $\quad e^{-\frac{t_2}{3RC}} = \frac{1}{2}$ or $\frac{t_2}{3RC} = \ln(2)$

or $\quad t_2 = 3CR \ln(2) \quad \ldots(2)$

From equations (1) and (2)

$$\frac{t_1}{t_2} = \frac{CR\ln(2)}{3CR\ln(2)} \quad \text{or} \quad \boldsymbol{\frac{t_1}{t_2} = \frac{1}{3}}$$

460. In steady state, a capacitor offers infinite resistance in the circuit. Hence the equivalent circuits can be drawn as follows

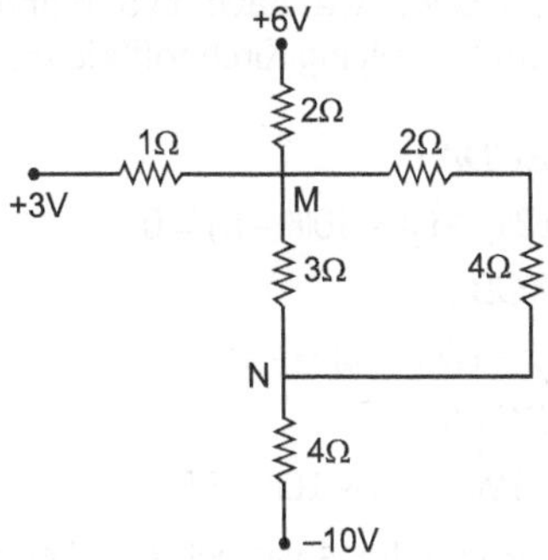

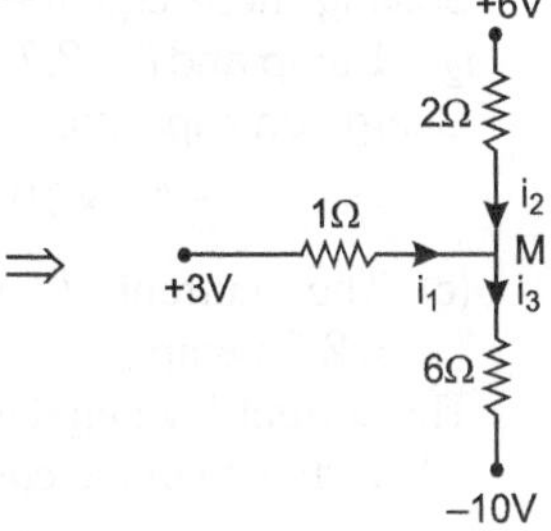

Let V be the potential of M. Applying Kirchhoff's first law

$$i_1 + i_2 = i_3$$

or $\quad \left(\frac{3-V}{1}\right) + \left(\frac{6-V}{2}\right) = \frac{V-(-10)}{6}$

or $\quad V = 2.6\,\text{V}$

Therefore, potential of point M is 2.6 V.

Now $\quad V_N - (-10) = 4i_3$

or $\quad V_N = -10 + 4i_3$

$$= -10 + 4\left(\frac{V+10}{6}\right)$$

$$= -10 + 4\left(\frac{2.6+10}{6}\right)$$

$$V_N = -1.6\,\text{V}$$

Hence $\quad \boldsymbol{V_M = +2.6\ \text{V}}$

and $\quad \boldsymbol{V_N = -1.6\ \text{V}}$

(b) Charge stored across 2μF capacitor is

$$q_2 = (2)(V_M - V_N)\,\mu\text{C} = (2)(2.6 + 1.6)$$

$$\boldsymbol{q_2 = 8.4\ \mu\text{C}}$$

Similarly charge stored across 1 μF capacitor is

$$q_1 = (1)(V_Q - V_P)\,\mu\text{C}$$

$$V_P = 0$$

Also $\quad V_M - V_N = 2.6 - (-1.6) = 4.2\,\text{V}$

Now $\quad V_M - V_Q = (4.2)\left(\frac{2}{2+4}\right) = 1.4\,\text{V}$

$\therefore V_Q = V_M - 1.4 = (2.6 - 1.4)\,\text{V} = 1.2\,\text{V}$

$\therefore q_1 = (1)(1.2 - 0)\,\mu\text{C}$

or $\quad \boldsymbol{q_1 = 1.2\ \mu\text{C}}$

Hence charge stored across 1μF and 2μF capacitors are 1.2μC and 8.4μC respectively.

461. (a) At time $t = 0$, circuit will be as shown below :

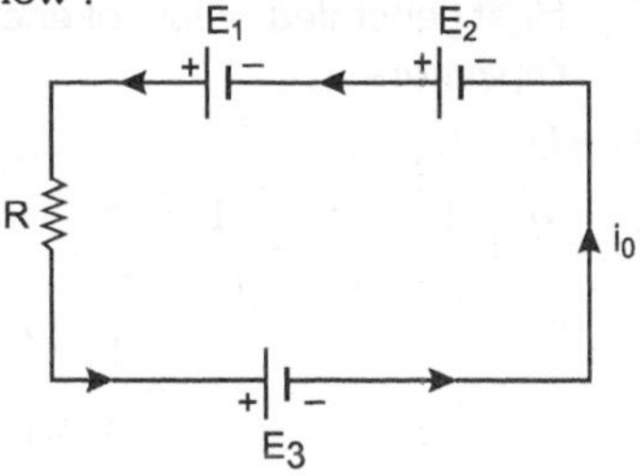

Here $\quad E_1 = \frac{q}{C_1} = \frac{30}{3} = 10\,\text{V}$

$$E_2 = \frac{q}{C_2} = \frac{30}{6} = 5\,\text{V}$$

and $\quad E_3 = \frac{q_3}{C_3} = \frac{30}{6} = 5\,\text{V}$

Therefore, initial current

$$i_0 = \frac{\text{net emf}}{\text{resistance}}$$

$$= \frac{E_1 + E_2 - E_3}{R} = \frac{10 + 5 - 5}{10}$$

or $\quad \boldsymbol{i_0 = 1\ \text{A}}$

(b) Let a charge q_0 flows through the circuit till steady state is reached again. Then charges on different capacitors will be as shown in figure.

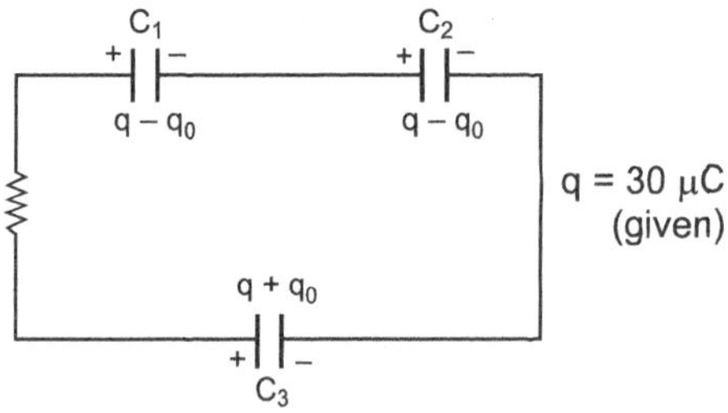

Applying Kirchhoff's voltage law in final state, we get

$$\frac{q-q_0}{C_1}-\frac{q+q_0}{C_3}+\frac{q-q_0}{C_2}=0$$

Substituting the values we get $q_0 = 15\,\mu C$.
Hence final charge on C_1 is $q - q_0$ or $15\,\mu C$ on C_2 is $q - q_0$ or $15\,\mu C$ and on C_3 is $q + q_0$ or $45\,\mu C$

$\therefore$ $\quad \mathbf{q_1 = 15\,\mu C,\ \ q_2 = 15\,\mu C}$

and $\quad \mathbf{q_3 = 45\,\mu C}$

(c) Heat generated = loss of energy stored in capacitors

$$H = U_i - U_f$$

$$H=\frac{q^2}{2}\left[\frac{1}{C_1}+\frac{1}{C_2}+\frac{1}{C_3}\right]-\frac{1}{2}\left[\frac{q_1^2}{C_1}+\frac{q_2^2}{C_2}+\frac{q_3^3}{C_3}\right]$$

$$= 300\,\mu J - 225\,\mu J \quad \text{or} \quad \boldsymbol{H = 75\,\mu J}$$

462. (a) When key K is open the circuit is a balanced Wheatstone bridge.

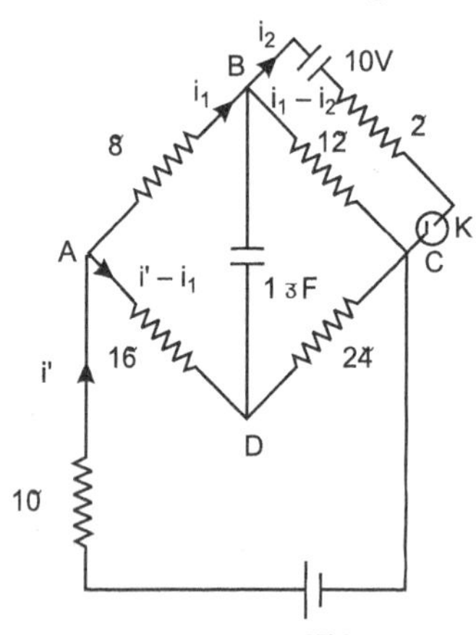

The equivalent resistance is given by :

$$R_{eq}=\left(\frac{20\times 40}{20+40}\right)=\left(\frac{40}{3}\right)\text{ohm.}$$

The total resistance

$$= R_{eq} + 10 = 23.3 \text{ ohm.}$$

The current $i=\left(\frac{55\times 3}{70}\right)$ amp. = 2.36 amp.

(b) When K is open, the bridge is balanced. $V_{BD} = 0$ *i.e.*, charge on capacitor = 0.

When K is closed, we have two different sources of e.m.f. Applying Kirchhoff's laws, we get,

For loop $ABCDA$:

$$8i_1 + 12(i_1 - i_2) - 40(i' - i_1) = 0$$

For loop $BXCB$:

$$2i_2 - 12(i_1 - i_2) = -10$$

For loop $ABCYA$:

$$8i_1 + 12(i_1 - i_2) + 10i' = 55$$

Solving these equations we get, $i_1 = 2$ amp, $i_2 = 1$ amp and $i' = 2.7$ amp.

Charge on capacitor

$$= 4.8\times 10^{-6}\text{ C.}$$

(c) The current I, when K was open = 2.36 amp.

The current i', when K was closed = 2.7 amp.

$\therefore$ The change in the current = 0.34 amp.

463. $$V_1 - V_2=\left(\frac{q_1}{4\pi\varepsilon_0}\right)\left(\frac{1}{a_1}-\frac{1}{a_2}\right)$$

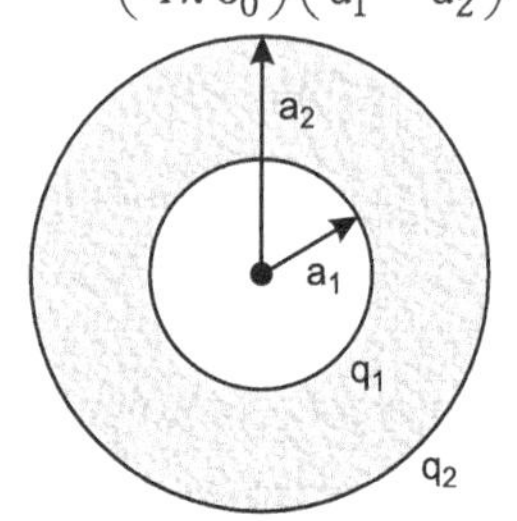

$$\therefore \quad q_1 = 4\pi\varepsilon_0\left(\frac{V_1 - V_2}{a_2 - a_1}\right)a_1 a_2$$

464. Interesting point to note in this question is that at a point distance $3R$ charge $-q$ will have repulsive force but on points close to the centre it will have a attraction force. One should be careful while applying the energy conservation equation in these type of problems.

First we have to find out a point where the electric field is zero because beyond that, charge will have attraction force and will itself reach the centre of the rings. Electric field at a point P distance x from the centre.

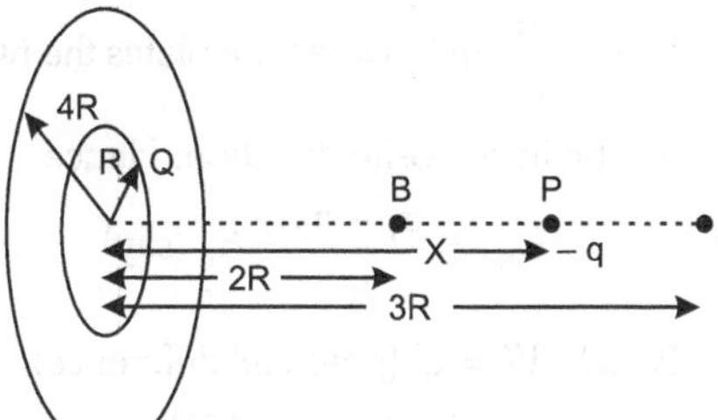

$$E_P = \frac{1}{4\pi \varepsilon_0} \frac{Qx}{(R^2 + x^2)^{3/2}} - \frac{1}{4\pi \varepsilon_0} \frac{8Qx}{(16R^2 + x^2)^{3/2}} = 0$$

$\Rightarrow$ $x = 2R$

So we have to give the charge enough KE so that it could reach this point because for $x < 2R$ field will be attractive. Now we can apply potential equation between these two points *i.e.*, distance $3R$ and $2R$.

Potential at A

$$= \frac{1}{4\pi\varepsilon_0} \left[\frac{Q}{(9R^2 + R^2)^{1/2}} - \frac{8Q}{(9R^2 + 16R^2)^{1/2}} \right]$$

$$= -\frac{1}{4\pi \varepsilon_0} \left(\frac{16 - \sqrt{10}}{10} \right) \frac{Q}{R}$$

Potential at point B

$$\frac{1}{4\pi\varepsilon_0} \frac{Q}{(4R^2 + R^2)^{1/2}} - \frac{1}{4\pi\varepsilon_0} \frac{8Q}{(4R^2 + 16R^2)^{1/2}}$$

$$= -\frac{1}{4\pi \varepsilon_0} \frac{3}{\sqrt{5}} \frac{Q}{R}$$

$$\text{Now } \frac{1}{2} mv^2 = \frac{1}{4\pi\varepsilon_0} \frac{Q.q}{R} \left[\frac{3}{\sqrt{5}} - \frac{16 - \sqrt{10}}{10} \right]$$

$$= \frac{1}{4\pi\varepsilon_0} \frac{Q.q}{R} \left[\frac{6\sqrt{5} + \sqrt{10} - 16}{10} \right]$$

$$\Rightarrow v_{min} = \left[\frac{2}{m} \cdot \frac{1}{4\pi\varepsilon_0} \frac{Qq}{R} \times \frac{\{6\sqrt{5} + \sqrt{10} - 16\}}{10} \right]^{1/2}$$

465. Charge stored in capacitor $q = CV$. It is the charge provided by battery. Thus the work done by battery

$$W = qE = CEV$$

Energy stored in capacitor $= \frac{1}{2} CV^2$.

Thus the total energy loss in the form of heat $Q = CEV - CV^2 = CV(2E - V)/2$

$$\therefore Q_2 = \frac{R_1}{R_1 + R_2} \times Q = \frac{R_1 CV}{2(R_1 + R_2)} (E - V)$$

466. At the surface of the charged sphere, whether it consists of a single piece or two pieces close together, the electric field stength is

$$E = \frac{1}{4\pi \varepsilon_0} \frac{Q}{R^2}$$

The electric charge per unit surface area is

$$\sigma = \frac{Q}{4\pi R^2}$$

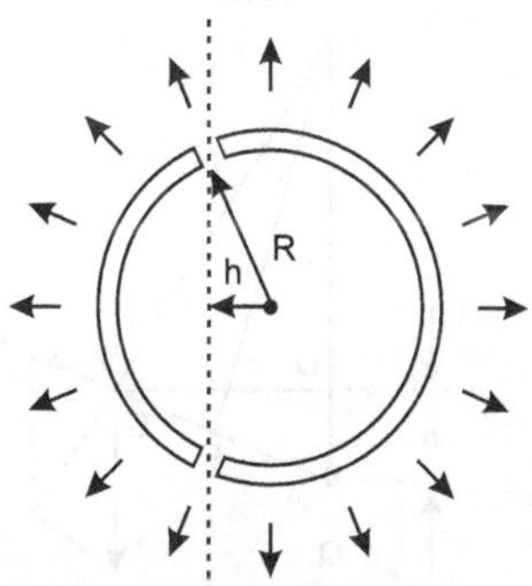

This electric field exerts a force $\Delta F = \frac{1}{2} E\, \Delta Q$ on the charge $\Delta Q = \sigma \Delta A$ which resides on a surface area ΔA, as illustrated in the figure. The reason for the factor of $\frac{1}{2}$ is that the electric field strength is E at the outer surface of the sphere and zero inside; its average value is therefore $\frac{E}{2}$.

The force per unit area exerted by the charges on the pieces of the sphere is therefore

$$\frac{\Delta F}{\Delta A} = \frac{Q^2}{32\pi^2 \varepsilon_0 R^4} = p.$$

The required force can be compared with the force with which a liquid at pressure p would push apart the two pieces of the sphere. As this force is also the product of p and the cross-sectional area of the intersection of the plane and sphere, *i.e.*, $p\pi(R^2 - h^2)$, it follows that the two parts of the sphere can be held together by a force

$$F = \frac{Q^2}{32\pi \varepsilon_0 R^4}(R^2 - h^2)$$

467. Using the notation in the figure, the equilbrium condition for the first ball is

$$\frac{mg}{F} = \frac{l}{x},$$

Where $$F = \frac{kqQ}{x^2}$$

is the Coulomb force action on the first ball and x is the distance between the balls carrying charges q and Q.

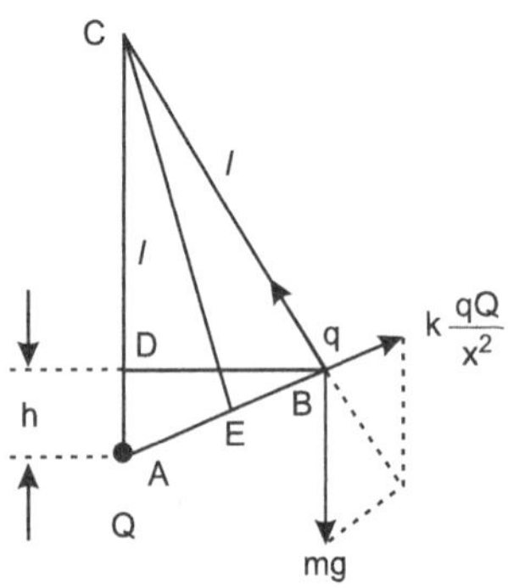

It is clear that the triangles ABD and CAE are similar and that consequently

$$\frac{x}{2} : \ell = h : x.$$

From the three equations above we can calculate the separation of the charges and the electrostatic energy of the system :

$$x = k\frac{qQ}{2mgh} \text{ and } E_{electro} = k\frac{qQ}{x} = 2mgh.$$

The workdone is the sum of the changes in electrostatic and gravitational potential energy,

$$W = 2mgh + mgh = 3mgh.$$

It is perhaps surprising that the work done does not depend on either the magnitudes of the charges or the length of the thread.

468. Electric field near a large metallic plate is given by $E = \frac{\sigma}{\varepsilon_0}$. In between the plates the two fields will be in opposite direction. Hence

$$E_{net} = \frac{\sigma_1 - \sigma_2}{\varepsilon_0} = E_0 \text{ (say)}$$

Now, $W = (q)$ (potential difference)

$$= (q)(E_0 a \cos 45°)$$

$$= (q)\left(\frac{\sigma_1 - \sigma_2}{\varepsilon_0}\right)\left(\frac{a}{\sqrt{2}}\right)$$

$$= \frac{(\sigma_1 - \sigma_2)\, qa}{\sqrt{2}\varepsilon_0}$$

469. For potential energy of the system of charges, total number of charge pairs will be ${}^8C_2$ or 28. Of these 28 pairs 12 unlike charges are at a separation 'a', 12 likes charges are at separation $\sqrt{2}a$ and 4 unlike charges are at separation $\sqrt{3}a$. Therefore the potential energy of the system

$$U = \frac{1}{4\pi\varepsilon_0}\left[\frac{(12)(q)(-q)}{a} + \frac{(12)(q)(q)}{\sqrt{2}a} + \frac{(4)(q)(-q)}{\sqrt{3}a}\right]$$

$$= -5.824\left(\frac{1}{4\pi\varepsilon_0}\cdot\frac{q^2}{a}\right)$$

The binding energy of this system is therefore $|U|$ or $5.824\left(\frac{1}{4\pi\varepsilon_0}\frac{q^2}{a}\right)$. So, work done by external forces in disassembling, this system of charges is

$$W = 5.824\left(\frac{1}{4\pi\varepsilon_0}\cdot\frac{q^2}{a}\right)$$

470. Applying energy conservation principle, increase in kinetic energy of the dipole
= decrease in electrostatic potential energy of the dipole

$\therefore$ kinetic energy of dipole at distance d from origin

$$= U_i - U_f$$

or $\quad KE = 0 - (-\vec{p}\cdot\vec{E}) = \vec{p}\cdot\vec{E}$

$$= (p\hat{i})\cdot\left(\frac{1}{4\varepsilon_0}\frac{q}{d^2}\hat{i}\right) = \frac{qp}{4\pi\varepsilon_0 d^2}$$

(b) Electric field at origin due to the dipole,

$$\vec{E} = \frac{1}{4\pi\varepsilon_0}\frac{2p}{d^3}\hat{i} \qquad (\vec{E}_{axis} \uparrow\uparrow \vec{p})$$

$\therefore$ Force on charge q

$$\vec{F} = q\vec{E} = \frac{pq}{2\pi\varepsilon_0 d^3}\hat{i}$$

471. Given $\quad q = 1\,\mu C = 10^{-6}\,C$

$m = 2\times 10^{-3}$ kg and $l = 0.8$ m

Let u be the speed of the particle at its lowest point and v its speed at highest point.
At highest point three forces are acting on the particle.

(i) Elecrostatic repulsion

$$F_e = \frac{1}{4\pi\varepsilon_0}\cdot\frac{q^2}{l^2} \qquad \text{(outwards)}$$

(ii) Weight $W = mg$ (inwards), and

(iii) Tension T (inwards)

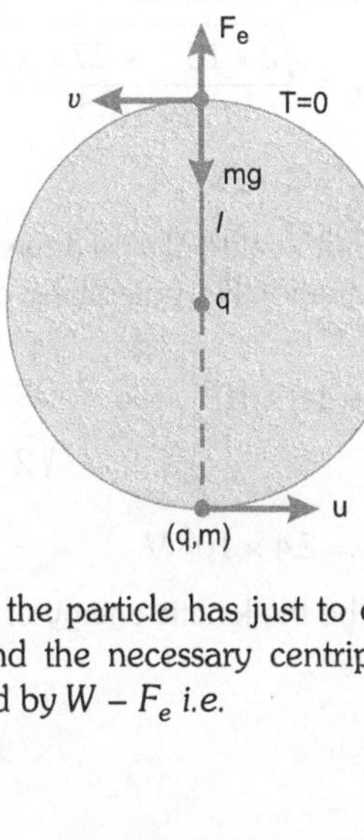

$T = 0$, if the particle has just to complete the circle and the necessary centripetal force is provided by $W - F_e$ *i.e.*

$$\frac{mv^2}{l} = W - F_e$$

or $\quad v^2 = \frac{l}{m}\left(mg - \frac{1}{4\pi\varepsilon_0}\cdot\frac{q^2}{l^2}\right)$

$$v^2 = \frac{0.8}{2\times 10^{-3}}\left(2\times 10^{-3}\times 10 - \frac{9.0\times 10^9\times(10^{-6})^2}{(0.8)^2}\right)\text{m}^2/\text{s}^2$$

or $\quad v^2 = 2.4\ \text{m}^2/\text{s}^2 \qquad \ldots(1)$

Now the electrostatic potential energy at the lowest and highest points are equal. Hence from conservation of mechanical energy :
increase in gravitational potential energy
= decrease in kinetic energy

or $\quad mg(2l) = \frac{1}{2}m(u^2 - v^2)$

or $\quad u^2 = v^2 + 4gl$

Substituting the values of v^2 from equation (1) we get

$$u^2 = 2.4 + 4(10)(0.8) = 34.4\ \text{m}^2/\text{s}^2$$

$\therefore \quad$ **u = 5.86 m/s**

Therefore, minimum horizontal velocity imparted to the lower ball, so that it can make complete revolution, is 5.86 m/s.

472. In the figure

$$q = 1\,\mu C = 10^{-6}\,C$$

$$q_0 = +\,0.1\,\mu C = 10^{-7}\,C$$

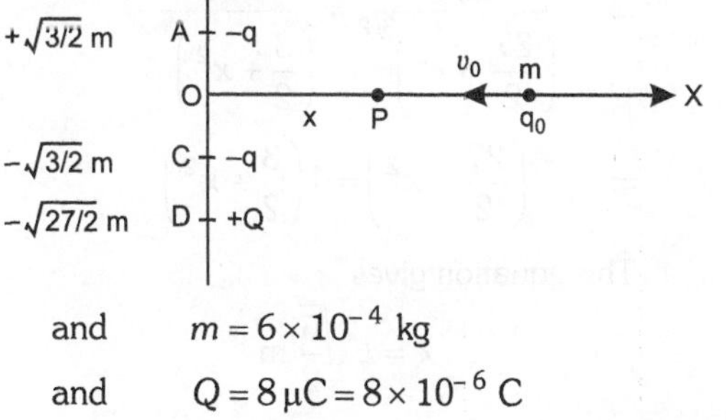

and $\quad m = 6\times 10^{-4}$ kg

and $\quad Q = 8\,\mu C = 8\times 10^{-6}\,C$

Let P be any point at a distance x from origin O. Then

$$AP = CP = \sqrt{\frac{3}{2} + x^2}$$

$$BP = DP = \sqrt{\frac{27}{2} + x^2}$$

Electric potential at point P will be

$$V = \frac{2KQ}{BP} - \frac{2Kq}{AP}$$

where $K = \frac{1}{4\pi\varepsilon_0} = 9 \times 10^9 \text{ Nm}^2/\text{C}^2$

$$\therefore V = 2 \times 9 \times 10^9 \left[\frac{8 \times 10^{-6}}{\sqrt{\frac{27}{2} + x^2}} - \frac{10^{-6}}{\sqrt{\frac{3}{2} + x^2}} \right]$$

$$V = 1.8 \times 10^4 \left[\frac{8}{\sqrt{\frac{27}{2} + x^2}} - \frac{1}{\sqrt{\frac{3}{2} + x^2}} \right] \quad \ldots(1)$$

$\therefore$ Electric field at P is

$$E = -\frac{dV}{dX}$$

$$= 1.8 \times 10^4 \left[(8)\left(-\frac{1}{2}\right)\left(\frac{27}{2} + x^2\right)^{-3/2} - (1)\left(-\frac{1}{2}\right)\left(\frac{3}{2} + x^2\right)^{-3/2} \right]$$

$E = 0$ on X-axis where

$$\frac{8}{\left(\frac{27}{2} + x^2\right)^{3/2}} = \frac{1}{\left(\frac{3}{2} + x^2\right)^{3/2}}$$

$$\Rightarrow \quad \frac{(4)^{3/2}}{\left(\frac{27}{2} + x^2\right)^{3/2}} = \frac{1}{\left(\frac{3}{2} + x^2\right)^{3/2}}$$

$$\Rightarrow \quad \left(\frac{27}{2} + x^2\right) = 4\left(\frac{3}{2} + x^2\right)$$

This equation gives

$$x = \pm\sqrt{\frac{5}{2}} \text{ m}$$

The least value of kinetic energy of the particle at infinity should be enough to take the particle upto $x = +\sqrt{\frac{5}{2}}$ m because

at $\quad x = +\sqrt{\frac{5}{2}}$ m

$E = 0 \;\Rightarrow\;$ Electrostatic force on charge q is zero or $F_e = 0$

for $\quad x > \sqrt{\frac{5}{2}}$ m

$\Rightarrow E$ is repulsive (towards positive X-axis)

and for $x < \sqrt{\frac{5}{2}}$ m

$\Rightarrow E$ is attractive (towards negative X-axis)

Now from equation (1), potential at

$$x = \sqrt{\frac{5}{2}} \text{ m}$$

$$V = 1.8 \times 10^4 \left[\frac{8}{\sqrt{\frac{27}{2} + \frac{5}{2}}} - \frac{1}{\sqrt{\frac{3}{2} + \frac{5}{2}}} \right]$$

$$V = 2.7 \times 10^4 \text{ volt}$$

Applying energy conservation at

$$x = \infty \quad \text{and} \quad x = \sqrt{\frac{5}{2}} \text{ m}$$

$$\frac{1}{2} m v_0^2 = q_0 V \quad \ldots(2)$$

$$\therefore \quad v_0 = \sqrt{\frac{2q_0 V}{m}}$$

Substituting the values

$$v_0 = \frac{\sqrt{2 \times 10^{-7} \times 2.7 \times 10^4}}{6 \times 10^{-4}}$$

$$\boldsymbol{v_0 = 3 \text{ m/s}}$$

$\therefore$ Minimum value of v_0 is 3 m/s

From equation (1), potential at origin ($x = 0$) is

$$V_0 = 1.8 \times 10^4 \left[\frac{8}{\sqrt{\frac{27}{2}}} - \frac{1}{\sqrt{\frac{3}{2}}} \right]$$

$$= 2.4 \times 10^4 \text{ V}$$

Let KE be the kinetic energy of the particle at origin.

Applying energy conservation at $x = 0$ and at $x = \infty$

$$KE + q_0V_0 = \frac{1}{2}mv_0^2$$

But $\frac{1}{2}mv_0^2 = q_0V$

from equation (2)

$$\therefore \quad KE = q_0(V - V_0)$$

$$KE = (10^{-7})(2.7 \times 10^4 - 2.4 \times 10^4)$$

$$\mathbf{KE = 3 \times 10^{-4}\ J}$$

Note: $E = 0$ or F_e on q_0 is zero at $x = 0$ and $x = \pm\sqrt{\frac{5}{2}}$ m.

Of these $x = 0$ is stable equilibrium position and $x = \pm\sqrt{\frac{5}{2}}$ is unstable equilibrium position.

473. Potential at a height *H* on the axis of the disc (V_P) :

The charge dq contained in the ring shown in figure

$$dq = (2\pi r dr)\,\sigma$$

Potential at P due to this ring,

$$dV = \frac{1}{4\pi\varepsilon_0}\cdot\frac{dq}{x}$$

where $x = \sqrt{H^2 + r^2}$

$$dV = \frac{1}{4\pi\varepsilon_0}\cdot\frac{(2\pi r dr)\sigma}{\sqrt{H^2 + r^2}} = \frac{\sigma}{2\varepsilon_0}\cdot\frac{r dr}{\sqrt{H^2 + r^2}}$$

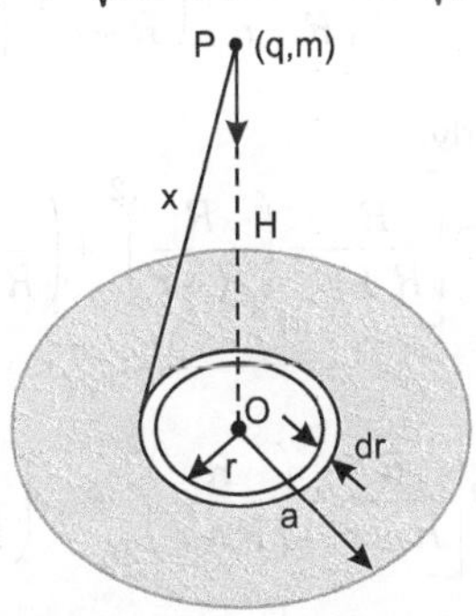

$\therefore$ Potential due to the complete disc

$$V_P = \int_{r=0}^{r=a} dV = \frac{\sigma}{2\varepsilon_0}\int_{r=0}^{r=a}\frac{r dr}{\sqrt{H^2 + r^2}}$$

$$= \frac{\sigma}{2\varepsilon_0}[\sqrt{a^2 + H^2} - H]$$

Potential at centre, (O) will be

$$V_O = \frac{\sigma a}{2\varepsilon_0} \qquad (H = 0)$$

(a) Particle is released from P and it just reaches point O. Therefore, from conservation of mechanical energy decrease in gravitational potential energy = increase in electrostatic potential energy

(Δ K.E. = 0 beacause $K_i = K_f = 0$)

$$\therefore \quad mgH = q[V_O - V_P]$$

or

$$gH = \left(\frac{q}{m}\right)\left(\frac{\sigma}{2\varepsilon_0}\right)[a - \sqrt{a^2 + H^2} + H] \quad \ldots(1)$$

$$\frac{q}{m} = \frac{4\varepsilon_0 g}{\sigma}$$

$$\therefore \quad \frac{q\sigma}{2\varepsilon_0 m} = 2g$$

Substituting in (1), we get

$$gH = 2g[a + H - \sqrt{a^2 + H^2}]$$

or $\frac{H}{2} = (a + H) - \sqrt{a^2 + H^2}$

or $\sqrt{a^2 + H^2} = a + \frac{H}{2}$

or $a^2 + H^2 = a^2 + \frac{H^2}{4} + aH$

or $\frac{3}{4}H^2 = aH$

or $H = \frac{4}{3}a$ and $H = 0$

$$\therefore \quad \boldsymbol{H = (4/3)\,a}$$

(b) Potential energy of the particle at height H = Electrostatic potential energy + gravitational potential energy

$$\therefore \quad U = qV + mgH$$

Here V = Potential at height H

$$\therefore \quad U = \frac{\sigma q}{2\varepsilon_0}[\sqrt{a^2 + H^2} - H + mgH] \quad \ldots(2)$$

At equilibrium position $F = \frac{-dU}{dH} = 0$

Differentiating (2) w.r.t. H

$$\text{or } mg + \frac{\sigma g}{2\varepsilon_0}\left[\left(\frac{1}{2}\right)(2H)\frac{1}{\sqrt{a^2+H^2}} - 1\right] = 0$$

$$\frac{\sigma q}{2\varepsilon_0} = 2mg$$

$$\therefore \quad mg + 2mg\left[\frac{H}{\sqrt{a^2+H^2}} - 1\right] = 0$$

$$\text{or} \quad 1 + \frac{2H}{\sqrt{a^2+H^2}} - 2 = 0$$

$$\frac{2H}{\sqrt{a^2+H^2}} = 1$$

$$\text{or} \quad \frac{H^2}{a^2+H^2} = \frac{1}{4} \quad \text{or} \quad 3H^2 = a^2$$

$$\text{or} \quad \boldsymbol{H = \frac{a}{\sqrt{3}}}$$

From equation (2) we can write U-H equation as

$$U = mg\,(2\sqrt{a^2+H^2} - H)$$

(Parabolic variation)

$$U = 2mga \text{ at } H = 0 \quad \text{and}$$

$$U = U_{\text{min}} = \sqrt{3}\, mga \text{ at } H = \frac{a}{\sqrt{3}}$$

Therefore, U-H graph will be as shown.

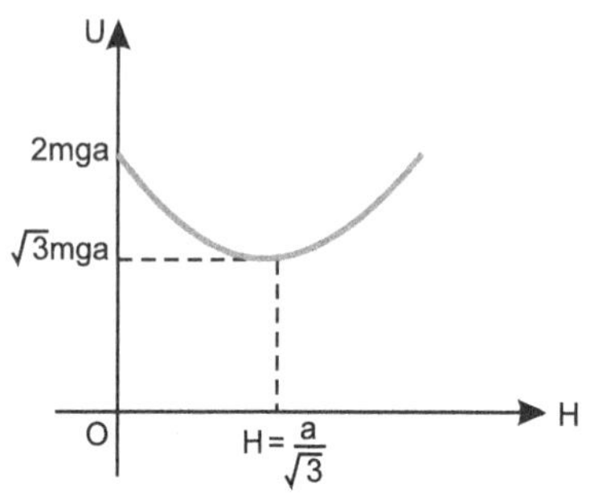

Note that at $H = \frac{a}{\sqrt{3}}$, U is minimum.

Therefore, $H = \frac{a}{\sqrt{3}}$ is stable equilibrium position.

474. Capacities of conducting spheres are in the ratio of their radii. Let C_1 and C_2 be the capacities of S_1 and S_2, then

$$\frac{C_2}{C_1} = \frac{R}{r}$$

(a) charges are distributed in the ratio of their capacities. Let in the first contact, charge acquired by S_2 is q_1. Therefore, charge on S_2 will be $Q - q_1$ say it is q'_1

$$\therefore \quad \frac{q_1}{q'_1} = \frac{q_1}{Q - q_1} = \frac{C_2}{C_1} = \frac{R}{r}$$

It implies that Q charge is to be distributed in S_1 and S_1 in the ratio of R/r.

$$\therefore \quad q_1 = Q\left(\frac{R}{R+r}\right) \qquad \ldots(1)$$

In the second contact, S_1 again acquires the same charge Q.

Therefore, total charge in S_1 and S_2 will be

$$Q + q_1 = Q\left(1 + \frac{R}{R+r}\right)$$

This charge is again distributed in the same ratio. Therefore, charge on S_2 in second contact,

$$q_2 = Q\left(1 + \frac{R}{R+r}\right)\left(\frac{R}{R+r}\right)$$

$$= Q\left[\frac{R}{R+r} + \left(\frac{R}{R+r}\right)^2\right]$$

Similarly,

$$q_3 = Q\left[\frac{R}{R+r} + \left(\frac{R}{R+r}\right)^2 + \left(\frac{R}{R+r}\right)^3\right]$$

and

$$q_n = Q\left[\frac{R}{R+r} + \left(\frac{R}{R+r}\right)^2 + \ldots \left(\frac{R}{R+r}\right)^n\right]$$

$$\text{or} \quad q_n = Q\frac{R}{r}\left[1 - \left(\frac{R}{R+r}\right)^n\right] \qquad \ldots(1)$$

$$\left[\because S_n = \frac{a\,(1-r^n)}{(1-r)}\right]$$

Therefore, electrostatic energy of S_2 after n such contacts

$$U_n = \frac{q_n^{\ 2}}{2C} = \frac{q_n^{\ 2}}{2(4\pi\varepsilon_0 R)}$$

or $$\boldsymbol{U_n = \frac{q_n^{\ 2}}{8\pi\varepsilon_0 R}}$$

where q_n can be written from equation (1).

(b) $$q_n = \frac{QR}{R+r}\left[1 + \frac{R}{R+r} + \ldots + \left(\frac{R}{R+r}\right)^{n-1}\right] \text{ as } n \to \infty$$

$$q_\infty = \frac{QR}{R+r}\left(\frac{1}{1-\frac{R}{R+r}}\right)$$

$$= \frac{QR}{R+r}\cdot\left(\frac{R+r}{r}\right) = Q\frac{R}{r}; \qquad \left[S_\infty = \frac{a}{1-r}\right]$$

$\because$ $$U_\infty = \frac{q_\infty^2}{2C} = \frac{Q^2R^2/r^2}{8\pi\varepsilon_0 R}$$

or $$\boldsymbol{U_\infty = \frac{Q^2 R}{8\pi\varepsilon_0 r^2}}$$

475. Let at any time 't' charge on capacitor C be Q and currents are as shown.

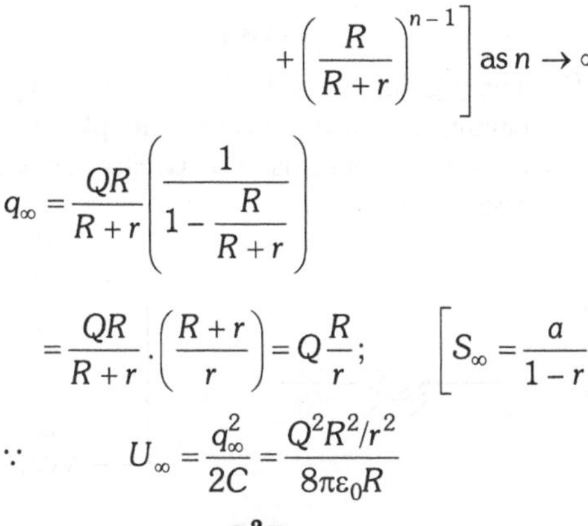

Since charge Q will increase with time 't', therefore,

$$i_1 = \frac{dQ}{dt}$$

(a) applying Kirchhoff's second law in loop $MNABM$

$$V = (i - i_1)R + iR$$

or $$V = 2iR - i_1R \qquad \ldots(1)$$

Similarly, applying Kirchhoff's second law in loop $MNSTM$ we have

$$V = i_1R + \frac{Q}{C} + iR \qquad \ldots(2)$$

Eliminating i from equation (1) and (2), we get

$$V = 3i_1r + \frac{2Q}{C}$$

$$\Rightarrow \quad 3i_1R = V - 2\frac{Q}{C}$$

$$\Rightarrow \quad i_1 = \frac{1}{3R}\left(V - \frac{2Q}{C}\right)$$

$$\Rightarrow \quad \frac{dQ}{dt} = \frac{1}{3R}\left(V - \frac{2Q}{C}\right)$$

$$\Rightarrow \quad \frac{dQ}{V - \frac{2Q}{C}} = \frac{dt}{3R}$$

$$\Rightarrow \quad \int_0^Q \frac{dQ}{V - \frac{2Q}{C}} = \int_0^t \frac{dt}{3R}$$

This equation gives

$$\boldsymbol{Q = \frac{CV}{2}(1 - e^{-2t/3RC})}$$

(b) $$i_1 = \frac{dQ}{dt} = \frac{V}{3R}e^{-2t/3RC}$$

From equation (1)

$$i = \frac{V + i_1R}{2R} = \frac{V + \frac{V}{3}e^{-2t/3RC}}{2R}$$

$\therefore$ Current through AB is

$$i_2 = i - i_1 = \frac{V + \frac{V}{3}e^{-2t/3RC}}{2R} - \frac{V}{3R}e^{-2t/3RC}$$

$$\boldsymbol{i_2 = \frac{V}{2R} - \frac{V}{6R}e^{-2t/3RC}}$$

$$\boldsymbol{i_2 = \frac{V}{2R}} \text{ as } t \to \infty$$

476. (i) Charge on capacitor A, before joining with an uncharged capacitor

$$q_A = CV = (100)(3)\mu C = 300\,\mu C$$

Similarly, charge on capacitor B

$$q_B = (180)(2)\,\mu C = 360\,\mu C$$

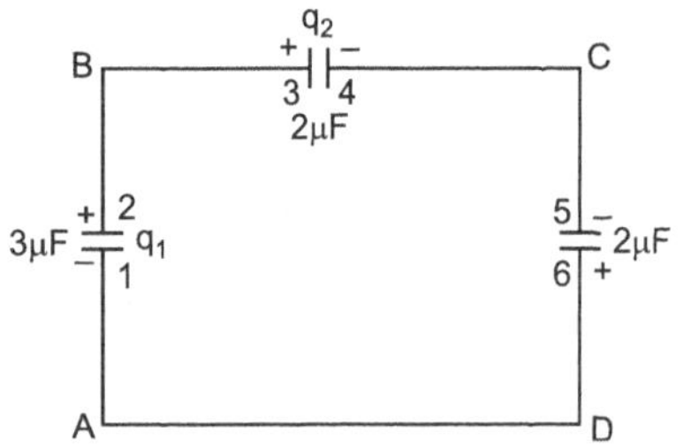

Let q_1, q_2 and q_3 be the charges on the three capacitors after joining them as shown in figure alongside

(q_1, q_2 and q_3 are in microcoulombs)

From conservation of charge

Net charge on plates 2 and 3 before joining = net charge after joining

$$\therefore \qquad 300 = q_1 + q_2 \qquad \dots(1)$$

Similarly, net charge on plates 4 and 5 before joining = net charge after joining

$$-360 = -q_2 - q_3$$

or
$$360 = q_2 + q_3 \qquad \dots(2)$$

Applying Kirchhoff's second law in closed loop *KLMNK*

$$\frac{q_1}{3} - \frac{q_2}{2} + \frac{q_3}{2} = 0$$

or
$$2q_1 - 3q_2 + 3q_3 = 0 \qquad \dots(3)$$

Solving equations (1), (2) and (3), we get

$$\mathbf{q_1 = 90\,\mu C}$$

$\mathbf{q_2 = 210\ \mu C}$ and $\mathbf{q_3 = 150\,\mu C}$

Therefore, final charges on the three capacitors are as shown in figure.

(ii) (a) Electrostatic energy stored before, completing the circuit

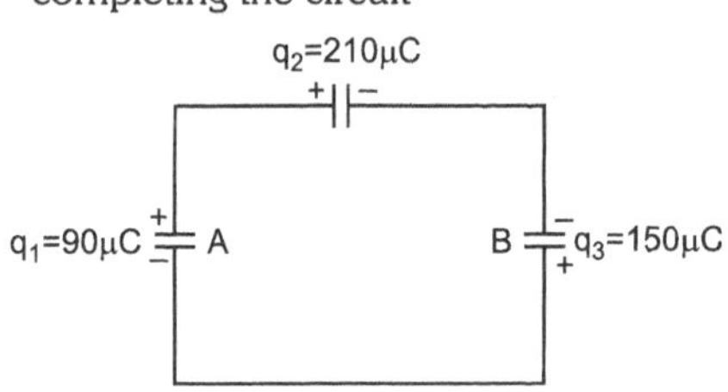

$$U_i = \frac{1}{2}(3 \times 10^{-6})(100)^2$$

$$+ \frac{1}{2}(2 \times 10^{-6})(180)^2 \left(U = \frac{1}{2}CV^2\right)$$

$$= 4.74 \times 10^{-2}\ \text{J}$$

or
$$U_i = \mathbf{47.4\ m\ J}$$

(b) Electrostatic energy stored after completing the circuit

$$U_f = \frac{1}{2}\frac{(90 \times 10^{-6})^2}{(3 \times 10^{-6})} + \frac{1}{2}\frac{(210 \times 10^{6})^2}{(2 \times 10^{-6})}$$

$$+ \frac{1}{2}\frac{(150 \times 10^{-6})^2}{(2 \times 10^{-6})} \left[U = \frac{1}{2}\frac{q^2}{C}\right]$$

$$= 1.8 \times 10^{-2}\ \text{J}$$

or
$$U_f = \mathbf{18\ mJ}$$

477. The problem is basically of discharging of *CR* circuit, because between the plates of the capacitor, there is a capacitor as well as resistance.

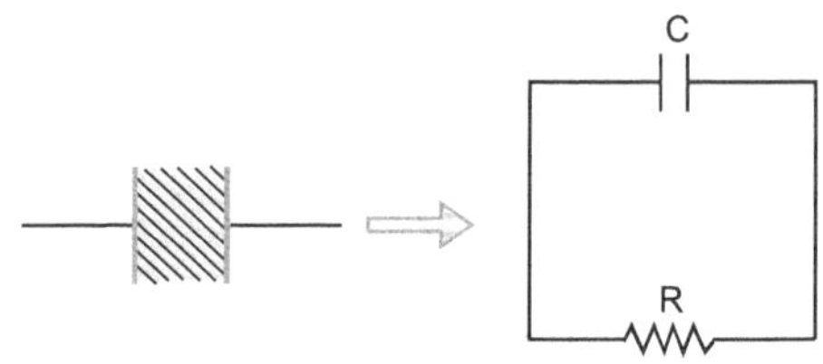

$$R = \frac{d}{\sigma A} \qquad \left(R = \frac{l}{\sigma A}\right)$$

and
$$C = \frac{K\varepsilon_0 A}{d}$$

$$\therefore \text{ Time constant, } \tau_C = CR = \frac{K\varepsilon_0}{\sigma}$$

Substituting the values, we have

$$\tau_1 = \frac{5 \times 8.85 \times 10^{-12}}{7.4 \times 10^{-12}} = 5.98\ \text{s}$$

Charge at any time *t* decreases exponentially as

$$q = q_0 e^{-t/\tau_C}$$

Here
$$q_0 = 8.85 \times 10^{-6}\ \text{C}$$

(Charge at time $t = 0$)

Therefore, discharging (leakage) current at time *t* will be given by

$$i = \left(-\frac{dq}{dt}\right) = \frac{q_0}{\tau_C} e^{-t/\tau_C}$$

or current at $t = 12$ s

$$i = \frac{(8.85 \times 10^{-6})}{5.98} e^{-12/5.98}$$

$$= 0.198 \times 10^{-6} \text{ A} = 0.198 \ \mu\text{A}$$

$$\boldsymbol{i = 0.198 \ \mu A}$$

478. (a) Let length and breadth of the capacitor be l and b respectively and d be the distance between the plates as shown in figure. Then consider a strip at a distance x of width dx

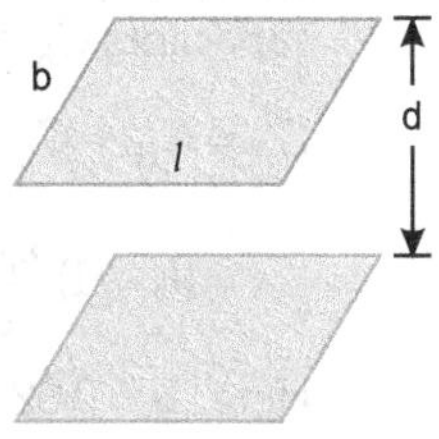

Now $\quad QR = x \tan\theta$

and $\quad PQ = d - x \tan\theta$

where $\tan\theta = d/l$ capacitance of PQ

$$C_1 = \frac{K_1\varepsilon_0(bdx)}{d - x\tan\theta} = \frac{K_1\varepsilon_0(bdx)}{d - \dfrac{xd}{l}}$$

$$C_1 = \frac{K_1\varepsilon_0 bldx}{d(l-x)} = \frac{K_1\varepsilon_0 A(dx)}{d(l-x)}$$

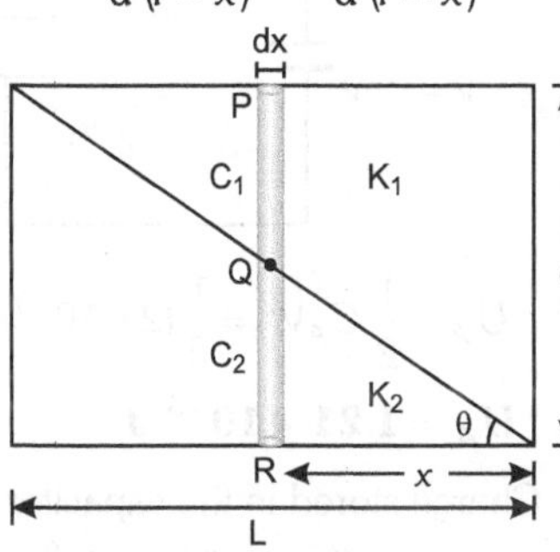

and C_2 = capacitance of QR

$$= \frac{K_2\varepsilon_0 b(dx)}{x\tan\theta}$$

$$C_2 = \frac{K_2\varepsilon_0 A(dx)}{xd} \qquad \left(\tan\theta = \frac{d}{l}\right)$$

Now C_1 and C_2 are in series. Therefore, their resultant capacity C_0 will be given by

$$\frac{1}{C_0} = \frac{1}{C_1} + \frac{1}{C_2}$$

Then $\quad \dfrac{1}{C_0} = \dfrac{1}{C_1} + \dfrac{1}{C_2}$

$$= \frac{d(l-x)}{K_1\varepsilon_0 A(dx)} + \frac{x.d}{K_2\varepsilon_0 A(dx)}$$

$$\frac{1}{C_0} = \frac{d}{\varepsilon_0 A(dx)}\left(\frac{l-x}{K_1} + \frac{x}{K_2}\right)$$

$$= \frac{d\{K_2(l-x) + K_1x\}}{\varepsilon_0 AK_1K_2(dx)}$$

$$\therefore \quad C_0 = \frac{\varepsilon_0 AK_1K_2}{d\{K_2(l-x) + K_1x\}}dx$$

$$C_0 = \frac{\varepsilon_0 AK_1K_2}{d\{K_2l + (K_1 - K_2)x\}}dx$$

Now the net capacitance of the given parallel plate capacitor is obtained by adding such infinitesimal capacitors placed parallel from $x = 0$ to $x = 1$

i.e. $\quad C_R = \displaystyle\int_{x=0}^{x=1} C_0$

$$= \int_0^l \frac{\varepsilon_0 AK_1K_2}{d\{K_2l + (K_1 - K_2)x\}}dx$$

Finally we get

$$C_R = \frac{K_1K_2\varepsilon_0 A}{(K_2 - K_1)d} \ln\frac{K_2}{K_1}$$

$$\boldsymbol{C_R = \frac{CK_1K_2}{K_2 - K_1} \ln\frac{K_2}{K_1}}$$

because $C = \dfrac{\varepsilon_0 A}{d}$

(b) Let q_1 and q_2 be the charges on the two spheres before connecting them.

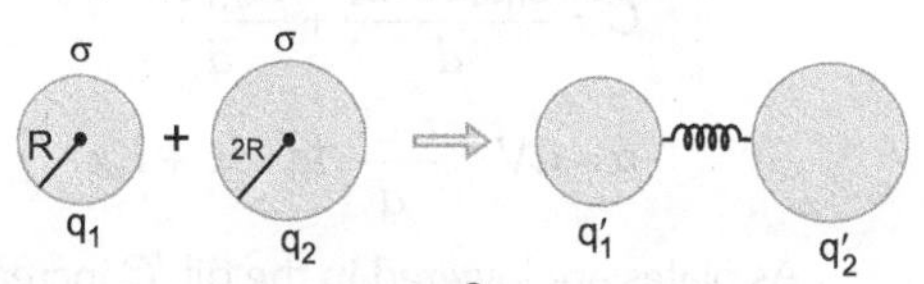

Then $\quad q_1 = \sigma(4\pi R^2)$,

and $\quad q_2 = \sigma(4\pi)(2R)^2 = 16\,\sigma\pi R^2$

Therefore, total charge (q) on both the spheres is

$$q = q_1 + q_2 = 20\,\sigma\pi R^2$$

Now after connecting, the charge is distributed in the ratio of their capacities, which in turn depends on the ratio of their radii $C = 4\pi\varepsilon_0 R$

$\therefore \qquad \dfrac{q'_1}{q'_2} = \dfrac{R}{2R} = \dfrac{1}{2}$

$\therefore \qquad q'_1 = \dfrac{q}{3} = \dfrac{20}{3}\sigma\pi R^2$

and $\qquad q'_2 = \dfrac{2q}{3} = \dfrac{40}{3}\sigma\pi R^2$

Therefore, surface charge densities on the spheres are :

$$\sigma_1 = \frac{q'_1}{4\pi R^2} = \frac{(20/3)\,\sigma\pi R^2}{4\pi R^2} = \frac{5}{3}\sigma$$

and $\sigma_2 = \dfrac{q'_2}{4\pi(2R)^2} = \dfrac{(40/3)\,\sigma\pi R^2}{16\pi R^2} = \dfrac{5}{6}\sigma$

Hence surface charge density on the bigger sphere is σ_2 *i. e.* $\frac{5}{6}\sigma$.

479. Let a be the side of the square plate. As shown in figure, C_1 and C_2 are in parallel. Therefore, total capacity of capacitor in the position shown is

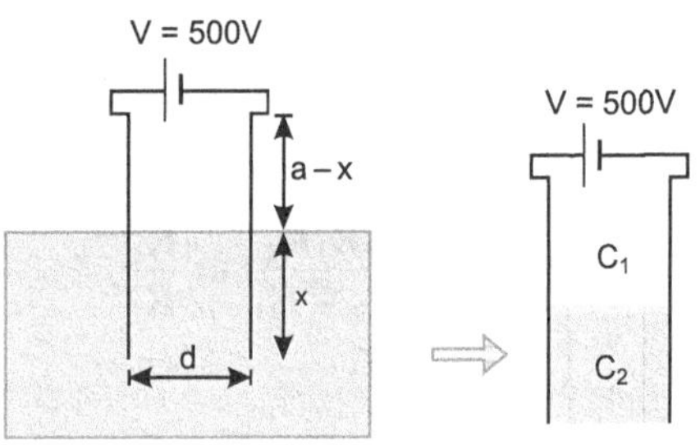

$$C = C_1 + C_2$$

$$C = \frac{\varepsilon_0 a(a-x)}{d} + \frac{K\varepsilon_0 ax}{d}$$

$$\therefore \qquad q = CV = \frac{\varepsilon_0 aV}{d}(a - x + Kx)$$

As plates are lowered in the oil, C increases or charge stored will increase.

Therefore, $\quad i = \dfrac{dq}{dt} = \dfrac{\varepsilon_0 aV}{d}(K-1)\,.\,\dfrac{dx}{dt}$

Substituting the values

$$\varepsilon_0 = 8.85 \times 10^{-12}\ \text{C}^2/\text{N - m}^2$$

$a = 1$ m, $V = 500$ volt and $d = 0.01$ m

$K = 11$ and $\dfrac{dx}{dt}$ = speed of plate

$= 0.001$ m/s

We get current

$$i = \frac{(8.85 \times 10^{-12})\,(1)\,(500)\,(11-1)\,(0.001)\ \text{A}}{(0.01)}$$

$$\boldsymbol{i = 4.43 \times 10^{-9}\ \text{A}}$$

480. (a) (i) Capacitor A is a combination of two capacitors C_K and C_O in parallel. Hence

$$C_A = C_K + C_O = \frac{K\varepsilon_0 A}{d} + \frac{\varepsilon_0 A}{d}$$

$$= (K+1)\frac{\varepsilon_0 A}{d} = (K+1)\frac{\varepsilon_0 A}{d}$$

Here $a = 0.02\ \text{m}^2$. Substituting the values, we have

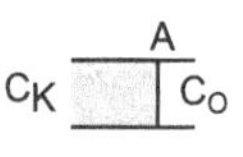

$$C_A = (9+1)\frac{(8.85 \times 10^{-12})\,(0.02)}{(8.85 \times 10^{-4})}\ \text{farad}$$

$$\boldsymbol{C_A = 2.0 \times 10^{-9}}\ \textbf{farad.}$$

Energy stored in capacitor A, when connected with a 110 V battery is

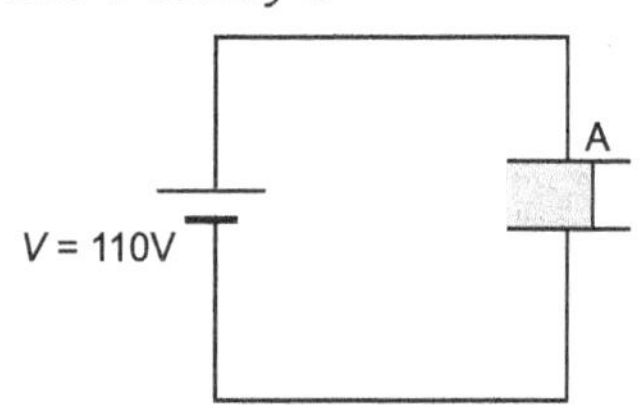

$$U_A = \frac{1}{2}C_A V^2 = \frac{1}{2}(2 \times 10^{-9})\,(110)^2$$

$$\boldsymbol{U_A = 1.21 \times 10^{-5}\ \text{J}}$$

(ii) Charge stored in the capacitor

$$q_A = C_A V = (2.0 \times 10^{-9})\,(110)$$

$$q_A = 2.2 \times 10^{-7}\ \text{C.}$$

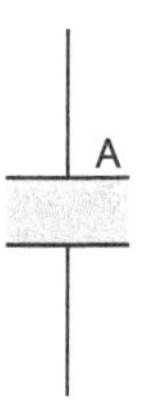

Now this charge remains constant even after battery is disconnected, But when the slab is removed, capacitance of A will get reduced. Let it be C'_A

$$C'_A = \frac{\varepsilon_0\,(2A)}{d} = \frac{(8.85 \times 10^{-12})\,(0.04)}{8.85 \times 10^{-4}}$$

and $\qquad C'_A = 0.4 \times 10^{-9}$ farad.

Energy stored in this case would be

$$U'_A = \frac{1}{2}\frac{q_A^{\ 2}}{C'_A} = \frac{1}{2}\frac{(2.2\times10^{-7})^2}{(0.4\times10^{-9})}\ \text{J}$$

$$U'_A = 6.05\times10^{-5}\ \text{J} > U_A$$

Therefore, work done to remove the slab would be

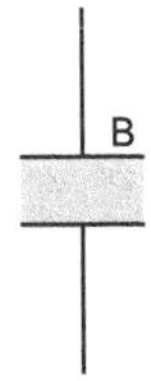

$$W = U'_A - U_A$$
$$= (6.05 - 1.21)\times10^{-5}\ \text{J}$$

or $\mathbf{W = 4.84\times10^{-5}\ J}$

(iii) Capacity of B when filled with dielectric is

$$C_B = \frac{K\varepsilon_0 A}{d} = \frac{(9)(8.85\times10^{-12})(0.02)}{(8.85\times10^{-4})}\ \text{farad}$$

$$C_B = 1.8\times10^{-9}\ \text{farad}$$

These two capacitors are in parallel. Therefore, net capacitance of the system is

$$C = C'_A + C_B = (0.4 + 1.8)\times10^{-9}\ \text{farad}$$

$$C = 2.2\times10^{-9}\ \text{farad.}$$

Charge stored in the system is

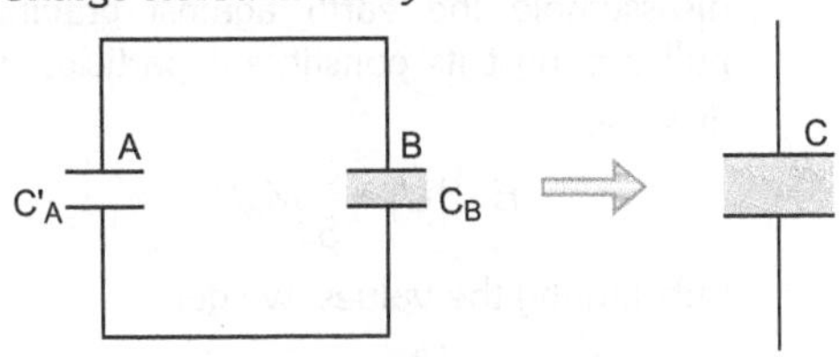

$$q = q_A = 2.2\times10^{-7}\ \text{C}$$

Therefore, energy stored,

$$U = \frac{1}{2}\frac{q^2}{C} = \frac{1}{2}\frac{(2.2\times10^{-7})^2}{(2.2\times10^{-9})}$$

or $\mathbf{U = 1.1\times10^{-5}\ J}$

(b) Total charge in the ring is $Q = (2\pi R)\lambda$

Potential due to a ring at a distance of x from its centre on its axis is given by

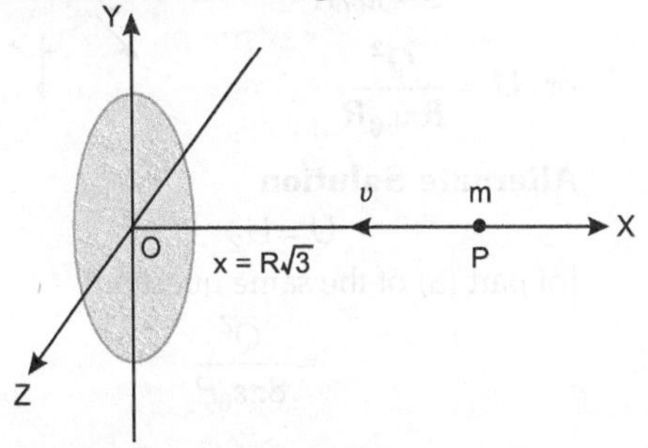

$$V(x) = \frac{1}{4\pi\varepsilon_0}\cdot\frac{Q}{\sqrt{R^2+x^2}}$$

and at the centre is

$$V_{\text{centre}} = \frac{1}{4\pi\varepsilon_0}\frac{Q}{R}$$

Using the above formulae

$$V_P = \frac{1}{4\pi\varepsilon_0}\cdot\frac{2\pi R\lambda}{\sqrt{R^2+3R^2}} = \frac{\lambda}{4\varepsilon_0}$$

and $V_0 = \frac{1}{4\pi\varepsilon_0}\cdot\frac{(2\pi R\lambda)}{R} = \frac{\lambda}{2\varepsilon_0}$ $\qquad V_0 > V_P$

P.D. between O and P is

$$V = V_O - V_P = \frac{\lambda}{2\varepsilon_0} - \frac{\lambda}{4\varepsilon_0} = \frac{\lambda}{4\varepsilon_0}$$

$$\therefore \quad \frac{1}{2}m\upsilon^2 \ge qV \quad \text{or} \quad \upsilon \ge \sqrt{\frac{2qV}{m}}$$

$$\text{or} \quad \upsilon \ge \sqrt{\frac{2q\lambda}{4\varepsilon_0 m}} \quad \text{or} \quad \upsilon \ge \sqrt{\frac{q\lambda}{2\varepsilon_0 m}}$$

Therefore, minimum value of speed υ should be

$$\boldsymbol{\upsilon_{\min} = \sqrt{\frac{q\lambda}{2\varepsilon_0 m}}}$$

481. (a) In this case the electric field exists from centre of the sphere to infinity. Potential energy is stored in electric field with energy density

$$u = \frac{1}{2}\varepsilon_0 E^2 \qquad \text{(energy/volume)}$$

(i) **Energy stored within the sphere (U_1)**

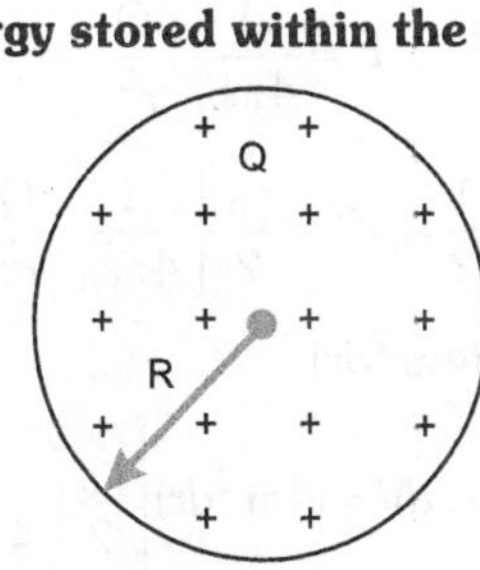

Electric field at a distance r is $(r \le R)$

$$E = \frac{1}{4\pi\varepsilon_0}\cdot\frac{Q}{R^3}\cdot r$$

$$\therefore \quad u = \frac{1}{2}\varepsilon_0 E^2 = \frac{\varepsilon_0}{2}\left[\frac{1}{4\pi\varepsilon_0}\frac{Qr}{R^3}\right]^2$$

Volume of element, $dV = (4\pi r^2)\, dr$

$\therefore$ Energy stored in this volume,

$dU = u\,(dV)$

$$dU = (4\pi r^2 dr)\frac{\varepsilon_0}{2}\left\{\frac{1}{4\pi\varepsilon_0}\cdot\frac{Q}{R^3}r\right\}^2$$

$$dU = \frac{1}{8\pi\varepsilon_0}\cdot\frac{Q^2}{R^6}\cdot r^4\,dr$$

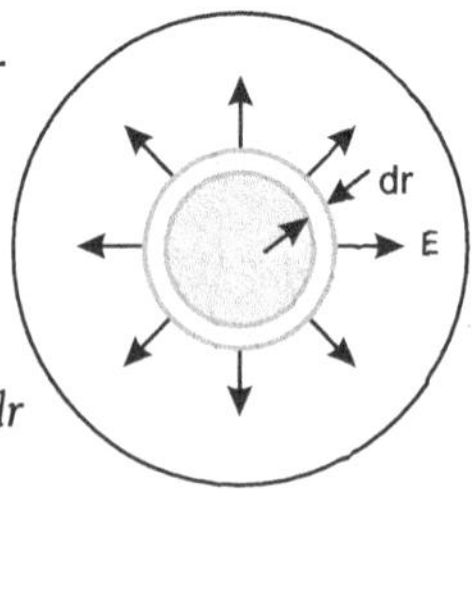

$$\therefore U_1 = \int_0^R dU$$

$$= \frac{1}{8\pi\varepsilon_0}\frac{Q^2}{R^6}\int_0^R r^4\,dr$$

$$= \frac{Q^2}{40\pi\varepsilon_0 R^6}[r^5]_0^R$$

$$U_1 = \frac{1}{40\pi\varepsilon_0}\cdot\frac{Q^2}{R} \qquad \ldots(1)$$

(ii) **Energy stored outside the sphere (U_2)**

Electric field at a distance r ($r \geq R$)

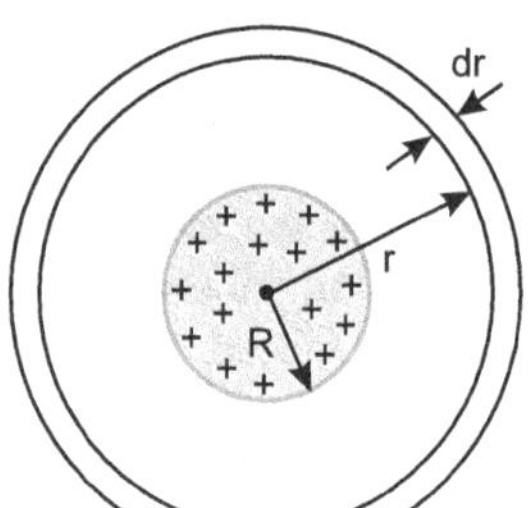

$$E = \frac{1}{4\pi\varepsilon_0}\cdot\frac{Q}{r^2}$$

$$\therefore \quad u = \frac{1}{2}\varepsilon_0 E^2 = \frac{\varepsilon_0}{2}\left\{\frac{1}{4\pi\varepsilon_0}\cdot\frac{Q}{r^2}\right\}^2$$

$$dV = (4\pi r^2 dr)$$

$$\therefore dU = u\,.\,dV = (4\pi r^2 dr)\left[\frac{\varepsilon_0}{2}\left(\frac{1}{4\pi\varepsilon_0}\cdot\frac{Q}{r^2}\right)^2\right]$$

$$dU = \frac{Q^2}{8\pi\varepsilon_0}\frac{dr}{r^2}$$

$$\therefore \quad U_2 = \int_R^\infty dU = \frac{Q^2}{8\pi\varepsilon_0}\cdot\int_R^\infty \frac{dr}{r^2}$$

$$U_2 = \frac{Q^2}{8\pi\varepsilon_0 R} \qquad \ldots(2)$$

Therefore, total energy of the system is

$$U = U_1 + U_2 = \frac{Q^2}{40\pi\varepsilon_0 R} + \frac{Q^2}{8\pi\varepsilon_0 R}$$

or $$\boldsymbol{U = \frac{3}{20}\frac{Q^2}{\pi\varepsilon_0 R}}$$

(b) Comparing this with gravitational forces, the gravitational potential energy of earth will be

$$U = -\frac{3}{5}\frac{GM^2}{R}$$

by replacing Q^2 by M^2 and $\frac{1}{4\pi\varepsilon_0}$ by G.

$$g = \frac{GM}{R^2} \quad \therefore \quad G = \frac{gR^2}{M}$$

$$\therefore \quad U = -\frac{3}{5}MgR$$

Therefore, energy needed to completely disassemble the earth against gravitational pull amongst its constituent particles will be given by

$$E = |U| = \frac{3}{5}MgR$$

Substituting the values, we get

$$E = \frac{3}{5}(10\text{ m/s}^2)(2.5\times 10^{31}\text{ kg - m})$$

$$\boldsymbol{E = 1.5\times 10^{32}}\text{ J}$$

(c) This is the case of a charged spherical conductor of radius R, energy of which is given by

$$U = \frac{1}{2}\frac{Q^2}{C}$$

or $$U = \frac{1}{2}\frac{Q^2}{4\pi\varepsilon_0 R}$$

or $$\boldsymbol{U = \frac{Q^2}{8\pi\varepsilon_0 R}}$$

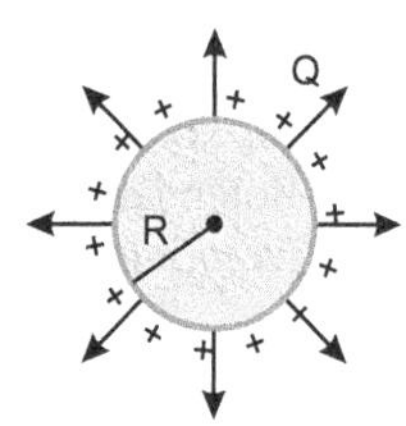

Alternate Solution

$$U = U_2$$

[of part (a) of the same question]

$$= \frac{Q^2}{8\pi\varepsilon_0 R}$$

CHAPTER 14 MAGNETICS

482. v is perpendicular to B

$$\therefore \quad \frac{mv^2}{r} = qvB$$

$$r = \left(\frac{mv}{qB}\right)$$

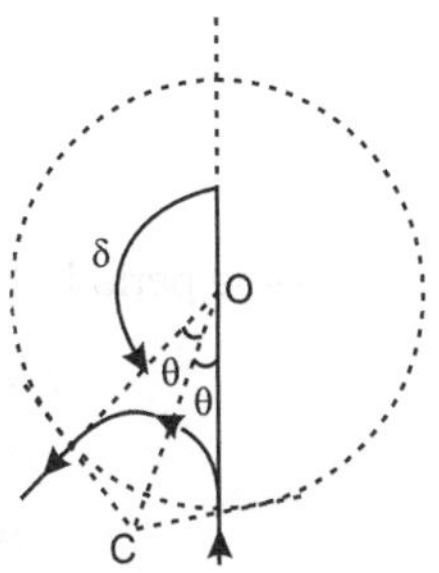

Where r is the radius of circular trajectory of the particle. Where c is the centre of its circular path.

$$\text{Deviation } \delta = (\pi - 2\theta) = \pi - 2\tan^{-1}\left(\frac{r}{R}\right)$$

$$= \pi - 2\tan^{-1}\left(\frac{mv}{qBR}\right)$$

483. (a) Magnetic moment of the ring

$$\vec{M} = \pi r^2 I_0 \hat{k}$$

$$\vec{B} = B_0 (2\hat{i} - 3\hat{j} + 5\hat{k})$$

$$\Rightarrow \text{Torque} = \vec{M} \times \vec{B} = I_0 B_0 \, \pi r^2 (3\hat{i} + 2\hat{j})$$

$$\Rightarrow \quad \tau_{AA'} = I_0 B_0 \, \pi r^2 \frac{(3\hat{i} + 2\hat{j}) \bullet (\hat{i} + \hat{j})}{\sqrt{2}}$$

$$\therefore \quad \alpha_{AA'} = \frac{\tau_{AA'}}{mr^2/2}$$

$$= \frac{5\sqrt{2} I_0 B_0 \pi}{m} \text{ rad/s}^2$$

(b) $U = -\vec{M} \bullet \vec{B}$

$$= -[(\pi r^2 I_0)\hat{k} \bullet B_0 (2\hat{i} - 3\hat{j} + 5\hat{k})]$$

$$= -\pi r^2 I_0 B_0 (5)$$

$$\Rightarrow U = -5\pi r^2 I_0 B_0 \text{ J}$$

(c) In uniform magnetic field force on a current carrying loop is zero.

484. Magnetic field is in negative Z-direction *i.e.*, perpendicular to paper inwards. Velocity vector is perpendicular to magnetic field, therefore, path of the particle will be a circle. Time period of which is given by

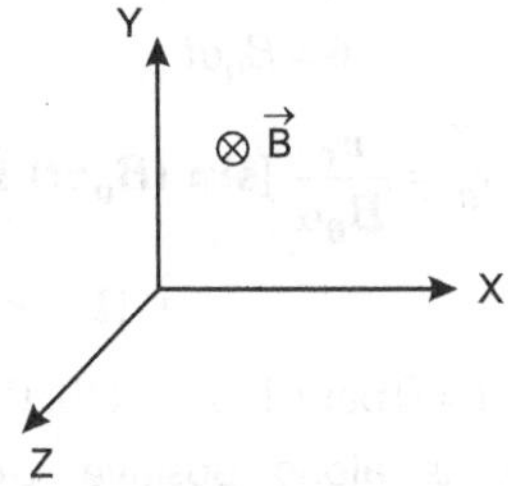

$$T = \frac{2\pi m}{B_0 q} = \frac{2\pi}{B_0 \alpha} \qquad \left(\frac{q}{m} = \alpha\right)$$

$$\text{or angular velocity } \omega = \frac{2\pi}{T} = B_0 \alpha \qquad \dots(1)$$

Radius of the circle is

$$r = \frac{mv_0}{B_0 q} = \frac{v_0}{B_0 \alpha} \qquad \dots(2)$$

In time t, the particle will rotate an angle $\theta = \omega t = B_0 \alpha t$ as shown in figure.

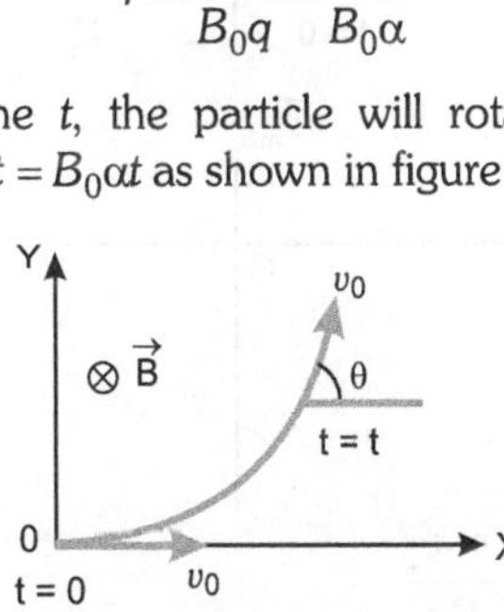

Hence velocity of particle at time t would be

$$\vec{v} = v_0 \cos\theta \, \hat{i} + v_0 \sin\theta \, \hat{j}$$

$$\text{or } \vec{\boldsymbol{v}} = \boldsymbol{v_0} \cos (\boldsymbol{B_0 \alpha t}) \, \hat{\mathbf{i}} + \boldsymbol{v_0} \sin (\boldsymbol{B_0 \alpha t}) \, \hat{\mathbf{j}}$$

Similarly position of particle can be found as :

$$\vec{r}_p = x\hat{i} + y\hat{j}$$

where $x = r\sin\theta$ and $y = r(1 - \cos\theta)$

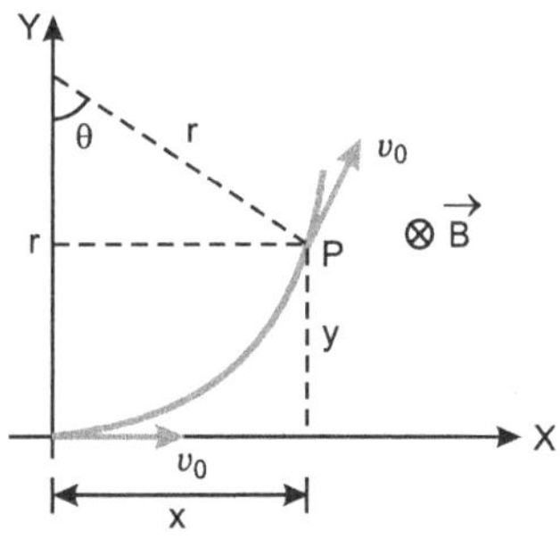

where $$r = \frac{v_0}{B_0\alpha}$$

and $$\theta = B_0\alpha t$$

Hence $$\vec{r}_p = \frac{v_0}{B_0\alpha}\,[\sin(B_0\alpha t)\,\hat{i} + \{1 - \cos(B_0\alpha t)\}\,\hat{j}]$$

485. At time $t = 0$ particle 1 is at (0, 0.4m, 0) and its velocity is along positive x-direction and magnetic force $\vec{F}_m$ is towards origin or along negative y-direction as shown. Hence according to Fleming's left hand rule, the magnetic field $\vec{B}$ should be along positive z-direction. Also

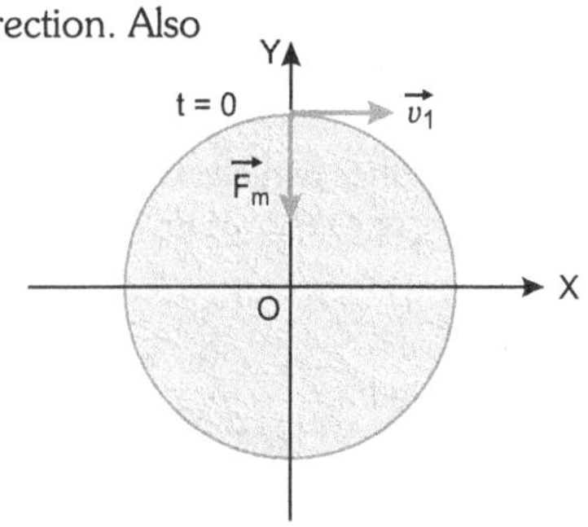

$$R = \frac{m_1 v_1}{B q_1}$$

$$\therefore \quad B = \frac{m_1 v_1}{R q_1} = \frac{(0.04)(5)}{(0.40(1)} = 0.5 \text{ tesla}$$

Now applying conservation of linear momentum before and after collision, we get

$$(m_1 + m_2)\,\vec{v} = m_1\vec{v}_1 + m_2\vec{v}_2$$

or $$(0.04 + 0.01)\,\vec{v} = (0.04)5\hat{i} + (0.01)(40)\hat{k}$$

or $$\vec{v} = 4\hat{i} + 8\hat{k}$$

Since velocity of the combined mass makes some angles ($\neq 0°, 180°$ or $90°$) with the magnetic field. Hence the path of the combined mass is helical. Radius of which is

$$R' = \frac{(m_1 + m_2)\,v_x}{(q_1 + q_2)\,B} = \frac{(0.05)(4)}{(2)(0.5)} = 0.2 \text{ m}$$

Time period of revolution is

$$T = \frac{2\pi(m_1 + m_2)}{(q_1 + q_2)\,B} = \frac{2\pi(0.05)}{(2)(0.5)} = \frac{\pi}{10} \text{ second}$$

The given time $t = \frac{\pi}{40}\text{ s} = \frac{T}{4}$

In this time the combined mass will rotate an angle $\theta = \frac{\pi}{2}$ or quarter circle in $x - y$ plane.

But now the radius will become 0.2 m

So $x = R = 0.2\text{ m} \Rightarrow y = R - R' = 0.2\text{ m}$

and $$z = v_z t = (8)\left(\frac{\pi}{40}\right) = 0.628\text{ m}$$

$\therefore$ Position of combined mass at $t = \frac{\pi}{40}$ **s** is (0.2 m, 0.2 m, 0.628 m)

486. Since the magnetic field is in positive z-direction, magnetic force will be in x-y plane. Electric field is in positive y-direction. Therefore, electrostatic force will be in y-direction *i.e.,* component of force along z-direction is always zero. Therefore, at any moment its velocity will have x and y components only.

Let at time t velocity of particle be

$$\vec{v} = v_x\,\hat{i} + v_y\,\hat{j}$$

Therefore, the Lorentz force on the particle

$$\vec{F} = q\{\vec{E} + (\vec{v} \times \vec{B})\}$$
$$= q\,[E_0\,\hat{j} + \{(v_x\,\hat{i} + v_y\,\hat{j}) \times (B_0\,\hat{k})\}]$$
$$= q\,[(E_0 - v_x B_0)\,\hat{j} + B_0 v_y\,\hat{i}]$$

or $\vec{a} = \frac{\vec{F}}{m} = \frac{q}{m}[B_0 v_y \hat{i} + (E_0 - v_x B_0)\hat{j}]$

or $\vec{a} = \alpha [B_0 v_y \hat{i} + (E_0 - v_x B_0)\hat{j}]$

i.e., $$a_x = \frac{dv_x}{dt} = \alpha B_0 v_y \qquad \ldots(1)$$

and $$a_y = \frac{dv_y}{dt} = \alpha (E_0 - v_x B_0) \qquad \ldots(2)$$

Differentiating equation (2) with respect to time

$$\frac{d^2 v_y}{dt^2} = -\alpha B_0 \frac{dv_x}{dt} = -\alpha^2 B_0^2 v_y \quad \text{(from 1)}$$

or $$\frac{d^2 v_y}{dt^2} = -\omega^2 v_y \qquad \text{where } \omega = \alpha B_0$$

i.e., v_y oscillates simple harmonically with angular frequency ω. Hence v_y can be written as

$v_y = v_0 \sin \omega t$, because $v_y = 0$ at $t = 0$

$$\therefore \quad \frac{dv_y}{dt} = v_0 \omega \cos \omega t \qquad \ldots(3)$$

At time $t = 0$, $\frac{dv_y}{dt} = v_0 \omega$ from equation (3) and $\frac{dv_y}{dt} = \alpha E_0$ from equation (2) because

$$v_x = 0 \quad \text{at} \quad t = 0.$$

$$\therefore \quad v_0 \omega = \alpha E_0$$

or $$v_0 = \frac{\alpha E_0}{\omega} = \frac{E_0}{B_0} \qquad (\omega = \alpha B_0)$$

Therefore, $$v_y = \frac{E_0}{B_0} \sin \omega t \qquad \ldots(4)$$

or $$\frac{dy}{dt} = \frac{E_0}{B_0} \sin \omega t$$

or $$\int_0^y dy = \frac{E_0}{B_0} \int_0^t \sin \omega t \, dt$$

or $$y = \frac{E_0}{B_0 \omega}(1 - \cos \omega t)$$

$$= \frac{E_0}{\alpha B_0^2}(1 - \cos \omega t)$$

$$y = a(1 - \cos \omega t) \qquad \text{where } a = \frac{E_0}{\alpha B_0^2}$$

Equation (1) can be written as

$$\frac{dv_x}{dt} = \alpha B_0 \left(\frac{E_0}{B_0}\right) \sin \omega t$$

or $$\int_0^{v_x} dv_x = \alpha E_0 \int_0^t \sin \omega t \, dt$$

or $$v_x = \frac{\alpha E_0}{\omega}(1 - \cos \omega t) = \frac{E_0}{B_0}(1 - \cos \omega t)$$

or $$\frac{dx}{dt} = \frac{E_0}{B_0}(1 - \cos \omega t)$$

or $$\int_0^x dx = \frac{E_0}{B_0} \int_0^t (1 - \cos \omega t)\, dt$$

or $$x = \frac{E_0}{B_0} t - \frac{E_0}{B_0 \omega} \sin \omega t$$

$$= \frac{E_0}{B_0 \omega}[\omega t - \sin \omega t]$$

$$= \frac{E_0}{\alpha B_0^2}(\omega t - \sin \omega t) \qquad \omega = \alpha B_0$$

or $$x = a(\omega t - \sin \omega t) \qquad a = \frac{E_0}{\alpha B_0^2}$$

therefore, $\boldsymbol{x = a(\omega t - \sin \omega t)}$

and $\boldsymbol{y = a(1 - \cos \omega t)}$

where $\boldsymbol{\omega = \alpha B_0}$ **and** $\boldsymbol{a = \frac{E_0}{\alpha B_0^2}}$

Magnetic force due to $B_0 \cos \theta$ component on the whole loop will be zero.

487. Magnetic force on a small current carrying element dl would be

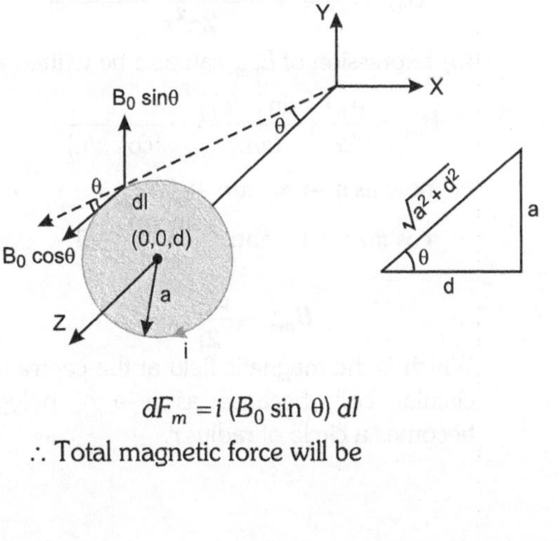

$$dF_m = i(B_0 \sin \theta)\, dl$$

$\therefore$ Total magnetic force will be

$$F_m = iB_0 \sin\theta \, \Sigma dl$$

$$= i\,B_0 \frac{a}{\sqrt{a^2+d^2}}(2\pi a) = \frac{\mathbf{2\pi a^2 iB_0}}{\sqrt{\mathbf{a^2+d^2}}}$$

488. One side of the polygon is $a = \dfrac{2\pi r}{n}$

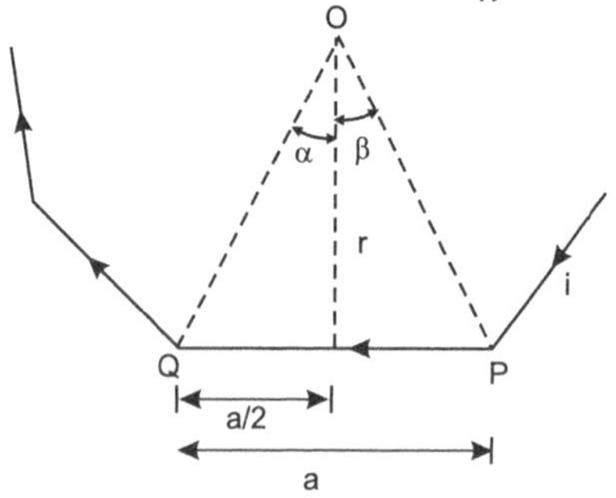

$$\alpha = \beta = \frac{1}{2}\left(\frac{2\pi}{n}\right) = \frac{\pi}{n}$$

$$r = \frac{a}{2}\cot\alpha = \left(\frac{\pi r}{n}\right)\cot\frac{\pi}{n}$$

Magnetic field due to one side PQ at centre is

$$B = \frac{\mu_0}{4\pi}\frac{i}{r}(\sin\alpha + \sin\beta)$$

$$B = \frac{\mu_0}{4\pi}\frac{i}{\left(\frac{\pi r}{n}\right)\cot\left(\frac{\pi}{n}\right)}\left(2\sin\frac{\pi}{n}\right)$$

$$= \frac{\mu_0 in \sin(\pi/n)\tan(\pi/n)}{2\pi^2 r}$$

$\therefore$ Net magnetic field due to all n sides will be

$$B_{net} = nB$$

$$\mathbf{B_{net}} = \frac{\boldsymbol{\mu_0 i n^2 \sin(\pi/n)\tan(\pi/n)}}{\mathbf{2\pi^2 r}}$$

(b) Expression of B_{net} can also be written as

$$B_{net} = \frac{\mu_0 i}{2r}\left\{\frac{\sin(\pi/n)}{(\pi/n)}\right\}^2 \cdot \frac{1}{(\cos \pi/n)}$$

now as $n \to \infty$, $\pi/n \to 0$,

$\cos\pi/n \to 1$ and $\dfrac{\sin\pi/n}{\pi/n} \to 1$

$$\therefore \quad \mathbf{B_{net}} = \frac{\boldsymbol{\mu_0 i}}{\mathbf{2r}}$$

Which is the magnetic field at the centre of a circular coil, because as $n \to \infty$, polygon becomes a circle of radius r.

489. (a) Magnetic force on dl will be

$$\mathbf{dF_m = iB\,(dl)}$$

From Fleming's left hand rule, direction of $d\vec{F}_m$ will be inwards (towards the centre).

(b) F is the force of compression in the wire. For equilibrium of element

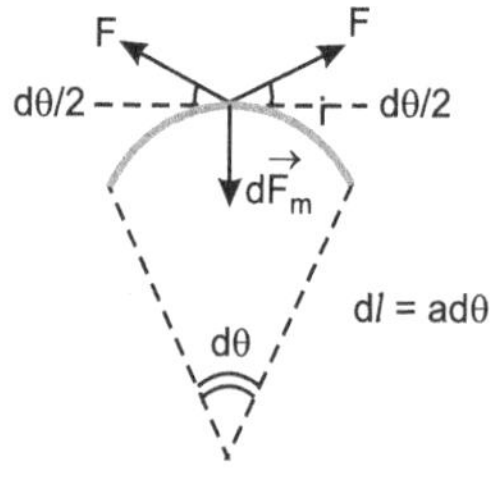

$$2F\sin\left(\frac{d\theta}{2}\right) = dF_m$$

or $\quad 2F\left(\dfrac{d\theta}{2}\right) = iB(ad\theta) \quad \left(\sin\dfrac{d\theta}{2} \approx \dfrac{d\theta}{2}\right)$

or $\quad F \cdot d\theta = iB\,(a\,d\theta)$

or $\quad \mathbf{F = iBa}$

(c) When the magnetic field is switched off, only unbalanced force is the compression force F. Due to which the length of the wire will increase and hence, radius will increase, length of wire is

$$l = 2\pi a$$

$$\therefore \quad dl = (2\pi)\cdot da$$

or increase in radius $da = \dfrac{dl}{2\pi}$ ...(1)

where dl = increase in length $= \dfrac{F \cdot l}{(\pi r^2)\cdot Y}$

$$\left(Y = \frac{Fl}{A\,\Delta l}\right)$$

$$= \frac{(iBa)(2\pi a)}{(\pi r^2)\,Y} = \frac{2iBa^2}{Yr^2}$$

Substituting this in equation (1), increase in radius

$$\mathbf{da} = \frac{\mathbf{ia^2B}}{\mathbf{\pi r^2 Y}}$$

490. First consider the force on 3 due to 1. B cos θ has no effect on the force onwire 3.

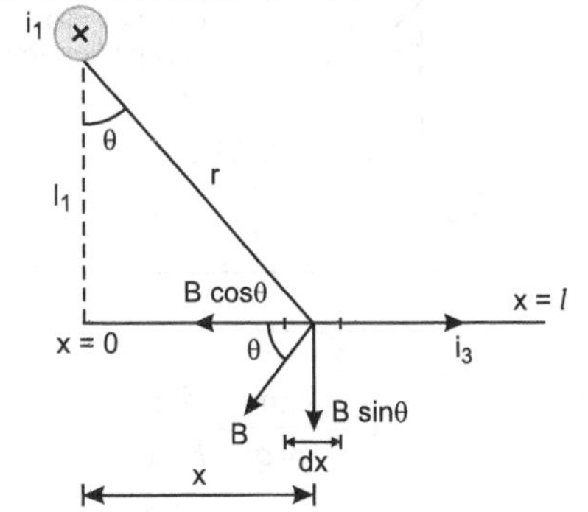

$$\therefore \quad F_1 = \int_0^l (i_3)(B \sin \theta)\, dx$$

$$= \int_0^l (i_3)\left(\frac{\mu_0 i_1}{2\pi r}\right)\left(\frac{x}{r}\right) \cdot dx$$

$$= \frac{\mu_0 i_1 i_3}{2\pi} \int_0^l \frac{x}{x^2 + l_1^2} \cdot dx$$

with proper substitutions we can find that

$$F_1 = \frac{\mu_0 i_1 i_3}{2\pi} ln \left(\sqrt{1 + \left(\frac{l}{l_1}\right)^2}\right) \qquad \text{(inwards)}$$

similarly

$$F_2 = \frac{\mu_0 i_2 i_3}{2\pi} ln \left(\sqrt{1 + \left(\frac{l}{l_2}\right)^2}\right) \qquad \text{(outwards)}$$

Therefore, net force will be

$\boldsymbol{F = F_1 - F_2}$

$$= \frac{\mu_0 i_3}{2\pi}\left[i_1 \ln \sqrt{1 + \left(\frac{l}{l_1}\right)^2} - i_2 \ln \sqrt{1 + \left(\frac{l}{l_2}\right)^2}\right] \qquad \text{(inwards)}$$

491. Magnetic force on the rod can be assumed to be acting at the centre of the rod, rightwards.

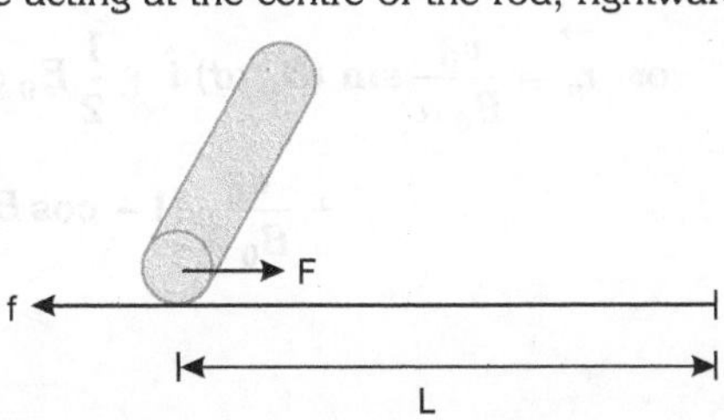

The magnitude of this force is :

$$F = idB = (48)(0.12)(0.24) \text{ N}$$

$$\text{or} \qquad F = 1.3824 \text{ N}$$

Now as there is no slipping, frictional force (f) will act in backward direction and instantaneous axis of rotation will be point of contact. Angular acceleration about which is eual to :

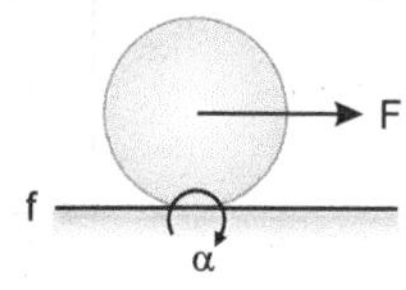

$$\alpha = \frac{\tau}{I} = \frac{F.R}{\left(\frac{1}{2}MR^2 + MR^2\right)}$$

$$= \frac{2F}{3MR} \qquad (R = \text{Radius})$$

and linear acceleration of COM of rod is :

$$a = R\alpha = \frac{2F}{3M}$$

$$= \frac{2\,(1.3824)}{3\,(0.72)}$$

$$= 1.28 \text{ m/s}^2$$

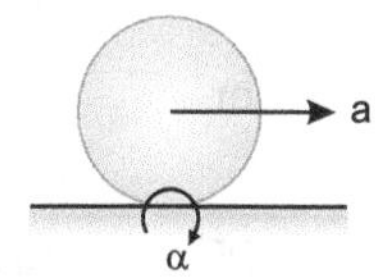

∴ speed of rod as it leaves the rails is :

$$v = \sqrt{2aL}$$

$$= \sqrt{2 \times 1.28 \times 0.45} = \mathbf{1.07 \text{ m/s}}$$

492. Magnetic field is in negative z-direction *i.e.*, perpendicular to paper inwards. Magnetic force $\vec{F_m}$ can not change the speed of a particle, therefore, the speed of the particle v_0 remains the same throughout. Let the particle makes an angle θ with the positive x-direction at x = x. Then

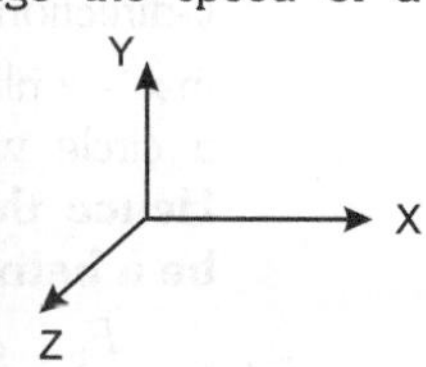

$$v_x = \frac{dx}{dt} = v_0 \cos \theta \qquad \text{...(i)}$$

$$\text{Now} \qquad a_m = \frac{F_m}{m} = \frac{(B_0 x)\, q v_0}{m} = B_0 v_0 x\alpha$$

$$a_y = a_m \cos \theta$$

$$\text{or} \qquad \left(\frac{dv_y}{dt}\right) = (B_0 v_0 x\alpha) \cos \theta$$

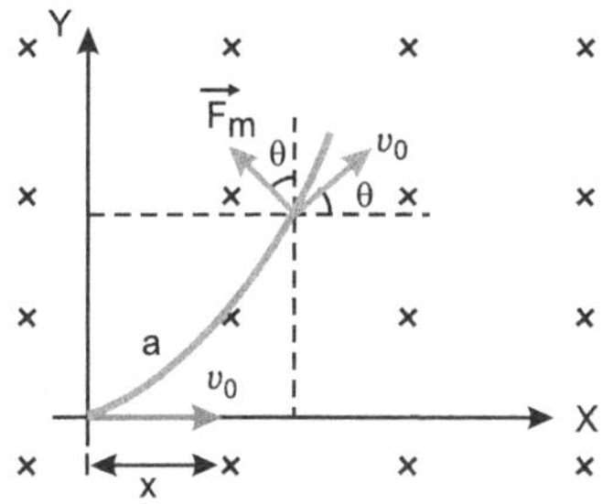

or $\left(\dfrac{dv_y}{dx}\right)\left(\dfrac{dx}{dt}\right) = (B_0 x\alpha)(v_0 \cos\theta)$

or $\dfrac{dv_y}{dx} = (B_0 x\alpha) \quad \left(\dfrac{dx}{dt} = v_0 \cos\theta\right)$

or $dv_y = B_0 x \alpha \cdot dx$

or $\displaystyle\int_0^{v_0} dv_y = B_0\alpha \int_0^{x_{max}} x\, dx$

or $v_0 = \dfrac{B_0 \alpha}{2} x_{max}^2$

or $\boldsymbol{x_{max}} = \sqrt{\dfrac{\boldsymbol{2v_0}}{\boldsymbol{B_0}\alpha}}$

493. (a) Electric field is in y-direction, therefore, electrostatic force $\vec{F}_e$ (which is also in y-direction) will provide an acceleration to the particle in y-direction. Therefore, y-component of its velocity will go on increasing. Magnetic field is also in y-direction so, magnetic force $\vec{F}_m$ will act in $x - z$ plane or the particle will rotate in a circle with its plane in $x - z$ plane. **Hence the path of the particle will be a helix with increasing pitch.**

(b) $a_y = \dfrac{F_y}{m} = \dfrac{qE_0}{m} = E_0\alpha \qquad \left(\dfrac{q}{m} = \alpha\right)$

$\therefore \quad v_y = a_y t = E_0 \alpha t \qquad \ldots(1)$

and $\quad y = \dfrac{1}{2} a_y t^2 = \dfrac{1}{2} E_0 \alpha t^2 \qquad \ldots(2)$

At time $t = 0$, particle is at origin. Its velocity is in x-direction so magnetic force will be in positive z-direction. The particle will rotate in $x - z$ plane with time period

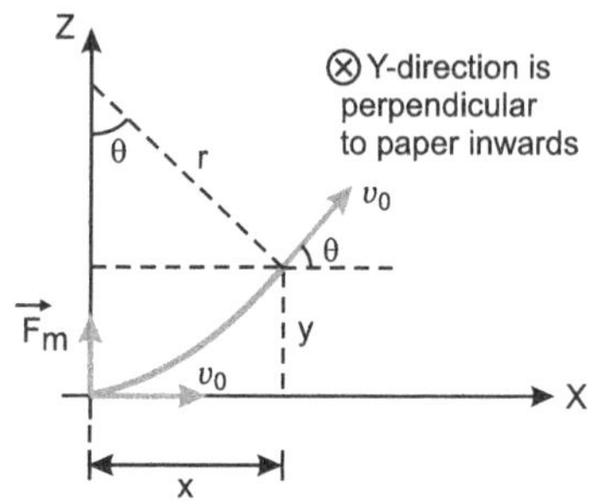

$$T = \frac{2\pi m}{B_0 q} = \frac{2\pi}{B_0\alpha}$$

or angular velocity $\omega = \dfrac{2\pi}{T} = B_0 \alpha$

Angle rotated by particle in time t is

$$\theta = \omega t = B_0 \alpha t \qquad \ldots(3)$$

$$v_x = v_0 \cos\theta = v_0 \cos B_0 \alpha t \qquad \ldots(4)$$

$$v_z = v_0 \sin\theta = v_0 \sin B_0 \alpha t \qquad \ldots(5)$$

$$x = r\sin\theta = \frac{mv_0}{B_0 q} \sin B_0 \alpha t$$

$$= \frac{v_0}{B_0 \alpha} \sin B_0 \alpha t \qquad \ldots(6)$$

and $\quad z = r(1 - \cos\theta)$

$$= \frac{v_0}{B_0 \alpha}(1 - \cos B_0 \alpha t) \qquad \ldots(7)$$

From equations (1), (4) and (5) velocity of particle at time t is

$$\vec{v} = v_x \hat{i} + v_y \hat{j} + v_z \hat{k}$$

or $\vec{\boldsymbol{v}}_{\boldsymbol{p}} = \boldsymbol{v_0} \cos (\boldsymbol{B_0} \alpha \boldsymbol{t})\, \hat{\boldsymbol{i}} + (\boldsymbol{E_0} \alpha \boldsymbol{t})\, \hat{\boldsymbol{j}} + \boldsymbol{v_0} \sin (\boldsymbol{B_0} \alpha \boldsymbol{t})\, \hat{\boldsymbol{k}}$

(c) From equations (2), (6) and (7) position of particle at time t is

$$\vec{r}_p = x\hat{i} + y\hat{j} + z\hat{k}$$

or $\vec{\boldsymbol{r}}_{\boldsymbol{p}} = \dfrac{\boldsymbol{v_0}}{\boldsymbol{B_0}\alpha} \sin (\boldsymbol{B_0} \alpha \boldsymbol{t})\, \hat{\boldsymbol{i}} + \dfrac{\boldsymbol{1}}{\boldsymbol{2}} \boldsymbol{E_0} \alpha \boldsymbol{t^2}\, \hat{\boldsymbol{j}} + \dfrac{\boldsymbol{v_0}}{\boldsymbol{B_0}\alpha}(\boldsymbol{1} - \cos \boldsymbol{B_0} \alpha \boldsymbol{t})\, \hat{\boldsymbol{k}}$

(d) Particle will touch the y-axis for n^{th} time after time $t = nT = \dfrac{2n\pi}{B_0\alpha}$.

So, y-co-ordinate at that moment will be

$$y_n = \frac{1}{2} E_0 \alpha t^2$$

$$= \frac{1}{2} E_0 \alpha \left(\frac{2n\pi}{B_0\alpha}\right)^2$$

or $$\boldsymbol{y_n = \frac{2n^2\pi^2 E_0}{B_0^2 \alpha}}$$

(e) y-compoment of velocity at that moment will be

$$v_y = E_0 \alpha t = (E_0 \alpha)\left(\frac{2n\pi}{B_0 \alpha}\right) = \frac{2n\pi E_0}{B_0}$$

$\sqrt{v_x^2 + v_z^2} = v_0$ *i.e.*, in $x - z$ plane speed v_0 remains constant.

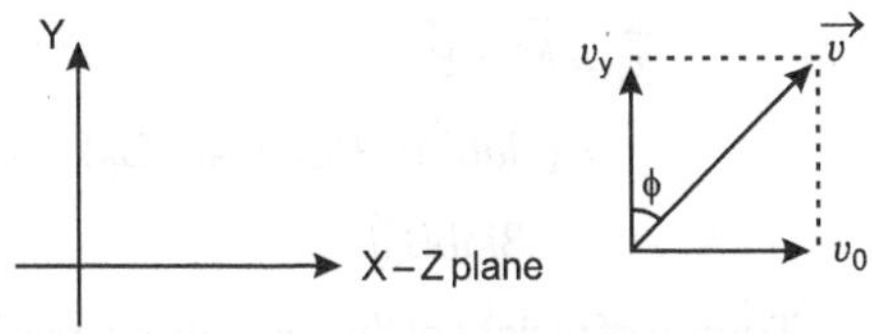

$$\tan\phi = \frac{v_0}{v_y}$$

or $\boldsymbol{\phi = \tan^{-1} \dfrac{v_0}{v_y} = \tan^{-1}\left(\dfrac{v_0 B_0}{2n\pi E_0}\right)}$

494. (a) Equal and opposite currents can be assumed in the cavities.

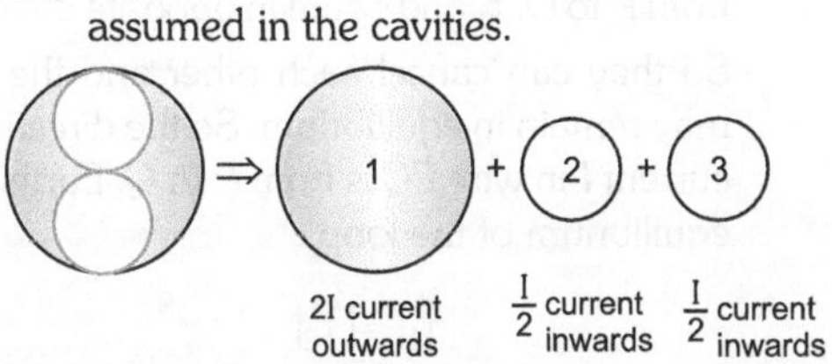

The current in cavity will be

$$i = \left\{\frac{I}{\pi a^2 - 2(\pi)\left(\dfrac{a}{2}\right)^2}\right\}\left\{\pi\left(\frac{a}{2}\right)^2\right\} = \frac{I}{2}$$

Total magnetic field at P_1 is

$$\vec{B} = \vec{B_1} + \vec{B_2} + \vec{B_3}$$

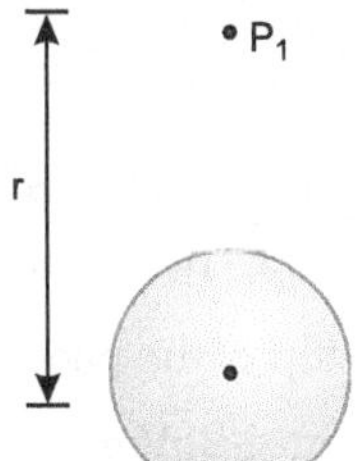

Here $\vec{B_1} = \left(\dfrac{\mu_0}{2\pi}\right)\left(\dfrac{2I}{r}\right)$ towards left

$\vec{B_2} = \left(\dfrac{\mu_0}{2\pi}\right)\left(\dfrac{I/2}{r - a/2}\right)$ towards right

and $\vec{B_3} = \left(\dfrac{\mu_0}{2\pi}\right)\left(\dfrac{I/2}{r + a/2}\right)$ towards right

Hence net magnetic field at P_1 is

$$B = \frac{\mu_0 I}{\pi}\left[\frac{1}{r} - \frac{1}{4(r - a/2)} - \frac{1}{4(r + a/2)}\right]$$

towards left

$$= \frac{\mu_0 I}{\pi}\left[\frac{1}{r} - \frac{1}{(4r - 2a)} - \frac{1}{(4r + 2a)}\right]$$

towards left

$$\boldsymbol{B = \frac{\mu_0 I}{\pi r}\left(\frac{2r^2 - a^2}{4r^2 - a^2}\right)}$$ **to the left**

(b) In the similar manner magnetic field at P_2 will be

$$B = B_1 - 2B_2 \cos\theta$$

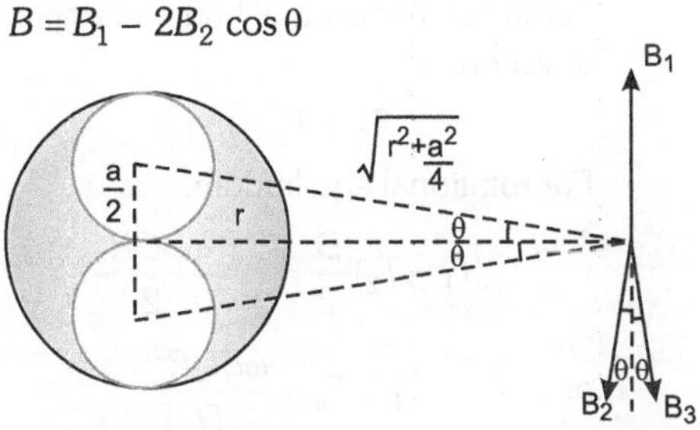

$$= \frac{\mu_0}{2\pi}\left(\frac{2I}{r}\right) - 2\left(\frac{\mu_0}{2\pi}\right)$$

$$\times \left(\frac{I/2}{\sqrt{r^2 + \dfrac{a^2}{4}}}\right)\left(\frac{r}{\sqrt{r^2 + \dfrac{a^2}{4}}}\right)$$

$$= \frac{\mu_0 I}{\pi}\left[\frac{1}{r} - \frac{2r}{4r^2 + a^2}\right]$$

$$= \frac{\mu_0 \boldsymbol{I}}{\pi \boldsymbol{r}} \frac{(\boldsymbol{2r^2} + \boldsymbol{a^2})}{(\boldsymbol{4r^2} + \boldsymbol{a^2})}$$

towards the top of the page.

495. $$r = \frac{\sqrt{2qVm}}{Bq}$$

or $$r \propto \sqrt{\frac{m}{q}}$$

$\therefore$ $$\frac{r_P}{r_\alpha} = \sqrt{\frac{m_P}{m_\alpha}}\sqrt{\frac{q_\alpha}{q_P}}$$

$$= \sqrt{\frac{1}{4}}\sqrt{\frac{2}{1}} = \frac{1}{\sqrt{2}}$$

496. In equilibrium,

$$2T_0 = mg$$

or $$T_0 = \frac{mg}{2} \qquad \ldots(1)$$

Magnetic moment,

$$M = iA = \left(\frac{\omega}{2\pi} Q\right)(\pi R^2)$$

$$\tau = MB \sin 90° = \frac{\omega BQR^2}{2}$$

Let T_1 and T_2 be the tensions in the two strings when magnetic field is switched on. ($T_1 > T_2$) For translational equilibrium of ring in vertical direction,

$$T_1 + T_2 = mg \qquad \ldots(2)$$

For rotational equilibrium,

$$(T_1 - T_2)\frac{D}{2} = \tau = \frac{\omega BQR^2}{2}$$

or $$T_1 - T_2 = \frac{\omega BQR^2}{D} \qquad \ldots(3)$$

Solving Eqs. (2) and (3) we have

$$T_1 = \frac{mg}{2} + \frac{\omega BQR^2}{2D}$$

As $T_1 > T_2$ and maximum values of T_1 can be $\frac{3T_0}{2}$, we have

$$\frac{3T_0}{2} = T_0 + \frac{\omega_{max} BQR^2}{2D} \qquad \left(\frac{mg}{2} = T_0\right)$$

$\therefore$ $$\omega_{max} = \frac{DT_0}{BQR^2}$$

497. (a) and (c) Let the direction of current in wire PQ is from P to Q and its magnitude be I.

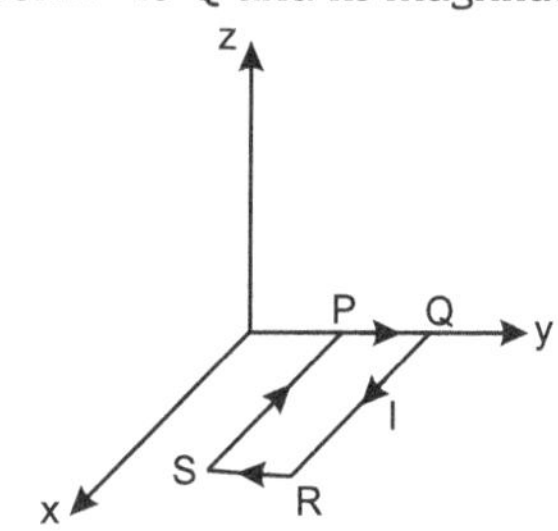

The magnetic moment of the given loop is :

$$\vec{M} = -Iab\hat{k}$$

Torque on the loop due to magnetic force is :

$$\vec{\tau_1} = \vec{M} \times \vec{B}$$

$$= (-Iab\hat{k}) \times \{(3\hat{i} + 4\hat{k}) B_0\}$$

$$= -3IabB_0\hat{j}$$

Torque of weight of the loop about axis PQ is :

$$\vec{\tau_2} = \vec{r} \times \vec{F}$$

$$= \left(\frac{a}{2}\hat{i}\right) \times (-mg\hat{k}) = \frac{mga}{2}\hat{j}$$

We see that when the current in the wire PQ is from P to Q, $\vec{\tau_1}$ and $\vec{\tau_2}$ are in opposite direction. So they can cancel each other and the loop may remain in equilibrium. So the direction of current I in wire PQ is from P to Q. Further for equilibrium of the loop :

$$|\vec{\tau_1}| = |\vec{\tau_2}|$$

or $$3IabB_0 = \frac{mga}{2}$$

$$\boldsymbol{I} = \frac{\boldsymbol{mg}}{\boldsymbol{6bB_0}}$$

(b) Magnetic force on wire RS is :

$$\vec{F} = I(\vec{l} \times \vec{B})$$

$$= I[(-b\hat{j}) \times \{(3\hat{i} + 4\hat{k}) B_0\}]$$

or $$\vec{F} = IbB_0(3\hat{k} - 4\hat{i})$$

498. (a) Given $i = 10$ A, $r_1 = 0.08$ m and $r_2 = 0.12$ m. Straight portions *i.e.*, *CD* etc. will produce zero magnetic field at the centre. Rest eight arcs will produce the magnetic field at the centre in the same direction *i.e.*, perpendicular to the paper outwards or vertically upwards and its magnitude is

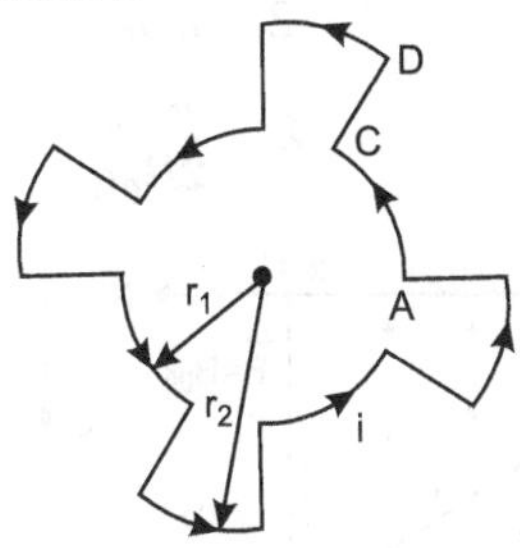

$$B = B_{\text{inner arcs}} + B_{\text{outer arcs}}$$

$$= \frac{1}{2}\left\{\frac{\mu_0 i}{2r_1}\right\} + \frac{1}{2}\left\{\frac{\mu_0 i}{2r_2}\right\}$$

$$= \left(\frac{\mu_0}{4\pi}\right)(\pi i)\left(\frac{r_1 + r_2}{r_1 r_2}\right)$$

Substituting the values, we have

$$B = \frac{(10^{-7})(3.14)(10)(0.08 + 0.12)}{(0.08 \times 0.12)} \text{ tesla}$$

$$\mathbf{B = 6.54 \times 10^{-5}\ tesla}$$

(Vertically upward or outward normal to the paper)

(b) **Force on AC :** Force on circular portions of the circuit *i.e.*, AC etc. due to the wire at the centre will be **zero** because magnetic field due to the central wire at these arcs will be tangential ($\theta = 180°$) as shown.

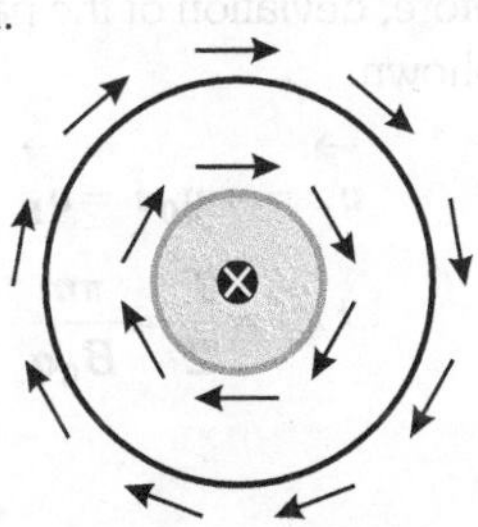

Force on CD :

Current in central wire is also $i = 10$ A

Magnetic field at *P* due to central wire,

$$B = \frac{\mu_0}{2\pi} \cdot \frac{i}{x}$$

$\therefore$ Magnetic force on element *dx* due to this magnetic field

$$dF = (i)\left(\frac{\mu_0}{2\pi} \cdot \frac{i}{x}\right) \cdot dx \qquad (F = ilB \sin 90°)$$

$$= \left(\frac{\mu_0}{2\pi}\right) i^2 \frac{dx}{x}$$

Therefore, net force on *CD* is

$$F = \int_{x=r_1}^{x=r_2} dF = \frac{\mu_0 i^2}{2\pi} \int_{0.08}^{0.12} \frac{dx}{x} = \frac{\mu_0}{2\pi} i^2 \ln\left(\frac{3}{2}\right)$$

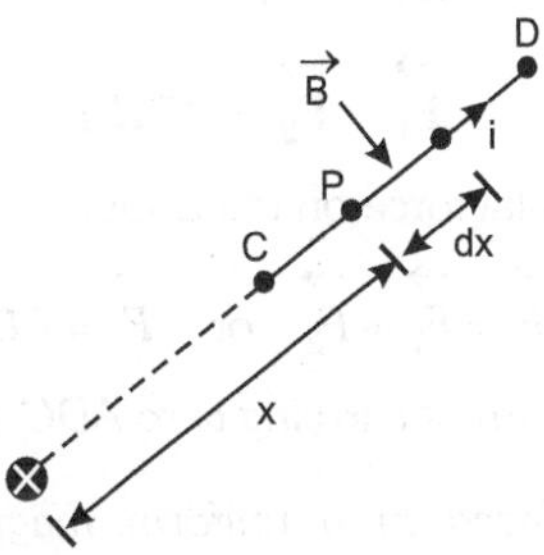

Substituting the values,

$$F = (2 \times 10^{-7})(10)^2 \ln(1.5)$$

or $\mathbf{F = 8.1 \times 10^{-6}\ N}$ (inwards)

Force on wire at the centre : Net magnetic field at the centre due to the circuit is in vertical direction and current in the wire in centre is also in vertical direction. Therefore, net force on the wire at the centre will be **zero**. ($\theta = 180°$). Hence

(i) Force acting on the wire at the centre is zero.

(ii) Force on arc $AC = 0$.

(iii) Force on segment *CD* is 8.1×10^{-6} N (inwards).

499. (a) Magnetic field ($\vec{B}$) at the origin

= Magnetic field due to semicircle *KLM*

+ Magnetic field due to other semicircle *KNM*

$$\therefore \quad \vec{B} = \frac{\mu_0 I}{4R}(-\hat{i}) + \frac{\mu_0 I}{4R}(\hat{j})$$

$$\vec{B} = \frac{\mu_0 I}{4R}\hat{i} + \frac{\mu_0 I}{4R}\hat{j} = \frac{\mu_0 I}{4R}(-\hat{i} + \hat{j})$$

$\therefore$ Magnetic force acting on the particle

$$\vec{F} = q(\vec{v} \times \vec{B})$$

$$= q\{(-v_0\hat{i}) \times (-\hat{i} + \hat{j})\}\frac{\mu_0 I}{4R}$$

$$\vec{F} = -\frac{\mu_0 q v_0 I}{4R}\hat{k}$$

(b) $\vec{F}_{KLM} = \vec{F}_{KNM} = \vec{F}_{KM}$

and $\vec{F}_{KM} = BI(2R)\hat{i} = 2BIR\hat{i}$

$$\therefore \quad \vec{F_1} = \vec{F_2} = \mathbf{2BIR\hat{i}}$$

or Total force on the Loop,

$$\vec{F} = \vec{F_1} + \vec{F_2} \quad \text{or} \quad \vec{F} = \mathbf{4BIR\hat{i}}$$

If a current carrying wire ADC (of any shape) is placed in a uniform magnetic field $\vec{B}$. perpendicular to the plane ADC.

Then $\vec{F}_{ADC} = \vec{F}_{AC}$

or $|\vec{F}_{ADC}| = I(AC)B$

From this we can conclude that net force on a current carrying loop in uniform magnetic field is zero. In the question, segments KLM and KNM also form a loop and they are also placed in a uniform magnetic field but in this case net force on the loop will not be zero. It would had been zero if, the current in any of the segments was in opposite direction.

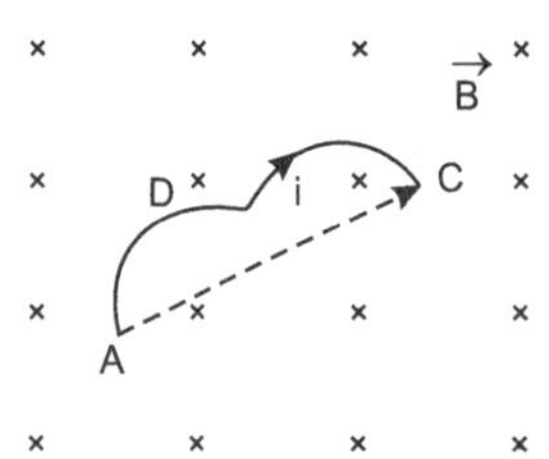

500. (a) $\theta = 30°$,

$$\sin\theta = \frac{L}{R}$$

Here $R = \dfrac{mv_0}{B_0 q}$

$$\therefore \quad \sin 30° = \frac{L}{\dfrac{mv_0}{B_0 q}}$$

or $\dfrac{1}{2} = \dfrac{B_0 q L}{m v_0}$

$$\therefore \quad \boldsymbol{L = \frac{mv_0}{2B_0 q}}$$

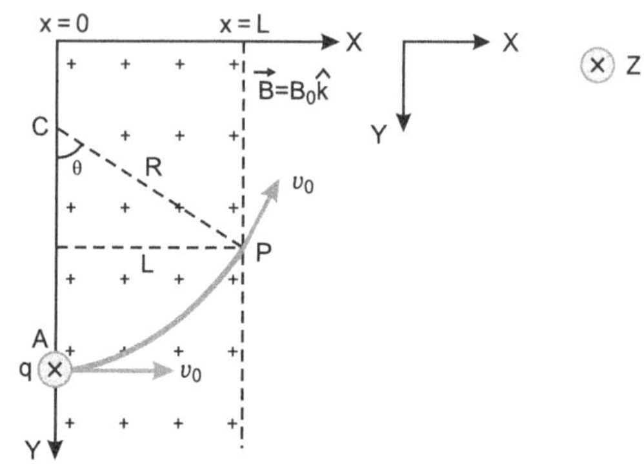

(b) In part (i)

$$\sin 30° = \frac{L}{R}$$

or $\dfrac{1}{2} = \dfrac{L}{R}$

or $L = \dfrac{R}{2}$

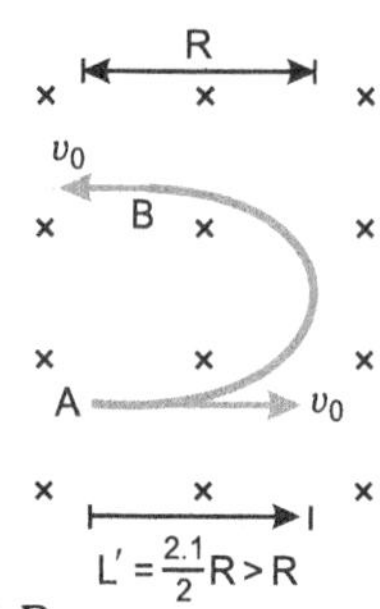

Now when $L' = 2.1\,L$ or $\dfrac{2.1}{2}R$

$$\Rightarrow \quad L' > R$$

Therefore, deviation of the particle is $\theta = 180°$ is as shown.

$$\therefore \quad \vec{v}_f = -v_0\hat{i} = \vec{v}_B$$

and $\boldsymbol{t_{AB} = \dfrac{T}{2} = \dfrac{\pi m}{B_0 q}}$

501. $\hat{j} = \frac{\vec{E}}{E}$ or $\frac{\vec{B}}{B} : \hat{i} = \frac{\vec{v}_0}{v_0}$ or $\hat{k} = \frac{\vec{v}_0 \times \vec{B}}{v_0 B}$

Force due to electric field will be along Y-axis. Magnetic force will not affect the motion of charged particle in the direction of electric field (or Y-axis). So,

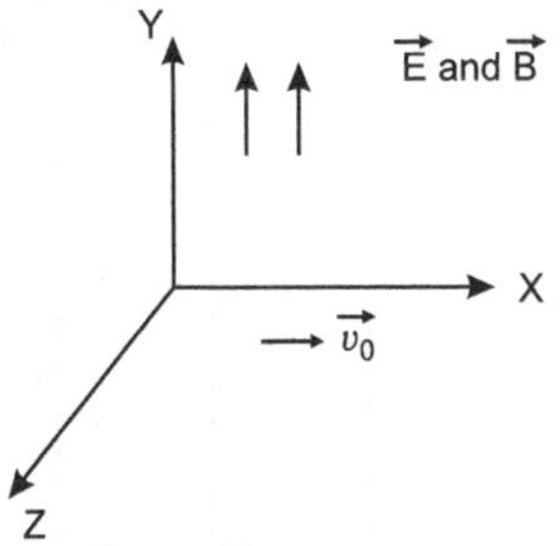

$$a_y = \frac{F_e}{m} = \frac{qE}{m} = \text{constant.}$$

Therefore, $v_y = a_y \quad t = \frac{qE}{m} . t$ ...(1)

The charged particle under the action of magnetic field describes a circle in x-z plane (perpendicular to $\vec{B}$) with

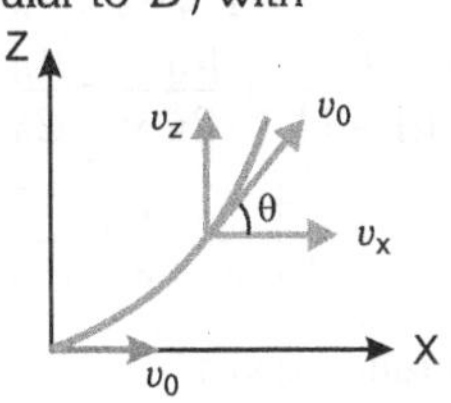

$$T = \frac{2\pi m}{Bq} \quad \text{or} \quad \omega = \frac{2\pi}{T} = \frac{qB}{m}$$

Initially $(t = 0)$ velocity was along X-axis. Therefore, magnetic force $(\vec{F}_m)$ will be along positive Z-axis $[\vec{F}_m = q(\vec{v}_0 \times \vec{B})]$. Let it makes an angle θ with X-axis at time t, then $\theta = \omega t$

$$\therefore \quad v_x = v_0 \cos \omega t = v_0 \cos\left(\frac{qB}{m} t\right) \quad ...(2)$$

$$v_z = v_0 \sin \omega t = v_0 \sin\left(\frac{qB}{m} t\right) \quad ...(3)$$

From (1), (2) and (3)

$$\therefore \quad \vec{v} = v_x \hat{i} + v_y \hat{j} + v_z \hat{k}$$

$$\therefore \quad \vec{v} = v_0 \cos\left(\frac{qB}{m} t\right)\left(\frac{\vec{v}_0}{v_0}\right) + \frac{qE}{m} t \left(\frac{\vec{E}}{E}\right) + v_0 \sin\left(\frac{qB}{m} t\right)\left(\frac{\vec{v}_0 \times \vec{B}}{v_0 B}\right)$$

$$\text{or } \vec{\boldsymbol{v}} = \cos\left(\frac{\boldsymbol{qB}}{\boldsymbol{m}} \boldsymbol{t}\right)(\vec{\boldsymbol{v}}_0) + \left(\frac{\boldsymbol{q}}{\boldsymbol{m}} \boldsymbol{t}\right)(\vec{\boldsymbol{E}}) + \sin\left(\frac{\boldsymbol{qB}}{\boldsymbol{m}} \boldsymbol{t}\right)\left(\frac{\vec{\boldsymbol{v}}_0 \times \vec{\boldsymbol{B}}}{\boldsymbol{B}}\right)$$

The path of the particle will be a helix of increasing pitch. The axis of the helix will be along Y-axis.

502. Magnetic moment of the loop,

$$\vec{M} = (iA)\hat{k} = (I_0 L^2)\hat{k}$$

Magnetic Field,

$$\vec{B} = (B \cos 45°)\hat{i} + (B \sin 45°)\hat{j} = \frac{B}{\sqrt{2}}(\hat{i} + \hat{j})$$

(a) Torque acting on the loop,

$$\vec{\tau} = \vec{M} \times \vec{B} = (I_0 L^2 \hat{k}) \times \left[\frac{B}{\sqrt{2}}(\hat{i} + \hat{j})\right]$$

$$\therefore \quad \vec{\boldsymbol{\tau}} = \frac{\boldsymbol{I_0 L^2 B}}{\sqrt{\boldsymbol{2}}}(\hat{\boldsymbol{j}} - \hat{\boldsymbol{i}})$$

$$\text{or} \quad |\vec{\tau}| = I_0 L^2 B$$

(b) Axis of rotation coincides with the torque and since torque is in $\hat{j} - \hat{i}$ direction or

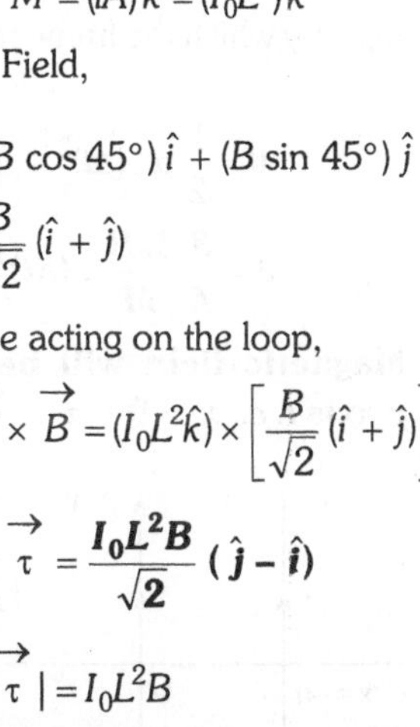

parallel to QS. Therefore, the loop will rotate about an axis passing through Q and S as shown alongside :

Angular acceleration, $\alpha = \dfrac{|\vec{\tau}|}{I}$

Where I = moment of inertia of loop about QS.

$$I_{QS} + I_{PR} = I_{ZZ}$$

(From theorem of perpendicular axis)

But $\qquad I_{QS} = I_{PR}$

$$\therefore \qquad 2I_{QS} = I_{ZZ} = \frac{4}{3} ML^2$$

$$\therefore \qquad I_{QS} = \frac{2}{3} ML^2$$

$$\therefore \qquad \alpha = \frac{|\vec{\tau}|}{I} = \frac{I_0 L^2 B}{\frac{2}{3} ML^2} = \frac{3}{2} \frac{I_0 B}{M}$$

$\therefore$ Angle by which the frame rotates in time Δt is

$$\theta = \frac{1}{2} \alpha \,.\, (\Delta t)^2$$

or $\qquad \boldsymbol{\theta = \dfrac{3}{4} \dfrac{I_0 B}{M} \,.\, (\Delta t)^2}$

503. (i) **Magnetic field will be zero on the y-axis *i. e.*, $x = 0 = z$**

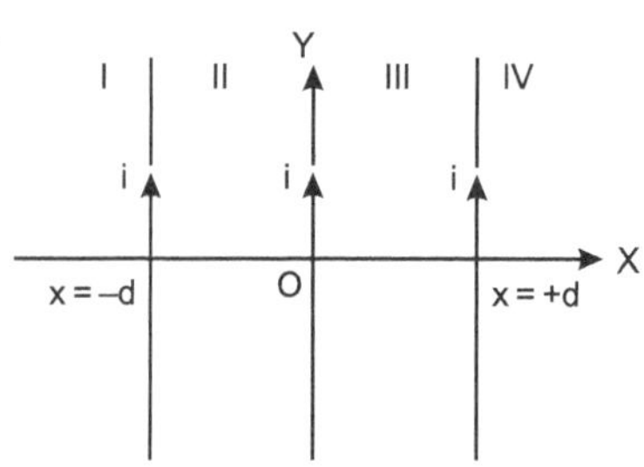

Magentic field can not be zero in region I and region IV because in region I magnetic field will be along positive z direction ⊙ due to all the three wires, while in region IV magnetic field will be along negative z-axis ⊗ due to all the three wires. It can be zero only in region II and III.

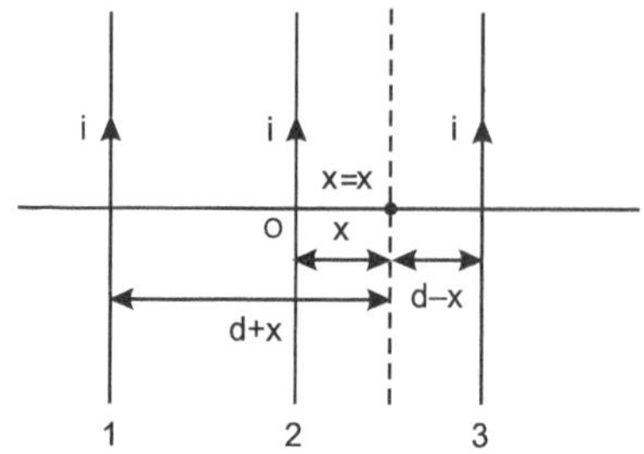

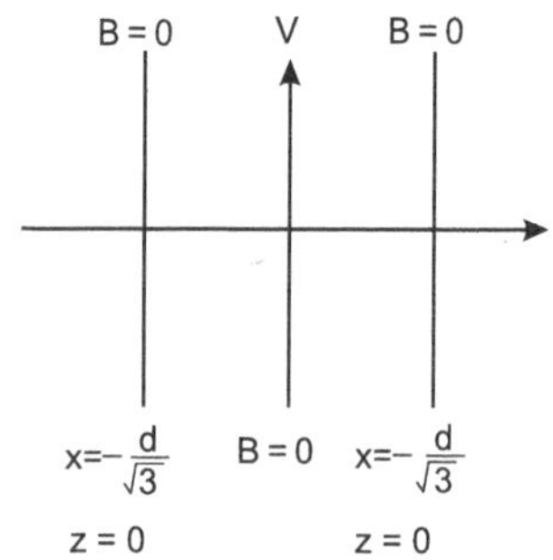

Let magnetic field is zero on line $z = 0$ and $x = x$. Then magnetic field on this line due to wires 1 and 2 will be along negative z-axis and due to wire 3 along positive z-axis. Thus

$$B_1 + B_2 = B_3$$

or $\qquad \dfrac{\mu_0}{2\pi} \dfrac{i}{(d+x)} + \dfrac{\mu_0}{2\pi} \dfrac{i}{x} = \dfrac{\mu_0}{2\pi} \dfrac{i}{(d-x)}$

or $\qquad \dfrac{1}{d+x} + \dfrac{1}{x} = \dfrac{1}{d-x}$

This equation gives

$$x = \pm \frac{d}{\sqrt{3}}$$

Hence there will be two lines

$$\boldsymbol{x = \frac{d}{\sqrt{3}}} \quad \text{and} \quad \boldsymbol{x = -\frac{d}{\sqrt{3}}} \qquad (z = 0)$$

where magnetic field is zero.

(ii) In this part we change our coordinate axes system, just for better understanding.

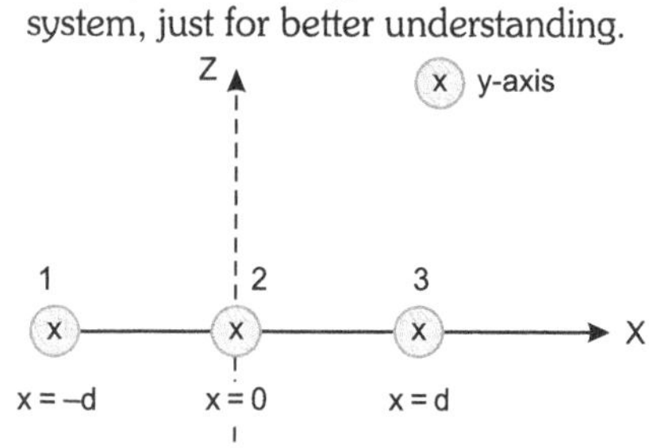

There are three wires 1, 2 and 3 as shown in figure. If we displace the wire 2 towards the z-axis, then force of attraction per unit length between wires (1 and 2) and (2 and 3) will be given by

$$F = \frac{\mu_0}{2\pi}\frac{i^2}{r}$$

The components of F along x-axis will be cancelled out. Net resultant force will be towards negative z-axis (or mean position) and will be given by

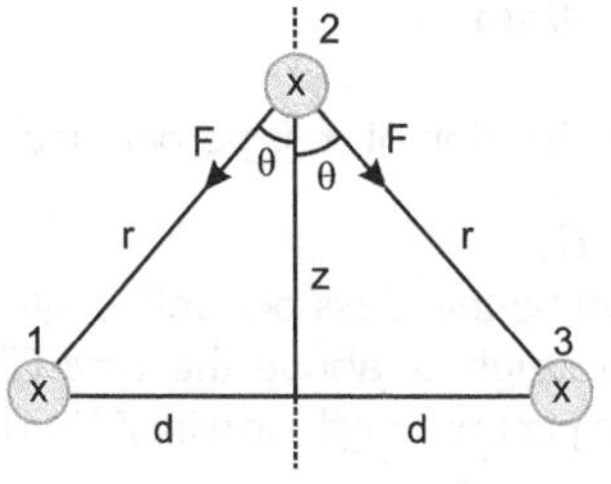

$$F_{net} = 2F\cos\theta = 2\left\{\frac{\mu_0}{2\pi}\frac{i^2}{r}\right\}\frac{z}{r}$$

$$F_{net} = \frac{\mu_0}{\pi}\frac{i^2}{(z^2+d^2)}.z$$

If $z << d$, then

$$z^2 + d^2 \approx d^2 \quad \text{and} \quad F_{net} = -\left(\frac{\mu_0}{\pi}\frac{i^2}{d^2}\right).z$$

Negative sign implies that F_{net} is restoring in nature.

Therefore, $\quad F_{net} \propto -z$

i.e., the wire will oscillate simple harmonically. Let a be the acceleration of wire in this position and λ is the mass per unit length of wire then

$$F_{net} = \lambda.a = -\left(\frac{\mu_0}{\pi}\frac{i^2}{d^2}\right)z$$

or $\quad a = -\left(\frac{\mu_0 i^2}{\pi\lambda d^2}\right).z$

∴ Frequency of oscillation

$$f = \frac{1}{2\pi}\sqrt{\frac{\text{Acceleration}}{\text{Displacement}}}$$

$$= \frac{1}{2\pi}\sqrt{\frac{|a|}{|z|}} = \frac{1}{2\pi}\frac{i}{d}\sqrt{\frac{\mu_0}{\pi\lambda}}$$

or $\quad f = \frac{i}{2\pi d}\sqrt{\frac{\mu_0}{\pi\lambda}}$

504. Given that $x >> a$.

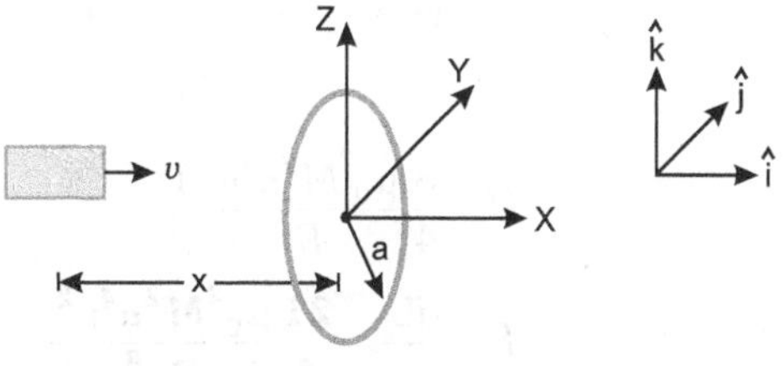

Magnetic field at the centre of the coil due to the bar magnet is,

$$B = \frac{\mu_0}{4\pi}\frac{2M}{x^3} = \frac{\mu_0}{2\pi}\frac{M}{x^3}$$

Due to this, magnetic flux linked with the coil will be,

$$\phi = B.S = \frac{\mu_0}{2\pi}\frac{M}{x^3}(\pi a^2) = \frac{\mu_0 Ma^2}{2x^3}$$

∴ Induced emf in the coil, due to motion of the magnet is

$$e = -\frac{d\phi}{dt} = -\left(\frac{\mu_0 Ma^2}{2}\right)\frac{d}{dt}\left(\frac{1}{x^3}\right)$$

$$= \frac{\mu_0 Ma^2}{2}\left(\frac{3}{x^4}\right).\frac{dx}{dt} = \frac{3}{2}\frac{\mu_0 Ma^2}{x^4}v \quad \left(\frac{dx}{dt} = v\right)$$

Therefore, induced current in the coil is,

$$i = \frac{e}{R} = \frac{3}{2}\frac{\mu_0 Ma^2}{Rx^4}v$$

Magnetic moment of the coil due to this induced current will be,

$$M' = iS = \frac{3}{2}\frac{\mu_0 Ma^2}{Rx^4}v(\pi a^2)$$

$$M' = \frac{3}{2}\frac{\mu_0\pi Ma^4 v}{Rx^4}$$

Potential energy of $\vec{M'}$ in $\vec{B}$ will be

$U = -M'B\cos 180°$

$$U = M'B = \frac{3}{2}\frac{\mu_0\pi Ma^4 v}{Rx^4}\left(\frac{\mu_0}{2\pi}\cdot\frac{M}{x^3}\right)$$

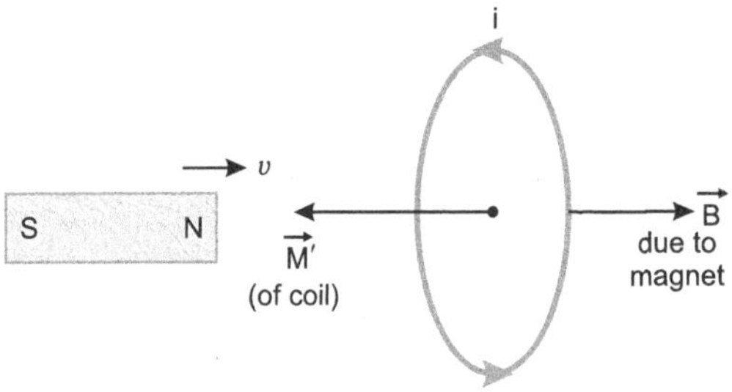

$$U = \frac{3}{4}\frac{\mu_0{}^2M^2a^4v}{R}\frac{1}{x^7}$$

$$\therefore \quad F = \frac{dU}{dx} = \frac{\mathbf{21}}{\mathbf{4}}\frac{\boldsymbol{\mu_0{}^2M^2a^4v^2}}{\boldsymbol{Rx^8}}$$

Positive sign of F implies that there will be a repulsion between the magnet and the coil. Note that here we can not apply

$$F = \frac{\mu_0}{4\pi}\frac{6MM'}{x^4} \quad \text{(directly)} \; \ldots(1)$$

because here M' is a function of x. However equation (1) can be applied where M and M' both are constants.

505. (i) In ground state ($n = 1$), according to Bohr's theory :

$$mvR = \frac{h}{2\pi} \quad \text{or} \quad v = \frac{h}{2\pi mR}$$

Now time period,

$$t = \frac{2\pi R}{v} = \frac{2\pi R}{h/2\pi mR} = \frac{4\pi^2 mR^2}{h}$$

Magnetic moment $M = iA$

where $i = \dfrac{\text{Charge}}{\text{Time period}}$

$$= \frac{e}{\dfrac{4\pi mR^2}{h}} = \frac{eh}{4\pi^2 mR^2}$$

and $A = \pi R^2$

$$\therefore \quad M = (\pi R^2)\left(\frac{eh}{4\pi^2 mR^2}\right)$$

$$\text{or} \quad \boldsymbol{M} = \frac{\boldsymbol{eh}}{\boldsymbol{4\pi m}}$$

Direction of magnetic moment $\vec{M}$ is perpendicular to the plane of orbit.

(ii) $\vec{\tau} = \vec{M} \times \vec{B}$

or $\tau = MB \sin\theta$

where θ is the angle between $\vec{M}$ and $\vec{B}$

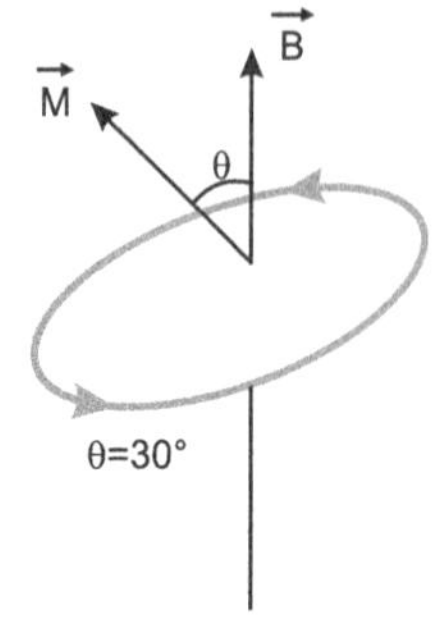

$$\theta = 30°$$

$$\therefore \tau = \left(\frac{eh}{4\pi m}\right)(B)\sin 30°$$

$$\therefore \tau = \frac{\boldsymbol{ehB}}{\boldsymbol{8\pi m}}$$

The direction of $\vec{\tau}$ is perpendicular to both $\vec{M}$ and $\vec{B}$.

506. Let m be the mass per unit length of wire AB. At a height x above the wire CD, magnetic force per unit length on wire AB will be given by

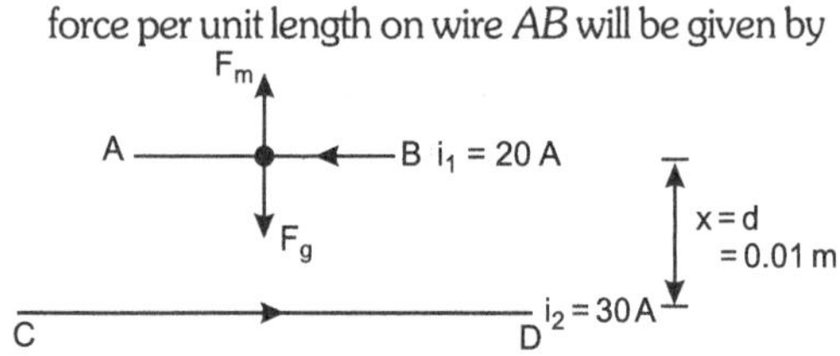

$$F_m = \frac{\mu_0}{2\pi}\frac{i_1 i_2}{x} \quad \text{(upwards)} \quad \ldots(1)$$

Weight per unit length of wire AB is

$$F_g = mg \quad \text{(downwards)}$$

Here m = mass per unit length of wire AB

At $x = d$, wire is in equilibrium *i.e.*,

$$F_m = F_g \quad \text{or} \quad \frac{\mu_0}{2\pi}\cdot\frac{i_1 i_2}{d} = mg$$

$$\text{or} \quad \frac{\mu_0}{2\pi}\frac{i_1 i_2}{d^2} = \frac{mg}{d} \quad \ldots(2)$$

When AB is depressed, x decreases therefore, F_m will increase, while F remains the same. Let AB is displaced by dx downwards. Differentiating (1) w.r.t. x, we get

$$dF_m = -\frac{\mu_0}{2\pi}\cdot\frac{i_1 i_2}{x^2}\cdot dx \quad \ldots(3)$$

i.e., restoring force,

$$F = dF_m \;\alpha\; - dx$$

Hence the motion of wire is simple harmonic. From equations (2) and (3), we can write

$$dF_m = -\left(\frac{mg}{d}\right).dx \qquad (x = d)$$

$\therefore$ Acceleration of wire $a = -\left(\frac{g}{d}\right)dx$

Hence period of oscillation

$$T = 2\pi\sqrt{\left|\frac{dx}{a}\right|} = 2\pi\sqrt{\left|\frac{\text{displacement}}{\text{acceleration}}\right|}$$

or $\quad T = 2\pi\sqrt{\frac{d}{g}} = 2\pi\sqrt{\frac{0.01}{9.8}}$

or $\quad \boldsymbol{T = 0.2\ \text{s}}$

507. Kinetic energy of electron,

$K = \frac{1}{2}mv^2 = 2\ \text{keV}$

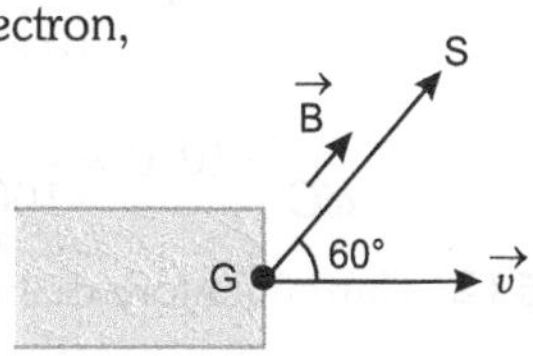

$\therefore$ speed of electron,

$v = \sqrt{\frac{2K}{m}}$

$v = \sqrt{\frac{2 \times 2 \times 1.6 \times 10^{-16}}{9.1 \times 10^{-31}}}\ \text{m/s}$

$v = 2.65 \times 10^7\ \text{m/s}$

Since the velocity $(\vec{v})$ of the electron makes an angle of $\theta = 60°$ with the magnetic field $\vec{B}$, the path will be a helix. So, the particle will hit S if

$GS = np \qquad$ Here $n = 1, 2, 3, \ldots$

$p = \text{pitch of helix} = \frac{2\pi m}{qB} v\cos\theta$

But for B to be minimum, $n = 1$

Hence $\quad GS = p = \frac{2\pi m}{qB} v\cos\theta$

$\therefore \qquad B = B_{\text{min}} = \frac{2\pi mv\cos\theta}{q(GS)}$

Substituting the values, we have

$$B_{\text{min}} = \frac{(2\pi)(9.1\times10^{-31})(2.65\times10^7)\left(\frac{1}{2}\right)}{(1.6\times10^{-19})(0.1)}\ \text{tesla}$$

or $\quad \boldsymbol{B_{\text{min}} = 4.73 \times 10^{-3}\ \text{tesla}}$

508. (a) Let us assume a segment of length dx at a point P, a distance x from the centre shown in figure.

Magnetic field at P due to current in wires A and B will be in the directions perpendicular to AP and BP respectively as shown.

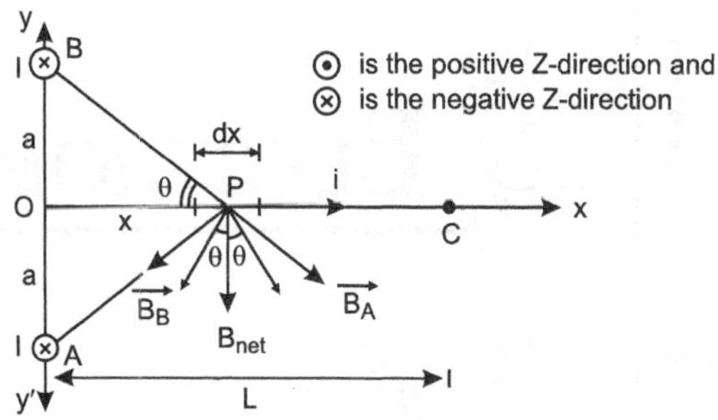

$$|\vec{B_A}| = |\vec{B_B}| = |\vec{B}| = \frac{\mu_0}{2\pi}\frac{i}{AP}$$

Therefore, net magnetic field at P will be along negative y-axis as shown

and $B_{\text{net}} = 2|\vec{B}|\cos\theta = 2\left(\frac{\mu_0}{2\pi}\right)\frac{i}{AP}\left(\frac{x}{AP}\right)$

$$B_{\text{net}} = \left(\frac{\mu_0}{\pi}\right)\frac{i.x}{(AP)^2} = \frac{\mu_0}{\pi}\cdot\frac{ix}{(a^2+x^2)}$$

Therefore, force on the element will be $(F = ilB)$

$dF = i\left\{\frac{\mu_0}{\pi}\frac{ix}{a^2+x^2}\right\}dx$ (in negative z-direction)

$\therefore$ Total force on the wire will be

$$F = \int_{x=0}^{x=L} dF = \frac{\mu_0 i^2}{\pi}\int_0^L \frac{x\,dx}{x^2+a^2}$$

$$= \frac{\mu_0 i^2}{2\pi} ln\left(\frac{L^2+a^2}{a^2}\right) \qquad \text{(in negative } z\text{-axis)}$$

Hence $\quad \boldsymbol{\vec{F} = -\frac{\mu_0 i^2}{2\pi} ln\left(\frac{L^2+a^2}{a^2}\right)\hat{k}}$

(b) If current in wire B is reversed, then magnetic fields due to A and B will be in the directions shown in figure. *i. e.*, net magnetic field $\vec{B}_{\text{net}}$ will be along positive x-axis and since current is also along positive x-axis, force on wire OC will be **zero**.

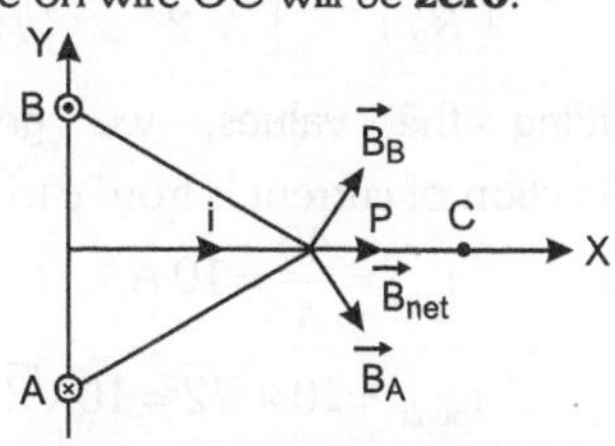

Note: $\vec{B}_A$ is not necessarily parallel to BP.

CHAPTER 15 ELECTROMAGNETIC INDUCTION

509. $$\left(\frac{1}{2}\right) = \frac{R}{\sqrt{R^2 + X_c^2}} \quad \ldots(1)$$

$$\left(\frac{\sqrt{3}}{2}\right) = \frac{R+10}{\sqrt{(R+10)^2 + X_c^2}} \quad \ldots(2)$$

Solving these two equations we get $R = 5$ ohm.

510. Applying Kirchhoff's law in loop abcdefa

$$E - L\left(\frac{di_1}{dt}\right) - i_1R_2 = 0$$

$$\int_0^{i_1} \frac{Ldi_1}{E - i_1R_2} = \int_0^t dt$$

$$\Rightarrow \quad i_1 = \frac{E}{R_2}\left\{1 - \exp\left(-R_2\frac{t}{L}\right)\right\}$$

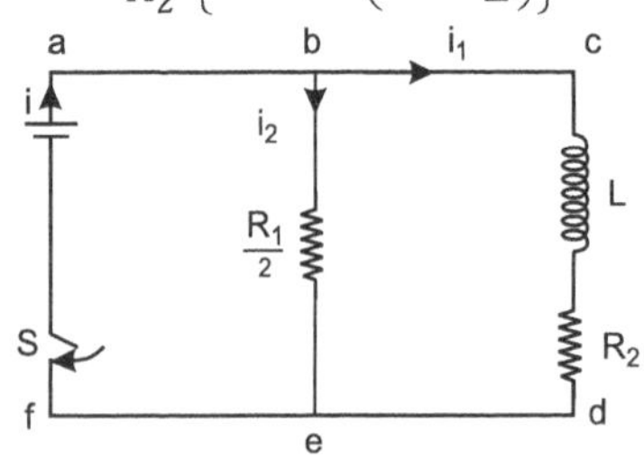

Potential drop across inductor is

$$L\frac{di_1}{dt} = E\exp\left(-R_2\frac{t}{L}\right) = 10e^{-5t} \text{ volts}$$

in steady state $i_1 = \dfrac{E}{R_2}$

When the switch S is opened,

$$i = \left(\frac{E}{R_2}\right)\exp\left[-\left(\frac{R_1}{2} + R_2\right)\frac{t}{L}\right]$$

Putting the values, we get, $i = 5\,e^{-10t}$ (direction of current is from c to d)

511. (a) $$i_{rms} = \frac{40}{4} = 10\text{ A}$$

$$i_{peak} = 10 \times \sqrt{2} = 10\sqrt{2}\text{ A}$$

(b) $$E_{rms} = \sqrt{40^2 + (40-10)^2} = 50\text{ V}$$

$$E_{peak} = E_0 = 50\sqrt{2}\text{ V}$$

(c) $$\omega L\, i_{rms} = 40$$

$$L = \frac{40}{10 \times 100\pi} = \frac{1}{25\pi}\text{ H}$$

$$\frac{i_{rms}}{\omega C} = 10,\ C = \frac{10}{100\pi \times 10} = \frac{1}{100\pi}\text{ F}$$

512. Current before closing the switch

$$i_0 = \frac{V}{R_0 + R}$$

Let current in the circuit is i at any time t. Applying KVL in the circuit

$$V = L\frac{di}{dt} + Ri$$

$$\Rightarrow \quad \frac{di}{dt} + \frac{Ri}{L} = \frac{V}{L}$$

On solving we get,

$$i = \frac{V}{R} + Ce^{-\frac{R}{L}t}$$

at $t = 0$ $$i = i_0 = \frac{V}{R_0 + R}$$

so, $$C = -V\frac{R_0}{R(R_0 + R)}$$

$$i = \frac{V}{R} - \frac{VR_0e^{\frac{-Rt}{L}}}{R(R + R_0)}$$

513. (a) $$I(t) = \frac{1}{R}(E_1 + E_2)$$

$$= \frac{25\sqrt{3}\,(1 + \sqrt{2}\sin\omega t)}{R}$$

Heat produced in one cycle of AC

$$= \int_0^{2\pi/\omega} I^2(t)\,Rdt$$

$$\frac{(25\sqrt{3})^2}{R}$$

$$= \int_0^{2\pi/\omega} (1 + 2\sin^2 \omega t + 2\sqrt{2} \sin \omega t)\, dt$$

$$= \frac{25 \times 25 \times 3}{50}\left[\frac{2\pi}{\omega} + \frac{2\pi}{\omega}\right] = \frac{3}{2}\text{J}$$

Number of cycles in 14 minute is

$$N = 14 \times 60 \times 50$$

Total heat produced

$$Q = \frac{3}{2} \times 14 \times 60 \times 50 \text{ J}$$

$$= 63000 \text{ J}$$

(b) $\because \quad Q = ms\,\Delta t$

$$\Rightarrow \quad T_f - T_i = \frac{Q}{ms} = \frac{63000}{3 \times 4200} = 5°\text{C}$$

$$T_f = 5 + T_i = 25°\text{C}$$

(c) $I_{DC}^2\, Rt = Q$

$$\Rightarrow I_{DC} = \sqrt{\frac{Q}{Rt}} = \sqrt{\frac{63000}{50 \times 14 \times 60}} = 1.22 \text{ A}$$

514. (a) $\cos\theta = \frac{R}{Z} = \frac{V_R}{V} = \frac{80}{V}$

$$0.8 = \frac{80}{V}$$

$$V = \frac{80}{0.8} = 100 \text{ volt}$$

(b) $\because \quad V^2 < V_R^2 + V_C^2$

(where $V_R = iR = 80$ volt)

$\therefore$Nature of element in box will be inductive.

$$\therefore \quad V^2 = V_R^2 + (V_C - V_L)^2$$

So, either $V_C - V_L = 60$ volt or $V_L - V_C = 60$ volt. Therefore V_L is either 40 volt or 160 volt. Since $i = 1$ A. Hence X_L is either 40 ohm or 160 ohm.

$$\therefore \quad 2\pi(50)L = 40 \text{ or } 160$$

$$\therefore \quad L = \left(\frac{0.4}{\pi}\right)\text{henry or}\left(\frac{1.6}{\pi}\right)\text{henry}$$

515. (a) Initial charge $= CV_0 = Q_0$

$$= 10 \times 10^{-3} \times 5 = 50 \text{ mc}$$

When capacitor is connected at position 1.

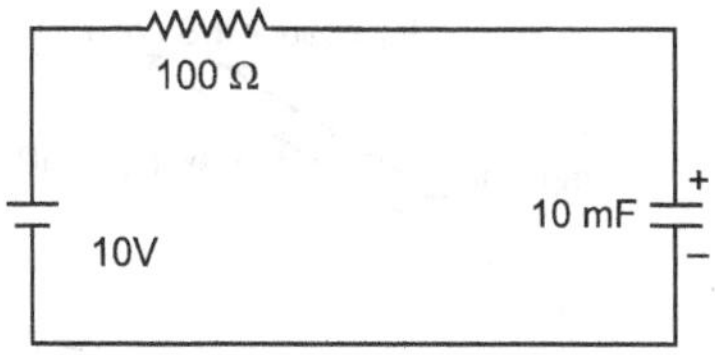

$$E - IR - \frac{q}{C} = 0$$

$$\Rightarrow \quad \int_0^t \frac{1}{RC}\, dt = \int_{Q_0}^q \frac{dq}{EC - q}$$

$q = 50\,(2 - e^{-1})$ mC, at $t = 1$ sec

$$q = 50\,(2 - e^{-1}) = 1.63 \times 50 = 81.5 \text{ mC}$$

Voltage across capacitor at that time

$$V = \frac{q}{C} = \frac{81.5 \times 10^{-3}}{10 \times 10^{-3}}$$

$$= 8.15 \text{ volt}$$

(b) $\frac{1}{2}Li_{max}^2 = \frac{1}{2}CV_{max}^2$

$$\therefore \quad i_{max} = \sqrt{\frac{C}{L}} \cdot V_{max}$$

$$= \sqrt{\frac{10}{2.5}} \times 8.15 = 16.3 \text{ A}$$

$$\text{Frequency} = \frac{1}{2\pi\sqrt{LC}} = \frac{1000}{2\pi \times 25}$$

$$= \frac{10^3}{50\pi}\text{ Hz} = \frac{20}{\pi}\text{ Hz}$$

516. Let v be the velocity of rod at any time t down the plane.

The motional emf $e = Bbv_\perp$

(here $v_\perp$ is the component of velocity perpendicular to magnetic field and length of conductor both, which is equal to $v\cos\alpha$ in this case)

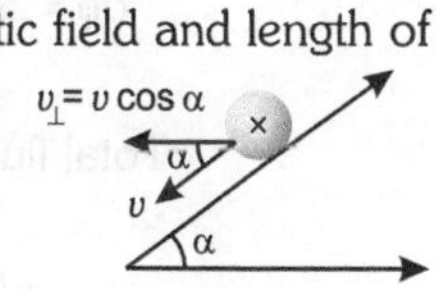

$$\therefore \quad e = Bbv\cos\alpha$$

current in the circuit $i = \frac{dq}{dt}$

where $\quad q = eC = CBbv\cos\alpha$

$$\therefore \quad i = (CBb\cos\alpha)\left(\frac{dv}{dt}\right) = (CBb\cos\alpha)\cdot a$$

where 'a' is the acceleration of the rod. This current i is perpendicular to paper inwards as shown in figure.

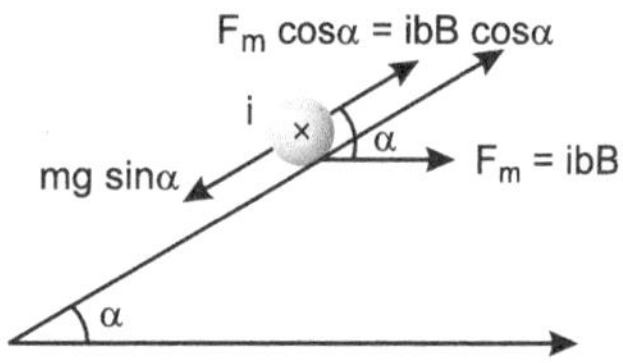

The rod is acted upon by the force of gravity and magnetic force. Let us write the equation of motion for the rod :

$$ma = mg \sin\alpha - ibB\cos\alpha$$
$$= mg\sin\alpha - (CB^2b^2\cos^2\alpha)\, a$$

or $$a = \frac{mg\sin\alpha}{m + CB^2b^2\cos^2\alpha} = \text{constant} \quad \ldots(1)$$

(a) $$t = \sqrt{\frac{2l}{a}} \qquad \left(l = \frac{1}{2}at^2\right)$$

or $$\boldsymbol{t = \sqrt{\frac{2l}{mg\sin\alpha}(m + Cb^2B^2\cos^2\alpha)}}$$

(b) $$v = \sqrt{2al} \qquad (v^2 = 2al)$$

or $$\boldsymbol{v = \sqrt{\frac{2\,mgl\sin\alpha}{m + Cb^2B^2\cos^2\alpha}}}$$

517. (a) Magnetic flux passing through an element shown in figure is

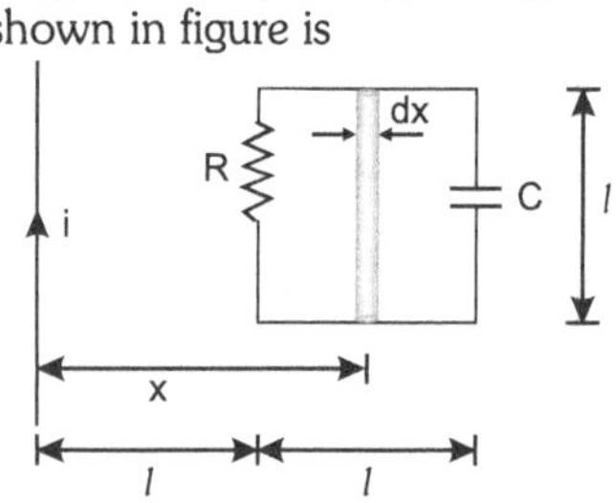

$$d\phi = Bldx = \frac{\mu_0 i}{2\pi x} l \cdot dx$$

$\therefore$ Total flux $$\phi = \int_l^{2l} \frac{\mu_0 il\,dx}{2\pi x}$$

$\therefore$ $$\phi = \frac{\mu_0 il}{2\pi}\ln(2)$$

$$= \left(\frac{\mu_0 l}{2\pi}\right)\ln(2)\frac{i_0}{t_0}\cdot t \qquad \left(i = \frac{i_0}{t_0}t\right)$$

Magnitude of induced emf

$$e_t = \left|\frac{d\phi}{dt}\right| = \frac{\mu_0 l\, i_0\ln(2)}{2\pi t_0} \quad \ldots(1)$$

$$= \text{constant}$$

So, it is a case of charging of a capacitor in $C - R$ circuit. Hence

$$q = q_0(1 - e^{-t/\tau_C}) \quad \ldots(2)$$

where $$q_0 = e_tC = \frac{\mu_0 li_0C\ln(2)}{2\pi t_0}$$

and $$\tau_C = CR$$

substituting in (2) we get

$$\boldsymbol{q = \frac{\mu_0\, li_0C\ln(2)}{2\pi t_0}(1 - e^{-t/CR})}$$

(b) Current at any time t is

$$i = \frac{dq}{dt} = \frac{e_t}{R}e^{-t/CR}$$

$\therefore$ Heat generated in resistance

$$H = \int_0^t i^2R\,dt = \int_0^t \frac{e_t^2}{R^2}e^{-2t/CR}R\,dt$$

$$H = \frac{e_t^2C}{2}(1 - e^{-2t/CR}) \quad \ldots(3)$$

Energy stored in capacitor

$$U = \frac{1}{2}\frac{q^2}{C} = \frac{1}{2}e_t^2C(1 - e^{-t/CR})^2 \quad \ldots(4)$$

$\therefore$ The desired ratio is

$$\boldsymbol{\frac{H}{U} = \frac{1 - e^{-2t/CR}}{(1 - e^{-t/CR})^2}}$$

(c) $$V_0 = \frac{q}{C} = e_t(1 - e^{-t/CR})$$

So $$\boldsymbol{t = CR\ln\left(\frac{e_t}{e_t - V_0}\right)}$$

where $$\boldsymbol{e_t = \frac{\mu_0 li_0\ln(2)}{2\pi t_0}}$$

518. In the figure $OA = x$,

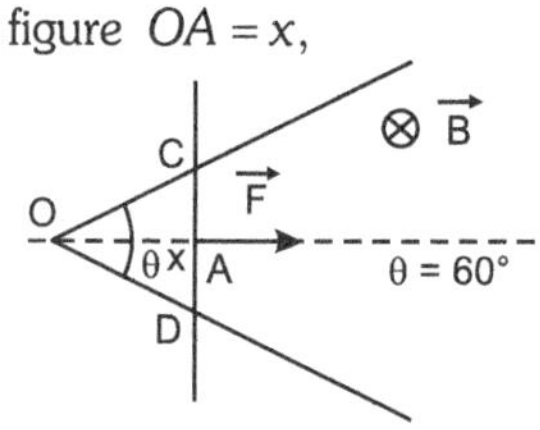

then $$OD = OC = \frac{2x}{\sqrt{3}}$$

and $$AD = AC = \frac{x}{\sqrt{3}}$$

So length of the conducting rod DC is

$$l = 2AD = \frac{2x}{\sqrt{3}}$$

$\therefore$ motional emf $e = Bvl = \frac{2}{\sqrt{3}} Bvx$

Resistance in the circuit $R = \rho\,(OD + OC)$

or $\quad R = 2\rho\,(OD) = \frac{4}{\sqrt{3}}\rho x$

$\therefore$ current in the circuit

$$i = \frac{e}{R} = \frac{Bv}{2\rho} \qquad \text{(anticlockwise)}$$

$\therefore$ magnetic force on the rod

$$F_m = ilB = \left(\frac{Bv}{2\rho}\right)\left(\frac{2x}{\sqrt{3}}\right)(B)$$

or $$F_m = \left(\frac{B^2 v}{\sqrt{3}\rho}\right)x \qquad \text{(leftwards)}$$

$\therefore$ External force applied is

$$\mathbf{F} = \left(\frac{\mathbf{B^2 v}}{\sqrt{\mathbf{3}}\rho}\right)\mathbf{x} \qquad \text{(rightwards)}$$

519. At any instant t, current in $L - R$ circuit is given as

$$i = i_0\,(1 - e^{-Rt/L}) = \frac{E}{R}(1 - e^{-Rt/L})$$

(E = applied emf)

At t = time constant (L/R)

$$i = \frac{E}{R}(1 - 1/e) = \frac{5}{10}\left(1 - \frac{1}{2.718}\right) = 0.316\text{ A}$$

$\therefore$ The rate at which energy is delivered by the battery is

$P_1 = Ei = (5)\,(0.316)$ watt = 1.58 watt ...(1)

At this time some energy will appear as joule heat in the resistance, given by

$$P_2 = i^2 R = (0.316)^2\,(10) = 0.998\text{ watt}$$

$$P_2 = 0.998\text{ watt} \qquad \text{...(2)}$$

Part of the energy is stored in magnetic field, linked with the inductor. It is given by the relation

$$P_3 = Li\left(\frac{di}{dt}\right)$$

$$\frac{di}{dt} = \frac{E}{L}e^{-Rt/L} = \frac{E}{L}\cdot\frac{1}{e} \qquad \left(\text{at } t = \frac{L}{R}\right)$$

$$\therefore \quad P_3 = (Li)\left(\frac{E}{L}\right)\left(\frac{1}{e}\right)$$

$$= \frac{Ei}{e} = \frac{5 \times 0.316}{2.718} = 0.582\text{ watt} \qquad \text{...(3)}$$

From equations (1), (2) and (3) we can see that

$$P_1 = P_2 + P_3$$

or the power delivered
= power dissipated as heat + power stored
It is same as required by the principle of conservation of energy.

520. Current in the circuit at time $t = 0$ is

$$i = \frac{E}{R} = \frac{20}{5} = 4A \qquad \text{(clockwise)}$$

$\therefore$ magnetic force on the rod is

$F_m = ilB = (4)\,(10)\,(0.5) = 20$ N (upwards)

weight of the rod

$W = mg = (1)\,(10) = 10$ N (downwards)

Since $F_m > W$, the rod will move upwards

Let v be the velocity of rod at any time t

motional emf $e = Bvl = (0.5)\,(v)\,(10) = 5\,v$

net emf $= E - e = 20 - 5\,v$

$$\text{current } i = \frac{20 - 5\,v}{R} = \frac{20 - 5\,v}{5} = (4 - v)$$

magnetic force

$$F_m = ilB = (4 - v)\,(10)\,(0.5) = (20 - 5\,v)$$

$\therefore \quad$ net force $= F_m - W \qquad$ (upwards)

or $$m\left(\frac{dv}{dt}\right) = (20 - 5\,v) - (10)$$

or $$(1)\left(\frac{dv}{dt}\right) = (10 - 5\,v) \qquad (m = 1\text{kg})$$

or $$\int_0^v \frac{dv}{10 - 5\,v} = \int_0^t dt$$

Solving this equation we get

$$\mathbf{v = 2\,(1 - e^{-5t})\ m/s}$$

521. Let v be the velocity of the rod at any time t.

Motional emf $\quad e = Bvl$

$$q = eC = BvlC$$

$$i_1 = \frac{e}{R} = \frac{Bvl}{R}$$

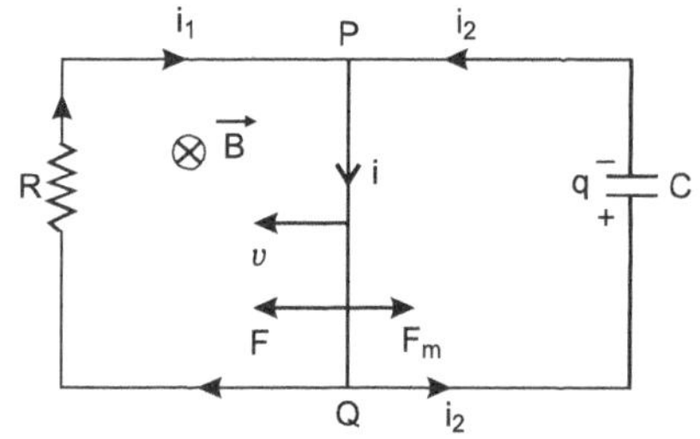

$$i_2 = \frac{dq}{dt} = BlC \cdot \frac{dv}{dt}$$

$$i = i_1 + i_2 = \frac{Bvl}{R} + BlC \cdot \frac{dv}{dt}$$

$$F_m = ilB = B^2l^2\left(\frac{v}{R} + C \cdot \frac{dv}{dt}\right)$$

Net force $F_{net} = F - F_m$

or $m \cdot \frac{dv}{dt} = F - B^2l^2\left(\frac{v}{R} + C \cdot \frac{dv}{dt}\right)$

or $\frac{dv}{dt}(m + B^2l^2C) = F - B^2l^2\frac{v}{R}$

or $\frac{dv}{F - \frac{B^2l^2v}{R}} = \frac{dt}{m + B^2l^2C}$

or $\int_0^v \frac{dv}{F - \frac{B^2l^2}{R} \cdot v} = \int_0^t \frac{dt}{m + B^2l^2C}$

This gives $\boldsymbol{v} = \frac{FR}{B^2l^2}\left[1 - e^{-\frac{B^2l^2}{R(m + B^2l^2C)}t}\right]$

522. (a) Magnitude of induced electric field due to change in magnetic flux is given by

$$\oint \vec{E} \cdot \vec{dl} = \frac{d\phi}{dt} = S \cdot \frac{dB}{dt}$$

or $E.l = \pi R^2(2B_0 t) \quad \left(\frac{dB}{dt} = 2B_0 t\right)$

Here E = induced electric field due to change in magnetic flux

or $E(2\pi R) = 2\pi R^2 B_0 t$

or $E = B_0 Rt$

Hence $F = QE = B_0QRt$

This force is tangential to ring. Ring starts rotating when torque of this force is greater than the torque due to maximum friction ($f_{max} = \mu mg$) or when

$$\tau_F \geq \tau_{f_{max}}$$

Taking the limiting case.

$\tau_F = \tau_{f_{max}}$ or $F.R = (\mu mg)R$

or $F = \mu mg$ or $B_0QRt = \mu mg$

It is given that ring starts rotating after 2 second. So, putting $t = 2$ second we get

$$\mu = \frac{\mathbf{2\,B_0RQ}}{\boldsymbol{mg}}$$

(b) After 2 second

$$\tau_F > \tau_{f_{max}}$$

Therefore, net torque is

$$\tau = \tau_F - \tau_{f_{max}} = B_0QR^2t - \mu mgR$$

Substituting $\mu = \frac{2B_0QR}{mg}$ we get,

$$\tau = B_0QR^2(t-2)$$

or $I\left(\frac{d\omega}{dt}\right) = B_0QR^2(t-2)$

or $mR^2\left(\frac{d\omega}{dt}\right) = B_0QR^2(t-2)$

or $\int_0^\omega d\omega = \frac{B_0Q}{m}\int_2^4 (t-2)\,dt$

or $\omega = \frac{2B_0Q}{m}$ ...(1)

Now magnetic field is switched off *i.e.*, only retarding torque is present due to friction. So, angular retardation will be

$$\alpha = \frac{\tau_{f_{max}}}{I} = \frac{\mu mgR}{mR^2} = \frac{\mu g}{R}$$

Therefore, applying

$$\omega^2 = \omega_0^2 - 2\alpha\theta$$

or $0 = \left(\frac{2B_0Q}{m}\right)^2 - 2\left(\frac{\mu g}{R}\right)\theta$

or $\theta = \frac{2B_0^2Q^2R}{\mu m^2 g}$

Substituting $\mu = \frac{2B_0RQ}{mg}$

we get $\theta = \frac{\boldsymbol{B_0Q}}{\boldsymbol{m}}$

523. (a) Voltage across the inductor

$$V_L = -L\frac{di}{dt}$$

$$= -(10^{-3})(20) \text{ volt}$$

$$= \mathbf{-20\ mV}$$

(b) Voltage across the capacitor

$$V_C = \frac{q}{C} = \frac{1}{C}\int_0^t i\,dt$$

$$= \left(\frac{1}{10^{-6}}\right)\int_0^t (20\,t)\,dt$$

$$= (10^6)\frac{20t^2}{2} \text{ volt} = \mathbf{(10^7\ t^2)\ volt}$$

(c) $\frac{1}{2}C V_C^2 > \frac{1}{2}Li^2$ or $CV_C^2 > Li^2$

or $(10^{-6})(10^7\,t^2)^2 > (10^{-3})(20\,t)^2$

or $t > 63.2 \times 10^{-6}$ s

or $\boldsymbol{t > 63.2\ \mu s}$

524. (a) Suppose at any time t, rod MN is at a distance x from the resistance R. Velocity of rod at that instant is

$$v = \frac{dx}{dt}$$

Motional emf $e = Bvl$ ...(1)

Total resistance R_{net} of the circuit at this instant is

$$R_{net} = R + 2\lambda x$$

$$\therefore \quad i = \frac{e}{R_{net}} = \frac{Blv}{R + 2\lambda x} \quad \text{(anticlockwise)}$$

or $$\boldsymbol{v = \frac{i\,(R + 2\lambda x)}{Bl}}$$

Instantaneous acceleration of the rod, is

$$a = \frac{dv}{dt} = \frac{2\lambda i}{Bl}\cdot\frac{dx}{dt} = \frac{2\lambda i}{Bl}\cdot v$$

or $$a = \frac{2\lambda i^2}{B^2l^2}(R + 2\lambda x) \quad \text{...(2)}$$

At this moment two forces are acting on the rod, applied force F (towards right) and magnetic force $F_m = ilB$ (towards left). So

$$F_{net} = F - F_m$$

or $$ma = F - ilB$$

$$\therefore \quad \mathbf{F = ilB + \frac{2m\lambda i^2}{B^2l^2}(R + 2\lambda x)}$$

(b) Work done by F per unit time is given by

$$P_1 = Fv = \frac{Fi\,(R + 2\lambda x)}{Bl} \quad \text{...(3)}$$

and heat generated in circuit per unit time is

$$P_2 = i^2 R_{net} = i^2(R + 2\lambda x) \quad \text{...(4)}$$

Therefore, fraction of work converted into heat is

$$f = \frac{P_2}{P_1} = \frac{i\,Bl}{F}$$

Substituting value of F we get

$$\boldsymbol{f = \left\{1 + \frac{2\,m\lambda i\,(R + 2\lambda x)}{B^3l^3}\right\}^{-1}}$$

Note: The remaining part of work done goes as the change in kinetic energy of rod per time *i.e.*,

$$\frac{dK}{dt} = \frac{d}{dt}\left(\frac{1}{2}mv^2\right) = mv\cdot\left(\frac{dv}{dt}\right) = (mv)(a)$$

Substituting the values

$$\frac{dK}{dt} = \frac{2\,m\lambda i^3}{B^3l^3}(R + 2\lambda x)^2 \quad \text{...(5)}$$

From equations (3), (4) and (5) we can see that

$$P_1 = P_2 + \frac{dK}{dt}$$

525. Let at time t angular velocity of rod be ω. Then induced emf in the circuit is

$$e = \frac{B\omega a^2}{2}$$

and therefore, current in the circuit will be

$$i = \frac{e}{R + r} = \frac{B\omega a^2}{2(R + r)} \quad \text{(radially outwards)}$$

Magnetic force on element dx, at a distance x from A is

$$F = (iB \cdot dx)$$

Torque on this element about A will be

$$d\tau = F \cdot x = (iBx)\,dx$$

$\therefore$ Total torque on the rod will be

$$\tau = \int_{x=0}^{x=a} d\tau \quad \text{or} \quad \tau = \frac{iBa^2}{2} \quad \text{(anticlockwise)}$$

Now angular retardation is

$$\alpha = \frac{\tau}{I}$$

or $$-\omega \cdot \frac{d\omega}{d\theta} = \frac{iBa^2}{2\,(ma^2/3)} \qquad \left(I = I_A = \frac{ma^2}{3}\right)$$

or $$-\omega \cdot \frac{d\omega}{d\theta} = \frac{3}{2}\frac{Bi}{m} = \frac{3}{2}\frac{B}{m}\frac{B\omega a^2}{2\,(R+r)}$$

or $$d\omega = -\frac{3}{4}\frac{B^2a^2}{m\,(R+r)} \cdot d\theta$$

or $$\int_{\omega_0}^{\omega} d\omega = -\frac{3}{4}\frac{B^2a^2}{m\,(R+r)}\int_0^{\theta} d\theta$$

or $$\omega = \omega_0 - \frac{3}{4}\frac{B^2a^2}{m\,(R+r)}\theta$$

But $$i = \frac{B\omega a^2}{2\,(R+r)}$$

or $$\boldsymbol{i = \frac{Ba^2}{2\,(R+r)}\left[\omega_0 - \frac{3}{4}\frac{B^2a^2\,\theta}{m\,(R+r)}\right]}$$

526. The circuit may be assumed to be equivalent to two circuits in parallel as shown below :

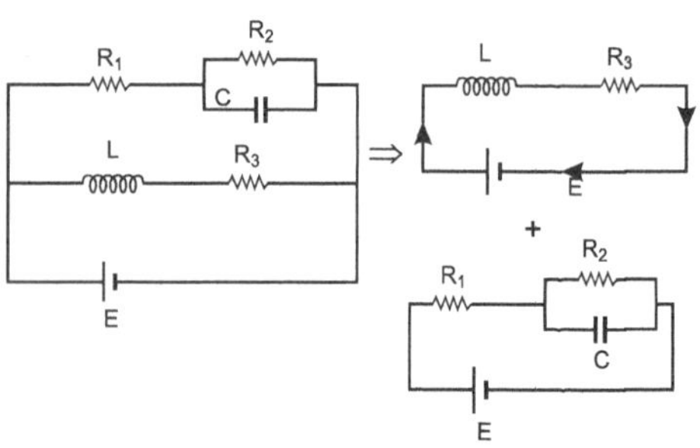

The first is a L-R circuit. Hence current i_L at any time t will be given by

$$i_L = i_0\,(1 - e^{-t/\tau_L})$$

Here $$i_0 = \frac{E}{R_3} \quad \text{and} \quad \tau_L = \frac{L}{R_3}$$

$$\therefore \qquad \boldsymbol{i_L = \frac{E}{R_3}\left(1 - e^{-\frac{R_3 t}{L}}\right)}$$

Current through capacitor at time t can be determined as follows :

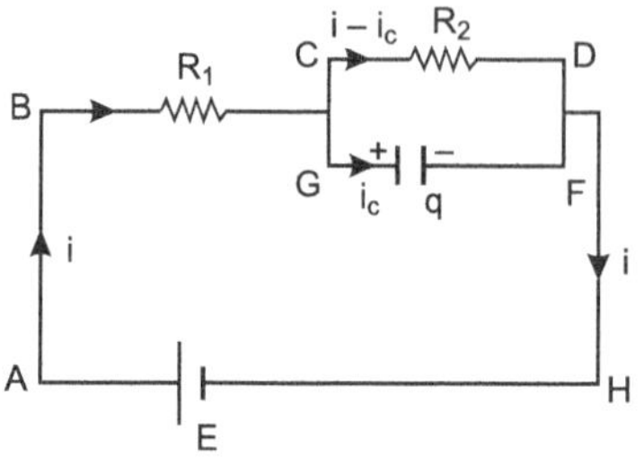

Let the currents in different branches be i, i_c and $i - i_c$ at any time t and q be the charge stored in the capacitor.

Applying Kirchhoff's loop law in $ABKDHA$

$$E - iR_1 - (i - i_c)\,R_2 = 0 \qquad \ldots(1)$$

in loop $KDFGK$

$$\frac{q}{C} - (i - i_c)\,R_2 = 0 \qquad \ldots(2)$$

Also $$i_c = \frac{dq}{dt} \qquad \ldots(3)$$

From equation (1)

$$i = \left(\frac{E + i_cR_2}{R_1 + R_2}\right)$$

Substituting this in equation (2) we get

$$\frac{q}{C} - \left(\frac{E + i_cR_2}{R_1 + R_2} - i_c\right)R_2 = 0$$

or $$\frac{q}{C} - \frac{ER_2}{R_1 + R_2} = i_c\left\{\frac{R_2^2}{R_1 + R_2} - R_2\right\}$$

or $$\left(\frac{dq}{dt}\right)\left\{\frac{R_2^2}{R_1 + R_2} - R_2\right\} = \frac{q}{C} - \frac{ER_2}{R_1 + R_2}$$

or $$\int_0^q \frac{dq}{\left(\frac{ER_2}{R_1 + R_2}\right) - \frac{q}{C}} = \frac{R_1 + R_2}{R_1 R_2}\int_0^t dt$$

Solving this we get

$$q = \frac{ECR_2}{R_1 + R_2}\left[1 - e^{-\left(\frac{R_1 + R_2}{CR_1R_2}\right)t}\right]$$

$$\therefore \qquad i_c = \frac{dq}{dt} = \frac{E}{R_1}e^{-\left(\frac{R_1 + R_2}{CR_1R_2}\right)t}$$

Hence currents through capacitor and inductor at any time t are

$$i_c = \frac{E}{R_1} e^{-\left(\frac{R_1 + R_2}{CR_1R_2}\right)t}$$

and $$i_L = \frac{E}{R_3}\left(1 - e^{-\frac{R_3}{L}t}\right)$$

527. Let v be the velocity of loop at time t and $v + dv$ in time $t + dt$.

Distance moved by the loop in time dt, down the plane is

$$ds = v \cdot dt$$

$$dh = (ds) \sin\theta = v \sin\theta \cdot dt$$

and $$dx = (ds)\cos\theta = v\cos\theta \cdot dt$$

or $$\frac{dx}{dt} = v\cos\theta$$

From conservation of mechanical energy :
Decreases in potential energy in time dt
= increase in kinetic energy in time dt
+ energy dissipated as heat in time dt

i.e. $$(mg)\, dh = \left\{\frac{1}{2}m(v + dv)^2 - \frac{1}{2}mv^2\right\} + i^2R \cdot dt$$

or $$(mgv \sin\theta)\, dt = \left\{\frac{1}{2}mv^2\left(1 + \frac{dv}{v}\right)^2 - \frac{1}{2}mv^2\right\} + i^2Rdt$$

or $$(mgv\sin\theta)\, dt = \left\{\frac{1}{2}mv^2\left(1 + 2\frac{dv}{v}\right) - \frac{1}{2}mv^2\right\} + i^2Rdt \quad (dv << v)$$

or $$(mgv\sin\theta)\, dt = (mv)\, dv + i^2Rdt \qquad \dots(1)$$

Here $$i = \frac{e}{R}$$

where $$e = \left|\frac{d\phi}{dt}\right| = \left|\frac{dB}{dt}\right| = \pi r^2 (B_0 \cdot a) \cdot \frac{dx}{dt}$$

or $$e = (\pi r^2 B_0 a)(v\cos\theta)$$

So $$i = \frac{\pi r^2 B_0 a v \cos\theta}{R}$$

Substituting this value in equation (1) we get

$$(mgv\sin\theta\, dt) = (mv\, dv) + \frac{\pi^2 r^4 B_0^2 a^2 v^2 \cos^2\theta}{R} \cdot dt$$

or $$\left(mg\sin\theta - \frac{\pi^2 r^4 B_0^2 a^2 v \cos^2\theta}{R}\right)dt = m \cdot dv$$

or $$\int_0^v \frac{dv}{mg\sin\theta - \frac{\pi^2 r^4 B_0^2 a^2 \cos^2\theta}{R} v} = \int_0^t \frac{dt}{m}$$

$$\therefore \quad v = \frac{mgR\sin\theta}{\pi^2 r^4 B_0^2 a^2 \cos^2\theta}\left(1 - e^{-\frac{\pi^2 r^4 B_0^2 a^2 \cos^2\theta\, t}{mR}}\right)$$

or $$\boldsymbol{v = \frac{g\sin\theta}{K}(1 - e^{-Kt})}$$

Where $$K = \frac{\pi^2 r^4 B_0^2 a^2 \cos^2\theta}{mR}$$

528. Power dissipated in the loop is

$$P = \frac{e^2}{r} \qquad \dots(1)$$

where e = induced emf in the loop

$$e = \left|\frac{d\phi}{dt}\right| = S\left|\frac{dB}{dt}\right| = (\pi a^2)\left|\frac{dB}{dt}\right|$$

Here $$B(x) = \frac{\mu_0}{2\pi}\frac{i_0}{x} = B$$

$$\therefore \left(\frac{dB}{dt}\right) = -\frac{\mu_0 i_0}{2\pi x^2} \cdot \frac{dx}{dt} = -\frac{\mu_0 i_0 v}{2\pi x^2} \quad \left(\frac{dx}{dt} = v\right)$$

$$\therefore \quad e = \frac{\pi a^2 \mu_0 i_0 v}{2\pi x^2} \qquad \dots(2)$$

From equations (1) and (2)

$$P = \frac{\pi^2 a^4 \mu_0^2 i_0^2 v^2}{4\pi^2 x^4 \cdot r}$$

or $$v = \sqrt{\frac{4Px^4 r}{\mu_0^2 i_0^2 a^4}}$$

or $$\boldsymbol{v = \frac{2x^2}{\mu_0 i_0 a^2}\sqrt{Pr}}$$

529. (a) Magnetic flux linked with the loop

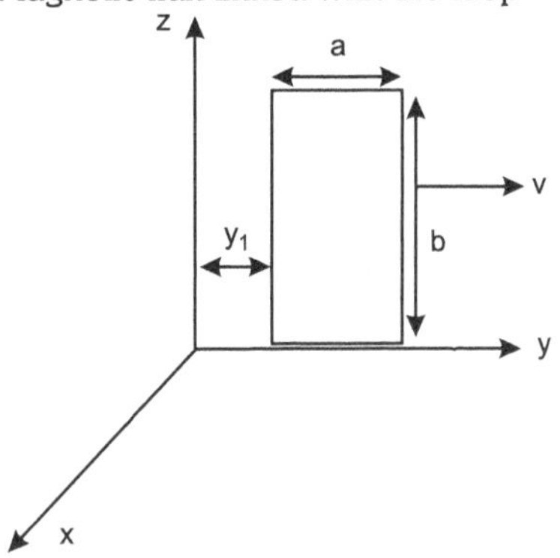

$$\phi = \int_{y_1}^{y_2} (6 - y)\, b dy$$

$$= 6b(y_2 - y_1) - \frac{b}{2}(y_2 - y_1)(y_2 + y_1)$$

$$= 6ab - \frac{ab}{2}[a + 2vt]$$

Induced e.m.f. $= -\dfrac{d\phi}{dt} = abv$

$$= (0.2)(0.5)(6.5)\text{ V}$$
$$= 0.65\text{ V}$$

(b) Here, $\phi = 6ab - \dfrac{ab}{2}[a + \omega t^2]$

where ω = acceleration

Hence, ε (induced e.m.f.)

$$= -\frac{d\phi}{dt} = ab\omega t$$

$$= (0.2)(0.5)\,2t = 0.2t\text{ V}$$

530. Flux linked with the loop at any instant

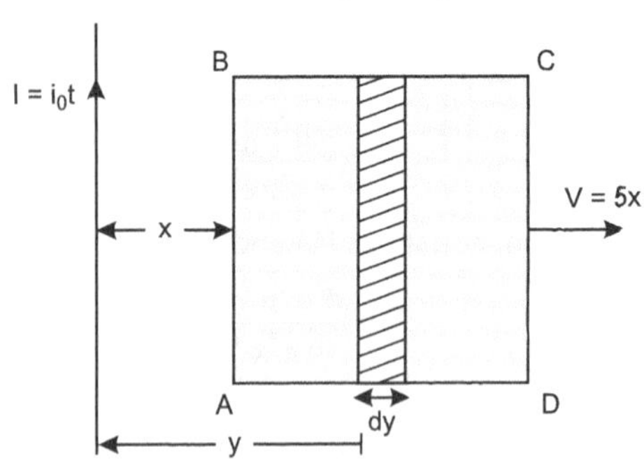

$$\phi = \int_x^{x+a} \frac{\mu_0 I}{2\pi y}\,.\, a dy$$

$$\phi = \frac{\mu_0 I_0 at}{2\pi}\ln\left(\frac{x+a}{x}\right) \quad \ldots(1)$$

$$\varepsilon = \frac{-d\phi}{dt}$$

$$= -\frac{\mu_0 I_0 a}{2\pi}\ln\frac{(x+a)}{x} + \frac{\mu_0 I_0 at}{2\pi}\left[\frac{1}{x+a} - \frac{1}{x}\right] \times \frac{dx}{dt}$$

$$I = \frac{\varepsilon}{R} = -\frac{\mu_0 I_0 a}{2\pi R}\ln\left(\frac{x+a}{x}\right) + \frac{\mu_0 I_0 a^2 t}{2\pi R x(x+a)}\frac{dx}{dt} \quad \ldots(2)$$

Now $\dfrac{dx}{dt} = 5x$

$$\Rightarrow \int_{x=\frac{a}{10}}^{10a} \frac{dx}{x} = 5\int_0^t dt$$

$$\Rightarrow t = \frac{\ln 100}{5}$$

Putting the value of t in equation (2)

$$I = 0.32\text{ A}$$

531. (a) Let at any moment the speed of the connector is v, current in the circuit is $i = \dfrac{dq}{dt}$ and charge on the capacitor is q. Then

$$m\frac{dv}{dt} = B\left(\frac{dq}{dt}\right)l$$

Integrating we get $v = \left(\dfrac{Bl}{m}\right)q \quad \ldots(1)$

Applying KVL in the loop we have,

$$\frac{q}{C} + \frac{dq}{dt}R = (E - Blv) \quad \ldots(2)$$

$$\Rightarrow \frac{dq}{dt} + \frac{q}{RC} = \frac{E}{R} - \frac{B^2l^2}{mR}q$$

$$\Rightarrow \int_0^q \frac{dq}{\frac{E}{R} - q\left(\frac{B^2l^2}{mR} + \frac{1}{RC}\right)} = \int_0^t dt$$

$$\Rightarrow \ln\left(\frac{\frac{E}{R} - q\lambda}{\frac{E}{R}}\right) = -\lambda t,$$

where $$\lambda = \left(\frac{B^2l^2}{mR} + \frac{1}{RC} \right)$$

$$q = \frac{E}{R}\left(\frac{1 - e^{-\lambda t}}{\lambda} \right)$$

$$\Rightarrow \quad q_{t \to \infty} = \frac{E}{R\lambda}$$

$$\Rightarrow \quad q_{t \to \infty} = \frac{EmC}{(CB^2l^2 + m)}$$

(b) From Eq. (1),

$$v_{t \to \infty} = \frac{BlCE}{(m + CB^2l^2)}$$

532. (a) Due to external force F, connector starts to accelerate and an e.m.f. is induced in it. Due to induced e.m.f., a current flows through it. According to Lenz's law, force Bil will (due to induced current) oppose motion of the connector.

Steady state velocity corresponds to zero acceleration. It means that force Bil must numerically be equal to F.

Let steady state velocity be v_0

$$e = Blv_0$$

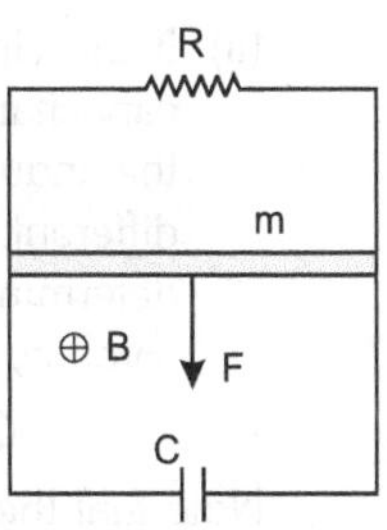

In case of steady state, charge on capacitor remains constant, therefore no current flows through it, hence current flows through resistance R alone.

∴ Induced current

$$i = \frac{e}{R} = \frac{Blv_0}{R}$$

Retarding force (Lorentz force)

$$= Bil = \frac{B^2l^2v_0}{R}$$

But $Bil = F$

$$\therefore \quad F = \frac{B^2l^2v_0}{R}$$

$$\Rightarrow \quad v_0 = \frac{FR}{B^2l^2} \quad \ldots(\text{i})$$

In steady state charge on the capacitor is $q_0 = Ce$

or $$q_0 = C(Blv_0) = \frac{CFR}{Bl} \quad \ldots(2)$$

(b) Let at some instant t, velocity of connector be v then induced e.m.f. $e = Blv$. At this instant, current through resistance is

$$i_R = \frac{e}{R} = \frac{Blv}{R}$$

and charge on capacitor is

$$q = Ce = CBlv$$

Since v is increasing, therefore q is also increasing or current is flowing through capacitor also. This current, $i_c = \frac{dq}{dt}$

or $$i_c = CBl\frac{dv}{dt} = CBla$$

where a is acceleration of the connector.

Total induced current through the connector is

$$i = i_R + i_C = \frac{Blv}{R} + CBla$$

Retarding force = Bil

∴ Net accelerating force on the connector

$$= F - Bil$$

$$\Rightarrow \quad ma = F - \left(\frac{B^2l^2v}{R} + CB^2l^2a \right)$$

$$\Rightarrow \quad (m + CB^2l^2)a = \frac{FR - B^2l^2v}{R}$$

But acceleration, $a = \frac{dv}{dt}$

$$\therefore \quad \frac{dv}{FR - B^2l^2v} = \frac{dt}{R(m + CB^2l^2)}$$

Integrating the above equation from ($t = 0, v = 0$) to (t, v)

$$v = \frac{FR}{B^2l^2}\left[1 - e^{-\left\{ \frac{B^2l^2}{R(m + CB^2l^2)} \right\}t} \right]$$

533.

$$mg - T = ma$$

$$T - F_m = ma$$

$$mg - F_m = 2ma \quad \ldots(1)$$

Let at any instant inductor is moving with velocity v

Hence $$E - L\frac{di}{dt} = 0$$

$$\frac{Blv dt}{L} = di$$

$$\int_0^x \frac{Bl}{L}\,dx = \int_0^i di \Rightarrow \frac{Bl}{L}x = i$$

Now $$F_m = Bil = B^2l^2\frac{x}{L}$$

Substituting in (1)

$$mg - \frac{B^2l^2x}{L} = 2ma = 2mv\frac{dv}{dx}$$

As acceleration is a function of displacement

$$\int_0^x \left(mg - \frac{B^2l^2x}{L}\right)dx = 2m\int_0^v v\,dv$$

$$\Rightarrow \quad mgx - \frac{B^2l^2x^2}{2L} = 2m\frac{v^2}{2}$$

$$\Rightarrow \quad v = \left(gx - \frac{B^2l^2x^2}{2mL}\right)^{1/2}$$

534. Let at distance x, velocity of slide wire be V. Then total resistance of the closed circuit at this instant will be

$$R = \lambda(2x + l)$$

Motional emf $e = Bvl$

Current $i = \frac{Bvl}{R} = \frac{Bvl}{\lambda(2x + l)}$

Magnetic force

$$F_m = ilB = \frac{B^2l^2v}{\lambda(2x + l)}$$

$$F_{net} = F - F_m$$

or $$mv \cdot \frac{dv}{dx} = F_0 v - \frac{B^2l^2v}{\lambda(2x + l)}$$

$$\therefore \quad m(dv) = \left[F_0 - \frac{B^2l^2}{\lambda(2x + l)}\right]dx$$

or $$m\int_{v_0}^v dv = \int_0^x \left[F_0 - \frac{B^2l^2}{\lambda(2x + l)}\right]dx$$

$$\therefore \quad v = v_0 + \frac{F_0x}{m} - \frac{B^2l^2}{2\lambda m}\ln\left(\frac{2x + l}{l}\right)$$

535. Let that the rod move with speed v and acceleration a along the inclined plane, while current I flows in it. The magnetic field brakes the rod in accordance with Lenz's law and its equation of motion is

$$ma = mg\sin\alpha - BIl.$$

This equation is the same in all three cases. The differences result from the different relationships between the induced voltage and the current flowing in the rod.

(i) If the circuit is closed by an ohmic resistor R, the current I and the induced voltage $V = Blv$ are connected by the relationship $I = \frac{V}{R} = \frac{Blv}{R}$. This shows that the braking force increases in proportion to the speed, with the result that the rod experiences decreasing acceleration and ultimately travels with uniform speed. This final maximum speed v_{max} can be found from the equation of motion by setting $a = 0$,

$$v_{max} = \frac{mgR\sin\alpha}{B^2l^2}$$

(ii) If the circuit is closed by a capacitor of capacitance C, the relationship between the induced voltage and the current is different. The charge on the capacitor is determined by the induced voltage and given by

$$Q = CV = CBlv.$$

Note that the current flowing through the rod is equal to the time derivative of this, *i.e.*,

$$I = \frac{dQ}{dt} = C\frac{dV}{dt} = CBl\frac{dv}{dt} = CBla.$$

In other words, the current flowing in the rod is directly proportional to the acceleration of the rod. If the above expression for the current is substituted into the equation of motion, the rod is found to move on the rails with uniform acceleration

$$a = \frac{mg\sin\alpha}{m + B^2l^2C}$$

Induction decreases the acceleration caused by gravity by, in effect, increasing the inertial mass of the rod. The speed of the rod and the charge on the capacitor are both directly proportional to the time elapsed.

(iii) If the circuit is closed by a coil of inductance L, the relationship between the induced voltage and the current is

$$L\frac{dI}{dt} = Blv = Bl\frac{dx}{dt}$$

We note that, since $I = 0$ and $x = 0$ at the start of the motion, the above formula implies that the current is proportional to the x-coordinate, $LI = Blx$. Substituting for the current, from this relationship into the equation of motion, gives

$$ma = mg \sin \alpha - \frac{B^2l^2}{L}x.$$

The force acting on the rod is therefore the sum of a constant term and a negative term proportional to the displacement. This is the same as the equation of motion of a body hung on a spring and then released. Thus, the rod makes harmonic oscillations about an equilibrium position

$$x_0 = \frac{mg\,L \sin \alpha}{B^2l^2}$$

The amplitude of the oscillation is $A = x_0$, and the dependence of the displacement of the rod on time is

$$x(t) = A(1 - \cos \omega t),$$

where $$\omega^2 = \frac{B^2l^2}{mL}$$

536. (i) At the instant when the capacitor is connected, a current $I = \dfrac{V_0}{R}$ starts flowing in the rod, which experiences a force $F = BIl$ and an initial acceleration $a = \dfrac{BlV_0}{mR}$. In accordance with Lenz's law, the voltage induced in the moving rod causes the current flowing in the rod to decrease. The charge Q on the capacitor decreases and consequently so does the voltage across it. Meanwhile the voltage induced in the rod increases, until the two voltages cancel each other out. The rod then continues with its maximum velocity given by

$$Blv_{max} = \frac{Q_{min}}{C} \qquad \ldots(1)$$

The equation of motion of the rod is

$$m\frac{dv}{dt} = ma = BIl = -Bl\frac{dQ}{dt} \qquad \ldots(2)$$

where the acceleration and the current have been expressed as the rates of change in velocity and charge, respectively. The proportionality between the two rates of change holds throughout. The speed of the rod increases from zero to v_{max}, whilst the charge on the capacitor decreases from $Q_0 = CV_0$ to Q_{min}. Equation (2) can therefore be rewritten as

$$mv_{max} = Bl(Q_0 - Q_{min})$$

The final velocity and the residual charge on the capacitor can be calculated using Eqs. (1) and (2),

$$v_{max} = \frac{BlCV_0}{m + B^2l^2C}$$

and $$Q_{min} = \frac{B^2l^2C^2V_0}{m + B^2l^2C}$$

(ii) The above relations show that the maximum velocity of the rod is proportional to the initial voltage V_0 across the capacitor. Thus, the final kinetic energy of the rod is proportional to V_0^2 (for given values of C and m), *i.e.*, proportional to the initial energy of the system. The coefficient of proportionality can be regarded as the efficiency η of the apparatus (considered as an electromagnetic gun), and can be written in the form

$$\eta = \frac{\frac{1}{2}mv_{max}^2}{\frac{1}{2}CV_0^2} = \frac{1}{\left(\dfrac{\sqrt{m}}{Bl\sqrt{C}} + \dfrac{Bl\sqrt{C}}{\sqrt{m}}\right)^2}$$

The product of the two terms in the brackets is 1, and from the inequality between arithmetic and geometric means, it follows that their sum is at least 2. This means that the efficiency of the electromagnetic gun cannot be more than 25 per cent.

Note: If the condition for maximum efficiency $m = CB^2l^2$ is substituted into the expression for the final charge on the capacitor we find that $Q_{min} = \frac{V_0C}{2}$, *i.e.*, only half of the initial charge on the capacitor is left. Thus, only one-quarter of the initial energy of the capacitor is left; one-quarter of it is transformed into the kinetic energy of the rod, and the other half is dissipated in the rod as Joule heat.

537. The current at time t is $I = kt$ in the outer coil, and $2I = 2kt$ in the inner one, where k is a constant. Because of these currents the magnetic field in the outer coil is $B = \mu_0 nkt$, whilst in the inner one it is $3B$, where n is the number of turns per unit length. The magnetic flux enclosed by the particle's trajectory of radius r is

$$\phi = \pi R^2 \times 2B + \pi r^2 \times B$$
$$= (2R^2 + r^2)\, \pi\mu_0 nkt$$

The (constant) magnitude of the induced electric field E can be calculated from the rate of change of magnetic flux with time:

$$E \times 2\pi r = \frac{d\phi}{dt}$$
$$= (2R^2 + r^2)\, \pi\mu_0 nk,$$

and so
$$E = \frac{(2R^2 + r^2)}{r} \frac{\mu_0 nk}{2}$$

The charged particle is held in its circular orbit by the magnetic field, and so, from the zero net radial component of the force acting on it, we obtain

$$\frac{mv^2}{r} = qvB \qquad \ldots(1)$$

The particle is accelerated along its circular orbit by the tangential component of the net force according to $ma_t = qE$, where m is the mass and q the electric charge of the particle. As the magnitude of the electric field is constant, the speed of the particle increases uniformly with time,

$$v = a_t t = \frac{qE}{m} t$$
$$= \frac{(2R^2 + r^2)}{r} \frac{\mu_0 nk}{2} \frac{q}{m} t$$

Inserting this and the value of B into equation (1), we get

$$\frac{m}{r} \frac{(2R^2 + r^2)}{r} \frac{\mu_0 nk}{2} \frac{q}{m} t = q\mu_0 nkt,$$

which is satisfied if

$$\frac{(2R^2 + r^2)}{2r^2} = 1, \qquad i.e., r = \sqrt{2}R$$

538. The total magnetic flux at the position of the ring is made up of that due to the external magnetic field and the effects of self-induction,

$$\phi = B_z \pi r_0^2 + LI$$

Any change in magnetic flux induces a current in the ring in accordance with

$$RI = \frac{\Delta\phi}{\Delta t}$$

However, this has to be zero since the ohmic resistance of a superconducting ring is zero. Accordingly, the magnetic flux through the ring has to be constant *i.e.*,

$$\phi = B_0(1 - \alpha z)\, \pi r_0^2 + LI = \text{constant}$$

From the initial conditions ($z = 0, I = 0$), the value of the constant is $\phi = B_0 \pi r_0^2$.

The current in the ring can be determined using the above equations which give

$$I = \frac{1}{L} B_0 \alpha \pi r_0^2 z$$

The Lorentz force acting on the ring (which can only be vertical, because of the symmetry of the assembly) can be expressed as

$$F_z = -B_r I(z)\, 2\pi r_0$$
$$= -B_0 \beta r_0 \frac{1}{L} B_0 \alpha \pi r_0^2\, 2\pi r_0 z = -kz$$

The Lorentz force is thus directly proportional to the vertical displacement of the ring, with the coefficient of proportionality calculable from the given data. (This result is only valid for small displacements, since the magnetic induction is not adequately described by the given formulae for large ones.)

The equation of motion of the ring is

$$ma_z = F_z - mg = -kz - mg$$

This means that the ring makes harmonic oscillations about the equilibrium position $z_0 = -\frac{mg}{k}$ with

$$z(t) - z_0 = A \cos \omega t,$$

where $\omega = \sqrt{\frac{k}{m}}\cdot$ From the initial conditions it follows that $A = -z_0$, and so

$$z(t) = \frac{g}{\omega^2}(\cos \omega t - 1).$$

The vertical z-coordinate is never positive, and it follows that the Lorentz force always points upwards, being zero at the topmost point of the oscillation. The current always flows in the same direction around the ring. Substituting the numerical data gives $\omega = 31.2\ s^{-1}$ and $A = 1$ cm. The time dependence of the current flowing in the ring can be expressed in terms of $z(t)$ as

$$I = \frac{1}{L} B_0 \alpha \pi r_0^2 z(t)$$

$$= \frac{1}{L} B_0 \alpha \pi r_0^2 A(\cos \omega t - 1).$$

The maximum value of the current, which flows at the bottom of the oscillation, is $I_{max} = 39$ A.

539. After a long time, resistance across an inductor becomes zero while resistance across capacitor becomes infinite. Hence net external resistance,

$$R_{net} = \frac{\frac{R}{2} + R}{2} = \frac{3R}{4}$$

current through the batteries,

$$i = \frac{2E}{\frac{3R}{4} + r_1 + r_2}$$

Given that potential across the terminals of cell A is zero.

$$\therefore \qquad E - ir_1 = 0$$

or $$E - \left(\frac{2E}{3R/4 + r_1 + r_2}\right) r_1 = 0$$

Solving this equation we get,

$$R = \frac{4}{3}(r_1 - r_2)$$

540. Inductive reactance

$$X_L = \omega L = (50)(2\pi)(35 \times 10^{-3}) \approx 11\ \Omega$$

Impedence $Z = \sqrt{R^2 + X_L^2}$

$$= \sqrt{(11)^2 + (11)^2} = 11\sqrt{2}\ \Omega$$

Given $V_{rms} = 220$ volt

Hence amplitude of voltage

$$V_0 = \sqrt{2} V_{rms} = 220\sqrt{2} \text{ volt}$$

$\therefore$ Amplitude of current $i_0 = \frac{V_0}{Z} = \frac{220\sqrt{2}}{11\sqrt{2}}$

or $$i_0 = 20 \text{ A}$$

Phase difference

$$\phi = \tan^{-1}\left(\frac{X_L}{R}\right) = \tan^{-1}\left(\frac{11}{11}\right) = \frac{\pi}{4}$$

In L-R circuit voltage leads the current. Hence instantaneous current in the circuit is,

$$i = (20 \text{ A}) \sin\left(\omega t - \frac{\pi}{4}\right)$$

Corresponding i-t graph is shown in figure.

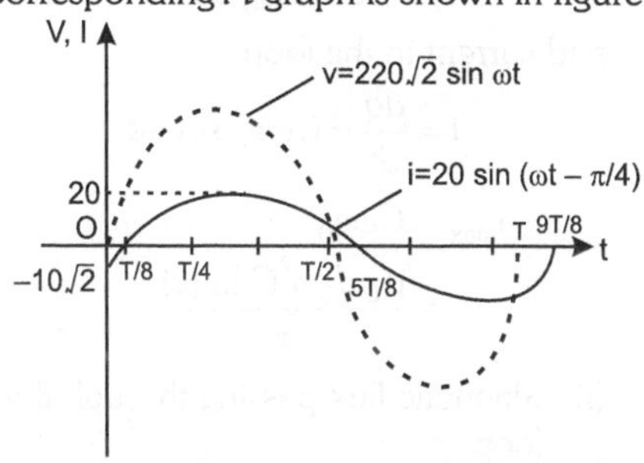

541. (i) For an elemental strip of thickness dx at a distance x from left wire, net magnetic field (due to both wires)

$$B = \frac{\mu_0}{2\pi}\frac{I}{x} + \frac{\mu_0}{2\pi}\frac{I}{3a - x} \quad \text{(outwards)}$$

$$= \frac{\mu_0 I}{2\pi}\left(\frac{1}{x} + \frac{1}{3a - x}\right)$$

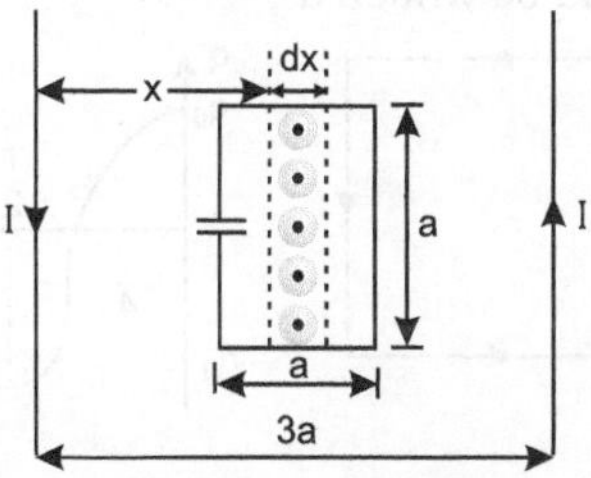

Magnetic flux in this strip,

$$d\phi = BdS = \frac{\mu_0 I}{2\pi}\left(\frac{1}{x} + \frac{1}{3a - x}\right) a\, dx$$

$\therefore$ total flux $\phi = \int_a^{2a} d\phi$

$$= \frac{\mu_0 I a}{2\pi} \int_a^{2a} \left(\frac{1}{x} + \frac{1}{3a - x}\right) dx$$

or $$\phi = \frac{\mu_0 I a}{\pi} \ln(2)$$

$$\phi = \frac{\mu_0 a \ln(2)}{\pi}(I_0 \sin \omega t) \quad \ldots(1)$$

Magnitude of induced emf,

$$e = \left| -\frac{d\phi}{dt} \right| = \frac{\mu_0 a I_0 \omega \ln(2)}{\pi} \cos \omega t$$

$$= e_0 \cos \omega t$$

where $$e_0 = \frac{\mu_0 a I_0 \omega \ln(2)}{\pi}$$

Charge stored in the capacitor,

$$q = Ce = Ce_0 \cos \omega t \quad \ldots(2)$$

and current in the loop

$$i = \frac{dq}{dt} = C\omega e_0 \sin \omega t \quad \ldots(3)$$

$$i_{max} = C\omega e_0$$

$$= \frac{\mu_0 a I_0 \omega^2 C \ln(2)}{\pi}$$

(i) Magnetic flux passing through the square loop

$$\phi \propto \sin \omega t \quad \text{[From Eq. (1)]}$$

i.e., ⊙ magnetic field passing through the loop is increasing at $t = 0$. Hence, the induced current will produce ⊗ magnetic field (from Lenz's law). Or the current in the circuit at $t = 0$ will be clockwise (or negative as per the given convention). Therefore, charge on upper plate could be written as,

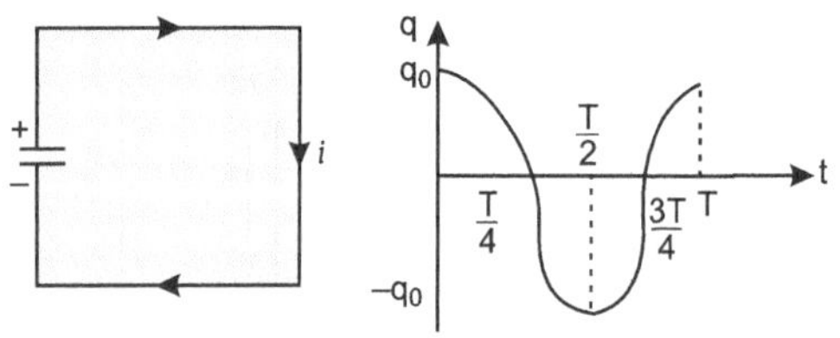

$$q = + q_0 \cos \omega t \quad \text{[from Eq. (2)]}$$

Here $q_0 = Ce_0$

$$= \frac{\mu_0 a C I_0 \omega \ln(2)}{\pi}$$

The corresponding *q-t* graph is shown in figures.

542. (a) Applying Kirchhoff's second law :

$$\frac{d\phi}{dt} - iR - L\frac{di}{dt} = 0$$

or $$\frac{d\phi}{dt} = iR + L\frac{di}{dt} \quad \ldots(1)$$

This is the desired relation between i, $\frac{di}{dt}$ and $\frac{d\phi}{dt}$.

(b) Equation (1) can be written as

$$d\phi = iRdt + Ldi$$

Integrating we get,

$$\Delta\phi = R\,\Delta q + Li_1$$

or $$\Delta q = \frac{\Delta\phi}{R} - \frac{Li_1}{R} \quad \ldots(2)$$

Here $$\Delta\phi_1 = \phi_f - \phi_i$$

$$= \int_{x = 2x_0}^{x = x_0} \frac{\mu_0}{2\pi}\frac{I_0}{x} l\, dx$$

$$= \frac{\mu_0 I_0 l}{2\pi} \ln(2)$$

So, from Eq. (2) charge flown through the resistance upto time $t = T$, when current is i_1, is

$$\boldsymbol{\Delta q = \frac{1}{R}\left[\frac{\mu_0 I_0 l}{2\pi} \ln(2) - Li_1\right]}$$

(c) This is the case of current decay in an *L-R* circuit. Thus

$$i = i_0 e^{-t/\tau_L} \quad \ldots(3)$$

Here $$i = \frac{i_1}{4}, \quad i_0 = i_1,$$

$$t = (2T - T) = T$$

and $$\tau_L = \frac{L}{R}$$

Substituting these values in Eq. (3) we get :

$$\boldsymbol{\tau_L = \frac{L}{R} = \frac{T}{\ln 4}}$$

543. (i) Given $R_1 = R_2 = 2\Omega$,

$E = 12V$

and $L = 400 \text{ mH} = 0.4 \text{ H}$

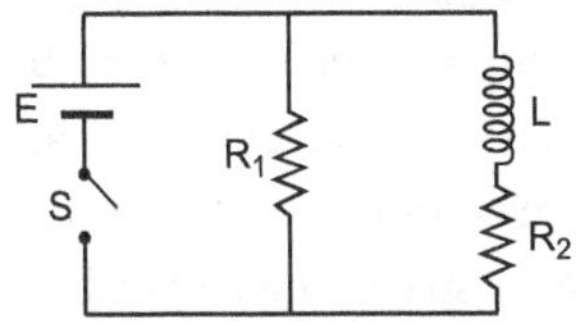

Two parts of the circuit are in parallel with the applied battery. So, the upper circuit can be broken as :

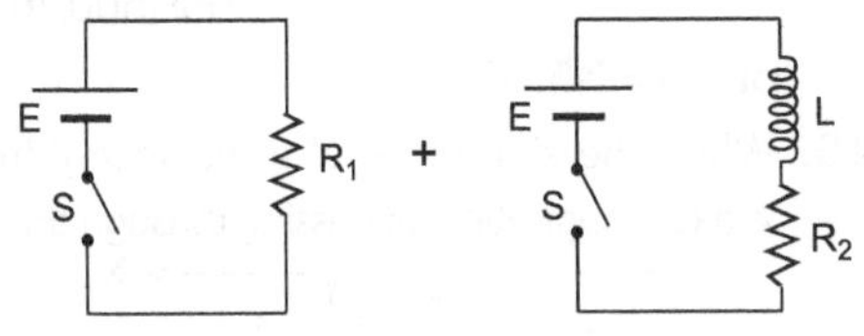

(a) (b)

Figure (b) is a simple *L-R* circuit, whose time constant is

$$\tau_L = \frac{L}{R_2} = \frac{0.4}{2} = 0.2 \text{ s}$$

and steady state current

$$i_0 = \frac{E}{R_2} = \frac{12}{2} = 6 \text{ A}$$

Therefore, if switch *S* is closed at time $t = 0$, then current in the circuit at any time *t* will be given by

$$i(t) = i_0 (1 - e^{-t/\tau_L})$$

or $i(t) = 6(1 - e^{-t/0.2}) = 6(1 - e^{-5t}) = i$ (say)

Therefore, potential drop across *L* at any time *t* is

$$V = \left| L \frac{di}{dt} \right|$$

$$= L(30 e^{-5t}) = (0.4)(30) e^{-5t}$$

or $\boldsymbol{V = 12 \, e^{-5t}}$ **volt**

(ii) The steady state current in L or R_2 is $i = 6 \text{ A}$

Now as soon as the switch is opened, current in R_1 is reduced to zero immediately. But in L and R_2 it decreases exponentially. The situation is as follows :

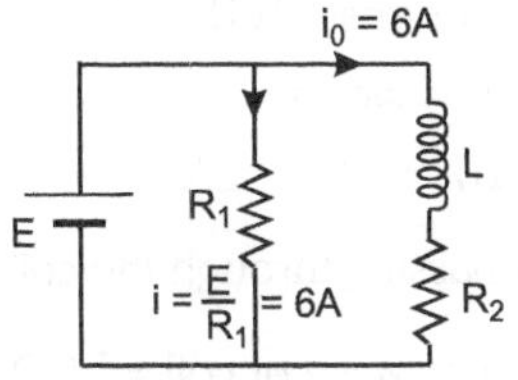

steady states condition

(c)

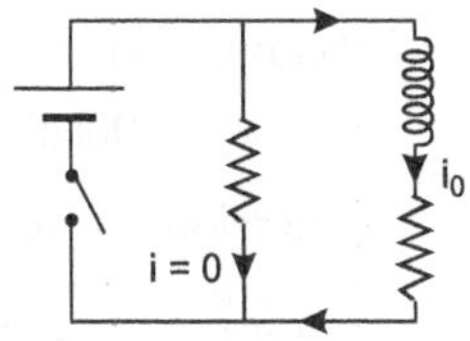

t = 0
S is open
(d)

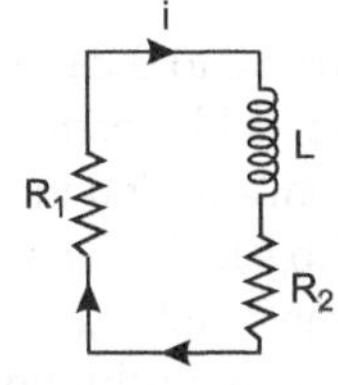

t = t
(e)

Refer Figure (e) : Time constant of this circuit would be

$$\tau_L' = \frac{L}{R_1 + R_2} = \frac{0.4}{(2+2)} = 0.1 \text{ s}$$

$\therefore$ Current through R_1 at any time *t* is

$$i = i_0 e^{-t/\tau_L'} = 6e^{-t/0.1}$$

or $\boldsymbol{i = 6e^{-10t}}$ **A**

Direction of current in R_1 is as shown in figure or clockwise.

544. Magnetic field (*B*) varies with time (*t*) as shown in figure

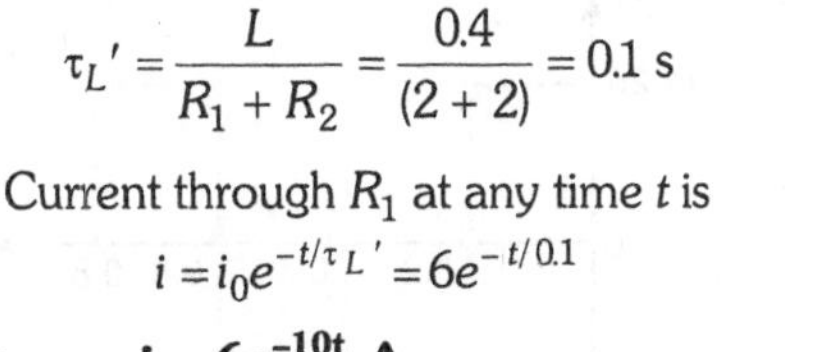

$$\left|\frac{dB}{dt}\right| = \frac{0.8}{0.2} = 4 \frac{T}{s}$$

Induced emf in the coil due to change in magnetic flux passing through it,

$$e = \left|\frac{d\phi}{dt}\right| = NA \left|\frac{dB}{dt}\right|$$

Here A = Area of coil $= 5 \times 10^{-3} \text{ m}^2$;

N = Number of turns = 100

Substituting the values we get

$$e = (100)(5 \times 10^{-3})(4) \text{ V} = 2 \text{ V}$$

Therefore, current passing through the coil

$$i = \frac{e}{R} \qquad (R = \text{resistance of coil} = 1.6\,\Omega)$$

or $$i = \frac{2}{1.6} = 1.25 \text{ A}$$

Note that from 0 to 0.2 s and from 0.4 s to 0.6 s, magnetic field passing through the coil increases, while during the time 0.2 s to 0.4 s and from 0.6 s to 0.8 s magnetic field passing through the coil decreases. Therefore, direction of current through the coil in these two time intervals will be opposite to each other. The variation of current (i) with time (t) will be as shown in the figure :

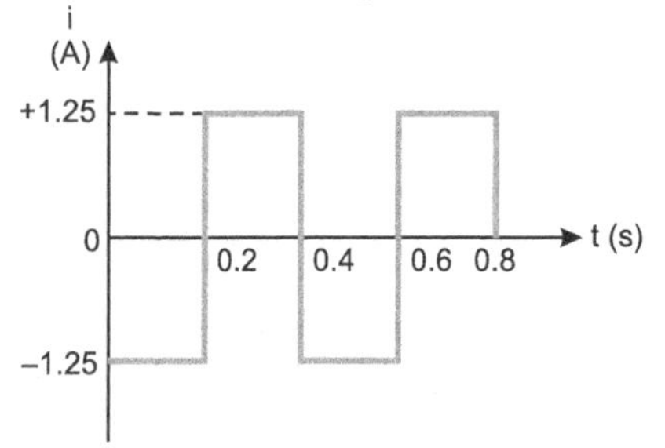

Power dissipated in the coil is

$$P = i^2R = (1.25)^2 (1.6) \text{ W} = 2.5 \text{ W}$$

Power is independent of the direction of current through the coil. Therefore, power (P) versus time (t) graph for first two cycles will be as follows

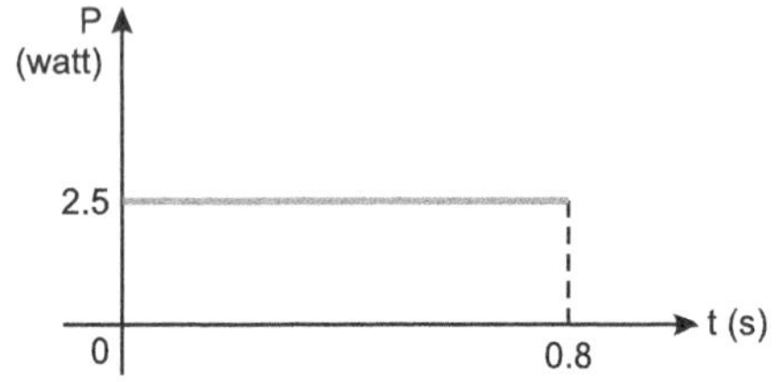

Total heat obtained in 12,000 cycles will be

$$H = P \cdot t$$
$$= (2.5)(12000)(0.4)$$
$$= 12000 \text{ J}$$

This heat is used in raising the temperature of the coil and the water. Let θ be the final temperature. Then

$$H = m_w S_w (\theta - 30°) + m_c S_c (\theta - 30°)$$

Here m_w = mass of water = 0.5 kg

S_w = Specific heat of water = 4200 J/kg - K

m_c = mass of coil = 0.06 kg

and S_c = specific heat of coil = 500 J/kg-K

Substituting the values we get

$$12000 = (0.5)(4200)(\theta - 30°) + (0.06)(500)(\theta - 30°)$$

or θ = **35.6°C**

545. When the side EF is at a distance y from the x-axis, magnetic flux passing through the loop is

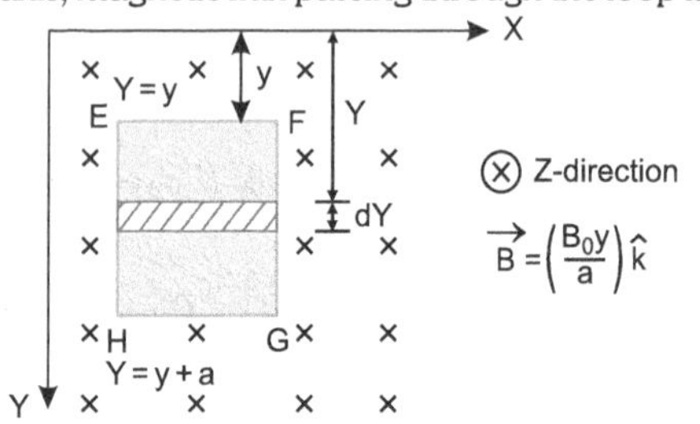

$$\phi = \int d\phi = \int_{Y=y}^{Y=y+a} \frac{B_0 Y}{a} (a dY)$$

$$\phi = \frac{B_0}{2}[(y+a)^2 - y^2]$$

(a) Induced emf is

$$e = \left|\frac{-d\phi}{dt}\right| = \left|-\frac{B_0}{2}[2(y+a) - 2y]\frac{dy}{dt}\right|$$

or $e = B_0 a \dfrac{dy}{dt}$

$$e = B_0 av$$

where $v = \dfrac{dy}{dt}$ = speed of loop

$\therefore$ **Induced current** $\boldsymbol{i = \dfrac{e}{R} = \dfrac{B_0 av}{R}}$

Direction : $|\vec{B}| \propto y$ i.e., as the loop comes down ⊗ magnetic field passing through the loop increases, therefore, the induced current will produce magnetic field or the induced current in the loop will be **counter-clockwise**.

Alternate Solution : (of part a)

Motional emf in EH and $FG = 0$ as $\vec{v} \parallel \vec{l}$

Motional emf in EF is

$$e_1 = \left(\frac{B_0 y}{a}\right)(a) v = B_0 y v \qquad (e = Blv)$$

Similarly motional emf in GH will be

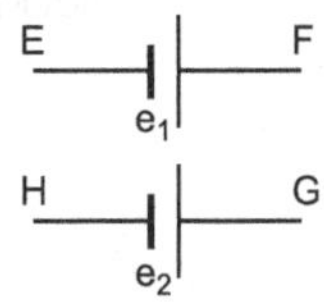

$$e_2 = \left\{\frac{B_0 (y + a)}{a}\right\}(a)(v)$$

$$= B_0 (a + y) v$$

Polarities of e_1 and e_2 are shown in adjoining figures.

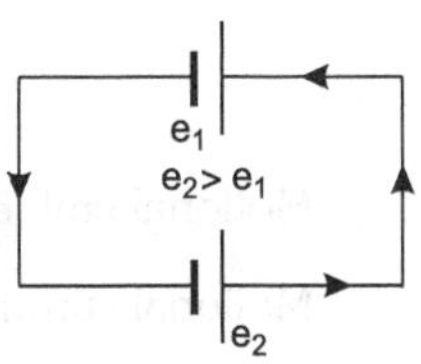

Net emf, $e = e_2 - e_1$

$$e = B_0\, a v$$

$$\therefore \qquad \boldsymbol{i = \frac{e}{R} = \frac{B_0 a v}{R}}$$

and direction of current will be counter-clockwise.

(b) Total Lorentz force on the loop : We have seen in part (a) that induced current passing through the loop (when its speed is v) is

$$i = \frac{B_0 a v}{R}$$

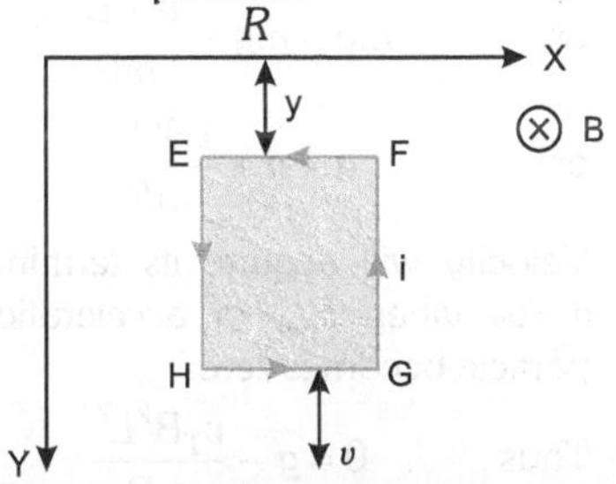

Now magnetic force on EH and FG are equal in magnitude and in opposite directions, hence they cancel each other and produce no force on the loop.

$$F_{EF} = \left(\frac{B_0 a v}{R}\right)(a)\left(\frac{B_0 y}{a}\right) \qquad (F = ilB)$$

$$= \frac{B_0^2 a v y}{R} \qquad \text{(downwards)}$$

and $$F_{GH} = \left(\frac{B_0 a v}{R}\right)(a)\left(\frac{B_0 (y + a)}{a}\right) \qquad \text{(upwards)}$$

$$= \frac{B_0^2 a v}{R}(y + a) \qquad F_{GH} > F_{EF}$$

$\therefore$ Net Lorentz force on the loop

$$= F_{GH} - F_{EF} = \frac{B_0^2 a^2 v}{R} \qquad \text{(upwards)}$$

or $$\boldsymbol{\vec{F} = -\frac{B_0^2 a^2 v}{R}\hat{j}}$$

(c) Net force on the loop will be

F = weight − Lorentz force (down wards)

or $$F = mg - \frac{B_0^2 a^2 v}{R}$$

or $$m\left(\frac{dv}{dt}\right) = mg - \left(\frac{B_0^2 a^2}{R}\right) v$$

$$\therefore \qquad \frac{dv}{dt} = g - \left(\frac{B_0^2 a^2}{mR}\right) v = g - Kv$$

where $$K = \frac{B_0^2 a^2}{mR} = \text{constant}$$

or $$\frac{dv}{g - Kv} = dt \quad \text{or} \quad \int_0^v \frac{dv}{g - Kv} = \int_0^t dt$$

This equation gives

$$v = \frac{g}{K}(1 - e^{-Kt})$$

Here $$\boldsymbol{K = \frac{B_0^2 a^2}{mR}}$$

i.e., speed of the loop is increasing exponentially with time t. Its terminal velocity will be

$$\boldsymbol{v_t = \frac{g}{K} = \frac{mgR}{B_0^2 a^2}} \qquad \text{at } t \to \infty$$

the v-t graph will be as shown.

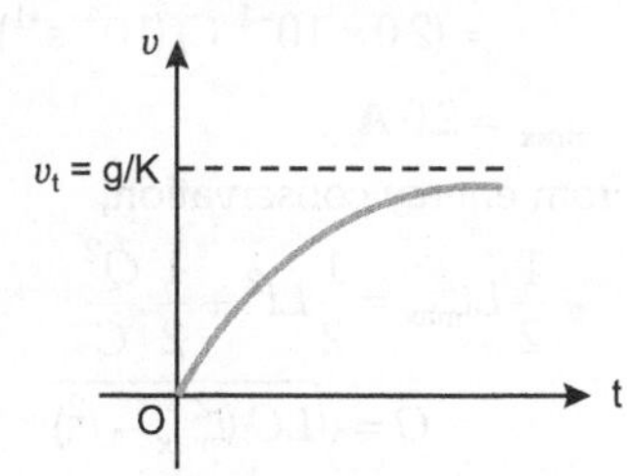

546. This is a problem of L-C oscillations.

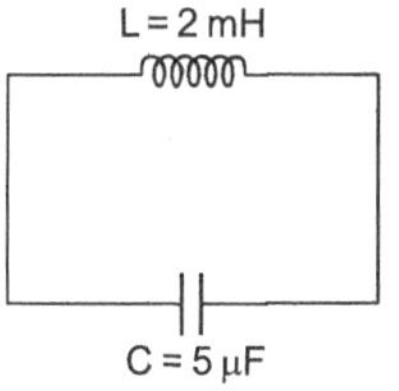

Charge stored in the capacitor oscillates simple harmonically as

$$Q = Q_0 \sin(\omega t \pm \phi)$$

Here Q_0 = maximum value of $Q = 200\,\mu C$
$= 2 \times 10^{-4} C$

$$\omega = \frac{1}{\sqrt{LC}} = \frac{1}{\sqrt{(2 \times 10^{-3}\text{ H})(5.0 \times 10^{-6}\text{ F})}}$$
$$= 10^4 \text{ s}^{-1}$$

Let at $t = 0$, $Q = Q_0$ then

$$Q(t) = Q_0 \cos \omega t \quad \dots(1)$$

$$i(t) = \frac{dQ}{dt} = -Q_0\omega \sin \omega t \quad \dots(2)$$

and $\frac{di(t)}{dt} = -Q_0\omega^2 \cos(\omega t) \quad \dots(3)$

(a) $Q = 100\,\mu C$ or $\frac{Q_0}{2}$

At $\cos(\omega t) = \frac{1}{2}$, from equation (3) :

$$\left|\frac{di}{dt}\right| = (2.0 \times 10^{-4} \text{ C})(10^4 \text{ s}^{-1})\left(\frac{1}{2}\right)$$

$$\left|\frac{di}{dt}\right| = \mathbf{10^4 \text{ A/s}}$$

(b) $Q = 200\,\mu C$ or Q_0 when $\cos(\omega t) = 1$ i.e., $\omega t = 0, 2\pi \dots$

At this time $i_t = -Q_0\,\omega \sin \omega t$

or $\qquad \boldsymbol{i(t) = 0} \qquad [\sin 0° = \sin 2\pi = 0]$

(c) $i(t) = -Q_0\,\omega \sin \omega t$

$\therefore$ Maximum value of i is $Q_0\omega$

or $\quad i_{max} = Q_0\omega$
$= (2.0 \times 10^{-4} \text{ C})(10^4 \text{ s}^{-1})$

$$\boldsymbol{i_{max}} = 2.0\,\mathbf{A}$$

(d) From energy conservation,

$$\frac{1}{2}Li_{max}^2 = \frac{1}{2}Li^2 + \frac{1}{2}\frac{Q^2}{C}$$

or $$Q = \sqrt{LC\,(i_{max}^2 - i^2)}$$

$$i = \frac{i_{max}}{2} = 1.0 \text{ A}$$

$$\therefore \quad Q = \sqrt{(2.0 \times 10^{-3})(5.0 \times 10^{-6})(2^2 - 1^2)}$$

$$Q = \sqrt{3} \times 10^{-4} \text{ C}$$

or $\quad \mathbf{Q = 1.732 \times 10^{-4} \text{ C}}$

547. (i) Let v be the velocity of the rod (as well as block) at any instant of time t.

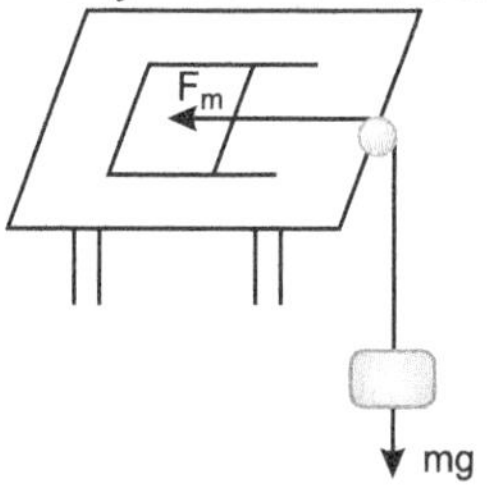

Motional emf, $e = BvL$

Motional current $= \frac{e}{R} = \frac{BvL}{R}$

and magnetic force on the rod

$$F_m = iLB = \frac{vB^2L^2}{R}$$

This magnetic force will be in the direction shown in figure.

Net force on the system at this moment will be

$$F_{net} = mg - F_m = mg - \frac{vB^2L^2}{R}$$

or $$ma = mg - \frac{vB^2L^2}{mR}$$

or $$a = g - \frac{vB^2L^2}{mR} \quad \dots(1)$$

Velocity will acquire its terminal value *i.e.*, $v = v_T$ when F_{net} or acceleration (a) of the particle becomes zero.

Thus $$0 = g - \frac{v_T B^2L^2}{mR}$$

or $$\boldsymbol{v_T = \frac{mgR}{B^2L^2}}$$

(ii) When $v = \frac{v_T}{2} = \frac{mgR}{2B^2L^2}$

then from equation (1), acceleration of the block.

$$a = g - \left(\frac{mgR}{2B^2L^2}\right)\left(\frac{B^2L^2}{mR}\right) = g - \frac{g}{2}$$

or $\quad \boldsymbol{a = \frac{g}{2}}$

548. (a) Consider a small element of length dx of the rod OA situated at a distance x from O.

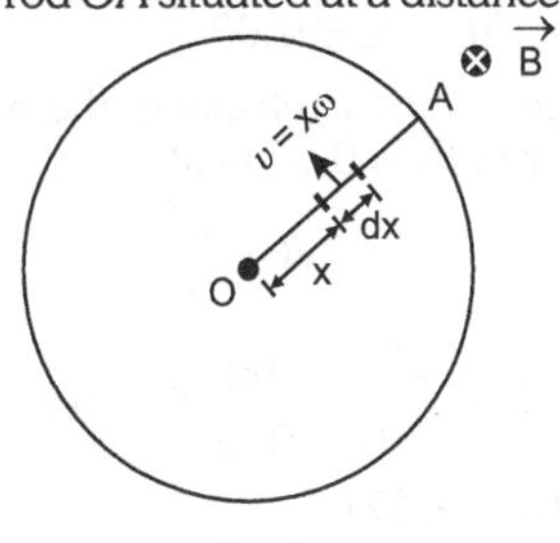

Fig. 1

Speed of this element,

$$\upsilon = x\omega$$

Therefore, induced emf developed across this element in uniform magnetic fied B

$$de = (B)(x\omega)\,dx \qquad (e = B\upsilon l)$$

Hence total induced emf across OA,

$$e = \int_{x=0}^{x=r} de = \int_0^r B\omega x\,dx = \frac{B\omega r^2}{2}$$

$$\therefore \qquad \boldsymbol{e = \frac{B\omega r^2}{2}}$$

(b) (i) At constant emf or P. D. $e = \dfrac{B\omega r^2}{2}$ is induced across O and A.

The equivalent circuit can be drawn as shown in the figure :

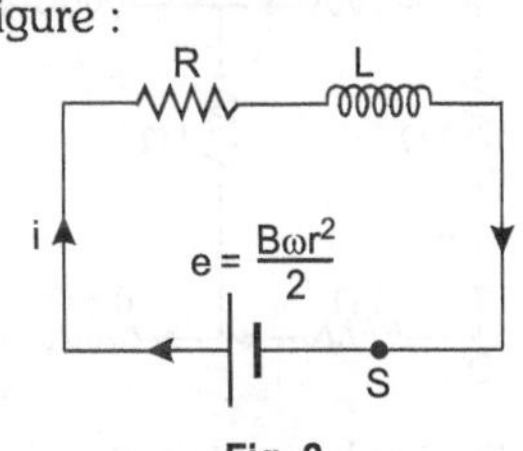

Fig. 2

Switch S is closed at time $t = 0$. Therefore, it is a case of growth of current in an L-R circuit. Current at any time t is given by $i = i_0\,(1 - e^{-t/\tau_L})$

Here $$i_0 = \frac{e}{R} = \frac{B\omega r^2}{2R}$$

$$\tau_L = \frac{L}{R}$$

$$\therefore \qquad \boldsymbol{i = \frac{B\omega r^2}{2R}\left[1 - e^{-\left(\frac{R}{L}\right)t}\right]}$$

The $i - t$ graph will be as follows :

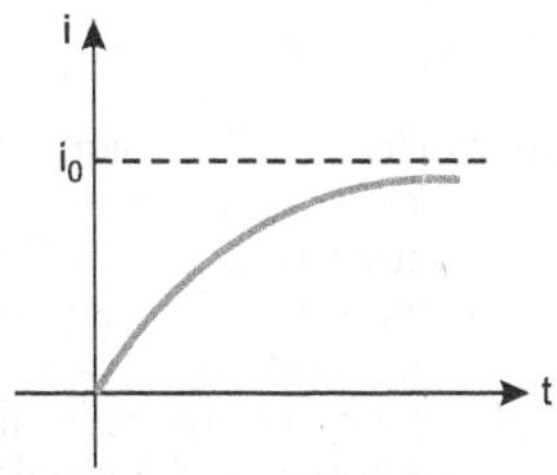

(ii) At constant angular speed, net torque = 0

The steady state current will be

$$i = i_0 = \frac{B\omega r^2}{2R}$$

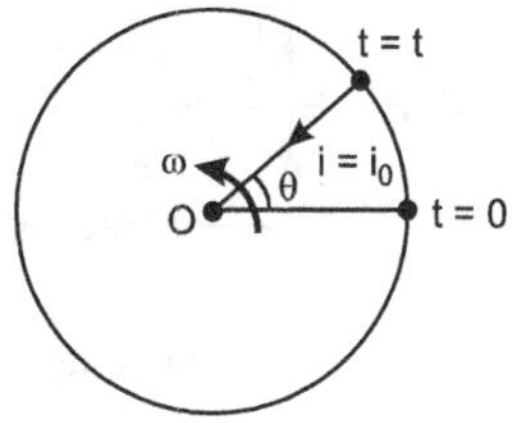

θ = ωt

Fig. 3

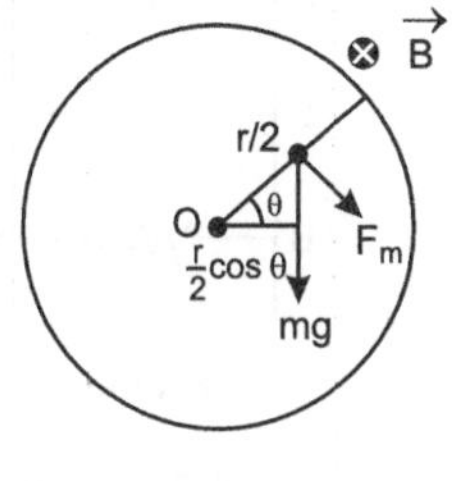

Fig. 4

From right hand rule we can see that this current would be inwards (from circumference to centre) and corresponding magnetic force (Fm) will be in the direction shown in figure and its magnitude is given by:

$$F_m = (i)(r)(B) = \frac{B^2\omega r^3}{2R} \qquad [F_m = ilB]$$

Torque of this force about centre O is

$$\tau F_m = F_m \cdot \frac{r}{2} = \frac{B^2\omega r^4}{4R} \qquad \text{(clockwise)}$$

Similarly, torque of weight (mg) about centre O is

$$\tau_{mg} = (mg)\frac{r}{2}\cos\theta = \frac{mgr}{2}\cos\omega t \quad \text{(clockwise)}$$

Therefore, net torque at any time t (after steady state condition is achieved) about centre O will be

$$\tau_{net} = \tau F_m + \tau_{mg}$$

$$= \frac{B^2\omega r^4}{4R} + \frac{mgr}{2}\cos\omega t \qquad \text{(clockwise)}$$

Hence the external torque applied to maintain a constant angular speed is

$$\boldsymbol{\tau_{ext} = \frac{B^2\omega r^4}{4R} + \frac{mgr}{2}\cos\omega t}$$

(but in anticlockwise direction)

Note that for $\frac{\pi}{2} < \theta < \frac{3\pi}{2}$, torque of weight will be anticlockwise the sign of which is automatically adjusted because $\cos\theta$ = negative for $\pi/2 < \theta < 3\pi/2$.

549. Let the magnetic field be perpendicular to the plane of rails and inwards ⊗. If v be the terminal velocity of the rails, then potential differences across E and F would be BvL with E at lower potential and F at higher potential. The equivalent circuit is shown in figure (2).

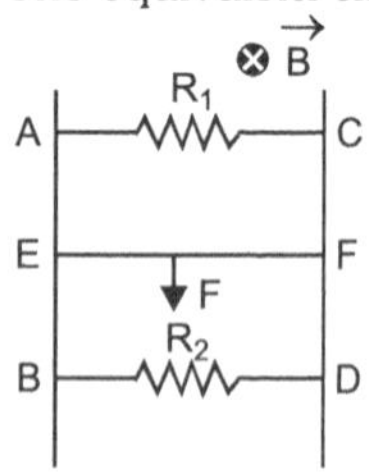

Figure-1

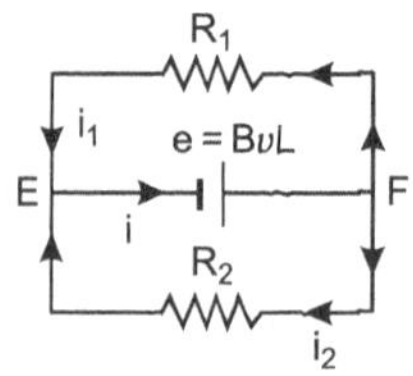

Figure-2

$$i_1 = \frac{e}{R_1} \quad \ldots(1) \qquad i_2 = \frac{e}{R_2} \quad \ldots(2)$$

Power dissipated in R_1 is 0.76 watt

Therefore, $ei_1 = 0.76$ watt ...(3)

Similarly $ei_2 = 1.2$ watt ...(4)

Now the total current in bar EF is

$$i = i_1 + i_2 \quad \text{(from } E \text{ to } F)\ldots5)$$

Under equilibrium condition,

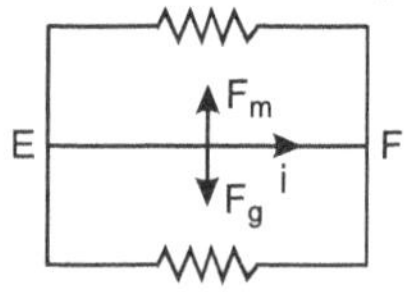

magnetic force (F_m) on bar EF = weight (F_g) of bar EF

i.e., $F_m = F_g$

or $iLB = mg$ ...(6)

From equation (6)

$$i = \frac{mg}{LB} = \frac{(0.2)(9.8)}{(1.0)(0.6)} \text{ A} \quad \text{or} \quad i = 3.27 \text{ A}$$

Multiplying equation (5) by e, we get

$$ei = ei_1 + ei_2 = (0.76 + 1.2) \text{ watt}$$

(From equations 3 and 4)

$$= 1.96 \text{ watt}$$

$$e = \frac{1.96}{i} \text{ volt}$$

$$e = \frac{1.96}{3.27} \text{ V} \quad \text{or} \quad e = 0.6 \text{ V}$$

But since $e = BvL$

$$\therefore \; v = \frac{e}{BL} = \frac{(0.6)}{(0.6)(1.00} \text{ m/s} = \mathbf{1.0 \text{ m/s}}$$

Hence terminal velocity of the bar is 1.0 m/s

Power in R_I is 0.76 watt

$$\therefore \qquad 0.76 = \frac{e^2}{R_1}$$

$$\therefore \quad R_1 = \frac{e^2}{0.76} = \frac{(0.6)^2}{0.76} \Omega = 0.47\,\Omega$$

$$\mathbf{R_1 = 0.47\Omega}$$

$$\text{Similarly} \quad R_2 = \frac{e^2}{1.2} = \frac{(0.6)^2}{1.2} \Omega = 0.3\,\Omega$$

$$\mathbf{R_2 = 0.3\ \Omega}$$

550. Induced emfs in two loops $AEFD$ and $EBCF$ would be

$$e_1 = \left|\frac{d\phi_1}{dt}\right| = S_1\left(\frac{dB}{dt}\right) = (1 \times 1)\,(1)V = 1V$$

$$\text{Similarly} \quad e_2 = \left|\frac{d\phi_2}{dt}\right| = S_2\left(\frac{dB}{dt}\right)$$

$$= (0.5 \times 1)\,(1)V = 0.5V$$

Now since the magnetic field is increasing, the induced current will produce the magnetic field in (•) direction. Hence e_1 and e_2 will be applied as shown in the figure :

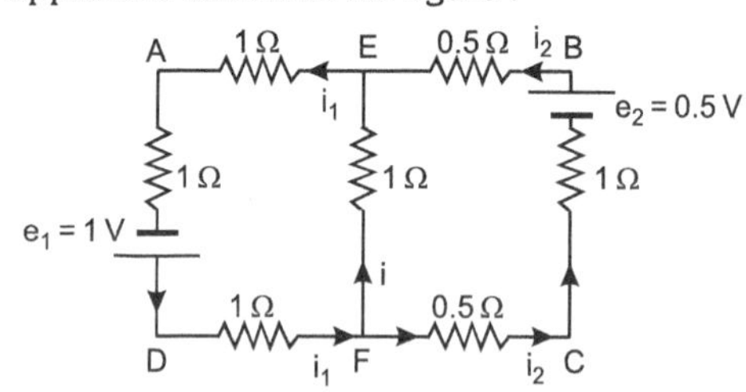

Kirchhoff's first law at junction F gives

$$i_1 = i + i_2 \qquad \ldots(1)$$

Kirchhoff's second law in loop $FEADF$ gives

$$3i_1 + i = 1 \qquad \ldots(2)$$

Kirchhoff's second law in loop $FEBCF$ gives

$$2i_2 - i = 0.5 \qquad \ldots(3)$$

Solving (1), (2) and (3), we get

$$i_1 = (7/22)A$$

and $i_2 = (6/22)A$ and $i = (1/22)A$

Therefore, current in segment AE is (7/22)A from E to A, current in segment BE is 6/22 A from B to E and current in segment EF is (1/22) A from F to E.

CHAPTER 16

Modern Physics

551. $E \propto T^4$

Since E has increased 81 times, the temperature has become 3 times. From Wein's displacement law $\lambda_m T$ = constant new wavelength corresponding to maximum intensity will become 3000 Å. Energy of photons corresponding to 3000 Å is,

$$E_0 = \frac{12375}{3000} \text{ eV} = 4.125 \text{ eV}$$

Maximum kinetic energy of photoelectrons,

$$K_{\text{max}} = E_3 - E_2$$

$$= \frac{(-13.6)}{(3)^2} - \frac{(-13.6)}{(2)^2} = 1.9 \text{ eV}$$

$$\therefore \quad W = E_0 - K$$

$$= (4.125 - 1.9) \text{ eV} = 2.225 \text{ eV}$$

552. (a) $$\frac{mv^2}{r} = Bev \quad \ldots(1)$$

$$mvr = \frac{nh}{2\pi} \quad \ldots(2)$$

Solving equations (1) and (2) we have

$$r = \sqrt{\frac{nh}{2\pi Be}} \quad \text{and} \quad v = \sqrt{\frac{Benh}{2\pi m^2}}$$

Thus, $$r_n = \sqrt{\frac{nh}{2\pi Be}}$$

(b) Kinetic energy,

$$K = \frac{1}{2} mv^2$$

$$= \frac{1}{2} m\left(\frac{Benh}{2\pi m^2}\right) = \frac{Benh}{4\pi m}$$

Thus, $$K_n = \frac{Benh}{4\pi m}$$

(c) $M = iA = (e)(2\pi f)(\pi r^2)$

$$= (e)(2\pi)\left(\frac{v}{2\pi r}\right)(\pi r^2) = \pi rev$$

Substituting the values we have,

$$M = \frac{nhe}{2m}$$

Now potential energy in n^{th} orbit,

$$U_n = MB \sin 90° = \frac{Benh}{2m}$$

(d) $$E_n = K_n + U_n = \left(\frac{1 + 2\pi}{4\pi}\right)\frac{Benh}{m}$$

(e) $$\phi_n = B(\pi r^2) = \frac{nh}{2e}$$

553. First line of Balmer series means that electron jumps from $n = 3$ to $n = 2$. Hence, the excitation energy emitted by two atoms in the process is

$$E_A = -13.6\, Z_A^2 \left[\frac{1}{2^2} - \frac{1}{3^2}\right],$$

where -13.6 eV is the energy of ground state of hydrogen atom and

Z_A = nuclear charge = Number of protons

or $$E_A = -13.6 \times \frac{5}{36} \times Z_A^2 \text{ eV}$$

Similarly, $$E_B = -13.6 \times Z_B^2 \left[\frac{1}{2^2} - \frac{1}{3^2}\right]$$

$$= -13.6 \times \frac{5}{36} \times Z_B^2$$

It is given that

$$E_B - E_A = 5.667 \text{ eV}$$

or $$13.6 \times \frac{5}{36}[Z_A^2 - Z_B^2] = 5.667$$

or $$Z_A^2 - Z_B^2 = \frac{5.667 \times 36}{13.6 \times 5} = 3 \quad \ldots(\text{i})$$

($Z_A^2 - Z_B^2$ should be an integer)

Since, the number of neutrons and protons in each atom is same.

$\therefore$ mass of atoms are

$$m_A = 2Z_A m_P$$

and $$m_B = 2Z_B m_P$$

[where M_p = mass of proton
= mass of neurton]

Momentum imparted by A to the target is

$$= \text{change in momentum of } A$$

$$= m_A v - (-m_A v) = 2m_A v$$

Momentum imparted by B, will similarly be $2m_B v$. Since, B imparts twice the momentum compared to A

$$\therefore \quad \frac{2m_B v}{2m_A v} = 2$$

or $$m_B = 2m_A$$

or $$Z_B = 2Z_A \quad \text{...(ii)}$$

Since, $$Z_B > Z_A,$$

equation (i) can be revised as

$$Z_B^2 - Z_A^2 = 3 \quad \text{...(iii)}$$

Solving (ii) and (iii)

$$Z_A = 1, \quad Z_B = 2$$

$\therefore A$ is ${}_1^2$H (Deuterium)

and B is ${}_2^4$He (Helium)

[Remember, it is given that number of neutrons is same as number of protons.]

554. The photon manifests any increase in its energy by increase in its frequency because its speed is already c which cannot increase any further. This is a fact which students must know. At times this assumption may not be written clearly in the problem.

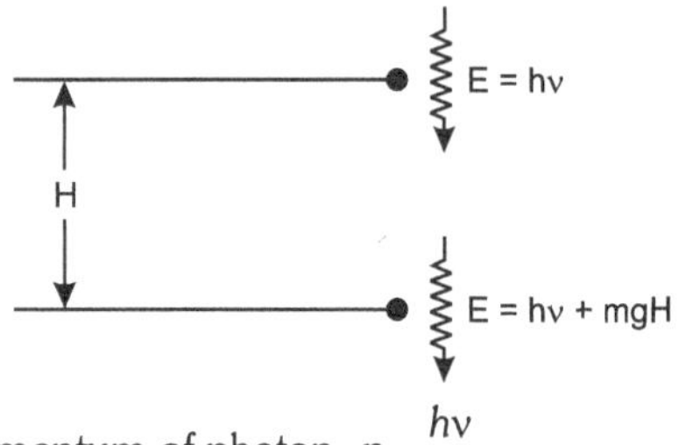

Momentum of photon, $p = \frac{h\nu}{c}$

Mass of photon, $m = \frac{p}{c} = \frac{h\nu}{c^2} \quad \text{...(1)}$

Though a photon has no rest mass, it acts as if it has a mass of $\frac{h\nu}{c^2}$. If the photon falls through a height H, its energy E shall increase by mgH. Since, the photon cannot have speed greater than c, the increase in energy is a consequence of increase in ν. Let the new frequency be ν'. Then

$$h\nu' = h\nu + mgH$$

or $$\nu' = \nu\left(1 + \frac{gH}{c^2}\right) \quad \text{[from (1)]}$$

555. The collision will be inelastic if a part of the kinetic energy is used to excite the atom. Let us assume that an energy ΔE is used in this way. Also, let the neutron and the hydrogen atom move at speeds v_1 and v_2 after collision.

(a) Using conservation of linear momentum

$$mv = mv_1 + mv_2$$

or $$v = v_1 + v_2 \quad \text{...(1)}$$

From energy conservation, we have

$$\frac{1}{2}mv^2 = \frac{1}{2}mv_1^2 + \frac{1}{2}mv_2^2 + \Delta E$$

or $$v^2 = v_1^2 + v_2^2 + \frac{2\Delta E}{m} \quad \text{...(2)}$$

Squaring (1) we have

$$v^2 = v_1^2 + v_2^2 + 2v_1 v_2$$

Comparing this with (2)

$$2v_1 v_2 = \frac{2\Delta E}{m}$$

Hence, $$(v_1 - v_2)^2 = (v_1 + v_2)^2 - 4v_1 v_2$$

$$= v^2 - \frac{4\Delta E}{m}$$

For $(v_1 - v_2)$ to be real

$$v^2 - \frac{4\Delta E}{m} \geq 0 \quad \text{or} \quad v^2 \geq \frac{4\Delta E}{m}$$

The minimum energy that can be absorbed by the hydrogen atom in ground state to go to an excited state is 10.2 eV. Thus, the minimum velocity of neutron for the collision to be inelastic is

$$v_{min}^2 = \frac{4 \times 10.2 \times 1.6 \times 10^{-19}}{1.67 \times 10^{-27}}$$

$$= 39.08 \times 10^8$$

or $\quad v_{min} = 6.25 \times 10^4$ m/s

(b) From the law of momentum conservation

$$mv - mv = mv_1 + mv_2$$

or $\quad v_1 = -v_2 \quad \ldots(3)$

From conservation of energy, we have

$$\frac{1}{2}mv^2 + \frac{1}{2}mv^2 = \frac{1}{2}mv_1^2 + \frac{1}{2}mv_2^2 + \Delta E$$

or $\quad 2v^2 = 2v_1^2 + \frac{2\Delta E}{m} \quad$ [using (3)]

or $\quad v_1^2 = v^2 - \frac{\Delta E}{m}$

Since v_1 is real $v^2 \geq \frac{\Delta E}{m}$

Minimum value of v is obtained when

$$\Delta E = 10.2 \text{ eV}$$

$\therefore \quad v_{min}^2 = \frac{10.2 \times 1.6 \times 10^{-19} \text{ J}}{1.67 \times 10^{-27} \text{ kg}}$

or $\quad v_{min} = 3.13 \times 10^4$ m/s

556. Pitch of the helix is given by

$$y = (v \cos \theta) . T = v(\cos 60°) T = \frac{vT}{2}$$

Where time period of circular motion

$$T = \frac{2\pi m}{eB}$$

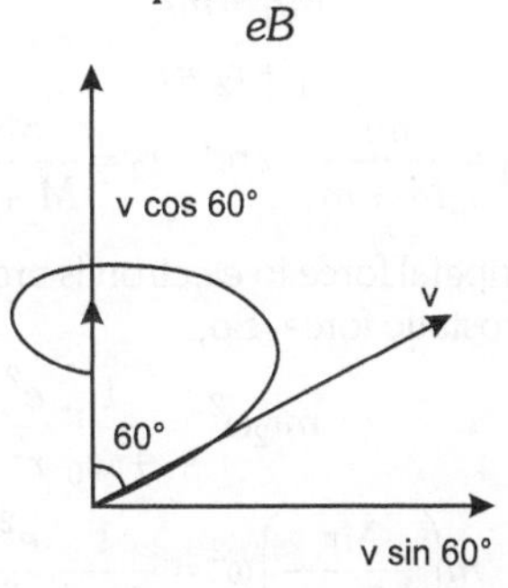

Pitch $y = \left(\frac{2\pi m}{Be}\right) v \cos 60° = \left(\frac{\pi m v}{Be}\right)$

$\therefore \quad KE = \frac{1}{2}mv^2 = \frac{1}{2}m\left(\frac{Bey}{\pi m}\right)^2$

Using Einstein's equation

$$\frac{1}{2}mv^2 = h\nu - \phi$$

$$\phi = h\nu - \frac{1}{2}m\left(\frac{Bey}{\pi m}\right)^2$$

$$= \left[4.9 - \frac{1}{2}\left(\frac{m}{e}\right)\left(\frac{Bey}{\pi m}\right)^2\right] \text{ electron-volts}$$

$$= \left[4.9 - \frac{1}{2}\frac{B^2y^2}{4\pi^2}\left(\frac{e}{m}\right)\right] \text{ electron-volt}$$

$$= (4.9 - 0.4) \text{ eV} = 4.5 \text{ eV}$$

Note: The maximum pitch will be corresponding to the electrons having maximum speed or maximum kinetic energy.

557. $\quad \lambda_{de} = \frac{h}{\sqrt{2m_e(E_{ph} - W)}}$

$$\lambda_{K\alpha} = \frac{4}{3R(Z-1)^2}$$

Given that

$$\lambda_{de} = (1.09)^2 \sqrt{10}\lambda_{K\alpha}$$

Substituting the values we have,

$$Z - 1 = 23$$

$\therefore \quad Z = 24$

558. We have for B

$$\frac{dN_B}{dt} = P - \lambda_2 N_B$$

$\Rightarrow \quad \int_0^{N_B} \frac{dN_B}{P - \lambda_2 N_B} = \int_0^t dt$

$\Rightarrow \quad \ln\left(\frac{P - \lambda_2 N_B}{P}\right) = -\lambda_2 t$

$\Rightarrow \quad N_B = \frac{P(1 - e^{-\lambda_2 t})}{\lambda_2}$

The number of nuclei of A after time t is

$$N_A = N_0 e^{-\lambda_1 t}$$

Thus $\dfrac{dN_c}{dt} = \lambda_1 N_A + \lambda_2 N_B$

$\Rightarrow \quad \dfrac{dN_c}{dt} = \lambda_1 N_0 e^{-\lambda_1 t} + P(1 - e^{-\lambda_2 t})$

$\Rightarrow N_c = N_0 (1 - e^{-\lambda_1 t}) + P\left(t + \dfrac{e^{-\lambda_2 t} - 1}{\lambda_2}\right)$

559. (i) Frequency of electron in n^{th} orbit is given by

$$\nu_n = \frac{K^2 m z^2 e^4}{2\pi n^3 \hbar^3} \quad \ldots(1)$$

Here $\hbar = \dfrac{h}{2\pi}$ and $K = \dfrac{1}{4\pi\varepsilon_0}$

Similarly $\nu_{n+1} = \dfrac{K^2 m Z^2 e^4}{2\pi (n+1)^3 \hbar^3} \quad \ldots(2)$

Energy of electron in n^{th} orbit is

$$E_n = -\frac{mK^2Z^2e^4}{2n^2\hbar^2}$$

and $E_{n+1} = -\dfrac{mK^2Z^2e^4}{2(n+1)^2\hbar^2}$

Therefore, frequency of photon ν when it jumps from $(n+1)^{th}$ energy state to n^{th} state will be given by

$$\nu = \frac{E_{n+1} - E_n}{2\pi\hbar} = \frac{mK^2Z^2e^4}{4\pi\hbar^3}\left[\frac{1}{n^2} - \frac{1}{(n+1)^2}\right]$$

$$= \frac{mK^2Z^2e^4}{2\pi\hbar^3}\,\frac{(2n+1)}{2n^2(n+1)^2}$$

or $\nu = \dfrac{K^2mZ^2e^4}{2\pi n^3\hbar^3}\left[\dfrac{\left(2 + \dfrac{1}{n}\right)}{2\left(1 + \dfrac{1}{n}\right)^2}\right] \quad \ldots(3)$

From equations (1), (2) and (3) we can see that

$$\nu_{n+1} < \nu < \nu_n$$

(ii) For large values of n

$$n + 1 \approx n \quad \text{and} \quad \frac{1}{n} \to 0,$$

So, from equations (1) (2) and (3) we see that

$$\nu = \nu_n = \nu_{n+1}$$

560. $F = -\dfrac{dU}{dr} = -kr$

Negative sign implies that force is acting towards centre. The necessary centripetal force to the particle is being provided by this force F. Hence

$$\frac{mv^2}{r} = kr \quad \ldots(1)$$

and $\quad mvr = n\hbar \quad \left(\text{where } \hbar = \dfrac{h}{2\pi}\right) \ldots(2)$

solving equations (1) and (2) we get

$$\boldsymbol{r = r_n = \sqrt{\frac{n\hbar}{m\omega}}} \quad \text{where} \quad \omega = \sqrt{\frac{k}{m}}$$

and total energy

$$E = U + K = \frac{kr^2}{2} + \frac{1}{2}mv^2$$

Substituting the values, we get

$$\boldsymbol{E = n\hbar\omega = E_n}$$

561. Given that M is the mass of the nucleus (proton) and m the mass of electron. Proton and electron both revolve about their centre of mass (COM) with same angular velocity (ω) but different linear speeds. Let r_1 and r_2 be the distances of COM from proton and electron. Let r be the distance between the proton and the electron. Then

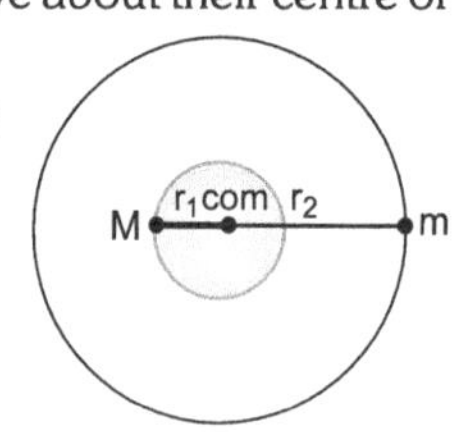

$$Mr_1 = mr_2$$
$$r_1 + r_2 = r$$

$\therefore \quad r_1 = \dfrac{mr}{M+m}$ and $r_2 = \dfrac{Mr}{M+m}$

Centripetal force to electron is provided by the electrostatic force. So,

$$mr_2\omega^2 = \frac{1}{4\pi\varepsilon_0}\frac{e^2}{r^2}$$

or $\quad m\left(\dfrac{Mr}{M+m}\right)\omega^2 = \dfrac{1}{4\pi\varepsilon_0}\dfrac{e^2}{r^2}$

or $\quad \left(\dfrac{Mm}{M+m}\right)r^3\omega^2 = \dfrac{e^2}{4\pi\varepsilon_0}$

Substituting $\dfrac{Mm}{M+m}=\mu$

(reduced mass of proton and electron)

$$\mu r^3\omega^2=\frac{e^2}{4\pi\varepsilon_0} \qquad \ldots(1)$$

Moment of inertia of the atom about COM is :

$$I=Mr_1^2+mr_2^2$$

or $$I=\left(\frac{Mm}{M+m}\right)r^2=\mu r^2 \qquad \ldots(2)$$

According to Bohr's theory

$$I\omega=n\hbar \qquad \text{where } \left(\hbar=\frac{h}{2\pi}\right)$$

or $$\mu r^2\omega=n\hbar \qquad \ldots(3)$$

solving equations (1) and (3) for r we get

$$r=\frac{4\pi\varepsilon_0 n^2\hbar^2}{\mu e^2} \qquad \ldots(4)$$

Electrical potential energy of the system is

$$U=-\frac{e^2}{4\pi\varepsilon_0 r}$$

and kinetic energy is

$$K=\frac{1}{2}I\omega^2=\frac{1}{2}\mu r^2\omega^2$$

But from equation (1)

$$\omega^2=\frac{e^2}{4\pi\varepsilon_0\mu r^3} \quad \therefore \quad K=\frac{e^2}{8\pi\varepsilon_0 r}$$

$\therefore$ Total energy of the atom is

$$E=K+U=-\frac{e^2}{8\pi\varepsilon_0 r}$$

Substituting value of r from equation (4) we get

$$E=-\frac{\mu e^4}{32\pi^2\varepsilon_0^2 n^2\hbar^2}$$

For ground state $n=1$

$$E_1=-\frac{\mu e^4}{32\pi^2\varepsilon_0^2\hbar^2}$$

or binding energy

$$\boldsymbol{E_b=|E_1|=\frac{\mu e^4}{32\pi^2\varepsilon_0^2\hbar^2}}$$

If motion of nucleus is not taken into account, in that case

$$E_b'=\frac{me^4}{32\pi^2\varepsilon_0^2\hbar^2},$$

$$m>\mu \quad \therefore \quad E_b'>E_b$$

$$\therefore \text{ percentage increase}=\frac{E_b'-E_b}{E_b}\times 100$$

$$=\left(\frac{m-\mu}{\mu}\right)\times 100=\left(\frac{m-\dfrac{mM}{M+m}}{\dfrac{mM}{M+m}}\right)\times 100$$

$$=\left(\frac{m}{M}\right)\times 100=0.00055\times 100=\mathbf{0.055\%}$$

562. The product nucleus ^{198}Hg is in excited state and possesses extra 1.088 MeV energy.

If ^{198}Hg would had been in ground state, the kinetic energy available to electron and antineutrino must have

$$Q=(m_{Au}-m_{Hg})\,931\text{ MeV}$$
$$=(197.968233-197.966760)\,931\text{ MeV}$$
$$=1.3714\text{ MeV}$$

Since ^{198}Hg is in excited state, actual kinetic energy available to electron and antineutrino is

$$K=(1.3714-1.088)\text{ MeV}$$
$$=\mathbf{0.2834\ MeV}$$

As β-ray and antineutrino has continuous spectrum starting form zero value, therefore, this is also the maximum kinetic energy of the electron emitted.

563. Let M be the total mass of Uranium mixture. Then the masses of the isotopes $_{92}U^{234}$ and $_{92}U^{238}$ in the mixture are

$$M_1=0.1\text{ M} \quad \text{and} \quad M_2=0.9\text{ M}$$

The mass number of isotopes are

$$A_1=234 \quad \text{and} \quad A_2=238$$

Number of molecules of these isotopes in the mixture are

$$N_1=\frac{M_1}{A_1}N_A \quad \text{and} \quad N_2=\frac{M_2}{A_2}N_A$$

Where N_A = Avogadro number.
Activity of a radioactive sample

$$R = \lambda N = \frac{ln\,2}{t_{1/2}} \cdot N \qquad (t_{1/2} = \text{half life})$$

$$\therefore \quad \frac{R_1}{R_2} = \left(\frac{N_1}{N_2}\right)\frac{(t_{1/2})_2}{(t_{1/2})_1}$$

$$= \left(\frac{M_1}{M_2}\right)\left(\frac{A_2}{A_1}\right)\frac{(t_{1/2})_2}{(t_{1/2})_1}$$

Substituting the values, we have

$$\frac{R_1}{R_2} = \left(\frac{0.1\text{ M}}{0.9\text{ M}}\right)\left(\frac{238}{234}\right)\left(\frac{4.5\times 10^5}{2.5\times 10^5}\right) = 0.2$$

$\therefore$ Percentage activity of

$$_{92}\mathbf{U}^{234} = \left(\frac{\mathbf{0.2}}{\mathbf{0.2+1}}\right)\times \mathbf{100} = \mathbf{16.67\%}$$

and percentage activity of

$$_{92}\mathbf{U}^{238} = \mathbf{(100 - 16.67)\% = 83.33\%}$$

564. In time interval dt, number of increase of daughter nuclei are

$$dN_2 = \lambda_1 N_1 dt - \lambda_2 N_2 dt$$

or $$dN_2 = \lambda_1 N_0 e^{-\lambda_1 t} dt - \lambda_2 N_2 dt$$

$$(N_1 = N_0\, e^{-\lambda_1 t})$$

or $$\frac{dN_2}{dt} + \lambda_2 N_2 = \lambda_1 N_0 e^{-\lambda_1 t} \qquad \ldots(1)$$

Case 1 : When $\lambda_1 >> \lambda_2$

i.e. $(t_{1/2})_1 << (t_{1/2})_2 \qquad (t_{1/2} = \text{half life})$

We can assume that $N_{20} \approx N_0$ so that

$$N_2 = N_0 e^{-\lambda_2 t}$$

(N_{20} = number of daughter atoms at time $t = 0$)
Physically this means that parent nuclei practically instantly transform into daughter nuclei, which then decay according to the law of radioactive decay with decay constant λ_2.

Case 2 : When $\lambda_1 << \lambda_2$ *i.e.*

$$(t_{1/2})_1 >> (t_{1/2})_2$$

In this case number of parent nuclei can be assumed to remain constant over a sizable time interval and is equal to N_0.

This transforms equation (1) into

$$\frac{dN_2}{dt} = -(\lambda_2 N_2 - \lambda_1 N_0)$$

or $$\int_0^{N_2} \frac{d\,N_2}{\lambda_1 N_0 - \lambda_2 N_2} = \int_0^t dt$$

which after integration gives

$$\boldsymbol{N_2 = \frac{\lambda_1}{\lambda_2} N_0(1 - e^{-\lambda_2 t})}$$

Note: To find the actual dependence of N_2 on t, we integrate (1). The solution has the form

$$N_2 = \frac{\lambda_1 N_0}{\lambda_2 - \lambda_1}(e^{-\lambda_1 t} - e^{-\lambda_2 t})$$

$$N_2 \approx N_0\, e^{-\lambda_2 t} \qquad \text{for } \lambda_1 >> \lambda_2$$

and $$N_2 \approx \frac{\lambda_1}{\lambda_2} N_0\,(1 - e^{-\lambda_2 t}) \qquad \text{for } \lambda_1 << \lambda_2$$

565. (a) Electric field is parallel to magnetic field and initial velocity is perpendicular to both the fields. Hence the resultant path will be a helix of increasing pitch. Velocity of α- particle at any time t, would be

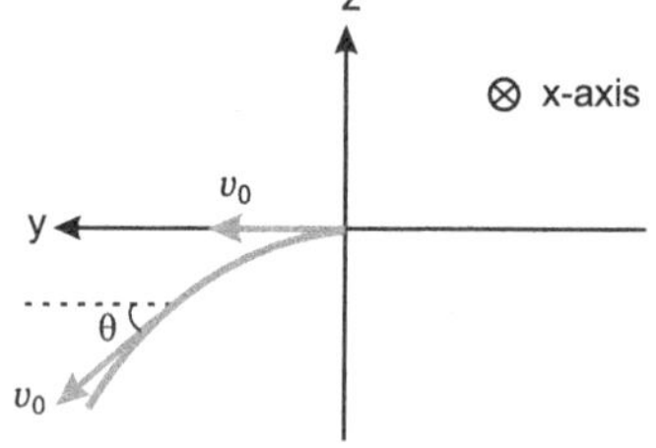

$$\vec{\boldsymbol{v}} = \left(\frac{\boldsymbol{q_\alpha E_0}}{\boldsymbol{m_\alpha}}\boldsymbol{t}\right)\hat{\mathbf{i}} + \boldsymbol{v_0}\,\mathbf{cos}\,\theta\,\hat{\mathbf{j}} - \boldsymbol{v_0}\,\mathbf{sin}\,\theta\,\hat{\mathbf{k}}$$

where $\theta = \omega t$ and $\omega = \dfrac{\boldsymbol{q_\alpha B}}{\boldsymbol{m_\alpha}}$

(b) Speed of α-particle at time t is

$$|\vec{v}| = \sqrt{\left(\frac{q_\alpha E_0}{m_\alpha}t\right)^2 + v_0^2}$$

Given $|\vec{v}| = 2v_0$ at,

$$t = \sqrt{3}\times 10^7\left(\frac{m_\alpha}{q_\alpha E_0}\right)\text{sec}$$

So, $4v_0^2 = (\sqrt{3}\times 10^7)^2 + v_0^2$

or $\boldsymbol{v_0 = 10^7}$ **m/s**

(c) When α-particle is emitted with velocity v_0 from a stationary nucleus X, daughter nucleus Y recoils with speed v which is given by

$$m_Y v = m_\alpha v_0$$

or $$v = \frac{m_\alpha v_0}{m_Y}$$

$$= \frac{4.003}{221.03} \times 10^7 \text{ m/s}$$

$$= 1.81 \times 10^5 \text{ m/s}$$

∴ Total energy released during α-decay of a nucleus X

$$E = \frac{1}{2} m_Y v^2 + \frac{1}{2} m_\alpha v_0^2$$

$$= \left\{\frac{221.03}{2} \times (1.81 \times 10^5)^2 + \frac{4.003}{2} \times (10^7)^2\right\} \left\{\frac{1.67 \times 10^{-27}}{1.6 \times 10^{-13}}\right\} \text{MeV}$$

$$= 2.126 \text{ MeV}$$

∴ Mass lost during α-decay

$$\Delta m = \frac{2.126}{931} \text{ u} = 0.00228 \text{ u}$$

∴ Mass of nucleus X

$$m_X = m_Y + m_\alpha + \Delta m$$

$$= 221.03 + 4.003 + 0.00228$$

$$m_X = 225.035 \text{ u}$$

Mass defect in nucleus X.

or $$\Delta m' = [92 m_p + (225 - 92) m_n - m_X]$$

$$= 1.9348 \text{ u}$$

∴ Binding energy per nucleon of nucleus

$$X = \frac{1.9348 \times 931}{225} \text{ MeV}$$

$$= \mathbf{8.00 \ MeV}$$

566. (a) Let R_{A_0} and R_{B_0} be the initial activities of A and B. Then

$$R_{A_0} + R_{B_0} = 10^{10} \text{ dps} \quad \ldots(1)$$

Activity of A after time $t = 20$ days (two half lives of A) is

$$R_A = \left(\frac{1}{2}\right)^2 R_{A_0} = 0.25 R_{A_0}$$

Similarly activity of B after $t = 20$ days (four half lives of B) is

$$R_B = \left(\frac{1}{2}\right)^4 R_{B_0} = 0.0625 R_{B_0}$$

Now it is given that

$$R_A + R_B = 20\% \text{ of } 10^{10}$$

or $$0.25 R_{A_0} + 0.0625 R_{B_0} = 0.2 \times 10^{10} \text{ dps} \quad \ldots(2)$$

Solving equations (1) and (2) we get

$$\boldsymbol{R_{A_0} = 0.73 \times 10^{10}} \textbf{ dps}$$

and $$\boldsymbol{R_{B_0} = 0.27 \times 10^{10}} \textbf{ dps}$$

(b) $$\frac{R_{A_0}}{R_{B_0}} = \frac{\lambda_A N_{A_0}}{\lambda_B N_{B_0}} = \frac{(t_{1/2})_B}{(t_{1/2})_A} \cdot \frac{N_{A_0}}{N_{B_0}}$$

$$\therefore \quad \boldsymbol{\frac{N_{A_0}}{N_{B_0}} = \left(\frac{R_{A_0}}{R_{B_0}}\right) \frac{(t_{1/2})_A}{(t_{1/2})_B}}$$

$$\boldsymbol{= \left(\frac{0.73}{0.27}\right)\left(\frac{10}{5}\right) = 5.4}$$

567. Let E_n and E_m be the energies of electron in nth and mth states. Then

$$E_n - E_m = h\nu_0 \quad \ldots(1)$$

In the second case when the atom is moving with a velocity v. Let v' be the velocity of atom after emitting the photon. Applying conservation of linear momentum,

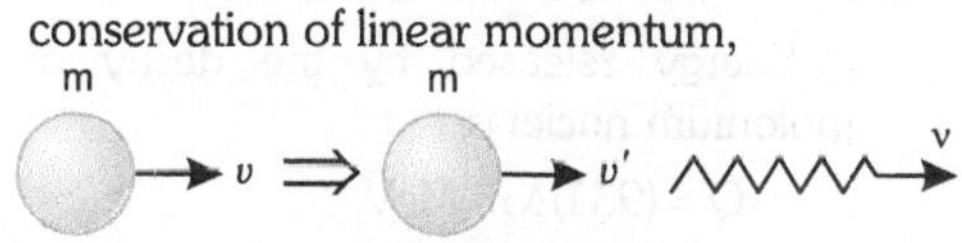

$$mv = mv' + \frac{h\nu}{c}$$

(m = mass of hydrogen atom)

or $$v' = \left(v - \frac{h\nu}{mc}\right) \quad \ldots(2)$$

Applying conservation of energy

$$E_n + \frac{1}{2} m v^2 = E_m + \frac{1}{2} m v'^2 + h\nu$$

or $h\nu = (E_n - E_m) + \frac{1}{2}m(v^2 - v'^2)$

$$= h\nu_0 + \frac{1}{2}m\left[v^2 - \left(v - \frac{h\nu}{mc}\right)^2\right]$$

$$= h\nu_0 + \frac{1}{2}m\left[v^2 - v^2 - \frac{h^2\nu^2}{m^2c^2} + \frac{2h\nu v}{mc}\right]$$

$$= h\nu_0 + \frac{h\nu v}{c} - \frac{h^2\nu^2}{2mc^2}$$

Here the term $\frac{h^2\nu^2}{2mc^2}$ is very small. So, can be neglected

$\therefore \quad h\nu = h\nu_0 + \frac{h\nu v}{c}$

or $\quad \nu\left(1 - \frac{v}{c}\right) = \nu_0$

or $\quad \nu = \nu_0\left(1 - \frac{v}{c}\right)^{-1}$

or $\quad \nu \approx \boldsymbol{\nu_0\left(1 + \frac{v}{c}\right)}$

as $\quad v << c$

568. (i) Nuclear reaction will be as follows

$$Po^{210} \longrightarrow Pb^{206} + He^4 + Q$$

mass defect in this equation is

$\Delta m = m_{Po} - m_{Pb} - m_{He}$

$= (209.98264 - 205.97440 - 4.00260)$ amu

$= 0.00564$

$\therefore$ Energy released by the decay of one polonium nuclei is

$Q = (931)\Delta m$ MeV

$= (931)(0.00564)$ MeV

$= 5.25084$ MeV

$= (5.25084 \times 1.6 \times 10^{-13})$ J

or $Q \approx (8.4 \times 10^{-13})$ J

693 days of polonium are equivalent to 5 half lives of polonium, because $\frac{693}{138.6} = 5$

So, number of nuclei left after 5 half lives are

$$N = N_0\left(\frac{1}{2}\right)^5 \qquad \ldots(1)$$

(N_0 = initial number of nuclei)

Number of nuclei disintegrated (activity) per day are

$$\left(-\frac{dN}{dt}\right) = \lambda N = \frac{ln(2)}{t_{1/2}} \times N$$

So energy released per day

$$= \left(\frac{ln(2)}{t_{1/2}}\right) N(8.4 \times 10^{-13}) \text{ J}$$

But only 10% of this energy is used as electric power. So,

$$\left(\frac{10}{100}\right)\left\{\frac{ln(2)}{t_{1/2}}\right\} N\{8.4 \times 10^{-13}\} = 1.2 \times 10^7 \text{ J}$$

$$\therefore \quad N = \frac{(1.2 \times 10^7)(100)(t_{1/2})}{(10)(ln\ 2)(8.4 \times 10^{-13})}$$

Substituting the values

$$N = \frac{(1.2 \times 10^7)(100)(138.6)}{(10)(\ln 2)(8.4 \times 10^{-13})}$$

$$N = 2.857 \times 10^{22}$$

$\therefore$ Initial number of nuclei required are

$N_0 = (2^5)N \qquad$ (from equation 1)

or $\quad N_0 = (2^5)(2.857 \times 10^{22})$

$= 9.1424 \times 10^{23} \qquad \ldots(2)$

$\therefore$ Mass of polonium required is

$$m = \left(\frac{9.1424 \times 10^{23}}{6 \times 10^{23}}\right)(210)$$

or $\quad$ **m = 319.984 gm**

(ii) Initial activity $= \lambda N_0 = \left(\frac{ln(2)}{t_{1/2}}\right) N_0$

$$= \left(\frac{0.693}{138.6}\right)(9.1424 \times 10^{23})$$

= **4.5712×10^{21} disintegrations per day**

569. Let N be the number of nuclei at any time t. The net rate of formation of radionucleide will be

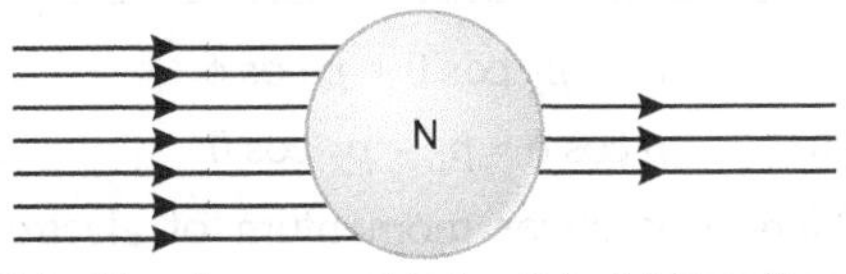

$$\frac{dN}{dt} = q - \lambda N$$

(λ = disintegration constant)

or $$\frac{dN}{q - \lambda N} = dt$$

or $$\int_0^N \frac{dN}{q - \lambda N} = \int_0^t dt$$

or $$N = \frac{q}{\lambda}(1 - e^{-\lambda t})$$

∴ Activity at time t is

$$A = \lambda N = q(1 - e^{-\lambda t}) \quad \ldots(1)$$

Given that $A = 10^8$ disintegrations per second. Substituting the values we get

$$(10^8) = (10^9)(1 - e^{-\lambda t})$$

So $1 - e^{-\lambda t} = 0.1$ or $e^{-\lambda t} = 0.9$

or $\lambda t = 0.105$ or $\left(\frac{ln\ 2}{T}\right)t = 0.105$

or $$t = \frac{(0.105)(T)}{ln\ (2)}$$

$$= \frac{(0.105)(14.3)}{0.693} \text{ days}$$

$$\boldsymbol{t = 2.167 \text{ days}}$$

From equation (1) we can see that activity (A) increases exponentially with time from 0 to q. So, $A - t$ graph will be as shown in figure.

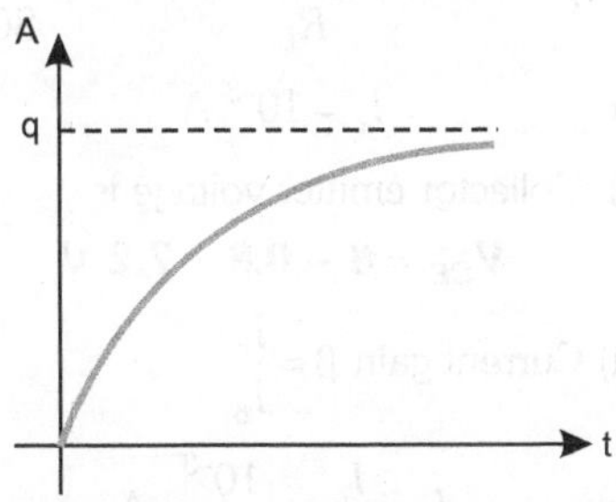

Here $q = 10^9$ disintegrations per second which is also the maximum value of activity of A.

570. Mass of the body left after time t is

$$m = m_0 e^{-\lambda t}$$

So $$\left(-\frac{dm}{dt}\right) = m_0 \lambda e^{-\lambda t}$$

and thrust force on the body is

$$F_t = u_r\left(-\frac{dm}{dt}\right) \quad \text{(in forward direction)}$$

or $$m\left(\frac{dv}{dt}\right) = u\,(m_0 \lambda e^{-\lambda t}) \quad (u_r = u)$$

or $$(m_0 e^{-\lambda t})\frac{dv}{dt} = m_0 u \lambda e^{-\lambda t}$$

or $$dv = u\lambda dt$$

or $$\int_0^v dv = u\lambda \int_0^t dt$$

or $$\boldsymbol{v = u\lambda t}$$

571. Let N be the number of radionuclei at any time t. Then

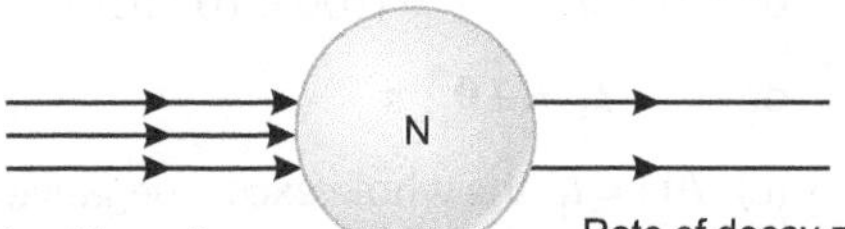

∴ net rate of formation of nuclei at time t is

$$\frac{dN}{dt} = \alpha - \lambda N \quad \text{or} \quad \int_0^N \frac{dN}{\alpha - \lambda N} = \int_0^t dt$$

or $$N = \frac{\alpha}{\lambda}(1 - e^{-\lambda t})$$

Number of nuclei formed in time $t = \alpha t$ and number of nuclei left after time

$$t = \frac{\alpha}{\lambda}(1 - e^{-\lambda t})$$

Therefore, number of nuclei disintegrated in time $t = \alpha t - \frac{\alpha}{\lambda}(1 - e^{-\lambda t})$

∴ energy released till time

$$t = E_0\left[\alpha t - \frac{\alpha}{\lambda}(1 - e^{-\lambda t})\right]$$

But only 20% of it is used in raising the temperature of water.

So $$0.2\,E_0\left[\alpha t - \frac{\alpha}{\lambda}(1 - e^{-\lambda t})\right] = Q$$

where $Q = ms\Delta\theta$

$\therefore$ $\Delta\theta$ = increase in temperature of water $= \dfrac{Q}{ms}$

$$\therefore \quad \Delta\theta = \frac{\mathbf{0.2}E_0\left[\alpha t - \frac{\alpha}{\lambda}(1 - e^{-\lambda t})\right]}{ms}$$

572. (a) Upto time t_1 a constant current $i = 10^{-12}$ A flows between the two plates.

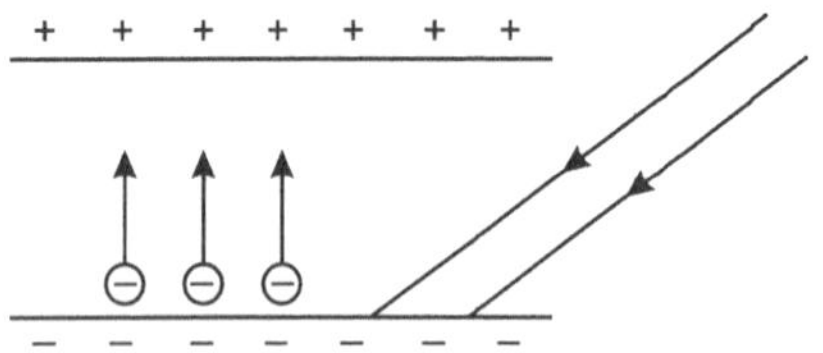

At time t_1, both the plates become neutral *i.e.*, potential difference $\boldsymbol{V = 0}$

(b) $Q = CV = \int_0^{t_1} I \cdot dt = \int_0^{t_1} (10^{-12})\, dt$

or $(10^{-12})t_1 = CV = (100 \times 10^{-6})(10)$

or $\quad \boldsymbol{t_1 = 10^9}$ **s**

(c) At $t = t_1$, the whole excess negative charge on lower plate reaches the upper plate and both the plates become neutral. For $t > t_1$. Some more negative charge reaches the upper plate which repels further movement of electrons from lower plate to upper plate. That is why current is now decreasing. At time $t = t_2$, the potential difference between the plates is equal to the stopping potential. Hence current becomes zero. Therefore, potential difference between the plates for $t > t_2$ is equal to the stopping potential or **1.5 V** (corresponding to maximum kinetic energy of electrons)

573. Energy of a photon is given by $E = pc$

Applying conservation of linear momentum, we get

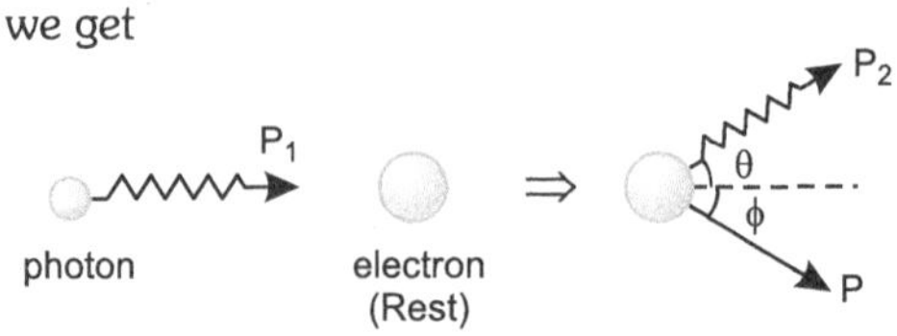

Initial position **Final position**

$$p_1 = p_2 \cos\theta + p\cos\phi$$

or $\quad p\cos\phi = p_1 - p_2\cos\theta \quad \ldots(1)$

Here p is linear momentum of electron at angle ϕ as shown in figure.

Also $\quad p\sin\phi = p_2\sin\theta \quad \ldots(2)$

Squaring and adding equation (1) and (2) we get

$$p^2 = p_1^2 + p_2^2 - 2p_1p_2\cos\theta \quad \ldots(3)$$

Applying conservation of energy, we get

$$p_1c + m_0c^2 = p_2c + c\sqrt{p^2 + m_0^2c^2} \quad \ldots(4)$$

Squaring this equation, we find :

$$p^2 = p_1^2 + p_2^2 - 2p_1p_2 + 2p_1m_0c - 2p_2m_0c \quad \ldots(5)$$

From equations (3) and (5) we can show that

$$\frac{1}{p_2} - \frac{1}{p_1} = \frac{1}{m_0c}(1 - \cos\theta)$$

574. The circuit is shown in figure

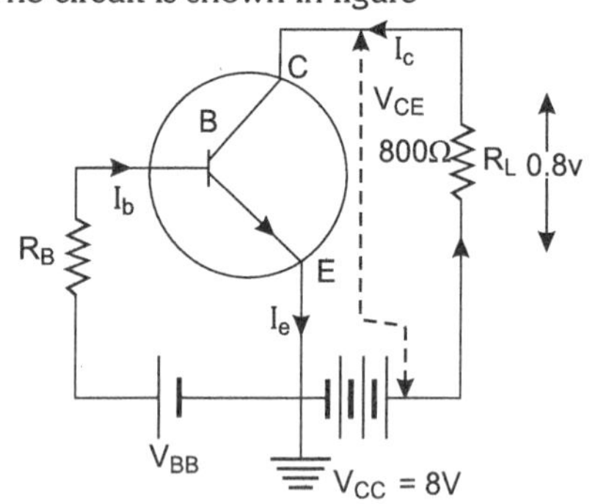

collector current

$$I_c = \frac{\text{voltage drop across } R_L}{R_L} = \frac{0.8}{800}$$

or $\quad I_c = 10^{-3}$ A

(i) Collector emitter voltage is

$$\boldsymbol{V_{CE} = 8 - 0.8 = 7.2}\ \textbf{V}$$

(ii) Current gain $\beta = \dfrac{I_c}{I_b}$

or $\quad I_b = \dfrac{I_c}{\beta} = \dfrac{10^{-3}}{25/26}$ A

or $\quad \boldsymbol{I_b = 1.04 \times 10^{-3}}$ **A**

(iii) Voltage gain

$$=\beta\cdot\frac{R_{\text{output}}}{R_{\text{input}}}=\frac{25}{26}\times\frac{800}{200}=\mathbf{3.846}$$

(iv) Power gain

= (current gain) (voltage gain)

$$=\left(\frac{25}{26}\right)(3.846)=\mathbf{3.698}$$

575. Potential difference across R_L is

$$I_cR_L=8-V_{CE}=(8-4)\text{ V}=4\text{ V}$$

$$\therefore\quad \boldsymbol{R_L}=\frac{4}{I_c}=\frac{4}{4\times10^{-3}}=\mathbf{10^3\,\Omega}$$

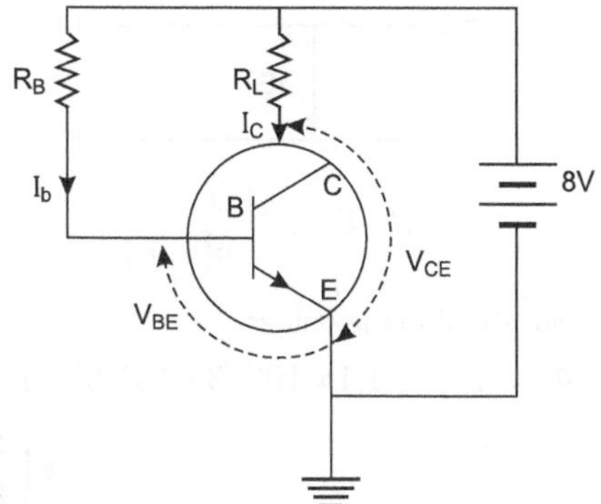

Further $\beta=\dfrac{I_c}{I_b}$

$$\therefore\quad I_b=\frac{I_c}{\beta}$$

$$=\frac{4\times10^{-3}}{100}=4\times10^{-5}\text{ A}$$

Now potential difference across R_B is

$$I_bR_B=8-V_{BE}=8-0.6=7.4\text{ V}$$

$$\therefore\quad R_B=\frac{7.4}{I_b}=\frac{7.4}{4\times10^{-5}}\,\Omega$$

or $\boldsymbol{R_B}=\mathbf{1.85\times10^5\,\Omega}$

576. Applying Kirchhoff's law between *ABMNP* we get

$$V_{BE}+i_bR_1=V_c$$

or $V_{BE}=V_c-i_bR_1=5.5-(10\times10^{-6})(500\times10^3)=0.5\text{ V}$ ...(1)

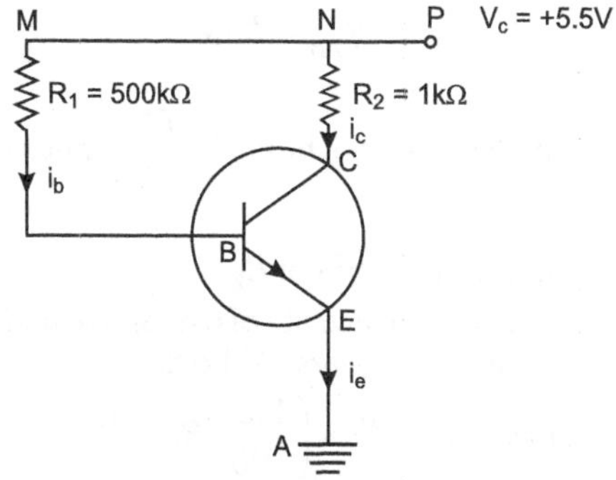

Similarly applying Kirchhoff's law between *AEBCNP* we get

$$V_{CE}+i_cR_2=V_c$$

or
$$V_{CE}=V_c-i_cR_2$$
$$=5.5-(5.2\times10^{-3})(10^3)$$
$$=0.3\text{ V}\qquad\text{...(2)}$$

From equation (1)

$$V_{BE}=V_B-V_E=0.5\text{ V}$$

and from equation (2)

$$V_{CE}=V_C-V_E=0.3\text{ V}$$

$$\therefore\quad V_{BC}=V_B-V_C=(0.5-0.3)\text{ V}=0.2\text{ V}$$

This is a NPN transistor, and V_{BE}, V_{BC} both are positive *i.e.*, $V_B>V_E$ and $V_B>V_C$ or emitter-base and collector-base both are forward biased. Hence **the circuit can not be used as an amplifier**. The transistor is in saturation mode.

577.
$$\lambda_{AB}=\lambda_A+\lambda_B$$

$$\therefore\quad \frac{1}{T_{AB}}=\frac{1}{T_A}+\frac{1}{T_B}\qquad\text{(Here } T=\text{half life)}$$

or
$$\frac{1}{T_{AB}}=2+4=6$$

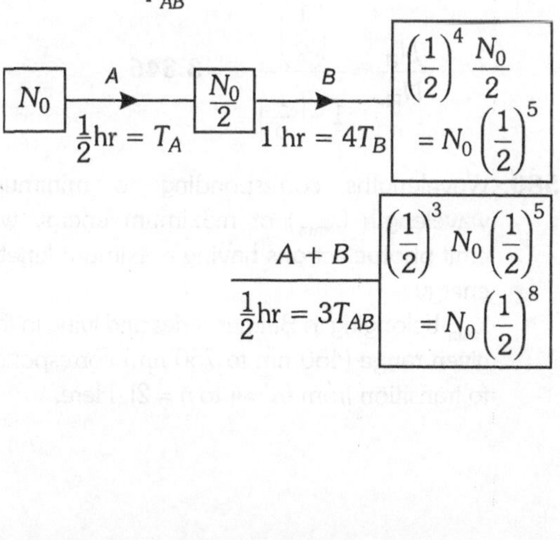

$$\therefore \quad T_{AB} = \frac{1}{6} \text{ hr}$$

Therefore, after 2 hours $N_0 \left(\frac{1}{2}\right)^8$ nuclei will be left.

578. Given, $R_1 = 2.5\%$ of R_2

Here R_1 is the activity of old sample and R_2 the activity of 10 years old bottle.

Now, $$R_0 e^{-\lambda t_1} = \left(\frac{2.5}{100}\right) R_0 e^{-\lambda t_2}$$

$$\therefore \quad e^{\lambda(t_1 - t_2)} = \frac{100}{2.5} = 40$$

$$\lambda (t_1 - t_2) = \ln (40) = 3.7$$

or $$\frac{0.693}{20} (t_1 - t_2) = 3.7$$

$$\therefore \quad t_1 - t_2 = 106.8$$

Given $$t_2 = 10 \text{ years}$$

$$\therefore \quad t_1 \approx 116.8 \text{ years}$$

579. Let N_0 be the initial number of nuclei of ^{238}U. After time t

$$N_U = N_0 \left(\frac{1}{2}\right)^n$$

Here n = number of half lives

$$= \frac{t}{t_{1/2}} = \frac{1.5 \times 10^9}{4.5 \times 10^9} = \frac{1}{3}$$

$$N_U = N_0 \left(\frac{1}{2}\right)^{1/3}$$

and $$N_{Pb} = N_0 - N_U = N_0 \left[1 - \left(\frac{1}{2}\right)^{1/3}\right]$$

$$\therefore \quad \frac{N_U}{N_{Pb}} = \frac{\left(\frac{1}{2}\right)^{1/3}}{1 - \left(\frac{1}{2}\right)^{1/3}} = \mathbf{3.846}$$

580. Wavelengths corresponding to minimum wavelength (λ_{min}) or maximum energy will emit photoelectrons having maximum kinetic energy.

λ_{min} belonging to Balmer series and lying in the given range (450 nm to 750 nm) corresponds to transition from ($n = 4$ to $n = 2$). Here,

$$E_4 = -\frac{13.6}{(4)^2} = -0.85 \text{ eV}$$

and $$E_2 = -\frac{13.6}{(2)^2} = -3.4 \text{ eV}$$

$$\therefore \quad \Delta E = E_4 - E_2 = 2.55 \text{ eV}$$

$$K_{max} = \text{Energy of photon} - \text{work function}$$
$$= 2.55 - 2.0 = \mathbf{0.55\ eV}$$

581. $$\Delta E = h\nu = Rhc (Z - b)^2 \left(\frac{1}{n_1^2} - \frac{1}{n_2^2}\right)$$

For K-series $b = 1$

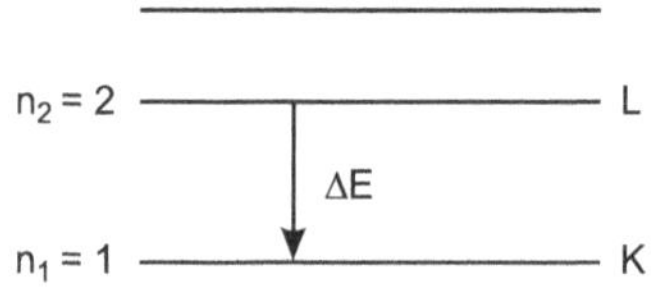

$$\nu = Rc (Z - 1)^2 \left(\frac{1}{n_1^2} - \frac{1}{n_2^2}\right)$$

Substituting the values

$$4.2 \times 10^{18} = (1.1 \times 10^7)(3 \times 10^8)(Z - 1)^2 \times \left(\frac{1}{1} - \frac{1}{4}\right)$$

$$\therefore \quad (Z - 1)^2 = 1697$$

or $$Z - 1 \approx 41 \quad \text{or} \quad Z = 42$$

582. Let n_0 the number of radioactive nuclei at time $t = 0$. Number of nuclei decayed in time t are given by $n_0 (1 - e^{-\lambda t})$, which is also equal to the number of beta particles emitted during the same interval of time. For the given condition,

$$n = n_0 (1 - e^{-2\lambda}) \qquad \ldots(1)$$

$$(n + 0.75n) = n_0 (1 - e^{-4\lambda}) \qquad \ldots(2)$$

Dividing (2) by (1) we get

$$1.75 = \frac{1 - e^{-4\lambda}}{1 - e^{-2\lambda}}$$

or $$1.75 - 1.75e^{-2\lambda} = 1 - e^{-4\lambda}$$

$$\therefore \quad 1.75e^{-2\lambda} - e^{-4\lambda} = \frac{3}{4}$$

Let us take $e^{-2\lambda} = x$

Then the above equation is,

$$x^2 - 1.75x + 0.75 = 0$$

or $$x = \frac{1.75 \pm \sqrt{(1.75)^2 - 4(0.75)}}{2}$$

or $$x = 1 \text{ and } \frac{3}{4}$$

$\therefore$ From Eq. (3) either

$$e^{-2\lambda} = 1 \quad \text{or} \quad e^{-2\lambda} = \frac{3}{4}$$

but $e^{-2\lambda} = 1$ is not accepted because which means $\lambda = 0$. Hence

$$e^{-2\lambda} = \frac{3}{4}$$

or $$-2\lambda \ln(e) = \ln(3) - \ln(4)$$
$$= \ln(3) - 2\ln(2)$$

$\therefore$ $$\lambda = \ln(2) - \frac{1}{2}\ln(3)$$

Substituting the given values,

$$\lambda = 0.6931 - \frac{1}{2} \times (1.0986)$$
$$= 0.14395 \text{ s}^{-1}$$

$\therefore$ Mean life $t_{mean} = \frac{1}{\lambda} = 6.947$ sec

583. Maximum kinetic energy of the photoelectrons would be

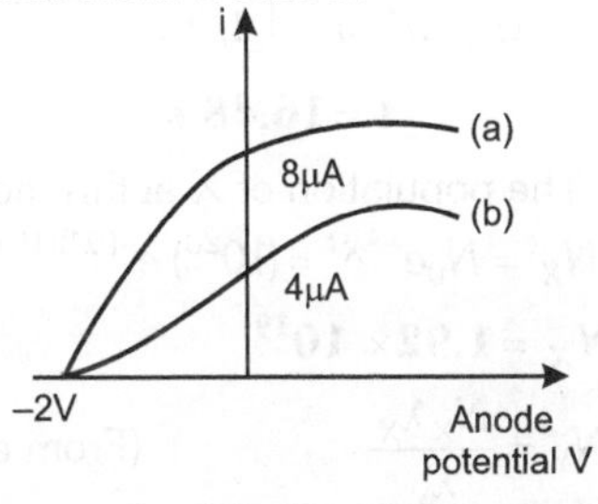

$$K_{max} = E - W = (5 - 3) \text{ eV} = 2 \text{ eV}$$

Therefore, the stopping potential is 2 volt. Saturation current depends on the intensity of light incident. When the intensity is doubled the saturation current will also become two fold. The corresponding graphs are shown in figure.

584. (a) Total 6 lines are emitted. Therefore,

$$\frac{n(n-1)}{2} = 6$$

or $$n = 4$$

So, transition is taking place between m^{th} energy state and $(m+3)^{th}$ energy state.

$\therefore$ $$E_m = -0.85 \text{ eV}$$

or $$-13.6\left(\frac{Z^2}{m^2}\right) = -0.85$$

or $$\frac{Z}{m} = 0.25 \quad \ldots(1)$$

Similarly $$E_{m+3} = -0.544 \text{ eV}$$

or $$-13.6\frac{Z^2}{(m+3)^2} = -0.544$$

or $$\frac{Z}{m+3} = 0.2 \quad \ldots(2)$$

Solving Eqs. (1) and (2) for Z and m we get,

$$m = 12 \quad \text{and} \quad \mathbf{Z = 3}$$

(b) Smallest wavelength corresponds to maximum difference of energies which is obviously $E_{m+3} - E_m$

$\therefore$ $\Delta E_{max} = -0.544 - (-0.85) = 0.306$ eV

$\therefore$ $$\lambda_{min} = \frac{hc}{\Delta E_{max}}$$
$$= \frac{1240}{0.306} = \mathbf{4052.3\ nm}$$

585. Area of plates $A = 5 \times 10^{-4}$ m^2

distance between the plates

$$d = 1 \text{ cm} = 10^{-2} \text{ m}$$

(a) Number of photoelectrons emitted upto $t = 10$ s are

$$n = \frac{\text{(number of photons falling in unit time)} \times \text{(Area} \times \text{time)}}{10^6}$$

$$= \frac{1}{10^6}[(10)^{16} \times (5 \times 10^{-4}) \times (10)]$$

$$= \mathbf{5.0 \times 10^7}$$

(b) At time $t = 10\text{ s}$

charge on plate A,

$$q_A = +ne$$
$$= (5.0 \times 10^7)(1.6 \times 10^{-19})$$
$$= 8.0 \times 10^{-12}\text{ C}$$

and charge on plate B,

$$q_B = (33.7 \times 10^{-12} - 8.0 \times 10^{-12})$$
$$= 25.7 \times 10^{-12}\text{ C}$$

$\therefore$ Electric field between the plates

$$E = \frac{(q_B - q_A)}{2A\varepsilon_0}$$

or $$E = \frac{(25.7 - 8.0) \times 10^{-12}}{2 \times (5 \times 10^{-4})(8.85 \times 10^{-12})}$$

$$= \mathbf{2 \times 10^3\ N/C}$$

(c) Energy of photoelectrons at plate A

$$= E - W = (5 - 2)\text{ eV} = 3\text{ eV}$$

Increase in energy of photoelectrons

$$= (eEd)\text{ joule} = (Ed)\text{ eV}$$
$$= (2 \times 10^3)10^{-2}\text{ eV} = 20\text{ eV}$$

Energy of photoelectrons at plate B

$$= (20 + 3)\text{ V} = \mathbf{23\ eV}$$

586. (i) Let at time $t = t$, number of nuclei of Y and Z are N_Y and N_Z. Then

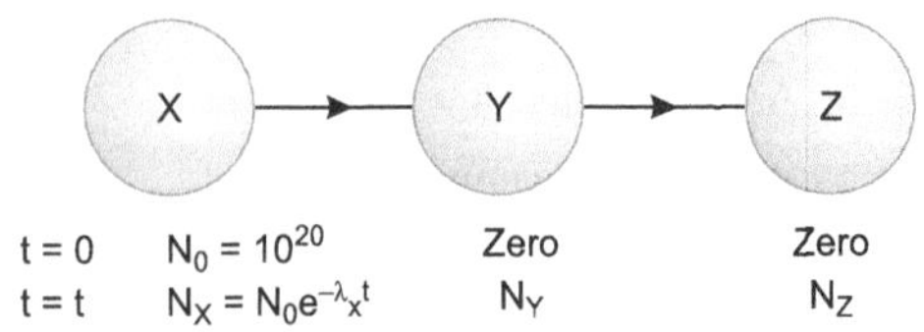

Rate equation of the populations of X, Y and Z are

$$\left(\frac{dN_X}{dt}\right) = -\lambda_X N_X \quad \ldots(1)$$

$$\left(\frac{dN_Y}{dt}\right) = \lambda_X N_X - \lambda_Y N_Y \quad \ldots(2)$$

and $$\left(\frac{dN_Z}{dt}\right) = \lambda_Y N_Y \quad \ldots(3)$$

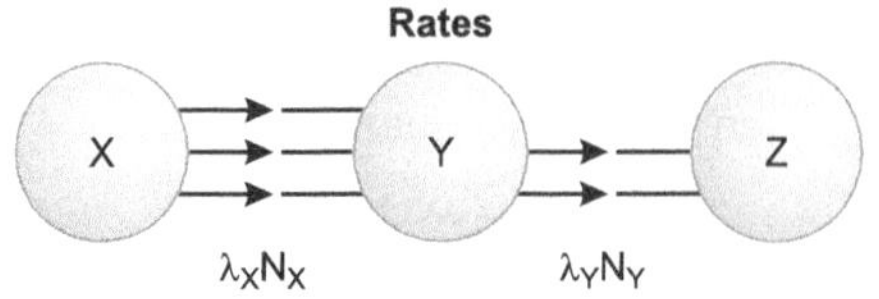

(ii) Given $N_Y(t) = \dfrac{N_0\lambda_X}{\lambda_X - \lambda_Y}[e^{-\lambda_Y t} - e^{-\lambda_X t}]$

For N_Y to be maximum

$$\frac{dN_Y(t)}{dt} = 0$$

i.e. $$\lambda_X N_X = \lambda_Y N_Y \quad \ldots(4)$$

[From equation 2]

or $$\lambda_X (N_0 e^{-\lambda_X t}) = \lambda_Y \frac{N_0\lambda_X}{\lambda_X - \lambda_Y}[e^{-\lambda_Y t} - e^{-\lambda_X t}]$$

or $$\frac{\lambda_X - \lambda_Y}{\lambda_Y} = \frac{e^{-\lambda_Y t}}{e^{-\lambda_X t}} - 1$$

or $$\frac{\lambda_X}{\lambda_Y} = e^{(\lambda_X - \lambda_Y)t}$$

or $$(\lambda_X - \lambda_Y)\, t \ln(e) = \ln\left(\frac{\lambda_X}{\lambda_Y}\right)$$

or $$t = \frac{1}{(\lambda_X - \lambda_Y)}\ln\left(\frac{\lambda_X}{\lambda_Y}\right)$$

Substituting the values of λ_X and λ_Y, we have

$$t = \frac{1}{(0.1 - 1/30)}\ln\left(\frac{0.1}{1/30}\right) = 15\ln(3)$$

or $$\mathbf{t = 16.48\ s}$$

(iii) The population of X at this moment

$$N_X = N_0 e^{-\lambda_X t} = (10^{20})\, e^{-(0.1)(16.48)}$$

$$\mathbf{N_X = 1.92 \times 10^{19}}$$

$$N_Y = \frac{N_X\lambda_X}{\lambda_Y} \quad \text{(From equation 4)}$$

$$= (1.92 \times 10^{19})\frac{(0.1)}{(1/30)}$$

$$= \mathbf{5.76 \times 10^{19}}$$

$$\therefore N_Z = N_0 - N_X - N_Y$$
$$= 10^{20} - 1.92 \times 10^{19} - 5.76 \times 10^{19}$$

or $$\mathbf{N_Z = 2.32 \times 10^{19}}$$

587. The reactor produces 1000 MW power or 10^9 W power or 10^9 J/s of power. The reactor is to function for 10 years. Therefore, total energy which the reactor will supply in 10 years is

$$E = \text{(Power) (time)}$$

$$= (10^9 \text{ J/s}) (10 \times 365 \times 24 \times 3600 \text{ s})$$

$$= 3.1536 \times 10^{17} \text{ J}$$

But since the efficiency of the reactor is only 10%, therefore, actual energy needed is 10 times of it or 3.1536×10^{18} J. One Uranium atom liberates 200 MeV of energy or $200 \times 1.6 \times 10^{-13}$ J or 3.2×10^{-11} J of energy. So, number of uranium atoms needed are

$$\frac{3.1536 \times 10^{18}}{3.2 \times 10^{-11}} = 0.9855 \times 10^{29}$$

or number of kg-moles of uranium needed are

$$n = \frac{0.9855 \times 10^{29}}{6.02 \times 10^{26}} = 163.7$$

Hence total mass of uranium required is

$$m = (n)M = (163.7)(235) \text{ kg}$$

or $m \approx 38470$ kg

or $\mathbf{m = 3.847 \times 10^4}$ **kg**

588. (i) Given mass of α-particle, $m = 4.002$ a.m.u. and mass of daughter nucleus $M = 223.610$ a.m.u. de-Broglie wavelength of α-particle,

$$\lambda = 5.76 \times 10^{-15} \text{ m}$$

So, momentum of α-particle would be

$$p = \frac{h}{\lambda} = \frac{6.63 \times 10^{-34}}{5.76 \times 10^{-15}} \text{ kg-m/s}$$

or $p = 1.151 \times 10^{-19}$ kg-m/s ...(1)

From law of conservation of linear momentum, this should also be equal to the linear momentum of the daughter nucleus (in opposite direction).

Let K_1 and K_2 be the kinetic energies of α-particle and daughter nucleus. Then total kinetic energy in the final state is :

$$K = K_1 + K_2 = \frac{p^2}{2m} + \frac{p^2}{2M} = \frac{p^2}{2}\left(\frac{1}{m} + \frac{1}{M}\right)$$

$$K = \frac{p^2}{2}\left(\frac{M+m}{Mm}\right)$$

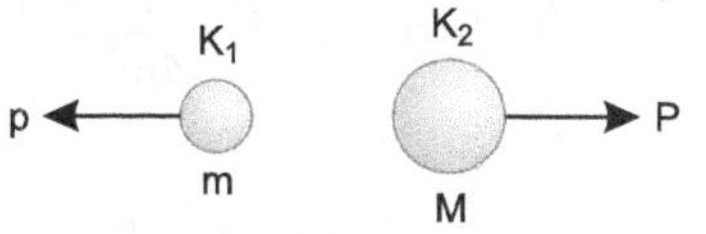

$$1 \text{ amu} = 1.67 \times 10^{-27} \text{kg}$$

Substituting the values, we get

$$K = \frac{(1.151 \times 10^{-19})^2}{2} \frac{(4.002 + 223.610)(1.67 \times 10^{-27})}{(4.002 \times 1.67 \times 10^{-27})(223.61 \times 1.67 \times 10^{-27})}$$

$$K = 10^{-12} \text{ J}$$

$$K = \frac{10^{-12}}{1.6 \times 10^{-13}} \text{ MeV} = 6.25 \text{ MeV}$$

or $\mathbf{K = 6.25}$ **MeV**

(ii) Mass defect,

$$\Delta m = \frac{6.25}{931.470} \text{ amu} = 0.0067 \text{ amu}$$

Therefore, mass of parent nucleus = mass of α-particle + mass of daughter nucleus + mass defect (Δm)

$$= (4.002 + 223.610 + 0.0067) \text{ amu}$$

$$= 227.62 \text{ amu}$$

Hence mass of parent nucleus is **227.62 amu.**

589. (a) Let ground state energy (in eV) be E_1

Then from the given condition

$$E_{2n} - E_1 = 204 \text{ eV}$$

or $$\frac{E_1}{4n^2} - E_1 = 204 \text{ eV}$$

or $$E_1\left(\frac{1}{4n^2} - 1\right) = 204 \text{ eV}$$

and $$E_{2n} - E_n = 40.8 \text{ eV} \quad \text{...(1)}$$

or $$\frac{E_1}{4n^2} - \frac{E_1}{n^2} = 40.8 \text{ eV}$$

or $$E_1\left(\frac{-3}{4n^2}\right)=40.8 \text{ eV} \quad \ldots(2)$$

From equation number (1) and (2)

$$\frac{1-\frac{1}{4n^2}}{\frac{3}{4n^2}}=5 \quad \text{or} \quad 1=\frac{1}{4n^2}+\frac{15}{4n^2}$$

or $$\frac{4}{n^2}=1 \quad \text{or} \quad \boldsymbol{n=2}$$

From equation (2)

$$E_1=-\frac{4}{3}n^2\,(40.8) \text{ eV}$$

$$=-\frac{4}{3}(2)^2\,(40.8) \text{ eV}$$

or $$\boldsymbol{E_1=-217.6 \text{ eV}}$$

$$E_1=-(13.6)Z^2$$

$$\therefore \quad Z^2=\frac{E_1}{-13.6}=\frac{-217.6}{-13.6}=16$$

$$\therefore \quad \boldsymbol{Z=4}$$

$$E_{min}=E_{2n}-E_{2n-1}$$

$$=\frac{E_1}{4n^2}-\frac{E_1}{(2n-1)^2} \qquad n=2$$

$$=\frac{E_1}{16}-\frac{E_1}{9}=-\frac{7}{144}E_1$$

$$=-\left(\frac{7}{144}\right)(-217.6) \text{ eV}$$

$$\therefore \quad \boldsymbol{E_{min}=10.58 \text{ eV}}$$

(b) Energy of incident photon

$$E_i=10.6 \text{ eV}=10.6\times 1.6\times 10^{-19} \text{ J}$$

$$=16.96\times 10^{-19} \text{ J}$$

Energy incident per unit area per unit time (intensity) $=2$ J

$\therefore$ Number of photons incident on unit area in unit time $=\dfrac{2}{16.96\times 10^{-19}}=1.18\times 10^{18}$

Therefore, number of photons incident per unit time on given area $(1.0\times 10^{-4} \text{ m}^2)$

$$=(1.18\times 10^{18})\,(1.0\times 10^{-4})$$

$$=1.18\times 10^{14}$$

But only 0.53% of incident photons emit photoelectrons

$\therefore$ Number of photoelectrons emitted per second (n)

$$n=\left(\frac{0.53}{100}\right)(1.18\times 10^{14})$$

or $$\boldsymbol{n=6.25\times 10^{11}}$$

$$\boldsymbol{K_{min}=0}$$

and $K_{max}=E_i-$ work function

$$=(10.6-5.6) \text{ eV}=5.0 \text{ eV}$$

$$\therefore \quad \boldsymbol{K_{max}=5.0 \text{ eV}}$$

(i) *From the past experience it has been observed that one question in modern physics is usually asked from Bohr's theory + photoelectric effect. Sometimes only one question is asked mixing both the theories and otherwise they are asked separately as two parts of the same question.*

(ii) *In Bohr's theory questions may be asked in near future based on dependence of r_n, v_n and E_n on mass of the electron. For this remember that*

$$r_n \propto \frac{1}{m}$$

$$E_n \propto m$$

and v_n is independent of mass of electron. Sometimes electron is replaced by some other particle say μ-meson, which is 210 times heavier than electron. Now again two cases are possible.

Case 1 : *When mass of nucleus >> mass of μ-meson or nucleus is assumed to be stationary. In that case r_n decrease 210 times, whereas $|E_n|$ increases 210 times and v_n remains the same.*

Case 2 : *When mass of nucleus is comparable to mass of μ-meson or any other particle. In that case m is replaced by reduced mass of nucleus and μ-meson; which will definitely be less than 210. Let us say it comes out to be 205. Then r_n will decrease by 205 times while $|E_n|$ will increase 205 times.*

590. Given : work function $W = 1.9$ eV

Wavelength of incident light

$$\lambda = 400 \text{ nm}$$

$\therefore$ Energy of incident light,

$$E = \frac{hc}{\lambda} = 3.1 \text{ eV}$$

(Substituting the values of h, c and λ)

Therefore, maximum kinetic energy of photoelectron

$$K_{max} = E - W = 3.1 - 1.9 = 1.2 \text{ eV}$$

Now the situation is as shown below :

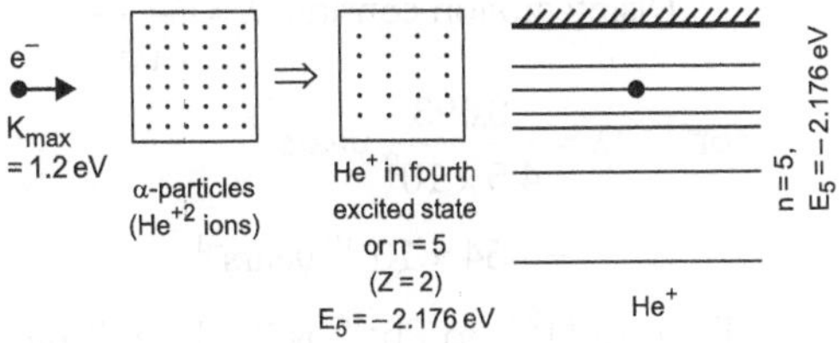

Energy of electron in 4th excited state of He^+ ($n = 5$) will be

$$E_5 = -13.6 \frac{Z^2}{n^2} \text{ eV}$$

$$E_5 = -(13.6)\frac{(2)^2}{(5)^2} = -2.176 \text{ eV}$$

Therefore, energy released during the combination $= 1.2 - (-2.176) =$ **3.376 eV**

Similarly energies in other energy states of He^+ will be

$$E_4 = -13.6 \frac{(2)^2}{(4)^2} = -3.4 \text{ eV}$$

$$E_3 = -13.6 \frac{(2)^2}{(3)^2} = -6.04 \text{ eV}$$

$$E_2 = -13.6 \frac{(2)^2}{2^2} = -13.6 \text{ eV}$$

The possible transitions are

$$\Delta E_{5\to4} = E_5 - E_4 = 1.3 \text{ eV} < 2 \text{ eV}$$

$$\Delta E_{5\to3} = E_5 - E_3 = \mathbf{3.94\ eV}$$

$$\Delta E_{5\to2} = E_5 - E_2 = 11.5 \text{ eV} > 4 \text{ eV}$$

$$\Delta E_{4\to3} = E_4 - E_3 = \mathbf{2.64\ eV}$$

$$\Delta E_{4\to2} = E_4 - E_2 = 10.2 \text{ eV} > 4 \text{ eV}$$

Hence the energy of emitted photons in the range of 2 eV and 4 eV are 3.376 eV during combination and 3.94 eV and 2.64 after combination.

591. (a) Let at time 't', number of radioactive nuclei are N.

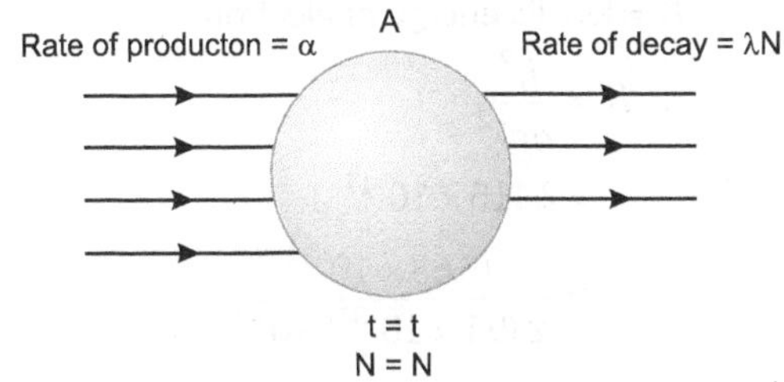

Net rate of formation of nuclei of A.

$$\frac{dN}{dt} = \alpha - \lambda N \quad \text{or} \quad \frac{dN}{\alpha - \lambda N} = dt$$

or $$\int_{N_0}^{N} \frac{dN}{\alpha - N} = \int_0^t dt$$

Solving this equation, we get

$$\boldsymbol{N = \frac{1}{\lambda}[\alpha - (\alpha - \lambda N_0)e^{-\lambda t}]} \qquad \ldots(1)$$

(b) (i) Substituting $\alpha = 2\lambda N_0$

and $$t = t_{1/2} = \frac{ln\,(2)}{\lambda}$$

in equation (1), we get

$$\boldsymbol{N = \frac{3}{2}N_0}$$

(ii) Substituting $\alpha = 2\lambda N_0$ and $t \to \infty$ in equation (1), we get

$$N = \frac{\alpha}{\lambda} = 2N_0 \quad \text{or} \quad \boldsymbol{N = 2N_0}$$

592. From the figure it is clear that

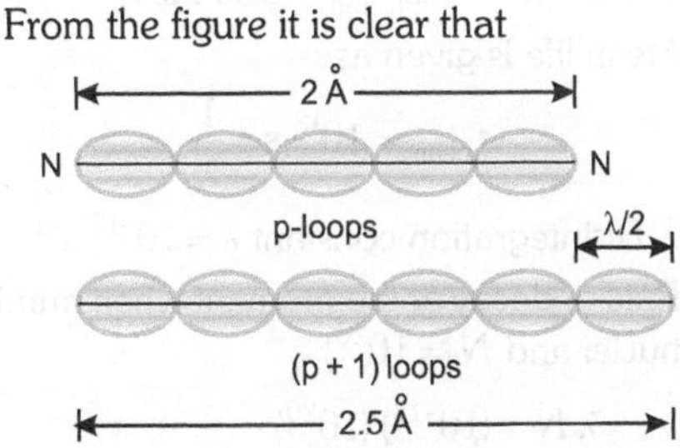

$$p \cdot (\lambda/2) = 2 \text{ Å}$$

$$(p+1)\cdot \lambda/2 = 2.5 \text{ Å}$$

$$\therefore \quad \lambda/2 = (2.5 - 2.0) \text{ Å} = 0.5 \text{ Å}$$

or $\lambda = 1\ \text{Å} = 10^{-10}$ m

(i) de Broglie wavelength is given by

$$\lambda = \frac{h}{p} = \frac{h}{\sqrt{2Km}}$$

K = kinetic energy of electron

$$\therefore\ K = \frac{h^2}{2m\lambda^2}$$

$$= 2.415 \times 10^{-17}\ \text{J}$$

$$= \frac{(6.63 \times 10^{-34})^2}{2(9.1 \times 10^{-31})(10^{-10})^2}$$

$$= \left(\frac{2.415 \times 10^{-17}}{1.6 \times 10^{-19}}\right)\text{eV}$$

$\therefore$ $\mathbf{K = 150.8\ eV}$

(ii) N  N The least value of d will be when only one loop is formed

$$\therefore\ d_{min} = \frac{\lambda}{2} \text{ or } \boldsymbol{d_{min} = 0.5\ \text{Å}}$$

593. The reaction involved in α-decay is

$$^{248}_{96}\text{Cm} \rightarrow {}^{244}_{94}\text{Pu} + {}^{4}_{2}\text{He}$$

Mass defect

$$\Delta m = \text{Mass of } {}^{248}_{96}\text{Cm} - \text{Mass of } {}^{244}_{94}\text{Pu} - \text{Mass of } {}^{4}_{2}\text{He}$$

$= (248.072220 - 244.064100 - 4.002603)$ u

$= 0.005517$u

Therefore, energy released in α-decay will be

$E_\alpha = (0.005517 \times 931)$ MeV $= 5.136$ MeV

Similarly, $E_{fission} = 200$ MeV (given)

Mean life is given as

$$t_{mean} = 10^{13}\ \text{s} = \frac{1}{\lambda}$$

$\therefore$ Disintegration constant $\lambda = 10^{-13}\ \text{s}^{-1}$

Rate of decay at the moment when number of nuclei and $N (= 10^{20})$

$= \lambda N = (10^{-13})(10^{20})$

$= 10^7$ disintegration per second.

Of these disintegrations, 8% are in fission and 92% are in α-decay.

Therefore, energy released per second

$= (0.08 \times 10^7 \times 200 + 0.92 \times 10^7 \times 5.136)$ MeV

$= 2.072 \times 10^8$ MeV

$\therefore$ Power output (in watt) = Energy released per second (J/s)

$= (2.072 \times 10^8)(1.6 \times 10^{-13})$ (J/s)

$\therefore$ **Power output = 3.32 × $\mathbf{10^{-5}}$ watt**

594. $\boldsymbol{t_{1/2} = 4.5 \times 10^9}$ **years**

$\therefore$ Disintegration constant $\lambda = \dfrac{0.693}{t_{1/2}}$

or $\lambda = \dfrac{0.693}{4.5 \times 10^9}\ \text{years}^{-1}$

$= 1.54 \times 10^{-10}\ \text{years}^{-1}$

Ratio of U^{238} to Pb^{206} is 3 : 1 i.e. 3 parts of U^{238} and 1 part of Pb^{206}. We may assume that at $t = 0$, all 4 parts were of U^{238}. Therefore, we can say that

$$\frac{N_0}{N} = \frac{4}{3} \quad (\text{of } U^{238}) \quad \ldots(1)$$

Here N_0 = Number of atoms of U^{238} at time $t = 0$

and N = Number of atoms of U^{238} at time $t = t$

But from $N = N_0 e^{-\lambda t}$ we have

$$\frac{N_0}{N} = e^{\lambda t} \quad \ldots(2)$$

Equating (1) and (2), we have

$$e^{\lambda t} = 4/3$$

or $\lambda t\, ln\,(e) = ln\,(4/3)$

or $\lambda t = 0.29$

$$\therefore\ t = \frac{0.29}{\lambda}$$

$$= \frac{0.29}{1.54 \times 10^{-10}}$$

$= \mathbf{1.88 \times 10^9}$ **years**

Therefore, age of ore is approximately 1.88×10^9 years.

595. (a) (i) Kinetic energy of electron in the orbits of hydrogen and hydrogen like atoms

= |Total energy|

$\therefore$ **Kinetic energy = 3.4 eV**

(ii) The de Broglie wavelength is given by

$$\lambda = \frac{h}{P} = \frac{h}{\sqrt{2Km}}$$

(K = kinetic energy of electron)

Substituting the values, we have

$$\lambda = \frac{(6.6 \times 10^{-34} \text{ J - s})}{\sqrt{2\,(3.4 \times 1.6 \times 10^{19} \text{ J})\,(9.1 \times 10^{-31} \text{ kg})}}$$

$$\lambda = 6.63 \times 10^{-10} \text{ m}$$

or $\lambda = \mathbf{6.63\ Å}$

(b) (i) In 10 second, number of nuclei has been reduced to half (25% to 12.5%). Therefore, its half life is

$$t_{1/2} = 10 \text{ s}$$

Relation between half life and mean life is

$$t_{mean} = \frac{t_{1/2}}{ln\ (2)} = \frac{10}{0.693} \text{ s}$$

$$t_{\mathbf{mean}} = \mathbf{14.43\ s}$$

(ii) From initial 100% to reduction till 6.25%, it takes four half lives.

$$100\% \xrightarrow[t_{1/2}]{} 50\% \xrightarrow[t_{1/2}]{} 25\%$$

$$\xrightarrow[t_{1/2}]{} 12.5\% \xrightarrow[t_{1/2}]{} 6.25\%$$

$\therefore$ $t = 4\ t_{1/2} = 4(10) \text{ s} = 40 \text{ s}$

$\mathbf{t = 40\ s}$

596. (a) Energy of emitted photons

$$E_1 = 5.0 \text{ eV} = 5.0 \times 1.6 \times 10^{-19} \text{ J}$$

$$= 8.0 \times 10^{-19} \text{ J}$$

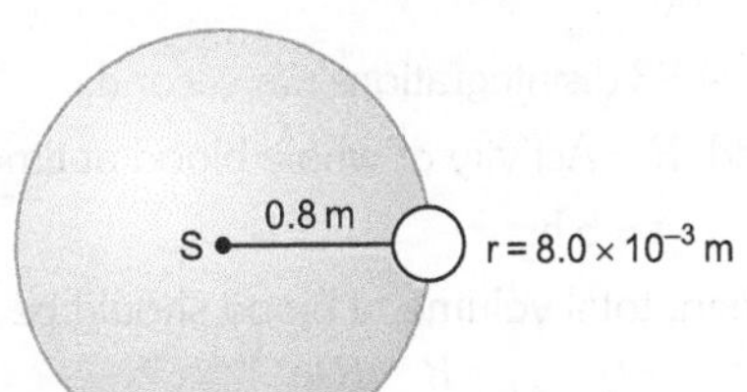

Power of the point source is 3.2×10^{-3} watt or 3.2×10^{-3} J/s

Therefore, energy emitted per second,

$$E_2 = 3.2 \times 10^{-3} \text{ J.}$$

Hence number of photons emitted per second

$$n_1 = \frac{E_2}{E_1} \quad \text{or} \quad n_1 = \frac{3.2 \times 10^{-3}}{8.0 \times 10^{-19}}$$

$$n_1 = 4.0 \times 10^{15} \text{ photons/sec.}$$

Number of photons incident on unit area at a distance of 0.8 m from the source S will be

$$n_2 = \frac{n_1}{4\pi(0.8)^2} = \frac{4.0 \times 10^{15}}{4\pi(0.64)}$$

$$\approx 5.0 \times 10^{14} \text{ photon/sec - m}^2$$

The area of metallic sphere over which photons will fall is :

$$A = \pi r^2 = \pi(8 \times 10^{-3})^2 \text{ m}^2$$

$$\approx 2.01 \times 10^{-4} \text{ m}^2$$

Therefore, number of photons incident on the sphere per second are

$$n_3 = n_2$$

$$A = (5.0 \times 10^{14} \times 2.01 \times 10^{-4})$$

$$\approx 10^{11} \text{ per second}$$

But since one photoelectron is emitted for every 10^6 photons hence number of photoelectrons emitted per second,

$$n = \frac{n_3}{10^6} = \frac{10^{11}}{10^6} = 10^5 \text{ per second}$$

or $\boldsymbol{n = 10^5}$ **per second**

(b) Maximum kinetic energy of photo-electrons

K_{max} = Energy of incident photons – work function

$$= (5.0 - 3.0) \text{ eV} = 2.0 \text{ eV}$$

$$= 2.0 \times 1.6 \times 10^{-19} \text{ J}$$

$$K_{max} = 3.2 \times 10^{-19} \text{ J}$$

The de-Broglie wavelength of these photoelectrons will be

$$\lambda_1 = \frac{h}{p} = \frac{h}{\sqrt{2K_{max}m}}$$

(Here h = Planck's constant and m = mass of electron)

$$\therefore \quad \lambda_1 = \frac{6.63 \times 10^{-34}}{\sqrt{2 \times 3.2 \times 10^{-19} \times 9.1 \times 10^{-31}}}$$

$$= 8.68 \times 10^{-10} \text{ m} = 8.68 \text{ Å}$$

Wavelength of incident light λ_2 (in Å)

$$= \frac{12375}{E_1 \text{ (in eV)}}$$

or $$\lambda_2 = \frac{12375}{5} \text{ Å} = 2475 \text{ Å}$$

Therefore, the desired ratio is

$$\frac{\boldsymbol{\lambda_2}}{\boldsymbol{\lambda_1}} = \frac{\mathbf{2475}}{\mathbf{8.68}} = \mathbf{285.1}$$

(c) As soon as electrons are emitted from the metal sphere, it gets positively charged and acquires positive potential. The positive potential gradually increases as more and more photoelectrons are emitted from its surface. Emission of photoelectrons is stopped when its potential is equal to the stopping potential required for fastest moving electrons.

(d) As discussed in part (c), emission of photoelectrons is stopped when potential on the metal sphere is equal to the stopping potential of fastest moving electrons.

Since $\quad K_{max} = 2.0$ eV

Therefore, stopping potential $V_0 = 2$ volt. Let q be the charge required for the potential on the sphere to be equal to stopping potential or 2 volt. Then

$$2 = \frac{1}{4\pi\varepsilon_0} \cdot \frac{q}{r} = (9.0 \times 10^9) \frac{q}{8.0 \times 10^{-3}}$$

$$\therefore \quad q = 1.78 \times 10^{-12} \text{ C}$$

Photoelectrons emitted per second $= 10^5$ [part a]

or charge emitted per second

$$= (1.6 \times 10^{-19}) \times 10^5 \text{ C}$$

$$= (1.6 \times 10^{-14}) \text{ C}$$

Therefore, time required to acquire the charge q will be

$$t = \frac{q}{1.6 \times 10^{-14}} \text{ sec} = \frac{1.728 \times 10^2}{1.6} \text{ sec}$$

or $$\boldsymbol{t \approx 111} \textbf{ second}$$

597. From the given conditions

$$E_n - E_2 = (10.2 + 17) \text{ eV} = 27.2 \text{ eV} \quad \ldots(1)$$

and $$E_n - E_3 = (4.25 + 5.95) \text{ eV} = 10.2 \text{ eV} \quad \ldots(2)$$

Equation (1) and (2) gives

$$E_3 - E_2 = 17.0 \text{ eV}$$

or $$Z^2 (13.6)(1/4 - 1/9) = 17.0$$

$$\Rightarrow \quad Z^2 (13.6)(5/36) = 17.0$$

$$\Rightarrow \quad Z^2 = 9 \Rightarrow \boldsymbol{Z = 3}$$

From equation (1)

$$Z^2 (13.6)(1/4 - 1/n^2) = 27.2$$

or $$(3)^2 (13.6)(1/4 - 1/n^2) = 27.2$$

or $$1/4 - 1/n^2 = 0.222$$

or $$1/n^2 = 0.0278$$

or $$n^2 = 36 \quad \therefore \quad \boldsymbol{n = 6}$$

598. λ = Disintegration constant

$$= \frac{0.693}{t_{1/2}} = \frac{0.693}{15} \text{ hrs}^{-1} = 0.0462 \text{ hr}^{-1}$$

Let R_0 = Initial activity = 1 microcurie
= 3.7×10^4 disintegrations per second.

r = Activity in 1 cm^3 of blood at $t = 5$ hrs

$$= \frac{296}{60} \text{ disintegration per second}$$

= 4.93 disintegrations per second,

and R = Activity of whole blood at time $t = 5$ hrs

Then, total volume of blood should be

$$V = \frac{R}{r} = \frac{R_0 e^{-\lambda t}}{r}$$

Substituting the values, we have

$$V = \left(\frac{3.7 \times 10^4}{4.93}\right) e^{-(0.0462)(5)}\ \text{cm}^3$$

$$V = 5.95 \times 10^3\ \text{cm}^3$$

or $\boldsymbol{V = 5.95}$ **litre**

599. (i) Let K_1 and K_2 be the kinetic energies of neutron and helium atom after collision and ΔE be the excitation energy.

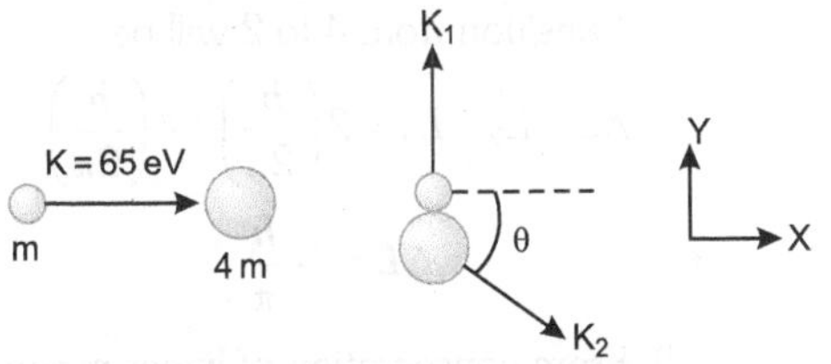

From conservation of linear momentum along X-direction

$$p_i = p_f$$

$$\Rightarrow \quad \sqrt{2Km} = \sqrt{2(4m)K_2}\cos\theta \quad \ldots(1)$$

$$(p = \sqrt{2Km})$$

Similarly, applying conservation of linear momentum in Y-direction, we have

$$\sqrt{2K_1 m} = \sqrt{2(4m)K_2}\sin\theta \quad \ldots(2)$$

Squaring and adding (1) and (2), we get

$$K + K_1 = 4K_2 \quad \ldots(3)$$

or $$4K_2 - K_1 = K = 65\ \text{eV} \quad \ldots(4)$$

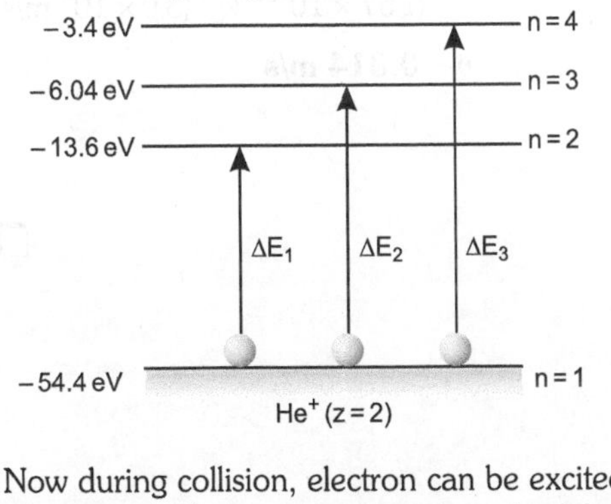

Now during collision, electron can be excited to any higher energy state. Applying conservation of energy, we get

$$K = K_1 + K_2 + \Delta E$$

or $$65 = K_1 + K_2 + \Delta E \quad \ldots(5)$$

ΔE can have the following values

$$\Delta E_1 = -13.6 - (-54.4)\ \text{eV} = 40.8\ \text{eV}$$

Substituting in (5), we get

$$K_1 + K_2 = 24.2\ \text{eV} \quad \ldots(6)$$

Solving (4) and (6), we get

$$K_1 = 6.36\ \text{eV} \quad \text{and} \quad K_2 = 17.84\ \text{eV}$$

Similarly, when we put

$$\Delta E = \Delta E_2 = \{-6.04 - (-54.4)\}\ \text{eV}$$

$= 48.36$ eV in equation (5), we get

$$K_1 + K_2 = 16.64\ \text{eV} \quad \ldots(7)$$

Solving (4) and (7), we get

$$K_1 = 0.312\ \text{eV} \quad \text{and} \quad K_2 = 16.328\ \text{eV}$$

Similarly, when we put

$$\Delta E = \Delta E_3 = \{-3.4 - (-54.4)\}$$

$= 51.1$ eV in equation (5), we get

$$K_1 + K_2 = 14\ \text{eV} \quad \ldots(8)$$

Now solving (4) and (8), we get

$$K_1 = -1.8\ \text{eV} \quad \text{and} \quad K_2 = 15.8\ \text{eV}$$

But since the kinetic energies can't have the negative values, the electron will not jump to third excited state or $n = 4$.

Therefore, the allowed values of K_1 (K.E. of neutron) are 6.36 eV and 0.312 eV and of K_2 (K.E. of the atom) are 17.84 eV and 16.328 eV and the electron can jump upto second excited state only ($n = 3$).

(ii) Possible emission lines are only three as shown in figure. The corresponding frequencies are

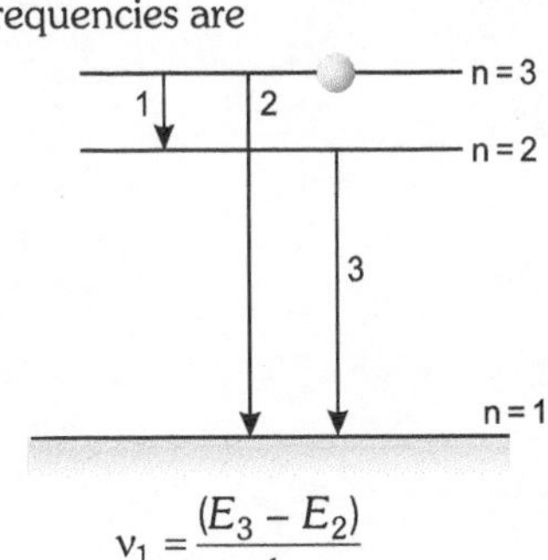

$$\nu_1 = \frac{(E_3 - E_2)}{h}$$

$$= \frac{\{-6.04 - (-13.6)\} \times 1.6 \times 10^{-19}}{6.63 \times 10^{-34}} \text{ Hz}$$

$$= 1.82 \times 10^{15} \text{ Hz}$$

$$\nu_2 = \frac{(E_3 - E_1)}{h}$$

$$= \frac{\{-6.04 - (-54.4)\} \times 1.6 \times 10^{-19}}{6.63 \times 10^{-34}} \text{ Hz}$$

$$= 11.67 \times 10^{15} \text{ Hz}$$

and $\nu_3 = \frac{(E_2 - E_1)}{h}$

$$= \frac{\{-13.6 - (-54.4)\} \times 1.6 \times 10^{-19}}{6.63 \times 10^{-34}} \text{ Hz}$$

$$= 9.84 \times 10^{15} \text{ Hz}$$

Hence the frequencies of emitted radiations are 1.82×10^{15} Hz, 11.67×10^{15} Hz and **9.84×10^{15} Hz**

600. (a) From Einstein's equation of photo-electric effect,

Energy of photons causing the photoelectric emision

= Maximum kinetic energy of emitted photons + work function

or $E = K_{max} + W = (0.73 + 1.82)$ eV

or $\mathbf{E = 2.55 \text{ eV}}$

(b) In case of a hydrogen atom,

$$E_1 = -13.6 \text{ eV}$$

$$E_2 = -3.4 \text{ eV}$$

$$E_3 = -1.5 \text{ eV}$$

and $E_4 = -0.85$ eV

Since $E_4 - E_2 = 2.55$ eV

Therefore, quantum numbers of the two levels involved in the emission of these photons are 4 and 2 (4 → 2)

(c) Change in angular momentum in transition from 4 to 2 will be

$$\Delta L = L_2 - L_4 = 2\left(\frac{h}{2\pi}\right) - 4\left(\frac{h}{2\pi}\right)$$

or $$\Delta \boldsymbol{L} = -\frac{\boldsymbol{h}}{\pi}$$

(d) From conservation of linear momentum

|Momentum of hydrogen atom|

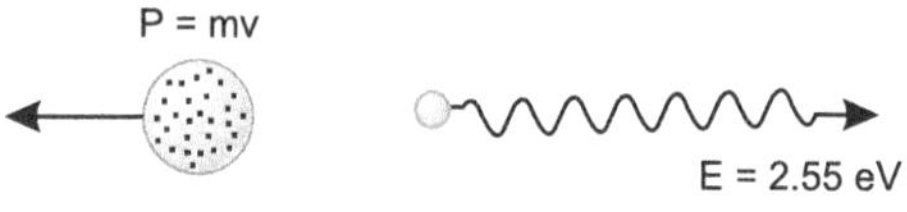

= |Momentum of emitted photon|

or $mv = \frac{E}{c}$ (m = mass of hydrogen atom)

or $v = \frac{E}{mc}$

$$= \frac{2.55 \times 1.6 \times 10^{-19} \text{ J}}{(1.67 \times 10^{-27} \text{ kg})(3.0 \times 10^8 \text{ m/s})}$$

$v = \mathbf{0.814 \text{ m/s}}$